Analytical Techniques for Atmospheric Measurement

Analytical Techniques for Atmospheric Measurement

Edited by

Dwayne E. Heard
University of Leeds

Blackwell
Publishing

Editorial Offices:
Blackwell Publishing Ltd, 9600 Garsington Road, Oxford OX4 2DQ, UK
Tel: +44 (0)1865 776868
Blackwell Publishing Professional, 2121 State Avenue, Ames, Iowa 50014-8300, USA
Tel: +1 515 292 0140
Blackwell Publishing Asia, 550 Swanston Street, Carlton, Victoria 3053, Australia
Tel: +61 (0)3 8359 1011

First published 2006 by Blackwell Publishing Ltd
2 2007

ISBN: 978–1–4051–2357–0

Library of Congress Cataloging-in-Publication Data
Analytical techniques for atmospheric measurement / edited by Dwayne E. Heard.
p. cm.
Includes bibliographical references and index.
ISBN-13: 978–1–4051–2357–0 (acid-free paper)
ISBN-10: 1–4051–2357–5 (acid-free paper)
1. Atmosphere—Measurement. 2. Meteorological instruments.
3. Weather forecasting. I. Heard, Dwayne E.
QC876.A53 2006
551.51′1′028—dc22

2005024526

A catalogue record for this title is available from the British Library

Set in 10/12pt Minion
by Integra Software Services Pvt. Ltd, Pondicherry, India
Printed and bound in India
by Replika Press Pvt. Ltd

The publisher's policy is to use permanent paper from mills that operate a sustainable forestry policy, and which has been manufactured from pulp processed using acid-free and elementary chlorine-free practices. Furthermore, the publisher ensures that the text paper and cover board used have met acceptable environmental accreditation standards.

For further information on Blackwell Publishing, visit our website:
www.blackwellpublishing.com

Contents

Preface

In almost all of the scientific journals where data from field measurements are presented, it is the results, and how they have advanced our understanding of our complex workings of the atmosphere, that have received the most attention. We must not forget, however, that the instruments making the measurements are themselves state-of-the-art and highly specialised, having been first developed in the laboratory by chemists, physicists or engineers, before being applied to the measurement of atmospheric composition. There is usually insufficient space in the journal articles for a thorough description of the instrument, and often only essential details are given. Although there are a number of journals that specialise in the description of instruments, for example *Review of Scientific Instruments*, there is no common theme to a given issue, and an instrument to measure an important atmospheric constituent will likely be sandwiched between articles covering instruments from unrelated fields. Published proceedings from conferences or symposia devoted to instrumentation do offer some coherence, but the articles tend to be specialised, and although appealing to the highly expert reader, they are, in the main, not accessible to scientists wishing to further their knowledge in a new area. There are a number of textbooks that contain sections on instruments for measurements of atmospheric composition, but these usually do not cover recent developments, nor do they give a broad or detailed coverage of the techniques involved.

New analytical techniques or detectors are often invented by researchers with little or no interest in atmospheric science and are then 'discovered' by the atmospheric field measurement community, who go on to utilise the method to make significant breakthroughs. More recently, there has been a stronger collaboration between atmospheric chemists, physicists, engineers and analytical scientists, to develop bespoke instruments with the sole intention of measuring new species in the atmosphere. Sometimes a scientist in a core discipline, for example a physical chemist, recognises the need for a particular measurement in the atmosphere, and realises that the techniques they have used in the laboratory for fundamental studies can equally be applied in the field. Once a rarity, this practice is now becoming more common.

Aims of the book

The major aim of this book is to take the focus away from the results and the advances resulting from field measurements (very important though they are), and to place the emphasis on the instruments themselves: how they work, their strengths and weaknesses for a particular task, the platforms on which they have been deployed, how they are calibrated etc. The book explains the fundamental physical principles upon which the instrumental techniques are based. For instance, what properties of molecules can be exploited to enable their detection? What limits the sensitivity and accuracy of a given instrument, and what information can be gained from its use? Instruments developed to make measurements of atmospheric composition are highly specialised, but often have evolved over many years via several incarnations. The book attempts to convey the excitement of the challenge to *quantitatively* measure trace atmospheric constituents that are at the heart of atmospheric chemistry, and responsible for many of the problems facing society today; for example, the warming of our atmosphere, the destruction of the ozone hole, and the formation of urban air pollution and acid rain. Measurements of the composition of our atmosphere range from the earth's surface to the edge of space. Unsurprisingly, the most important species to measure, those that exert special control on our atmosphere, are often the most difficult to measure – only being present in miniscule concentrations as a result of being so reactive. For some species, for example the hydroxyl radical (OH), it has taken over twenty years of instrument development to enable a reliable and accurate measurement. In some cases, just a single measurement has opened up new areas for study – for example chlorofluorocarbon (CFC) measurements pioneered by Jim Lovelock using the electron capture detector.

Intended audience

The book is designed to appeal to two major types of audience. One class of readers are those who wish to gain a general understanding of instrumentation for measurement of atmospheric composition, the fundamental principles upon which techniques depend, their major capabilities, together with highlights of the important results and the advances in understanding that have resulted – but without wanting a detailed discussion of the underlying atmospheric chemistry or physics. The chapters have been written so that this information is easily retrievable, and is accessible to the non-expert. These readers will be final year undergraduates or postgraduate students starting out on a research project, or postdoctoral fellows or faculty (or indeed anyone) who wish to move into or learn more about composition measurements in the atmosphere. The other class of readers are field scientists or instrument developers who are more experienced, and who will be interested in the finer detail of specific instruments, and latest developments, and perhaps wish to discover if a particular technique were suitable for a new measurement. The individual chapters within this book contain a high level of instrumental detail not normally found in the literature due to lack of space in regular journal articles, and also include a discussion of the strengths and weaknesses of a particular method.

Intended outcomes

Upon completion of this book, it is hoped that you will be able to describe the analytical techniques that are used to make quantitative measurements in the atmosphere, and understand the physical principles upon which the techniques are based. You will be able to discuss, for each analytical method, issues such as species selectivity, calibration, sensitivity (lower detection limit), time-response, uncertainties (accuracy and precision), interferences and the method of data analysis. You will also be able to give examples of species that can be detected using the different techniques, with details of their detection, and describe the experimental platforms in the field on which instruments have been deployed, for example from the ground, ships, aircraft, balloons, rockets, satellites and spacecraft. You will be able to describe applications of the techniques used in the field to further understand the key chemical processes in the atmosphere facing our society today, and will be able to give specific highlights. Finally, it is hoped that you will be able to discuss why measurement of atmospheric composition in the field is such an important component of atmospheric science.

Acknowledgements

I am indebted to the authors of individual chapters who agreed so readily to contribute to this book, and for writing authoritative accounts of their specialist areas enriched by years of experience of measurements in the field. I would also like to thank Paul Sayer and Sarahjayne Sierra, of Blackwell Publishing, for their encouragement and advice during the preparation of this book.

Dwayne Heard
Leeds, November 2005

Contributors

James D. Allan	School of Earth, Atmospheric and Environmental Sciences, Sackville Street Building, The University of Manchester, Sackville Street, PO Box 88, Manchester, M60 1QD, UK
Hugh Coe	School of Earth, Atmospheric and Environmental Sciences, Sackville Street Building, The University of Manchester, Sackville Street, PO Box 88, Manchester, M60 1QD, UK
Ronald C. Cohen	Department of Chemistry, University of California, Latimer Hall, Berkeley, CA 94720-1460, USA
Alan Fried	Atmospheric Chemistry Division, National Center for Atmospheric Research, 3450 Mitchell Lane, Boulder, Colorado 80301, USA
Jacqueline F. Hamilton	University of York, Department of Chemistry, Heslington, York, YO19 5DD, UK
Dwayne E. Heard	School of Chemistry, University of Leeds, Leeds, LS2 9JT, UK
Andreas Hofzumahaus	Forschungszentrum Juelich GmbH, Institut fuer Chemie und Dynamik der, Geosphaere, ICG-II: Troposphaere, Leo Brandtstr., D-52425 Juelich, Germany
Alastair C. Lewis	University of York, Department of Chemistry, Heslington, York, YO19 5DD, UK
John M.C. Plane	University of East Anglia, School of Environmental Sciences, Norwich NR4 7TJ, UK
Dirk Richter	Atmospheric Chemistry Division, National Center for Atmospheric Research, 3450 Mitchell Lane, Boulder, Colorado 80301, USA
Alfonso Saiz-Lopez	School of Chemistry, University of Leeds, Leeds, LS2 9JT, UK
Andrew J. Weinheimer	Atmospheric Chemistry Division, National Center for Atmospheric Research, PO Box 3000, Boulder, Colorado, CO 80305-3000, USA

Jonathan Williams Max-Planck-Institut für Chemie, Abt. Luftchemie, J.J.-Becher-Weg 27, D-55128 Mainz, Germany

Ezra C. Wood Department of Chemistry, University of California, Latimer Hall, Berkeley, CA 94720-1460, USA

Chapter 1

Field Measurements of Atmospheric Composition

Dwayne E. Heard

1.1 The role of field measurements in atmospheric science

1.1.1 Our changing atmosphere

It is difficult to switch on the radio or television, or open a newspaper, without being exposed to accounts of environmental concern resulting from a change in the composition of our atmosphere; for example, increased instances of drought or flooding as a result of the warming of our atmosphere through the increased emissions of greenhouse gases, or the increase in deaths attributable to a deterioration of urban air quality as a result of increased emissions of harmful gases and particles and their secondary oxidation products. Sensational documentaries based on the possible occurrence of some future apocalyptic event are becoming more widespread, and although in many cases they are not based on any scientific fact, the awareness of the general public on environmental issues is raised considerably. In a highly technological and information-rich age, it is becoming generally accepted that our atmosphere is changing as a result of mankind's activities, and if current trends are left unchecked, there may be dire consequences ahead. The key questions are, how quickly is our atmosphere changing and what can be done to mitigate the consequences? In order to predict the future changes in atmospheric composition, and hence the degree of expected climate change through global warming, numerical models are used extensively. The models require three types of input:

(1) An estimation of the future rates of emissions of trace gases. For natural emissions, for example isoprene from trees and plants, the rate of emission is strongly temperature dependent and hence there is a positive feedback with any future warming. For anthropogenic emissions, various scenarios for future use of fossil fuels are estimated, and many assumptions have to be made.
(2) The model must be able to disperse and mix trace gases away from emission sources, for example via advection by the wind or via convective uplift in weather fronts, and on a larger scale through global circulation.
(3) Following their emission, trace gases are removed from the atmosphere by solar-induced photolysis, chemical transformation or by physical deposition to surfaces, forming a large number of chemical intermediates and secondary species, which

themselves undergo further transformation. It is often the secondary products that are most harmful to the atmosphere, and it is crucial that the model contains an accurate description of the chemical mechanisms that describe the atmospheric degradation of trace gases emitted into the atmosphere.

1.1.2 The importance of atmospheric chemistry

The earth's atmosphere is oxidising not because of reaction of trace species with O_2 but owing to the presence of several key oxidising intermediates, which initiate the atmospheric removal of trace gases in the troposphere, eventually forming CO_2 and water vapour. The most important of these is the hydroxyl radical (OH), which is mostly generated in the daytime as a result of ozone photolysis to form electronically excited oxygen atoms, $O(^1D)$, which react rapidly with water vapour to form OH. The hydroxyl radical reacts with virtually all trace gases, including CO, hydrocarbons, oxygenated volatile organic compounds (VOCs), hydrochlorofluorocarbons (HCFCs), used as replacements for chlorofluorocarbons (CFCs) following their ban after the Montreal Protocol, and SO_2. Under normal circumstances, OH concentrations are very low at night, and the nitrate radical (NO_3) replaces OH as the major oxidising species. NO_3 is generated by the reaction of NO_2 with ozone, and reacts either by hydrogen atom abstraction or by addition to double bonds. Ozone itself is the third major oxidising species, reacting, for example, with unsaturated molecules through addition to double bonds forming ozonides which decompose to form a variety of unstable intermediates. Common to the oxidation by OH, NO_3 or O_3 is the formation of intermediate peroxy radicals, RO_2, where R is an organic fragment, including R = H, which reacts with nitric oxide, emitted following the burning of fossil fuels, to form nitrogen dioxide (NO_2). NO_2 gas is brown, as a result of strong absorption of sunlight in the blue and green parts of the spectrum, and is rapidly photolysed by sunlight ($\lambda < 400\,nm$) to form ground-state oxygen atoms, $O(^3P)$, which almost instantaneously combine with O_2 to form O_3, which is harmful to humans and plants in high concentrations.

The conversion of primary emissions, such as VOCs, eventually to CO_2 and water vapour can be extremely complex, involving several reactions. To give an example, *the Master Chemical Mechanism* (Jenkin *et al.*, 2003), http://mcm.leeds.ac.uk/MCM/, describes the complete oxidation pathways for the top 135 VOC emissions in the UK, and consists of 13 600 chemical species and 5900 chemical reactions. An important input to the model is the rate coefficient for each of these reactions over a range of conditions of temperature and pressure encountered in the atmosphere, and although many have been measured in the laboratory, the majority are not known and have to be estimated. In addition, solar-induced photodissociation and deposition to surfaces (the ground, ocean or aerosols) must be included to completely describe the chemistry of the atmosphere, and it is necessary in the laboratory to measure absorption cross-sections as a function of wavelength and temperature, and to measure photodissociation quantum yields as a function of wavelength, temperature *and* pressure.

The focus of this book is not about the chemistry in the atmosphere that is examined via field measurements, but the techniques by which the field measurements are made. The interested reader is referred to a number of excellent textbooks and review articles to discover more about the chemistry of the atmosphere, which in certain locations can be relatively simple involving a handful of species, but in polluted environments requires thousands of chemical species to adequately describe its complexities (Ehhalt, 1998; Finlayson-Pitts & Pitts, 2000; Ravishankara, 2003; Seinfeld & Pandis, 1998; Wayne, 2000). Suggestions for further reading are given in Section 1.11.

1.1.3 Why field measurements of atmospheric composition are important

The composition of the air we breathe was unknown until the second half of the eighteenth century, when carbon dioxide, nitrogen and oxygen were all discovered and isolated, with constituents at lower concentrations, such as methane and ozone, not being discovered until almost a century later. Since the first atmospheric measurements of ozone at ground level were made by Schönbein in 1858, scientists have continued to develop ever more sensitive instruments to discover and measure thousands of trace gases in the atmosphere. However, it is only in the last forty years or so that the link between concentrations of trace gases and global environmental issues has been made. Figure 1.1 shows the change in the mean global radiative forcings due to a number of agents in the period 1750–2000 (Houghton, 2005; IPCC, 2001). It can be seen that several trace gas species, and also

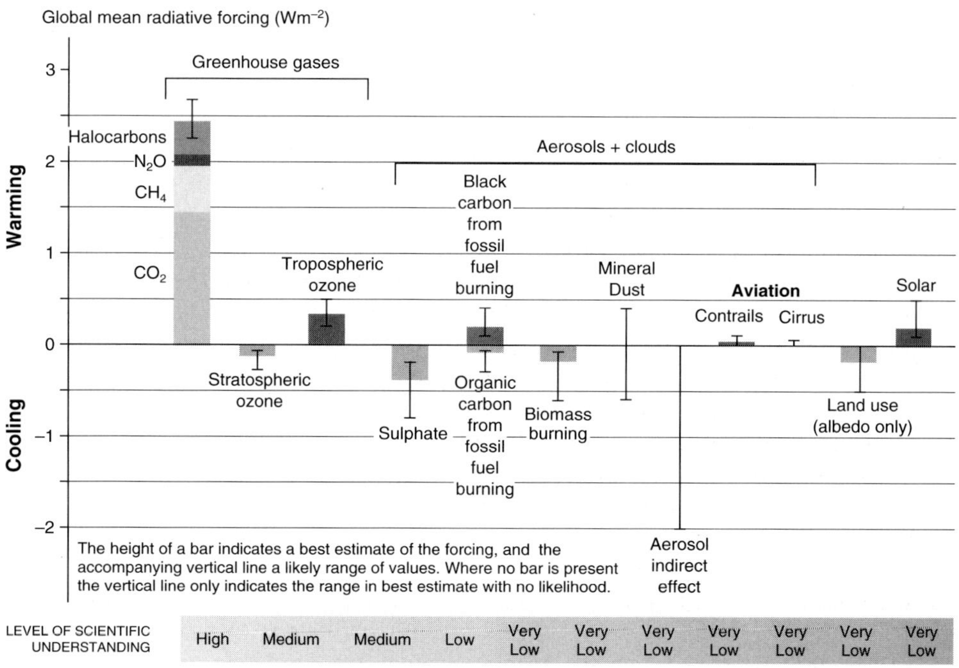

Figure 1.1 Anthropogenic and natural forcing of the climate for the year 2000, relative to 1750. Note that water vapour is also a greenhouse gas, but is not shown. Taken with permission from the 2001 report of the Intergovernmental Panel on Climate Control (Houghton, 2005; IPCC, 2001).

particulate matter (aerosols and clouds), are able to exert a global warming or cooling effect, or both. The graph is based upon our knowledge of the concentrations of these species, and how they interact with incoming or scattered solar radiation. Long-term and quantitative determination of these species in the atmosphere is crucial to calculate the individual contributions to global warming and hence climate change, and to inform policy makers. The concentration of greenhouse gases is increasing all the time, and their measurement is crucial to calculate future changes in atmospheric temperature (see Section 1.7.2). Understanding the trends in chemically and radiatively important gases and particles, and how they can be controlled in the future, is perhaps the most important challenge facing society today.

Field measurements are necessary over a wide range of temporal and spatial scales in order to record any long-term trends, and also to test how well models can predict the composition of the current atmosphere. Although a complete understanding of the complex process within our atmosphere requires an integration of field measurements, computer modelling and laboratory studies, almost all of the major breakthroughs have been initiated by field observations. Without the development of a suite of sensitive and accurate field instrumentation we would not be aware of the links between greenhouse gases/aerosols and global warming, the formation of ozone holes in the stratosphere, the deterioration in air quality on our cities, the changes in the oxidising capacity of the atmosphere, or other threats to our well-being.

An example is the discovery of the ozone hole, first observed in the mid-1980s as a reduction in the overhead column of ozone measured above Antarctica, measured from the ground using a Dobson spectrophotometer that relies on the absorption of ultraviolet (UV) light from the sun (UV absorption is covered in Chapter 3). These measurements were later confirmed using satellite measurements (see Section 1.4.8), with the ozone 'hole' demonstrating a marked spatial, altitudinal and seasonal dependence. An interesting aside is that the more sophisticated satellite instruments had recorded all the necessary data to observe the formation of the ozone hole, but the data analysis software had been programmed not to include data below a certain value! The ozone hole continued to grow in size and intensity, and the observations led to a flurry of activity amongst atmospheric chemists, who eventually postulated that heterogeneous chemistry occurring on the surface of polar stratospheric clouds that formed during the long Antarctic winter was able to generate active forms of chlorine that destroyed ozone when the sun returned. The work led to the Nobel Prize for Chemistry being awarded in 1995 to three scientists for their elucidation of the mechanism of formation of the ozone hole (but not to the instrument developers!). The postulated mechanism was proven beyond doubt by further field observations, as shown in Figure 1.2, that indicated a clear anti-correlation between concentrations of ozone and chlorine monoxide intermediates, measured by in situ instruments aboard the ER-2 research aircraft that flew into the stratospheric polar vortex from Patagonia.

Another example is the invention of the electron capture detector by Jim Lovelock, which enabled the detection of CFCs, present only at the part per trillion level (ppt, 1 part in 10^{15}), which led to the realisation that mankind was releasing into the atmosphere chemical species for which there were no natural removal mechanisms, and a species which if left unregulated could lead to disastrous consequences for stratospheric ozone. These field measurements of CFCs, followed by the understanding of the

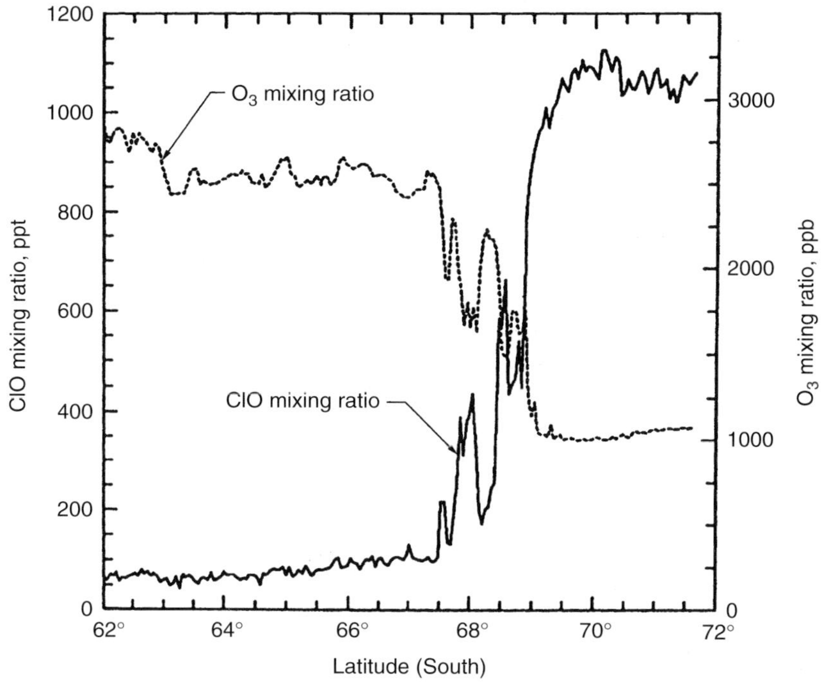

Figure 1.2 Concentrations of O_3 (UV absorption, Chapter 3) and ClO radicals (resonance fluorescence detection of Cl atoms following conversion of ClO by reaction with NO, Chapter 4) measured simultaneously aboard the NASA ER-2 aircraft as it flew into the polar vortex during a flight originating in Punta Arenas, Chile. There is almost perfect anticorrelation between rising ClO levels and the destruction of ozone. (Taken with permission from the American Geophysical Union, Anderson *et al.*, 1989.)

destructive nature of their degradation in the atmosphere, led to the implementation of the Montreal Protocol in 1987 and subsequent amendments. Concentrations of CFCs are now beginning to level-off, or are falling (dependent upon their atmospheric lifetime), as shown in Figure 1.3, but it will be at least several decades before critical chlorine loadings in the stratosphere fall below the threshold for formation of the Antarctic ozone hole.

During the heat wave experienced during the summer of 2003, when temperatures in the UK rose for the first time above 100°F (37.8°C), concentrations of ozone rose in the SE of England to 150 parts per billion (ppb, 1 part in 10^9), levels not observed for 20 years, yet at relatively low concentrations of nitrogen oxides (a result of stringent controls on exhaust pipe emissions). Field measurements of isoprene, a biogenic emission from plants, demonstrated concentrations that rose exponentially with temperature, with concentrations at the ppb level during the heat wave, sufficient to generate large quantities of ozone following its atmospheric oxidation. These measurements highlight the importance of including changes in the rate of biogenic emissions in any global climate change model.

Measurements by satellites have played a crucial role in understanding the effects of major volcanic eruptions. Stratospheric aerosols affect the atmospheric energy balance by scattering and absorbing solar and terrestrial radiation, and perturb stratospheric chemical cycles by catalysing heterogeneous reactions which markedly perturb nitrogen, chlorine

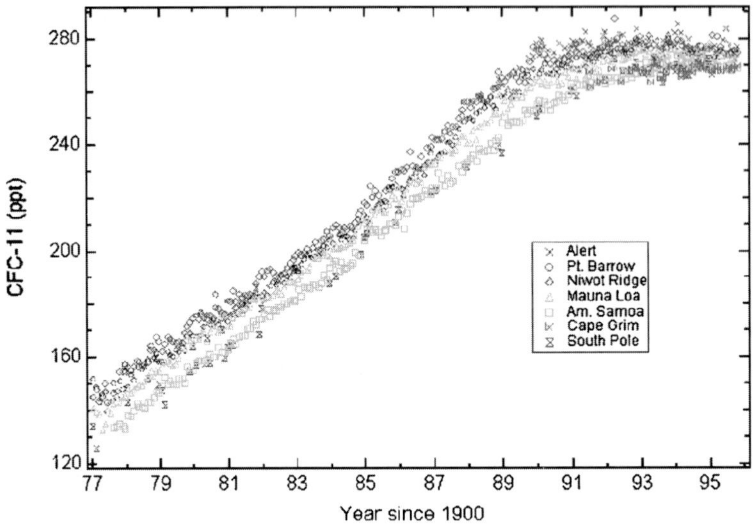

(a)

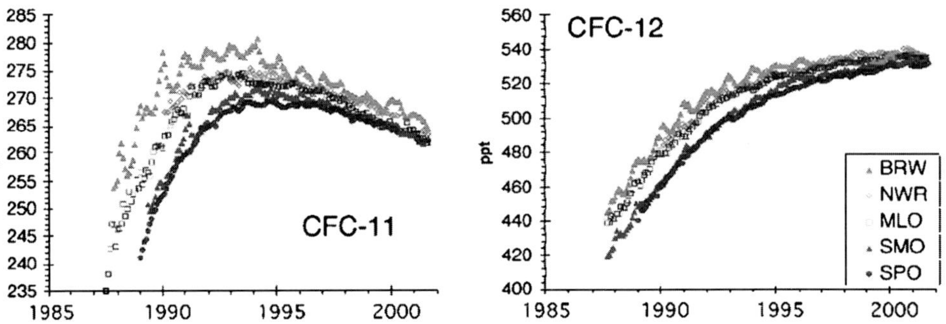

(b)

Figure 1.3 (a) Long-term trend (mixing ratios in pptv) of CFC -11 ($CFCl_3$) measured by ground-based instruments of the NOAA CMDL network since 1977. In the stratosphere the maximum is reached about 5 years later. (Data courtesy of Jim Butler and Jim Elkins, NOAA.) (b) More recent measurements of CFC-11 ($CFCl_3$) and CFC-12, from the same CMDL network. The atmospheric lifetimes of CFC-11 and CFC-12 are 45 and 100 years, respectively. The values for the northern hemisphere are slightly higher, because hemispherical mixing is always faster than the transport across the equator. The measurements are made using gas-chromatographs equipped with an electron capture detector. (Data courtesy of Jim Butler and Jim Elkins, NOAA.)

and ozone levels. These aerosols are small (0.1 μm radius) sulphuric acid solution droplets produced primarily through oxidation of sulphur dioxide injected into the stratosphere by volcanic eruptions. The Stratospheric Aerosol and Gas Experiment II (SAGE II) instrument, http://www-sage2.larc.nasa.gov/, launched on the Earth Radiation Budget Satellite (ERBS) in 1984, monitored the long-term global effects of the Mt. Pinatubo volcanic eruption in June–July 1991, which is estimated to have deposited 3.6×10^7 tonnes of aerosol material into the stratosphere. The instrument measured the stratospheric optical depth in the near-infrared ($\lambda = 1.02\,\mu m$). Measurements just after the eruption and again a year later showed that the volcanic material initially concentrated in the Tropics had spread across the entire globe. The Pinatubo aerosol layer warmed the local subtropical stratosphere by about 2.5–3°C within three months of the eruption, and a statistically significant global average surface cooling was predicted by the end of 1992.

There is a need to continually develop new instruments to measure the spatial and temporal distributions of trace gases, be these harmful pollutants or short-lived intermediates, as these observations are required to monitor trends and as input to models and as target species to test the performance of computer models. There has been a phenomenal improvement in the measurement of many trace gases. Only 10 or 15 years ago, measurement accuracy for some trace species was limited to a factor of two, often with quite long averaging periods. An accuracy of 10% or less is now common, with averaging times in some cases of less than a second. New trace gases are being discovered all the time. The most powerful greenhouse gas ever discovered, SF_5CF_3, was measured by gas chromatography with detection via a magnetic sector mass spectrometer (see Chapter 5 for details) with mixing ratios of $\sim$0.005 pptv, trapped in air within snow in Antarctica (Sturges *et al.*, 2000). SF_5CF_3 is thought to be a by-product of the use of SF_6 by the electronics industry.

1.1.4 The challenges of field measurements in the atmosphere

Many field instruments have been developed to measure a very large number of trace species, whose mole fractions (or mixing ratios) in the atmosphere vary from the per cent level down to parts per quadrillion (ppqv $= 10^{-15}$). A crucial property of a trace gas is its atmospheric lifetime, which is defined as the time taken for the concentration to decay to $1/e$ (37%) of its initial value once its source is removed. The range of atmospheric lifetimes is truly enormous, varying from less than a second, as is the case for free-radicals such as the hydroxyl radical (OH) which mediates virtually all of atmospheric chemistry, to hundreds of years, for example CFCs. Figure 1.4 shows the atmospheric lifetime for a range of species together with the timescales for mixing processes across a range of spatial scales. The atmospheric lifetime determines the degree of transport away from the source region, and whether a trace gas is well mixed globally or exhibits significant spatial structure. The atmospheric lifetime of a trace gas is one of the most critical parameters when choosing the sampling strategy and detection method adopted for its measurement in the atmosphere. The atmospheric lifetime, τ, of a trace gas, X, is the reciprocal of the rate of removal, k', from the atmosphere, $\tau = 1/k'$ (Kurylo & Orkin, 2003; Ravishankara & Lovejoy, 1994). The rate of removal of X is the sum of the rates of photolysis (J, see Chapter 9), dry or wet deposition to surfaces and reactions in the gas phase. For some species photolytic removal or removal on surfaces can be dominant, but for the majority of trace species, k' is controlled by the rate of gas phase reactions, in particular with the hydroxyl radical, OH, and is given by:

$$k' = k_{OH+X}[OH] \qquad (1.1)$$

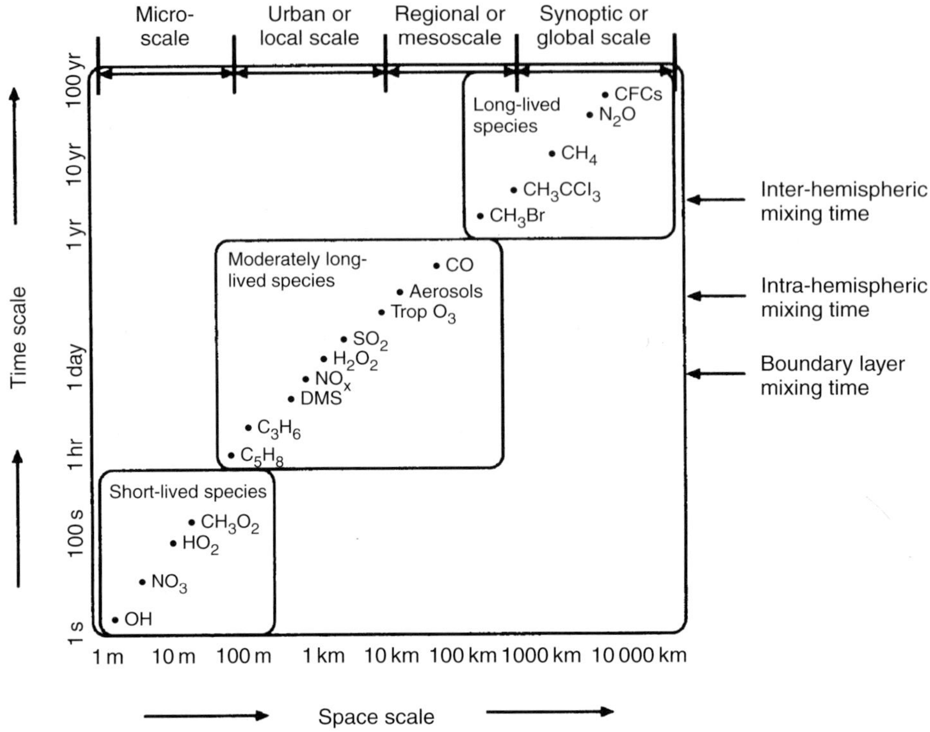

Figure 1.4 Lifetimes and distribution of atmospheric trace gases. The graph shows the very wide range of temporal and spatial scales over which atmospheric trace gases demonstrate variability in the atmosphere. (From Brasseur *et al.*, 1999, with credits to W.L. Chameides.)

where k_{OH+X} is the second-order rate constant for the reaction of OH with X and [OH] is the OH concentration. Thus it is critically important to know the concentration of OH, the quantitative measurement of which illustrates the serious challenges of field measurements. The importance of OH as the dominant oxidant in the atmosphere was recognised in the late 1960s/early 1970s, from which time a considerable effort worldwide has been invested in the development of field instruments for its detection. OH concentrations are very low, a typical daily maximum (at solar noon) being a few 10^6 molecule cm^{-3} ($\sim$0.1 pptv); OH reacts very quickly on surfaces, and will not be transmitted through sample lines; OH concentrations vary significantly on short spatial scales; OH lifetimes are very short ($\sim$1 s), and thus any measurement technique must be in situ, have excellent sensitivity, temporal and spatial resolution, and not be subject to interference from other species. For OH, in common with many other field determinations, it is this last criterion, namely the elimination of interferences, that is the most difficult to achieve. Laser-induced fluorescence (LIF) was suggested in 1972 as a suitable method for OH detection (see Chapter 4 for further details). However, while using this method an interference plagued OH measurements for many years, namely the laser photolysis of O_3 to generate $O(^1D)$, with subsequent reaction of $O(^1D)$ with atmospheric water vapour to generate OH. The first results in 1974 gave a maximum [OH] of 1.5×10^8

molecule cm^{-3}, with nighttime values of 5×10^6 molecule cm^{-3}, and at the time these were accepted, as model calculations of OH were subject to very large errors because kinetic data for key reactions relevant to OH were not accurately known. In the mid-1980s measurements by several instruments using LIF during the NASA GTE/CITE campaign were later discredited, with an evaluation panel concluding that OH concentrations had not been measured! The laser-generated interference is now well recognised, and current OH field instruments using LIF rely on a different excitation scheme and pressure regime, and are free from such interferences. Some thirty years or more after the first OH measurements, several methods are now routinely used to measure OH (Table 1.1), with reported accuracies as good as 10%. An important lesson was learnt by the entire community of the dangers of artefact signals, but this has led to the emergence of very rigorous acceptance criteria for new instruments to measure OH.

The concentration of trace species varies with latitude, longitude, altitude and time of day, sometimes on very short scales, and a wide range of instruments deployed on a variety of measurement platforms is required. The atmosphere displays a wide range of temperatures (from +50℃ (323 K) at the surface down to −130℃ (143 K) in the mesosphere), pressures (decreasing exponentially from 1 atm at sea level to 10^{-6} atm at 100 km above sea level), solar intensity and meteorological conditions (wind, snow, hail), and many practical and logistical barriers must be surmounted even to attempt a measurement.

1.1.5 Comparison with calculations from numerical models

Our understanding of atmospheric chemistry is manifested through atmospheric models which describe the main physical and chemical processes that control how trace gases and particulate matter are distributed spatially on local, regional and global scales, and also as a function of time. It is natural therefore to compare field measurements of atmospheric composition on a variety of temporal and spatial scales, using a plethora of instrumental techniques and platforms that are the subject of this book, with the calculations of numerical models. The field measurement alone of a trace species provides only limited information. Atmospheric models are used for a variety of purposes, for example to calculate future trends in greenhouse gases, such as methane, that control future changes in atmospheric temperature, or substances harmful to health, such as ozone, to define future air quality. Policy makers make extensive use of model predictions to define emission reduction legislation, which can cost the economy billions of pounds to implement. Whatever the application, the output of the desired model is only as good as the input ('garbage in garbage out'), and as one of the model inputs is the chemical mechanism used to describe atmospheric transformation, it must be as accurate as possible.

Often the major degradation pathways are not known due to an absence of kinetic data on rate coefficients or product branching ratios, and these parameters have to be estimated using structure–activity relationships or numerical calculation (for which thermodynamic data are often required). In order to be confident about the ability of a model to calculate future atmospheric composition, it is important to design ways to test the accuracy of the mechanism used, and to identify which parts of it require

Table 1.1 Analytical techniques for selected classes of neutral atmospheric trace constituents

Class of constituent	Species	Analytical technique (chapter number)
Hydrogen compounds	OH	Laser-induced fluorescence (4), UV-visible DOAS (3), chemical ionisation mass spectrometry (5), far-IR emission spectroscopy (1)
	HO_2	Fluorescence (after conversion) (4), matrix isolation electron-spin resonance (1), peroxy radical chemical amplifer (7), far-IR emission spectroscopy (1), microwave absorption spectroscopy (1)
	H_2O vapour	Fluorescence after photolysis (4), IR absorption spectroscopy (2), surface acoustic wave (1), cavity ring-down spectroscopy (3), dew/frost point hygrometer (1), Lyman-α UV absorption (3)
	H_2O_2	IR absorption spectroscopy (2), liquid chromatography (after conversion, 8), chemiluminescence and fluorescence (after conversion, 7)
	H_2	Gas chromatography, reaction with hot HgO bed, Hg determined photometrically (8), mass spectrometry (5)
Nitrogen compounds	NO	IR absorption spectroscopy (2), UV-visible DOAS (3), fluorescence (4), chemiluminescence (7), solid-state sensors (1)
	NO_2	IR absorption spectroscopy (2), UV-visible DOAS (3), laser-induced fluorescence (4), chemiluminescence (after conversion, 7), sensors (after conversion, 1)
	NO_3	UV-visible DOAS (3), matrix isolation ESR (1), cavity ring-down spectroscopy (3), laser-induced fluorescence (4)
	N_2O_5	Cavity ring-down spectroscopy (3) or LIF after thermolysis to NO_3 and NO_2 (4)
	HONO	UV-visible DOAS (3), fluorescence (after conversion, 7)
	HNO_3	IR absorption spectroscopy (2), fluorescence (after conversion, 4), chemical ionisation mass spectrometry (5)
	Sum NO_y	Fluorescence (after conversion, 4), chemiluminescence (after conversion, 7)
	HO_2NO_2	Fluorescence (after conversion, 4), chemical ionisation mass spectrometry (after conversion, 5)
	Alkyl nitrates	Fluorescence (after thermolysis and conversion, 4), Chemical ionisation mass spectrometry (5), chromatography (8)
	PAN	Chemical ionisation mass spectrometry (5), gas chromatography with electron-capture detection (8)
	Other PANs	Chromatography (8)
	N_2O	IR absorption spectroscopy (2), gas chromatography (8)
	NH_3	IR absorption spectroscopy (2), UV-visible DOAS (3)
Halogenated compounds	HCl, HBr, HF	IR absorption spectroscopy (2)
	IO, I_2, OIO	UV-visible DOAS (3)
	BrO	UV-visible DOAS (3), fluorescence (after conversion, 4)
	OBrO	UV-visible DOAS (3)
	ClO	UV-visible DOAS (3), far-IR absorption spectroscopy (1), fluorescence (after conversion, 4)

Table 1.1 (Continued)

	OClO	UV-visible DOAS (3)
	$ClONO_2$	Fluorescence (after thermolysis and conversion, 4)
	ClOOCl	Fluorescence (after thermolysis and conversion, 4)
	Organic halocarbons	Gas chromatography, mass spectrometry (5, 8)
Volatile organic compounds	CH_4	IR absorption spectroscopy (2), chromatography (8), cavity ring-down spectroscopy (2, 3)
	Non-methane hydrocarbons	IR absorption spectroscopy (2), UV-visible DOAS (3), chemical ionisation mass spectrometry (5), chemiluminescence (alkenes, 7), gas chromatography (8), cavity ring-down spectroscopy (2)
	CH_2O	IR absorption spectroscopy (2), UV-visible DOAS (3), fluorescence after conversion (4), chromatography (8)
	Oxygenated compounds (carbonyl, alcohols, carboxylic acids)	UV-visible DOAS (3), chemical ionisation mass spectrometry (including proton-transfer mass spectrometry) (5), chromatography (8)
	Organic peroxides	Liquid chromatography (after conversion, 8), fluorescence (after conversion, 7)
	CO	IR absorption spectroscopy (2), fluorescence (4), chromatography, reaction with hot HgO bed, Hg determined photometrically (8), solid-state sensors (1)
	CO_2	IR absorption spectroscopy (2), cavity ring-down spectroscopy (2) chromatography (8)
Polyaromatic hydrocarbons	PAH	Chromatography (after extraction, 8), fluorescence (naphthalene, 4)
Sulphur compounds	HCN	IR absorption spectroscopy (2)
	OCS	IR absorption spectroscopy (2), gas chromatography (8)
	CS_2	UV-visible DOAS (3), gas chromatography (8)
	SO_2	UV-visible DOAS (3), fluorescence (4)
	H_2SO_4	Chemical ionisation mass spectrometry (5)
	DMS (dimethyl sulphide)	Chemical ionisation mass spectrometry (5), chromatography (8)
Peroxy radicals	Sum RO_2	Peroxy radical chemical amplifer (7), matrix isolation ESR (1), chemical ionisation mass spectrometry (after conversion, 5, 7)
	$CH_3CO{\cdot}O_2$	Matrix isolation ESR (1)
Ozone	O_3	IR absorption spectroscopy (2), UV-visible DOAS (and non-dispersive) (3), chemiluminescence (7), solid-state sensors (1), electrochemical (7)
Elemental mercury	Hg	Fluorescence (4), cavity ring-down spectroscopy (3)
Aerosols	Chemical composition	Filter samples (6), volatility (6), mass spectrometry (6, 5, 8)

updating, through improved kinetic measurements in the laboratory. A powerful method is to use the model to calculate atmospheric composition for our current atmosphere or past trends in key species, and compare these to field measurements. Care must be taken to choose an appropriate model to compare with an appropriate measurement (usually defined by the lifetime of the trace gas or aerosol).

A commonly used molecule to test the accuracy of chemical mechanisms in a variety of environments is the hydroxyl radical, OH. Its short chemical lifetime, due to its high reactivity, means that its budget (and hence concentration) is only controlled by local in situ chemistry, and not by transport processes. Hence zero-dimensional or box models, which consider a box in which the sample is well mixed, can be used to describe the chemistry of OH under specific conditions without having to incorporate transport into or out of the box. The rate of change of OH concentration is given by:

$$\mathrm{d[OH]/dt} = \mathrm{P(OH)} - \mathrm{L(OH)} = \mathrm{P(OH)} - [\mathrm{OH}]\Sigma_i k_{i+\mathrm{OH}}[i] \tag{1.2}$$

where $[i]$ is the concentration of a given sink (photolysis or heterogeneous losses of OH are not important loss terms for OH). OH is generated and destroyed quickly and so its concentration rapidly assumes that of a steady state ($\mathrm{d[OH]}/\mathrm{d}t = 0$) with a balance between production and loss, and hence

$$\begin{aligned} \mathrm{P(OH)} &= \mathrm{L(OH)} \\ [\mathrm{OH}] &= \mathrm{P(OH)}/\Sigma_i k_{i+\mathrm{OH}}[i] \end{aligned} \tag{1.3}$$

Comparison of measured values in the field with OH calculated from Equation 1.3 will provide a sensitive test of how well a mechanism has described the production and loss terms (i.e. the chemistry!). The lifetime of OH is less than 1 second, and hence is made within a few metres of the inlet of an instrument designed to measure OH. As input, the model requires the concentration of all sinks, $[i]$, to be measured, as well as all terms contributing to the production term P(OH) (including radiation), ideally within a few metres of the OH instrument inlet. In reality many of the species required as input either cannot be or are not measured (e.g. $O(^1D)$, HO_2), and must be calculated, but themselves are coupled closely to the concentrations of many other species. In general, if the number of chemical species in the mechanism is N, then N differential equations must be solved simultaneously to calculate [OH]. There are now many examples of model comparisons with OH measurements from a variety of platforms. The agreement is variable, depending on the environment under study, pointing to inadequacies in our understanding of the underlying chemistry. For remote environments, where the air is clean and relatively few reactions and species are required to describe the chemistry, there is now good agreement between measured OH and concentrations predicted using box models, as shown in Figure 1.5 for the data taken during a field campaign at Mace Head, Ireland, in 2002. If the chemical mechanism were to be used in a chemistry-transport model (CTM) to predict the future composition in remote areas (e.g. of ozone), then we can be more confident that the results are accurate.

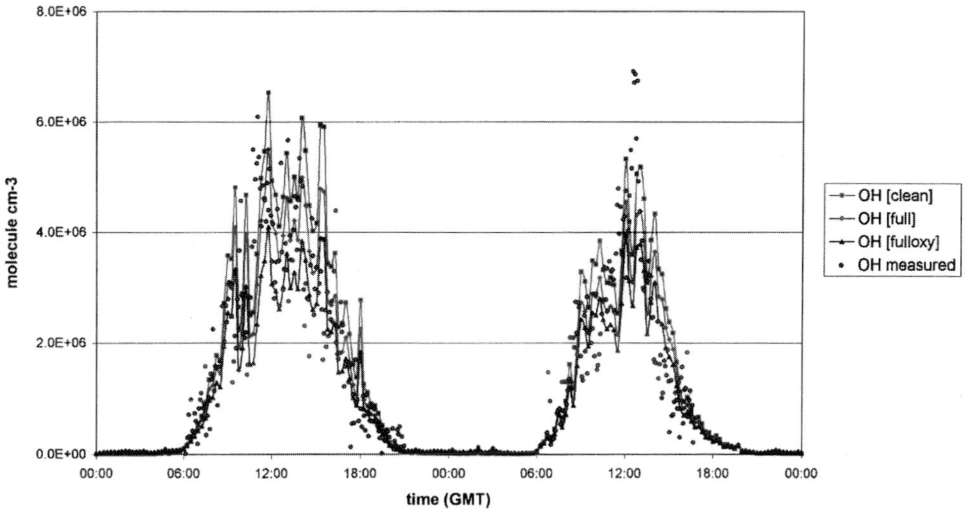

Figure 1.5 Model–measurement comparison for OH radicals for 9–10 August 2002 during the NAMBLEX campaign at Mace Head, Ireland. The measurements were made in clean westerly air using the Leeds FAGE instrument (laser-induced fluorescence at low-pressure), and have been averaged up to 15 min time intervals to facilitate comparison. For the model calculation, mechanisms with different degrees of chemical complexity were used. The 'clean' model was constrained to reaction of OH with CO and CH_4 only, the 'full' model also included 23 hydrocarbons and chloroform, and the 'fulloxy' model also included 23 hydrocarbons, chloroform, acetone, acetyaldehyde and methanol. The calculated OH concentration decreases as more and more of its sinks are included in the mechanism.

1.2 Scope, structure and content of this book

1.2.1 Scope and structure of the book

This book will primarily address instruments used to make quantitative measurements of the chemical composition of the atmosphere. Instruments to measure the structure of the atmosphere – for example, micro-meteorological determinations of the instantaneous three-dimensional wind velocity (although important in order to allow interpretation of chemical data) or the measurement of deposition velocities to the earth's surface – are not covered. However, as the majority of atmospheric processes in the atmosphere are initiated by the absorption of sunlight, at a rate dependent upon the rate of photolysis of certain trace gases, the book includes a chapter on techniques to measure photolysis frequencies directly or to measure radiative properties of sunlight to enable the calculation of photolysis frequencies. Also, the book will concentrate on real-time in situ methods, or those sampling a long path through the atmosphere, and that have short integration times on the order of minutes or less. Methods that require the accumulation of a sample, perhaps via a denuder, will typically not be covered.

The chapters are organised chapter-wise according to experimental technique (e.g. absorption, fluorescence, chromatography, mass spectrometry, etc.), rather than by class of species measured. It is realised that instruments may be highly specific for the measurement of one species only – and under a particular technique there may be a large number of instrument types. Clearly it is not possible to describe all of these, and examples selected will be the most commonly used variants. Each chapter will contain the following common elements:

(1) The fundamental physical principles of each technique will be covered, together with a description of apparatus commonly deployed in the field, which may include a variety of platforms to probe different regions of the atmosphere. Sampling is given particular attention, as reactive or labile species may change their chemical composition during sampling and prior to detection.
(2) The detection limits, accuracy, precision, sources of noise, degree of selectivity and possible interferences for monitoring of different species using a given technique will be discussed.
(3) A flavour of field-data obtained using the technique, including examples of time-series, and key breakthroughs that have resulted from field observations of key speices. In some cases comparison with model calculations are included to demonstrate how our understanding of the atmosphere can be probed via the use of field measurements.
(4) Use of the technique, where appropriate, in routine air quality monitoring, perhaps as a network of instruments that are relatively inexpensive to set up and operate, and which can be run continuously with little or no user interaction.
(5) A discussion of quality control, which is of paramount importance as field observations are increasingly used to support legislation controlling emission of substances harmful to health or the environment. Calibration of instruments and intercomparison of methods are important components of quality control and are covered extensively in each chapter.
(6) A description of the future requirements and the latest developments involving a given technique.
(7) Suggestions for further reading in general texts, as well as a comprehensive bibliography of the research literature covering the historical development of the technique, descriptions of state-of-the-art instruments used worldwide, and examples of their deployment and results obtained. It is hoped that the book will provide a one-stop shop to quickly obtain information, both basic and in detail, for all the commonly used analytical methods for atmospheric measurement.

Of course, a particular atmospheric species can sometimes be measured by several methods, with some being more fit-for-purpose than others for a particular application or deployment platform. The organisation of the book via technique is less convenient for readers who are interested in learning about alternative methods for detection of a given species. In Section 1.3.3, atmospheric measurements are organised according to class of species, with detailed cross-referencing to sections of various chapters of the book that cover detection of that species via various methods. A brief discussion of the pros and cons of the methods available is given for some key species.

1.2.2 Previous texts describing methods for determining atmospheric composition

There are very few books dedicated to the measurement of atmospheric composition, and even fewer that are current at the time of writing. A few books in edited series volumes cover measurements in the atmosphere with an emphasis on the techniques used. For example, *Measurement Challenges in Atmospheric* Chemistry (Newman, 1993) is the result of a symposium at an American Chemical Society meeting held in 1990, and contains a series of technically detailed chapters organised according to the measurement of a variety of key trace species. *Selective Detectors* (Sievers, 1995), applicable to the determination of atmospheric composition is also rather specialised. *Handbook of Air Pollution Analysis* (Perry & Young, 1977) is one of the few books dedicated to the measurement of atmospheric trace gases. The structure is largely organised by class of species, but there are chapters on general issues such as inlets and sampling methods, as well as remote monitoring techniques and the planning and execution of an air pollution study. However, the book was published almost 30 years ago, and hence the methods described are somewhat antiquated, and many species at that time had not been recognised as important to measure in the atmosphere. In 2003, a synthesis of a decade of tropospheric chemistry research was published by the International Global Atmospheric Chemistry Project (IGAC, report completed in 2001) (Brasseur *et al.*, 2003), and contains a succint discussion of advances in field measurements in both the gas and aerosol phases, together with the results from field campaigns in the context of atmospheric photooxidants and tropospheric aerosols. Methods for measurements of atmospheric trace gases are reviewed by several authors for selected species in Calvert (1994).

There are also a number of advanced undergraduate textbooks that contain very useful descriptions of the fundamental physical methods of techniques used to make composition measurements, together with some details on the instruments and platforms themselves, and examples of field data and their interpretation (Brasseur *et al.*, 1999; Finlayson-Pitts & Pitts, 2000). More specialised review articles are cited in individual chapters of this book, but there are a number of articles that review a wide range of instrumental methods rather than just for individual techniques or class of species. A review of a very wide range of instrumentation and measurement techniques for ground-based and airborne field studies of gas-phase tropospheric chemistry can be found in Clemitshaw (2004). The coverage of techniques and their deployment is very comprehensive indeed, but due to the nature of the review, each method is covered fairly briefly and does not discuss the fundamental principles upon which the methods are based. In a short review, Roscoe and Clemitshaw provide a selective view of the past, present (1997) and future measurement techniques in gas-phase tropospheric chemistry (Roscoe & Clemitshaw, 1997). As discussed in Section 1.8, there have been a number of large intensive field campaigns, and for some the detailed results are disseminated in the open literature via special issues of journals, which may contain up to 20–30 papers in succession, many of which contain quite detailed accounts of field instruments for composition measurements (see Further reading). *Encyclopaedia of Atmospheric Sciences* (Holton *et al.*, 2003) (2780 pages) is a comprehensive collection of review articles that covers all of atmospheric science, and several of the articles discuss methods for atmospheric composition measurements (for example, (Clemitshaw, 2003; Heard, 2003).

1.2.3 Content of the book: Summary of individual chapters

Chapter 1 provides an introduction to the determination of atmospheric composition, and an overview of the various techniques used to measure trace gases, aerosols and radiation. As each chapter covers the commonly used techniques, it is inevitable that some, although important, have not been covered, and this chapter (Section 1.5) provides a brief description of methods not covered elsewhere in the book. In order to assist navigation through the remainder of the book, and for reference, detailed tables are provided according to experimental technique and by class of trace species measured. The various platforms available for the deployment of instruments are described, including satellites, with examples. Other issues, which are recurring themes in later chapters, are also briefly introduced, for example the calibration of instruments, detection limits, instrument accuracy and precision, and intercomparison of various techniques. The general requirements when designing a major field campaign are also discussed, examples given of major field campaigns mounted around the time of writing, and a case study is provided of the instrument portfolio deployed during a campaign at Mace Head, Ireland. A discussion is also given of long-term measurements, as well as monitoring networks, which provide the backbone of data which regulatory bodies rely on heavily to direct future policy and that provide evidence about changes in our atmosphere and climate. There is also discussion of the use of simulated environments in outdoor environmental chambers, where the composition of the sampled air can be carefully controlled. Finally, the chapter concludes with a brief discussion of the future directions for field measurements of atmospheric composition.

Chapters 2–4 cover measurement techniques that involve excitation of a molecule to an excited state through absorption of a photon of light. The source of radiation can be natural, for example the sun, moon, stars, or the earth itself, or artificial, for example lamps or lasers. In Chapters 2 and 3 the concentrations of atmospheric species are derived using the Beer–Lambert law from measurement of the amount of light absorbed. In Chapter 2, Alan Fried and Dirk Richter describe how the excitation of vibrations within molecules via the absorption of infrared (IR) radiation is used to quantitatively measure the concentration of atmospheric molecules. The method is widely applicable over a range of spatial scales. In Chapter 3, John Plane and Alfonso Saiz-Lopez describe how the excitation of electrons within molecules via the absorption of visible and UV radiation is used to measure the concentration of the molecules. Although similar in concept, measurement of the absorption of IR or UV/visible radiation requires very different approaches. The converse of absorption is emission of radiation from excited states, and, in Chapter 4, Ezra Wood and Ronald Cohen describe the use of fluorescence from electronically excited atomic or molecular states to monitor atmospheric species. A major advantage is the observation of photons on a null background, rather than the measurement of a very small absorption of radiation as in some cases, but set against this is the absence of a convenient fundamental law to relate signal to concentration, and calibrations must be carried out. The use of lasers as radiation sources is commonplace, their narrow spectral linewidths and wavelength tunabilty making them ideal for highly selective molecular excitation. The observation of IR or microwave emission from vibrationally or rotationally excited molecules as a method of monitoring is used from space-borne or high altitude platforms, and is covered briefly in Section 1.5.4. Furthermore, Section 1.4.8 discusses the use of satellites for global measurements of trace species, using remote sensing techniques based on differential optical absorption spectroscopy of scattered or transmitted sunlight, or the detection of radiation in the IR and far-infrared/microwave regions emitted by excited molecules.

Chapter 4 is restricted to the generation of fluorescence via direct optical excitation, rather than fluorescence (chemiluminescence) generated as a result of an electronically excited state being produced as the product of an exothermic chemical reaction or an inelastic collision involving energy transfer from another molecule. The chapter does not consider fluorescence detection following a previous derivatisation or chromatographic separation of the desired analyte (covered in Chapters 7 and 8, respectively), but does discuss fluorescence detection following simple chemical conversion, for example HO_2 detection via $HO_2 + NO \rightarrow OH + NO_2$ with detection of OH by LIF, or thermolysis, for example $N_2O_5 \xrightarrow{\Delta} NO_3 + NO_2$ with detection of NO_2 by LIF. N_2O_5 has also been monitored in the atmosphere via detection of the NO_3 product, using cavity ring-down spectroscopy (Chapter 3). Free-radical species, for example BrO and ClO, have been quantified in the atmosphere through their conversion to Br and Cl atoms, with detection by resonance fluorescence at vacuum UV wavelengths (Chapter 4).

Chapter 5 by Jonathan Williams is dedicated to the monitoring of atmospheric species following their ionisation, with selectivity achieved using subsequent separation by mass prior to detection. Ions can be spatially controlled and detected with exquisite sensitivity, and mass spectrometry is the most widely applicable of all methods. The chapter is primarily concerned with methods of ionisation and the subsequent separation by mass, and the detection of ions, rather than with any prior separation, for example via chromatographic methods (Chapter 8), that can lead to additional selectivity. Ionisation can be achieved through collisions with electrons or ionising radiation, or through ion–molecule reactions with positively or negatively charged reagent ions, with detection of either positively or negatively charged product ions. An important reagent ion is the hydronium ion (H_3O^+), which can be used to protonate a wide range of species. Very high resolving power mass spectrometers can distinguish molecules containing different isotopes within a given molecule, and measurement of isotope ratios enables much to be learnt about the sources and chemistry of that molecule. Indeed isotope ratios for molecules trapped within ice-cores are used as proxies for the temperature of the atmosphere going back millions of years.

In Chapter 6, by James Allan and Hugh Coe, methods for the measurement of the physical and chemical properties of atmospheric aerosols are covered. Aerosols come in a wide variety of sizes, shapes and chemical composition, and in order to understand their toxic properties, their ability to absorb or scatter radiation, or their ability to act as sinks (via surface uptake) for gaseous intermediates, it is crucial to have a detailed knowledge of the number density and chemical composition as a function of size. Understanding, and hence being able to predict, how aerosols are formed, and subsequently grow, with associated changes in composition, is crucial to understanding their role in, for example, global warming and the toxicity of urban air pollution. A focus of Chapter 6 is the discussion of mass spectrometric methods to determine the chemical composition of size-segregated aerosols. A major difference compared with the detection of gaseous species (Chapter 5) is the requirement to volatilise the aerosols prior to or at the same time as ionisation, and to take account of the different efficiencies of volatisation as a function of composition.

As mentioned above, electronically excited species can be generated as a result of exothermic chemical reactions, in either the gas or the liquid phase, or occurring upon surfaces, or via energy transfer from collisions with other molecules, and Chapter 7, by Andrew Weinheimer, describes methods relying on the observation of chemiluminescence. As for directly excited fluorescence, the method is very sensitive on account of a null background, but has the advantage of not requiring a light source, rather the mixing of another reagent with the analyte of interest. Chemiluminescence detectors are very common, due to their relative simplicity and low expense, and are deployed very widely, for example in the detection of nitric oxides. Chapter 7 also covers techniques that convert the analyte to a species that can be detected via other methods, but mainly restricted to chemiluminesence or fluorescence. The conversion may be chemical, either in the gas phase (e.g. peroxy radicals via a chemical amplifier) or in the liquid phase following dissolution in a scrubber (e.g. HCHO), or following photolytic conversion (e.g. $NO_2 \rightarrow NO$). Chemical conversion of analytes is also covered in other chapters; for example, HO_2 can be converted easily to OH via the addition of NO, and thus HO_2 detection is considered together with OH in Chapter 4 on fluorescence methods.

A well-established method of separating analytes is chromatography, and when combined with a variety of detectors, is widely used for measurements of atmospheric composition. In Chapter 8, Jacqueline Hamilton and Alastair Lewis describe a number of gas and liquid chromatographic systems in common use, giving a comprehensive account of practical issues for their operation, including sampling methods, separation strategies, types of detectors, instrument calibration and quality control. For some detection methods, for example mass spectrometry, the details of the fundamental principles and field instrumentation are covered in other chapters, and the emphasis in Chapter 8 is on the separation of analytes and the coupling between chromatographic column and detection system. As well as the sampling from the gas phase, the composition of aerosols can also be analysed using chromatographic methods, with collection of material on filter paper by high-volume sampling, followed by extraction or sample volatisation. For polluted environments, containing many thousands of volatile organic compounds, conventional single-column chromatography is only able to separate and identify a small fraction of species. Developments in the use of multidimensional chromatography (the use of more than one column) to further separate complex mixtures, are given, in particular when coupled to a time-of-flight mass spectrometer, which gives an excellent time response and enables a further dimension of separation to be achieved. Chromatographic methods have been applied widely to study a range of atmospheric problems, and a selection of examples is given.

The chemistry of the atmosphere is driven to a large extent by reactive species photochemically generated following the interaction of atmospheric constituents with radiation of sufficient energy to break chemical bonds. An example is the hydroxyl radical, the most important atmospheric oxidant, which is mainly generated by the photolysis of ozone to generate electronically excited oxygen atoms, followed by reaction with abundant hydrogen atom donors (e.g. H_2O vapour). The composition of the atmosphere is greatly influenced by photochemical processes, and in Chapter 9, Andreas Hofzumahaus describes techniques used to determine the rate (or frequency) at which molecules are photodissociated in the atmosphere. The photolysis frequency of a molecule at a given wavelength (to generate a specific product) is dependent upon the product of three main parameters: (1) the absorption coefficient, (2) the photodissociation quantum yield to generate that product, and (3) the amount of light available to excite the molecule in the first place. It is necessary to know these for all wavelengths over which the molecule absorbs light, and both (1) and (2) are also pressure and temperature dependent. For many molecules (1) and (2) have been determined in the laboratory, and the amount of direct radiation from the sun can be calculated under clear sky conditions for any location/altitude/time on the planet. However, molecules are photolysed not only by direct solar radiation but also by radiation scattered by air molecules, aerosols and clouds, or by radiation reflected from the ground, and (3) is therefore very difficult to calculate. In Chapter 9 several methods to determine the rate of photolysis are described. Measurement is either taken directly using an actinic flux filter radiometer, equipped with a suitable optical filter/detector combination, or by a chemical actinometer, whereby the concentration of a photoproduct is measured as a result of photodissociation of a known precursor. Alternatively, the solar flux can be measured as a function of wavelength (parameter (3) above) and values of (1) and (2) – measured in the laboratory – are used to calculate the rate of photolysis. A number of practical systems are described, and where appropriate, their calibration, together with their deployment on various platforms in atmospheric chemistry field experiments, is also described. The temporal and spatial variations of photolysis frequencies are illustrated for a number of important photolabile species, together with the controlling influence these rates have on the concentration of photochemically generated intermediates, for example the hydroxyl radical.

1.3 The measurement of atmospheric composition

1.3.1 Units of concentration

Concentrations of the most important atmospheric trace gases range over more than 10 orders of magnitude. Concentrations are expressed in several units, and it is useful to define their relationship. The most common method to report levels is the mixing ratio, which is the ratio of the number of molecules of the species of interest to the number of molecules in the volume of air of interest. An advantage of using mixing ratios is that even if the temperature or pressure of the air changes, the mixing ratio is conserved. Examples of common units of mixing ratio are the following: ppmv (parts per million by volume) – 1 in 10^6 molecules of air; ppbv (parts per billion by volume) – 1 in 10^9 molecules of air; pptv (parts per trillion by volume) – 1 in 10^{12} molecules of air; and ppqv (parts per quadrillion by volume) – 1 in 10^{15} molecules in air.

Another unit of concentration, often used by legislators, is micrograms of species per cubic metre of air ($\mu g\,m^{-3}$), but for comparisons it is necessary to correct values to a standard temperature and pressure. For NO_2, at 101.3 kPa (1 atm at sea level), and 293 K, 1 ppb is equivalent to 1.913 $\mu g\,m^{-3}$, whereas for NO it is 1.248 $\mu g\ m^{-3}$ as it is a lighter molecule. It is not possible to express particulate matter concentrations as a mixing ratio, and their measurements are given in $\mu g\,m^{-3}$.

The number density of an atom or molecule (X) in the gas phase, commonly referred to as concentration, [X], is usually reported in units of molecules cm^{-3}, and is linked to the total pressure, p (in Pa), and its mixing ratio, χ_X, by the expression:

$$[X] = N_a \chi_X (p/RT) \times 10^{-6} \tag{1.4}$$

where $N_a = 6.022 \times 10^{23}$ mol^{-1} is the Avogadro constant, the factor 10^{-6} is to convert from the SI unit of m^{-3} to cm^{-3} and R is the gas constant 8.314 J mol^{-1} K^{-1}. At sea level, using air as an example (i.e. mixing ratio $\chi_{air} = 1$), $p = 10^5$ Pa, $T = 298$ K, gives [air] $= 2.43 \times 10^{19}$ molecule cm^{-3}. As another example, a measurement of HO_2 radicals at high altitude ($p = 0.25$ atm $= 2.5 \times 10^4$ Pa, $T = 200$ K) yielded a mixing ratio of $\chi_{HO_2} = 4$ pptv, which corresponds to a concentration of:

$$[HO_2] = 4 \times 10^{-12} \times N_a (p/RT) \times 10^{-6} = 3.62 \times 10^7 \text{ molecules cm}^{-3}$$

1.3.2 Selection criteria for instruments

There are a number of selection criteria to consider when it comes to choosing the most suitable instrument for a given task.

(a) The universality of the method is. It is a clear advantage to be able to measure several species simultaneously.
(b) What is the sensitivity of the instrument? Is this significantly less than the atmospheric concentration of the desired analyte?
(c) Is the instrument specific for a given species, and are there interferences from other species? Optical methods have an advantage in that spectra can be used to fingerprint a given molecule, but it needs to be proven that the spectra are due to ambient sources of the species and not generated within the instrument.
(d) Sampling efficiency needs to be good, and well understood, as some highly reactive molecules (e.g. free-radicals like OH) are destroyed on surfaces, and others (e.g. NH_3, HNO_3) are sticky, and transmission may vary according to the age and condition of the inlet. The measurement of HNO_3 is an example of where specialised inlets have had to be designed to tackle this problem, especially for aircraft sampling (Ryerson *et al.*, 1999).
(e) The accuracy and precision of the instrument is important (see Section 1.6.1), and there are different requirements depending on the application. High accuracy is required in order to bring together data from a global network of monitoring stations, whereas high precision is required to measure fluxes of gases using very short integration periods or to monitor long-term trends for an individual trace species.
(f) Integration or averaging period (response time) is another important factor, and the requirements vary widely, depending on the chemical lifetime and hence temporal variability in the atmosphere. Hydroxyl radical measurements ideally require time responses of a few seconds, whereas measurements of CFCs can have much longer averaging periods.

(g) Spatial resolution is one of the most important criteria, again because of the widely differing atmospheric lifetimes and hence spatial variability away from source. If one is interested in understanding the formation and destruction of the OH radical, for comparison with a constrained box-model, then ideally all the sources and sinks should be measured as close to the OH instrument inlet as possible. If, however, the aim is to compare with the calculations of a global CTM, with a grid-square size of 100 km × 100 km, then a measurement representative of the average within that footprint is required, perhaps provided from a satellite or a remote sensing device, a mobile laboratory or aircraft, or a network of ground-based sensors. The two major classes of methods are in situ and remote sensing, the former sampling from a single point in the atmosphere (normally an inlet but this could be a small volume defined by a folded optical path), the latter normally using a distant source of light, for use in absorption spectroscopy. As a general rule, remote sensing methods have inferior spatial resolution. In the limit, one only obtains a total column measurement of the species of interest between the light source (e.g. the sun, or the earth's surface for scattered light) and the detector. However, in some cases the spatial resolution of remote sensing methods can be very good, for example in Light Detection and Ranging (LIDAR) methods (Section 1.5.1) when a pulsed source of light is used. Care must be taken when comparing measurements of the same species taken with instruments that sample with different spatial resolutions, or deciding upon the optimum dataset to constrain a photochemical model. For example, during the NAMBLEX field campaign (Section 1.8.2), NO_3 radicals were measured using an in situ method (broadband cavity ring-down spectroscopy, see Chapter 3), with the instrument housed on the roof of a container near the shore, and also by differential optical absorption spectroscopy DOAS (Chapter 3), using a path length of 2 × 4.2 km, with the retro-reflector on an offshore island, and the light source and spectrometer housed in a building on the shore close to the CRDS instrument. At night, concentrations of NO_3 from the DOAS instrument were considerably higher than from the CRDS instrument, but this observation could be explained by the emission of small concentrations of NO, locally, probably from the soil, that react quickly with NO_3, reducing its concentration, but which are not present over the open ocean.

(h) The method should be non-intrusive with respect to the operation of nearby instruments, not blocking other inlets from the prevailing wind. The structure supporting the instrument should also be taken into consideration. Remote sensing methods (e.g. with lasers or other light sources) are ideal in this regard but suffer from problems of spatial resolution.

(i) Weight, size, requirement of calibration, electrical power requirements, autonomy (does it need to be left for months unattended?) and ease of operation (are skilled operators always available?) are other operational criteria, and will vary mostly according to the deployment platform.

(j) Ease of data reduction is a key criterion. Having to wait months for validated data is a disadvantage.

(k) Is the instrument available commercially? The cost of an instrument is not just derived from the hardware but also from the work-years of effort required in its development, and in its upkeep and maintenance. If the running costs and requirement of a skilled operator are taken into account, a commercial instrument may be a better long-term solution, even if the performance in some of the other criteria is not quite as good. Commercial instruments will probably be more reliable than research-based state-of-the-art instruments, but this is not always the case! The detection limits for commercial instrumentation are improving all the time, although for some species, especially free-radicals, there are no commercial packages available. For monitoring networks, where high numbers of devices are needed, low cost, ease of use and reliability are necessary.

1.3.3 Instruments organised by classification of trace species

This book is primarily concerned with the analytical methods and instruments used to make measurements of atmospheric composition. The chapters are organised according to analytical technique, and cover many examples of species measured in the atmosphere. However, if it is desired to determine how to measure a given species in the atmosphere (or photolysis rates), the structure of the book does not allow for immediate comparison of the various methods available. The purpose of this section is to provide a single look-up table, with cross-referencing to individual chapters, to enable such a comparison to be made quickly. Also, for a given technique, Section 1.3.4 provides in a single table the range of species that have been measured in the atmosphere, again with cross-referencing to individual chapters. No attempt is made to suggest whether one method or another is better suited than another for measurement of a particular species, as the choice is driven by several criteria (see Section 1.3.2), and the optimum technique may vary according to the specific application. Individual chapters do provide examples of intercomparisons for a given species, and so a judgement can be made, for a given platform (e.g. ground-based), of the merits of particular methods. Section 1.5 discusses methods that are not covered in other chapters.

Table 1.1 lists analytical methods that have been used to measure individual species in the atmosphere, organised according to chemical family. The table is not intended to be exhaustive, and does not contain all the species measured in the atmosphere, and for each species, may not cover all techniques. It is hoped, however, that all the major species and how they are normally measured are covered. Satellite measurements and photolysis frequencies are not included in this table. The optimum technique for a given species is dependent upon the platform used, conditions encountered (e.g. altitude) and desired sensitivity, temporal and spatial resolution. The table is restricted to molecules that have actually been measured under ambient conditions rather than under artificial conditions in the laboratory, and does not cover methods analysing samples collected onto filters. Likewise, the methods listed for each molecule are only those that have the required sensitivity for ambient measurements. Finlayson-Pitts & Pitts (2000) also provide a listing organised by species measured and cover details of the major methods used.

1.3.4 Instruments organised by analytical technique

Table 1.2 separates analytical techniques used to make composition measurements in the atmosphere into their major classifications. For each overall classification (e.g. optical), the individual analytical methods are listed, together with the chemical species that have

Table 1.2 Analytical techniques for measurement of atmospheric composition and photolysis frequencies

Classification	Technique	Chapter	Species measured
Optical	Infrared absorption spectroscopy	2	CO_2, O_3, H_2O, N_2O, CO, CH_4, NO, NO_2, HNO_3, hydrocarbons, NH_3, CH_2O, H_2O_2, HCl, HBr, HF, HCN, OCS
	Differential optical absorption spectroscopy (DOAS)	3	NO_3, O_3, NO, NO_2, HONO, NH_3, SO_2, HCHO, CS_2, benzene, toluene, naphthalene, phenol, p-cresol, OH, ClO, OClO, BrO, OBrO, IO, OIO, I_2
	Cavity ring-down spectroscopy	2, 3	OIO, I_2, NO_3, CH_4, hydrocarbons N_2O_5 (after conversion), elemental Hg, H_2O, O_2
	Fluorescence methods including laser-induced fluorescence (LIF)	4	Direct OH, NO, NO_2, NO_3, SO_2, CH_2O, CO After conversion: ClO, BrO, $ClONO_2$, ClOOCl, HO_2, H_2O, NO_y, peroxy nitrates, alkyl nitrates, hydroxyl-alkyl nitrates, HNO_3, N_2O_5, elemental Hg
	Far-infrared emission spectroscopy	1	OH, HO_2
	Microwave/far infrared absorption spectroscopy	1	ClO, HO_2
Mass spectrometric	Ion impact	5	Universal technique, but in particular all types of VOCs
	Chemical ionisation mass spectrometry (including proton-transfer-MS)	5	OH, HNO_3, H_2SO_4, aldehydes, ketones, DMS, non-methane hydrocarbons, HO_2NO_2, alkyl nitrates, PAN, HO_2, sum RO_2, halocarbons
	Aerosol mass spectrometry	6	Chemical composition of aerosols as function of size
Chemical	Peroxy radical chemical amplifier (PERCA)	7	Sum $HO_2 + RO_2$
	Chemiluminescence	7	O_3, NO, NO_2, NO_y, isoprene
	Derivatization	7	H_2O_2, ROOH, CH_2O, HONO
Chromatographic	Gas chromatography	8	Permanent gases, CO, CO_2, CH_4, saturated and unsaturated hydrocarbons (aliphatic and aromatic), C_2H_2, DMS, CH_2O, aldehydes, ketones, alcohols, CFCs, halocarbons, N_2O, PAN, PPN, organic nitrates
	Liquid chromatography	8	PAH, peroxides, aldehydes, ketones, CH_2O, organic acids, biomolecules, proteins, inorganic ions

Table 1.2 (Continued)

Classification	Technique	Chapter	Species measured
Photolysis frequencies	Filter radiometry	9	$J(O^1D)$, $J(NO_2)$
	Spectral radiometry	9	$J(O^1D)$, $J(NO_2)$, J(HONO), $J(CH_2O)$, J value of any species subject to knowledge of absorption cross-section and quantum yield
	Chemical actinometry	9	$J(O^1D)$, $J(NO_2)$
Measurements from space	Satellites (various techniques)	1	Aerosols, O_3, N_2O, NO, NO_2, HNO_3, CF_3Cl, CF_2Cl_2, $CFCl_3$, other CFCs, O_3, $ClONO_2$, CO_2, CH_4, H_2O, BrO, OClO, SO_2, CH_2O, HF, HCl, CO, N_2O_5, ClO, O_3, O_2, $O_2(a^1\Delta)$, O_4 (O_2 dimer), HOCl
	Space shuttle, SpaceLab-3	1	O_3, NO, NO_2, N_2O_5, $ClONO_2$, HCl, HF, CH_4, CFCs, ClO, H_2O, OH
Other	Matrix isolation ESR	1	HO_2, sum of all RO_2, $CH_3CO \cdot O_2$, NO_3, NO_2
	Solid-state sensors	1	CO, O_3, hydrocarbons, NO_x, H_2O
	Electrochemical	7	O_3

been measured. For some measurements, only column amounts are measured, rather than vertically resolved or in situ measurements. Once again, the table is not intended to be exhaustive, rather indicating commonly used techniques and measured species. Individual chapters (indicated in the table) should be consulted for more comprehensive tables of equipment used and species covered. Clemitshaw (2004) provides a very comprehensive listing of methods organised according to type of analytical technique. For some species, several methods are available for their detection, with different sensitivities and accuracies, and a listing in this table does not imply a technique as the optimum method for detection. The table is restricted to techniques used to make ambient measurements in the atmosphere, and does not cover techniques that have only been demonstrated to have excellent sensitivity in the laboratory for specific species.

1.4 Instrument platforms

For many aspiring field scientists (the author included), the transition from 'the laboratory' to 'the field' comes as somewhat of a shock. Suddenly, a prized instrument that has proven itself in the comfortable confines of the laboratory must be exposed to the harsh reality of the atmosphere in order to make a real measurement. Instruments require a platform on which to operate, to provide physical support, electrical power, protection from the elements (in particular the ingress of water), protection from overheating or distortion due to changes in pressure, and must enable unimpeded sampling which does not perturb the composition of the surrounding atmosphere. In this section, various types of platform are discussed, ranging from the ground to space, and both fixed and mobile.

1.4.1 Ground-based platforms, including vehicle-based mobile laboratories

The simplest platform is a building or fixed structure on the ground, with an inlet protruding from the structure or atop a neighbouring tower via a sampling line. There are many ground-based stations situated around the world. Some are well-established, highly instrumented, permanent stations, with well-documented meteorological conditions, whereas others are simply a source of power and possibly a few buildings that provide basic facilities for small, intensive field measurements. There is a trend for larger instruments to be housed within custom-fitted, air-conditioned, standard size (20 feet, 6 m, in length) shipping containers, with attendant towers or sampling structures, and the requirements of the site are only a flat area of land together with electrical power (and for some instruments, running water). This type of arrangement has the added advantage of faster set-up and tear-down times. The UK Universities Facility of Atmospheric Measurements (UFAM), http://www.env.leeds.ac.uk/ufam/, consists of a distributed set of ground-based and specialised airborne instruments designed to make in situ measurements of chemical composition as well as small- and meso-scale measurements of physical parameters (e.g. wind profiles, turbulence) in the atmosphere. Many of the instruments are housed in shipping containers for ease of transportation. Electrical power is usually not a limiting factor at ground-based sites, with the exception of extremely remote locations where all power has to be provided through diesel-powered (or possibly wind- or solar-powered) generators. Often a common sampling height for instrument inlets (e.g. 10 m above the ground), or a common sampling line (with several 'Tees' going to different instruments), is used during intensive campaigns involving several instruments, although this is not possible for some bulky instruments with a requirement for short inlets (e.g. OH measurements).

It is important that there is a good record of the local meteorology, including micrometeorological measurements of local mixing or internal boundary layers. Detailed meteorological measurements are essential during all campaigns, in particular if there are power-generating diesel generators, whose exhaust must be downwind. To sample very clean air, which is typical of the background atmosphere free of human influence, one normally has to travel considerable distances to very remote locations where field stations are situated. Examples of remote monitoring sites are given in Section 1.7.2.2. Environments include semi-rural, suburban, urban, desert, coastal, polar, oceanic (island observatories), forested, mountainous, from sea level up to 3580 m (e.g. the Jungfraujoch Sphinx Observatory in Switzerland). High-altitude sites are ideal to avoid interference from water vapour or boundary layer pollution. Figure 1.6 shows photographs of the Mace Head field site, Ireland, and the Clean Air Sector Laboratory at the Halley Station, Antarctica, and Table 1.3 contains examples of locations where intensive field campaigns have been mounted.

A number of ground-based mobile laboratories have been developed for the in situ measurement of atmospheric composition whilst on the move. There is a requirement of onboard power to supply to instruments, sufficient air-conditioning to keep instruments cool, a sampling system able to sample the surrounding air (with the relative velocity of the vehicle) without contamination from the vehicle's exhaust, and fast-response ($\sim$1 s) instruments to provide adequate spatial resolution at higher speeds. Applications include

(a)

(b)

Figure 1.6 (a) The Mace Head Atmospheric Research Station, situated on the west coast of Ireland, and photographed during the NAMBLEX field campaign (July–September 2002). Many of the instruments brought to the site specifically for the campaign are housed inside shipping containers. The retro-reflecting mirror for a DOAS system is positioned on the island in the distance. (b) The Clean Air Sector Laboratory (right), operated by the British Antarctic Survey, and positioned approximately 1.5 km from the main Halley Base on the Brunt Ice Shelf, Antarctica. The shipping container on the left houses the FAGE instrument from the University of Leeds, which measures OH and HO_2 radicals, and which was deployed next to CASLAB during the CHABLIS (Chemistry in the Antarctic Boundary Layer and the Interface with snow) summer intensive campaign, January–February 2005. (Photograph courtesy of Dr W.J. Bloss.)

Table 1.3 Major field campaigns for the measurement of atmospheric composition (highly selective, since 1991)

Campaign or acronym	Year	Dates	Location			Type of environment	Platform	Comments
			Place	Latitude	Longitude			
MLOPEX-2	1992	April–May	Mauna Loa, Hawaii, USA	20N	156W	Free-troposphere	Ground	3400 m altitude
TOHPE	1993	Aug–Oct	Fritz Peak and Idaho Hill, Colorado, USA	40N	105W	Forested	Ground	2687–3070 m altitude
POPCORN	1994	August	Pennewit, Germany	54N	12E	Continental, rural	Ground	
ACE-1	1995	Nov–Dec	Southwest Pacific Ocean (South of Australia)				Aircraft, ground	
ALBATROSS	1996	Oct–Nov	North and South Atlantic	68N–50S	various	Open ocean MBL	Ship	Aboard ship R/V Polarstern
EASE97	1997	April–May	Mace Head, Ireland	53N	10W	Coastal MBL	Ground	
SONEX	1997	Autumn	North Atlantic	various	various	Air-flight corridor	Aircraft	
BERLIOZ	1998	July–August	Pabstthum, Germany	53N	13E	Rural and urban plume	Ground	
PARFORCE	1998 1999	September June	Mace Head, Ireland	53N	10W	Coastal MBL	Ground	
SOAPEX-2	1999	Jan–Feb	Cape Grim, Tasmania	41S	142E	Coastal MBL	Ground	

Table 1.3 (Continued)

Campaign or acronym	Year	Dates	Location			Type of environment	Platform	Comments
			Place	Latitude	Longitude			
PEM-Tropics B	1999	March–April	Tropical pacific	various	various	MBL and free/upper troposphere	Aircraft P-3B and DC-8	
SOS (Nashville)	1999	June–July	Cornelia Fort, Nashville, Tennessee, USA	36N	88W	Polluted urban	Ground	
SOLVE	1999–2000	Winter	Polar vortex (from Kiruna, Sweden)	various	various	Polar stratosphere	Aircraft	
INDOEX	1999	Feb–March	Indian Ocean	various	various		Ship, aircraft	
Rishiri Island	2000	June	Rishiri Island, Japan	45N	141E	Coastal MBL	Ground	
PROPHET-2	2000	July–August	Pellston, Michigan, USA	46N	85W	Forested	Ground	
TEXAQS	2000	Aug–Sept	Houston, Texas, USA	29N	95W	Polluted urban	Ground	
TOPSE	2000	Feb–May	USA, Canada, Arctic	various	various	Above remote continental and ice sheet	Aircraft	
ISCAT	2000–2001	Nov–Jan	South Pole	90S		Ice sheet	Ground	2800 m altitude

TRACE-P/ ACE-Asia	2001	March–April	Western Pacific	various	various	mainly above Pacific ocean	Aircraft P-3B and DC-8	
PMTACS-NY	2001	June–August	New York City, USA	41N	74W	Polluted urban	Ground	
MINOS	2001	August	Finokalia Station, Crete, Greece	35N	26E	Coastal MBL	Ground	
NAMBLEX	2002	July–Sept	Mace Head, Ireland	53N	10W	Coastal MBL	Ground	
ANTCI	2003	Nov–Dec	South Pole	90S		Ice sheet	Ground	2800 m altitude
ECHO	2003		Julich, Germany	51N	6E	Forested	Ground	
TORCH	2003	July–August	Writtle, Essex, UK	51N	1E	Semi-polluted rural, London plume	Ground	
	2004	May–June	Weybourne, Norfolk, UK	53N	1E	Semi-polluted rural	Ground	
INTEX	2004	June–Aug	NE USA, Atlantic	various	various		Aircraft, ground, ship	
ITOP	2004	July–Aug	Based at Faial, Azores	38N	28W	Mid-Atlantic ocean	Aircraft	

Notes: Special issues of journals were dedicated to the presentation of the results from some of these campaigns. A selected listing of special issues is given in the 'Further Reading' section.

Glossary of field campaign acronyms used in Table 1.3

Acronym	Full name
ACE	Aerosol Characterisation Experiment
ANTCI	Antarctic Tropospheric Chemistry Investigation
BERLIOZ	Berliner Ozone Experiment
EASE 97	Eastern Atlantic Spring Experiment 1997
ECHO	Emission and CHemical Transformation of Biogenic Volatile Organic Compounds
INDOEX	Indian Ocean Experiment
ISCAT	Investigation of Sulphur Chemistry in the Antarctic Troposphere
INTEX	Intercontinental Chemical Transport Experiment
ITOP	Intercontinental Transport of Pollution
MINOS	Mediterranean Intensive Oxidant Study
MLOPEX	Mauna Loa Observatory Photochemistry Experiment
NAMBLEX	North Atlantic Marine Boundary Experiment
PARFORCE	New Particle Formation and Fate in the Coastal Environment
PEM	Pacific Exploratory Mission
PMTACS-NY	PM2.5 Technology Assessment and Characterization Study – New York
POPCORN	Photochemistry of Plant Emitted Compounds and OH Radicals in Northeastern Germany
PROPHET	Program for Research on oxidants: Photochemistry, Emissions and Transport
SOAPEX	Southern Ocean Atmospheric Photochemistry Experiment
SOLVE	SAGE III Ozone Loss Validation Experiment
SONEX	Subsonic Assessment, Ozone and Nitrogen Oxide Experiment
SOS	Southern Oxidants Study
TEXAQS	Texas Air Quality Study
TOHPE	Tropospheric OH Photochemistry Experiment
TOPSE	Tropospheric Ozone Production about the Spring Equinox
TORCH	Tropospheric Organic Chemistry Experiment
TRACE-P	Transport and Chemical Evolution over the Pacific

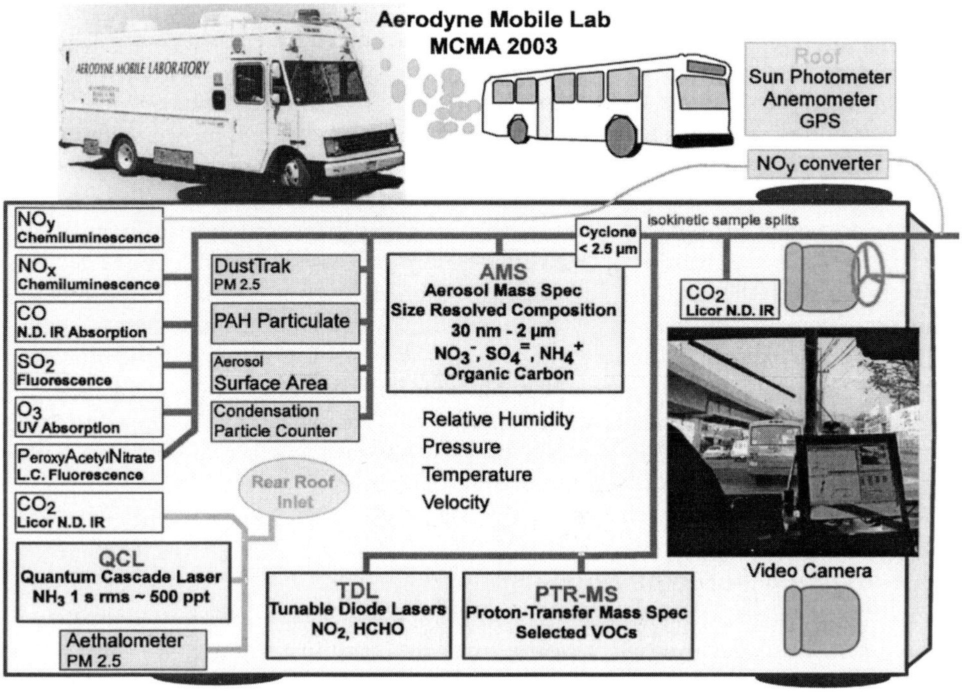

Figure 1.7 Aerodyne Research Mobile laboratory as deployed in the Mexico City Metropolitan Area 2003 Field Campaign. The facility is equipped to monitor a wide range of trace gas and particle pollutant concentrations and on-road measurement of in-use vehicle emission ratios (Reprinted with permission from Kolb *et al.*, 2004, © American Chemical Society).

the location of emission plumes, quantifying greenhouse gas emissions (e.g. from natural gas and landfills), and the spatial distribution of regulated pollutants (O_3, NO_X, CO, SO_2 and particulate matter) in urban centres. Figure 1.7 shows a schematic of a system operated by Aerodyne Inc., together with the instruments deployed, which is described in detail in Kolb *et al.* (2004). An alternative monitoring strategy is to use the laboratory to temporarily sample at a number of fixed sites (using a battery pack to power the instruments when the engine is not running) to establish pollutant profiles across an urban centre (Seakins *et al.*, 2002).

1.4.2 Ship-borne platforms

Ocean-going research vessels are operated by a number of national agencies for the measurement of atmospheric composition, including ice-breakers for polar regions. Examples include the *Ronald H. Brown*, operated by the US National Oceanic and Atmospheric Administration, the *RV Polarstern*, operated by the Alfred Wegener Institute, and the *RRS Discovery* and *Charles Darwin*, operated by the UK Natural Environmental Research Council. Details of the entire worldwide fleet of research vessels, including their specifications and cruise schedules, are maintained on-line by the University of Delaware,

College of Marine Studies (http://www.researchvessels.org/). Ship-borne platforms have the advantage that heavy and power-hungry instruments can be accommodated, although space in laboratories is limited. Some ships have deck space for shipping containers, enabling larger instruments to be deployed. Cruises may last for long periods, several months at a time, and thus very long time-series can be measured, providing latitudinal gradients of key species from polar to equatorial regions. Care must be taken that the sampling is forward of the exhaust stack, and that local wind-fields are well understood. Pseudo-Langragian experiments have been performed in which an instrumented ship is positioned either upwind or downwind of a ground-based site, with connected flow between the sites allowing the chemical evolution of a given airmass to be followed from measurements in changes in composition. Alternatively, measurements along a ship-track away from an ocean upwelling hotspot will enable the atmospheric processing of marine emissions to be studied. It is possible for ice-breakers to remain locked in the ice during polar winters, and to obtain seasonal changes in composition during the transition from winter to spring to summer as the sea-ice cracks, breaks up and disperses. Ship measurements are also useful for satellite validation.

1.4.3 Balloon-borne platforms

In situ instruments deployed on the platforms considered in Sections 1.4.1 and 1.4.2 do not allow for vertically resolved measurements, and although remote sensing techniques can provide some information of this type, it is often limited by scattering and absorption of light in the boundary layer. Balloons represent a relatively simple and inexpensive solution to this problem. Small (weather) balloons carrying a lightweight radiosonde (up to a few kg) have been used in their thousands to obtain profiles of meteorological parameters (pressure, temperature, humidity, wind direction and speed). An ozonesonde is also commonly deployed on balloons as part of an international network, and consists of an I_2/I^- (aq) electrochemical cell (see Section 1.5.3) to measure ozone, and aerosols have also been measured using small instruments. Many stations worldwide release a fixed number of balloons each day to obtain a long-term record. The balloons are constant-pressure devices (the pressure inside the balloon is the same as ambient pressure), containing a fixed mass of helium, with thin membranes that expand as the balloon ascends, according to the ideal gas equation:

$$pV = nRT \tag{1.5}$$

where p and T are ambient pressure and temperature, n is the (fixed) number of moles of helium (1 mole contains 6.02×10^{23} molecules) and V is the balloon volume. Data are sent back to the ground via telemetry, with the balloons eventually bursting and the sonde (usually not recovered) falling back to earth.

To lift heavier instruments, a larger balloon is required, as more lift can be generated by a larger volume of the lighter helium. Scientific balloons can be very large (up to 10^6 m^3) and are able to lift payloads of up to $\sim$ 1000 kg to altitudes of 30–45 km. There are severe constraints on the instruments – they must be automatic and capable of withstanding wide ranges of pressures and temperatures. Instrument packages are often contained with gondolas that may be up to 100 m or more below the balloon, so that the instruments are not shaded by the balloon. Weather conditions have to be perfect for launch, and this limits the deployment of balloon instruments. Long duration balloons that fly for several weeks are able to circumnavigate the globe one or more times, and drift with the wind at an almost constant altitude. Such balloons have been used in the tropics to measure trace species involved in stratospheric ozone depletion that have been transported from the surface to the tropopause by deep convection, and are ideal for satellite validation. The history of constant-level balloons as observational platforms for atmospheric research has been reviewed by Businger *et al.* (1996). Much reliance is placed on global positioning systems to record the precise flight track, and instruments include lightweight instruments to measure O_3 and H_2O vapour.

Tethered balloons are also quite widely used, with payloads of a few kg, although their altitude ceiling is limited to about 1500 m because of the weight of the tethering cable, but with a winching device the height of the instrument payload can be precisely changed.

1.4.4 Aircraft-borne platforms

Aircraft are an excellent means to obtain a three-dimensional distribution of trace gases and aerosols in the atmosphere, on account of their high speed, long range and ability to sample a range of altitudes quickly. Space is usually at a premium and hence instruments must be lightweight and compact, must not generate radio-frequency emissions that may interfere with aircraft avionics, and ideally should use small amounts of electrical power, although larger aircraft can generate significant power (but initially at a rather inconvenient frequency of 400 Hz, and this has to be converted to 50 Hz 240 V a.c. or 28 V d.c.). There are greater demands on instruments on aircraft, which should be designed to withstand high stress loadings. Research aircraft vary from the small, for example a Cessna single-engined turboprop aircraft, with a relatively low ceiling (13 500 feet), low range (~1000 km) and short duration (~6 h), all the way to large civil airliners, for example the highly-modified four jet-engined McDonnell Douglas DC-8 that is operated by NASA. The NASA DC-8 has a very large instrument payload (13 600 kg), has a flight duration of 12 h, can fly from 1000 to 42 000 feet (13.7 km), and has a range of 10 000 km. It is beyond the scope of this introductory chapter to list in detail the performance specification of all the research aircraft in use today, but further sources of information can easily be found on the Internet. Some of the larger fully instrumented aircraft, in addition to the NASA DC-8, are the NASA WB-57 (18 km ceiling), the UK BAe-146 (Facility for Airborne Atmospheric Measurements, FAAM), the M-55 Geophysica (initially Russian Air Force, 21 km ceiling), the NASA Lockheed ER-2 aircraft (converted U-2 series reconnaissance platform, 21 km ceiling, 1200 kg payload), the Lockheed P3-B (operated by several agencies), the Lockheed Hercules C-130 series transport aircraft (12 hr endurance, 17 000 kg scientific payload, capable of flying at 50 feet over water), the German DLR Falcon 20 and French Mystere Falcon 20.

There have been many aircraft-based field campaigns, and some of these are listed in Table 1.3. Co-ordination with other research aircraft is a feature of field campaigns, and it has been possible to follow the chemical evolution and photochemical processing of identified plumes over very large distances. An example is the ICARTT (International Consortium for Atmospheric Research on Transport and Transformation) project, which monitored the export of polluted layers from North America across the North Atlantic towards Europe. Four NASA and NOAA research aircraft were based on the US eastern seaboard (as was the *Ronald H. Brown* research vessel), the UK BAe-146 was based on the mid-Atlantic island of Faial in the Azores, and the German Falcon aircraft was based at Creil, France.

Essential to the flight planning and scientific design of aircraft sorties are forecast modelling tools; for example, reverse domain filling trajectories, which predict at any time for a given flight path the origin of the sampled air. Inlet systems should ideally be close to the front of the aircraft, as the skin boundary layer increases in width along the length of the fuselage. Satellite validation is becoming a more important part of aircraft field campaigns as the range of species monitored by satellites in the troposphere increases (see Section 1.4.8).

1.4.5 Commercial passenger or freight aircraft platforms

Analysing atmospheric trace gases and aerosols using commercial passenger aircraft represents a cost-effective solution for recording many thousands of hours of flight data. The main advantage being that near-global coverage is possible, as opposed to limited coverage using a limited number of research aircraft. Other applications include comparison with satellite measurements along the flight-paths or land-based observatories close to take-off and landing. There are currently three passenger aircraft systems operating worldwide, which have been reviewed in 2005 (Brenninkmeijer *et al.*, 2005), and from which much of the material in this section has been taken. The certification of equipment for use on passenger aircraft by aviation authorities can be a lengthy process. Most of the data are taken at cruising altitudes of 9–12 km, and for mid-latitudes, stratospheric air is sometimes sampled.

The EU-funded Civil Aircraft for the Regular Investigation of the Atmosphere Based on an Instrument Container (CARIBIC, www.caribic-atmospheric.com) project began in 1997; it used an automated container deployed on intercontinental flights using a Boeing 767, but now a more powerful instrument package is deployed using a new Airbus A340-600 operated by Lufthansa. There are 1–2 flights per month, and measurements (technique and chapter number in brackets) include the following:

- O_3 (chemiluminescence, 7, and UV absorption, 3)
- CO (Vacuum UV fluorescence, 4; gas chromatography and a reducing gas detector, 8) H_2O vapour (diode laser photoacoustic detection, 2)
- NO and NO_y (chemiluminescence, 7) mercury vapour (atomic fluorescence, 4)
- CO_2 (non-dispersed IR absorption, 2)
- O_2 (electrochemical cells, 1)
- methanol, acetone, acetaldehyde (proton transfer mass spectrometry, 5)
- aerosols (>4nm, >12nm, >18nm, condensation particle counter, 6)
- aerosol size distribution (15–5000 nm, optical particle counter, 6)
- aerosol elemental composition (filter impactor, 6)
- particle morphology (impactor, electron microscopy, 8)

- non-methane hydrocarbons, halocarbons, CO_2, CH_4, N_2O, SF_6 (whole air sampler with 28 glass flasks, analysis by GC and GC-MS, 5 and 8)
- oxygenated VOCs (adsorption tubes followed by GC analysis, 8)
- BrO, HCHO, OClO, O_4 (O_2 dimer, all remote sensing DOAS, 3)
- cirrus clouds.

Measurement of ozone and water vapour by Airbus In-service Aircraft (MOZAIC, www.aero.obs-mip.fr/mozaic) is an EU-funded network of five Airbus 340–300 aircraft that contain in situ instruments for the measurement of O_3, H_2O vapour, NO_y ($= NO_X + HNO_3 + HONO +$ alkyl nitrates $+$ PAN) and CO. Measurements of aerosols are planned. Between September 1994 and June 2004 there have been 20 960 flights and 160 000 flight hours made over the continents of Europe, North America, Asia, South America, Africa and the Atlantic Ocean, and hence there is a fantastic dataset for use by the atmospheric community.

The JAL Foundation (www.jal-foundation.or.jp/) has funded the collection of air flask samples from a Japan Airlines Boeing 747, mainly on routes between Japan and Australia, and expansion is underway to involve seven 747-400 and 777 aircraft (flask sampling and in situ CO_2 measurements).

1.4.6 Uninhabited aerial vehicles

Uninhabited aerial vehicles (UAVs) have undergone considerable development, driven primarily for military use. They offer considerable advantages where manned craft are dangerous or difficult to deploy, or for very long missions when pilot fatigue is a limiting factor. UAVs have also been used for a variety of atmospheric monitoring applications. In situ measurements of vertical profiles of the composition of the atmosphere in regions where current aircraft cannot reach are needed to validate and calibrate satellite measurements. The US National Oceanic and Atmospheric Administration (NOAA) have tested (May 2005) an Altair UAV, which is a high altitude version of the Predator UAV, whose payload includes a gas chromatograph for measurements every 70 seconds of long-lived gases – SF_6, nitrous oxide (N_2O), CFC-11 ($CFCl_3$), CFC-12 (CF_2Cl_2) and H-1211 ($CBrClF_2$) – and a UV absorption instrument for measurement of O_3. The vertical distribution of water vapour has been remotely measured with passive microwave sensors. The altitude ceiling of this UAV is ~15 km; it has a combined internal and external payload of ~1500 kg, and a duration of 30 hours. Further information on this demonstration project can be found at http://uav.noaa.gov/index.html.

1.4.7 Rocket platforms

The maximum altitude for balloons is about 45 km, and in situ measurements for vertical profiles at higher altitudes have been achieved with rockets, which have payloads of up to few 100 kg and can reach 200–300 km. The rockets are recovered after use, and hence must be launched from remote locations; for example, from Andoya, on the Norwegian Coast (rockets reovered from the ocean), Wallops Island, Virginia, USA and

White Sands, New Mexico, USA. Rocket-based measurements of the hydroxyl radical have been made between 45 and 70 km altitude in the mesosphere using solar-induced fluorescence spectroscopy, with ~ 2 km vertical resolution. Measurements can be during the rocket ascent (lasting only a few minutes) or during a slower descent below a parachute.

1.4.8 Satellites and other space-borne platforms

The use of satellites represents one of the most important advances in the monitoring of our atmosphere. Measurements of atmospheric composition from space have revolutionised our understanding of the three-dimensional distribution of trace gases in the atmosphere and of the impact humans have had on it. Satellites have enabled direct observation of intercontinental long-range transport of biogenic and anthropogenic emissions and the global observation of regulatory species over periods of several years. An example is ozone, whose tropospheric background levels have steadily risen as a result of the oxidative processing of VOCs in the presence of NO_X during long-range transport. Sources of emission (urban areas) may be thousands of kilometres away from the regions of resulting elevated ozone. Satellite data have shown unequivocally that pollution is a global issue, and provide a direct comparison with the output from CTMs, enabling their realistic evaluation. Satellite data will allow the following questions to be answered: Will the recovery of the ozone layer follow the model predictions, or will it be altered because of cooling in the stratosphere caused by greenhouse gases? What is controlling the concentrations of tropospheric pollutants? What is the impact of changes in atmospheric composition, for example upper tropospheric aerosols, ozone and water vapour, on climate?

This book is organised according to analytical technique, each of which has been deployed on a number of platforms, and hence there is no chapter dedicated to the use of satellite platforms. In this section, no attempt is made to provide a comprehensive treatment of satellite instruments and methods, rather a summary is given of the satellite platforms available, their orbits and viewing geometries, the temporal coverage and vertical resolution, a description of some of the instruments and the species that have been measured. The subject of satellite measurements is now very large indeed, with numerous satellites in orbit equipped with arrays of instruments, and the volume of data becoming available is truly enormous. Further information can be found in Brasseur *et al.* (1999, 2003), Burrows (1999), Calvert (1994), Platt & Stutz (2006), Scott (2005), and from a number of excellent Web pages, for example:

http://jwocky.gsfc.nasa.gov/ (TOMS)
http://www-iup.physik.uni-bremen.de/gome/ (GOME)
http://www-iup.physik.uni-bremen.de/sciamachy/index.html (SCIAMACHY)
http://envisat.esa.int/instruments/tour-index/mipas/ (MIPAS)
http://www.eos.ucar.edu/mopitt/ (MOPITT)
http://www-sage2.larc.nasa.gov/ (SAGE II)
http://eos-aura.gsfc.nasa.gov/ (AURA)

Additional references are given below and further information on satellites and their data products can be obtained from the Internet.

Satellite instruments rely on passive remote sensing of scattered or transmitted sunlight, or emitted thermal or microwave radiation (essentially DOAS from space). Chapters 2 and 3 describe the techniques of IR absorption spectroscopy and UV-visible DOAS, respectively. An excellent resource for the use of DOAS from satellites can be found in Platt & Stutz (2006). Naturally, there are extreme constraints on satellite instruments, for example in power consumption, weight, size, reliability, autonomy, resistance to cosmic radiation. The development costs are very high and there may be many years from instrument concept to measurement.

Several modes of observation are possible from satellites, which depend both on the orbit of the satellite and on the viewing angle of the instrument. There are two types of orbit, geostationary orbit (GEO) and low-earth orbit (LEO). At present, instruments that use DOAS are all aboard satellites in LEO. The satellite in GEO is 35 790 km from the earth's surface, and maintains a constant position with respect to the rotating earth. Hence the orbital period (time for one orbit) is exactly equal to the period of rotation of the earth (23 hr, 56 min, 4.09 s). By orbiting at the same rate, in the same direction as earth, the satellite appears stationary (synchronous with respect to the rotation of the earth). GEO satellites give the 'big picture' view, enabling coverage, for example, of weather events, and is especially useful for monitoring severe local storms and tropical cyclones. Being positioned in the equatorial plane, distorted images are provided of polar regions with poor spatial resolution.

Low-earth orbit satellites enable observation of the entire surface of the earth, but only for a short period each day. Many satellites are launched into an almost polar orbit, passing over the North and South Poles, at an altitude of 700–800 km. The satellite must travel quickly ($\sim$27 000 km hr^{-1}) relative to the earth so that the centrifugal force balances the significant gravitational force. The orbit period is 90–100 min, and provides a thorough coverage of the earth's surface, which spins $\sim$ 24° of longitude about its axis between successive orbits (the observation is akin to peeling an orange in one piece). In a sun-synchronous orbit, the satellite passes over a given longitude at intervals of 12 hours (once in the dark), and enables regular data collection at consistent times and long-term comparisons. The majority of satellites used for compositional measurements use this type of orbit so that variation with location can be distinguished from variation with time of day. Inclined orbits have an inclination somewhere between equatorial (0°) and polar (90°) orbit, with a period of a few hours, and are not sun synchronous, passing over a given location at different times. An example is the Upper Atmosphere Research Satellite (UARS) which contains six instruments for studies of the chemistry of the stratosphere, mesosphere and lower thermosphere. Care has to be taken when positioning a satellite. NASA is tracking more than 8000 objects larger than a cricket ball, all orbiting the earth at 27 000 km hr^{-1}.

Limb-viewing along a horizontal path or nadir-viewing looking directly below the satellite are both used and provide complementary information. Limb observations of sunlight scattered at the horizon provide excellent vertical resolution but poor horizontal resolution (400 km). Cloud, aerosol and spectral interference make measurements below 8 km very difficult using limb-viewing. Nadir observations of sunlight reflected by earth's surface or atmosphere (earthshine) provide better horizontal resolution and are capable of detecting down to the earth's surface. However, the vertical resolution is poor (often only a column measurement is possible). Other types of observation are solar, lunar and stellar occultation, where the light source is viewed directly, and absorption through the atmosphere is measured. In addition thermal emission from excited molecules in the atmosphere is observed rather than scattered sunlight. The main disadvantages of satellite measurements are the poor time resolution for a given location (once per day) and the bias of the measurements towards clear sky conditions.

Some of the species observed from satellites are listed in Table 1.2. A very comprehensive listing of species, together with the satellite platform and instrument used, the the type of orbit, and the altitude range of the measurement, can be found in Platt & Stutz (2006). Examples of instruments (satellite in brackets) include the Global Ozone Monitoring Experiment – 2, GOME-2 (ESA ERS-2), Stratospheric Aerosol and Gas Experiment II, SAGE II (Earth Radiation Budget Satellite), Solar Backscatter Ultraviolet Ozone Experiment, SBUV (Nimbus-7), Scanning Imaging Absorption Spectrometer for Atmospheric Cartography, SCIAMACHY (ESA-Envisat), and Total Ozone Monitoring Spectrometer, TOMS (Earth Probe). The GOME instrument (Burrows *et al.*, 1999) makes trace gas measurements using four medium resolution spectrometers, in the range of 290–790 nm, each with 1024 channels. The reflected sunlight from the atmosphere is analysed in nadir mode using the DOAS technique (Chapter 3), and column densities of NO_2, BrO, OClO, SO_2, H_2O, O_3 and O_4 (the weakly bound O_2 van der Waals dimer) are reported every 1.5 s. A scanning mirror moving perpendicular to the direction of travel of the satellite directs light into the spectrometer, and for each spectrum the 'ground pixel' (or footprint on the ground) is 320 km × 40 km. The swathe is 960 km wide (3 pixels). Measurements of stratospheric composition began with O_3 back in 1970 (with the SBUV instrument on the Nimbus 4 satellite) and NO_2 starting in 1979 (Platt & Stutz, 2006). One of the workhorses over the years has been the TOMS instrument, which measures the total column of O_3, and which first demonstrated clearly the extent of the ozone hole over Antarctica. The instrument (Heath *et al.*, 1975) consists of a grating UV-visible spectrometer, and six wavelengths are monitored between 213 and 380 nm, where few other molecules absorb. A disadvantage is that no measurements are possible for the polar night, but the instrument can also measure SO_2 (strong absorption features in the same spectral region), and has tracked plumes emanating from volcanoes and reaching the stratosphere. Another example is stratospheric OClO, which is a good indicator of chlorine activation.

Measurements of trace gases and aerosols in the troposphere from satellites are a more recent phenomenon, and are much more difficult than those in the upper atmosphere. The presence of clouds means that virtually all measurements are made using the nadir-viewing geometry, with poor vertical resolution. Obtaining the tropospheric contribution to the total column is especially difficult for species that are highly abundant in the stratosphere, for example O_3 (only ~10% of the total column is from the troposphere) and NO_2. For other species, for example HCHO, water vapour and SO_2, the column absorption is dominated by tropospheric contributions. There are several procedures (retrieval methods) for separating out the 'stratospheric' signal for species abundant in the stratosphere. The multi-step reference sector method (used for NO_2) compares signals from regions where it is known that there are no tropospheric contributions to the total NO_2 column with signals from polluted regions, with the difference giving the tropospheric NO_2 column. In the cloud slicing method the difference between columns measured above clouds and in cloud-free conditions is used to get the tropospheric contribution. The spectral line shape of an absorption feature is a function of temperature and pressure, and hence its careful observation can be used to extract a vertical profile. Finally, the difference in column amounts between two satellite instruments viewing the same point, but with one sensitive only to stratospheric altitudes, has been used to obtain troposphere columns, for example for O_3 (Fishman *et al.*, 1990). The retrieval algorithms are extremely complex and time-consuming, and must be developed for each separate species of interest. The absorption pathlength in the troposphere will depend upon the height above sea level, and for high-precision measurements of long-lived species which are well mixed (e.g. CO_2 and CH_4) and which have fairly uniform concentrations, the presence of mountain ranges must be taken into account.

Some of the first measurements of tropospheric CO, produced as a result of incomplete combustion during burning processes, and a good indicator of atmospheric pollution, were made by the Measurement of Air Pollution from Satellites (MAPS) instrument in 1984 from the space shuttle, whereas tropospheric O_3 was first determined in 1977 as a difference of measurements from SAGE and TOMS instruments. The Measurements of Pollution in the Troposphere (MOPITT) instrument aboard the Terra satellite has been used for tropospheric CO and CH_4 measurements in the near-IR and IR regions. The MOPITT instrument operates both at short wavelengths, where absorption of reflected sunlight provides the total CO column, and in the fundamental CO vibrational 1–0 band (4.8 μm), where middle and upper tropospheric amounts of CO are determined (Drummond & Mand, 1996). Subtraction allows determination of concentrations near the surface. Gas filters containing CO at different pressures (for selective absorption of radiation from different altitudes) are used to achieve some vertical resolution, so very high spectral resolution is not required. Retrieval methods in the near-IR and IR have to take into account the temperature and pressure dependence of the spectral lines, and there are many overlapping bands from multiple species (in particular water vapour).

Newer instruments such as GOME, SCIAMACHY and TES (Tropospheric Emission Spectrometer, deployed on the NASA EOS-CHEM Aura satellite) make high quality tropospheric measurements. The SCIAMACHY instrument makes measurements in limb, nadir and occultation modes, and uses spectral bands all the way from 240 nm to 2380 mm to measure tropospheric column amounts of O_3, CH_2O, SO_2, BrO, NO_2, H_2O, CO, CH_4 and N_2O (Bovensmann *et al.*, 1999). Vertically resolved concentrations of O_3, H_2O, N_2O and CH_4 will also be measured. The spectrometers aboard GOME only extend to 790 nm, and the ground-footprint of SCIAMACHY for nadir-viewing is also superior (30 km × 60 km). Figure 1.8 shows a map of the tropospheric NO_2 column for Europe in August 2002 retrieved from SCIAMACHY DOAS measurements. NO_2 is emitted directly and can also be generated through the reactions of HO_2 and RO_2 radicals (R is an organic group, for example CH_3) with NO, which is emitted by industrial and transport sources, and there is a clear correlation between industrial regions and elevated levels of NO_2. Global maps of tropospheric NO_2 show peaks over industrialised regions and cities, with the highest levels globally observed over China. It is also possible to observe ship tracks from satellites through NO_2 signatures. Fuel oil used by ships is largely unregulated and contains high levels of nitrogen and sulphur, resulting in high levels of NO_x and SO_2 being emitted.

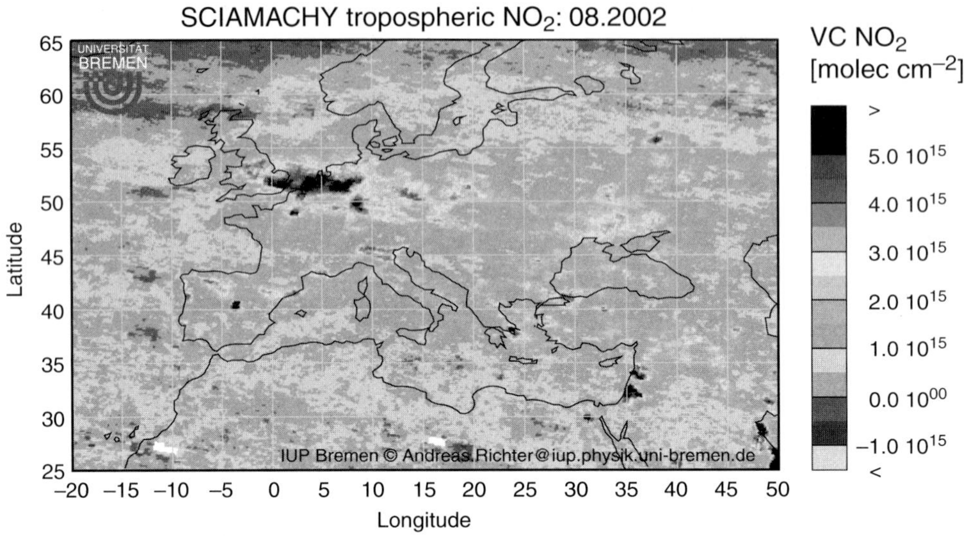

Figure 1.8 Tropospheric vertical column for NO_2 over Europe measured by the SCIAMACHY instrument aboard the ENVISAT satellite. The pixel size is 30 × 60 km, and the retrieval used the reference sector method. (Reproduced with permission from Richter *et al.*, 2004, 2005, University of Bremen.) (Reproduced in colour as Plate 1 after page 270.)

Although at the moment it is not possible to measure global distributions of short-lived free-radicals (e.g. OH, HO_2, RO_2, NO_3) in the troposphere from space, it is possible to observe markers of recent photochemical activity. An example is formaldehyde, CH_2O, which is formed from the OH radical-initiated oxidation of methane and other hydrocarbons, and also during biomass burning. High levels of CH_2O have been observed over forests or regions where there are high levels of biogenic hydrocarbons, such as isoprene and monoterpenes.

Finally, in polar regions, enhanced levels of BrO, a short-lived halogen species formed in the oxidation of Br atoms by ozone, have been observed by GOME (Richter *et al.*, 2002; Wagner & Platt, 1998) and SCIAMACHY. The levels are consistent with concentrations measured on the ground, and concentrations which occur at the same time as ozone depletion events (Bottenheim *et al.*, 1990). The BrO levels peak following the long polar winter in regions containing sea-ice, and disappear once the sea-ice melts. The area of the BrO clouds is very large, and an autocatalytic mechanism (the so-called 'bromine explosion') has been postulated which involves the oxidation on the sea-ice of Br^-(aq) in sea salt, followed by the release of active bromine compounds.

The NASA space shuttle has also been used to provide measurements from space, although limited to short durations. The MAPS instrument made measurements of CO in the 1980s using IR absorption. The Atmospheric Trace Molecular Spectroscopy (ATMOS) instrument, a high-resolution (0.015 cm^{-1}) Fourier transform spectrometer, has also flown on four missions on the Space Shuttle and on Spacelab-3, making measurements in occultation of solar radiation after it has passed through the atmosphere at sunrise or sunset (Brasseur *et al.*, 1999). The interferograms are radioed to a receiving station on the ground where a fast Fourier transformation is performed to yield absorption spectra, which are normalised using a spectrum taken above the atmosphere to eliminate structure from the solar spectrum and instrumental effects. Vertical resolution is possible, and profiles of almost 40 species have been measured, including in the upper troposphere. The hydroxyl radical has also been detected using solar-induced fluorescence in the $A^2\Sigma^+(v' = 0) - X^2\Pi_i(v'' = 0)$ band near 309 nm, and measured as a function of latitude (50°S to 61°N) and altitude (40–93 km) by the Middle High Resolution Spectrograph Investigation (MAHRSI) instrument (limb scanning with ~0.02 nm resolution) on the Space Shuttle.

Determination of atmospheric composition is not limited to our own atmosphere, and measurements have been made in the atmospheres of other planets and moons, using both remote sensing and in situ instruments. An example is the Huygens probe, released from the Cassini spacecraft, which measured vertically resolved concentrations of trace gases and aerosols (from 160 km to the surface) in the atmosphere of Titan, a moon of Saturn, on 25 December, 2004. The composition of aerosols was determined by pyrolysis and a gas chromatograph with mass spectrometric detection. Initial results confirmed that the atmosphere is composed mainly of CH_4 and N_2, with CH_4 concentrations increasing close to the surface, with clouds of CH_4 and C_2H_6 observed, and the presence of ^{40}Ar suggests volcanic activity. However, further description of these instruments is beyond the scope of this book.

1.5 Analytical methods not covered elsewhere in this book

Inevitably there are analytical methods which do not neatly fall under the headings used to structure the remainder of this book, but are important to cover nonetheless, as they provide unique information about the composition of the atmosphere. In this section, methods not covered elsewhere in this book to make composition measurements in the atmosphere are discussed, but not at the level of detail found in other chapters. Novel methods only tested in the laboratory are not covered, but brief mention of promising methods is made in Section 1.10. References are provided for further information.

1.5.1 LIDAR methods

The power of Light Detection and Ranging (LIDAR) methods is the ability to provide altitude profile information. Differential Absorption Lidar (DIAL) operates in a similar manner to DOAS (Chapter 3). Laser light is transmitted at two wavelengths into the atmosphere, one wavelength tuned to an absorption feature of the species under interrogation, and one wavelength tuned slightly off-resonance with there is little absorption. Both wavelengths are backscattered by Rayleigh scattering, but the return signal is stronger for the wavelength that is not absorbed. By using a pulsed laser and the fact that light travels 300 m in a microsecond, measuring the time delay between the outgoing laser radiation and the return signal, a vertical profile can be built up of the absorbing species. LIDAR has been used to measure NO_2, H_2O vapour, O_3, aerosol and SO_2 (e.g. in stack emission plumes) in the troposphere, as well as measurements in the lower stratosphere, in particular of O_3, from ground and airborne LIDARs. For example, DIAL measurements from an aircraft flying at 9 km were able to show clear destruction of O_3 in the lower stratosphere between 17 and 22 km as the aircraft entered the polar vortex at latitudes greater than 65°S (Browell, 1989). A long-term record of the vertical distribution of O_3 extending from 3 to 38 km, and with spatial and temporal resolutions of 0.36–3.6 km and 30 min, has been reported above the Hohenpeissenberg Observatory in southern Germany, and agrees with profiles measured using O_3 sondes. A LIDAR system operated by the University of Illinois has made measurements of metal layers (e.g. Na, Fe) in the mesosphere and lower thermosphere around 80–90 km. The source of the metals is the ablation from meteorites entering the atmosphere, and over polar regions these metallic species are taken up on the surfaces of polar mesospheric clouds and meteoric smoke particles. The occurrence of polar mesospheric clouds is increasing in frequency, and is thought to be an early indicator of climate change. The major limitation of LIDAR preventing wider applicability is sensitivity and the complexity (and cost) of the instrumentation and data reduction procedures. Current developments include active remote sensing from space, for example the NASA Lidar In-space Technology Experiment (LITE) project involves a three-wavelength backscatter LIDAR that has flown on the space shuttle, and development is underway for future systems for free-flying satellite platforms.

1.5.2 Matrix isolation electron spin resonance (MIESR)

A method that enjoyed early success (1978) (Mihelcic *et al.*, 1978) for tropospheric measurements of free-radicals was that of matrix isolation electron spin resonance (MIESR). Still the *only* direct method for atmospheric detection of HO_2 is MIESR, and it is highly selective. Air is expanded through a supersonic nozzle into a low-pressure region where it impinges on a cold finger containing a solid matrix (D_2O-ice) at −196°C. After sufficient material has been deposited, the sample is kept frozen at cryogenic temperatures, and is transferred to the laboratory for analysis. Although the method is direct, the sampling times are long (30 min) and the number of measurements is limited by the number of cold fingers available. Other species can also be analysed simultaneously from the ESR spectrum, for example NO_3 radicals, NO_2, $CH_3C(O)O_2$ (acetylperoxy radical) and the sum of peroxy radicals ($HO_2 + CH_3O_3 + C_2H_5O_2 + \ldots$, or more simply $HO_2 + \Sigma RO_2$) (Mihelcic *et al.*, 1990). The latter quantity is a useful proxy for the rate of in situ free-radical initiated formation of O_3, as peroxy radicals convert NO to NO_2, which leads to ozone after photolysis. The sum of peroxy radicals can also be measured using a peroxy radical chemical amplifier (PERCA), first developed in the early 1980s (Cantrell & Stedman, 1982), and also by chemical ionisation mass spectrometry following conversion to OH (both are described further in Chapter 7). The MIESR method is calibrated with a known concentration of HO_2 radicals generated from the photolysis of water vapour in air (see Section 4.3). The uncertainty of the MIESR method is given as $\pm 2.5 \times 10^7\,cm^{-3}$ (1 pptv) (Platt *et al.*, 2002). The future of the MIESR method is unclear, as the pioneer of this method, Dr Djuro Mihelcic, will shortly retire from KFA Julich, but it is vitally important for the community that either this method continues to operate, or that another, capable of directly measuring speciated peroxy radicals, is developed.

1.5.3 Solid-state and electrochemical sensors

There is a considerable demand for instruments that have very low weight, are highly compact and use very small amounts of electrical power, yet offer high sensitivity, are selective and have good temporal resolution. If these challenging specifications can be met, one can imagine some highly novel new applications for measurements of atmospheric composition, for example deployment on balloons, commercial aircraft, satellites or intelligent UAVs. In addition, if the devices are cheap to manufacture, then large numbers can be deployed in networks to determine concentration fields, either on the street canyon scale (sensors at junctions, pedestrian crossings) across cities, or on the regional scales. Using telemetry, or the Internet, many hundreds or even thousands of devices can be linked electronically and used to make time-synchronised, simultaneous measurements of key atmospheric indicators, which may be correlated with sources (e.g. traffic movement) or meteorological conditions. One of the problems is that of quality control, as the inherent sensitivity of individual sensors may differ significantly, and although calibrations are performed, sensitivities may vary with time in an unknown fashion. Hence a crucial parameter is the ability of the sensor to maintain its sensitivity over long periods. Data-rich networks with excellent temporal and spatial coverage could be used to design optimum traffic flow patterns or calculate personal exposure to pollutants at the kerbside.

In the troposphere measurements of key pollutants such as CO, NO, NO_2 and O_3, volatile organic compounds and particulates are ideal target species for such devices, whereas at the higher altitudes of the upper troposphere–lower stratosphere, accurate H_2O vapour measurements are crucial. Sensors have been developed for some of these species, largely based on solid-state devices that rely on a change of surface or electrochemical property upon surface contact with the required species. Small sensors for O_3 or NO_2 based on chemiluminescence are discussed in Chapter 7, and NO_2 measurements have also been made using small diffusion tubes (requiring no power supply or calibration), but with poor temporal resolution (several weeks). NO_2 has also been measured using electrochemical cells, which involves the electrochemical reduction of NO_2 between two electrodes immersed in an electrolyte reservoir, with the migration of electrons produced by the reaction being measured as a current flow. However, the sensitivity is limited and ozone may produce an interference, and these devices are not yet widely used. The majority of our information on the vertical distribution of ozone (in the troposphere and stratosphere) has been obtained with ozonesondes suspended from balloons, which are small, disposable instruments based on changes in electrochemical potential or conduction when O_3 oxidises solutions of potassium iodide. When ozone is pumped through the electrochemical cell, which consists of iodine–iodide redox electrodes in an aqueous solution of KI, I_2 is formed and is reconverted to iodide by the cell, causing electrons to flow through the cell's external circuit. In order to calculate the O_3, only the current and the amount of air pumped through the cell are required, although the instrument must be calibrated and corrections made for changes in temperature and flow-rate. Interferences from high levels of SO_2 are a problem with this type of instrument.

Solid-state, thick film sensors based on semi-conducting oxides have been developed for monitoring of trace gases (Carotta *et al.*, 2000; Williams, 1999). The sensors, which essentially are resistors that respond to a specific gas, are manufactured from nano-structured semi-conducting metal oxides, which are maintained at a high surface temperature. As current is passed through the sensor, an electrical response proportional (but not necessarily in a linear fashion) to the concentration of a specific trace gas is generated. CO, O_3, NO and NO_2 have been detected in the atmosphere using thick film sensors, and there are now some commercial products becoming available. In the case of $NO_x(NO+NO_2)$ there was reasonable agreement with chemiluminescence analysers.

A solid-state tungsten oxide (WO_3) sensor has been used for making ground-based O_3 measurements and profile measurements from balloon platforms (Hansford *et al.*, 2005). The device is small, lightweight, has low power requirements and is inexpensive. At high temperatures, vacancies in the oxide lattice at the gas–sensor interface are generated by thermal ejection of oxygen atoms, releasing electrons into the conduction band of the semiconductor. Vacancies can be filled by reaction with O_2 or O_3, and this changes the resistivity of the sensor. Reaction with O_3 is much faster and can be detected at ppbv concentrations. The resisitivity of the WO_3 surface is also temperature dependent and hence the surface is maintained at a constant ~530°C when in operating mode, using feedback control loops for stability. For deployment on large research balloons, data are logged to flash memory cards, whereas for flights of sondes on meteorological balloons, data are telemetered back to the ground as the balloons are not recovered. The device operates using rechargeable batteries, and only weighs 0.7 kg, with dimensions of 135 mm × 225 mm × 260 mm, costs ~£200, and uses 1 W of power. The resistance of the sensor is not linear with ozone concentration, is carefully calibrated in the laboratory, and is able to determine O_3 at the 1 ppbv level. The performance of the solid-state device relative to a commercially available ozonesonde (see Chapter 7) was tested during a balloon flight, with results agreeing to within 25% for most of the profile. On the ground, measurements have been made with response times down to ~20 sec.

Although chilled mirror hygrometers offer the primary standard for measurement of water vapour in the atmosphere, a relatively new solid-state water vapour sensor has emerged which offers advantages with size, weight, power usage and cost. Surface acoustic wave (SAW) devices are metal transducers transmitting ultrasonic waves across the surface of a piezoelectric material, and are very sensitive to surface effects, for example adsorption. The electrical output can be used to detect changes on the surface of the device. The SAW device operates at a higher frequency than a quartz microbalance and so is more sensitive to changes at the surface, for example in mass. Commercial SAW devices for water vapour are now available in a new generation of condensation hygrometers. The temperature-controlled quartz sensor surface is cooled until dew or frost collects on the surface, and changes in the phase of the acoustic wave is observed rather than reduction of reflected laser light using a chilled mirror.

1.5.4 Far-infrared and microwave absorption and emission spectroscopy

Transfer of population between rotational energy levels in a molecule is associated with the emission or absorption of radiation in far-IR (wavelengths of 20–1000 μm) and microwave regions (wavelengths of 0.1–10 cm), although a molecule must have a permanent dipole to be detectable by this method. Detectors are less sensitive in the far-IR region of the spectrum and hence it is difficult to achieve a high signal-to-noise

(S/N) ratio, but it is easier to work at high spectral resolution. Far-IR detection from balloon-borne platforms has been used to measure vertical profiles of several species, for example OH and HO_2 radicals. These instruments measure thermal emission spectra whilst observing various angles above the earth's limb (i.e. at different slant columns) and are able to measure vertical profiles and diurnal profiles for a given altitude. This method is a longer-wavelength version of Fourier transform infrared (FTIR) spectroscopy (Chapter 2). These methods have reduced sensitivity in the lower stratosphere (and cannot be used in the troposphere) because of strong water vapour absorption. The altitude of the balloon is usually fixed, and a complex retrieval procedure is required to generate altitude profiles. The average sampling latitude for a given telescope pointing angle can be quite different from that of the balloon itself. Heard has reviewed measurements of OH and HO_2 radicals in the stratosphere, which include absorption and emission spectroscopy at long wavelengths (Heard, 2003).

For heavier molecules with smaller rotational constants, emission and absorption are in the microwave region, and a very sensitive radio receiver can be used to detect radiation in a narrow spectral range surrounding a single rotational transition. The radio-frequency radiation is generated by a mixer that generates the difference in frequency between the atmospheric radiation and radiation from an oscillator in the instrument. The spectral shape of the line is measured at very high resolution, and can be used to obtain a vertical profile of the molecule being observed. Due to the pressure broadening, the width of the line is proportional to pressure, and hence signal in the wings of the line originates from molecules at higher pressures, or lower altitudes, compared to molecules responsible for signal close to line centre. The altitude profile in the stratosphere and mesosphere can be obtained from fitting the line, with a vertical resolution of 5–10 km (depending on altitude). Examples include chlorine monoxide (ClO) measured between 10–50 km in the stratosphere above Antarctica, and also ozone (up to 80 km). Satellite measurements in the microwave region to measure vertical profiles of many species were discussed briefly in Section 1.4.8.

1.5.5 Measurement of fluxes of trace gases and aerosols

In addition to the use of analytical methods to measure atmospheric composition, which is the main focus of this book, it is also possible to measure the rate at which a molecule is emitted into the atmosphere (emission flux), or deposited to the earth's surface, including the oceans, (deposition flux) by combining concentration measurements with micrometeorological data. The vertical flux is the average over time and area of the product of the vertical velocity and the mixing ratio of the species. It is very important to know, for example, the rate at which plants release CO_2 into the atmosphere, or take up O_3 through stomatal or surface uptake. The measurement of fluxes, including aerosol fluxes, is a very wide research area in its own right, and so only the bare essentials are provided here of the methods used. Further information can be obtained from Brasseur *et al.* (1999), Dabberdt *et al.* (1993), Seinfeld & Pandis (1998). Fluxes in general are more difficult to measure than concentrations, and, for a given molecule, the deposition rate may change with type of surface; for example, SO_2 deposition is enhanced for moist surfaces.

Enclosure methods involve putting a bag or box around vegetation, soil, or snow-block, and pumping air through it, and measuring the difference in concentration of the trace gas (e.g. isoprene) in the air as it enters and leaves the enclosure. However, the rate of emission or deposition may be changed by the presence of the enclosure (solar radiation may be attenuated), and the piece of vegetation or soil chosen may not be representative of the whole, necessitating further measurements. The eddy correlation method uses the covariance of fluctuations in ambient concentrations and vertical wind speed, measured using a sonic anemometer, which measures the difference in the velocity of sound in the upward and downward direction. Transport of a given trace gas takes place via turbulent eddies, and the differences in concentrations of gases in air moving downward to the surface and air that is moving upward from the surface (where the species may have been emitted or deposited to the surface) are measured. A variant of this method is to use a sampling system that is controlled automatically by the sign of the vertical wind velocity. Using switching values, the trace gas is directed onto one of two detectors depending on the direction of the vertical wind. Finally, deposition or emission at the surface results in a concentration gradient with height above the surface, and if the concentration and wind speed are measured at different heights, the turbulence of the atmosphere and the rate of deposition can be obtained. A disadvantage, of course, is that the instrument must either have multiple inlets displaced vertically or be capable of movement without disturbing the sampling in any way.

Analytical methods used for eddy correlation methods must have very fast response times, of the order of 0.1s, and precision ($<5\%$) is more important than accuracy. Fast-response IR absorption methods, combined with three-dimensional determinations of wind speed, have been used to measure fluxes of CH_4, CO_2 and H_2O from a variety of surface types. Deposition (dry or wet) velocities and emission fluxes have been measured using a variety of techniques for CO_2, NO, NO_2, O_3, NH_3, HONO, HNO_3, H_2O_2, SO_2, DMS and isoprene.

The measurement of the rate of exchange of aerosols between the atmosphere and surfaces is difficult, but have been made using the eddy correlation and the gradient methods, particularly over forests. It is particularly difficult to measure deposition of aerosols into bodies of water, and surrogate surfaces which contain water have been used. For urban surfaces rapid response particle counters have been used in conjunction with ultrasonic anemometers to measure deposition rates and emission rates of particles, the former being linked with degradation of building materials.

1.6 Quality assurance and quality control

Quality assurance concerns the data measurement process itself, whilst quality control refers to post-collection activities for optimisation of data accuracy and precision. Under the heading of quality assurance one would include choice of measurement site, the choice of where individual instruments are deployed at a given site, taking account of wind direction, obstructions to air flow and so on, selection of instruments for a particular measurement (often there is a choice), calibration of instruments, and maintenance and repair of instruments. Training of instrument operators and providing comprehensive instruction manuals with agreed protocols are also important aspects of quality assurance. On the other hand, quality control activities would include the following:

(i) Management of data sets, including procedures for submission to databases;
(ii) Ratification of data, using flags to indicate levels of data quality, and filtering out data where there are known instrumental problems or if records of meteorological conditions indicate sampling conditions at a particular time were unsuitable; and
(iii) Providing the means to adjust data; for example, a subsequent calibration indicating a drift in instrument sensitivity.

All data should be used unless there is good reason for rejection.

1.6.1 Precision and accuracy

Measurement by its nature is not an exact science, and all measurements will be subject to some error. For a scientist involved in field measurements, precision and accuracy have very different meanings. For every measurement there is a right answer, which is the aim of a given instrument to obtain. An analogy is useful here – shooting with a gun at a target, for which the bullseye is the perfect result. The operation here is the firing of the gun, whereas in field work it is the operation of the analytical instrument, and the result is the score on the target, whereas in fieldwork it is the concentration measured. If the marksman fires several shots at the target, and they are bunched closely together, but a long way from the bullseye, then the precision, or the uniformity, or reproducibility, is very good, and the quality of the operation is very good. However, the quality of the result, the score, the accuracy, is not so good. A uniform measurement has been achieved which is inaccurate. If the true result is known (obvious for the marksman), then an adjustment can be made (perhaps to the sights or to the aim) to improve the accuracy. In fieldwork the perfect result is not known in advance, and calibration using a known concentration can help to improve accuracy (but there are errors in the preparation of the calibration source) in addition to the good precision that the analytical method has achieved. Another scenario is that the marksman has hit the target close to the bullseye, but the individual shots are quite scattered, indicating good accuracy but poor precision. For fieldwork, this indicates that the methodology associated with a particular technique needs to be changed if greater precision is needed, as a fundamental limit has been reached with the existing setup. An advantage of using an instrument with high precision is that a 'rogue' measurement (through failure of the instrument at that time) can easily be identified, and rejected. However, if the precision is poor, it makes the elimination of rogue points more difficult, and if this point is included, the overall accuracy of the measurement (perhaps averaged over a longer time period) would be reduced. In the world of computer-generated data, one must be careful not to report data with a precision that is much higher than the methodology can provide.

Higher precision normally is associated with higher costs, and so one must carefully ask what level of precision is necessary. For example, if a tracer species, such as N_2O or CH_4, is measured, and one is interested in very small changes over an extended time period, then a precision of 1% or less may be necessary. High accuracy is also necessary if data from several monitoring stations are being compared. Hydroxyl radical measurements are often compared with the predictions of atmospheric numerical models, which are constrained by measurements of longer-lived species, and which contain a large number of kinetic and photochemical parameters. It is difficult to ascertain model uncertainties given their complexities, but methods have been developed (Sommariva *et al.*, 2004), and at the 95% confidence level these are usually ~25–40%. Thus, it is not necessary to have OH measurements at the 1% level of accuracy (although this would often help with trying to interpret any systematic differences between models and measurements!).

For any instrument, it is possible, given a thorough knowledge of the underlying fundamental principles of operation and all the instrumental parameters, to calculate the sensitivity of the instrument, in terms counts of signal (or other observable) for a given concentration of species in the atmosphere. However, with very few exceptions, the result is unreliable as many assumptions need to be made regarding the instrumental parameters. Usually one does not know the precise physical state of key components, for example the reflectivity of a mirror, which could degrade in time once exposed to the atmosphere, or the gain of a particular amplifier circuit, which could change with temperature. Instrumental parameters, and hence the sensitivity of instruments, change with time. It is necessary, therefore, for the majority of instruments, to perform a calibration using a known concentration of the trace gas or aerosol in question. One notable exception is the method of optical absorption spectroscopy (Chapters 2 and 3), which uses the Beer–Lambert law to link measured absorption with atmospheric concentration, if the path-length is known, and also the absorption cross-section at wavelengths used in the measurement. Even for absorption instruments though, it may not be possible to measure in the open atmosphere, and the trace species of interest may have to be introduced into a multipass cell, and there may be associated sampling losses, which have to be understood through calibration.

1.6.2 Calibration of instruments

It is often the calibration of instruments that represents the major challenge in performing an accurate and reliable trace gas measurement. Hence methods that do not require calibration offer a significant advantage. Calibration involves measuring the response of an instrument when a known concentration, ideally close to a typical ambient value, of the trace gas or aerosol in question is introduced to the sampling inlet of the instrument, ideally the same sampling geometry as for ambient air sampling. For instruments that rely on scrubbing into the aqueous phase prior to analytical detection, it is crucial that the efficiency of transfer into the aqueous phase is included in the calibration. Calibration hence involves the preparation of a known concentration of trace gases or aerosols. Calibration of instruments is considered in detail for each technique in subsequent chapters.

In some cases it is possible to purchase a compressed cylinder from a commercial supplier that contains either the pure trace gas (at a certified level of purity), a certified mixture of trace gas diluted in some buffer gas (although if this is not air then any change in sensitivity between the buffer gas in question and air must be understood), or a mixture of several trace gases with certified concentrations. An example of the latter is a mixture of hydrocarbons for the calibration of gas chromatographs that are able to measure many species during a single elution through the column (Chapter 8). If the concentrations provided are higher than ambient levels, then a dilution system must be designed, and mass flow-controllers should be calibrated carefully. It is often prudent to use in-line scrubbing agents to remove unwanted impurities, even if the manufacturer does not list them, as degradation in the cylinder can occur over time. An example is the removal of other oxides of nitrogen from a cylinder of nitric oxide (NO).

Unfortunately, some species are unstable, and decompose shortly after preparation, either in the gas-phase or are catalysed by surfaces. Examples include oxygenated VOCs and HNO_3. In some cases, it is possible to use permeation tubes containing liquid samples held at a known temperature that generate a gaseous sample at a known rate into a flow of air. The permeation tube can be weighed at intervals to establish the amount of material released. Examples are formaldehyde, HNO_3, Cl_2 and HCHO, for which commercial devices are available. For others, it is not possible to purchase the trace gas at all, and it is must be prepared carefully in the laboratory followed by purification. Examples include ozone, which can be prepared using a silent discharge in O_2, or UV photolysis of O_2 (typically using the 184.9 nm line of a mercury lamp), and nitrous acid, HONO, which can be prepared using several methods, usually involving the reaction of a strong acid with a salt containing the nitrite ion, NO_2^-.

The calibration of instruments to measure free-radical concentrations in the atmosphere represents a difficult challenge, as free-radicals are very short-lived and are removed rapidly at surfaces. No calibration is necessary for free-radical measurements by open-path differential optical absorption spectroscopy or cavity ring-down spectroscopy (e.g. NO_3, IO, see Chapter 3), as the methods are self-calibrating, but absorption methods that use enclosed multipass cells still need to be calibrated to assess any wall losses. It is necessary to generate the free-radicals at the total atmospheric pressure sampled by the instrument (ideally in air). Methods normally involve a chemical titration reaction, or the photolysis of a suitable precursor molecule, sometimes followed by conversion to the radical of interest. For OH radicals, a known concentration in a flow at a total pressure of a few Torr (suitable for in situ instruments that measure in the stratosphere) can be generated from the reaction $H + NO_2 \rightarrow OH + NO$, with H atoms generated in a microwave discharge of H_2. NO_2 is the limiting reagent, and its partial pressure in a flow is calculated from the ratio of the flow of NO_2 (measured using a calibrated mass flow controller) to the total flow ($NO_2 + H_2 +$ any buffer gas). This method is not suitable for the calibration of instruments to measure OH in the troposphere at close to one atmosphere total pressure. In this case, the most commonly used method is the photolysis of water vapour at 184.9 nm, using radiation from a mercury lamp, in a flow of synthetic air which generates OH and HO_2 radicals in a 1:1 ratio. If only a source of HO_2 radicals is required, addition of CO to the calibration gas mixture rapidly converts OH to HO_2 radicals. For CH_3O_2, the photolysis of CH_3I at 254 nm followed by the fast reaction of CH_3 with O_2 has been used to generate CH_3O_2 radicals. Using OH as an example, the OH concentration following the photolysis of water vapour is given by:

$$[OH] = [H_2O]\ \sigma_{H_2O,184.9\,nm}\ \phi_{OH}\ F_{184.9\,nm}\ t \qquad (1.6)$$

where σ is the water vapour absorption cross-section, ϕ is the photodissociation quantum yield of OH and F is the photon flux of the lamp, all at 184.9 nm, and t is the photolysis exposure time. $[H_2O]$ can be measured using a commercial device (e.g. dew-point hygrometer), and the cross-section and quantum yield are well established. However, it is difficult to measure F and t, although this has been done by several groups, using photodiodes that are absolutely calibrated against a national standard to measure F, and by using pitot tubes to measure flow-velocity and hence t (Stevens *et al.*, 1994). The product of F and t can also be measured using a chemical actinometer. In the example above, O_2 present in the air flowing together with H_2O vapour is also photolysed at 184.9 nm to generate two O atoms, which recombine rapidly with O_2 to form two molecules of O_3, which can be measured using a commercial analyser based on UV absorption spectroscopy (Schultz *et al.*, 1995). There are other complications for calibration of OH instruments that are peculiar to the individual method used. Further details on the calibration of OH instruments using laser-induced fluorescence spectroscopy can be found in Chapter 4 (Section 4.3.2).

For aerosols, commercial instruments are available for the generation of a range of aerosol diameters, and their size and number distribution can be determined using particle mobility analysers. Hence for an aerosol time-of-flight mass spectrometer, the response of the instrument for a given aerosol diameter (determined by the efficiency of sampling, volatisiation and ionisation) can be calibrated by using the previously characterised aerosol source. Calibration of the response for aerosol composition can be determined by generating aerosols from solutions with known concentrations of trace species (e.g. sulphate, nitrate, ammonium, chloride, volatile organic compounds), and monitoring a given mass-to-charge ratio (e.g. $m{:}z = 44$ is often used to monitor the sum of organics).

1.6.3 Intercomparison of instruments

Confidence in a field measurement is increased if the same result can be obtained with two or more instruments that utilise independent methodologies whilst sampling at the same location. Following the development of a novel analytical method or instrument, the community expects at some stage an intercomparison with another accepted and thoroughly tested method (if of course one exists). Each of the subsequent chapters contains sections describing intercomparisons between the analytical method in question and other types. Intercomparisons are an important part of quality assurance, and should be built into the design of field experiments, but unfortunately they are not as commonplace as perhaps they should be. Care must be taken to take into account any differences in spatial resolution of the instruments that may lead to sampling of slightly different air masses, and also differing integration periods. The alternative to positioning instruments side-by-side and performing ambient measurements with the same air mass is to circulate calibration standards around several instruments, which may be part of networks on different continents. The latter is more common for chromatography instruments using calibrated mixtures of volatile organic compounds. Formal blind intercomparisons between instruments is often facilitated by an impartial referee, who collects data from different participants for comparison, and who may also operate calibration equipment to provide concentrations of species to the various instruments known only to the referee. Simulation chambers are an ideal environment to provide stable conditions over long periods for intercomparison studies (Section 1.9).

The very low uncertainties that are now available for some techniques are illustrated by the results from an intercomparison of vertical column amounts of HCl, HF, N_2O, HNO_3, CH_4, O_3, CO and N_2 derived from the spectra recorded by two ground-based FTIR spectrometers (Chapter 2) operated side-by-side using the sun as a light source (Meier *et al.*, 2005). For most gases, the differences were typically 1% or less, the main exceptions were HNO_3 and CO_2 (2–3% difference) when spectra were measured using Mercury Cadmium Telluride (MCT) detectors, and due to effects of detector non-linearity.

1.6.3.1 Intercomparison case study: The hydroxyl radical (OH)

Intercomparisons have been crucial for the development of techniques for the atmospheric detection of the OH radical, on account of the early problems with its atmospheric detection using LIF spectroscopy, when signals were almost entirely due to laser-generated artefacts (Heard & Pilling, 2003). Measurements of OH at the same time and location using two or more completely different methods are rare, yet a comparison of the data in different environments is the *best* method to validate the methodology used, and to demonstrate to the atmospheric chemistry community the reliability of OH measurements. Accurate measurements of OH are essential to provide robust modelling of atmospheric chemistry. There have been several intercomparisons of instruments to measure OH under field conditions, and these have been reviewed by Heard & Pilling (2003).

During the 1993 Tropospheric OH Photochemistry Experiment (TOHPE), an extensive intercomparison was carried out at Fritz Peak in the Rocky Mountains, Colorado (Mount & Eisele, 1992). OH concentrations were measured using the in situ chemical ionisation mass spectrometry technique (see Chapter 5) and the open-path DOAS technique (over a long-path of 2 km × 10 km, see Chapter 3), and agreed within error limits (±30%, at the 95% confidence limit) over half of the time; and over a quarter of the time, disagreement could be explained by differences in concentrations of trace gases, for example NO_x, that control the OH abundance over the long-path and at the in situ site. In situ OH measurements by LIF at low pressure (FAGE, see Chapter 4) and folded long-path absorption (DOAS, 38.5 m between mirrors, see Chapter 3) were carried out during the POPCORN campaign held in a clearing in a cornfield in rural North-East Germany in August 1994 (Hofzumahaus *et al.*, 1998). An extensive OH dataset allowed an intercomparison of relative diurnal profiles and simultaneously measured absolute concentrations, as shown in Figure 1.9. The mean detection limit for the LIF instrument was 5×10^5 molecule cm^{-3} ($S/N = 2$, measurement time of 60 s), with a calibration accuracy of ~16%, compared to a mean DOAS detection limit of 1.5×10^6 molecule cm^{-3} ($S/N = 2$, integration time of 200 s), with an uncertainty of only 7%. A detailed statistical analysis yielded a correlation coefficient of $r^2 = 0.81$ and a weighted linear fit with a gradient of 1.09 ± 0.04, with LIF concentrations being lower than average by 9%.

In March and April 2001, the NASA DC-8 and P-3B aircraft flew within ~1 km of each other on three occasions during the TRACE-P campaign in the western Pacific. (Eisele *et al.*, 2003). OH was measured on the DC-8 and P-3B using the FAGE and CIMS techniques, respectively. Although both OH instruments tracked changes in OH production associated with changes in NO_x and $J(O^1D)$, a plot of the CIMS versus FAGE

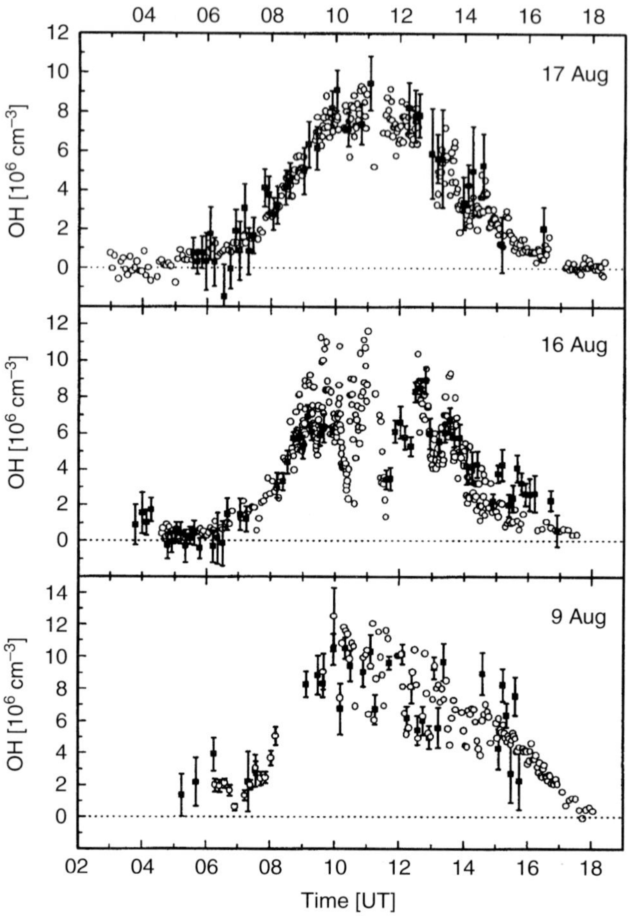

Figure 1.9 OH intercomparison during the 1994 POPCORN field campaign. Diurnal concentration profiles of OH measured by FAGE (open circles; scaled by a factor of 1.09) and DOAS (solid squares) on 9, 16 and 17 August 1994. Local noon is 11:17 UT. (Taken with permission from Hofzumahaus *et al.*, 1998.)

OH data gave a constrained fit of slope 1.5 (within combined uncertainties). A calibration problem associated with these measurements was suggested – not surprisingly, given the difficulty of calibrating these instruments in flight.

1.7 Atmospheric chemistry and policy

1.7.1 Health effects and environmental policy

For some trace species, emissions and concentrations are subject to a number of national and international agreements, and are primarily based on health effects. In the UK three species are most closely targeted and regulated: NO_2, O_3 and particulate matter. At high concentrations, NO_2 causes inflammation of the respiratory airways, and there is evidence that long-term exposure may affect lung function and that for some individuals exposure enhances the response to allergens (Pilling, 2004). The European Union (EU) Directive has an hourly averaged objective of 200 $\mu g\,m^{-3}$, with 18 exceedances per year, and an annual average objective of 40 $\mu g\,m^{-3}$. NO_2 also contributes to the formation of ozone and of secondary particles and is a contributing factor for acidification and eutrophication. Under the EU Directive, the UK is committed to reducing its national annual emission of NO_x from 1728 kTonnes as NO_2 in 2000 to below 1167 kTonnes (Pilling, 2004) by 2010.

Ozone is harmful to vegetation, with a critical level of 40 ppb over a growing season, and has a damaging effect on certain materials, for example polymers, rubber and textiles, and is the most common irritant of the common air pollutants and exposure to high concentrations produces inflammation of the respiratory tract, and at even higher concentrations is associated with chronic decline in lung function. For O_3, the standard is 50 ppb for an 8-hour running average.

For particulate matter, the diameter is the key parameter controlling the penetration into the lungs. The metric used in the UK is PM_{10}, which describes the mass of particles with a size less than 10 μm. Health effects resulting from both short-term and long-term exposure are linked mainly to respiratory and cardiovascular disorders. The EU limit for PM_{10} is a concentration of 50 $\mu g\,m^{-3}$ for a 24-hour day with up to 7 exceedances per year by 1 January, 2010, with the annual limit being 40 $\mu g\,m^{-3}$ (by 1 January, 2005) decreasing to an indicative limit value of 20 $\mu g\,m^{-3}$ by 1 January, 2010 (Pilling, 2005).

1.7.2 Monitoring networks

In order to monitor trends and exceedances in air quality standards, and also in long-term trends in ozone-depleting and greenhouse gases, a variety of ground-based measurement networks have been set up worldwide on a range of spatial scales. Good accuracy and especially good precision are important. With the advent of measurements of trace species in the troposphere from satellites (Section 1.4.8), global coverage is now becoming possible, albeit with large grid-sizes at the surface (although this is improving dramatically), and with limited temporal coverage during a diurnal cycle because of a single overpass time and the presence of clouds.

1.7.2.1 Air quality networks

Automatic networks are operated at both the national and the regional scale. In the USA, the Environmental Protection Agency (EPA) http://www.epa.gov/ operates monitoring networks for several trace species, including CO, O_3, NO_x, SO_2, dioxins, and PM_{10} and $PM_{2.5}$ particulates, and UV radiation, in some cases from hundreds of sites. For $PM_{2.5}$, there are anion, cation and carbon analyses using ion chromatography and other methods from filter samples. Some data are available on-line (e.g. for O_3 and particles) together with a national outlook, and the EPA is legally obliged to issue final designations for all areas of the country stating that they either meet or do not meet the 8-hour ozone standards. The EPA funds the Clean Air Status and Trends Network (CASTNET), which has operated since 1987, to measure dry acidic deposition and rural, ground-level ozone, and together with other monitoring networks it is able to evaluate the effectiveness of emission control strategies. The CASTNET consists of over 80 sites, and includes weekly measurements of sulphate, nitrate, ammonium, sulphur dioxide, and nitric acid, and hourly measurements of O_3 concentrations.

In the UK, funding comes from the Department of the Environment, Food and Rural Affairs (DEFRA), as well as from local authorities, for monitoring networks to measure CO, ozone, NO_2 (and NO_x), particulate matter (PM_{10}), SO_2 and hydrocarbons. CO preferentially combines with haemoglobin in the blood, reducing its oxygen-carrying capacity, and leading to headaches, depressed heart and breathing rates and death at high levels.

A list of monitoring sites in the UK Automatic Urban and Rural Network can be found on the website http://www.stanger.co.uk/siteinfo/Default.asp, and also on the air quality archive http://www.airquality.co.uk/archive/index.php, with up-to-date information on current and historic air quality. The number of sites increased from 74 at the end of 1996 to 120 by March 2004. Rural sites mainly monitor O_3 but also SO_2, NO_x and PM_{10} in some cases. The DEFRA also funds a hydrocarbon network (35 sites) monitoring hourly concentrations of 25 volatile organic compounds, including two carcinogenic compounds, benzene and 1,3 butadiene, in urban roadsides, urban background and rural locations across the country (Fowler *et al.*, 1997). Hourly data on ambient SO_2, NO_x, O_3, CO and particulate matter concentrations are used to provide air pollution information to the public each day. NO_2 concentrations are measured using diffusion tubes and chemiluminescence analysers (Chapter 7). For a study in 2001, the UK Air Quality Expert Group considered NO_2 measurements from about 200 automatic monitoring sites (Pilling, 2004). The number of sites had expanded significantly in order to meet the requirements of EU directives. Many of these sites also have co-located O_3 chemiluminescence analysers. In addition, there is a large-scale survey of NO_2 concentrations using passive diffusion tubes, involving over 1300 sites operated by more than 300 local authorities.

The Tapered Element Oscillating Microbalance (TEOM, Chapter 6) is the most commonly used instrument used for PM_{10} monitoring. This continuous sampler provides real-time data with a short time resolution ($<1\,hr$) and a precision of $\pm 2\,\mu g\,m^{-3}$, but a disadvantage is that it incorporates a heated manifold, which may lead to losses of semi-volatile species such as ammonium nitrate. Hence a correction is necessary when comparing with measurement of PM_{10} using other methods. Inside the instrument is a tapered glass tube, whose frequency of oscillation depends upon its mass. Any change in the mass of the tube, due to the deposition of particles onto a small filter affixed to one end, will result in a change in the resonant frequency, with the change being proportional to the additional mass. The instrument uses a size-selective inlet (see Chapter 6 for details of inlets used for aerosol measurements). There are 240 PM_{10} monitoring sites in the UK, 15 of which have co-located $PM_{2.5}$ analysers, and measurements are also made of black smoke, polycyclic aromatic hydrocarbons (PAHs), heavy metals (lead, nickel, arsenic and cadmium) and major ions (such as sulphate, nitrate and chloride).

Measurement uncertainty is very important if data are to be used to monitor health damaging pollutants and to enforce/influence legislation. The objectives set down by EU directives for NO_2 are 15% for continuous measurements (e.g. using chemiluminescence analysers) and 25% for indicative measurements (e.g. using diffusion tubes), at the 95% confidence level for individual measurements averaged over the limit period (e.g. 8 hr). Quality control arrangements ensure that measurement precision standards of at least $\pm 10\,\mu g\,m^{-3}$ for SO_2, $\pm 8\,\mu g\,m^{-3}$ for NO_x and O_3, $\pm 0.5\,mg\,m^{-3}$ for CO and $\pm 4\,\mu g\,m^{-3}$ for PM_{10} are achieved at automatic monitoring sites. Further details on uncertainties of particular techniques are given in later chapters. Industry is under increasing pressure to clean up its exhausts, under penalty of heavy fines for non-compliance, and many gas and particulate analysers are employed to monitor stack emissions.

1.7.2.2 Networks for long-term measurements of stratospheric ozone depleting substances, greenhouse gases and tropospheric ozone precursors

Air quality networks tend to be regional and within the same country. There are also a number of monitoring networks that transcend continental boundaries and which have been operating for over thirty years. These networks document global change, provide data to evaluate ozone depletion and climate change, enable the predictions of models to be tested, and inform policy makers. Finally, with the advent of long-term measurements of tropospheric trace species from satellites, the ground-based networks provide an important validation service.

The World Meteorological Organization has set up the Global Atmospheric Watch (GAW) programme, which coordinates networks to measure ozone (total column, tropospheric, vertical profiles), greenhouse gases, solar radiation, aerosol characteristics and tropospheric ozone precursors (CO, NO_x). The NOAA Climate Monitoring and Diagnostics Laboratory (CMDL, http://www.cmdl.noaa.gov/) programme supports measuring sites all around the world, both on the ground and using tall towers and aircraft platforms. Sites are located at Point Barrow, Alaska (71°N, 156°W), Trinidad Head, California (41°N, 124°W), Mauna Loa (19.5°N, 156°W, 3397 m above sea level), American Samoa (14°S, 171°W) and the South Pole (90°S, 102°W). As well as in situ instruments, whole air samples taken using flasks at a range of sites, all away from local influences, are shipped back to the USA for analysis. Surface measurements have revealed distinctive seasonal and latitudinal patterns, as well as long-term trends, for several species, for example CO, CO_2, CH_4, N_2O, O_3 (tropospheric and stratospheric), hydrocarbons, halogenated VOCs, aerosols and radiation. Another network consisting mainly of in situ instruments is the Atmospheric Lifetime Experiment/Global Atmospheric Gases Experiment ALE/GAGE, which has been measuring CH_4, N_2O, several CFCs and other chorine compounds using gas chromatography (Chapter 8) since 1978 [*Prinn et al.*, 1983]. The advanced GAGE (AGAGE, http://agage.eas.gatech.edu/) now uses fully automated instruments. The AGAGE monitoring sites are Mace Head (Ireland, 53°N, 10°W), Cape Grim (Tasmania, 41°S, 145°E), Trinidad Head (California, 41°N, 124°W), Ragged Point (Barbados, 13°N, 59°W) and Cape Matatula (American Samoa, 14°S, 171°W), with data also obtained at Ny-Ålesund (Svalbard, 78.9°N, 11.9°E) and the Jungfraujoch (Switzerland, 46°N, 8°E, 3580 m above sea level). Another ground-based network is the Network for Detection of Stratospheric Change (NDSC), which monitors stratospheric change (temperature, O_3, other trace species and radiation) using a range of remote sensing instruments. A further example is the GAW ozonesonde and CO_2 network.

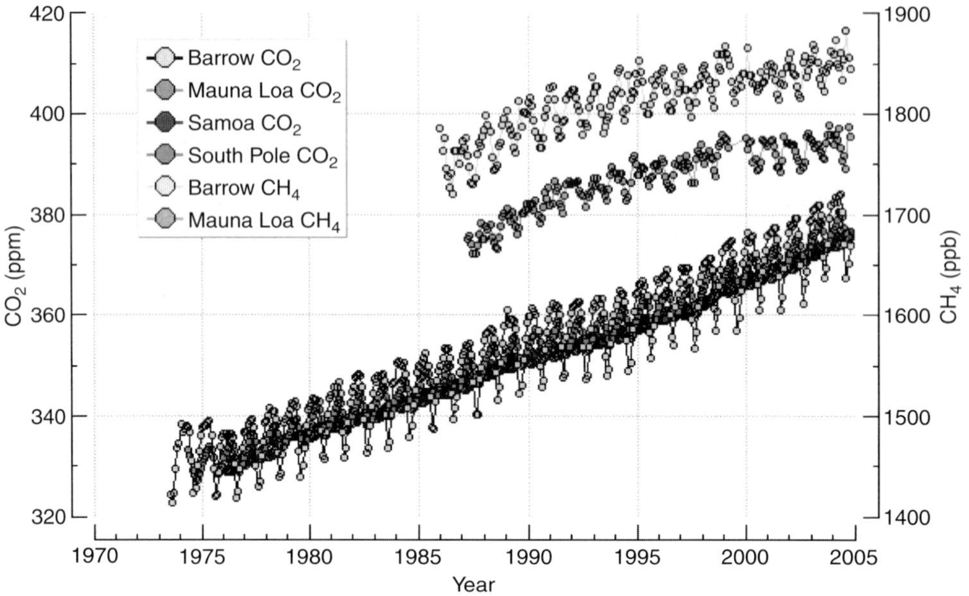

Figure 1.10 Concentrations of the greenhouse gases methane and CO_2 measured by the NOAA CMDL monitoring network since 1973. Methane is measured using a gas chromatograph with a flame ionisation detector and CO_2 is measured using a non-dispersive infrared (NDIR) analyser. (Data courtesy of Pieter Tans, NOAA.) (Reproduced in colour as Plate 2 after page 270.)

Figure 1.3 shows the long-term trend in concentrations of CFC-11 ($CFCl_3$) and CFC-12 (CF_2Cl_2), and Figure 1.10 shows measurements of CO_2 and CH_4 from 1973 to 2005, taken at a variety of remote, background sites. $CFCl_3$ (atmospheric lifetime of 45 years) was banned following the 1987 Montreal Protocol, and as shown concentrations are now beginning to fall off. However, concentrations of CF_2Cl_2 have been slower to respond since their ban following the Montreal Protocol, due to the longer atmospheric lifetime (100 years) of this species. However, the loading of chlorine in the stratosphere will not fall below the critical value for formation of the ozone hole for at least another fifty years or so, according to predictions from a range of global climate models.

Although not a monitoring network, Berresheim *et al.* (2005) have made a study of the long-term behaviour of the OH radical over five years at the Hohenpeissenberg observatory in Southern Germany, using a chemical ionisation mass spectrometer (see Chapter 5). In this way a seasonal study of the behaviour of this radical has been possible. The results indicate a very close correlation between OH radical concentrations and the rate of OH production (P(OH)) from the rate of ozone photolysis to generate $O(^1D)$ atoms (referred to as $J(O^1D)$, see Chapter 9). A linear relationship is expected from Equation 1.3, with the slope of the correlation changing as the chemical lifetime of OH (by reaction with its sinks) changes when there is a change in the sampled air mass, perhaps bringing in more polluted air from urban areas.

Monitoring of long-lived anthropogenically emitted species (e.g. methyl chloroform, CH_3CCl_3), whose loss rate is controlled almost totally by OH radicals has been used to infer globally averaged OH concentrations, although the value obtained

(1.1×10^6 molecule cm^{-3}) is weighted towards the tropics, where most of the CH_3CCl_3 is oxidised. Measurements of CH_3CCl_3 between 1978 and 1994 did not suggest any significant trend in [OH], which suggests that production rates of OH during this period must have increased (e.g. UV levels due to reduced ozone in the stratosphere), as the concentrations of two key sinks of OH – CO and CH_4 – increased during this period. However, concentrations of CH_3CCl_3 are falling year-on-year, as it is a banned ozone depleting substance following the Montreal Protocol, and it will not be easy to infer any future long-term trends in OH concentrations, and hence the oxidising capacity of the atmosphere, from future measurements of CH_3CCl_3.

1.8 Major field campaigns for measurement of atmospheric composition

Major intensive field campaigns are normally mounted over a period of 1–3 months. The duration maximises the probability of experiencing the desired meteorological conditions (fine weather) to achieve the scientific objectives. For a ground-based experiment, the past history of the sampled air mass, the range of emissions the air has encountered over the previous five days or so (viewable using back trajectories), is the most critical parameter, as it controls the mix of trace gases. Often particular wind directions are undesirable as they allow local sources to interfere with the experiment. For a Lagrangian study, when it is desired to study the evolution of composition within a moving air mass, perhaps between a land-based observatory and a ship or aircraft, or between two or more aircraft over a very wide geographical area, the opportunity for connected flow is quite limited, and much use is made of forecast modelling. Process studies are often best studied using intensive field campaigns involving large number of instruments and scientists from national or international consortia. The advantage of this approach is that models can be constrained with a large number of input parameters, and it is possible to study several research strands at the same site. One missing measurement can endanger an entire science objective, and instrument redundancy should be built into field campaigns, which are expensive to operate. Even if only two instruments for measurement of the same species work perfectly, confidence in that measurement is higher as an intercomparison can be performed.

1.8.1 The design of field campaigns

It is now accepted that the concerted deployment of a suite of field instrumentation is necessary to provide a detailed understanding of the chemical composition for a given environment. Naturally, it is not possible to measure every conceivable trace species, and campaigns must be designed to cover the necessary measurements to deliver a particular science objective. Modelling plays a key role over a range of length scales, from processes at local scale to long-range transport. There are examples of field sites where multiple intensive campaigns have been mounted over a period of several years, with each successive campaign becoming more and more comprehensive in terms of species coverage, as technology evolves and new laboratory techniques make it into the field.

During the Eastern Atlantic Spring Experiment at Mace Head in 1997, OH and HO_2 measurements were made using LIF (see Chapter 4), and compared with the predictions of zero-dimensional model. Despite the model being constrained with a large number of sources and sinks of HO_x, and the most up-to-date kinetics and photochemical data being used, the model consistently overpredicted both species, sometimes by up to a factor of 1.5 to 2, suggesting that important sinks of each species were not being measured. Upon return to Mace Head in 2002 for the North Atlantic Marine Boundary Layer Experiment (NAMBLEX), a wider range of species was measured, notably oxygenated volatile organic compounds (OVOCs) – methanol, acetone and acetyaldehyde – and halogen oxides (IO and BrO). The concentrations of OVOCS were found to be removing a larger fraction of OH molecules than non-methane hydrocarbons, and indeed the agreement of the model with measured OH was much better once OVOCs were included. For HO_2, the levels of measured IO and BrO were found to remove a significant fraction of HO_2, and model–measurement agreement was better when these reactions were included. With the benefit of hindsight, it became clear that measurements of key species necessary to fully describe the chemistry were missing in the previous campaign.

The requirements of a measurement platform and instrument portfolio depend on the precise aims of the study. If CO_2 or methane concentrations representative of the background atmosphere are required to assess any trends in their contribution to global warming, then a ground-based field site remote from any major emission sources of these compounds is needed. Ideally, in order to provide consistency checks and to rule out local sources, a network of such stations is needed (see Section 1.7.2). If the objective is to understand the destruction of ozone in the lower stratosphere over a range of geographical locations, then probably one would choose a combination of remote sensing methods with good vertical resolution (e.g. airborne LIDARs or satellite instruments) with in situ instruments aboard high flying aircraft, (e.g. the ER-2) to measure ozone and a range of short-lived free-radical species and their longer-lived precursors, together with aerosol species (e.g. polar stratospheric clouds).

The number of field campaigns mounted is now very large indeed, and it is out of the scope of this book to try to attempt a comprehensive compilation or discuss them in any detail. In a 2003 review by Heard & Pilling, 46 major field intensive campaigns, between 1991 and 2001, which involved tropospheric measurements of OH and HO_2 radicals were listed. Table 1.3 provides a very selective list of some of the major field campaigns that have taken place in the last five years, and is included to illustrate the wide variety of platforms and geographical locations where measurements have been taken. There are a number of special issues in journals dedicated to the results from intensive field campaigns, and some of these are listed in 'Further reading'. For illustrative purposes, the details of one field campaign are given in Section 1.8.2.

The volumes of data created during a comprehensive field campaign are truly enormous, and if the project is to succeed, there must be an agreed mechanism, and deadline, for depositing field data. All large field campaigns should have a central database. In the UK, all fieldwork data for atmospheric composition are managed by the British Atmospheric Data Centre (BADC). The BADC is centrally involved at all stages of the field project from its inception, and facilitates the agreement of data protocols to protect intellectual property rights, with data normally released to the general public after some agreed delay following the end of the project. The Centre also provides the tools, technical support and secure environment for on-line deposition of data, using an agreed, common format, usually NASA Ames or Net-CDF, and is responsible for setting up password-protected workspace areas where data can quickly be deposited by project participants and shared with other users prior to the formal delivery of validated data to the central archived database.

1.8.2 Case study of a field campaign: The 2002 NAMBLEX campaign

In the summer of 2002, ca. 50 scientists from 12 institutions participated in the NAMBLEX campaign, at Mace Head (53°N, 10°W), on the West Coast of Ireland. Mace Head experiences mainly unpolluted air from the Atlantic Ocean brought in by the prevailing westerly winds. The site is part of AGAGE and is a GAW Station (Simmonds *et al.*, 1996). The aims of NAMBLEX were to test quantitatively the basic understanding of oxidation processes in clean and moderately polluted air using comparisons of measured and model-predicted free-radicals, to study extensively the chemistry of halogen species in the marine boundary layer (MBL) through observation of reactive intermediates and their sources and sinks, to examine the origins and role of reactive hydrocarbons in the MBL, and to investigate the size-distributed composition of aerosols and the processes involved in new particle creation. To achieve these aims a wide range of gas phase and aerosol species was measured using a range of state-of-the-art instrumentation, as shown in Table 1.4, which also lists averaging times and detection limits. Wherever possible, inlets were positioned at the same height above the ground (10 m is often used), and their horizontal displacement was minimised. Figure 1.6 shows a photograph of the main field site at Mace Head during NAMBLEX. The majority of the instruments were deployed in bespoke converted shipping containers, of standard size for convenience of transportation by road or sea. These mobile laboratories are self-sufficient, requiring only a source of electricity. Some instruments were also housed in buildings adjacent to the containers. Measurements of several free-radical species were made, namely OH, HO_2, RO_2, NO_3, BrO and IO (see also Table 1.1 for methods of their detection), together with their sources and sinks, and instrument inlets were placed as close as possible. For some species (e.g. CH_2O, peroxides, NMHC) measurements were made using more than one technique, enabling an intercomparison to be performed (see Section 1.6.3).

Chemists, physicists and meteorologists worked together closely during NAMBLEX. Detailed chemical compositional measurements ran alongside physical measurements of the structure of the boundary layer, using, for example, a small tethered balloon equipped with a basic met package (with telemetry), a SODAR (sonic detection and ranging) and a wind-profiler (1290 MHz Doppler radar) for vertical distributions of wind speed and direction, and several sonic anemometers mounted at various heights on masts, to measure the u, v and w wind components in real time. Combining the measurements from the wind profiler, SODAR and sonic anemometers allow the validity of trajectory calculations to be tested. There were several events where the local wind direction at the site was decoupled from the synoptic long-range flow, as represented by five-day air mass back-trajectories, which indicate the trajectory of the air sampled at the site during the previous five days. Air mass trajectories were calculated using wind fields modelled by

Table 1.4 Species measured and instruments deployed during the North Atlantic Marine Boundary Layer Experiment (NAMBLEX), Mace Head, Ireland, held during the summer of 2002

Species measured	Analytical technique	Temporal resolution	Detection limit	Chapter
OH	Fluorescence Assay by Gas Expansion (FAGE)	30 s	2×10^5 molecule cm^{-3} (150 s), 0.002 pptv	4
HO_2	Addition of NO to convert to OH, then FAGE	30 s	1×10^6 molecule cm^{-3} (150 s) 0.04 pptv	4, 7
$\Sigma RO_2 + HO_2$	Peroxy Radical Chemical Amplifier	1 min	0.4 pptv (10 min)	7
NO_3, IO, OIO, BrO, I_2	Differential Optical Absorption Spectroscopy, optical path 2×4.2 km	5 min	0.5, 0.8, 5, 2, 5 pptv	3
NO_3, OIO, I_2	Broad-band cavity ring-down spectroscopy	100 s–10 min	1, 4, 20 pptv	3
Vertical profiles of NO_3	Zenith-pointing DOAS and altitude retrieval	1 profile/day	$(1–5) \times 10^{13}$ molecule cm^{-2}	3
Photolysis frequencies, including $J(O^1D)$, $J(NO_2)$	Calibrated filter (2π sr and 4π sr) radiometers and scanning spectral-radiometer	1 s		9
O_3	UV absorption spectroscopy	1 s	0.6 ppbv	3
NO	NO/O_3 chemiluminescence detectors	10 s	1 pptv	7
NO_2	Photochemical convertor $NO_2 \rightarrow NO$		2 pptv	7
ΣNO	Gold tube/CO converter $NO_y \rightarrow NO$		5 pptv	7
ΣNO_y-HNO_3	Gold tube convertor/denuder to remove HNO_3		10 pptv	7
H_2O_2, CH_3OOH	Scrubbing/derivatisation, high performance liquid chromatography	8 min	10 pptv	7
H_2O_2, Σ ROOH	Dual-channel fluorometric detector	20 s	10 pptv	7
H_2O vapour	IR absorption Broad band cavity ring-down spectroscopy	10 s	5 ppmv	2
CH_2O	Fluorometric detection (Hantzsh reaction)	1 min	20 pptv	7
	Selective trapping and gas chromatography / helium ionization detection	15 min	50 pptv	8

Table 1.4 (Continued)

CH_4, speciated C_2-C_7 non-methane hydrocarbons, including isoprene, DMS	Automated GC/flame ionisation detection using programmed temperature injection Seawater sampling using head-space gas chromatography	40 min	1 pptv	8
CH_3OH, $(CH_3)_2CO$, CH_3CHO	Gas chromatography with flame ionisation detection, selective column for oxygenates	40 min	1 pptv	8
Peroxy acetyl nitrate (PAN)	Gas chromatography/electron capture detection	5 min	10 pptv	5
Alkyl nitrates	Negative ion chemical ionisation gas chromatography/mass spectrometry	45 min	<1 pptv	8
Reactive halocarbons	Gas chromatography/mass spectrometry	40 min	<1 pptv	5
Aerosol number and size distributions (3 nm to 20 μm)	Scanning mobility particle sizer (3–500 nm), range of optical particle counters (0.1–300 μm), CN counters, >3, 5, 10 nm)	1 min		6
Hygroscopic properties (growth factors) for 20–300 nm aerosol particles	Hygroscopic tandem differential mobility analyser	5 min		6
Aerosol chemical composition	Aerosol mass spectrometer (Aerodyne Inc.)	0.05 s	100 particles s^{-1} (0.25 μg m^{-3} for a 3 min average	6
	Aerosol time of flight mass spectrometer (TSI Inc.)			6
	Cascade impactors (filters) with ion chromatography (NH_4^+, Na^+, Cl^-, SO_4^{2-})	12 hr		6,8
Vertically resolved wind-speed and direction *u*, *v*,	Wind profiler (Doppler radar), Sodar, tethered balloon	15 min		1
w wind velocity	Sonic anemometers	0.1 s		1
CO	Gas chromatography/hot mercuric oxide reduction gas detector coupled with UV detection	30 min		8

Table 1.4 (Continued)

Species measured	Analytical technique	Temporal resolution	Detection limit	Chapter
CO	Gas chromatography/flame ionisation detector	40 min		8
CO_2 (in situ)	Non-dispersive infrared spectroscopy	2 hr		2
CFCs, HCFCs, HFCs	Gas chromatography/electron capture detector/mass spectrometer	2 hr		5, 8
H_2	Gas chromatography/hot mercuric oxide reduction gas detector coupled with UV detection	40 min		8
CH_4	Gas chromatography/Flame ionisation detector	40 min		8

the European Centre for Medium Range Weather Forecasting (ECMWF), which can be downloaded from the website http://badc.nerc.ac.uk/data/ecmwf-trj/. It was possible to observe the formation of an internal boundary layer during the transition from the ocean to the land (where the surface roughness increases), and also to identify when sea-breeze effects dominated, periods when trajectory calculations used to deduce air mass origins are unreliable. This information is critical for the correct positioning of instruments relative to one another, and to interpret long time-series of composition measurements.

1.9 Instrumented chambers for the study of simulated atmospheres

The atmosphere is not very accommodating. Extended bad weather or simply not sampling air masses with the desired history during a campaign can be highly frustrating, and the scientific impact of what is usually an expensive operation is seriously compromised. Being able to choose the composition of the atmosphere that is sampled offers many advantages, especially when stable conditions are required over long periods, perhaps in order to perform instrument intercomparisons or test the zero response of an instrument. In more polluted environments, it is not possible to study the oxidative degradation of a single VOC via field measurements, in order to compare with model calculations, as many thousands of VOCs may be present. Atmospheric chambers with simulated atmospheres are ideal if it is desired to study the mechanism for degradation of one species at a time. It is possible to systematically vary individual parameters under controlled conditions, and to remove the vagaries of the meteorology.

There are a number of highly instrumented outdoor atmospheric chambers worldwide, which use the sun's radiation to initiate photochemistry. Excluded in this section are indoor smog-chambers which use artificial light sources, such as black lamps or xenon lamps, to initiate photochemistry. Indoor environmental chambers are discussed in some detail by Finlayson-Pitts & Pitts (2000). Two examples of large outdoor chambers are the EUPHORE (European Photoreactor Program) community facility, which consists of two hemispherical chambers at CEAM (The Mediterranean Centre for Environmental Studies), Valencia, Spain, and the SAPHIR (Simulation of Atmospheric Photochemistry in a Large Reaction) chamber at Forschungszentrum Jülich, Germany. Photographs of the chambers are shown in Figure 1.11. Both chambers are highly instrumented, containing a wide variety of in situ instruments, using most of the techniques described in this book. A feature of both chambers is the ability to measure free-radical species, for example OH and HO_2 radicals. The chambers are equipped with motorised shutters and hence can also be operated in the dark. A useful feature is the ability to quickly switch on or off the solar radiation, and monitor the rise or fall of radical precursors or the free-radicals themselves. It is found that the walls of the chamber are sources of free-radicals, and this must be characterised thoroughly. The Jülich SAPHIR chamber can accommodate several standard-sized shipping containers underneath it, with sampling probes entering from beneath, and has been used for a number of intercomparisons, providing stable conditions that can be altered at will over long periods.

An EU-funded project EUROCHAMP (Integration of European Simulation Chambers for Investigating Atmospheric Processes) aims to provide a grid of environmental trace gas and aerosol chambers designed for the scientific investigation of atmospheric chemical processes.

1.10 Future directions

In recent years there has been an increase in the number of species monitored, the type of platform on which measurements are possible, and, importantly, an improvement in the accuracy and precision of reported measurements. Important questions are the following: Which species do we really need to measure but are not currently being measured? Which species already measured need to be measured better? Are there emerging and promising techniques in the laboratory, that may enable new or better measurements? Which regions of the atmosphere are still poorly characterised and merit further study?

In the following chapters the authors have each highlighted current developments and future needs for specific analytical techniques, and these points are not repeated here. Research grant funding for future development of analytical techniques is largely driven by the scientific needs with respect to the individual species in the atmosphere. Thus, although in individual chapters there are discussions regarding the future directions for that particular technique (e.g. in Chapter 2, discussions that an improvement in IR laser sources will lead to a better measurement of formaldehyde, CH_2O), the future direction for a particular species may not involve that type of technique at all. Common to all techniques, there is a need to make instruments much smaller and portable, requiring advances in electronics. Intercomparisons and the sharing of calibration standards have no doubt led to improved confidence in atmospheric measurements, but further work is needed. The flow of air into sampling inlets varies from technique to technique, and so a calibration method that works well with one instrument may be unsuitable for another. A possible way forward is to agree a standard inlet type so that calibration sources can be circulated amongst several groups.

(a)

(b)

Figure 1.11 (a) One of the two EUPHORE atmospheric simulation chambers at Valencia, Spain. Further information on this facility, including instruments, can be obtained from http://www.gva.es/ceam/index_i.htm. (b) The SAPHIR atmospheric simulation chamber at Jülich, Germany. For further information, see http://www.fz-juelich.de/icg/icg-ii/saphir.

In a chapter on field measurements within the IGAC Synthesis Book (Tyndall *et al.*, 2003), which was completed in April 2001, the authors (25 in total) speculated about the future of measurement techniques, and considered (a) the evolution of the techniques themselves, and (b) the needs of the atmospheric community. A summary is included here. For (a), the anticipated trends included a wider application of mass spectrometric methods (see Chapter 5), improvements in optical spectrometry (see Chapters 2–4), miniaturised, automated gas chromatographs (see Chapter 8), and continued development of satellite instrumentation (see also below). For (b), there was seen to be a need for more compact, universal instruments that can be readily calibrated. The measurement techniques for some free-radicals were seen to be unsatisfactory, for example RO_2 radicals, or too difficult for routine measurements. A shortlist of requirements included (Tyndall *et al.*, 2003) the following: techniques for high-frequency hydrocarbon and other VOC measurements; rapid, reliable measurements of water vapour, NH_3 and NO_2; an international OH/HO_2 instrument intercomparison; instrumentation for unstable N reservoirs (N_2O_5, HO_2NO_2); the development of single-particle instruments with molecular speciation (see Chapter 6); measurement techniques for X, XO, OXO, HOX (where X = Cl, Br and I); additional monitoring stations in the Tropics; sondes and LIDARS to measure vertical profiles of a wider variety of trace (especially short-lived) gases; and satellite-based instruments for global three-dimensional distributions of trace gases in the troposphere (see also below).

Since 2001, some of these requirements have been realised. For example, an international OH/HO_2 intercomparison involving German, British and Japanese groups was performed in the summer of 2005 at the Jülich SAPHIR chamber (see Section 1.9); aerosol mass spectrometers capable of analysing single particles have been developed and deployed (see Chapter 6); further short-lived halogen species have been detected, including I_2, using DOAS (see Chapter 3); and new satellite instruments, for example SCIAMACHY and those on the AURA satellite, are providing high quality tropospheric measurements. The use of mass spectrometry is certainly on the increase, for example there are significant numbers of proton-transfer-reaction mass spectrometers (using quadrupole and time-of-flight mass spectrometers, see Chapter 5) being used by the community for real-time measurement of a wide range of VOCs (in particular oxygenated species), and two-dimensional chromatography combined with flame-ionisation detection or mass spectrometry (see Chapter 8) is being used to speciate and quantify ever larger numbers of VOCs in polluted air. Future improvements in sensitivity and time-resolution are still necessary, and there is a need for the detection of water soluble, thermally labile and polar organic species which are readily taken up into the condensed phase.

At present, free-radical concentrations in the troposphere are only measured using in situ techniques at selected locations, and for short periods, usually during intensive field campaigns (although there is a five-year study of OH concentrations at a rural site in southern Germany (Berresheim *et al.*, 2005)). There is a need for long-term trends for short-lived reactive molecules in the troposphere, such as ozone and free-radicals, in particular OH (which controls the oxidising of the daytime troposphere), and also global-scale measurements of a wide range of tropospheric free radicals, including OH, HO_2, RO_2, XO (X = I, Br and Cl) and NO_3. Continuous measurements of tropospheric OH from a geographically diverse range of ground-based instruments, or from satellites would, for the first time, enable any long-term trends in global or regional OH levels to be monitored directly. Satellite measurements of free-radicals in the troposphere would be a major advance, but are probably not likely in the near future. Instruments must become more reliable, be capable of unattended operation over longer periods, and require lower levels of operator expertise. There are some very compact instruments undergoing development in the laboratory, for example cavity ring-down spectrometers or locked cavity absorption techniques using small continuous diode lasers, that may become suitable for measurements of free-radicals (Bakowski *et al.*, 2002; Brown, 2003). The idea of intracavity laser spectroscopy for ultra-sensitive detection of atmospheric OH was first suggested in 1991 (McManus & Kolb, 1991) but only around 2000 have suitable continuous wave laser sources at 308 nm become available (Barry *et al.*, 2000).

One of the main reasons for the measurement of short-lived free-radicals is to provide a target species to test the accuracy of chemical mechanisms. However, in order to adequately constrain the model, it is necessary to know the co-measured concentrations of longer species whose reactions are either sources or sinks for the free-radical being investigated. In the case of OH, there are thousands of species in polluted air, mainly VOCs, which remove OH, and despite developments in instrumentation to measure ever larger number of VOCs, it will never be possible to measure all the OH sinks, and therefore completely constrain the model. In polluted environments, modelled OH levels sometimes exceed the measurements, consistent with missing sinks. Another approach is to measure the atmospheric lifetime of OH, τ_{OH}, which is the reciprocal of the total loss rate by reaction with its sinks, and is given by

$$\tau_{OH} = \frac{1}{\Sigma_i k_i[X_i]} \qquad (1.7)$$

where X_i represents a sink for OH, and k_i is the rate coefficient for the reaction $X_i + OH$. One measurement therefore defines the rate of removal of OH by reaction with all its sinks, and several groups worldwide have now developed field instruments for the measurement of τ_{OH} directly (see Chapter 4). Measurement of the lifetime of other short-lived intermediates could also aid in the quantification of their budgets.

Considerable improvements in satellite products can be expected in the near future (Brasseur *et al.*, 2003). The global distribution of aerosols is highly variable and vertically resolved distributions with good sensitivity are necessary to quantify their contribution to the radiative budget of the atmosphere and for comparison with models. New methods under development include the LIDAR in-space technology experiment (LITE), which uses lasers at several wavelengths operating from satellites. Improvements in laser technology may allow more reliable operation of lasers in space, and differential absorption LIDAR (DIAL), which is well proven on the ground, may prove feasible from satellites.

Observations from geostationary orbits (~36 000 km) will enable horizontal resolutions of 10 km or less, and temporal resolutions of ~10–15 min. In contrast to instruments on satellites in low earth orbit, it will be possible to view the same spot continuously, and obtain a full diurnal profile with good temporal resolution. With good spatial *and* temporal resolution it will then be possible to perform process studies, monitoring the evolution of shorter-lived species within pollution events, in a similar way to ground-based intensive field campaigns. Several satellites in geostationary orbit would provide a full coverage of the earth, except for the polar regions. New technology, in the form of detectors, communication and data handling, is being developed for geostationary satellites. It is important to validate any new satellite product via carefully planned ground- or aircraft-based measurements. However, it is not always straightforward to compare in situ measurements with satellite observations.

Acknowledgements

I would like to thank Professor Ulrich Platt from the University of Heidelberg for supplying information about the measurement of atmospheric composition from space using DOAS instruments on satellite platforms (Platt & Stutz, 2006).

Further reading

Books and other general reviews

Brasseur, G., Orlando, J.J. & Tyndall, G. (1999) *Atmospheric Chemistry and Global Change*, OUP, Oxford.

Brasseur, G.P., Prinn, R.G. & Pszenny, A.A.P. (2003) *Atmospheric Chemistry in a Changing World. An Integration and Synthesis of a Decade of Tropospheric Chemistry Research*, 300 pp., Springer-Verlag, Berlin.

Calvert, J.G. (1994) The chemistry of the atmosphere: Its impact on global change, *Chemistry for the 21st Century*, Blackwell Scientific Publications, Oxford.

Clemitshaw, K.C. (2004) A review of instrumentation and measurement techniques for ground-based and airborne field studies of gas-phase tropospheric chemistry, *Critical Reviews in Environmental Science and Technology*, **34**, 1–108.

Finlayson-Pitts, B.J. & Pitts, J.N. (2000) *Chemistry of the Lower Atmosphere. Theory, Experiments and Applications*, Academic Press, San Diego.

Holton, J.R., Pyle, J.A. & Curry J.A. (2003) *Encyclopaedia of Atmospheric Sciences*, Elsevier.

Newman, L. (1993) Measurement challenges in atmospheric chemistry, in *Advances in Chemistry*, M.J. Comstock (ed.), 407 pp., American Chemical Society, Washington DC.

Platt, U. & Stutz, J. (2006) *Differential Optical Absorption Spectroscopy*, Springer, Heidelberg.

Ravishankara, A.R. (2003) Atmospheric chemistry: Long-term issues, *Chemical Reviews*, **103 (12)**, 4505–5262.

Roscoe, H.K. & Clemitshaw, K.C. (1997) Measurement techniques in gas-phase tropospheric chemistry: A selective view of the past, present and future, *Science*, **276**, 1065–1072.

Sievers, R.E. (1995) Selective detectors. Environmental, industrial and biomedical applications, *Chemical Analysis: A Series of Monographs on Analytical Chemistry and Its Applications*, J.D. Winefordner (ed.), 261 pp., John Wiley & Sons, New York.

Wayne, R.P. (2000) *Chemistry of Atmospheres*, Oxford University Press Inc., New York.

Journal special sections for some major field campaigns (selective)

(in chronological order of the campaign dates)

The Tropospheric OH Photochemical Experiment (TOHPE, 1993) *J. Geophys. Res.*, 102 (D5), 1997.

Photochemistry of Plant-emitted Compounds and OH Radicals in Northeastern Germany (POPCORN, 1994) *J. Atmos. Chem.*, 31 (1–2), 1998.

Aerosol Characterisation Experiment. Radiative Effects of Aerosols in the Remote Marine Atmosphere (ACE 1, 1995) *J. Geophys. Res.*, 103 (D13), 1998, 104 (D17), 1999.
Subsonic Assessment Ozone and Nitrogen Oxides Experiment (SONEX, 1997) *Geophys. Res. Lett.*, 26 (20), 1999.
North Atlantic Regional Experiment (NARE, 1993, 1996, 1997) *J. Geophys. Res.*, 103 (D22), 1998, 104 (D11), 1999.
International Photolysis Frequency Measurement and Modelling Intercomparison Study (IPMMI, 1998) *J. Geophys. Res.*, 108 (D16), 2003.
Berliner Ozone Experiment (BERLIOZ, 1998) *J. Geophys. Res.*, 108 (D4), 2003.
New Particle Formation and Fate in the Coastal Environment (PARFORCE, 1998, 1999) *J. Geophys. Res.*, 107 (D19), 2002.
Southern Oxidants Study – Atlanta Supersite Project (SOS, 1999) *J. Geophys. Res.*, 108 (D7), 2003.
Indian Ocean Experiment (INDOEX, 1999) *J. Geophys. Res.*, 107 (D19), 2002.
Pacific Exploratory Mission Tropics B (PEM-Tropics B, 1999, 2000) *J. Geophys. Res.*, 101 (D23), 2001, 108 (D2), 2002.
Southern Africa Regional Science Initiative (SAFARI, 2000) *J. Geophys. Res.*, 108 (D13), 2003.
SAGE III – Ozone Loss and Validation Experiment and Third European Stratospheric Experiment on Ozone (SOLVE-THESEO, 1999, 2000) *J. Geophys. Res.*, 107 (D20), 2002, 108 (D5), 2003.
Investigation of Sulfur Chemistry in the Antarctic Troposphere (ISCAT, 2000) *Atmos. Environ.*, 38 (32), October 2004, pp. 5363–5373.
Tropospheric Ozone Production about the Spring Equinox (TOPSE, 2000) *J. Geophys. Res.*, 108 (D4), 2003.
Intercontinental Transport and Chemical Transformation 2002 (ITCT 2K2, 2000) *J. Geophys. Res.*, 109, 2004.
Mediterranean Intensive Oxidant Study (MINOS, 2001) *Atmos. Chem. Phys.*, 3, 2003.
Transport and Chemical Evolution over the Pacific – Aerosol Characterisation Experiment Asia (TRACE-P, ACE-Asia, 2001) *J. Geophys. Res.*, 108 (D20 and D21), 2003, 109 (D16), 2004.
Characterization of Asian Aerosols and Their Radiative Impacts on Climate (ACE-Asia, 2002 *J. Geophys. Res.*, 108 (D23), 2003, 109 (D19), 2004.
North Atlantic Marine Boundary Layer Experiment (NAMBLEX, 2002) *Atmos. Chem. Phys.*, 5, 2005.

References

Anderson, J.G., Brune, W.H., Lloyd, S.A., Toohey, D.W., Sander, S.P., Starr, W.L., Loewenstein, M. & Podolske J.R. (1989) Kinetics of O_3 destruction by ClO and BrO within the Antarctic Vortex – An analysis based on in situ ER-2 data, *J. Geophys. Res. Atmos.*, 94 (D9), 11 480–11 520.
Bakowski, B., Corner, L., Hancock, G., Kotchie, R., Peverall, R., & Ritchie, G.A.D. (2002) Cavity-enhanced absorption spectroscopy with a rapidly swept diode laser, *Appl. Phys. B-Lasers and Opt.*, 75 (6–7), 745–750.
Barry, H.R., Bakowski, B., Corner, L., Freegarde, T., Hawkins, O.T.W., Hancock, G., Jacobs, R.M., Peverall R. & Ritchie G.A.D. (2000) OH detection by absorption of frequency doubled diode laser radiation at 308 nm, *Chem. Phys. Lett.*, 319, 125.
Berresheim, H., Plass-Dulmer, C., Elste, T., & Rohrer, F. (2005) Five years of OH measurements at the Global Atmosphere Watch Observatory Hohenpeissenberg, *Geophys. Res. Abstr.*, 7, 01902.
Bottenheim, J.W., Barrie, L.A., Atlas, E., Heidt, L.E., Niki, H., Rasmussen R.A. & Shepson, P. (1990) Depletion of lower tropospheric ozone during Arctic spring – The polar sunrise experiment, *J. Geophys. Res.*, 95, 18 535–18 568.

Bovensmann, H., Burrows, J.P., Buchwitz, M., Frerick, J., Noel, S., Rozanov, V.V., Chance, K.V., & Goede, A.P.H. (1999) SCIAMACHY – Mission objectives and measurement modes, *J. Atmos. Sci*, 56, 127–150.

Brasseur, G., Orlando, J.J., & Tyndall, G. (1999) *Atmos. Chem. Global Change*, Oxford University Press, Oxford.

Brasseur, G.P., Prinn, R.G., & Pszenny, A.A.P. (2003) *Atmospheric Chemistry in a Changing World. An Integration and Synthesis of a Decade of Tropospheric Chemistry Research*, pp. 300, Springer-Verlag, Berlin.

Brenninkmeijer, C.A.M., *et al.* (2005) Analyzing atmospheric trace gases and aerosols using passenger aircraft, *Eos Trans.*, 86 (8), 77.

Browell, E.V. (1989) Differential absorption lidar sensing of ozone, *Proc. IEEE*, 77, 419–432.

Brown, S.S. (2003) Absorption Spectroscopy in High-Finesse Cavities for Atmospheric Studies, *Chem. Rev.*, 103 (12), 5219–5238.

Burrows, J.P. (1999) Current and future passive remote sensing techniques used to determine atmospheric constituents, in *Developments in Atmospheric Sciences 24: Approaches to Scaling Trace Gas Fluxes in Ecosystems*, Bouwman, A.F. (ed.), pp. 315–347, Elsevier, Amsterdam.

Burrows, J.P., Weber, M., Buchwitz, M., Rozanov, V., Ladstätter-Weißenmayer, A., Richter, A., DeBeek, R., Hoogen, R., Bramstedt, K., Eichmann, K.-U., Eisinger, M., & Perner, D. (1999) The Global Ozone Monitoring Experiment (GOME): Mission concept and first scientific results, *J. Atmos. Sci*, 56, 151–171.

Businger, S., Chiswell, S.R., Ulmer, W. & Johnson R. (1996) Balloons as a Lagrangian measurement platform for atmospheric research, *J. Geophys. Res. Atmos.*, 101 (D2), 4363–4376.

Calvert, J.G. (1994) The chemistry of the atmosphere: Its impact on global change, *Chemistry for the 21st Century*, Blackwell Scientific Publications, Oxford.

Cantrell, C.A., & Stedman, D.H. (1982) A possible technique for the measurement of atmospheric peroxy radicals, *Geophys. Res. Lett.*, 9, 846.

Carotta, M.C., Martinelli, G., Crema, G., Gallana, M., Merli, M., Ghiotti, M., & Traversa, E. (2000) Array of thick film sensors for atmospheric pollutant monitoring, *Sensors and Actuators*, B68, 1–8.

Clemitshaw, K.C. (2003) Tropospheric Chemistry and Composition/Hydroxyl Radical, in *Encyclopedia of Atmospheric Sciences*, Holton, J.R., Pyle, J.A. & Curry, J.A. (eds) Elsevier.

Clemitshaw, K.C. (2004) A review of instrumentation and measurement techniques for ground-based and airborne field studies of gas-phase tropospheric chemistry, *Critical Reviews in Environmental Science and Technology*, 34, 1–108.

Dabberdt, W.F., Lenschow, D.H., Horst, T.W., Zimmerman, P.R., Oncley, S.P. & Delany, A.C. (1993) Atmosphere-Surface Exchange Measurements, *Science*, 260 (5113), 1472–1481.

Drummond, J.R. & Mand, G.S. (1996) The measurement of pollution in the troposphere (MOPITT) instrument: Overall performance and calibration requirements, *J. Atmos. Oceanic Tech.*, 13, 314.

Ehhalt, D.H. (1998) Atmospheric chemistry – Radical ideas, *Science*, 279 (5353), 1002–1003.

Eisele, F., Mauldin, L., Cantrell, C., Zondlo, M.A., Apel, E., Fried, A., Walega, J., Shetter, R., Lefer, B., Flocke, F., Weinheimer, A., Avery, M., Vay, S., S. G., H. Singh, Brune, W.H., Harder, H., Martinez-Harder, M., Bandy, A., Thornton, D., Heikes, B., Kondo, Y., Riemer, D., Sandholm, S., Tan, D., Dibb, J. & Talbot, R. (2003) TRACE-P Intercomparison, *J. Geophys. Res.-Atmos.*, 108, 8791.

Finlayson-Pitts, B.J. & Pitts, J.N. (2000) *Chemistry of the Lower Atmosphere. Theory, Experiments and Applications*, Academic Press, San Diego.

Fishman, J.F., Watson, C.E., Larsen, J.C. & Logan, J.A. (1990) Distribution of tropospheric ozone determined from satellite data, *J. Geophys. Res.*, 95, 3599.

Fowler, D., *et al.* (1997) Ozone in the United Kingdom. Fourth report of the Photochemical Oxidants Review Group.

Hansford, G.M., Freshwater, R.A., Bosch, R.A., Cox, R.A., Jones, R.L., Pratt, K.F.E. & Williams, D.E. (2005) A low cost instrument based on a solid state sensor for balloon-borne atmospheric O_3 profile sounding, *J. Environ. Monit.*, 7 (2), 158–162.

Heard, D.E. (2003) Stratospheric chemistry and composition: Hydroxyl radical, in *Encyclopaedia of Atmospheric Sciences*, Holton, J.R., Pyle, J.A. & Curry, J.A. (eds) pp. 2185–2193, Elsevier, Amsterdam.

Heard, D.E. & Pilling, M.J. (2003) Measurement of OH and HO_2 in the troposphere, *Chem. Rev.*, 103 (12), 5163–5198.

Heath, D.F., Krueger, A.J., Roeder, H.A., & Henderson, B.D. (1975) The solar backscatter ultraviolet and total ozone mapping spectrometer (SBUV/TOMS) for NIMBUS G, *Opt. Eng.*, 14, 323.

Hofzumahaus, A., Aschmutat, U., Brandenburger, U., Brauers, T., Dorn, H.P., Hausmann, M., Hessling, M., Holland, F., Plass-Dulmer, C. & Ehhalt, D.H. (1998) Intercomparison of tropospheric OH measurements by different laser techniques during the POPCORN campaign 1994, *J. Atmos. Chem.*, 31 (1–2), 227–246.

Holton, J.R., Pyle, J.A. & Curry, J.A. (2003) *Encyclopaedia of Atmospheric Sciences*, Elsevier, London.

Houghton, J. (2005) Global warming, *Rep. Prog. Phys.*, 68, 1343–1403.

IPCC (2001) Third Assessment Report: Climate Change 2001, Synthesis report, Watson, R.T. (ed.), 184pp., Geneva.

Jenkin, M.E., Saunders, S.M., Wagner, V. & Pilling, M.J. (2003) Protocol for the development of the Master Chemical Mechanism, MCM v3 (Part B): Tropospheric degradation of aromatic volatile organic compounds, *Atmos. Chem. Phys.*, 3, 181–193.

Kolb, C.E., Herndon, S.C., McManus, B., Shorter, J.H., Zahniser, M.S., Nelson, D.D., Jayne, J.T., Canagaratna, M.R. & Worsnop, D.R. (2004) Mobile laboratory with rapid response instruments for real-time measurements of urban and regional trace gas and particulate distributions and emission source characteristics, *Environmental Science & Technology*, 38 (21), 5694–5703.

Kurylo, M.J. & Orkin, V.L. (2003) Determination of atmospheric lifetimes via the measurement of OH radical kinetics, *Chem. Rev.*, 103 (12), 5049–5076.

McManus, J.B. & Kolb, C.E. (1991) Long-path intracavity laser for the measurement of atmospheric trace gases, *SPIE*, 1433, 340.

Meier, A., Paton-Walsh, C., Bell, W., Blumenstock, T., Hase, F., Goldman, A., Steen, A., Kift, R., Woods, P. & Kondo, Y. (2005) Evidence of reduced measurement uncertainties from an FTIR instrument intercomparison at Kiruna, Sweden, *J. Quant. Spectroscopic Radiative Transfer*, 96, 75–84.

Mihelcic, D., Ehhalt, D.H., Klomfass, J., Kulessa, G.F., Schmidt, U. & Trainer, M. (1978) Measurements of free radicals in the atmosphere by matrix isolation and electron paramagnetic resonance, *Berichte Der Bunsen-Gesellschaft-Physical Chemistry Chemical Physics*, 82, 16.

Mihelcic, D., Volzthomas, A., Patz, H.W. & Kley, D. (1990) Numerical-analysis of Esr-spectra from atmospheric samples, *J. Atmos. Chem.*, 11 (3), 271–297.

Mount, G.H. & Eisele, F.L. (1992) An Intercomparison of tropospheric Oh measurements at Fritz-Peak-Observatory, Colorado, *Science*, 256 (5060), 1187–1190.

Newman, L. (1993) Measurement challenges in atmospheric chemistry, in *Advances in Chemistry*, edited by M.J. Comstock, 407pp., American Chemical Society, Washington DC.

Perry, R. & Young, R.J. (1977) *Handbook of Air Pollution Analysis*, 503pp., John Wiley & Sons, New York.

Pilling, M.J., *et al.* (2004) Air quality expert group. Nitrogen dioxide in the United Kingdom, in *DEFRA*, Pilling, M.J., (ed.).

Pilling, M.J., *et al.* (2005) Air quality expert group. Particulate matter in the united kingdom, in *DEFRA*, Pilling, M.J. (ed.).

Platt, U., Alicke, B., Dubois, R., Geyer, A., Hofzumahaus, A., Holland, F., Martinez, M., Mihelcic, D., Klupfel, T., Lohrmann, B., Patz, W., Perner, D., Rohrer, F., Schafer, J. & Stutz, J. (2002) Free radicals and fast photochemistry during BERLIOZ, *J. Atmos. Chem.*, 42 (1), 359–394.

Platt, U. & Strutz, J. (2006) *Differential Optical Absorption Spectroscopy*, Springer, Heidelberg.

Prinn, R., Simmonds, P.G., Rasmussen, R.A., Rosen, R.D., Alyea, F.N., Cardenlino, C.A., Crawford, A.J., Cunnold, D.M., Fraser, P.J. & Lovelock, J.E. (1983) The atmospheric lifetime experiment, 1. Introduction, instrumentation and overview, *J. Geophys. Res.*, 88, 8353.

Ravishankara, A.R. (2003) Atmospheric chemistry: Long-term issues, *Chem. Rev.*, 103 (12), 4505–5262.

Ravishankara, A.R. & Lovejoy, E.R. (1994) Atmospheric lifetime, its application and its determination – CFC-substitutes as a case-study, *J. Chem. Society-Faraday Trans.*, 90 (15), 2159–2169.

Richter, A., Wittrock, F., Ladstätter-Weißenmayer, A. & Burrows, J.P. (2002) GOME measurements of stratospheric and tropospheric BrO, *Adv. Space Res.*, 29 (11), 1667–1672.

Richter, A., Eyring, V., Burrows, J.P., Bovensmann, H., Lauer, A., Sierk, B., & Crutzen, P.J. (2004) Satellite measurements of NO_2 from International Shipping Emissions, *Geophys. Res. Lett.*, 31, L23110, doi:10.1029/2004GLO20822.

Richter, A., Burrows, J.P., Nüß, H., Granier, C., & Niemeier, V. (2005) Increase in tropospheric nitrogen dioxide over China observed from space, *Nature*, 437, 129–132, doi:10.1038/nature 04092.

Roscoe, H.K. & Clemitshaw, K.C. (1997) Measurement techniques in gas-phase tropospheric chemistry: A selective view of the past, present and future, *Science*, 276, 1065–1072.

Ryerson, T.B., Huey, L.G., Knapp, K., Nueman, J.A., Parris, D.D., Sueper, D.T. & Fehsenfeld, F.C. (1999) Design and initial characterization of an inlet for gas-phase NOy measurements from aircraft, *J. Geophys. Res.-Atmos.*, 104 (D5), 5483–5492.

Schultz, M., Heitlinger, M., Mihelcic, D. & Volz-Thomas, A. (1995) Calibration source for peroxy-radicals with built-in actinometry using H_2O and O_2 photolysis at 185 nm, *J. Geophys. Res.-Atmos.*, 100 (D9), 18 811–18 816.

Scott, A. (2005) A different perspective on the world, *Chemistry World*, pp. 41–45.

Seakins, P.W., Lansley, D.L., Hodgson, A., Huntley, N. & Pope, F. (2002) New directions: Mobile laboratory reveals new issues in urban air quality, *Atmos. Environ.*, 36 (7), 1247–1248.

Seinfeld, J.H. & Pandis, S.N. (1998) *Atmospheric Chemistry and Physics*, John Wiley & Sons, New York.

Sievers, R.E. (1995) Selective detectors: Environmental, industrial and biomedical applications, in *Chemical Analysis: A Series of Monographs on Analytical Chemistry and Its Applications*, Winefordner J.D. (ed.), 261pp., John Wiley & Sons, New York.

Simmonds, P.G., Derwent, R.G., McCulloch, A., O'Doherty, S. & Gaudry, A. (1996) Long-term trends in concentrations of halocarbons and radiatively active trace gases in Atlantic and European air masses monitored at Mace Head, Ireland from 1987–1994, *Atmos. Environ.*, 30 (23), 4041–4063.

Sommariva, R., Haggerstone, A.-L., Carpenter, L.J., Carslaw, N., Creasey, D.J., Heard, D.E., Lee, J.D., Lewis, A.C., Pilling, M.J. & Zador, J. (2004) OH and HO_2 chemistry in clean marine air during SOAPEX-2, *Atmos. Chem. Phys.*, 4, 839–856.

Stevens, P.S., Mather, J.H. & Brune, W.H. (1994) Measurement of tropospheric OH and HO_2 by laser-induced fluorescence at low-pressure, *J. Geophys. Res.-Atmos.*, 99 (D2), 3543–3557.

Sturges, W.T., Wallington, T.J., Hurley, M.D., Shine, K.P., Sihra, K., Engel, A., Oram, D.E., Penkett, S.A., Mulvaney, R. & Brenninkmeijer, C.A.M. (2000) A potent greenhouse gas identified in the atmosphere: SF_5CF_3, *Science*, 289 (5479), 611–613.

Tyndall, G.S., Winker, D.M., Anderson, T.L., Eisele, F.L. *et al.* (25 authors) (2003) Advances in laboratory and field measurements, in *Atmospheric Chemistry in a Changing World: An Integration and Synthesis of a Decade of Tropospheric Chemistry Research*, Brasseur, G.P., Prinn, R.G. & Pszenny, A.A.P. (eds) 300pp., Springer-Verlag, Berlin.

Wagner, T. & Platt, U. (1998) Observation of tropospheric BrO from the GOME Satellite, *Nature*, 395, 486–490.

Wayne, R.P. (2000) *Chemistry of Atmospheres*, Oxford Science Publications.

Williams, D.E. (1999) Semiconducting oxides as gas-sensitive resistors, *Sensors and Actuators B-Chemical*, 57 (1–3), 1–16.

Chapter 2
Infrared Absorption Spectroscopy

Alan Fried and Dirk Richter

2.1 Introduction

Absorption spectroscopy is the study of the absorption of electromagnetic radiation by atoms and molecules. In this approach, one measures the absorbed light intensity as a function of wavelength. Spectroscopic data can be used to study molecular structure such as the size, shape and nature of the absorbing molecules as well as the nature of the chemical bonds and the interactions among energy levels involved in the transitions. Since different molecules absorb electromagnetic radiation at different specific wavelengths, absorption spectroscopy has also found wide-ranging applications as a tool for measuring trace concentrations of various molecules in (1) laboratory-based kinetic and mechanistic studies, (2) combustion studies, (3) medical applications, (4) automotive studies, (5) cigarette emission studies, (6) studies of flux to and from various surfaces, (7) detection of explosives and bio-agents, and (8) measurements of trace atmospheric constituents for atmospheric studies, to name a few.

This chapter will focus on the last application, with specific emphasis on measurements employing electromagnetic radiation in the infrared (IR) spectral region. Such measurements are critical for understanding a number of important atmospheric processes, including the build-up of greenhouse gases and their role in future climatic change as well as detailed mechanistic information regarding chemical processing in the atmosphere, to name a few examples. With the exception of noble gases and homonuclear diatomic molecules such as O_2, N_2, and H_2, nearly all atmospheric constituents exhibit moderate to strong absorptions in the IR, making IR absorption spectroscopy an extremely versatile tool for atmospheric studies. Although liquid and solid aerosols have been shown to play major roles in atmospheric chemistry, this chapter will only focus on IR absorption from gas-phase molecular species. Techniques employing absorption of radiation in the ultraviolet (UV) and visible regions of the electromagnetic spectrum, which are also valuable tools for atmospheric research, will be discussed in Chapter 3. Approaches involving measurements of the subsequent re-emission of the absorbed energy in the form of fluorescence will be discussed in Chapter 4.

This chapter will introduce many of the important concepts commonly used in absorption spectroscopy, including the fundamentals of electromagnetic radiation and its interaction with molecules, absorption cross-sections and selection rules, Beer–Lambert law, vibrational bands, vibrational-rotational features, lineshapes, as well as many other basic concepts. The important fundamental mathematical formulations will be presented along with information regarding how such expressions can be derived and the underlying assumptions. However, in no case do we endeavor to present rigorous derivations or rigorous quantum theory, both of which can be found in numerous textbooks (see 'Further reading' in the 'Reference' section). Many of the details regarding rotational and vibrational energy levels and the various interactions will also not be presented. Instead, the focus is on the important conclusions and inferences that one should extract from the expressions that are presented. After discussing the fundamental theory, we will highlight the types of IR absorption approaches that have commonly been employed to study trace atmospheric constituents approximately since 1990 as well as the specific atmospheric gases that have been studied. The strengths and weaknesses of these measurements will be discussed and, wherever possible, comparisons with other approaches and other wavelength regions will be given. This chapter will employ SI units wherever possible. However, in many cases expressions and important variables are denoted in the *cgs* system of units, and this chapter will adopt this philosophy when these expressions are presented.

2.2 Fundamentals of infrared absorption spectroscopy

2.2.1 Electromagnetic radiation in the infrared

We first present some of the fundamentals related to the terms used throughout this chapter. All electromagnetic radiation is classified in terms of its wavelength. Figure 2.1 depicts a classical electromagnetic wave with orthogonal electric and magnetic fields, both of which are orthogonal to the direction of light propagation. The distance between the wave peaks is the wavelength and is denoted by λ, while the number of waves per unit time is the frequency ν in units of cycles per second or Hertz. These parameters are

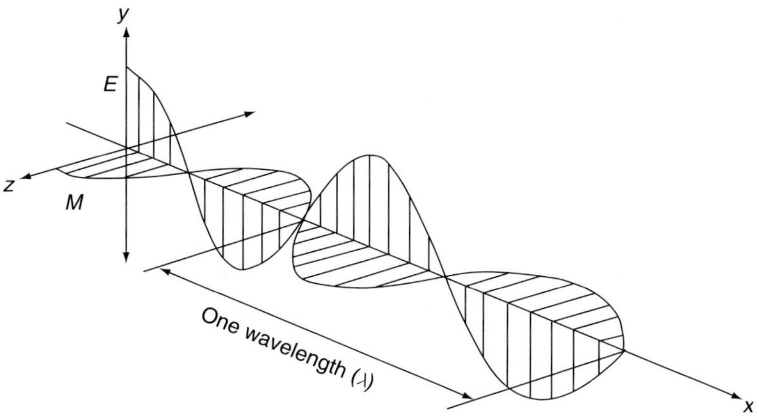

Figure 2.1 Electromagnetic wave with orthogonal electric and magnetic fields.

related through the speed of light in vacuum c ($c = 2.99792 \times 10^8$ m s^{-1}) by the following relationship:

$$\nu = c/\lambda \tag{2.1}$$

The energy (E) of a photon for this wave can be expressed in terms of this frequency via a proportionality constant, h (where $h = 6.626 \times 10^{-34}$ J s), known as Planck's constant:

$$E = h\nu \tag{2.2}$$

It is important to note that applications for each spectral region of the electromagnetic spectrum have developed independent of one another, and hence each spectral region has acquired its own preferred set of units. In the IR spectral region, the units of wavelength are frequently given in terms of μm(10^{-6} m) and nm(10^{-9} m). In the mid-IR spectral region (to be defined shortly) one almost exclusively works in units of wave numbers, $\bar{\nu}$, which is the reciprocal of the wavelength in cm and has units of cm^{-1}:

$$\bar{\nu} = 1/\lambda \quad (\text{cm}^{-1}) \tag{2.3a}$$

and thus,

$$E = h\nu = hc/\lambda = h\bar{\nu}c \tag{2.3b}$$

A wavelength of 1 μm is 1000 nm, 10 000 Å, 1×10^{-4} cm, and 10 000 cm^{-1}.

The parameters of wavelength and wave numbers in all the above expressions refer to vacuum, whereas most fundamental spectroscopic measurements employ a spectrometer operated in a non-vacuum environment. The wavelength in air (λ_{air}) is related to the wavelength in vacuum by the refractive index n via:

$$\lambda_{vac} = n\lambda_{air} = \lambda_{air} + (n-1)\lambda_{air} \tag{2.3c}$$

At IR wavelengths in the 1–16 μm range considered in this chapter, the term $(n-1)$ takes on values around 2.73×10^{-4} for standard air (dry air at 15°C at 1 atm pressure and 0.03% CO_2). Values for other conditions are given by Coleman *et al.* (1960). In trace atmospheric measurements the distinction between vacuum and air is automatically dealt with, since one typically compares their observed spectrum to reference spectra given in vacuum units.

Figure 2.2 depicts the different regions of the electromagnetic spectrum employed in atmospheric studies, which extends from the vacuum UV to the microwave spectral regions. It is important to note that the exact delineations between spectral regions are approximate and differ somewhat from reference to reference. As shown, the IR spectral region spans the approximate wavelength range from the end of the visible region at 0.7 μm to the start of the microwave region at 500 μm. It will be more useful for the discussions of this chapter to further characterize the IR spectral region in terms of three sub-regions, where again the exact delineations vary somewhat from text to text. These three regions are: (1) the near-IR, which extends from approximately 0.7 to 2.5 μm; (2) the mid-IR, which extends from 2.5 to 25 μm; and (3) the far-IR, which extends from

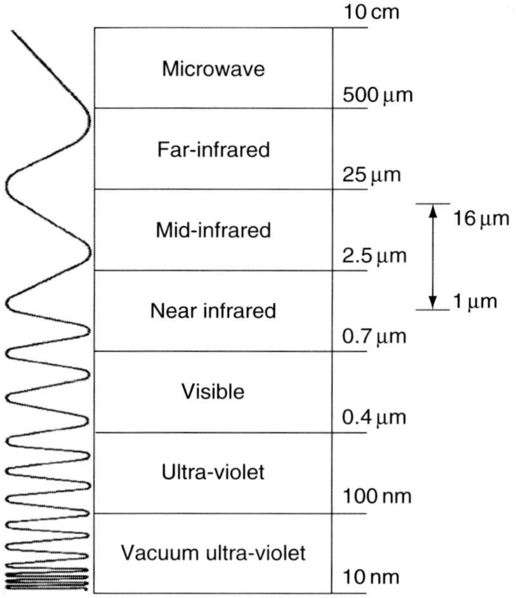

Figure 2.2 Regions of the electromagnetic spectrum with approximate wavelength delineations on the right-hand axis.

25 to 500 µm. This chapter will only consider the 1–16 µm spectral region where most IR studies of atmospheric constituents tend to focus.

The near- and mid-IR spectral regions are distinguished from other spectral regions not only by wavelength but also by the nature of the molecular interactions involved, the sheer number of gas-phase atmospheric constituents exhibiting absorption features, and the characteristics of the recorded spectra in terms of absorption strengths and linewidths. We will examine each of these aspects in further detail in the following sections.

2.2.2 Molecular interactions in the IR

When electromagnetic radiation is incident upon a molecule, the oscillating electric field, shown in Figure 2.1, under the right set of conditions, interacts with the dipole moment (μ) of the molecule to produce a change in energy of the molecule from some lower state with energy E_1 to an upper state with energy E_2. One can find a detailed mathematical treatment of such a system in most books on quantum mechanics and spectroscopy, and two such excellent references are Barrow (1962) and Levine (1970). In these and most references, one treats the molecule quantum-mechanically and the radiation field as a classical wave employing time-dependent perturbation theory. Considering only the interactions involving the electric field, since the magnetic force is much smaller, one arrives at an expression for the probability of a transition that is proportional to the square of the product of three terms: the maximum amplitude of the electric field E^0, the transition moment $\mu_{1,2}$, and an expression that contains two fractions – one term that becomes important for absorption and the other term that becomes important for stimulated emission. Considering only absorption from state 1 to state 2, the second fraction drops out and the transition probability is proportional to:

$$\left(E^0\mu_{1,2}\left[\frac{1-e^{(2\pi i/h)(\Delta E-h\nu)t}}{\Delta E-h\nu}\right]\right)^2 \tag{2.4}$$

The term ΔE represents the difference in energy between the two molecular energy levels ($\Delta E = E_2 - E_1$) and $h\nu$ represents the photon energy from Equation 2.2. The transition moment is given by:

$$\mu_{1,2} = \int \Psi_2{}^*\mu\Psi_1\,\mathrm{d}\tau \tag{2.5}$$

Here, Ψ_1 and Ψ_2 represent the wavefunctions for the two energy levels, and μ is the dipole moment operator along a specific direction τ of the electric field. One can write Expressions 2.4 and 2.5 for each of the x, y, and z components of the electric field. The dipole moment operator for a distribution of i charges is determined from the summation of the product of the electric field charge (q_i) times a position vector:

$$\mu = \mathbf{i}\Sigma_i(q_i x_i) + \mathbf{j}\Sigma_i(q_i y_i) + \mathbf{k}\Sigma_i(q_i z_i) \tag{2.6}$$

The terms **i**, **j**, and **k** represent unit vectors in the x, y, and z directions and the terms x_i, y_i, and z_i are the Cartesian coordinates in these directions.

Equation 2.4 contains the important result that in order for a electric-dipole moment–allowed transition allowed by the electric-dipole moment to be non-negligible, each of the three terms must be non-zero. One obtains a non-negligible value for the fractional term in the bracket only when the energy difference between the two energy levels is equal to the photon energy, namely $\Delta E = h\nu$. In this case the denominator approaches zero and the quotient becomes large but not infinite. Under these circumstances, the square of the transition moment, in equation 2.5, determines the magnitude of the electric dipole–allowed absorption strength for a given field strength. Since only certain molecular transitions will result in non-zero transition moments, spectroscopists devise *selection rules* to determine which transitions are allowed. However, this analysis only refers to electric dipole–allowed transitions, and electric quadrupole transitions orders of magnitude weaker may still occur.

Absorption of electromagnetic radiation in the IR spectral region, with a few exceptions, involves interactions with molecular vibrations and rotations. The resulting absorption spectra involve transitions between rotational levels of some lower vibrational state (generally the ground vibrational level) and rotational levels of some higher vibrational state. For this reason these transitions are referred to as vibrational-rotational transitions. In order for such transitions to occur, the square of the transition moment expression of equation 2.5 using vibrational wavefunctions must be non-zero. One can further determine conditions for IR transitions by examining the variation of the dipole moment during a vibration. This is accomplished by expanding each of the dipole-moment Cartesian coordinates in terms of normal coordinates (Q_i), a $3N$ coordinate system (where N is the number of atoms in the molecule) that translates and rotates with the molecule. For small vibrational displacements, the transition moment of equation 2.5 becomes the sum of two terms, one involving the permanent dipole moment in the equilibrium position (μ_0) and the other involving the summation for the derivative of the dipole moment as a function of each of the i normal coordinates:

$$\mu_{1,2} = (\mu)_0 \int \Psi_2{}^* \Psi_1 \, d\tau + \sum_i \left(\frac{\delta\mu}{\delta Q_i} \right)_0 \int \Psi_2{}^* Q_i \Psi_1 \delta\tau \qquad (2.7)$$

The first integral vanishes since the vibrational wavefunctions are orthogonal to each other for different vibrational states. The second term, which represents the transition moment for vibrational transitions, is non-zero only when the dipole moment derivative in front of the integral is non-zero. A change in either the magnitude or direction of the dipole moment will satisfy this criterion. Since noble gases and homonuclear diatomic molecules such as O_2, N_2, and H_2 do not satisfy this condition, they do not possess allowed IR transitions. In addition, many vibrational modes of polyatomic molecules are also absent because this condition is not satisfied. However, N_2 exhibits weak quadrupole transitions in the 3.8–5.0 μm spectral range, with integrated line intensities 6–15 orders of magnitude weaker than typical trace gases of atmospheric interest.

2.2.3 *Vibrational bands and rotational lines in the IR*

Having assessed the general conditions necessary for IR absorption, we now further examine vibrational-rotational spectral features and their general properties. As in the previous section, we present only a very brief overview, which is necessary for subsequent discussions of trace gas detection employing IR absorption spectroscopy. More detailed treatments of this subject can be found in classic textbooks such as Herzberg (1945) and King (1964), and in books by Banwell and McCash (1994) and Duxbury (2000).

An IR absorption spectrum consists of a series of bands due to transitions where the vibrational quantum number changes in the ground electronic state. Under higher resolution one observes a series of closely spaced absorption lines within each vibrational band due to simultaneous changes in the rotational quantum number. In the case of diatomic molecules there is only one vibrational mode, which can take on various vibrational quantum numbers. Dipole moment–allowed selection rules indicate that the change in rotational quantum number ΔJ may take on values of ± 1, and in certain electronic states a value of 0. Rotational lines where $\Delta J = +1$ during the transition from the ground to the excited vibrational state are designated R-branch lines, and where $\Delta J = -1$, as P-branch lines. Q-branch lines occur when $\Delta J = 0$, and result in a series of closely spaced lines near the band origin ω_0, the transition frequency in cm^{-1} between the two vibrational levels in the absence of rotation. In the absence of centrifugal distortion (a changing internuclear distance as the molecule vibrates) and other interactions, one may write the following simple general expression for the frequencies of such rotational lines:

$$\overline{\nu}_{1,2} = \omega_0 + B'_v J'(J'+1) - B_v{}'' J''(J''+1) \qquad (2.8)$$

Here, B_v' and B_v'' are the rotational constants in the upper and lower vibrational states, respectively (where B is proportional to the inverse of the square of internuclear distance), while J' and J'' represent the rotational quantum numbers of these states.

Like diatomic molecules, there is one principal moment of inertia in linear polyatomic molecules, and thus rotational spectra can be described by Equation 2.8 for the vibrational mode (to be discussed) under consideration. There are, however, subtle differences between diatomic and linear polyatomic spectra. In linear polyatomic molecules, like CO_2, the two bending vibrations (to be discussed) are degenerate, resulting in the presence of l-doubling of the vibrational levels. A second difference relates to differences in the statistical weights, factors that give the multiplicity of the rotational energy levels. Finally, as the rotational B values for polyatomic molecules are significantly smaller than diatomics due to the increased mass (by one or more orders of magnitude), the rotational features of polyatomic molecules are closer in spacing than diatomics. Generally as the molecule gets larger the rotational features become denser.

For illustrative purposes, we show in Figure 2.3 the band structure and associated rotational features for the asymmetric stretching vibration of the linear molecule CO_2. As can be seen, the P-branch features form a progression at lower frequencies from the band origin, while the R-branch features form a progression at higher frequencies. As in all symmetric triatomic molecules, where the O atoms are equivalent, in CO_2 alternate rotational lines are absent. We will further discuss the rotational lines of Figure 2.3 in relation to line intensities in Section 2.2.5. In nonlinear polyatomic molecules, one encounters more complicated rotational structures since one must consider more than one axis of rotation. Further discussion of this topic will not be considered here.

In polyatomic molecules one resolves the various vibrational motions into a sum of normal vibrations where all atoms in the molecule oscillate with the same frequency and simultaneously pass through their equilibrium positions. In linear polyatomic molecules, where each atom possesses 3 translational and 2 rotational degrees of freedom, there are $3N-5$ vibrational modes. In nonlinear polyatomic molecules, there are $3N-6$ vibrational modes due to the 3 translational and 3 rotational degrees of freedom. Therefore IR absorption spectra in polyatomic molecules consist of a series of different vibrational–rotational features for each mode of vibration. Each distinct mode of vibration is called a *band*. Different vibrational bands for a given molecule often appear in different regions of the IR, and the larger the molecule the greater the number of vibrational bands.

In some texts, the actual vibrational quantum numbers for the transitions are given. For example, for nonlinear triatomic molecules like H_2O (3 normal modes of vibration) one designates the vibrational transitions by the following notation:

$$(\nu_1, \nu_2, \nu_3)\text{upper state} \leftarrow (\nu_1, \nu_2, \nu_3)\text{lower state}$$

where ν here is not to be confused with the frequency in Equation 2.2. By convention, the highest frequency vibration within each symmetry type (see Chapter 3 of Banwell and McCash, 1994) is designated by the lowest subscript and this designation progresses to higher subscripts as the frequency decreases. In the case of H_2O, symmetric stretching transitions involve the quantum number ν_1, and the fundamental band (one quantum of excitation) has a band origin at 3657 cm^{-1} and is designated by the notation 100 $\leftarrow$ 000. Frequently, one simply uses the shorthand notation ν_1 to designate such a transition.

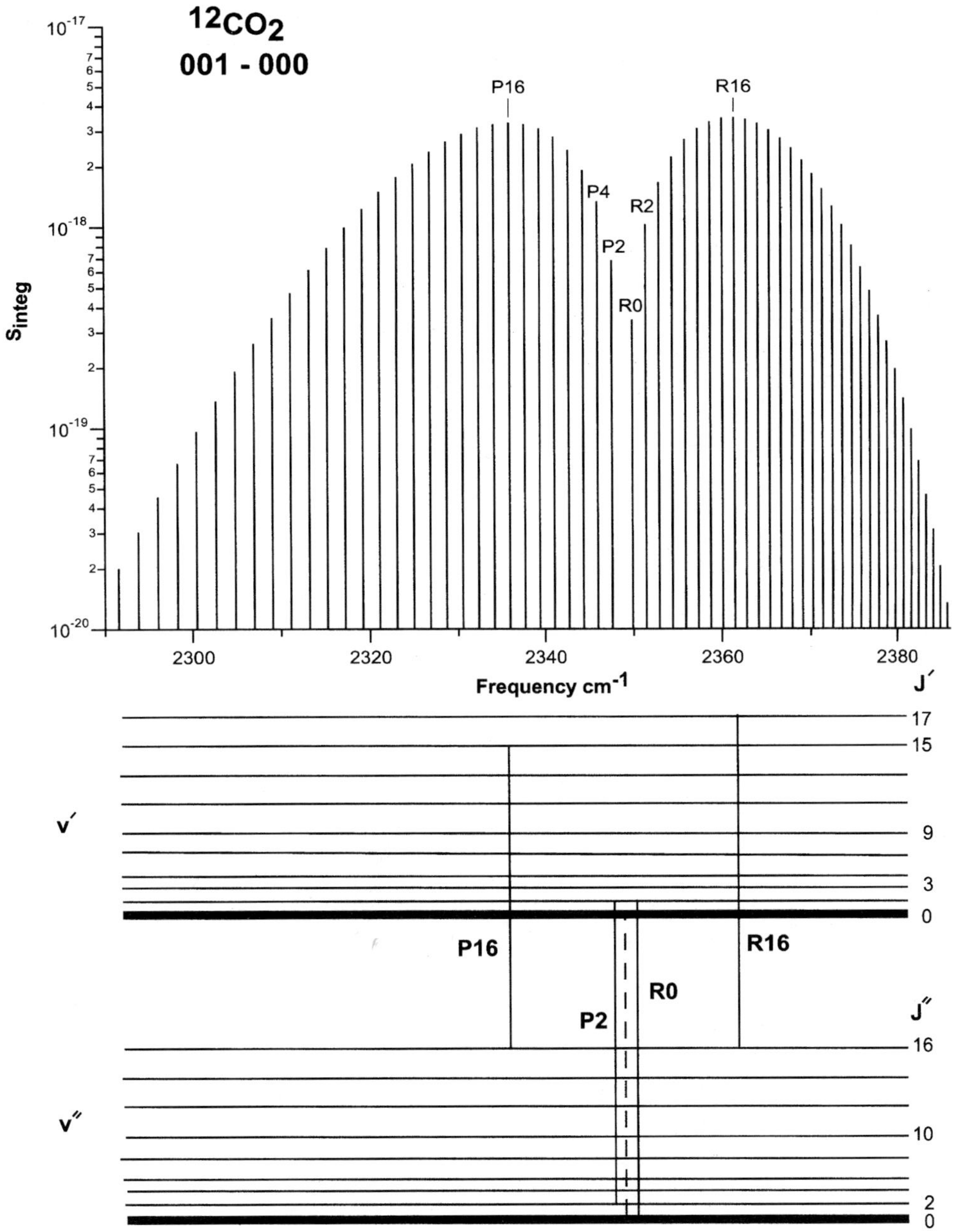

Figure 2.3 Vibrational-rotational transitions and energy levels for the asymmetric vibrational mode, ν_3, of the $^{12}CO_2$ molecule at a temperature of 296 K. The top panel displays the integrated line intensities (S_{integ} in units of $cm^2\,cm^{-1}molecule^{-1}$) and the lower panel displays the energy levels, the spacings for which are not drawn to scale. The band origin is displayed by the dashed line.

The bending vibrations of H_2O (same symmetry as above) involve the quantum number ν_2, and the fundamental band has an origin at 1595 cm^{-1} and a designation 010 ← 000. The asymmetric stretching vibration (different symmetry) involves the quantum number ν_3 and the fundamental band has an origin at 3756 cm^{-1} and a designation 001 ← 000.

As discussed previously, not all vibrational modes appear in the IR (i.e. are IR active). Figure 2.4 shows the three IR active bands of H_2O and the two IR active bands of CO_2; the ν_1 symmetric stretch of CO_2 does not produce a change in dipole moment, and hence this band is IR inactive. By contrast, all normal vibrations in asymmetrical polyatomic molecules, like the ν_1 symmetric stretch of HCN, are IR active. In degenerate bending vibrations of linear polyatomic molecules (e.g. the ν_2 bending vibration of CO_2), a superscript appears above the bending quantum number, and this signifies the presence of l-doubling, which splits the vibrational levels.

One can further analyse the integral in the second term of Equation 2.7 to qualitatively assess the strength of vibrational transitions. For non-zero values of the dipole moment derivative, the integral on the right hand side of Equation 2.7 is non-zero when the vibrational quantum number changes by $\pm1, \pm2, \pm3\ldots$ The most intense absorption transitions occur when the vibrational state for one vibrational mode at a time changes by 1 quantum number, that is by +1. Such transitions, which are called *fundamental transitions* or *fundamental bands*, occur in the mid-IR spectral region. The ν_1 transition above for H_2O is an example of this. Since vibrational motions are anharmonic, combination and difference bands may also appear in the mid-IR spectral region as long as there is a change in dipole moment. In this case more than one vibrational mode at a time is excited. In a combination band, one can observe transitions involving $\nu_1 + \nu_2$, for example, while transitions involving $\nu_1 - \nu_2$ exemplify a difference band. Combination

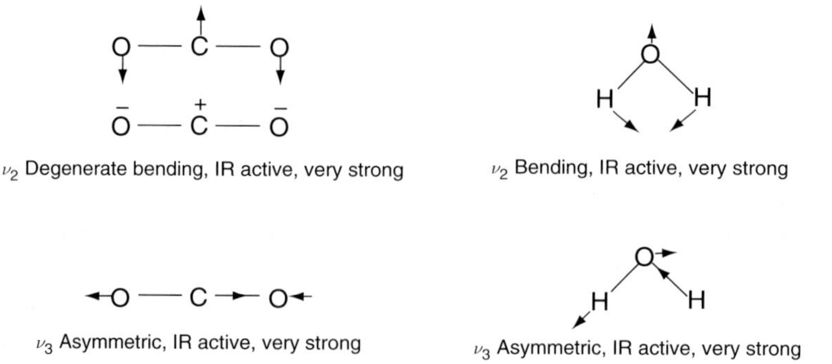

Figure 2.4 Normal vibrational modes for the linear molecule CO_2 and the non-linear molecule H_2O. The arrows show the atomic displacements during the vibration. The − and + on the atoms of CO_2 show these displacements in and out of the plane of the page.

and difference bands are generally significantly weaker than fundamental bands. For example, the strongest line in the ν_3 fundamental of CO_2 shown in Figure 2.3 is the R(16) rotational line, which has an integrated intensity of 3.5×10^{-18} cm^2 cm^{-1}molecule^{-1} at 296 K. This same rotational line in the combination band 101 $\leftarrow$ 000, where the symmetric and asymmetric stretch each have one quanta of excitation, has an integrated intensity weaker by a factor of 60 times. Overtone vibrational transitions may also occur. In this case the vibrational quantum number changes by +2, +3, . . . , and these appear at shorter wavelengths in the near-IR and in some cases in the visible region. Such transitions are significantly weaker than fundamental transitions, typically by an order of magnitude or more for each change in vibrational quantum number above +1. The R(16) line in the 201 $\leftarrow$ 000 overtone-combination band of CO_2, which occurs at 1.96 μm in the near-IR, has an integrated line intensity weaker by a factor of 139 times than the 101 band. When detecting gases like CO_2 and H_2O, working with near-IR overtones is generally not a limitation, since the ambient concentrations are so large and the laser sources and detectors are significantly better than those for the mid-IR spectral region. However, detection of most trace gas species still requires the significantly stronger mid-IR fundamental absorption lines.

Two other types of vibrational bands may also appear in the IR: hot bands and isotopic bands. Hot bands (where the lowest vibrational state is not the ground state) only become important at high temperatures such as in flames and in combustion stacks where there is a significant population of molecules in high vibrational levels. The relationship between absorption strength and ground vibrational population will be presented shortly in a later section. Isotopic bands, like that of $^{13}CO_2$, are very similar in nature to the main isotope of $^{12}CO_2$, shown in Figure 2.3, only displaced at lower frequencies. The band origin occurs at 2284 cm^{-1}. However, atmospheric absorption lines from such bands are significantly weaker than the main isotope due to the lower isotopic abundance ($\sim$ by a factor of 99 in this case).

2.2.4 Vibrational bands of atmospheric spectral features in the mid-IR

Figure 2.5 shows a plot of the fundamental absorption bands in a narrow window of the mid-IR spectral region for four major atmospheric constituents, and Table 2.1 tabulates the band origins and integrated band intensities for these and other species from Smith *et al.* (1985) and Pugh and Rao (1976). This is only a select list of atmospheric absorbers, and the interested reader is referred to the references above as well as the 2000 HITRAN database (Park *et al.*, 1987; Goldman *et al.*, 2002; Rothman *et al.*, 2003) for more comprehensive lists and spectral atlases. In addition, Maki and Wells (1991) and Guelachvili and Rao (1986) provide spectral maps and transitions of select absorption lines used in the wavenumber calibration of IR spectrometers. Figure 2.5 plots these four atmospheric constituents using the mixing ratios shown, which are typical of values found in the troposphere. As can be seen, CO_2 absorption features in the 4 μm spectral region are extremely strong, and, under the conditions given, could mask much weaker absorptions by trace species in this region. Water vapor absorptions near the fundamental band origins given in Table 2.1 could also present similar problems for trace gas measurements

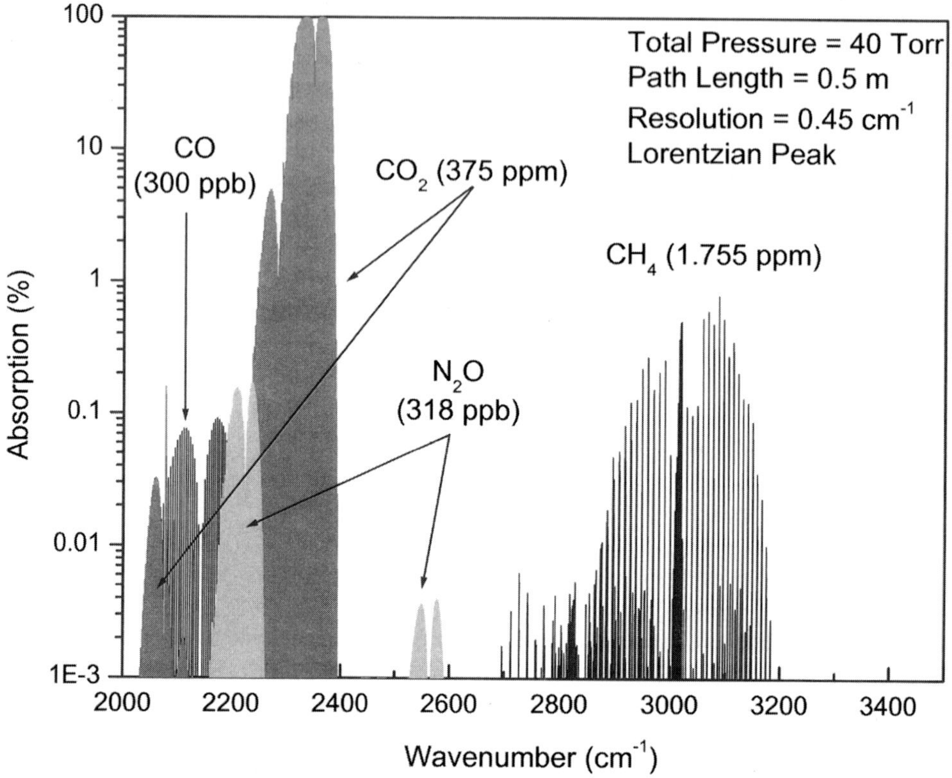

Figure 2.5 A low-resolution display of vibrational-rotational lines of four major atmospheric constituents in the 2000 to 3200 cm^{-1} window under the conditions displayed in the plot.

of other species in the lower troposphere. High resolution spectra for the specific regions of interest, however, are further required to determine if these abundant gases will present a problem.

Table 2.1 and the references cited above illustrate the sheer number of atmospheric constituents that absorb in the mid-IR spectral region. This is in contrast to the visible and UV spectral regions, where significantly fewer atmospheric constituents exhibit strong absorptions. This is one aspect that makes mid-IR absorption measurements a particularly attractive tool for atmospheric research. However, this is not without unique measurement challenges. Since many mid-IR atmospheric studies employ laser sources of one kind or another, advances in mid-IR measurement performance are closely tied to advances in laser source development. Although there has been a great deal of activity in this area, progress in mid-IR laser source development has not kept pace with progress in other regions of the electromagnetic spectrum in terms of performance and ease of use. In fact, as of this writing, the optimum laser sources reside in the near-IR spectral region, more particularly in the 0.9–1.8 μm spectral region, where advances from the telecommunications industry have helped push the development. Unfortunately, the number of atmospheric gases that can be studied in this spectral region are somewhat limited due

Table 2.1 Band origins (rounded to nearest cm^{-1}) and integrated band strengths for some of the important trace gases from Smith *et al.* (1985) unless otherwise noted. Use strengths as a guide, since in many cases multiple values are presented by Smith *et al.* (1985). The bands are for the major isotopes

Molecule	Band	Band origin cm^{-1}	S_v cm^{-2} atm^{-1} at 300 K
CO	1-0	2143	235
HBr	1-0	2559	55[a]
HCl^{35}	1-0	2886	121[b]
HF	1-0	3962	408[b]
NO	1-0	1876	113
OH	1-0	3570	
CO_2	$\nu_2{}^1$	667	200
	ν_3	2349	2230
HCN	ν_1	2097	0.1
	ν_2	712	234
	ν_3	3312	219
H_2O	ν_1	3657	10[a]
	ν_2	1595	254
	ν_3	3756	192[a]
N_2O	ν_1	1285	277
	$\nu_2{}^1$	589	29
	ν_3	2224	1173
NO_2	ν_1	1320	2
	ν_2	750	13
	ν_3	1617	1400
	$\nu_1 + \nu_3$	2906	70
O_3	ν_1	1103	10[a]
	ν_2	701	18[a]
	ν_3	1042	361
OCS	ν_1	859	27
	ν_2	521	12
	ν_3	2062	2184
CH_2O	ν_1	2782	194
	ν_2	1746	301
	ν_3	1500	45
	ν_4	1167	26
	ν_5	2843	356
CH_4	ν_1	2917	0.03
	ν_2	1534	2[a]
	ν_3	3019	276
	ν_4	1306	136

[a] Band strengths from Pugh and Rao (1976).
[b] Band strengths taken from Pine *et al.* (1985).

to the significantly weaker line strengths. We will further discuss in Section 2.5.1.2 the performance, attributes, and drawbacks of various laser sources throughout the IR.

The second main challenge of working in the mid-IR spectral region concerns H_2O vapor. Water vapor mixing ratios can be very high and highly variable; typical mixing ratios can vary from 0.04 (40 000 parts per million by volume, ppmv) to the 10^{-6} range (1 ppmv). Although one has to worry about overtone and combination bands of H_2O in the visible and UV spectral regions (electronic transitions occur in the vacuum UV), absorption line strengths here are typically very weak (in the 10^{-26} range with some lines as strong as 10^{-24}). By contrast, discrete strong lines of H_2O appear throughout the mid-IR spectral region. Absorption lines in the very strong fundamental bending vibration shown in Figure 2.4 (band origin $1595\,cm^{-1}$), for example, are as strong as $2.9 \times 10^{-19}\,cm^2\,cm^{-1}\,molecule^{-1}$ at 296 K. This coupled with the very high mixing ratios and the reasonably large air and self-broadening coefficients (to be discussed) adds to the challenges of quantitative mid-IR absorption measurements. One has to worry about direct spectroscopic interference not only from H_2O (i.e. overlap of H_2O lines with those from the species of interest), but also from the broad tails of H_2O vapor absorption, which may lead to changes in background spectral structure. Open path IR measurements, like those from a ground-based Fourier transform infrared (FTIR) spectrometer, have to pay particular attention to atmospheric H_2O vapor. In many cases, such measurements have difficulty accessing certain spectral regions, since the broad wings of H_2O vapor lines totally obscure weaker neighboring absorptions.

2.2.5 Vibrational–rotational spectral line intensities

We have presented in previous sections the general factors associated with absorption line intensities without further expansion. Since the sensitivity with which one can measure an atmospheric trace gas is directly related to these intensities, we will now present quantitative expressions with associated units (in parentheses). It is important to keep in mind that one may encounter many different names for the term 'spectral line intensities'. For example, absorption line strengths, absorption cross-sections, and absorption coefficients, to name a few, have all been used synonymously. However, strictly speaking, spectral line intensity and absorption cross-section are given on a per molecule basis, while many of the other terms generally refer to the aggregate absorption for a group of molecules. To be sure, one should be cautious and careful while examining the units. This is important to further ensure that the absorption pathlength has not been folded into one's definition. Finally, for reasons that will become clear when discussing the Beer–Lambert law, all references to spectral line intensities in atmospheric studies as well as many other disciplines are given in terms of the logarithmic base *e*. This is in contrast to absorption in solutions, where the logarithmic base 10 is typically used.

This section is very useful for understanding the various factors which comprise spectral line intensities, and their temperature dependencies, a potentially important aspect when considering measurements at very low temperatures in the upper atmosphere. However, since various databases such as the HITRAN spectroscopic database (Rothman *et al.*, 2003) tabulate line intensities for most small molecules of atmospheric interest and provide a means to calculate these values at different temperatures, a comprehensive reading of this section is not necessary for understanding the other sections of this chapter.

As before, we consider here absorption for a two-state system between a lower state vibrational-rotational energy level E_1 and an upper state E_2. Absorption between the two states is induced by a uniform electromagnetic field of wavelength exactly equal to the energy difference of the two states ($\Delta E = h\nu$). It is useful to keep in mind throughout the discussion that follows that the strength of an absorption feature is intimately related to the population of the lower absorbing state; many of the terms that will be given express the fraction of the total number of molecules that are present in this lower absorbing state.

The spectral line intensity, $S_{1,2}$, integrated over an individual vibrational-rotational absorption line at a temperature, T(K), for a single molecule can be expressed by the product of a vibrational band strength, S_v^0, a rotational term, $R_{1,2}$, and a Herman–Wallis vibration–rotation interaction factor, F, as (see for example, Rothman *et al.* (1992) and Pine *et al.* (1985)):

$$S_{1,2} = S_v^0 R_{1,2} F \tag{2.9}$$

The rotationless vibrational band strength is defined at a temperature T by:

$$S_v^0(T) = \left(\frac{8\pi^3}{3hc}\right) \omega_0 L \left(\frac{T_0}{T}\right) I_a [Q_v(T)]^{-1} \exp\left(\frac{-hcG_v''}{kT}\right) |R_v|^2\, 10^{-36} \tag{2.10}$$

Here h is Planck's constant; c is the speed of light; k is Boltzmann's constant ($k = 1.381 \times 10^{-16}$ erg mole^{-1} K^{-1}, for simplicity of calculation, $hc/k = 1.4388$ cm K); ω_0 is the band origin frequency (cm^{-1}); L is the Loschmidt's number, which is equal to 2.68678×10^{19} molecules cm^{-3} atm^{-1} at a temperature of $T_0 = 273.15$ K (here 1 atm is defined as 760 torr = 101.325 kPa = 1013.25 mb); I_a is the isotopic abundance; Q_v is vibrational partition function at a temperature of T; and R_v is the vibrational transition moment $\mu_{1,2}$, defined in equation 2.5. The vibrational partition function, which is very close to unity near room temperature, is calculated from the summation of all vibrational energies G″ in cm^{-1}. This function, and the rotational partition function (to be discussed), gives the population of the absorbing level. The factor 10^{-36} converts the dipole transition moment from units of Debye to ergs cm^3 ($1\text{D}^2 = 10^{-36}$ ergs cm^3).

The rotational term of the line strength, $R_{1,2}$ (not to be confused with the dipole transition moment R), is given by:

$$R_{1,2}(T) = (HL)\left(\frac{\bar{\nu}_{1,2}}{\omega_0}\right) \exp\left(\frac{-hcE_r''}{kT}\right) [Q_r(T)]^{-1} \left[1 - \exp\left(\frac{-hc\nu_{1,2}}{kT}\right)\right] \tag{2.11}$$

Like the vibrational partition function, the last exponential factor in brackets is near unity for all but very high temperatures at IR wavelengths less than ~10 μm (frequencies >1000 cm^{-1}); at 1000 K this term takes on a value of 0.987 for transitions near 3000 cm^{-1} compared to a value of 1.000 at 296 K. This term, which accounts for induced emission, also becomes important at room temperature for wavelengths >10 μm (frequencies <1000 cm^{-1}); for example, at the 25 μm (400 cm^{-1}) end of the mid-IR spectral region, this term takes on a value of 0.857 at 296 K.

In Equation 2.11, the term *HL* is the Hönl–London factor, which gives the intensity distribution for the rotational lines in the different rotational branches, and is a function of the rotational and vibrational quantum numbers for each vibrational band type and rotational branch. For example, in the case of the 001 ← 000 band of CO_2 in Figure 2.3, the $R(16)$ line is a factor of 1.063 stronger than the corresponding $P(16)$ line, and this is primarily dictated by the ratio of Hönl–London factors for the $J'' = 16$ rotational energy level ($17/16 = 1.0625$) in this case. The frequencies $\nu_{1,2}$ and ω_0 in Equation 2.11 represent the vibrational-rotational transition and vibrational band centre frequencies in cm^{-1}, respectively, E_r'' is the rotational energy level in cm^{-1}, and $Q_r(T)$ is the rotational partition function, and is given by the sum over all rotational levels by:

$$Q_r(T) = \sum_i g_i \exp\left(\frac{-hcE_i}{kT}\right) \tag{2.12}$$

Here the summation is over all the i degenerate ground states with g_i being the degeneracy of the state i. This factor, which depends upon the particular molecule in question, includes the nuclear spin statistical weights. As discussed by Rothman *et al.* (1998) caution must be exercised when calculating the statistical weights for degenerate states.

The Herman–Wallis vibration–rotation interaction factor F is calculated from a power series expansion of the rotational quantum number, and the expansion coefficients are generally determined experimentally for a specific band of the molecule in question. In some representations for the integrated spectral line intensity, the Herman–Wallis and Hönl–London factors are folded into the square of the dipole transition moment, R^2. The HITRAN database and the associated write-up (Rothman *et al.*, 1998) is one such example.

The units for the spectral line intensity $S_{1,2}$ are governed by the units for the vibrational band strength, since the rotational and Herman–Wallis terms are unitless. Vibrational band strengths are typically expressed in units of $cm^{-2}\,atm^{-1}$, and therefore the resulting vibrational-rotational line intensities are in these same units. These units are typically employed in most atmospheric absorption calculations. However, many of the databases like the 2000 HITRAN database (Rothman *et al.*, 2003), give the vibrational-rotational line intensities in molecular units of $cm^2\,cm^{-1}molecule^{-1}$. To convert from molecular units to $cm^{-2}\,atm^{-1}$ at a temperature T (K), one simply multiplies the HITRAN values by the Loschmidt's number times the ratio ($273.15/T$).

As the vibrational-rotational spectrum of CO_2 in Figure 2.3 shows, the integrated line intensities for the R- and P-branches exhibit a distribution with maximum intensities at R(16) and P(16) for the two branches at 296 K. As discussed above, the Hönl–London factor, which increases for increasing rotational quantum number, is one factor dictating this distribution. The other major factor is the rotational population, which is described by a Boltzmann distribution (see the first exponential term in Equation 2.11). As can be seen in Equation 2.11, as the rotational energy increases (increasing the rotational quantum number) the exponential factor decreases. These two factors, which oppose one another, maximize at a rotational lower state quantum number of $J'' = 16$ in this case. Other molecules will maximize at other values. Integrated spectral line intensities in the HITRAN database span the range from $10^{-18}\,cm^2\,cm^{-1}$ $molecule^{-1}$, for strong absorbers such as from select lines of CO_2, OCS, CS_2, HF, to name a few examples, to $10^{-28}\,cm^2\,cm^{-1}$ $molecule^{-1}$, for weak absorbers. Integrated spectral line intensities for formaldehyde (CH_2O), which will be discussed in Examples 1 and 2, are moderate in value (strong lines are in the $5–7 \times 10^{-20}\,cm^2\,cm^{-1}$ $molecule^{-1}$ range).

The final aspect regarding vibrational-rotational line intensities is temperature, which has an influence on the number of absorbing molecules in the ground state. As can be seen, there are a number of temperature-dependent factors in Equations 2.10–2.12. In addition to the molecular number density factor (T_0/T), the vibrational (Q_v) and rotational (Q_r) partition functions along with the various exponential terms all have temperature dependencies. The net effect of all these terms generally results in increased absorption line intensities with decreasing temperature. The exact magnitude depends upon the particular molecule under study. In the case of the R(16) line of CO_2 in Figure 2.3, the absorption line intensity increases from a value of $3.52 \times 10^{-18}\,cm^2\,cm^{-1}\,molecule^{-1}$ at 296 K to a value of $4.34 \times 10^{-18}\,cm^2\,cm^{-1}\,molecule^{-1}$ at 200 K. In addition, the peak of the distribution shifts toward a lower rotational quantum number; the peak line now occurs at R(12) with a line intensity of 4.59×10^{-18}. One can utilize this increased line intensity to realize increased sensitivity when studying trace atmospheric gases in the upper troposphere and lower stratosphere, where temperatures can be as low as 190 K. However, absorption lineshapes, which also exhibit temperature-dependent factors (to be discussed), must also be considered.

2.3 Quantitative trace gas measurements employing IR absorption spectroscopy

2.3.1 IR absorption lineshapes and linewidths

In the previous discussion, vibrational-rotational spectral features were treated as infinitely sharp widthless transitions between a pair of energy levels. However, all transitions are spread over a finite range of wavelengths or frequencies with a maximum at a frequency of ν_0 at the line centre. Throughout the rest of this chapter, we will use the terms ν_0 and ν to represent frequencies (in cm^{-1} units) at the line centre and at other positions over the vibrational-rotational absorption feature, respectively. In this context, the term ν is not to be confused with its other usages previously discussed. We will also shorten our previous designation of $S_{1,2}$ for the integrated vibrational–rotational line intensity to just S. Consideration of absorption linewidth and its associated shape are very important factors when carrying out quantitative absorption measurements. In this section we present a brief discussion of three of the typical absorption lineshapes one encounters in atmospheric measurements. For simplicity we assume only one absorbing species, and that broadening due to finite energy level lifetimes (natural broadening) and field broadening (due to strong applied electric and magnetic fields) are negligible compared to broadening by two other mechanisms: the random thermal motion of absorbing molecules relative to the analyzing and receiving device, and collisions between molecules.

Figure 2.6 shows the two extremes that one may encounter in the absorption lineshape $\Gamma(\nu)$ in atmospheric measurements. These lineshapes are Doppler and Lorentz. A third lineshape not shown in Figure 2.6 (Voigt) will also be considered in this chapter. The lineshape factor $\Gamma(\nu)$ describes how the vibrational-rotational spectral line intensities of the previous section are distributed with frequency. In this figure and the following equations, the maximum lineshape factor $\Gamma(\nu_0)$ occurs at line centre for the three distributions that will be discussed, and the symbols γ_D, γ_L, and γ_V, respectively, denote the Doppler,

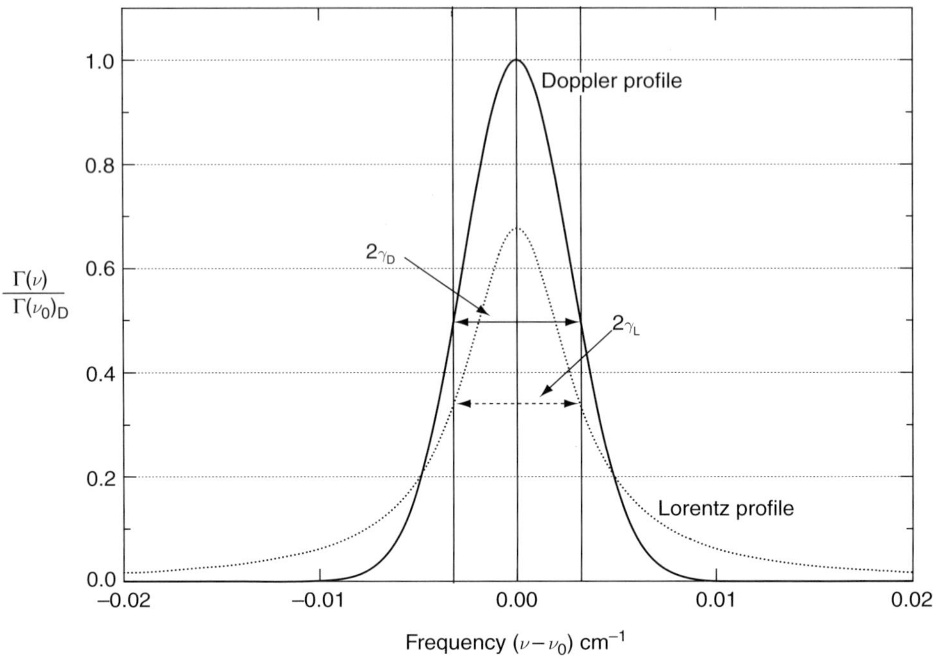

Figure 2.6 Doppler and Lorentz profiles $\Gamma(\nu)$ for the same integrated intensity and halfwidth values ($\gamma_D = \gamma_L$). Both profiles are normalized to the peak of the Doppler $\Gamma(\nu_0)_D$ at line center. The peak of the Doppler profile is 1.48 times the peak of the Lorentz profile.

Lorentz, and Voigt distribution halfwidths in cm^{-1} at half the maximum line centre values.

At low pressures (typically less than 1 torr), the random thermal motions of the molecules follow a Maxwell–Boltzmann velocity distribution, resulting in a Gaussian absorption lineshape whose frequency intensity distribution $\Gamma(\nu)$ in cm units is given by the following:

$$\Gamma(\nu)_D\,\mathrm{cm} = \frac{1}{\gamma_D}\sqrt{\frac{\ln 2}{\pi}}\exp\left[-\left(\frac{\nu-\nu_0}{\gamma_D}\right)^2\ln 2\right] \tag{2.13}$$

The pre-exponential term is the peak line centre value $\Gamma(\nu_0)_D$. Equation 2.13 is the general form of the Gaussian distribution with an integrated value of 1. The Doppler halfwidth is related to the molecular weight M and temperature T (K) of the gas by:

$$\gamma_D\,\mathrm{cm}^{-1} = 3.581\times 10^{-7}\nu_0\sqrt{\frac{T}{M}} \tag{2.14}$$

Since IR databases deal with integrated line intensities, while the sensitivity of real-world instruments typically rely on peak line centre absorptions, it is very useful to integrate Equation 2.13 (over all frequencies spanning the absorption line) to relate the peak line centre value to the integrated Doppler linewidth and halfwidth. This results in the following very useful expression:

$$\Gamma(\nu_0)_D\,\text{cm} = \frac{\int \Gamma(\nu)}{\gamma_D}\sqrt{\frac{\ln 2}{\pi}} = \frac{0.4697}{\gamma_D}\int \Gamma(\nu) \tag{2.15}$$

Since most atmospheric trace gas measurements are not carried out using Doppler-limited pressures, one must also consider two other lineshapes. At high pressures, typically at several hundred Torr, collisions between molecules dominate the lineshape, which is described by the following Lorentzian distribution:

$$\Gamma(\nu)\,\text{cm} = \frac{1}{\pi}\left[\frac{\gamma_L}{(\nu-\nu_0)^2+\gamma_L^2}\right] \tag{2.16}$$

Here the Lorentz halfwidth $\gamma_L(\text{cm}^{-1})$ at temperature T is related to the broadening coefficient $\gamma_L^0(\text{cm}^{-1}\,\text{atm}^{-1})$ by:

$$\gamma_L(T)\,\text{cm}^{-1} = \gamma_L^0(T_0)\left(\frac{T_0}{T}\right)^n P \tag{2.17}$$

In this expression, P is the total sample pressure in atm, the exponent n gives the temperature dependence for the broadening and is typically around 0.5, from simple collision theory. However, small deviations from this value have been measured or inferred in many cases. The superscript on the broadening coefficient refers to the reference temperature T_0, which usually is 296 or 298 K. Values for γ_L^0 and n are given for various molecules in spectral databases such as HITRAN. It is worth noting that the temperature dependence for the Lorentz broadening is opposite to that for Doppler broadening, and the net overall effect of temperature will depend upon the exact Lorentz and Doppler contributions to the absorption lineshape. As with the Doppler profile, one can derive the following useful expression that relates the peak line centre Lorentz value $\Gamma(\nu_0)$ to the integrated value and halfwidth:

$$\Gamma(\nu_0)\,\text{cm} = \frac{1}{\pi\gamma_L}\int \Gamma(\nu) \tag{2.18}$$

A comparison of the peak line centre values for the Doppler and Lorentzian profiles of Equations 2.15 and 2.18 indicates that the Doppler profile yields an absorption 1.48 times larger than the Lorentz profile for equivalent line halfwidths (i.e. a pressure where $\gamma_L = \gamma_D$). This result is shown in Figure 2.6. One should keep in mind that this is a special case at a unique sampling pressure for illustrative purposes, only to show the comparison of the two functions. At such a sampling pressure one cannot simply choose to operate on one function or the other, since both functions become important in describing the lineshape, as will now be discussed.

At intermediate sampling pressures of approximately several torr to ~100 torr (the exact range of which depends upon the particular molecule and the values of γ_D and γ_L), both Doppler and Lorentz lineshapes contribute to the overall lineshape, and results in a new function described by a Voigt function. Since in atmospheric studies most instruments operate in this pressure regime, the Voigt function is most frequently the function of interest. The reader is referred to Mitchell and Zemansky (1961) and Penner (1959) for some of the original discussions of the Voigt profile and to Armstrong (1967) and Humlicek (1982) for more discussions and solutions. The Voigt profile is generated by the convolution of the Doppler and Lorentzian lineshape functions above. In such a convolution process, one assumes that the two lineshape functions are independent of one another, and that the new function is generated from the integral of sliding one function across the other over the entire absorption line (i.e. limits of the integral from $-\infty$ to $+\infty$). The convolution process results in the following expression:

$$\Gamma(\nu)\,\mathrm{cm} = \Gamma(\nu_0)_\mathrm{D}\left[\frac{y}{\pi}\int_{-\infty}^{\infty}\frac{\mathrm{e}^{-t^2}}{y^2+(x-t)^2}\,\mathrm{d}t\right] \tag{2.19a}$$

with

$$x = \left(\frac{\nu-\nu_0}{\gamma_\mathrm{D}}\right)\sqrt{\ln 2};\quad y = \left(\frac{\gamma_\mathrm{L}}{\gamma_\mathrm{D}}\right)\sqrt{\ln 2};\quad t = \left(\frac{\delta}{\gamma_\mathrm{D}}\right)\sqrt{\ln 2} \tag{2.19b}$$

The parameter δ, which is incorporated in the parameter of integration t, is used to express the Doppler and Lorentz frequency differences during the convolution process in terms of a single variable. This convolution is carried out by keeping one function fixed at a value of x and sliding the other function across in $\delta\,\mathrm{cm}^{-1}$ steps over the entire line. The function of Equation 2.19a shows that the Voigt profile can be described by a unitless fraction (the term in the brackets) times the peak line centre Doppler function, $\Gamma(\nu_0)_\mathrm{D}$. This fraction approaches 1.0 at low pressures (Doppler profile) and takes on values of 0.02 and lower at high pressures (Lorentz profile), depending upon the broadening coefficient and the pressure.

The frequency dependence of the various spectral line intensity profiles just introduced can be related to the integrated spectral line intensity S of the previous section through the following:

$$S(\nu) = \Gamma(\nu)S \tag{2.20a}$$

$$S(\nu_0) = \Gamma(\nu_0)S \tag{2.20b}$$

Since the shape factor has the unit of cm, the unit of $S(\nu)$ becomes cm^2 $\mathrm{molecule}^{-1}$.

Unfortunately, unlike the Doppler and Lorentz functions, there is no simple relationship between the integrated and peak Voigt functional values. One must solve Equation 2.19 explicitly for each molecule at the sampling pressures and temperatures employed. This is not a problem since Equation 2.19 was derived in such a manner as to approximate the real part of the complex probability function, for which there are a number of efficient polynomial expansion solutions (Humlicek, 1982 and Armstrong, 1967, to name a few). Efficient computer code can thus be written to routinely solve the Voigt function under different conditions. However, one can also solve Equation 2.19 numerically using a simple spreadsheet program, and it is very instructive to do so for one set of conditions at the line centre, as shown by the following example.

Example 1

Conditions for example

Gas is CH_2O, $M = 30.026\,\text{g mole}^{-1}$
$T = 296\,\text{K}$, $P = 50.0\,\text{torr} = 0.0658\,\text{atm}$
$\nu_0 = 2831.6417\,\text{cm}^{-1}$ (from HITRAN)
$S = 5.04 \times 10^{-20}\,\text{cm}^2\,\text{cm}^{-1}\,\text{molecule}^{-1}$ at 296 K (from HITRAN)
$\gamma_L = (0.108\,\text{cm}^{-1}\,\text{atm}^{-1}$ at 296 K from HITRAN)* $(0.0658\,\text{atm}) = 0.0071\,\text{cm}^{-1}$
$\gamma_D = 0.0032\,\text{cm}^{-1}$ (Equation 2.14)
$y = 1.858$ (Equation 2.19b)
$x = 0$ (Equation 2.19b, for the Voigt value at line centre where $\nu = \nu_0$)
Range of convolution $= \pm 0.05\,\text{cm}^{-1}$ over the line, number of steps in convolution $= 1000$
Resolution of convolution $= 0.1\,\text{cm}^{-1}$ in 1000 steps $= 1 \times 10^{-4}\,\text{cm}^{-1}\,\text{step}^{-1}$

Steps in convolution

(1) Calculate values of t and ultimately the entire term in the integral of Equation 2.19a for each $\delta\,\text{cm}^{-1}$ step in sliding one function over the other. We start at a δ value $= -0.05\,\text{cm}^{-1}$, and t becomes -13.072.
(2) Calculate e^{-t^2} and ultimately the integrand of Equation 2.19a, which results in negligibly small values of 6.2×10^{-75} and 3.6×10^{-77}, respectively, for this particular convolution step.
(3) Repeat this procedure for each of the 1000 steps in the δ convolution and store the resultant value of the integrand each time. At line centre where δ is 0, we calculate a value of 0.2898 for the integrand.
(4) Integrate all the above individual determinations using a trapezoidal approach, where the average value of two adjacent determinations is multiplied by the step size and summed over all the steps.
(5) Multiply the integral by $y/\pi = 0.591$ to arrive at a value of 0.272. The Voigt line centre value is thus 27.2% of the peak of the Doppler profile. After multiplying by the Doppler peak ($7.43 \times 10^{-18}\,\text{cm}^2\,\text{molecule}^{-1}$) one arrives at a final Voigt line centre spectral intensity of $2.02 \times 10^{-18}\,\text{cm}^2\,\text{molecule}^{-1}$.
(6) One can repeat the entire procedure using a different value for v in the function x to determine the Voigt function at some other frequency.

Figure 2.7(a) shows the results of calculating ($T = 296\,\text{K}$) the Voigt function at line centre for three different molecules. Both methods of calculation (the simple spreadsheet method in Example 1 or the Humlicek (1982) polynomial expansion) yield the same results. In each case we plot the individual Voigt results for the molecule in question normalized to its specific Doppler value at line centre, $\Gamma(\nu_0)_V/\Gamma(\nu_0)_D$ (this is the fractional term in the brackets of Equation 2.19a). The main plot shows the relationship as a function of the parameter y, while the inset shows the relationship as a function of the pressure in torr. The three molecules shown – which have vastly different Doppler widths at the frequencies selected: OH ($\gamma_D = 0.0053\,cm^{-1}$), CH_2O ($\gamma_D = 0.0032\,cm^{-1}$) and HBr ($\gamma_D = 0.0018\,cm^{-1}$) – were chosen to represent the range of Doppler halfwidths that one would typically encounter in atmospheric studies. As can be seen in the main plot, all three molecules yield the same behavior as a function of the parameter y, as expected from Equation 2.19a; for each fixed value of x in this equation the y parameter governs the value of the fractional term in the brackets. Hence, in addition to the example calculation above, one may use the main plot of Figure 2.7(a) as a convenient means to quickly determine the value for the normalized line centre Voigt function. In this procedure, one (1) determines the y value for the specific molecule of interest at the sampling pressure employed, and (2) the normalized line centre Voigt function is determined utilizing the curve.

The inset in Figure 2.7(a) shows the normalized line centre Voigt function versus sampling pressure, and this indicates that equivalent y values occur at different sampling

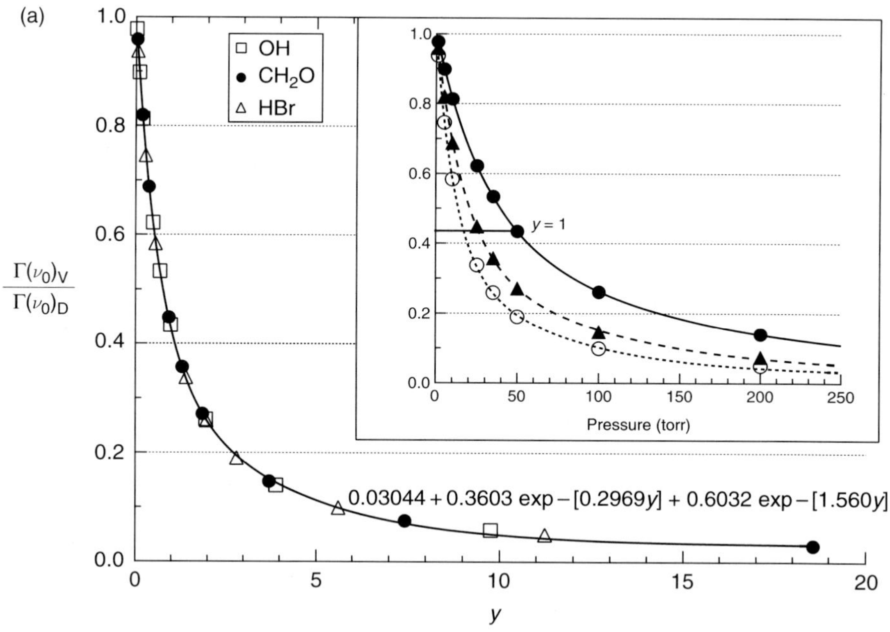

Figure 2.7 (a) Calculated (at 296 K) line center Voigt function $\Gamma(\nu_0)_V$ normalized to the line center Doppler function $\Gamma(\nu_0)_D$ versus the parameter y for three different molecules with vastly different Doppler widths and comparable pressure broadening coefficients ($\sim 0.1\,cm^{-1}\,atm^{-1}$) : OH ($\nu_0 = 3568.4169\,cm^{-1}$, $\gamma_D = 0.0053\,cm^{-1}$), CH_2O($\nu_0 = 2831.6417\,cm^{-1}$, $\gamma_D = 0.0032\,cm^{-1}$), and HBr ($\nu_0 = 2606.1991\,cm^{-1}$, $\gamma_D = 0.0018\,cm^{-1}$). The curve is a double exponential fit of the data and the resulting fit equation is given above the curve for facility in calculating the normalized function. The inset shows this same normalized Voigt function versus sampling pressure in Torr. The horizontal line intersects the three curves at y=1 at different sampling pressures, and this yields approximately the same normalized ratio value of 0.44 for the three molecules. (b) Plots of the Voigt line center spectral intensities in units of $cm^2 molecule^{-1}$ versus sampling pressure for the three molecules in Figure 2.7a.

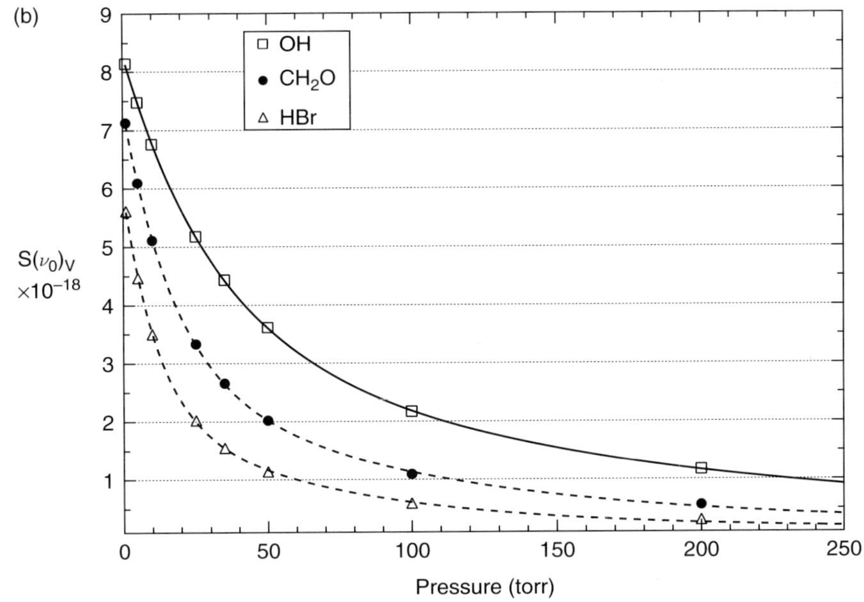

Figure 2.7 (Continued).

pressures for the three molecules. As can be seen, pressure broadening has a smaller effect on the normalized ratio for larger Doppler halfwidths, and this effect is most pronounced at intermediate pressures in the 10–50 torr range. The value for the normalized Voigt function from either plot or the spreadsheet calculations can be used to determine the Voigt function at line centre from a simple multiplication of this normalized value with the peak of the Doppler function at line centre. Equation 2.20b is then used to calculate the Voigt spectral line intensity at line centre, $S(\nu_0)_V$, and Figure 2.7(b) plots these values as a function of pressure for the three molecules. The values in this plot, which show similar behavior as the inset of Figure 2.7(a), are the final values needed in the Beer–Lambert Law absorbance calculations of the next section.

The Voigt profile just presented assumes that the Doppler distribution is unaltered by collisions between molecules. However, at low pressures approximately in the 1–10 torr range, the Maxwell–Boltzmann velocity distribution can be affected by collisions, and this results in the phenomenon known as *collisional-narrowing* or *Dicke-narrowing* (see Pine and Fried, 1985, and references therein). This effect is most pronounced in light molecules (larger Doppler widths), cases where the pressure-broadening coefficients are small, and where the mismatch in rotational energy levels between collisional partners is small (i.e. where the collisional partner is the gas of interest such as in self-broadening measurements). In such cases the Voigt profile underestimates the peak absorbance and compensates with too much absorbance in the wings of the profile. Various collision models and profiles have been developed to better describe the resulting lineshape, but these will not be presented here. At typical sampling conditions, where the pressures are of several 10s of torr and the broadening gas is air, the errors in the Voigt profile are generally less than a few percent.

The IR absorption lineshapes just presented play an important role not only on measurement sensitivity but also on selectivity. In addition, these lineshapes dictate whether or not one can employ frequency modulation techniques, which rely on modulating on and off absorption features. The relatively narrow absorption features of Figure 2.7 all have Voigt linewidths less than $0.01\,\mathrm{cm}^{-1}$ for sampling pressures of 50 torr. This is a major advantage of working in the IR spectral region compared to the visible and ultraviolet regious, particularly when in situ measurements are carried out employing a narrow spectral bandwidth laser light source and an absorption cell operated at reduced sampling pressures of typically several torr to ~100 torr. Under these conditions, the IR Voigt absorption lineshapes for individual or small groups of lines are generally well resolved from other absorption features, including features from the same molecule and/or interference molecules. Even in cases where spectral overlapping occurs, high selectivity can still be achieved in many cases since the interfering lines are generally well characterized and small in number, and the absorption lineshapes are well known. As a result, spectral deconvolution methods are frequently successful in minimizing such problems. By contrast, in the visible and UV spectral regions the absorption features often appear as dense clumps of unresolvable features, even under reduced sampling pressures. The significantly increased Doppler halfwidths in the UV compared to the IR are partially responsible for this congestion. Equation 2.14 indicates that γ_D is directly proportional to the line centre frequency (ν_0), which in the UV is approximately one order of magnitude larger than the IR.

2.3.2 Beer–Lambert absorption law and absorbance

The discussion in the previous section relating line centre Voigt spectral line intensities to pressure does not give the complete picture for the instrument sensitivity one might expect from an IR spectrometer. As we will now discuss, one must also fold in the number density of the absorbing species and the sample pathlength, l. The absorber number density is expressed by the product of three terms, P (the sample pressure in atm units), the sample-volume mixing ratio χ (typically in ppmv, 10^{-6}; in ppbv, 10^{-9}; or in pptv, 10^{-12}), and the total number density for the bath gas in the probed sampling volume, N (molecules $\mathrm{cm}^{-3}\,\mathrm{atm}^{-1}$). These terms are used in the well-established Beer–Lambert Law, which relates the above molecular parameters to measured light intensities:

$$I(\nu) = I_0(\nu)\exp[-S(\nu)(P\chi N)l] = I_0(\nu)\exp[-A(\nu)] \qquad (2.21)$$

This expression provides the basis upon which many IR absorption instruments rely, directly or indirectly, for converting measured light intensities to quantitative mixing ratio results. In this expression, $I(\nu)$ and $I_0(\nu)$ are the frequency-dependent transmitted and incident intensities, respectively, $S(\nu)$ is the spectral line intensity discussed in the previous section, and the term $A(\nu)$ is the absorbance. The ratio I/I_0 is referred to as the *transmittance* or sometimes simply as the *transmission*. Strictly speaking, according to the International Union of Pure and Applied Chemistry, the term 'absorbance' is reserved for Equation 2.21 written in base 10 logarithmic units, and the term $A(\nu)$ in Equation 2.21 is properly called the *optical depth*, *optical density*, or *optical mass*. However, it is common practice in atmospheric studies to designate $A(\nu)$ in base e units as the absorbance, and we shall adopt this common usage here. In remote measurements (to be discussed) the exponent is oftentimes referred to as the *extinction*, τ, which has contributions from molecular absorption and scattering, as well as aerosol scattering and absorption.

Equation 2.21 is written for a single absorbing species; multiple absorbers are written as a summation of terms in the exponent. Equation 2.21 applies equally well to intensities integrated over the absorption feature or at line centre. In the case of the former, one employs the integrated spectral line intensity, S, while in the latter one employs the line centre spectral intensity, $S(\nu_0)$, and the line centre absorbance $A(\nu_0)$. In either case, it is important to emphasize that the Beer–Lambert law relationship provides quantitative results based on fundamental parameters and fundamental measured quantities, which are independent of instrumental factors. Hence, direct absorption measurements using this approach are considered absolute and, as will be discussed, do not require calibration standards. This is in contrast to other measurement techniques such as fluorescence. The lower panel of Figure 2.8 depicts the resulting relationship between the incident and the transmitted intensities for a hypothetical absorption.

We now examine in more detail the unitless exponential term of Equation 2.21 at line centre. In most cases in atmospheric studies, the unitless absorbance is considerably less than 0.2, and the exponential term e^{-A} can be replaced by its linear form $(1 - A)$. This is referred to as the 'optically thin' absorption region. Using this approximation, one can easily derive the following expression between the absorbance and the fractional change in light intensity, $\Delta I/I_0$:

$$A(\nu_0) = \frac{\Delta I(\nu_0)}{I_0(\nu_0)} \tag{2.22}$$

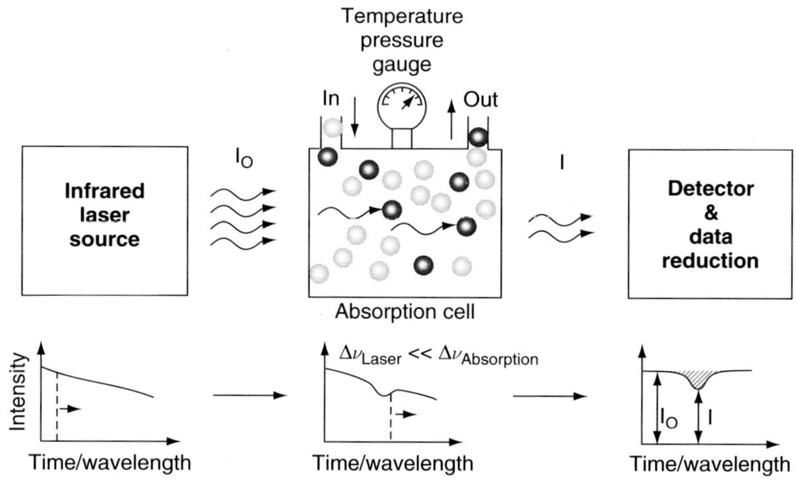

Figure 2.8 The top panel depicts the fundamental components of an IR absorption setup. The lower panel shows the resulting recorded direct absorption profile along with the incident (I_0) and transmitted (I) intensities employed in Beer–Lambert Law calculations. Since the wavelength of the laser source is scanned in time, the x-axis can be represented by either variable.

The smallest resolvable fractional change in light intensity at line centre, and hence the minimum detectable line centre absorbance, is frequently employed in many atmospheric studies as one means of characterizing instrument performance. This term is designated $A(\nu_0)_{min}$ or A_{min} for short, and will be used throughout the rest of this chapter. Rewriting Equation 2.21 for line centre conditions we arrive at the following expression with associated units:

$$A(\nu_0) = [S(\nu_0)\,\mathrm{cm}^2\,\mathrm{molecule}^{-1}][(P\,\mathrm{atm})(\chi)(N\,\mathrm{molecules\ cm}^{-3}\,\mathrm{atm}^{-1})][l\,\mathrm{cm}] \quad (2.23)$$

This expression, which is the culmination of many of the previous expressions, is highlighted to indicate its importance and the fact that it will be employed throughout the remainder of this chapter when discussing instrument performance. Mostly, the instrument performance figure of merit is given by A_{min} normalized to absorption pathlength and an effective detection bandwidth, typically in units of $\mathrm{Hz}^{-1/2}$:

$$\frac{A_{min}}{l(\mathrm{cm})\sqrt{\text{effective bandwidth (Hz)}}} \quad (2.24)$$

The following example illustrates how Equation 2.23 is used to calculate line centre absorbance. We employ in this example the molecule CH_2O, the conditions discussed in the Example 1, and realistic sampling conditions we employ with our airborne tunable diode laser system.

Example 2

Conditions of example

Gas is CH_2O
$M = 30.026\,\mathrm{g\,mole}^{-1}$
$T = 296\,\mathrm{K}$
$\nu_0 = 2831.6417\,\mathrm{cm}^{-1}$ (from HITRAN)
$S = 5.04 \times 10^{-20}\,\mathrm{cm}^2\,\mathrm{cm}^{-1}\,\mathrm{molecule}^{-1}$ at 296 K (from HITRAN)
$P = 50.0\,\mathrm{torr} = 0.0658\,\mathrm{atm}$
$S(\nu_0)_V = 2.02 \times 10^{-18}\,\mathrm{cm}^2\,\mathrm{molecule}^{-1}$ (from Example 1, at $P = 50$ torr, or from Figure 2.7b)
Pathlength $l = 100\,\mathrm{m} = 10\,000\,\mathrm{cm}$
$\chi = [CH_2O] = 1\,\mathrm{ppbv} = 1 \times 10^{-9}$
$N = L(T_0/T) = 2.68678 \times 10^{19}(273.15/296) = 2.479 \times 10^{19}$ molecules $\mathrm{cm}^{-3}\,\mathrm{atm}^{-1}$
$(P\,\mathrm{atm})(\chi)(N\,\mathrm{molecules\ cm}^{-3}\,\mathrm{atm}^{-1}) = (0.0658\,\mathrm{atm})(1 \times 10^{-9})(2.479 \times 10^{19}$ molecules $\mathrm{cm}^{-3}\,\mathrm{atm}^{-1}) = 1.6 \times 10^9$ molecules cm^{-3}

$A(\nu_0) = (2.02 \times 10^{-18}\,\mathrm{cm}^2\,\mathrm{molecule}^{-1})(1.6 \times 10^9\,\mathrm{molecules\ cm}^{-3})(10\,000\,\mathrm{cm}) = 3.3 \times 10^{-5}$

Figure 2.9 shows the resulting $A(\nu_0)$ calculations for CH_2O as well as for OH and HBr at other sampling pressures, using the data of Figure 2.7 and Equation 2.23. The three profiles of Figure 2.9 reveal the results of two opposing terms with pressure: the decreasing spectral line centre intensities with increasing pressure (Figure 2.7b) and the increasing absorber number density N with increasing pressure. At low pressures, $A(\nu_0)$ linearly increases with pressure (Doppler regime), while at high pressures $A(\nu_0)$ asymptotically approaches its pressure-independent Lorentz value. In the case of the latter, the increase in pressure in Equation 2.23 is cancelled by the P^{-1} dependence; the Lorentz line centre spectral intensity of Equation 2.23 is proportional to the Lorentz line centre shape factor of Equation 2.18, which from Equation 2.17 indicates a P^{-1} dependence. At intermediate sampling pressures both pressure dependencies are operable. The profiles of Figure 2.9, which essentially reflect the inverse behavior of Figure 2.7b, indicate a different asymptotic roll-off region for the three molecules shown. The shaded points represent the $A(\nu_0)$ value equal to ~90% of the asymptotic value at high pressure for each profile, and these occur at different pressures. The larger the Doppler influence, the higher the roll-off pressure (e.g. compare the largest Doppler profile of OH versus the smallest Doppler profile for HBr).

Figure 2.9 reveals the optimum pressure regime for the three example molecules shown using mid-IR detection. The highest pressure yields the largest line centre absorbance, and

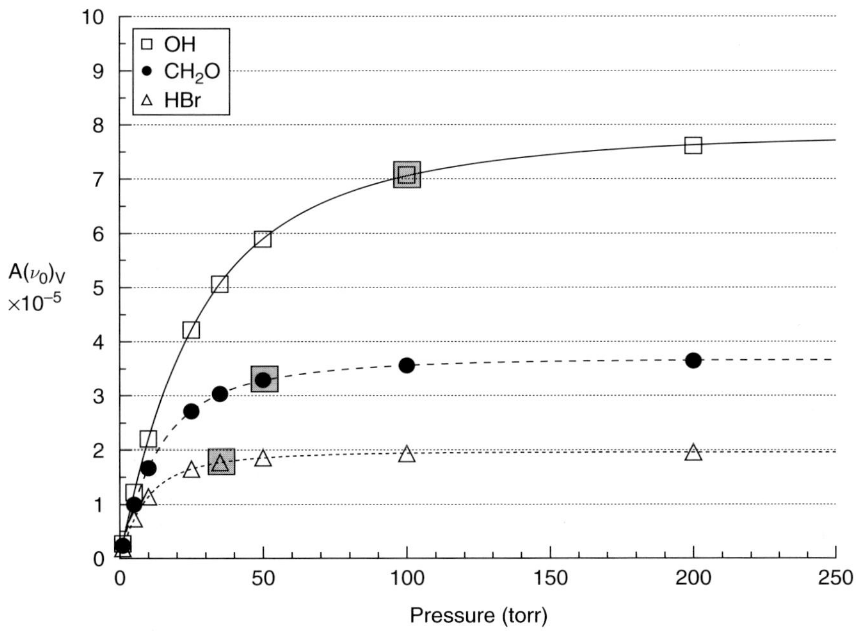

Figure 2.9 Plot of Voigt line center absorbance, $A(\nu_0)_V$, calculation for OH, CH_2O, and HBr for the conditions of Figure 2.7a as a function of sampling pressure for conditions of: T = 296 K, mixing ration = 1 ppbv, pathlength = 100 m. The curves are double exponential fits of the data, and the three points highlighted by shading on the three profiles represent ~90% of the maximum line center absorbance value for each profile. The relative magnitudes of the three profiles largely reflect the differences in the integrated spectral line intensities for the three molecules. At 50 torr, the Voigt halfwidths are 9.4×10^{-3}, 8.3×10^{-3}, and $6.5 \times 10^{-3}\ cm^{-1}$ respectively for the three molecules.

hence the highest sensitivity at the expense of measurement selectivity. Conversely, sampling in the linear low-pressure regime yields the best selectivity at the expense of sensitivity and the precision with which the sampling pressure must be controlled (i.e. a fluctuation in the sampling pressure here results in a larger sensitivity change compared to the roll-off regime). Given these two extremes, one should operate at sampling pressures somewhere around the shaded points for the three profiles. One can use Figure 2.9 to estimate the optimum pressure regime for other molecules. However, Equation 2.23 and the calculations of Example 1 for the particular conditions at hand should be utilized for precise determinations.

2.4 Trace gases in the atmosphere

With the exception of nitrogen, which exhibits the weak IR quadrupole transitions discussed earlier, the mixing ratios for IR-active atmospheric constituents vary from the percent range in the case of H_2O near the surface to the pptv range in the case of various trace species.

Table 2.2 Typical mixing ratios of various trace atmospheric species[a]

Constituent	Chemical formula	Volume mixing ratio	Data source
Carbon dioxide	CO_2	375 ppmv	b
Ozone (troposphere)	O_3	10–500 ppbv	f
Ozone (stratosphere)	O_3	0.5–10 ppmv	f
Nitrous oxide	N_2O	318 ppbv	b
Carbon monoxide (background)	CO	50–160 ppbv	c
Carbon monoxide (Airborne Bkg levels to pollution plumes)	CO	17–1113 ppbv	d
Methane	CH_4	1755 ppbv	b
Sum of Non-methane hydrocarbons	Σ NMHCs	1–25 ppbv	e
Nitric oxide	NO	0–1450 pptv	d
Nitrogen dioxide	NO_2	0–3349 pptv	d
Nitric acid	HNO_3	3–7412 pptv	d
Ammonia	NH_3	10 pptv–1 ppbv	f
Hydroxyl radical	OH	0–0.6 pptv	d
Hydroperoxy radical	HO_2	0.2–34 pptv	d
Hydrogen peroxide	H_2O_2	50 pptv–10 ppbv	d
Formaldehyde	CH_2O	25 pptv–11 ppbv	d
Sulfur dioxide	SO_2	3 pptv–31 ppbv	d
Sum of halocarbons	Σ HCs	1.4–5 ppbv	d
Carbonyl sulfide	OCS	438–1219 pptv	d

[a] Mixing ratios vary considerably depending upon season, altitude, latitude, year, air mass type, proximity to local sources and sinks as well as many other variables. The last column indicates the source of the data for the mixing ratio range.
[b] 2003 global average background levels from the NOAA Climate Monitoring & Diagnostic Laboratory (CMDL), www.cmdl.noaa.gov
[c] 2003 range of background levels found in the southern and northern hemispheres from NOAA CMDL.
[d] Range of airborne measurements during TRACE-P airborne campaign (Jacob *et al.*, 2003, data available from: www-gte.larc.nasa.gov) spanning the 0–12 km altitude range.
[e] Airborne measurements of the sum of alkanes and alkenes during the TOPSE 2000 airborne campaign (Atlas *et al.*, 2003) spanning the 0–8 km altitude range.
[f] Brasseur and Schimel (1999).

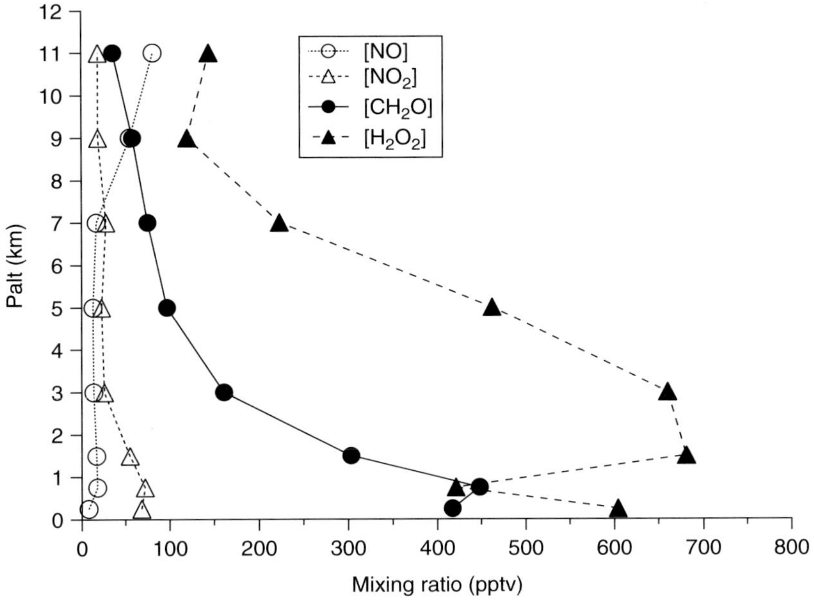

Figure 2.10 Pressure altitude median mixing ratios for eight different altitude bins for four different trace constituents measured (1 minute measurements) in the Western Pacific region (110–150° longitude) during the TRACE-P study. In the case of NO and NO_2 these measurements were obtained from the Georgia Institute of Technology Laser-Induced Fluorescence Instrument as described in Sandholm *et al.* (1997); the CH_2O measurements were obtained from our tunable diode laser absorption spectrometer as described in Fried *et al.* (2003b); and the H_2O_2 measurements were obtained from the University of Rhode Island Instrument as described in Snow *et al.* (2003).

During the NASA Transport and Chemical Evolution over the Pacific (TRACE-P) airborne campaign (Jacob *et al.*, 2003, data available from: www-gte.larc.nasa.gov), for example, H_2O vapor mixing ratios spanned the range from 1 to 3% (10 000–31 000 ppmv) near the ocean surface to 3–50 ppmv at altitudes around 12 km. Other atmospheric constituents measured during this campaign and/or measured by global monitoring networks are tabulated in Table 2.2. As can be seen, with the exception of the abundant atmospheric gases H_2O, CO_2, O_3, N_2O, CO, and CH_4, all other trace species typically exhibit atmospheric mixing ratios below 50 ppbv and more frequently below 1 ppbv. In fact, in many instances during this and many other campaigns, the mixing ratios of most trace atmospheric constituents are frequently below 100 pptv, particularly when sampling remote background air at high altitudes. This behavior is shown in Figure 2.10 for four different gases measured during the TRACE-P campaign. Such ultra-low mixing ratios require not only long absorption pathlengths but also very low A_{min} values of less than 10^{-5}.

2.5 Measurement approaches employing IR absorption spectroscopy

In all the IR absorption methods that will be discussed, an absorption spectrum of one form or another is ultimately generated, and quantitative results are obtained by measuring the change in light intensity as a function of wavelength or frequency (in wavenumbers). The methods that will be discussed differ with respect to the means employed in acquiring absorption spectra as well as the methods used in converting the acquired spectra into absorber mixing ratios. However, all methods are linked by the fundamental requirement to wavelength disperse either the source or the received radiation. In this section, we will present an overview of the various approaches that have been employed, with specific emphasis on the broad distinctions between approaches. Further details on each technique will then be presented in subsequent sections. The various applications that will be discussed are only meant to provide a broad overview of the field and not an exhaustive review of all applications.

One can characterize IR absorption methods in two broad categories: in situ and remote. Figures 2.8 and 2.11 further depict two different variants of in situ measurements. In both cases, the probed sample, which is in close proximity to the IR source and detector, captures an in situ, spatially localized snapshot of the mixing ratio for the target gas of interest. In most cases, the probed sample is continuously drawn through an absorption cell at reduced pressure, and this is shown by the fundamental components depicted in Figure 2.8. This approach is designated as extractive in situ sampling and has the advantage that one may control the sampling conditions of pressure and temperature to optimize the measurement performance. In practice, the absorption cell is typically a multipass cell and a number of transfer optics are required to image the IR beam into and out of the multipass cell and onto an IR detector.

The same basic components shown in Figure 2.8 are also applicable to the second type of in situ system: open path in situ airborne measurements. Only in this case, the sampling cell is replaced by the open atmospheric path, one example of which is depicted in Figure 2.11. This example is described in detail by Diskin *et al.* (2002) for ambient water vapor measurements on NASA's DC-8 aircraft. In this approach, in situ water vapor is probed in the 28.5 m round trip open path between the laser transceiver mounted in the aircraft cabin and a panel of retroreflecting road-sign material mounted on the outboard engine of the DC-8 aircraft. Unlike extractive in situ measurements, one can neither control the sampling conditions in this open path arrangement nor add calibration standards and/or zero air (air scrubbed of the gas of interest) to the sampling stream while in flight. Instead one must employ a calibration-calculation matrix, which is derived in the laboratory, to calibrate the instrument for various sampling conditions of pressure and temperature. Despite these issues, open path in situ measurements are ideally suited for reactive and 'sticky' gases, where the sample does not contact inlet and cell sampling surfaces.

The IR light source schematically depicted in Figure 2.8 is an extremely important component of all in situ measurement systems, and one can further define two broad categories of IR light sources. The first category includes tunable solid-state laser sources, as well as semi-continuously tunable gas laser sources, where accidental coincidences with the molecular absorption features of interest are exploited. The study by Nelson *et al.* (2002) is one of many examples of the former. The second category utilizes broadband IR emission from hot filaments sources, which are ultimately wavelength dispersed before or after the sampling cell using various types of monochromators, FTIR spectrometers, or gas correlation cells and filter wheels. Both types of IR light sources yield high temporal resolution and have produced very important flux measurements of various trace gases.

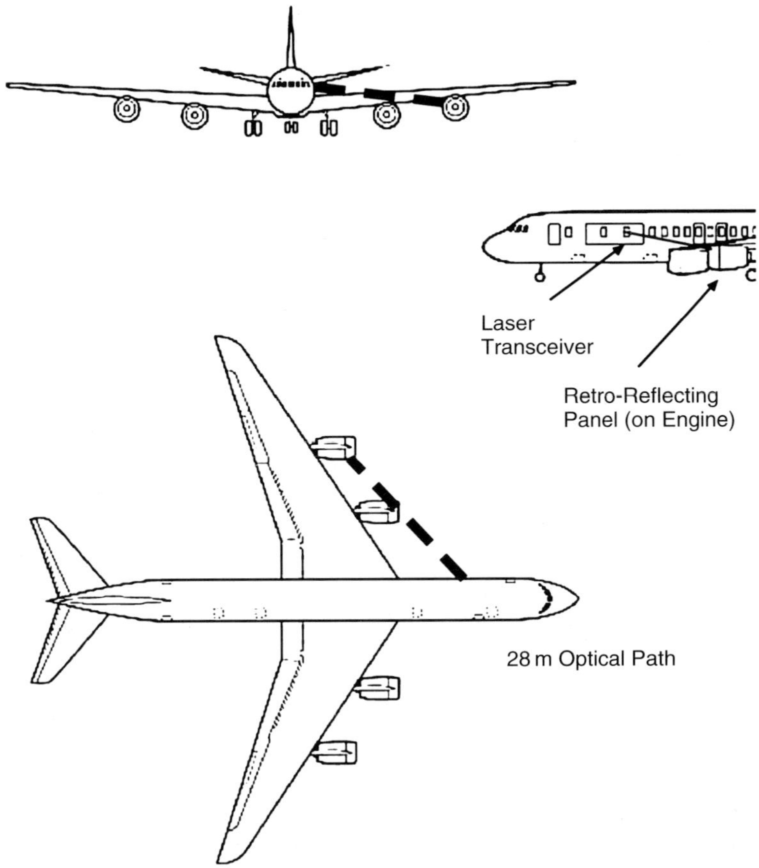

Figure 2.11 An example of an in situ open path system for airborne measurements of ambient water vapor on NASA's DC-8 aircraft (Diskin *et al.*, 2002). Ambient water is probed in the 28.5 m round trip open path between the laser transceiver mounted in the aircraft cabin and a panel of retroreflecting road-sign material mounted on the outboard engine of the DC-8 aircraft. This figure was graciously supplied by Diskin, Sachse, and colleagues at NASA.

For example, Ham and Heilman (2003) employed a commercial LI-COR instrument using a broadband IR emission source and a chopper filter wheel assembly to measure the eddy covariance flux of CO_2 and H_2O from a tallgrass prairie and a cedar forest. The study by Griffith *et al.* (2002) represents an example of FTIR flux measurements of CO_2, CH_4, N_2O, and H_2O over agricultural fields using the flux-gradient technique. Section 2.6 will highlight a third example where IR absorption measurements employing a tunable solid-state laser source have been utilized to measure trace gas fluxes. The studies of Grutter (2003) and Yokelson *et al.* (2003) represent examples of extractive in situ FTIR measurements to investigate pollution and smoke emissions on ground-based and airborne platforms, respectively. Although one can cite many other examples and approaches, the discussions in the rest of this chapter will focus exclusively on the two most widely employed in situ IR sources: tunable solid-state laser sources and broadband hot filament emission sources, followed by wavelength dispersion using an FTIR spectrometer.

In remote IR absorption measurements, the probed sample is spatially separated from the IR radiation source and detector, and such measurements can be either active or passive in nature. Active remote measurements typically employ broadband hot filament IR light sources similar to those employed for in situ measurements, followed by an FTIR spectrometer to measure trace gas concentrations and/or emission fluxes over a surface or a region under study. In many cases, transmitting and receiving telescopes are employed with the FTIR system to probe several hundred meters of open path. The studies by Griffith *et al.* (1991) and Grutter *et al.* (2003) are two such examples of this approach employed to study emissions from fires and pollution over a major metropolitan area, respectively. Balloon-borne tunable diode laser measurements employing a 500 m open path retroreflector array is yet another example where remote IR absorption measurements have been successfully carried out using an active light source (Webster *et al.*, 1994a).

In passive remote measurements, an absorption instrument receives IR radiation from some other light source, usually the sun, and the IR absorption in the intervening atmosphere or sample in a plume is measured (see Figure 2.12). Many studies, at the time of writing, employ an FTIR instrument using a solar tracker for such studies, and such

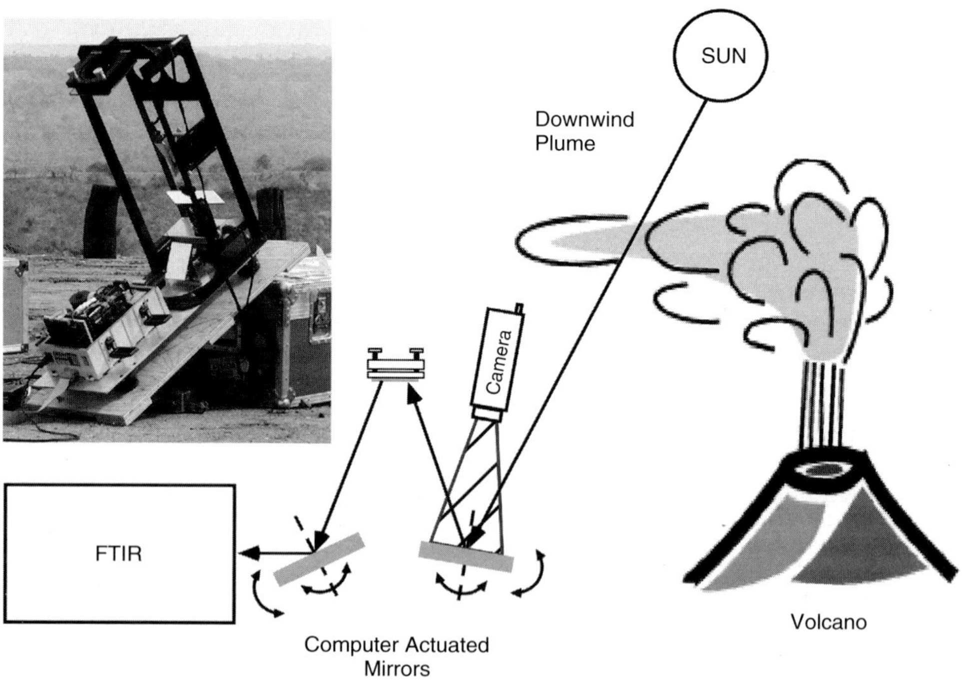

Figure 2.12 An example of passive remote FTIR measurements using the sun as a source, employing a solar tracker to study emission plumes from volcanoes (Duffell *et al.*, 2001).

measurements have been carried out on a wide range of platforms, including platforms that are ground-based (Rinsland *et al.*, 1992, 2002, 2003 and Griffith *et al.*, 2003), airborne (Mankin *et al.*, 1979, Mankin and Coffey, 1989, and Coffey *et al.*, 2003), balloon-borne (Toon *et al.*, 2002), and satellite-borne (Persky, 1995 and references therein), to cite a few examples. Duffell *et al.* (2001) have utilized remote solar absorption FTIR measurements to study volcanic emissions from a truck platform. Chapter 1 contains a brief discussion of absorption measurements from satellites and so will not be further discussed here.

Both active and passive remote IR absorption measurements can suffer from the same problems as in situ open path measurements: namely, absorption due to ambient H_2O vapor may take out so much IR radiation that much weaker trace absorptions become totally obscured. This can be a particular problem in passive remote measurements where absorption pathlengths are typically many tens of kilometer. In these cases one must work either at higher altitudes or in 'atmospheric windows' where H_2O vapor absorption is much weaker.

The above introduction and brief examples of in situ and remote techniques serve to illustrate the extreme flexibility and power of IR absorption measurements in atmospheric studies in terms of the variety of platforms and number of molecules that have been studied. In the sections that follow, we will present specific examples of in situ measurements employing tunable solid-state laser sources as well as in situ and remote FTIR measurements. As we will show, these various measurement approaches yield important complimentary data about the atmosphere. In situ measurements, particularly those employing tunable solid-state lasers, yield very sensitive measurements for a select number of atmospheric gases, which are important for furthering our understanding of localized chemical processes and transformations. Such measurements, however, are limited in space, time, and the total number of atmospheric species that can be studied by one instrument at any given time. Remote FTIR measurements, on the other hand, provide much broader temporal and spatial coverage on a much larger set of atmospheric species, and these measurements are critical for detecting tropospheric and stratospheric trends. It is important to re-emphasize that the data reduction approaches used in both types of approaches are based on the Beer–Lambert Law relationships presented earlier. However, extractive in situ measurements can also employ calibration standards.

2.5.1 *In situ measurements*

2.5.1.1 Multipass absorption cells

It is clear from the very low mixing ratios indicated in Section 2.4 for various atmospheric constituents that in situ absorption measurements in many cases require (1) very high measurement sensitivity (low values of A_{min}), and simultaneously (2) long absorption pathlengths. Since the spectral line intensity for the CH_2O absorption line shown in Example 2 is typical of many atmospheric gases, these calculations provide an excellent example of the required sensitivity and pathlength needed to successfully measure ambient levels of typical trace gases under remote background conditions. As the example shows, one can detect ambient CH_2O mixing ratios of 1 ppbv with a signal-to-noise ratio (SNR) of ~ 33 employing an absorption pathlength of 100 m and an absorbance sensitivity (A_{min}) of $\sim 1 \times 10^{-6}$. As Equation 2.23 shows, degraded A_{min} values, lower ambient

CH_2O mixing ratios and/or lower spectral line intensities in the case of other molecules will reduce the measured SNR. This example and the mixing ratios of Table 2.2 clearly indicate that absorption pathlengths of tens of meters to several hundred meters are necessary for the measurement of mixing ratios of many trace gases under moderate to remote background conditions. By contrast, pathlengths of several kilometers are typically necessary for DOAS measurements, and this will be discussed in Chapter 3.

Such long optical pathlengths have been readily achieved in atmospheric studies employing multipass absorption cells where the optical beam is reflected back and forth between a set of mirrors. Three fundamental designs have traditionally been employed to achieve pathlengths up to ~252 m in physical base lengths equal to or less than 1.5 m. Figure 2.13 depicts the three multipass cells, which are based upon the designs of White (1942), Herriott *et al.* (1964), and the astigmatic design of Herriott and Schulte (1965). Chernin (1993) further describes a whole series of multiple element multipass cells, but these have not been widely employed in atmospheric studies and will not be further discussed here.

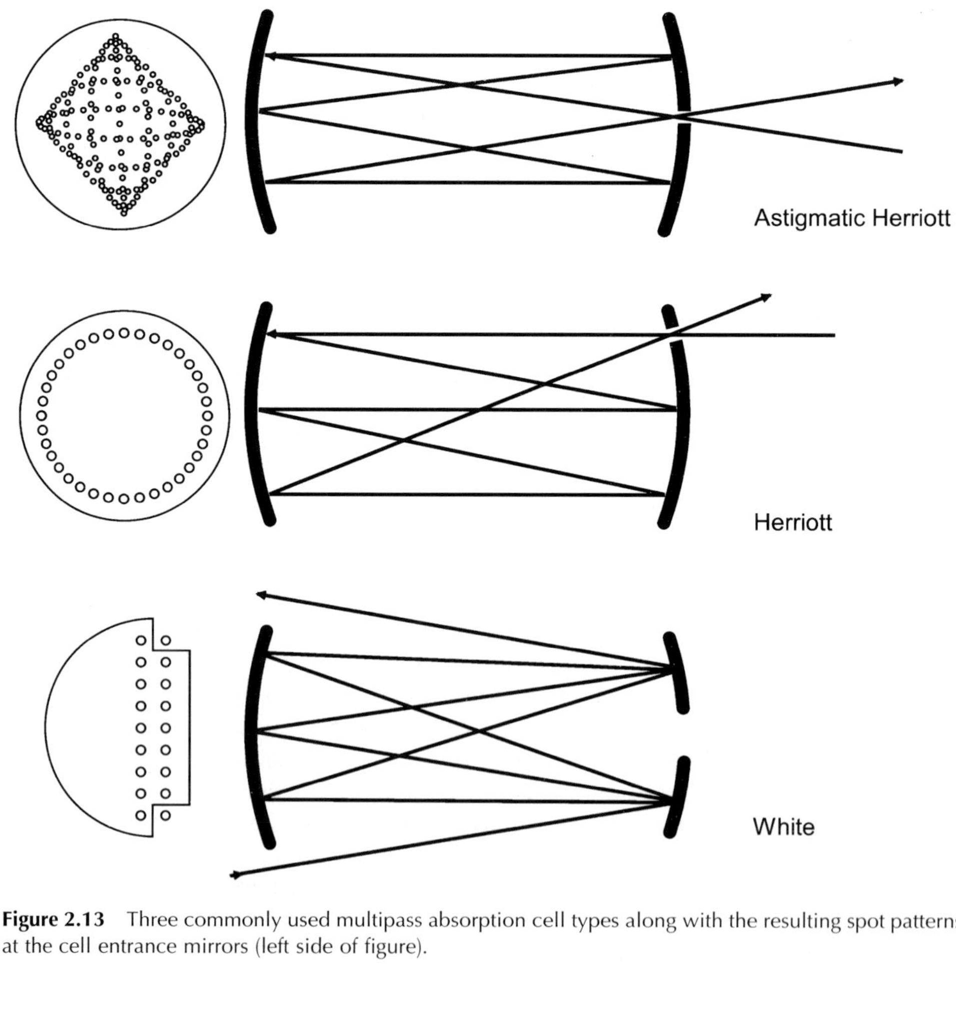

Figure 2.13 Three commonly used multipass absorption cell types along with the resulting spot patterns at the cell entrance mirrors (left side of figure).

In all three designs, the optical beam is re-imaged using curved mirrors. The White optical design consists of three spherical mirrors of identical radii of curvature, and the front mirror is placed at its confocal distance from the two identical D-shaped rear mirrors (i.e. spaced at their common radii of curvature). One focuses the input beam at the position of the front mirror and this re-images the beam back on the front mirror. As shown, the re-imaged beam forms a line of spots in two rows on the front mirror. In a modified design published by Horn and Pimentel (1971) and deployed on an aircraft platform by Schiff *et al.* (1990), the exit spot in the traditional White cell hits a corner reflector (two flat mirrors), which translates the beam vertically forming an additional two rows of spots. This arrangement reduces the output beam astigmatism; more effectively utilizes the front mirror space; and reduces the sensitivity to misalignment of the input beam (Schiff *et al.*, 1990). The sample volume of this cell is 28 liters, and pathlengths up to 213 m have been achieved by adjusting the angle for one of the back D-mirrors. Both ground-based (Mackay *et al.*, 1996) and airborne measurements (Schiff *et al.*, 1990) have successfully been acquired using this 5-mirror optical design.

Since these early implementations, newer White cells have achieved moderate pathlengths (l) in smaller sampling volumes, and these include studies by: Werle and Slemr (1991) ($l = 100$–200 m, 6 liter volume); Roths *et al.* (1996) ($l = 126$ m, 6.4 liter volume); and Wienhold *et al.* (1998) ($l = 64$ m, 2.7 liter volume). Small sampling volumes are critical not only for reducing instrument response times but also in maximizing measurement duty cycle within the instrument stability period. This will be presented when we discuss the concept of Allan Variance.

In off-axis resonator cells based upon the simple and astigmatic Herriott designs (1964, 1965), there are only two mirrors, which are spaced at nearly their common radii of curvature. In the simple Herriott cell, both mirrors are spherical, and the input beam is injected through a hole in one mirror slightly off axis. The beam recirculates around the mirrors in an elliptical pattern before exiting the cell through the same coupling hole. The number of passes is adjustable by changing mirror separation, and the condition where the beam closes on itself at the coupling hole is known as the *re-entrant condition*. In the astigmatic Herriott cell (Herriott and Schulte, 1965) the mirrors are cylindrically deformed such that the radii of curvature of both mirrors are different in the horizontal and vertical planes. The beam spots trace out sinusoidal patterns on both mirrors with different frequencies in the horizontal and vertical planes. The resulting Lissajous pattern, which is shown in the top panel of Figure 2.13, utilizes nearly the entire surface of both mirrors. As a result, the optical path optimally fills the cell absorption volume, which in turn allows for small cell volumes for a given absorption pathlength. The original design of Herriott and Schulte (1965) accomplished this by distorting the mirrors in their mounts. McManus *et al.* (1995) and Zahniser *et al.* (1995) developed and commercialized a much more rugged design (Aerodyne Research Incorporated) for field use where the astigmatic curvature is built permanently into both mirrors. In their design, McManus *et al.* (1995) implemented a novel approach in which one mirror is rotated about its optic axis during alignment to compensate for manufacturing errors in mirror radii. This reduces mirror tolerances and fabrication costs.

All three multipass cell designs have found widespread usage in atmospheric studies. However, many studies, at the time of writing, employing solid-state laser sources have taken advantage of the simpler two-mirror design of the astigmatic Herriott cell, coupled with the fact that the mirrors are not adjustable once aligned. Although this aspect reduces flexibility, it is a requisite for very high sensitivity on vibrating aircraft platforms. By contrast, White cells are still employed for in situ extractive FTIR measurements (Yokelson *et al.*, 1999, 2003; Grutter, 2003). Such cells can accommodate larger input and output beams, which are characteristic of FTIR spectrometers (~8 mm beam diameters, Yokelson private communications). However, as a result, the ratios pathlength to cell sampling volume are not as high as with laser sources. Grutter (2003) achieved a pathlength of 32 m in a 10 liter sampling volume, while Yokelson *et al.* (1999, 2003) achieved a pathlength of 97.5 m in a 16 liter sampling volume.

There are, however, two areas where precautions are necessary when employing astigmatic Herriott cells with solid-state laser sources. Since the entrance and exit beams are only separated by several millimeters at the cell entrance, precautions are needed to avoid imaging scattered light (from the cell coupling hole and entrance window) onto the sample detector. Secondly, since the required input mode matching *f*/# is at least *f*/45 in astigmatic Herriott cells compared to values ~*f*/22 for White cells, the transfer optics of the former must be mechanically more stable than the latter.

Since the original pioneering work of McManus *et al.* (1995), there have been several variants to the original astigmatic design. Richard *et al.* (2002) describe an ultra stable version of the astigmatic Herriott employing four carbon fiber stabilizing rods in place of the single aluminium bar currently on top of the commercial cell. In collaboration with Aerodyne researchers, our laboratory has implemented a similar design (Dyroff *et al.*, 2004), which results in a factor of approximately two improvement in stability upon changes in the differential pressure across the cell. In the design by Richard *et al.* (2002), pathlengths up to 252 m were achieved in a 3.6 liter sampling volume (55.6 cm base path and 453 reflections). The maximum pathlength that one can achieve in realistic Herriott cells for airborne measurements depends upon factors such as the available laser power, laser spatial beam quality, mirror reflectivity, the size of the coupling hole, maximum base length that can be accommodated, and the cell stability. Although we cannot precisely place an upper limit on the maximum pathlength achievable, we estimate an optimum pathlength for airborne platforms somewhere in the ~100–400 m range. As will be discussed in Section 2.5.1.3.5, one can achieve significantly longer pathlengths of up to about 10 km employing cavity enhanced absorption methods, but at the expense of approximately one to two orders of magnitude poorer resolvable fractional change in light intensity. As will be discussed, these techniques only yield improved performance relative to conventional absorption methods for a select group of abundant atmospheric gases that absorb in the near-IR spectral region.

2.5.1.2 Tunable solid-state laser sources

The performance characteristics of tunable mid-IR laser sources are particularly important for sensitive and selective absorption measurements. Tittel *et al.* (2003) present a detailed discussion of the various mid-IR sources, and only a brief overview will be presented here. Figure 2.14 shows various laser sources with output wavelengths in the near- and mid-IR spectral regions. Typical electrical power consumption for these lasers extends up to several amps at voltages of up to 6 volts.

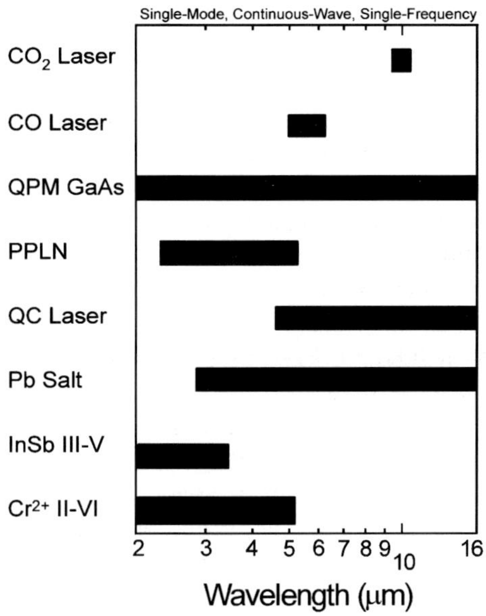

Figure 2.14 Various laser sources and their output wavelengths. With the exception of the CO_2 and CO lasers, which are gas lasers, all the devices are continuously tunable solid-state devices. QPM and PPLN denote quasi-phase matched and periodically poled lithium niobate, respectively, while QC stands for quantum cascade laser.

Numerous criteria should be reviewed before selecting a specific IR laser source. Sometimes the choices are limited and predetermined by the specific spectroscopic technique applied, the desired sensitivity and selectivity of the IR absorption measurement, and the wavelength region. Laser characteristics to consider include the operating wavelength, optical power, frequency-stability and tunability, linewidth, beam quality, cooling requirements (Peltier or liquid nitrogen), and commercial availability of the desired wavelength.

One can easily achieve moderate spectroscopic absorbances down to $A_{min} \sim 10^{-3}$ with minimal effort employing almost all laser types and techniques, and the selection criteria is mostly driven by cost and availability. To achieve a medium absorbance of $A_{min} \sim 10^{-3}$ to 5×10^{-5}, one must take reasonable care in selecting the laser source and in setting up the instrument. However, there are many options and a great deal of flexibility in choosing the laser source and spectroscopic technique. In order to build instruments, which can detect very low values of A_{min} of 10^{-6}–10^{-7}, one must take the utmost care in selecting, implementing, and matching the laser source with the spectroscopic technique. Most often minute effects such as small beam pointing instability can be the limiting technical noise for a given spectroscopic method.

We review here four different tunable solid-state laser sources that are presently being employed for high measurement performance. With the exception of the pulsed quantum cascade (QC) laser, all the laser sources discussed in this section have typical linewidths (half width at half maximum) in the 10^{-5}–$10^{-3}\,cm^{-1}$ region, with most devices typically in the $10^{-4}\,cm^{-1}$ (3–30 MHz) range. The pulsed QC laser has a typical linewidth of $\sim 0.005\,cm^{-1}$ (150 MHz). Thus, with the exception of the latter, all the laser energy from the various devices that will be discussed is efficiently coupled into vibrational-rotational absorption features under typical in situ sampling pressures in the 10–50 torr range. The instrument or 'slit' function in this case is negligible. Hence the line center absorption and absorption profiles are not perturbed. At sampling pressures of 50 torr, for example, the typical laser linewidths for the four devices considered here are about two orders of magnitude smaller than the absorption linewidths for the three example molecules shown in Figure 2.9. Here, the three absorption features have linewidths in the 0.007–0.009 cm^{-1} range. For comparison, one experiences a reduction in line centre absorption by ~29% when the laser and absorption features have identical Gaussian halfwidths.

2.5.1.2.1 LEAD SALT AND InSb LASERS

Lead salt (~3–30 μm), InSb (2–5 μm) and other heterojunction, interband diode lasers are directly electronically tunable mid-IR lasers (Joulie, 2003). Lead salt lasers are Fabry-Perot-type devices and have a typical 300–500 micron chip length, and 250 microns width and height. The end faces of the chip are typically uncoated and lasing is achieved by applying a forward current through the pn-junction, which is grown orthogonal to the optical axis. Since the photon energy approximates the band gap energy for these semiconductors, lead salt lasers are required to operate at cryogenic temperatures, typically at liquid nitrogen temperatures. The tuning range of lead salt lasers can extend over several tens to several hundreds cm^{-1}(~5 cm^{-1}/K). However, this is not continuous as one often experiences various mode jumps, which are governed by the Fabry-Perot cavity and the temperature- and current-dependent refractive index and laser gain functions. Continuous and high speed (kHz to MHz) mode hop-free tuning over 1–2 cm^{-1} by current modulation (~0.05 cm^{-1}/mA) is normally possible and sufficient to tune over an atmospherically broadened absorption line in the mid-IR spectral region. Output powers of 0.1–0.5 mW are characteristic and normally sufficient with the use of long pathlength multipass absorption cells. New developments to improve the performance of lead salt diode lasers and to increase their operating temperatures have slowed significantly as the market for such devices has diminished. The operating characteristics of these lasers can be susceptible to temperature cycling and to slight manipulations of the laser power lead. These perturbations can sometimes induce slight but intolerable changes in the laser operating characteristics, particularly the operating wavelength. In addition, the output beam from lead salt lasers is highly astigmatic and divergent. This requires transfer optics with high magnification to mode match the laser to multipass absorption cells. Typically such systems are fairly large, requiring an optical bench of ~2–3 feet on a side. This makes such a system susceptible to small mechanical perturbations, particularly on airborne platforms, which may lead to changes in the background spectral structure. Despite such challenges, spectrometers based on lead salt diode lasers have proven to produce excellent data in ground and airborne deployments; with care, minimal detectable fractional absorbances of $A_{min} = 7 \times 10^{-7}$ have been achieved (Fried *et al.*, 1998a; Werle, 1998).

2.5.1.2.2 NEAR-IR, DFB, FIBER LASER, AND FIBER AMPLIFIERS FOR PARAMETRIC CONVERSION

Tunable near-IR diode and fiber laser sources that have been developed for the optical telecommunication industry possess many of the desired properties to perform highly sensitive absorption spectroscopy without the need for cryogens. Such lasers have many important attributes such as they (1) are operated at room temperature; (2) are frequency stable over time; (3) are of single mode and single frequency; (4) are narrow in linewidth (<10 MHz); (5) are either widely tunable or wavelength selectable to within ~50 GHz; (6) have high output powers (1–50 mW); (7) are capable of high frequency current modulations (MHz regime) and are tunable with rates of ~0.02 cm^{-1}/mA; and (8) possess a low diverging Gaussian intensity distribution if coupled through a single mode optical fiber. These lasers cover the near-IR spectral region between 1 and 2 μm and have been successfully applied to cavity absorption spectroscopy and photoacoustic spectroscopy (Baer *et al.*, 2002; Webber *et al.*, 2003). The use of optical rare earth–doped fiber amplifiers allows one to further increase the output powers of fiber and diode lasers to several Watts, only limited by the onset of nonlinear effects such as stimulated Brillouin scattering (SBS) and the material damage threshold intensity of the fiber facet (Ghatak & Thyagarajan, 1998; Agrawal, 2001).

2.5.1.2.3 DIFFERENCE FREQUENCY GENERATION (DFG)

Parametric frequency down conversion employing nonlinear optical crystals can be used to directly transfer these properties of fiber-based near-IR diode and fiber laser sources to the mid-IR wavelength region (Richter *et al.*, 2002). Two co-aligned laser beams focused into a nonlinear crystal comprise a single pass cw-DFG system. The conversion efficiency is determined by the beam quality, spatial overlap, focusing condition of the two beams, as well as the figure of merit and length of the crystal employed. To ensure that pump and generated beams do not disperse with respect to one another and to conserve momentum, phase matching has to be an integral part of the parametric frequency conversion process and incorporated into the crystal. Traditionally, this has been achieved using nonlinear crystals with birefringent properties. With the advent of optically engineered quasi-phase matched frequency converters such as periodically poled lithium niobate (PPLN), frequency conversion to the mid-IR has become a more efficient and flexible tool. Phase matching is engineered into the crystal by incorporating a grating along the optical axis. Such grating structure can be permanently written into ferroelectric crystals. First, a linear array of electrodes is attached to the surface of the crystal by means of a lithographic process. A well-defined series of high voltage pulses is then applied through the cross-section of the crystal, which permanently creates an inversion domain every coherence length and effectively introduces a correcting phase shift for the three interacting waves. As a result, quasi-phase matching offers the following advantages over birefringent phase matching; (1) parametric conversion no longer depends on the birefringent properties of the crystal to conserve the momentum and energy; (2) no optical walk-off effects; (3) access to larger diagonal nonlinear optical coefficients; and (4) the use of non-birefringent materials with much larger nonlinear coefficients, such as GaAs. With quasi-phase matched, optical conversion crystals, the possible range of generated mid-IR wavelengths is only limited by the transparency of the particular crystal material used. For example, PPLN features a practical transparency from 0.4 μm up to 4.6 μm.

Modest input power levels of several hundred mW of pump power in both input lasers are required to generate 10–100 μW of tunable mid-IR power. Mid-IR liquid nitrogen–cooled indium antimonide (InSb) and Peltier-cooled mercury cadmium telluride (MCT) detectors typically require minimum incident powers of 1–5 μW to resolve an A_{min} of

10^{-6}. Hence the required pump powers for DFG are determined by the losses expected in the multipass sampling cell and by the required A_{min} detection levels. Any mid-IR wavelengths of up to ~4.6 μm can be generated with DFG using the available range of near-IR pump sources (diode laser, fiber laser and/or other comparable performing solid-state laser sources). Mid-IR DFG measurements in our laboratory using CH_2O as a test gas have demonstrated limits of detection that are at least a factor of two lower than achievable using lead-salt diode lasers; as of this writing 1-second measurements yield minimum detectable line center absorbances in the $1.8–2.6 \times 10^{-6}$ range (1 σCH_2O precisions of 55–77 pptv). Although such high performance awaits more rigorous tests on airborne platforms, these results are extremely encouraging and even lower limits of detection are expected in the future. The unique ability to integrate such a DFG source directly to a multipass absorption cell and the far superior optical beam quality are important factors in this regard (Dyroff *et al.*, 2004). These same factors will also make such a system significantly more compact and rugged than a lead salt system.

2.5.1.2.4 QUANTUM CASCADE (QC) LASER

Faist *et al.* (1994) demonstrated the first working intersubband QC laser in 1994. Since then, the development of various QC types and approaches covering the wavelength regions in the mid-to-far-IR have been demonstrated (Hofstetter and Faist, 2003). The operating principle of a QC device is unique. Photons are emitted when electrons make discrete transitions along quantum wells created by alternating layers of semiconductor materials (InGaAs or AlInAs on InP substrates). The width of these layers determines the lasing wavelength. This is in contrast to traditional interband semiconductors in which the potential energy difference between the valence and conduction bands (bandgap) governs the emission wavelength. Cascading the stages of alternating layers through which a single electron passes permits the generation of multiple photons and higher optical output powers (Faist *et al.*, 1994).

The focus of QC device research thus far has emphasized the development of a laser that can be built to cover a wide wavelength range with reliable single mode and single frequency operation (by incorporation of a DFB structure) and a device that is capable of high fidelity frequency modulation. In addition, these efforts have also emphasized the maximization of output power and an increase in the operating temperature from liquid nitrogen to near-room temperature operation using Peltier temperature control. While near-room temperature QC lasers have been realized in pulsed mode, as of this writing, cw operation with a few exceptions still requires liquid nitrogen cooling (Beck *et al.*, 2002). The Beck *et al.* (2002) publication provides an excellent recent review of QC devices. Similar to any side emitting semiconductor laser, QCs have a highly divergent and astigmatic output beam and require the use of high magnification for mode matching to a multipass absorption cell.

Since the inception and development of QC laser technology, there has been very high expectation that such devices would ultimately become the preferred mid-IR laser source. Although the QC device, since its invention in 1994, has indeed improved upon many of the drawbacks of lead salt diode lasers presented earlier, the performance of QC devices has still not surpassed that achieved by lead-salt diode lasers. Of the various laser sources presented, as of this writing, there still is no clear-cut ideal laser source in terms of ultimate spectroscopic performance and practical usability to routinely achieve

A_{min} values better than the 1×10^{-5} to 1×10^{-6} range currently available. Many research and commercial groups around the world ardently pursue such goals.

Current improvements for cw-QC devices are focusing on various issues, including: (1) lowering the high compliance voltage and designing low noise current driver electronics; (2) lowering current threshold densities and operating currents; (3) improving overall wall-plug efficiency (presently <1%) and heat removal management; (4) increasing operating temperatures; (5) increasing limited tunability; and (6) enhancing the ability to manufacture device structures in a predictable way to hit precise wavelengths. The latter is important, since as of this writing, it typically requires lead times of up to one year to obtain QC devices that hit specific desired wavelengths.

In 2004, Evans *et al.* reported a 400 mW output power QC device emitting single mode 6 μm radiation at room temperature. Such developments and more accessible QC devices in the future will certainly fuel the development toward a universal mid-IR laser source. However, until then, scientists who wish to employ mid-IR laser sources based on QC, lead salt, DFG, or other tunable mid-IR sources not mentioned in this chapter, will need to carefully assess all the attributes of each device so that one can measure the particular trace species of interest with the required sensitivity, selectivity, and speed.

2.5.1.3 Measurement techniques using tunable solid-state lasers, noise reduction, and signal enhancement

The solid-state laser sources just discussed have been employed in atmospheric studies using one of four primary techniques: (1) direct absorption; (2) wavelength modulation; (3) frequency modulation; and (4) cw cavity absorption methods. In this section, we briefly discuss the measurement principles of each technique, restricting our discussions to wavelengths in the 1–16 μm spectral region. Since this field has, and will continue, to grow exponentially, it is only possible to highlight a few select examples of each approach. It is very instructive to first briefly examine the different noise sources in tunable solid-state lasers, with specific emphasis on the most widely employed lasers to date for atmospheric studies, lead-salt diode lasers.

2.5.1.3.1 NOISE SOURCES

The sensitivity of any IR spectrometer can be expressed in terms of its (SNR). For a fixed line centre absorbance, one can write the SNR in terms of the laser power incident on the detector (P_D), and the quadrature addition of the root-mean-square (rms) currents from detector/preamplifier thermal noise (N_{TH}), detector shot noise (N_{SN}), and the laser excess source noise (N_{EX}) in accordance with the analysis of Werle *et al.* (1989) and Werle and Slemr (1991):

$$SNR = \frac{P_D}{\sqrt{(N_{TN})^2 + (N_{SN})^2 + (N_{EX})^2}} \tag{2.25}$$

This analysis does not include optical noise (to be discussed), which cannot be quantified in such general terms since it is very specific to a given optical setup and can change from minute to minute even for a fixed setup. In addition, electrical noise on the laser controller and/or detection circuitry is also not included in this expression. The laser excess source noise considered here is the laser amplitude modulation (AM) noise that is within the measurement bandwidth and centered on the measurement detection frequency. This noise source is laser dependent, but is influenced by conditions of operation. For example, laser-operating conditions where multiple lasing modes are simultaneously present will experience intermode partitioning noise. This noise source, along with optical feedback caused by scattering from various optical components, adds to the laser excess noise. Solid-state lasers also exhibit frequency modulation (FM) noise, which, although in many cases is quite small, causes amplitude noise vis-à-vis an absorption feature. Residual amplitude modulation (RAM) is another potential noise source not considered in Equation 2.25. This noise source, which depends on the particular laser and operating conditions, results from the fact that the laser amplitude changes as the frequency is scanned and/or modulated. It can also produce a large background signal, which can totally mask absorption features of interest.

The AM noise from both the laser and detector fall off with a $1/f$ 'pink noise' frequency dependence, which in the case of lead-salt diode lasers can plateau at frequencies as high as 200 MHz (Werle *et al.* 1989) or at frequencies as low as 1 MHz (Carlisle *et al.*, 1989). Near-IR diode lasers also show this $1/f$ source dependence but roll off at different frequencies (Hollberg *et al.*, 1998). The shot noise, which originates from the discrete nature of the generation of photoelectrons (Werle *et al.*, 1989), is frequency independent (White noise) but is proportional to the square root of the laser power incident on the detector. The thermal noise, which arises from the Johnson noise of the detector and preamplifier components and the inherent thermal noise of the detector, is independent of both laser power and frequency. Hence, both the White-type shot noise and the thermal noise are proportional to the square root of the detection bandwidth.

From the discussion above, one observes that at high enough frequencies the spectrometer noise will no longer be dominated by the $1/f$ laser excess noise. In this case, the frequency-independent thermal and shot noise contributions will determine the SNR, and optimum quantum-limited performance is achieved whenever the shot noise is equal to or greater than the thermal noise (Werle *et al.*, 1989). Thus, in order to achieve the highest measurement sensitivities with IR spectrometers employing solid-state laser sources, the detection frequency should be as high as possible. However, additional noise sources than those considered above typically dominate most IR spectrometers, and, as we will see, the optimum detection frequency becomes a tradeoff between excess source noise and the stability of the spectral background structure (Werle, 1995).

In all IR spectrometers, optical noise ultimately plays a dominant role in determining the stability of this background structure. As in all lasers, the light from cw tunable solid-state laser sources is highly temporally coherent, and, as a consequence, light scattering from any optical element in the beam path may ultimately impinge on the detector along with the primary beam. Since the detector output is responsive to all such beams, the primary signal of interest can reside on top of an undulating background structure. In many cases, this structure contains multiple frequencies, amplitudes, and time constants from multiple scattering sources. In many cases, optical windows and beam splitters may be the cause of such multiple beam paths. Every optical system contains at least one window, and most likely several windows mounted on laser dewars, multipass absorption cells, and detectors. Even in a perfect system with no scattering, small movements of any optical element in the beam path will translate to small movements of the primary beam on the detector, which can produce phase and amplitude noise fluctuations. Such optical sources of noise are frequently referred to in the literature as 'etalon', 'etalon fringe noise', or 'optical interference fringe noise', even though a classic fringe structure may not be present. Throughout the rest of this chapter, we collectively designate all such noise sources as 'optical noise'. In some cases, variable background structure may be caused by variable ambient absorption from molecules such as H_2O-vapor in the open absorption path external to the multipass cell. Here the pressure-broadened wings of H_2O-vapor could significantly change the background curvature for the analyte gas of interest, even when the H_2O absorption feature is $0.2\,cm^{-1}$ away from the feature of interest (Fried *et al.*, 1998a). It is also entirely possible that unidentified, very weak absorption features from other atmospheric molecules may play a role in changing the background structure. Depending upon the experimental particulars, optical noise from the above sources yields, with very few exceptions, equivalent A_{min} values somewhere in the 10^{-4}–10^{-7} range. We will now present the various measurement approaches employed in atmospheric studies, which attempt to minimize these various noise sources.

2.5.1.3.2 DIRECT ABSORPTION

The first approach is based on direct absorption. In this method, the laser wavelength (frequency in cm^{-1}) is rapidly scanned over absorption features of the target molecule, and mixing ratios are deduced from measurements of incident and transmitted intensities, sample pressure and temperature, and pathlength, employing the Beer–Lambert absorption relationship presented earlier. The scan is typically accomplished by scanning the injection current to the laser source while holding the laser temperature fixed. Since the current scan also affects the laser temperature, both tuning mechanisms become important whenever the laser is slowly scanned, resulting in a changing laser current–temperature relationship from scan to scan (Jennings, 1980). Jennings and subsequently Fried *et al.* (1988) have shown that this effect can be eliminated by rapidly sweeping the laser tuning current (at frequencies $\geq\sim$20Hz) over the absorption feature in a time that is fast compared to the thermal time constant of the laser crystal. This technique is known as *sweep integration.*

Direct absorption spectra acquired using sweep integration are carried out by applying a sweep ramp, usually a saw tooth ramp, to the laser injection current, and the detector signal thus acquired is co-averaged in a computer or signal averager as the laser is repetitively swept over the same absorption features. In some implementations of this approach, a small fraction of the main IR beam is split off and sent through a reference cell containing the target gas of interest. This reference channel is used to co-align spectral absorption features before co-averaging so as to avoid smearing and degradation of the resultant spectrum. Zahniser *et al.* (1995) have employed sweep integration with a lead-salt diode laser using 100 discrete ramp steps at a 300 kHz rate, resulting in sweep frequencies of 3 kHz. Nelson *et al.* (2002) from the same laboratory have extended this approach to pulsed non-cryogenically cooled QC lasers for detection of NO. These researchers employed a pulse rate of 1 MHz, which produced a 10 kHz sweep rate for a 100-point spectrum. Richard *et al.* (2002) employed a 100 Hz sweep ramp to acquire absorption features of CH_4 in the upper troposphere/lower stratosphere. In all the direct absorption measurements just discussed, absorption features are fit to Voigt profiles using nonlinear least-squares fitting routines together with the Beer–Lambert absorption expression.

The most popular routine for this purpose is the Levenberg–Marquardt nonlinear least-squares algorithm published by Marquardt (1963). Since the direct absorption approach is based upon fundamental principles, no calibration standards are required. This approach can be further used to measure and/or verify the concentration of calibration standards where such standards are employed. However, based upon our experience in the field, it is highly desirable in all cases to periodically introduce calibration standards near the system inlet entrance. This minimizes potential systematic errors from inlet/cell surface losses, particularly on airborne platforms where one samples over a wide range of temperature, pressure, and relative humidity conditions.

Employing direct absorption, Nelson *et al.* (2002), Hori *et al.* (1999), and Zahniser *et al.* (1995) routinely achieve minimum detectable line centre absorbance values, A_{min}, (at the 1σ precision level) of $1–3 \times 10^{-5}$ using a 210 m pathlength for 1 second of averaging. This corresponds to minimum detectable mixing ratios of 120 pptv $Hz^{-1/2}$, 80 pptv $Hz^{-1/2}$, and 55 pptv $Hz^{-1/2}$, for NO, NO_2, and NH_3, respectively. Using Equation 2.24, this corresponds to pathlength-normalized absorbance values of $5 \times 10^{-10}\,cm^{-1}\,Hz^{-1/2}$ to $14 \times 10^{-10}\,cm^{-1}\,Hz^{-1/2}$. It is important to emphasize that these performance estimates represent the true instrument figure of merit, since they were obtained from the rms noise of repetitive background measurements. As discussed by Tittel *et al.* (2003, and references therein), the true figure of merit of a real world instrument is the ability to replicate the same result when one samples a constant input concentration. This is in contrast to high frequency baseline noise or measurement precision of a limited number of scans. Since low frequency noise sources from a variety of causes may also be important, the ability to replicate the same answer from many successive measurements can be far worse than the inherent sensitivity of a single or small group of scans. The study by Fried *et al.* (1998a) further illustrates this fact. Unfortunately, in many cases instrumental figures of merit are too optimistic since they are not based upon such replicate measurements from a time series, particularly under field conditions. This all too frequently obfuscates the true comparison between techniques.

As with the examples above, it is essential when discussing instrument performance to present the complete set of conditions under which such determinations are carried out. An ideal comprehensive assessment should contain the following detailed information: the sampling pathlength, the level of precision, the measurement time, the measurement bandwidth, the minimum detectable mixing ratio, the corresponding line centre absorbance, the measurement platform and sampling conditions, and finally an assessment as to the performance reproducibility (i.e. a statement as to whether or not such performance can be routinely obtained on a day-to-day basis or every once in a while). A performance histogram is ideal for this purpose (see Figure 11 of Tittel *et al.*, 2003). Although the measurement time and measurement bandwidth are inter-convertible, this conversion is complicated by the fact that different techniques employ different frequency selective filters, different number of scans and fitting points. This opens up the possibility that the inter-conversion may not be treated the same for each study reported in the literature. For this reason it is highly desirable to report both A_{min} normalized to the bandwidth as well as A_{min} for 1 second of averaging, which in many cases are equivalent. It is also highly desirable in this regard to report the A_{min} value for the optimum averaging time using an Allan Plot, and this will be presented in Section 2.5.1.3.6. We will attempt to indicate such comprehensive assessments wherever possible in the remainder of this chapter.

The balanced beam and balanced ratiometric detection (BRD) methods are two variants of the direct absorption approach successfully employed by a number of researchers. In these approaches, the laser beam is split into a signal beam and a reference beam, and both are detected using similar IR detectors. The signal beam passes through the multipass sampling cell while the reference beam does not. The common mode noise residing on both beams from laser excess noise, optical noise up to the beam splitter, and laser power supply noise cancel under the right circumstances when the detector outputs are either subtracted or ratioed in the logarithmic domain. For complete common mode noise cancellation, the reference beam must be accurately scaled so as to approximate the sample beam over the full scan range at the time of measurement. Hobbs (1997) developed an electronic feedback circuit based upon the d.c. detector levels to adjust and maintain the detector scalings without the requirement for an exact balance between the two arms. In this BRD approach the net output signal depends upon the ratio of the signal to reference arm currents, which changes when the signal beam is partially absorbed by the sample gas of interest. Allen *et al.* (1995) have demonstrated noise-equivalent absorbances as low as $2 \times 10^{-7}\,\mathrm{Hz}^{-1/2}$ in the near-IR spectral region. This is close to the shot noise limit of their study. Sonnenfroh *et al.* (2001) employed this technique with a quasi-cw, room-temperature, QC laser emitting in the near-IR spectral region. However, in this study the noise-equivalent absorbance (9×10^{-4}) was degraded considerably from the near-IR value due to large detector background currents and low average laser power. A similar approach has also been employed with second harmonic detection (to be discussed) by Zhu and Cassidy (1995) using two different near-IR lasers in the 1.3–1.4 μm spectral region. Based upon the replicate precision over 10 seconds of scans, these researchers achieved a 1σ rms baseline precision of 1.4×10^{-6} using an equivalent noise bandwidth of 1.25 Hz. These measurements were carried out in the laboratory employing a White cell with a 4 m pathlength and $2f$ detection at 10 kHz.

2.5.1.3.3 WAVELENGTH MODULATION TECHNIQUES

The second and third primary measurement techniques commonly employed with solid-state laser sources are based upon the closely related techniques of wavelength modulation spectroscopy (WMS) and frequency modulation spectroscopy (FMS). As we have discussed, sweeping the laser injection current rapidly changes the laser output frequency, and typical sweep frequencies in the 25–3000 Hz range have been employed for this purpose. One can also apply much higher frequencies in the kHz to MHz regime to modulate the laser injection current as the frequency is simultaneously swept across the absorption line or locked to the line centre. As discussed previously, it is desirable to employ high measurement detection frequencies to minimize laser source noise as well as other $1/f$ noise sources, and the modulation process enables detection at such optimum frequencies.

Most atmospheric studies at the time of writing employ modulation frequencies in the kHz regime, typically between 1 and 100 kHz, and this technique is known as *wavelength modulation spectroscopy* (WMS), *derivative spectroscopy*, or *harmonic detection*. In most applications, one typically detects the modulated absorption at twice the modulation frequency ('second harmonic' or '$2f$ detection') using a lock-in amplifier set to $2f$. As shown in Figure 2.15, which depicts the resulting second harmonic spectrum as well as the first harmonic ($1f$) and direct absorption spectra, the second harmonic signal exhibits a

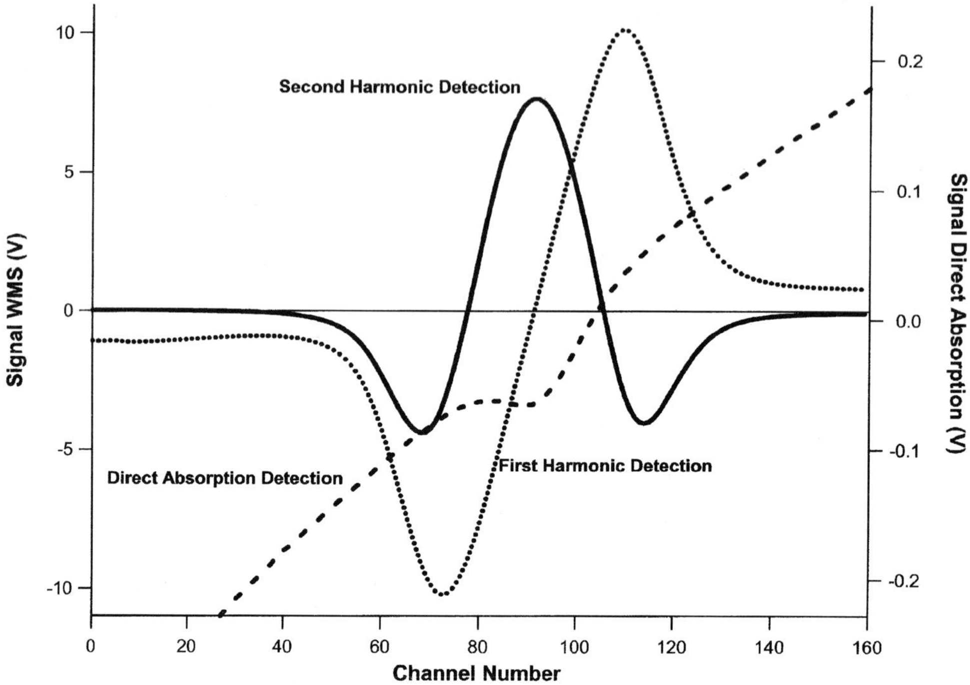

Figure 2.15 First and second harmonic wavelength modulation spectroscopy (WMS) signals in volts (left-hand axis) along with direct absorption profile (right hand axis) for pure CO at 44.2 torr in a 16 cm absorption cell (absorbance = 0.018) using a the near-IR DFB laser at 1.566 μm. The harmonic and direct absorption detection voltages have different gains (see text for an explanation of the different shapes). (All spectra were acquired in our laboratory by Dr Petter Weibring.)

maximum at line centre and a baseline near 0 volts. The first harmonic signal, by contrast, is zero at line centre, and the baseline, although flat, is nonzero. Higher harmonics may also be used, but with each higher harmonic the signal is significantly decreased (compare the $1f$ and $2f$ spectra of Figure 2.15). The $2f$, as shown in Figure 2.15, reveals both positive and negative lobes.

In addition to up-converting the detection frequency, both WMS and FMS techniques produce a signal that is directly proportional to the mixing ratio of the target gas of interest rather than a small change in a large signal, which is detected in direct absorption. In both modulation cases, the background structure in theory is flat and at 0 d.c. Hence, the amplification and digitization process, in the case of digital lock-in amplifiers, is effectively centered around the absorption signal of interest rather than the large laser intensity with its highly sloping background structure prevalent in direct absorption measurements (see Figure 2.15). Furthermore, using modulation techniques, one can choose the modulation frequency and/or amplitude to minimize optical noise structure that may be present in direct absorption (Reid *et al.*, 1980).

Finally, since modulation spectroscopy relies on a large change in the absorption coefficient with wavelength (i.e. derivative-type response), modulation approaches can help to discriminate against broad featureless absorptions, such as those from the wings of atmospheric pressure-broadened water lines and those from big unresolved organic molecules. As discussed in an earlier section, the larger the molecule the closer the vibrational-rotational line spacings. Molecules, such as large organic gases, typically containing five or more atoms will in most cases exhibit unresolved absorption features. Although such features can change the modulated signal background structure and curvature, they will generally not produce a modulated signal per se, even when spectrally overlapped with the target molecule of interest. This aspect, which is in contrast to direct absorption, becomes more important as an added degree of selectivity when measuring trace gases in the atmosphere at levels below 100 pptv. However, this same aspect also limits modulation spectroscopy to small molecules generally containing up to five atoms and/or very narrow Q-branch transitions.

There have been many reviews of second harmonic detection, and three excellent sources can be found in the book chapter by Brassington (1995), the detailed discussion of the theory by Reid and Labrie (1981) and the application for the detection of HCl by Fried *et al.* (1984). We only summarize the results here. For a sinusoidal modulation at angular frequency ω and amplitude a (half modulation amplitude in cm^{-1}), one can write the time variation of the laser frequency (wavelength) about a mean value of ν as:

$$\nu(t) = \nu + m\gamma_v \cos\theta \qquad (2.26)$$

One should not confuse the laser frequency ν in cm^{-1} with the modulation frequency. In Equation 2.26 the modulation phase angle ωt is replaced by θ and the modulation amplitude a is normalized by the Voigt halfwidth at half maximum (γ_v) to yield the dimensionless modulation parameter m via:

$$m = a/\gamma_v \qquad (2.27)$$

Incorporating the laser time dependence in Equation 2.26 in a cosine Fourier series, one can write the second harmonic Voigt absorption coefficient, $H_2(\nu, m)$, as:

$$H_2(\nu, m) = \frac{2}{\pi S(\nu_0)_v} \int_0^{\pi} S_v(\nu + m\gamma_V \cos\theta, y) \cos 2\theta \, d\theta \qquad (2.28)$$

Here S_v and $S(\nu_0)$ are the Voigt absorption coefficients at the frequency determined by the modulation and at line centre, respectively, and the term y has been defined in Equation 2.19b. Equation 2.28, which is used to calculate the modulated absorption coefficient centered around a specific frequency ν, reveals that a large number of frequencies actually contribute to each value of the modulated absorption coefficient. At line centre for example, the modulated absorption coefficient has contributions extending from both wings of the absorption line, and as a consequence the maximum positive second harmonic absorbance, which occurs at $m = 2.2$ (See Reid & Labrie, 1981), is lower than the direct absorbance. The exact ratio depends upon the modulation waveform applied (Iguchi, 1986). For the Voigt lineshape of Examples 1 and 2, the positive lobe of the modulated absorbance at line centre, when normalized to that from direct absorption, yields values of 0.36, 0.40, and 0.53, for sinusoidal, triangle, and quasi-square modulation waveforms, respectively. The corresponding normalized peak-to-peak second harmonic absorbance (sum of positive and negative lobes) yields values of

0.57, 0.62, and 0.86, respectively, for the three waveforms mentioned above. One obtains slightly different ratios for Gaussian and Lorentzian lineshapes. However, in all cases, the reduced modulated line centre absorbance relative to direct absorption is one of the slight disadvantages of modulation spectroscopy.

Under optical thin conditions and a Voigt lineshape, Fried *et al.* (1984) and Reid and Labrie (1981) have shown that the processed second harmonic signal $S_{2f}(\nu)$ can be expressed by the following expression:

$$S_{2f}(\nu) = CI_0 H_{2v}(\nu, m) P\chi Nl \tag{2.29}$$

In this expression C is a dimensionless proportionality constant, which represents the second harmonic transfer function of the detector/preamplifier, lock-in amplifier, and associated cabling. The a.c. gain is included in this term. As before, the term I_0 is the incident laser intensity, which we assume in this expression does not change as we modulate over the absorption line. This assumption is a good approximation when small to moderate modulation amplitudes are employed (Reid and Labrie, 1981). The other terms have been defined in Equation 2.23. Since the modulated absorption coefficient at a particular wavelength has the same units (cm^2 $molecule^{-1}$) as the direct absorption coefficient, the second harmonic signal has the same units as I_0 (usually volts).

The proportionality term C and the dependence of the modulated absorption coefficient on both the particular modulation waveform and modulation amplitude in Equation 2.29 indicate that the second harmonic signal is not a simple function of the absorber mixing ratio. In contrast to direct absorption, where the absorber-mixing ratio is related to fundamental instrument-independent quantities, second harmonic measurements contain instrument-dependent factors. However, one can relate the absorber-mixing ratio to fundamental quantities via indirect means for a fixed instrumental setup. May and Webster (1993) described one procedure to determine the above proportionality constant for a fixed modulation amplitude and waveform, which they designate as the 'response number'. In this analysis, the proportionality constant is determined from the ratio of the observed to calculated peak-to-peak second harmonic signal normalized by the absorbance. The d.c. and a.c. gains are included in this determination. An alternative approach successfully employed by Fried *et al.* (1984, 1998b) uses a calibration standard for another gas (reference gas, *r*) that absorbs in the same spectral region as the sample gas (*s*) of interest. In this approach, using the same sampling pressure and pathlength, one can express the ratio of the peak-to-peak second harmonic signals at line centre as:

$$\frac{S_{2f}(\nu_0)_s}{S_{2f}(\nu_0)_r} = \frac{(I_0)_s}{(I_0)_r}\frac{(H_{2\nu})_s}{(H_{2\nu})_r}\frac{[\chi]_s}{[\chi]_r} \tag{2.30}$$

Here the terms *P*, *l*, *N* and *C* in Equation 2.29 all cancel out, and one can solve for the sample mixing ratio $[\chi]_s$ from the ratio of second harmonic signals, laser powers, and calculated second harmonic absorption coefficients. A third approach, which is employed in most field studies, compares the sample second harmonic signal to that from a known calibration standard (cal) for the same gas, in accordance with:

$$[\chi]_s = \frac{S_{2f}(\nu_0)_s}{S_{2f}(\nu_0)_{cal}}[\chi]_{cal} \tag{2.31}$$

This approach assumes identical sampling conditions of laser power, sample pressure and temperature, and pathlength, for both sampling modes. This approach has the advantage in that it corrects to first order for any sample losses in the inlet line and sampling cell during field operation. This approach, however, requires an accurate knowledge of the calibration-mixing ratio, which can be obtained using direct absorption in the laboratory. Our airborne measurements of CH_2O, for example, employ both direct absorption and the second harmonic ratioing approach given by Equation 2.30 (using methane standards as the reference gas) in the laboratory before and after field studies. In this manner, we realize the advantages of both an absolute calibration and a field calibration under a variety of sampling temperatures, pressures, and relative humidities.

A range of A_{min} values have been reported in the literature for 2*f* WMS detection employing sweep integration. This range reflects many particulars of the studies, including such factors as the pathlength employed, the specific type of laser used, and whether or not the performance estimate is based upon laboratory or actual field measurements. Since it is impossible to cite each and every reference, we only cite a few representative examples. As discussed previously, Zhu and Cassidy (1995) report $A_{min}(1\sigma)$ values of 1.4×10^{-6} for laboratory studies using near-IR diode lasers, an equivalent bandwidth of 1.25 Hz, 2*f* detection at 10 kHz, and an electronic noise subtracting circuit. Bomse *et al.* (1992) carried out very extensive laboratory studies comparing various measurement approaches employing a single pass absorption cell. They report A_{min} values of $\sim 8 \times 10^{-7}$ using WMS at frequencies of 100 kHz for 1 Hz equivalent detection bandwidths.

Our group has carried out extensive airborne measurements of CH_2O employing 2*f* WMS detection using a 100 m multipass absorption cell and a 3.5 μm lead-salt diode laser (Fried *et al.*, 1998a, 1999a, 2003a,b; Wert *et al.*, 2003a,b). These measurements thus provide a rather extensive database on instrument performance during actual airborne measurements. Over three different airborne campaigns (TOPSE 2000, TexAQS 2000, and TRACE-P), Wert *et al.* (2003a) using various approaches report $CH_2O(1\sigma)A_{min}$ values based on replicate measurements that fall within the range between 4×10^{-6} and 13×10^{-6} when extrapolated to 1-second averaging times. The lower limit is a factor of 2.5 improvement relative to the direct absorption values presented earlier. When integrated over 1 minute, our results yielded A_{min} values of 5×10^{-7} to 1.8×10^{-6} and minimum detectable CH_2O mixing ratios (1σ) of 15–55 pptv, with most values falling within the range of 20–40 pptv. Normalizing to pathlength, these 60-second averages correspond to $5 \times 10^{-11}\,cm^{-1}$ to $2 \times 10^{-10}\,cm^{-1}$. A histogram for the entire spread of our airborne CH_2O performance during the TOPSE 2000 campaign can be found in Figure 11 of Tittel *et al.* (2003).

In a variant of the 2*f* WMS approach just discussed, the laser is 'line-locked' to the centre of the absorption feature rather than swept across the feature. As discussed by Loewenstein (1988), this approach is advantageous for applications where rapid time response is required and/or transient species are being observed in laboratory kinetic studies. For this purpose, the zero-crossing signal of either the first or third harmonic at line centre (see Figure 2.15) is used in a feedback loop to keep the laser locked to line centre. Sachse *et al.* (1987) have successfully employed this approach on their DACOM (Differential Absorption of CO Measurement) instrument to carry out fast airborne measurements of CO, N_2O, and CH_4 on time periods of ~ 1 second. Typical airborne line centre absorbance precisions were $\sim 2 \times 10^{-5}$ for a 0.03 Hz electronics bandwidth.

Loewenstein (1988) and Loewenstein *et al.* (1989) employed this same technique with their ATLAS (Airborne Tunable Diode Laser Absorption Spectrometer) system to carry out fast measurements of CO and N_2O on NASA's ER-2 high altitude research aircraft. Optical noise represents the main performance limitation with this approach. However, in contrast to sweeping over the absorption line, line-fitting analysis has not been utilized, at least in past studies, to identify and minimize the effects of optical noise. Rapid background subtraction (to be discussed), which can effectively capture and minimize such noise sources, would certainly improve the measurement performance of line-locking techniques further. However, at the time of writing, this approach has not been typically employed with line-locking methods.

2.5.1.3.4 FREQUENCY MODULATION TECHNIQUES

In frequency modulation spectroscopy (FMS), one employs modulation frequencies that are equal to or greater than both the laser and absorption linewidths. Typical frequencies in this case fall within the range of several MHz to several GHz. As a result, one can employ significantly higher modulation and sweep frequencies than WMS, and as discussed previously, one can in principle achieve quantum-limited performance where excess laser noise no longer dominates. The high frequency modulation (ν_m) produces multiple sidebands about the laser centre frequency ($\nu_c \pm n\nu_m$), and the absorption line of interest is probed by only one or two of these sidebands. In WMS by contrast, where the modulation frequency is much less than the laser and absorption linewidths, one can consider the absorption feature as being probed by a very large number of closely spaced sidebands. The two main sidebands in FMS are equal in intensity and 180° out of phase with one another, and as a result the signal at ν_m is 0. The presence of an absorber in the sample cell will negate this cancellation, producing a signal at ν_m. As discussed in the review by Brassington (1995), one obtains a positive absorption signal as the first sideband passes through the absorption feature as the laser centre frequency is swept. This is followed by a negative signal as the second sideband passes through the absorption feature. This technique is called *single tone FMS*, or STFMS, and the detected signal is processed by a double balanced mixer. This approach requires detectors with bandwidths as high as several hundred MHz. Because such detectors in the mid-IR are expensive, the technique of two tone FMS, or TTFMS, has been developed. In this case, two closely spaced high frequency modulations, separated by ~1–10 MHz, are applied to the laser, and the detection is carried out at the 1–10 MHz beat frequency. Although the laser excess noise is not completely eliminated, TTFMS significantly reduces detector costs. Comprehensive theoretical discussions of FMS and comparisons with WMS can be found in Silver (1992) and Bomse *et al.* (1992), and the reviews by Brassington (1995) and Werle (2004), as well as the publications by Werle *et al.* (1989), Werle and Slemr (1991), Werle (1995) and Werle and Jänker (1996), provide excellent background on FMS techniques and their limitations.

Since FMS employs high frequency RF modulation, the experimental setup is more complicated than that for WMS. The ultimate detection sensitivity of FMS spectrometers is particularly dependent upon various laser parameters such as the laser power, noise spectrum, and amplitude modulation index as well as the laser beam quality. The latter is important since the optical radiation must ultimately be focused onto small area detectors in order to achieve such high detection bandwidths in the case of STFMS. In addition, RF pickup in the signal processing chain can also limit performance, and this further requires careful shielding and impedance matching of the RF modulation frequency imposed on the laser d.c. injection current. As discussed by Silver (1992), such impedance matching is especially difficult to achieve with mid-IR lead-salt diode lasers in cryogenic housings. The FMS techniques also exhibit additional practical difficulties. As discussed in the above references by Werle and colleagues, optimum FMS signals are typically achieved at much lower pathlengths than WMS (Werle and Slemr, 1991), thus placing restrictions on minimal detectable mixing ratios that can be achieved. If both techniques are operated with their optimal number of passes through their respective multipass cells, FMS can improve the sensitivity only by approximately one order of magnitude relative to WMS instead of two or more orders of magnitude anticipated in early studies (Werle and Slemr, 1991).

There are additional drawbacks using FMS, which further minimize the performance disparity with WMS in realistic field systems. High frequency modulation is more susceptible to alignment instabilities (Werle, 1995) than lower frequency WMS techniques. At modulation frequencies of 100–200 MHz employing STFMS, for example, Werle (1995) calculates that short-term optical alignment stability must be less than 10 μm to achieve minimum detectable absorbances less than 10^{-6} on a consistent basis. This compares to optical stabilities in the mm range for WMS techniques. Werle and Jänker (1996) also uncovered another drawback using FMS. Employing modulation frequencies of 100 MHz, these researchers found that refractive index fluctuations generated by turbulent gas flow through a multipass absorption cell produce phase and optical pathlength fluctuations that limit performance. At flow rates of 10 liters min^{-1}, such phase fluctuations limited their A_{min} to values in the range between 5×10^{-7} and 7×10^{-7} using integration times of up to 10-seconds (Werle and Jänker 1996).

Bomse *et al.* (1992) describe a variant of the STFMS technique, which they term *high frequency WMS*, to circumvent the drawbacks just discussed. In this case, the laser is modulated at frequencies in the 1–20 MHz range and $2f$ detection is achieved using RF mixers instead of lock-in amplifiers. Aside from this difference and the further reduction in laser excess noise, high frequency WMS is identical to $2f$ WMS discussed earlier. Bomse *et al.* (1992) report A_{min} values in the low- to mid-10^{-7} range for 1 Hz detection bandwidths. Although this is a slight improvement over the WMS results presented earlier, the laser can be swept over the absorption lines of interest at much higher rates than with WMS. Typical WMS rates are 50–100 Hz, while rates for high frequency WMS are $\geq$500 Hz. This has two advantages: the measurement time is significantly decreased, an important attribute for flux studies; and secondly, low frequency perturbations during the laser sweep are minimized. Such perturbations can arise from such sources as mechanical vibrations during aircraft operation and/or electrical perturbations on the laser sweep ramp. However, the ultimate performance of WMS and FMS techniques are in most instances dominated by optical noise. As a consequence, early optimism of exceedingly low minimum detectable absorbances in the 10^{-7}–10^{-8} range for 1 Hz bandwidths in select laboratory setups using single pass absorption cells (Carlisle and Cooper, 1989; Carlisle *et al.*, 1989; Werle *et al.*, 1989) employing FMS has not been realized in realistic field systems employing multipass absorption cells.

At the time of writing, the number of field studies employing FMS techniques has been somewhat limited, with most of these studies using near-IR laser sources.

Hovde *et al.* (1993) describe two different FMS systems for airborne measurements of H_2O-vapor and a system for CH_4 measurements. Wienhold *et al.* (1993) describe a two-channel TTFMS system for the fast (1 Hz rates) detection of CO and other species. This instrument, which employed mid-IR lead-salt diode lasers and was operated in both airborne and ground-based studies, achieved A_{min} values of $\sim 10^{-5}$ for 1-second averaging times. Wienhold *et al.* (1998) further describe this instrument and its deployment on three airborne campaigns for measurements of CO, N_2O, and CH_4. Werle and Kormann (2001) exploited the fast response times achievable with FMS techniques to carry out eddy-correlation flux measurements of CH_4 at 10 Hz from rice paddy fields.

2.5.1.3.5 CONTINUOUS-WAVE CAVITY ABSORPTION SPECTROSCOPY

Cavity Absorption Spectroscopy (CAS) takes advantage of long path absorption spectroscopy by coupling tunable laser radiation to a high-finesse cavity. Regardless of the specific type of coupling of the radiation to the cavity, long, effective path lengths of several thousand meters can be achieved. Numerous variants of cw-CAS have been developed and derived from the pulsed cavity ring-down spectroscopy (CRDS) technique, first demonstrated in 1988 by O'Keefe and Deacon (1988). In CRDS, a laser pulse is coupled into a high-finesse cavity formed by two highly reflective ($R > 99.9\%$) spherical mirrors and the radiation that leaves the cavity through one mirror is observed as it exponentially 'rings' down the optical cavity. The exponential decay rate is determined by the mirror losses (scatter, absorption, transmission) and absorption by the sample gas present in the cavity while traversing between the two mirrors. Comparing the cell decay rate when the cell is empty and when an absorber is present, directly relates to the mixing ratio of the absorber. We will only further discuss cavity absorption measurements based upon cw-laser sources in this chapter, since such measurements yield much higher sensitivity than pulsed sources in the IR.

Two principal forms of cw-CAS have been derived from the original pulsed CRDS study: namely, cw-CRDS and Cavity Enhanced Absorption Spectroscopy (CEAS) or Integrated Cavity Output Spectroscopy (ICOS). While in cw-CRDS one observes individual ring-down events, in CEAS and ICOS one observes time integrated ring-down events similar to traditional absorption spectroscopy.

Continuous wave CRDS features a high spectroscopic sensitivity. However, this approach requires active laser-to-cavity locking with fast electronics. Good laser beam quality is also important in this approach, as the excitation of higher order modes leads to multi-order exponential ring-down functions (Hodges *et al.*, 2004). Overall, cw-CRDS has demonstrated sensitivities in excess of $10^{-10}\,cm^{-1}$, but at the cost of a complex and alignment sensitive setup that is difficult to implement for non-laboratory applications. The CEAS and ICOS techniques and its advanced variant, off-axis ICOS (OA-ICOS, Paul *et al.*, 2001), have tremendously simplified CAS, as the cavity is passively operated without the need of locking. The variant OA-ICOS uses essentially the same optical mirror arrangement as an astigmatic multipass Herriott cell without a coupling hole, and couples the laser beam to the cavity at a small angle with the added advantage of avoiding direct back reflections to the laser source. By following a Lissajous figure pattern across the mirror cross-section, the free spectral range is significantly decreased and forms a very dense cavity mode spectrum, making the cavity transmission almost independent of wavelength. This approach also no longer requires dithering the laser wavelength or the cavity by a piezoelectric transducer, as required by CEAS and ICOS, respectively, to effectively couple the laser into a large number of transverse cavity modes.

Both OA-ICOS and CEAS have been most successfully applied in the near-IR spectral region, where Baer *et al.* (2002) report A_{min} values of $3 \times 10^{-11}\,cm^{-1}\,Hz^{-1/2}$. Such high performance, which is approximately one order of magnitude better than our field second harmonic measurements in the mid-IR, needs to be further verified by repetitive measurements. In the near-IR spectral region, these approaches take advantage of the convenience of commercially available high power (10–50 mW) tunable telecom diode lasers, low noise detectors ($D^* \sim 10^{12}\,cm\,Hz^{1/2}\,W^{-1}$), and high quality reflective mirror coatings (> 99.99%). Translation of this performance to the mid-IR spectral region is limited by: (1) higher noise detectors ($D^* \sim 10^{11}\,cm\,Hz^{1/2}\,W^{-1}$); (2) adequate tunable laser sources of sufficient output power (tens of mW) and high beam quality; and (3) similar-performing highly reflective coated mirrors. However, if one could ultimately circumvent these issues and translate this spectroscopic performance to the mid-IR spectral region, one could achieve extremely low mixing ratio detection limits for a whole series of atmospherically important gases. As discussed previously, the fundamental vibrational-rotational transitions are several orders of magnitude stronger in the mid-IR than the near-IR. For example, spectral line intensities for CO_2 in the mid-IR at 4.3 μm are ~20 000 times stronger than near-IR transitions at 1.6 μm where OA-ICOS has successfully been employed. In such instances, medium pathlength-normalized A_{min} values of $10^{-8}\,cm^{-1}$ ($A_{min} = 10^{-4}$, $l = 100\,m$) in the mid-IR would require an equivalent A_{min} of $5 \times 10^{-13}\,cm^{-1}$ to achieve the same sensitivity in terms of number density in the near-IR. Despite this disadvantage, near-IR OA-ICOS achieves respectable mixing ratio detection limits for the more abundant atmospheric gases (Baer *et al.*, 2002).

2.5.1.3.6 BACKGROUND STRUCTURE AND MEASUREMENT PERFORMANCE

Most, if not all, IR spectrometers employing solid-state laser sources are ultimately limited by the stability of the background structure on which the absorption resides. This is the case for direct absorption, WMS and FMS approaches just presented. An example of such background structure is displayed in Figure 2.16 along with ambient, calibration, and fit absorption profiles for CH_2O. The background structure, which is due to optical noise, plays a significant role in the determination of the ultimate mixing ratio one deduces. This is particularly true here where the ambient CH_2O mixing ratios are near the detection limit. In this case, our airborne CH_2O results would be in error by ~100 pptv if the background spectrum is not first removed. Figure 2.17 shows this even more dramatically in the case of our laboratory CH_2O measurements where an input-mixing ratio of 3.714 ppbv was sampled. Backgrounds were acquired for 10 seconds before and after each 20 second sample, and including the sample, background, and intervening flushing periods, each data point represents a 1 minute acquisition. In the upper panel the backgrounds were ignored, while the average of the two backgrounds was subtracted before fitting in the lower panel. In both cases, the sample data were fit to a CH_2O standard acquired previously using a real-time multiple linear regression approach (Sewell *et al.*, 1993). The retrieved CH_2O mixing ratio averaged 4.036 ± 0.305 ppbv in the case of the upper panel and 3.714 ± 0.055 ppbv using background subtraction in the lower

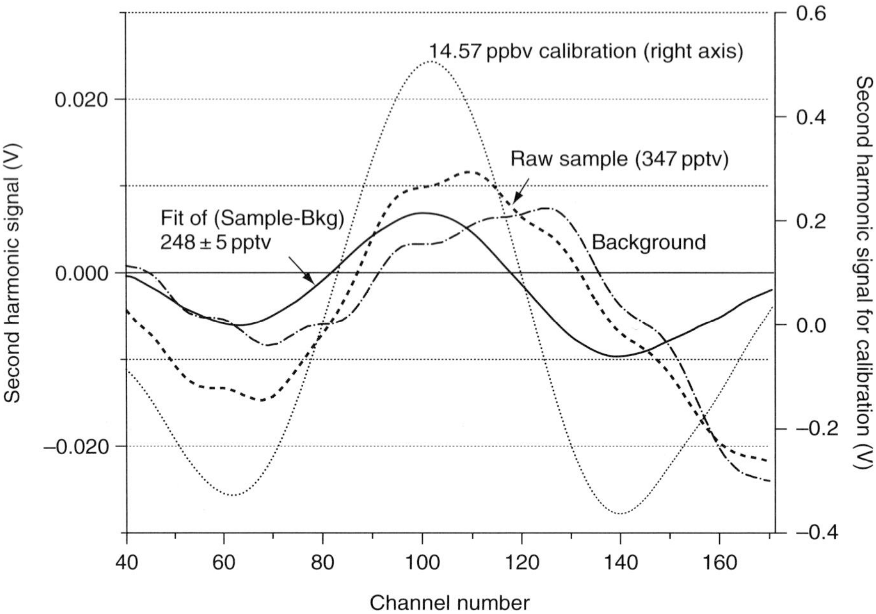

Figure 2.16 Second harmonic profiles of ambient CH_2O, calibration, background and fit spectra acquired while in flight (4 km altitude) during a recent airborne campaign (Wert *et al.*, 2003b). The raw ambient and background spectra were acquired for 10 seconds, and these spectra along with the fit of ambient-background use the left axis. The ambient-background was fit employing the 14.57 ppbv CH_2O spectrum (right axis).

panel. Not only does the background structure yield erroneously high CH_2O results, but also it degrades the measurement precision by a factor of 5.5 in this case.

Fried *et al.* (1997, 1998a), Zahniser *et al.* (1995), and Werle *et al.* (1993) among others have presented the merits of such rapid background subtraction for WMS, direct absorption, and FMS, respectively. If carried out correctly, rapid background subtraction effectively captures and removes optical noise. As addressed by Werle *et al.* (1993), acquisition of sample and background spectra as well as the associated cell/inlet flushing times needs to be accomplished within a characteristic stability period, t_{opt}, in order for background subtraction to be effective. Werle *et al.* (1993) popularized the use of Allan Variance Plots, first introduced by Allan in 1966 for characterization of frequency standards, for assessment of the stability of diode laser optical systems. As shown in Figure 2.18, an Allan Plot is constructed by repetitively sampling a constant input-mixing ratio in very small time increments (0.5 seconds in this case). The resulting average measurement variance between adjacent groups of data is displayed on a log–log plot as the number of elements in each group is successively increased (i.e. increasing the integration time). As shown, the measurement variance decreases linearly with integration time (White noise regime) until t_{opt} is reached ($\sim$100 seconds in this case), whereupon instrument drifts start to dominate. In this particular measurement study, which was carried out in the laboratory employing a temperature-stabilized optical system, the square root of the Allan Variance yields an estimate of the instrument performance.

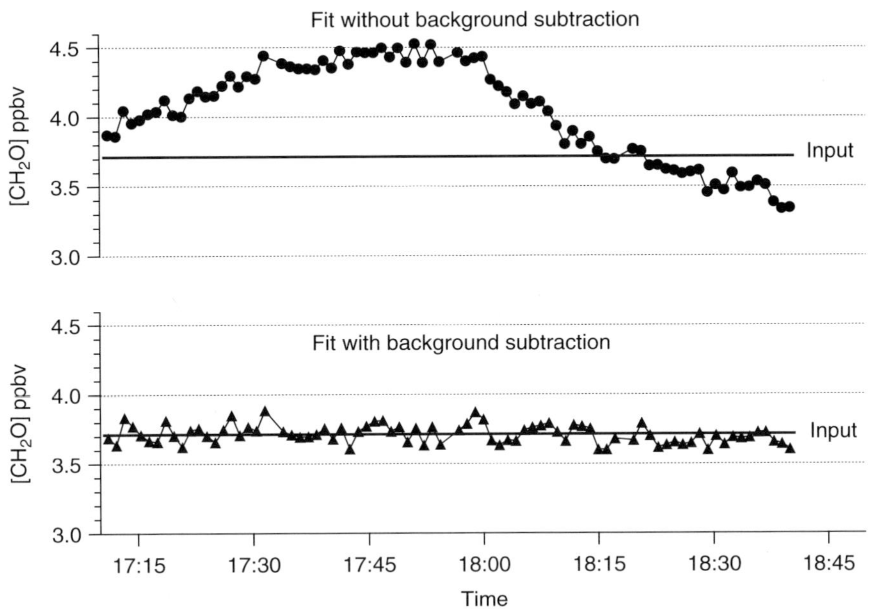

Figure 2.17 The importance of employing rapid background subtraction. In both cases, 20-second CH_2O measurements in the laboratory were obtained by sampling a constant 3.714 ppbv CH_2O in zero air mixture at the entrance of our inlet system. The input concentration (solid horizontal line in both panels) are indicated along with the retrieved CH_2O mixing ratios without background subtraction (upper panel) and with the background spectrum subtracted (lower panel). Backgrounds were acquired for 10 seconds before and after each 20 second sample and the average of the two backgrounds was subtracted from each intervening sample before fitting in the case of the lower trace.

At 60 seconds, this results in a CH_2O mixing ratio of 28 pptv, which corresponds to $A_{min} = 9 \times 10^{-7}$ employing 2*f* WMS. This compares to a similar A_{min} value of 5×10^{-7} achieved by Werle *et al.* (1993) for NO_2 in the laboratory using STFMS in 60 seconds.

2.5.1.3.7 SUMMARY OF THE PERFORMANCE OF DIFFERENT APPROACHES

In the previous sections, we have described various measurement approaches based upon sweep integration direct absorption, 2*f* WMS, STFMS, TTFMS, and CAS, to achieve high measurement performance using solid-state laser sources coupled with multipass cells. At the time of writing, there is no clear-cut optimum measurement approach that stands out when considering real-world atmospheric measurements operating in the mid-IR spectral region; each technique exhibits unique strengths and weaknesses. The OA-ICOS technique appears to yield the highest performance of $3 \times 10^{-11}\,\text{cm}^{-1}\,\text{Hz}^{-1/2}$ (Baer *et al.*, 2002) due to the very long optical pathlengths achievable (4.2 km pathlength). However, these measurements need to be further confirmed by replicate precision measurements, and at present such performance was achieved in the near-IR, where one can only take advantage of this performance for a select group of more abundant atmospheric gases. As discussed, it would be highly desirable to translate this performance into the mid-IR spectral region. The present near-IR OA-ICOS performance is limited by residual optical interference

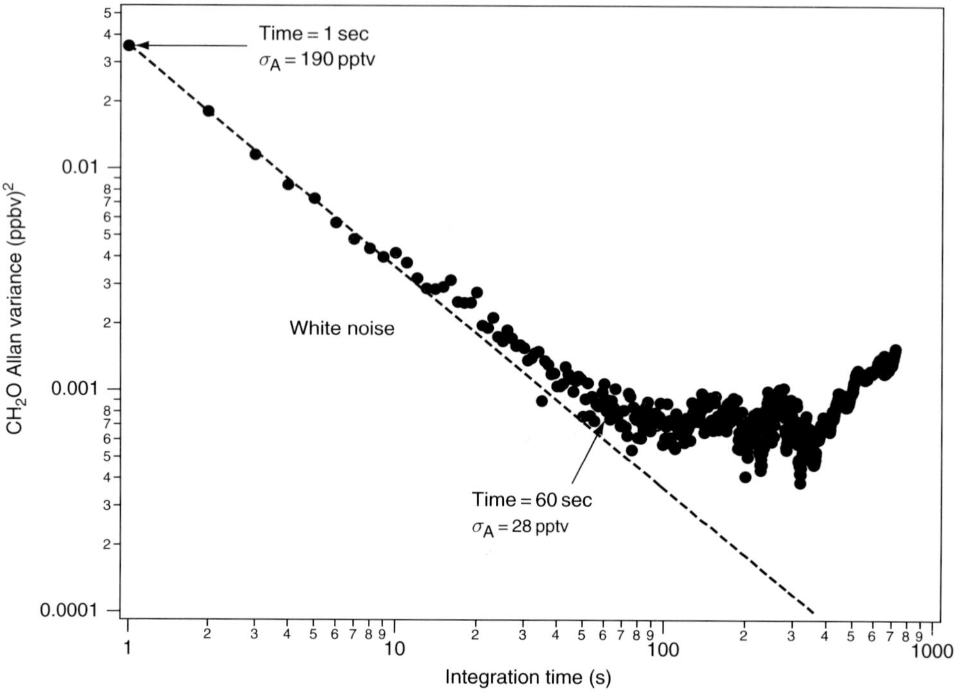

Figure 2.18 Formaldehyde Allan Variance plot carried out in the laboratory using the TDLAS setup shown in Figure 2.19 and 2*f* WMS coupled with sweep integration. The square root of the Allan Variance, σ_A, yields the limit of detection at the time indicated.

effects in the absorption cell. The literature-reported A_{min} values for direct absorption (1×10^{-5} to 3×10^{-5} in 1 second) are slightly less sensitive ($\sim$ factor of 2.5) than those achieved with the various modulation techniques. The magnitude and stability of the background structure and the effectiveness of capturing and removing such structure during the measurement ultimately dictate the performance of these various modulation techniques. The comparative merits of 2*f* WMS and the various FMS approaches, which are very nicely summarized by Silver (1992) and Bomse *et al.* (1992), indicate that comparable laboratory A_{min} values in the 10^{-7}–$10^{-6}\,\mathrm{Hz}^{-1/2}$ range are achievable with all modulation techniques. These researchers indicate that high frequency WMS (HFWMS) techniques, employing modulation frequencies around 10 MHz, offer the best compromise between lower frequency WMS techniques and higher frequency FMS approaches. The FMS technique has the advantage over WMS in that the laser can be swept much faster over the absorption features of interest, yielding higher time-response data and fewer perturbations over each scan ramp. However, the simplicity of 2*f* WMS has thus far favored this approach for most field studies. Realistic A_{min} values of 4×10^{-6} to 13×10^{-6} for 1-second averaging times have been achieved by our group (Wert *et al.*, 2003a) on airborne platforms using an absorption pathlength of 100 m. Clearly additional studies, particularly those in real field environments, are needed to further assess the ultimate merits of the various approaches. Nevertheless, it is clear that for moderate IR absorbers like CH_2O($S_{integ} = 5 \times 10^{-20}\,\mathrm{cm^2\,cm^{-1}\,molecule^{-1}}$), A_{min} values in the 10^{-6}–10^{-7} range using 100–200 m sampling pathlengths are required to further improve our understanding of chemical processes in the background atmosphere. Less stringent detection sensitivities are needed for stronger IR absorbers and/or higher ambient mixing ratios.

2.5.1.3.8 *EXAMPLE OF A SOLID-STATE LASER FIELD INSTRUMENT*

We present in this section one example of an in situ field instrument, using a solid-state laser source, which has been employed in numerous atmospheric studies. Although the particulars for each measurement approach described previously may be different, many of the fundamental concepts are common to all techniques and are illustrated by the following discussion. Upon completion of this discussion, we present in Table 2.3 typical examples of each measurement approach, the molecules studied, and the platforms employed.

Figure 2.19 depicts an optical schematic of the airborne tunable diode laser system developed and deployed by our laboratory on numerous airborne campaigns for measurements of CH_2O (Fried *et al.*, 1998a, 1999a, 2003a,b; Wert *et al.*, 2003a,b). As shown,

Table 2.3 Trace gas measurements using a variety of approaches employing tunable solid-state laser sources on a wide range of platforms

Gas measured	Sampling approach	Technique	Reference
	Ground based measurements		
CH_4	In situ extractive MPC	STFMS & TT FMS	Werle and Kormann (2001)
H_2O, CH_4	In situ extractive MPC	HFWMS	Hovde *et al.* (1993)
$NO_2, HNO_3, CH_2O, H_2O_2$	In situ extractive MPC	SH	Sauer *et al.* (2003)
NO	In situ extractive MPC	DA	Nelson *et al.* (2002)
CH_2O, H_2O_2	In situ extractive MPC	SH	Mackay *et al.* (1996)
NO_2, HNO_3	In situ extractive MPC	DA	Hori *et al.* (1999)
NO, NO_2, O_3, NH_3, CH_2O	In situ extractive MPC	DA	Nelson (2004)
CH_2O	In situ extractive MPC	SH	Fried *et al.* (1997)
	Ship based measurements		
NO_2, CH_2O, H_2O_2	In situ extractive MPC	SH	Harris *et al.* (1992)
CH_2O, H_2O_2, CO	In situ extractive MPC	SH	Wagner *et al.* (2001)
	Low-to-mid-altitude airborne measurements in the troposphere		
NO_2, HNO_3	In situ extractive MPC	SH	Schiff *et al.* (1990)
CO, N_2O, CH_4	In situ extractive MPC	SH line locking	Sachse *et al.* (1987) Vay *et al.* (1998)
CO, H_2O_2, CH_2O, NO_2	In situ extractive MPC	SH	Roths *et al.* (1996)
CH_2O	In situ extractive MPC	SH	Fried *et al.* (1998a, 1999a, 2003a,b) Wert *et al.* (2003a,b)
H_2O	In situ extractive MPC	DA	Silver and Stanton (1987)
H_2O	In situ open path	SH line locking	Diskin *et al.* (2002)

Table 2.3 (Continued)

Gas measured	Sampling approach	Technique	Reference
High-altitude airborne measurements in the upper troposphere and lower stratosphere			
CH_4	In situ extractive MPC	DA	Richard *et al.* (2002)
CO, N_2O	In situ extractive MPC	SH and SH line locking	Loewenstein (1988)
N_2O, CH_4	In situ extractive MPC	SH	Loewenstein *et al.* (2002)
CO, N_2O, CH_4	In situ extractive MPC	TTFMS	Wienhold *et al.* (1998)
HCl, NO_2, HNO_3, CH_4, N_2O	In situ extractive MPC	SH	Webster *et al.* (1994b)
CH_4	In situ extractive MPC	TTFMS	Amato *et al.* (2002)
H_2O, CH_4, CO_2	Open path MPC	DA	Durry *et al.* (2004)
Balloon-borne measurements			
NO, NO_2, HNO_3, O_3, HCl, H_2O, CH_4, N_2O	Remote open path	SH, slow scanning	Webster *et al.* (1990, 1994a) May and Webster (1993)

TTFMS and STFMS: Two-tone and single-tone frequency modulation spectroscopy, respectively.
MPC: Multipass absorption cell.
SH: 2*f* Second harmonic measurements using sweep integration unless otherwise indicated.
DA: Direct absorption.
HFWMS: High frequency wavelength modulation spectroscopy.

the IR radiation from a lead-salt diode laser, operating at 3.5 μm, is directed through a multipass astigmatic Herriott cell (Aerodyne Research Inc.). For this purpose, the beam is collimated, focused, and refocused using two parabolic and one ellipsoidal mirror. After 182 passes through the cell (optical pathlength of 100 m), the output IR beam is directed onto sample and reference indium-antimonide photovoltaic detectors. The optical system employs a minimum number of components, only two of which are adjustable. The entire optical system, including the optical bench, is mounted in a thermally controlled enclosure maintained at 30℃. Both of these aspects are critical for high performance aircraft measurements and are essential for extending the system stability period out to 1 minute and longer (see Figure 2.18).

Absorption data are acquired using second harmonic WMS (2*f* WMS) coupled with sweep integration. In this approach the diode laser is repetitively swept over an isolated CH_2O absorption feature (2831.6417 cm^{-1}) using a sawtooth ramp applied to the laser-tuning current at a frequency of 50 Hz. A 50 kHz quasi-square wave modulation waveform is simultaneously applied to the laser-tuning current and the 2*f* signal at 100 kHz is detected in both the sample and the reference arms using lock-in amplifiers. The lock-in amplifier outputs are then digitized and co-averaged by computer. The reference arm, which contains a low-pressure cell of pure CH_2O, serves as a wavelength reference. In this process, the line centre of the reference arm is determined on every scan, and each scan is then appropriately shifted in memory to align the peak centers before co-averaging. This fast spectral shifting is very important for high measurement performance.

During airborne operation, ambient air is continuously drawn through the inlet shown in Figure 2.19 and through the Herriott cell at controlled pressures of around 50 torr using typically flow rates of 8–10 standard liters per minute. Background spectra are

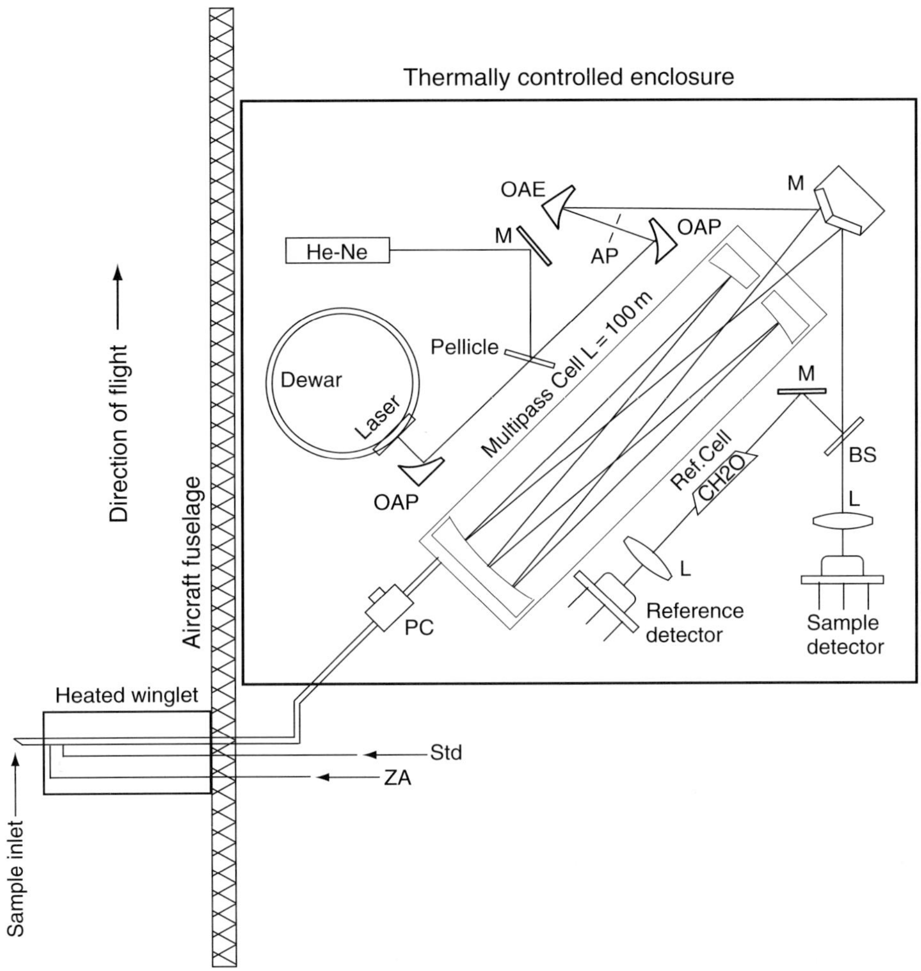

Figure 2.19 Schematic of airborne tunable diode laser system for measurements of CH_2O. The components are as follows: PC (pressure controller); OAP (off axis parabolic mirror); OAE (off axis ellipsoidal mirror); M (flat mirror); AP (aperture); L (lens); BS (beam splitter); Std (CH_2O gas-phase calibration standard injection line); and ZA (zero air addition line). Cell flow is maintained using a vacuum pump (not shown) attached to the cell outlet port located beneath the input/output beam on the right-hand side of the figure.

acquired by passing ambient air from a second inlet through a heated scrubber system (not shown) and this flow stream is re-routed back into the inlet line a few centimeters from the tip at flow rates higher than the inlet flow using the zero-air addition line (see ZA flow in Figure 2.19). Typically, twelve 5 second ambient spectra are acquired during a 1 minute acquisition block, and this is preceded and followed by 10 second background acquisitions. The background pairs are then averaged (time weighted) and subtracted point by point from each of the ambient spectra. This rapid background subtraction not only removes optical noise, as discussed previously, but also reduces the effects of sample line outgassing (see Wert *et al.*, 2002). Since two backgrounds are acquired with every 1 minute ambient sample, it is essential to employ sampling cells with low sampling volumes (3 liters in our case) in order to stay within the stability period and to avoid low ambient duty cycles. Calibration spectra are typically acquired every 30–60 minutes by adding a CH_2O calibration standard to the scrubbed background air, and this generates calibration-mixing ratios in the 12–14 ppbv range at the Herriott cell entrance. As discussed previously, each background-subtracted ambient spectrum is fit in real time to a background-subtracted calibration spectrum employing a multiple linear regression approach. Many other additional details regarding the instrument and its operation, data acquisition, and fitting procedures can be found in the references given above.

Table 2.3 lists some examples of other trace gases measured by other groups using a variety of the approaches employing tunable solid-state laser sources discussed previously on a wide range of platforms. Some instruments employ multiple diode laser sources with multiple liquid nitrogen dewars, collection optics, and multiple detectors for simultaneous measurements of trace gases that absorb in different spectral regions. The three- and four-channel airborne TRISTAR and FLAIR instruments respectively described by Wienhold *et al.* (1998) and Roths *et al.* (1996) are examples of this type of instrument. However, the basic detection and performance principles are similar to the single channel system just described. Table 2.3, which is not all-inclusive, shows a representative set of instruments using solid-state IR laser sources employed: (1) on ground and shipboard platforms; (2) on low- to mid-altitude airplanes such as NASA's DC-8, NCAR's C-130, and NCAR's and NASA's Electra aircraft for tropospheric measurements; (3) on high altitude airplanes, such as NASA's ER-2 and WB-57 and Russia's Geophysica aircraft for measurements in the upper troposphere and lower stratosphere (UT/LS); and (4) on balloon-borne platforms for stratospheric measurements. The latter two platforms are the most demanding since they require autonomously operated instruments without an onboard operator. We only include in this table those instruments actually employed in atmospheric studies.

2.5.1.4 *In situ measurements employing FTIR*

The technique of Fourier Transform Infrared (FTIR) spectroscopy is an extremely versatile technique for measuring atmospheric constituents. As indicated in Section 2.5, FTIR instruments can be operated to acquire in situ measurements from aircraft platforms using an extractive sampling cell or to acquire remote measurements on ground-based and aircraft platforms. Comprehensive treatises on FTIR spectroscopy can be found in the textbooks by Griffiths (1975) and Beer (1992), and the review article by Persky (1995), and only a brief overview of the technique will be presented here. Although this discussion is somewhat abbreviated compared to the discussion of tunable laser sources, this by no means is meant to convey any reduced importance such measurements have on atmospheric studies. In fact, as will be shown, both in situ and remote FTIR measurements provide unique atmospheric measurements that are not available from studies employing tunable laser sources.

Figure 2.20 illustrates the basic concept of the FTIR approach, which has its roots in the work of Michelson in 1880. A broadband light source, from either an incandescent light source mounted with the instrument or the sun or some other source of radiation, enters the instrument and is split into two optical paths, one of which contains a moving mirror. The combined beams are directed onto a detector and the intensity variation due

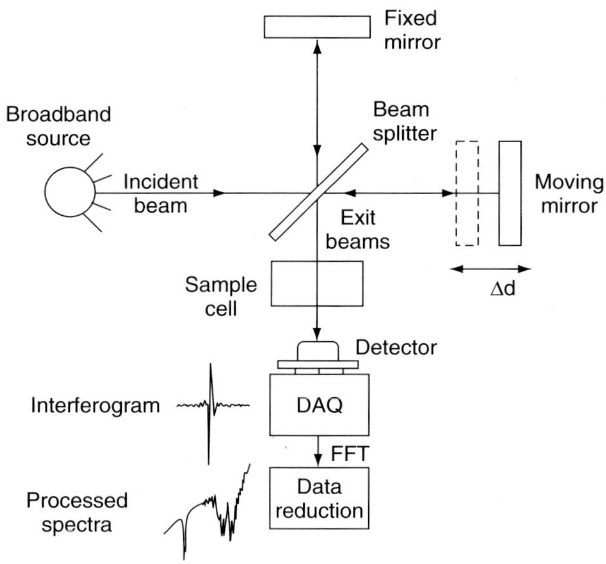

Figure 2.20 Schematic of the basic optical concept employed in FTIR instruments. The DAQ represents a data acquisition system and FFT stands for the process of carrying out a fast Fourier Transform.

to optical interference is measured as a function of the path difference or retardation. For in situ measurements, the source IR beam passes through a multipass absorption cell, which samples ambient air, and this is shown in Figure 2.21. In modern FTIR instruments, the recorded intensity variation, which is known as an *interferogram*, is recorded as the moving mirror is continuously scanned in a linear manner. In the case of a perfectly collimated, infinitely narrow bandwidth, monochromatic, light source, impinging on a perfect 50/50 beam splitter, the detected signal is a maximum when the two paths are exactly equal. At this zero retardation, the two beams perfectly constructively interfere. Conversely, the two beams perfectly destructively interfere when the retardation of the moving mirror is at $\lambda/2$, and this produces zero light at the detector. The signal at the detector will thus vary sinusoidally for a constant velocity–moving mirror, producing a maximum each time the retardation is an integral multiple of λ. One can write the following simple expression that relates the time variation of the detected intensity (I) to the retardation (δ in cm):

$$I(\delta) = B(\overline{\nu})\cos(2\pi\overline{\nu}\delta) \tag{2.32}$$

The term $B(\overline{\nu})$ gives the detected intensity of the light source in frequency units of wavenumbers modified by various instrumental factors. This term thus represents a recorded spectrum similar to that acquired by a scanning solid-state laser source. Mathematically, $I(\delta)$ can be thought of as the cosine Fourier transform of $B(\overline{\nu})$, and the spectrum is determined by taking the Fourier transform of the interferogram $I(\delta)$. In the case of an infinitely narrow bandwidth, monochromatic light source, the interferogram is a perfect cosine wave. Continuous radiation will produce an interferogram that is the composite

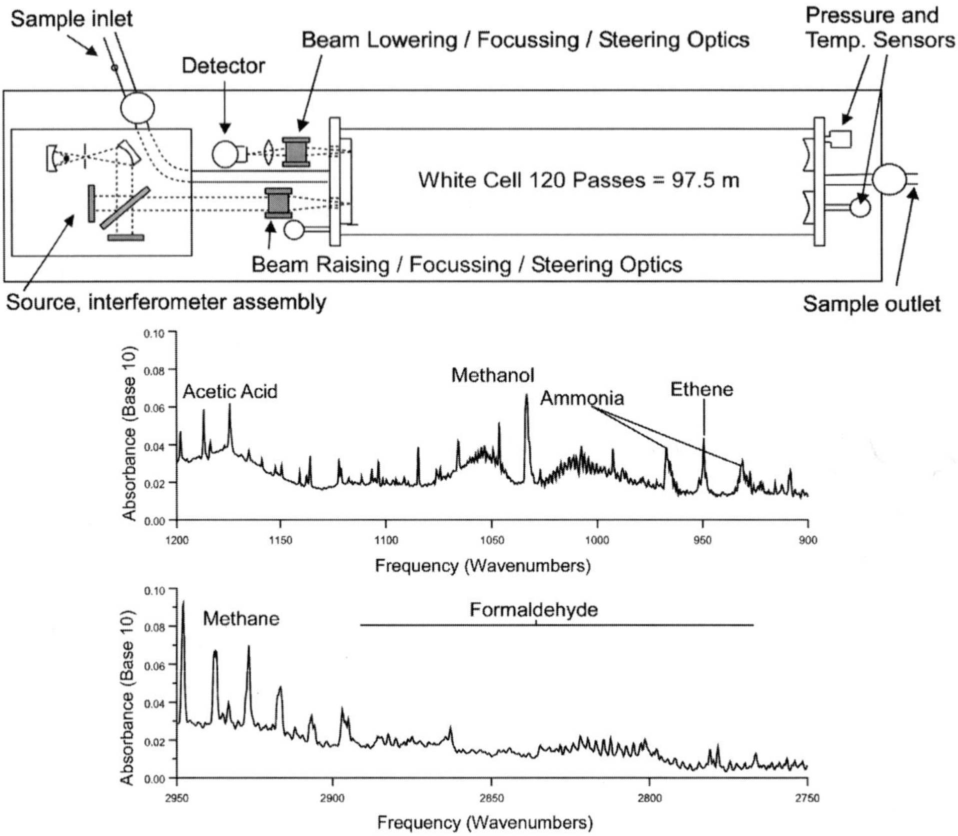

Figure 2.21 Optical and flow schematic of Airborne Fourier Transform Infrared (AFTIR) instrument (top panel) and frequency spectrum (lower panels) acquired by this instrument in fresh smoke plumes during the laboratory fire studies. These figures were graciously supplied by Robert Yokelson of the University of Montana, and more details regarding the AFTIR system can be found in Yokelson (1999 and 2003).

of interferograms for each frequency and this results in the following Fourier integral pair:

$$I(\delta) = \int_0^\infty B(\overline{\nu}) \cos(2\pi\overline{\nu}\delta)\, d\overline{\nu} \tag{2.33}$$

and

$$B(\overline{\nu}) = 2\int_0^\infty I(\delta) \cos(2\pi\overline{\nu}\delta)\, d\delta \tag{2.34}$$

The desired spectrum is thus obtained by the inverse Fourier transform process of Equation 2.34. Since in practice one cannot measure out to $+\infty$ retardation, FTIR instruments have finite spectral resolution determined by the practical mirror travel. Examples of resolution employed are 0.001 cm^{-1} in ground-based and some larger airborne instruments (Worden *et al.*, 1997), through to 0.06 cm^{-1} (airborne or ground-based, Coffey *et al.*, 2003), 0.5 cm^{-1} (airborne or ground-based, Yokelson *et al.*, 1999), and 1 cm^{-1} for airborne measurements (McMillan *et al.*, 2003). Individual vibrational-rotational lines of molecules have pressure-broadened absorption halfwidths of $\sim$0.1 cm^{-1} at atmospheric pressure and $<$ 0.01 cm^{-1} at stratospheric pressures of $\sim$50 torr. Therefore, measurements of these features using lower resolution instruments must be convolved with an accurate instrument function to retrieve meaningful mixing ratios. In addition, the problem of overlapping features due to different molecules can be worse at lower resolution and tend to decrease selectivity.

However, some selectivity can be recovered by using multi-component fits or independent information from other spectral regions where the molecule of interest absorbs in a different band. This latter option is often feasible since FTIR instruments can cover thousands of wavenumbers in $\sim$1 second at moderate (0.5 cm^{-1}) resolution. This same aspect also allows the simultaneous measurement of a large number of atmospheric species. The in situ FTIR spectrum displayed in Figure 2.21 is one example of this. The ability of FTIR measurements to probe large spectral regions is a significant advantage over measurements acquired by tunable solid-state laser sources, where measurements are acquired in very select narrow spectral windows.

The textbook by Griffiths (1975) discusses the various noise sources in FTIR measurements, and typically the variation of the photon flux incident on the detector is ultimately the limiting noise source. This noise, which is directly related to the light source stability in the case of in situ measurements employing an incandescent light source, typically yields A_{min} values for 1 minute averages of $\sim 2 \times 10^{-4}$ for airborne instruments, and values of $\sim 10^{-5}$ in laboratory settings where the light source has been temperature stabilized. Since the noise is proportional to the inverse of the resolution, A_{min} values for high resolution ground-based instruments are typically less sensitive than the values above. For example, the high resolution ground-based instruments used to acquire remote FTIR measurements from solar spectra at the Kitt Peak National Observatory and the International Scientific Station at the Jungfraujoch in the Swiss Alps (Rinsland *et al.*, 1992) achieve A_{min} values in the range between 10^{-2} and 2.5×10^{-4} for integration times up to 33 minutes. By comparison with the results from previous sections, FTIR instruments yield measurements that are $\sim$1–3 orders of magnitude lower in sensitivity than achievable with comparable pathlengths using tunable solid-state laser sources.

As an example, Yokelson *et al.* (2003) report CH_2O detection limits of 15–20 ppbv with an airborne FTIR instrument (AFTIR) using a pathlength of 97.5 m for 1 minute averages. This compares to $\sim$3 orders of magnitude (30–50 pptv) lower limits of detection with our airborne lead-salt diode laser system (Wert *et al.*, 2003a) employing a comparable pathlength of 100 m. This performance enhancement, however, is achieved at the expense of reduced versatility; our lead-salt measurements at the time of writing just report CH_2O mixing ratios, whereas the AFTIR measurements by Yokelson *et al.* (2003) report mixing ratios for H_2O, CO_2, CO, CH_4, NO_2, NO, C_2H_4, CH_3COOH, CH_2O, CH_3OH, HCN, NH_3, HCOOH, and C_2H_2. Such a large suite of gases opens up possibilities to explore unanticipated species and/or new interpretations after the fact. As can be seen, both types of measurements are complementary, each yielding unique and valuable information about the atmosphere.

2.5.2 Remote measurements

As stated earlier, remote measurements employing IR absorption spectroscopy have been carried out employing a wide variety of approaches and platforms. However, in this section, we will only briefly highlight some of the unique measurement aspects and considerations associated with passive ground-based and airborne FTIR measurements using solar and lunar light sources.

The same FTIR fundamentals just discussed in the Section 2.5.1.4 apply to these measurements, with the exception that the open atmospheric path replaces the extractive in situ sampling cell. Total observation pathlengths up to ~100 km are achieved. However, even if the trace gas of interest is uniformly distributed throughout the troposphere, such as in the case of OCS, mixing ratios typically decrease rapidly above ~20 km, due to photolysis. Hence the effective absorption pathlength in this case is ~200 times that achieved with typical in situ techniques (~100 m). After consideration of the inherent sensitivity differences discussed in the last section, one achieves nearly comparable absorption precisions. However, lower precisions are achieved for trace gases distributed in highly localized layers, where the pathlengths are considerably smaller than above.

In contrast to in situ measurements, where local mixing ratios are determined, remote FTIR measurements do not readily yield detailed mixing ratio distributions but rather total vertical column amounts in units of molecules cm^{-2} above the ground-based or airborne platform (Coffey *et al.*, 1981). However, as we will see, this is a distinct advantage for temporal trend analysis. Another distinction between in situ and remote measurements is the issue of spectroscopic interferences. In contrast to in situ measurements, where in many cases one can minimize the effects of broad absorption tails due to H_2O vapor, CO_2, and other abundant gases by working at reduced sampling pressures, remote measurements do not have this luxury. In many cases complete interference-free lines cannot be found, and the resulting data retrieval must take account of such interferences.

These issues notwithstanding, remote FTIR spectroscopy remains an extremely powerful tool for measuring a multitude of important trace atmospheric gases simultaneously. Moreover, such measurements can be carried out on a sustained basis compared to periodic in situ airborne measurements, and these measurements can provide larger spatial coverage using a network of instruments. Both aspects make remote ground-based FTIR measurements ideally suited for long-term spatial and temporal trends. In recognition of this fact, the Network for the Detection of Stratospheric Change (NDSC) has included a series of high quality ground-based FTIR instruments in their suite of measurements to study the temporal and spatial variability of a number of important stratospheric gases, including the column abundances of O_3, HCl, NO, NO_2, $ClONO_2$, and HNO_3. At present this network consists of 20 primary stations and a larger number of complementary stations located around the globe. Comprehensive information on this network can be found at: http://www.ndsc.ncep.noaa.gov.

As discussed in detail by Coffey *et al.* (1981), vertical column amounts are determined from remote FTIR measurements from the integrated absorption along the line of sight to the sun, and then this quantity is converted to a vertical column amount based on the viewing geometry (see Chapter 3 for details). Since both depend upon the absorber distribution, a relative absorber distribution profile, based upon measurements and/or model mixing ratios, is assumed. A multiplicative scaling factor to this assumed profile is iteratively determined by a least-squares fit so as to produce results that best match the observations. The total vertical column is then determined by summing up all the scaled vertical absorber layers. A more detailed discussion of these procedures can be found in Coffey *et al.* (1981).

Although errors in the assumed relative vertical distribution systematically affects the final retrieved vertical column amounts (since the retrieved amounts depend upon the pressure of the absorber), such errors are generally smaller than measurement imprecision and errors in the line parameters (Coffey *et al.*, 1981). Generally, a single or several individual vibrational-rotational lines are used in these determinations. In total vertical columns retrieved with newer high resolution ground-based FTIR instruments, individual absorption features in some cases are fully resolved. Since the measured absorption feature(s) is (are) comprised of features with different shapes for different temperature and pressure layers (refer to Section 2.3.1 on lineshapes and the dependence with T and P), one can incorporate this information in the fitting process to allow the assumed profile to change somewhat in the retrieval. This is not possible with lower resolution airborne spectrometers. However, in this case, one has the added benefit of fitting retrievals only above the aircraft flight altitude, which can be changed to provide additional information.

2.6 Advances in atmospheric studies employing IR absorption measurements and future directions

As can be seen from the previous sections, IR absorption spectroscopy utilizing various approaches yields tremendous flexibility in studying the atmosphere, and numerous advances have indeed been achieved employing such tools. This section will briefly outline some of these advances. Our goal is to present a few representative examples employing the various measurement techniques discussed in this chapter, recognizing the fact that is impossible to highlight all such important studies.

2.6.1 *Measurements of CH_2O mixing ratios and model comparisons*

Ground-based and airborne in situ IR absorption measurements have provided significant advances to our knowledge of atmospheric processing. The extensive CH_2O measurement–model comparisons from our laboratory and many other research groups represent one of many such examples. This gas is produced from the oxidation of methane and non-methane volatile organic compounds (VOCs) in the atmosphere. Measurement–model comparisons of CH_2O yield critical information regarding our overall knowledge of various atmospheric processes, including cycling among various trace gas families, VOC oxidation mechanisms in the troposphere, and an assessment of hydrogen radical sources in the upper troposphere (Fried *et al.*, 2002 and references therein and Frost *et al.*, 2002). As discussed by Fried *et al.* (2003a,b and references therein), CH_2O measurement–model comparisons have had a rather long history, yielding both positive (model underpredictions) and negative deviations as well as good agreement. This points to uncertainties in the measurement of this gas in the background atmosphere, where mixing ratios can be as low as 50 pptv, and in our knowledge of VOC processing and atmospheric transport.

To address these issues, our laboratory as well as other research groups have carried out highly sensitive measurements of CH_2O and other trace gases since 1997, employing tunable diode laser absorption spectroscopy and other methods on ground-based, airborne, and ship-borne platforms. Numerous CH_2O measurement comparisons have been carried out between the TDLAS results and those acquired by other methods on the various platforms (Gilpin *et al.*, 1997; Harder *et al.*, 1997; Fried *et al.*, 2002; Wert *et al.*, 2003a,b). On average, most of the comparisons revealed agreement within the mutual measurement uncertainties. However, unexplained differences, particularly for some point-to-point comparisons on airborne platforms, have also been observed in some instances. These differences notwithstanding, TDLAS measurements of CH_2O acquired during various studies have been systematically compared with the results from photochemical box models (Wagner *et al.*, 2001; Fried *et al.*, 2002, 2003a,b; Frost *et al.*, 2002). These studies carefully assessed both measurement and model uncertainties. During the time periods when such measurement–model comparisons are valid, these studies have demonstrated for the first time consistent overall agreement, suggesting that additional VOC oxidation processes are generally not operative. This is an important finding since other studies (Ayers *et al.*, 1997; Jaeglé *et al.*, 2000) have postulated the need for new CH_2O production mechanisms to explain model underpredictions in many regions of the atmosphere. The reasons for these differences are still unclear.

However, airborne measurements from our laboratory also identified a number of time periods of significant measurement–model disagreement, suggesting that important unknown CH_2O sources and/or production mechanisms may indeed be operative (Fried *et al.*, 2002, 2003a; Frost *et al.*, 2002). For example, during very cold and low light sampling conditions during the TOPSE 2000 study at high northern latitudes, CH_2O observations were consistently elevated relative to box models, sometimes by as much as 392 pptv (Fried *et al.*, 2003a). As discussed by these researchers, such additional CH_2O sources can dramatically affect the hydrogen radical budget in the upper troposphere, which in turn has an impact on the oxidizing capacity of the atmosphere. Conversely, comparisons in clouds and in the lower marine troposphere in the presence of marine aerosols during the 2001 NASA TRACE-P campaign (Fried *et al.*, 2003b), indicated for the first time rather significant CH_2O uptake. Such uptake can dramatically affect mid-tropospheric ozone production rates and the budgets of radical species being transported to the upper atmosphere. All these discrepancies need to be further investigated.

The complementary in situ extractive FTIR measurements of CH_2O and other gases by Yokelson in biomass burning plumes (Tabazadeh *et al.*, 2004) indicated the importance of oxygenated volatile organic compounds (OVOCs) and for the first time additional production mechanisms of CH_2O. Both results, which were unexpected, lead to higher reactivity and faster ozone formation in biomass smoke plumes and an increase of hydrogen radicals in the global troposphere from biomass plumes.

2.6.2 Flux measurements of CH_4

Werle and Kormann (2001) developed and deployed a ground-based instrument, based on the FMS technique discussed previously, to carry out eddy correlation flux measurements of methane emissions from rice paddy fields. Methane is an important greenhouse gas, and rice paddies have been identified as a major methane emission source. These researchers carried out a detailed flux comparison using their instrument to sample ambient air above a rice paddy by the eddy correlation technique and by a closed chamber technique employing a gas chromatograph. The latter technique, which has been used almost exclusively to extrapolate the results of local measurements to global methane flux estimates from rice paddies, needed to be confirmed by independent methods. Werle and Kormann (2001) arrived at the very important finding that the resulting chamber results were 60–90% higher than their eddy correlation measurements. These researchers speculated that strong artificial turbulence from a fan used to rapidly mix the air within the closed chamber might be responsible for part of this discrepancy. These important results need to be further confirmed and the discrepancies need to be fully resolved. Nevertheless, this study suggests that global methane estimates from rice paddies may be overestimated, and chamber flux measurements of other trace gases may have similar errors.

2.6.3 Measurements in the Antarctic polar vortex

in situ tunable diode laser measurements of N_2O, CO, and CH_4, on low and high altitude airborne platforms have provided important information as to the origin of the air mass being sampled. The N_2O measurements from the ATLAS spectrometer (Loewenstein *et al.*, 1989) during the Airborne Antarctic Ozone Experiment, for example, provided critical tracer information regarding the age of air in the polar vortex. The N_2O measurements in particular revealed that the Antarctic polar vortex was occupied by old air during the polar winter and that upwelling of newer air from lower altitudes did not occur, as suggested by several theories.

Another in situ tunable diode laser instrument, the Aircraft Infrared Absorption Spectrometer (ALIAS) developed by Webster *et al.* (1994b), provided the first in situ measurements of HCl and ClO on NASA's ER-2 high altitude aircraft over the southern hemisphere. Both of these gases are intimately involved in ozone destruction, and the new airborne measurements allowed a systematic study of the processes governing chlorine activation (Jaeglé *et al.*, 1997).

2.6.4 Isotopic ratio measurements of water in the UT/LS

As was shown in Section 2.2.3, different isotopic bands of the same molecule are generally well resolved, and one can take advantage of this fact to investigate sources and sinks of various trace gases using IR absorption spectroscopy. Although efforts to apply high-resolution IR absorption measurements to this endeavor are still in their formative stages of development, one can find many applications in the literature including high precision measurements of $^{13}CO_2$ to $^{12}CO_2$ ratios and HDO to H_2O ratios. As one example, we cite the study of Webster and Heymsfield (2003). These researchers measured water isotopes (D/H, $^{18}O/^{16}O$, and $^{17}O/^{16}O$) by IR absorption spectroscopy in and out of cirrus clouds formed in situ and convectively generated in anvils over subtropical regions of the upper troposphere/lower stratosphere (UT/LS). These measurements provided direct observational evidence for elucidating the mechanisms responsible for the growth of water vapor in the lower stratosphere, an important aspect to further our understanding of ozone depletion mechanisms (Webster and Heymsfield, 2003).

2.6.5 Remote FTIR measurements

Remote FTIR measurements of HCl, HF, NO_2, and HNO_3 acquired aboard NASA's DC-8 aircraft and on balloons as part of SAGE III Ozone Loss and Validation Experiment (SOLVE), provided important information about chlorine activation and denitrification in the Arctic stratosphere (Coffey *et al.*, 2003). Within the polar vortex, these researchers observed markedly reduced columns of HCl and NO_2 and elevated columns of HF and HNO_3. These observations are consistent with heterogeneous reactions of HCl and NO_2 on polar stratospheric clouds to form HNO_3 and active forms of chlorine.

Ground-based FTIR measurements of carbonyl sulfide (OCS) total column abundances from two of the NDSC network sites (the National Solar Observatory at Kitt Peak and the International Scientific Station of the Jungfraujoch in Switzerland) provided new information regarding long-term trends and variability of this gas (Rinsland *et al.*, 1992). The relatively unreactive and insoluble sulfur gas OCS can be transported from its source regions in the troposphere to the stratosphere where it ultimately gets converted to fine sulfuric acid aerosols. This process was first postulated by Crutzen to explain the existence of the background aerosol sulfate layer known as the Junge layer. It has been postulated that long-term enhancements to this layer may influence the Earth's radiation budget and climate through increased solar scattering and enhance the global destruction of ozone from heterogeneous reactions on the sulfate aerosols (see Fried *et al.*, 1999b and references therein). The remote FTIR measurements, recorded between 1977 and 1991 at the two sites, indicated no significant change in the OCS total column abundance at northern mid-latitudes.

One can find more details regarding the use of remote FTIR measurements in various monitoring applications on the United States EPA Technology Innovation Program Web site (http://clu-in.org/programs/21m2/openpath/op-ftir/). This reference in addition provides further details about the different configurations that have been employed using open path FTIR instruments as well as the gases detected.

2.6.6 Future directions

Advances in atmospheric chemistry from various types of IR absorption measurements will continue to grow at an exponential rate. We have every confidence that slow but steady improvements in tunable laser sources, data acquisition and reduction techniques, sampling cells, and more stable optical/mechanical designs will eventually lead to smaller autonomously operated field instruments with routine limits of detection at and below 10 pptv for various atmospheric trace gases. Such performance will undoubtedly open up many new avenues of research and lead to new findings regarding atmospheric processes.

Future technological advances in laser sources most likely will have the largest impact on trace atmospheric measurements. It would be particularly important in this regard to develop new laser sources that (1) are user-friendly; (2) can access all the desired mid-IR wavelengths with devices that are readily available with powers of many hundreds of mW; (3) have improved spatial beam qualities; (4) exhibit reduced residual technical noise; and (5) exhibit increased laser frequency and polarization stabilities. Implementation of such desirable characteristics will not only improve measurement detection limits but will also enhance the versatility of any given instrument. Furthermore, we anticipate that realization of such attributes will ultimately lead to the development of simple, elegant, and straightforward trace gas absorption systems that can be operated by personnel with significantly less training than at present.

Acknowledgements

The authors would like to thank Drs. Petter Weibring, Steve Massie, and Aaron Goldman for providing critical reviews of this chapter and to Drs. Mike Coffey and Robert Yokelson for critical reviews of the FTIR sections.

Further reading

Banwell, C.N. & McCash, E.M. (1994) *Fundamentals of Molecular Spectroscopy*, Fourth Edition, McGraw-Hill International (UK) Limited.

Barrow, G.M. (1962) *Molecular Spectroscopy*, McGraw-Hill, New York.

Beer, R. (1992) *Remote Sensing by Fourier Transform Spectrometry*, in the Chemical Analysis Series, Volume 120, John Wiley & Sons, New York.

Brassington, D.J. (1995) Tunable diode laser absorption spectroscopy for the measurement of atmospheric trace species, in *Advances in Spectroscopy*, 24, R.E. Hestler (ed.), John Wiley, pp. 83–148.

Duxbury, G. (2000) *Infrared Vibration-Rotation Spectroscopy – From Free Radicals to the Infrared Sky*, John Wiley & Sons Ltd, England.

Griffiths, P.R. (1975) Chemical Infrared Fourier Transform Spectroscopy, in *Chemical Analysis Series*, 43, John Wiley & Sons, New York.

Herzberg, G. (1945) Molecular Spectra and Molecular Structure II. Infrared and Raman Spectra of Polyatomic Molecules, Van Nostrand Reinhold Company, New York.

King, G.W. (1964) *Spectroscopy and Molecular Structure*, Holt Rinehart and Winston, Inc., New York.

Levine, I.N. (1970) *Quantum Chemistry Volume II: Molecular Spectroscopy*, Allyn and Bacon, Boston.

Rothman, L.S., Rinsland, C.P., Goldman, A. *et al.* (1998) The HITRAN molecular spectroscopic database and HAWKS (HITRAN Atmospheric Workstation): 1996 edition, *J. Quant. Spectrosc. Radiat. Transfer*, 60, 665–710.

Rothman, L.S., Barbe, A., Benner, D.C. *et al.* (2003) The HITRAN molecular spectroscopic database: edition of 2000 including updates through 2001, *J. Quant. Spectrosc. Radiat. Transfer.*, 82, 5–44.

Smith, M.A.H., Rinsland, C.P., Rao, K.N. (1985) Intensities and collision broadening parameters from infrared spectra, in *Molecular Spectroscopy: Modern Research*, Vol. III. K.N. Rao (ed.), Academic Press, New York.

Tittel, F.K., Richter, D. & Fried, A. (2003) Mid-infrared laser applications in spectroscopy, in *Topics in Applied Physics: Solid-State Mid-Infrared Laser Sources*, I.T. Sorokina & K.L. Vodopayanov (eds), 89, 445–510, Springer-Verlag, Berlin, Heidelberg.

United States Environmental Protection Agency, Technology Innovation Program, *Measurement and Monitoring Technologies for the 21st Century, Open Path Technologies: Measurement at a Distance*, http://clu-in.org/programs/21m2/openpath/op-ftir/

References

Agrawal, G. (2001) *Nonlinear Fiber Optics*, Third Edition, Academic Press.

Allen, M.G., Carleton, K.L., Davis, S.J. *et al.* (1995) Ultra-sensitive dual beam absorption and gain spectroscopy: Applications for near-IR and visible diode laser sensors, *Appl. Opt.*, 34 (18), 3240–3249.

Amato, F.D., Mazzinghi, P. & Castagnoli, F. (2002) Methane analyzer based on TDL's for measurements in the lower stratosphere: Design and laboratory tests, *Appl. Phys.* B, 75, 195–202.

Armstrong, B.H. (1967) Spectrum line profiles: The Voigt function, *J. Quant. Spectrosc. Radiat. Transfer*, 7, 61–88.

Atlas E.L., Ridley, B.A. & Cantrell, C. (2003) The Tropospheric Ozone Production about the Spring Equinox (TOPSE) Experiment: Introduction, *J. Geophys. Res.*, 108 (D4), 8353, doi:10.1029/2002JD003172.

Ayers, P.G., Gillett, R.W., Granek, H. *et al.* (1997) Formaldehyde production in clean marine air, *Geophys. Res. Lett.*, 24, 401–404.

Baer, D.S., Paul, J.B., Gupta M. & O'Keefe, A. (2002) Sensitive absorption measurements in the near-infrared region using off-axis integrated-cavity-output spectroscopy, *Appl. Phys.* B. 75, pp. 261–265.

Beck, M., Hofstetter, D., Aellen, T., Faist, J., Oesterle, U., Ilegems, M., Gini, E. & Melchior, H. (2002) Continuous Wave Operation of a Mid-Infrared Semiconductor Laser at Room Temperature, *Science*, 295, 301–305.

Bomse, D.S., Stanton, A.C. & Silver, J.A. (1992) Frequency modulation and wavelength modulation spectroscopies: comparison of experimental methods using a lead-salt diode laser, *Appl. Opt.*, 31 (6), 718–731.

Brasseur, G.P. & Schimel, D. (1999) Chapter 1: Atmospheric chemistry and the earth system, in *Atmospheric Chemistry and Global Change*, G.P. Brasseur, J.J. Orlando & G.S. Tyndall (eds), Oxford University Press, New York.

Carlisle, C.B. & Cooper, D.E. (1989) Tunable-diode-laser frequency-modulation spectroscopy using balanced homodyne detection, *Opt. Lett.*, 14 (23), 1306–1308.

Carlisle, C.B., Cooper, D.E. & Preiser, H. (1989) Quantum noise-limited FM spectroscopy with a lead-salt diode laser, *Appl. Opt.*, 28 (13), 2567–2576.

Chernin, S.M. (1993) New generation of multipass systems, in *Tunable Diode Laser Spectroscopy, Lidar, and DIAL Technique for Environmental and Industrial Measurements*, A Fried, D.K. Killinger & H.I. Schiff (eds), pp. 99–108, Proceedings of SPIE, 2112.

Coffey, M.T., Mankin, W.B. & Goldman, A. (1981) Simultaneous spectroscopic determination of the latitudinal, seasonal, and diurnal variability of stratospheric N_2O, NO, NO_2, and HNO_3, *J. Geophys. Res.*, 86 (C8), 7331–7341.

Coffey, M.T., Mankin, W.G., Hannigan, J.W. & Toon, G.C. (2003) Airborne spectroscopic observations of chlorine activation and denitrification of the 1999/2000 winter Arctic stratosphere during SOLVE, *J. Geophys. Res.*, 108 (D5), 8303, doi: 10.1029/2001JD001085.

Coleman C.D., Bozman, W.R. and Meggers, W.F. (1960) *Table of Wavenumbers*, U.S. Department of Commerce, National Bureau of Standards, NBS monograph 3, Volumes 1, 2.

Diskin, G.S., Podolske, J.R., Sachse, G.W. & Slate, T.A. (2002) Open-path airborne tunable diode laser hygrometer, in *Diode Laser and Applications in Atmospheric Sensing*, A. Fried (ed.), pp. 198–204, Proceedings of SPIE, 4817.

Duffell, H., Oppenheimer, C. & Burton, M. (2001) Volcanic gas emission rates measured by solar occultation spectroscopy, *Geophys. Res. Lett.*, 28 (16), 3131–3134.

Durry, G., Amarouche, N., Zéninari, V. *et al.* (2004) In situ sensing of the middle atmosphere with balloon-borne near-infrared laser diodes, *Spectrochimica Acta Part A – Molecular and Biomolecular Spectroscopy*, 60(14), 3371–3379.

Dyroff, C., Fried, A., Richter, D., Walega, J.G., Zahniser, M.S. & McManus, J.B. (2004) Design and performance assessment of a stable astigmatic Herriott cell for trace gas measurements on airborne platforms, *Proceedings of Laser Applications to Chemical and Environmental Analysis*, Topical Meeting, February 2004, Annapolis, MD.

Evans, A., Yu, J.S., David, J., Doris, L., Mi, K., Slivken, S. & Razeghi, M. (2004) High-temperature, High-power, continuous-wave operation of buried heterostructure quantum-cascade lasers, *Appl. Phys. Lett.* 84, 314–316.

Faist, J., Capasso, F., Sivco, D.L., Sitori, C., Hutchinson, A.L. & Cho, A.Y. (1994) Quantum Cascade Laser, *Science*, 264, 553.

Fried, A., Sams, R. & Berg, W.W. (1984) Application of tunable diode laser absorption for trace stratospheric measurements of HCl: Laboratory results, *Appl. Opt.*, 23 (11), 1867–1880.

Fried, A., Sams, R., Dorko, W., Elkins, J.W. & Cai, Z.-T. (1988) Determination of nitrogen dioxide in air compressed gas mixtures by quantitative tunable diode laser absorption spectrometry and chemiluminescence detection, *Anal. Chem.*, 60 (5), 394–403.

Fried, A., Sewell, S., Henry, B., Wert, B.P., Gilpin, T. & Drummond, J.R. (1997) Tunable diode laser absorption spectrometer for ground-based measurements of formaldehyde, *J. Geophys. Res.*, 102 (D5), 6253–6266.

Fried, A., Henry, B., Wert, B., Sewell, S. & Drummond, J.R. (1998a) Laboratory, ground-based, and airborne tunable diode laser systems: Performance characteristics and applications in atmospheric studies, *Appl. Phys.* B, 67, 317–330.

Fried, A., Henry, B. & Sewell, S. (1998b) Potential calibration errors in carbonyl sulfide permeation devices: Implications for atmospheric studies, *J. Geophys. Res.*, 103 (D15), 18 895–18 906.

Fried, A., Wert., B.P., Henry, B. & Drummond, J.R. (1999a) Airborne tunable diode laser measurements of formaldehyde, *Spectrochimica Acta.*, A 55, 2097–2110.

Fried, A., Tyndall, G. and Barth, M. (1999b) Sulfur compounds, Chapter 10, in *Atmospheric Chemistry and Global Change*, G.P. Brasseur, J.J. Orlando & G.S. Tyndall (eds), Oxford University Press, New York.

Fried, A., Lee, Y.-N., Frost, G. *et al.* (2002) Airborne CH_2O measurements over the North Atlantic during the 1997 NARE campaign: Instrument comparisons and distributions, *J. Geophys. Res.*, 107 (D13), 10.1029/2000JD000260.

Fried, A., Wang, Y., Cantrell, C. *et al.* (2003a) Tunable diode laser measurements of formaldehyde during the TOPSE 2000 study: distributions, trends, and model comparisons, *J. Geophys. Res.*, 108 (D4), 8365, doi: 10.1029/2002JD002208.

Fried, A., Crawford, J., Olson, J. *et al.* (2003b) Airborne tunable diode laser measurements of formaldehyde during TRACE-P: distributions and box model comparisons, *J. Geophys. Res.*, 108 (D20), 8798, doi: 10.1029/2003JD003451.

Frost, G.J., Fried, A., Lee, Y.-N., *et al.* (2002) Comparisons of box model calculations and measurements of formaldehyde from the 1997 North Atlantic Regional Experiment, *J. Geophys. Res.*, 107 (D13), 10.1029/2001JD000896.

Ghatak, A. & Thyagarajan, K. (1998) *An Introduction to Fiber Optics*, Cambridge University Press.

Gilpin, T., Apel, E., Fried, A., Sewell, S., Wert, B.P., Calvert, J.G., Genfa, Z., Dasgupta, P., Harder, J.W., Heikes, B., Hopkins, B., Westberg, H., Kleindienst, T., Lee, Y.-N., Zhou X., & Lonneman W. (1997) Intercomparison of Six Ambient [CH_2O] Measurement Techniques. *J. Geophys. Res.*, 102, 21161–21188.

Goldman, A., Vanallen, J.W., Blatherwick, R.D. *et al.* (2002) *New Atlas of Stratospheric IR Absorption Spectra Volume II. The Spectra*, Department of Physics University of Denver, Denver, Colorado, 80208.

Griffith, D.W.T., Mankin, W.G., Coffey, M.T., Ward, D.E. & Riebau, A. (1991) FTIR remote sensing of biomass burning emissions of CO_2, CO, CH_4, CH_2O, NO, NO_2, NH_3, and N_2O, in *Global Biomass Burning: Atmospheric, Climatic, and Biospheric Implications*, J.S. Levine (ed.), pp. 230–239, MIT Press, Cambridge, Mass.

Griffith, D.W.T., Leuning, R., Denmead, O.T. & Jamie, I.M. (2002) Air-land exchanges of CO_2, CH_4, and N_2O measured by FTIR spectroscopy and micrometeorological techniques, *Atmos. Environ.*, 36, 1833–1842.

Griffith, D.W.T., Jones, N.B., McNamara, B. *et al.* (2003) Intercomparison of NDSC ground-based solar FTIR measurements of atmospheric gases at Lauder, New Zealand. *J. Atmos. Ocean Technol.*, 20 (8), 1138–1153.

Grutter, M., Flores, E., Basaldud, R. & Ruiz-Suárez, L.G. (2003) Open-path FTIR spectroscopic studies of the trace gases over Mexico City. *Atmos. Oceanic Opt.*, 16 (3), 232–236.

Grutter, M., (2003) Multi-gas analysis of ambient air using FTIR spectroscopy over Mexico City, *Atmósfera*, 16, 1–13.

Guelachvili, G. & Rao, K.N. (1986) Handbook of Infrared Standards with Spectral Maps and Transition Assignments between 3 and 2600 μm, Academic Press, Inc., Harcourt Brace Jovanovich, Publishers, Orlando, Florida.

Ham, J.M. and Heilman, J.L. (2003) Experimental test of density and energy-balance corrections on carbon dioxide flux as measured using open-path eddy covariance, *Agron. J.*, 95, 1393–1403.

Harder, J.W., Fried, A., Sewell, S. & Henry, B. (1997) Comparison of tunable diode laser and long-path ultraviolet/visible spectroscopic measurements of ambient formaldehyde concentrations during the 1993 OH Photochemistry Experiment, *J. Geophys. Res.*, 102 (D5), 6267–6282.

Harris, G.W., Klemp, D., Zenker, T. & Burrows, J.P. (1992) Tunable diode laser measurements of trace gases during the 1988 Polarstern cruise and intercomparisons with other methods, *J. Atm. Chem.*, 15, 315–326.

Herriott, D., Kogelnik, H. & Kompfner, R. (1964) *Off-axis paths in spherical mirror interferometers*, *Appl. Opt.*, 3 (4), 523–526.

Herriott, D.R. & Schulte, H.J. (1965) Folded optical delay lines, *Appl. Opt.*, 4 (8), 883–889.

Hobbs, P.C.D. (1997) Ultrasensitive laser measurements without tears, *Appl. Opt.*, 36, 903–920.

Hodges, J.T., Layer, H.P., Miller, W.W. & Scace, G.E. (2004) Frequency-stabilized single-mode cavity ring-down apparatus for high-resolution absorption spectroscopy, *Rev. Sci. Inst.*, 75, 849–863.

Hofstetter, D. and Faist, J. (2003) High performance quantum cascade lasers and their applications, in *Topics in Applied Physics: Solid-State Mid-Infrared Laser Sources*, I.T. Sorokina & K.L. Vodopayanov (eds) 89, 61–96, Springer-Verlag, Berlin, Heidelberg.

Hollberg, L., Fox, R., Waltman, S. & Robinson, H. (eds) (1998) *Precision Spectroscopy, Diode Lasers, and Optical Frequency Measurement Technology*, United States Dept. of Commerce NIST Tech. Note 1504.

Horn, D. and Pimentel, G.C. (1971) 2.5-km low temperature multiple-reflection cell, *Appl. Opt.*, 10, 1892–1898.

Hori, C.V., Zahniser, M.S., Nelson, D.D. *et al.* (1999) Nitric acid and nitrogen dioxide flux measurements: A new application of tunable diode laser absorption spectroscopy, in *Application of Tunable Diode and Other Infrared Sources for Atmospheric Studies and Industrial Processing Monitoring II*, A. Fried (ed.), pp. 152–161, Proceedings of SPIE, 3758.

Hovde, D.C., Silver, J.A. & Stanton, A.C. (1993) Measuring atmospheric methane and water vapor using near-infrared diode lasers, in *Tunable Diode Laser Spectroscopy, Lidar, and DIAL Technique for Environmental and Industrial Measurements*, A. Fried, D.K. Killinger and H.I. Schiff (eds), pp. 110–117, Proceedings of SPIE, 2112.

Humlicek, J. (1982) Optimized computation of the Voigt and complex probability functions, *J. Quant. Spectrosc. Radiat. Transfer*, 27, 437–444.

Iguchi, T. (1986) Modulation waveforms for second-harmonic detection with tunable diode lasers, *J. Opt. Soc. Am.*, B 3 (3), 419–423.

Jacob, D.J., Crawford, J.H., Kleb, M.M. *et al.* (2003) Transport and Chemical Evolution over the Pacific (TRACE-P) aircraft mission: Design, execution, and first results, *J. Geophys. Res.*, 108 (D20), 9000, doi:10.1029/2002JD003276.

Jaeglé, L., Webster, C.R., May, R.D. *et al.* (1997) Evolution and stoichiometry of heterogeneous processing in the Antarctic stratosphere, *J. Geophys. Res.*, 102 (D11), 13 235–13 253.

Jaeglé, L., Jacob, D.J., Brune, W.H. *et al.* (2000) Photochemistry of HOx in the upper troposphere at northern midlatitudes, *J. Geophys. Res.*, 105, 3877–3892.

Jennings, D.E. (1980) Absolute line strengths in ν_4 $^{12}CH_4$: A dual beam diode laser spectrometer with sweep integration, *Appl. Opt.*, 19 (16), 2695–2700.

Joulie, A. Christol, P., Baranov, A.N. & Vicet, A. (2003) Mid-infrared 2–5 μm heterojunction laser diodes, in *Topics in Applied Physics: Solid-State Mid-Infrared Laser Sources*, I.T. Sorokina & K.L. Vodopayanov (eds), 89, 1–59.

Loewenstein, M. (1988) Diode laser harmonic spectroscopy applied to in situ measurements of atmospheric trace molecules, *J. Quant. Spectrosc. Radiat. Transfer*, 40 (3), 249–256.

Loewenstein, M., Podolske, J.R., Chan, K.R. & Strahan, S.E. (1989) Nitrous oxide as a dynamical tracer in the 1987 airborne Antarctic ozone experiment, *J. Geophys. Res.*, 94 (D9), 11 589–11 598.

Loewenstein, M., Jost, H., Grose, J. *et al.* (2002) Argus: A new instrument for the measurement of the stratosphere dynamical tracers, N_2O and CH_4, *Spectrochimica Acta.*, A 58, 2329–2345.

Mackay, G.I., Karecki, D.R. & Schiff, H.I. (1996) Tunable diode laser absorption measurements of H_2O_2 and HCHO during the Mauna Loa Observatory Photochemistry Experiment, *J. Geophys. Res.*, 101 (D9) 14 721–14 728.

Maki, A.G. & Wells, J.S. (1991) *Wavenumber Calibration Tables from Heterodyne Frequency Measurements*, NIST Special Publication 821, US Government Printing Office, Washington, DC 20402.

Mankin, W.G., Coffey, M.T. & Griffith, D.W.T. (1979) Spectroscopic measurement of carbonyl sulfide (OCS) in the stratosphere, *Geophys. Res. Lett.* 6 (11), 853–856.

Mankin, W.G. & Coffey, M.T. (1989) Airborne measurements of stratospheric constituents over Antarctica in the austral spring 1987, 1, method and ozone observations, *J. Geophys. Res.*, 94, 11 413–11 421.

Marquardt, D.W. (1963) Peak fitting by nonlinear least squares, *J. Soc. Ind. Appl. Math.*, 11, 431–441.

May, R.D. & Webster, C.R. (1993) Data processing and calibration for tunable diode laser harmonic absorption spectrometers, *J. Quant. Spectrosc. Transfer*, 49, 335–347.

McManus, J.B., Kebabian, P.L. & Zahniser, M.S. (1995) Astigmatic mirror multipass absorption cells for long path-length spectroscopy. *Appl. Opt.*, 34 (18), 3336–3348.

McMillan, W.W., McCourt, M.L., Revercomb, H.E. *et al.* (2003) Tropospheric carbon monoxide measurements from the scanning high-resolution interferometer sounder on 7 September 2000 in southern Africa during SAFARI 2000, *J. Geophys. Res.*, 108, 8492, doi: 10.1029/2002JD002335.

Mitchell, A.C.G. & Zemansky, M.W. (1961) *Resonance Radiation and Excited Atoms.* Cambridge University Press, Cambridge, Mass.

Nelson, D.D., Shorter, J.H., McManus, J.B. & Zahniser, M.S. (2002) Sub-part-per-billion detection of nitric oxide in air using a thermoelectrically cooled mid-infrared quantum cascade laser spectrometer, *Appl. Phys.* B, 75, 343–350.

Nelson, D.D. (2004) *Environmental monitoring with pulsed quantum cascade lasers: Seeking sensitivity and stability*, Invited talk at the 2004 Laser Applications to Chemical and Environmental Analysis, February 2004.

O'Keefe, A. and Deacon, D.A.G. (1988) Cavity-ring down optical spectrometer for absorption measurements using pulsed laser sources, *Rev. Sci. Instrum.* 59, 2544.

Park, J.H., Rothman, L.S., Rinsland, C.P. *et al.* (1987) *Atlas of Absorption Lines from 0 to 17900 cm^{-1}*, NASA Reference Publication 1188.

Paul, J.B., Lapson, L. & Anderson, J.G. (2001) Ultrasensitive absorption spectroscopy with a high finesse optical cavity and off-axis alignment, *Appl. Opt.*, 40, 4904–4910.

Penner, S.S. (1959) *Quantitative Molecular Spectroscopy and Gas Emissivities.* Addison-Wesley Publishing Company, Reading, Mass.

Persky, M.J. (1995) A review of spaceborne infrared Fourier transform spectrometers for remote sensing, *Rev. Sci. Instrum.*, 66 (10), 4763–4796.

Pine, A.S., Fried, A. & Elkins, J.W. (1985) Spectral intensities in the fundamental bands of HF and HCl, *J. Mol. Spectrosc.* 109, 30–45.

Pine, A.S. & Fried, A. (1985) Self-broadening in the fundamental bands of HF and HCl, *J. Mol. Spectrosc.*, 114, 148–162.

Pugh, L.A. & Rao, K.N. (1976) Intensities from infrared spectra, in *Molecular Spectroscopy: Modern Research Volume II*, K.N. Rao (ed.), Academic Press, New York.

Reid, J., El-Sherbiny, M., Garside, B.K. & Ballik, E.A. (1980) Sensitivity limits of a tunable diode laser spectrometer, with application to the detection of NO_2 at the 100-ppt level, *Appl. Opt.*, 19 (19), 3349–3353.

Reid, J. & Labrie, D. (1981) Second-harmonic detection with tunable diode lasers-comparison of experiment and theory, *Appl. Phys.* B, 26, 203–210.

Richard, E.C., Kelly K.K., Winkler, R.H. *et al.* (2002) A fast-response near-infrared tunable diode laser absorption spectrometer for in situ measurements of CH_4 in the upper troposphere and lower stratosphere, *Appl. Phys.* B, 75, 183–194.

Richter, D., Fried, A., Wert, B.P., Walega, J.G. & Tittel, F.K. (2002) Development of a tunable mid-IR difference-frequency laser source for highly sensitive airborne trace gas detection, *Appl. Phys.* B, 75, 281–288.

Rinsland, C.P., Zander, R., Mahieu, E., *et al.* (1992) Ground-based infrared measurements of carbonyl sulfide total column abundances: Long-term trends and variability, *J. Geophys. Res.*, 97 (D5), 5995–6002.

Rinsland, C.P., Goldan, A., Mahieu, E. *et al.* (2002) Ground-based infrared spectroscopic measurements of carbonyl sulfide: free tropospheric trends from a 24-year time series of solar absorption measurements, *J. Geophys. Res.*, 107 (D22), 4657, doi:10.1029/2002JD002522.

Rinsland, C.P., Mathieu, E., Zander, R. *et al.* (2003) Long-term trends of inorganic chlorine from ground-based infrared solar spectra: past increases and evidence for stabilization, *J. Geophys. Res.*, 108 (D8), 4252, doi:10.1029/2002JD003001.

Rothman, L.S., Hawkins, R.L., Wattson, R.B. & Gamache, R.R. (1992) Energy levels, intensities, and linewidths of atmospheric carbon dioxide bands, *J. Quant. Spectrosc. Radiat. Transfer*, 48, 537–566.

Roths, J., Zenker, T., Parchatka, U. *et al.* (1996) Four-laser airborne infrared spectrometer for atmospheric trace gas measurements, *Appl. Opt.*, 35 (36), 7075–7084.

Sachse, G.W., Hill, G.F., Wade, L.O. & Perry, M.G. (1987) Fast-response, high-precision carbon monoxide sensor using a tunable diode laser absorption technique, *J. Geophys. Res.*, 92 (D2), 2071–2081.

Sandholm, S., Smmyth, S., Bai, R. & Bradshaw, J. (1997) Recent and future improvements in two-photon laser-induced fluorescence NO measurement capabilities, *J. Geophys. Res.*, 102 (D23), 28 651–28 661.

Sauer, C.G., Pisano, J.T. & Fitz, D.R (2003) Tunable diode laser absorption spectrometer measurements of ambient nitrogen dioxide, nitric acid, formaldehyde, and hydrogen peroxide in Parlier, California, *Atm. Environ.*, 37, 1583–1591.

Schiff, H.I., Karecki, D.R., Harris, G.W., Hastie, D.R. & Mackay, G.I. (1990) A tunable diode laser system for aircraft measurements of trace gases, *J. Geophys. Res.*, 95 (D7), 10 147–10 153.

Sewell, S., Fried, A., Henry, B. & Drummond, J. (1993) A field diode laser spectrometer employing an astigmatic Herriott cell, in *Tunable Diode Laser Spectroscopy, Lidar, and DIAL Technique for Environmental and Industrial Measurements*, A. Fried, D.K. Killinger & H.I. Schiff (eds), pp. 72–80, Proceedings of SPIE, 2112.

Silver, J.A. & Stanton, A.C. (1987) Airborne measurements of humidity using a single-mode Pb-salt diode laser, *Appl. Opt.*, 26 (13), 2558–2566.

Silver, J.A. (1992) Frequency-modulation spectroscopy for trace species detection: Theory and comparison among experimental methods, *Appl. Opt.*, 31 (6), 707–717.

Snow, J.A., Heikes, B.G., Merrill, J.T. *et al.* (2003) Winter-spring evolution and variability of HO_x reservoir species, hydrogen peroxide, and methyl hydroperoxide, in the northern middle to high latitudes, *J. Geophys. Res.*, 108 (D4), 8362, doi: 10.1029/2002JD002172.

Sonnenfroh, D.M., Rawlins, W.T., Allen, M.G. *et al.* (2001) Application of balanced detection to absorption measurements of trace gases with room-temperature, quasi-CW quantum-cascade lasers, *Appl. Opt.*, 40 (6), 812–820.

Tabazadeh, A., Yokelson, R.J., Singh, H.B. *et al.* (2004) Heterogeneous chemistry methanol in tropospheric clouds, *Geophys. Res. Lett.*, 31, L06114, doi: 10.1029/2003GRL018775.

Toon, G., Sen, B. & Blavier, J.F. (2002) Comparison of improved limb atmospheric spectrometer and MkIV profiles of atmospheric trace gases measured above Alaska in May 1997. *J. Geophys. Res.*, 107 (D24), 8211, doi:10.1029/2001JD000640.

Vay, S.A. *et al.* (1998) DC-8 observations of aircraft CO, CH_4, N_2O, and H_2O(g) emission studies during SUCCESS, *Geophys. Res. Lett.*, 25, 1717–1720.

Wagner, V., Schiller, C. & Fischer, H. (2001) Formaldehyde measurements in the marine boundary layer of the Indian Ocean during the 1999 INDOEX cruise of the R/V Ronald H. Brown, *J. Geophys. Res.*, 106 (D22), 28, 529–28, 538.

Webber, M.E., Pushkarsky, M. & Patel, C.K.N. (2003) Fiber-amplifier-enhanced photoacoustic spectroscopy with near-infrared tunable diode lasers, *Appl. Opt.*, 42, 2119–2126.

Webster, C.R., May, R.D., Toumi, R. & Pyle, J. (1990) Simultaneous in situ measurements of NO, NO_2, HNO_3, O_3, N_2O, and jNO_2, using the BLISS diode laser spectrometer, *J. Geophys. Res.*, 95, 13 851–13 866.

Webster, C.R., May, R.D., Allen, M. & Jaeglé, L. (1994a) Balloon profiles of stratospheric NO_2, and HNO_3 for testing the heterogeneous hydrolysis of N_2O_5 on sulfate aerosols, *Geophys. Res. Lett.*, 21, 53–56.

Webster, C.R., May, R.D., Trimble, C.A. *et al.* (1994b) Aircraft (ER-2) Laser Infrared Absorption Spectrometer (ALIAS) for in-situ stratospheric measurements of HCl, N_2O, CH_4, NO_2, and HNO_3, *Appl. Opt.* 33, 454–472.

Webster, C.R. & Heymsfield, A.J. (2003) Water isotopes ratios D/H, $^{18}O/^{16}O$, $^{17}O/^{16}O$ in and out of clouds map dehydration pathways, *Science*, 302, 1742–1745.

Werle, P., Slemr, F., Gehrtz, M. & Bräuchle, C. (1989) Wide-band noise characteristics of a lead-salt diode laser: Possibility of quantum limited TDLAS performance, *Appl. Opt.*, 28, 1638–1642.

Werle, P. & Slemr, F. (1991) Signal-to-noise ratio analysis in laser absorption spectrometers using optical multipass cells, *Appl. Opt.*, 30 (4), 430–434.

Werle, P., Mucke, R. & Slemr, F. (1993) The limits of signal averaging in atmospheric trace-gas monitoring by tunable diode-laser absorption spectroscopy (TDLAS), *Appl. Phys.* B, 57, 131–139.

Werle, P. (1995) Laser excess noise and interferometric effects in frequency-modulated diode-laser spectrometers, *Appl. Phys.* B, 60, 499–506.

Werle, P. & Jänker, B. (1996) High-frequency-modulation spectroscopy: Phase noise and refractive index fluctuations in optical multipass cells, *Opt. Eng.*, 35 (7), 2051–2057.

Werle, P. (1998) A review of recent advances in semiconductor laser-based gas monitors, *Spectrochim. Acta*, 54, 197–236.

Werle, P. & Kormann, R. (2001) Fast chemical sensor for eddy-correlation measurements of methane emissions from rice paddy fields, *Appl. Opt.*, 40 (6), 846–858.

Werle, P. (2004) Diode-laser sensors for in-situ gas analysis, in *Lasers in Environmental and Life Sciences-Modern Analytical Methods*, J.P. Hering, P. Lay & S. Stry (eds), Springer Verlag, Heidelberg.

Wert, B.P., Fried, A. & Henry, B. (2002) Evaluation of inlets used for the airborne measurement of formaldehyde, *J. Geophys. Res.*, 107 (D13), 10.1029/2001JD001072.

Wert, B.P., Fried, A., Rauenbuehler, S., Walega, J. & Henry, B. (2003a) Design and performance of a tunable diode laser absorption spectrometer for airborne formaldehyde measurements, *J. Geophys. Res.*, 108 (D12), 4350, doi: 10.1029/2002JD002872.

Wert., B.P., Trainer, M., Fried, A. *et al.* (2003b) Signatures of terminal alkene oxidation in airborne formaldehyde measurements during TexAQS 2000, *J. Geophys. Res.*, 108 (D3), 4104, doi: 10.1029/2002JD002502.

White, J.U. (1942) Long optical path of large aperture. *J. Opt. Soc. Am.*, 32, 285–288.

Wienhold, F.G., Zenker, T. & Harris, G.W. (1993) A dual channel two tone frequency modulation tunable diode laser spectrometer for ground-based and airborne trace gas measurements, in *Tunable Diode Laser Spectroscopy, Lidar, and DIAL Technique for Environmental and Industrial Measurements*, A. Fried, D.K. Killinger and H.I. Schiff (eds), pp. 31–44, Proceedings of SPIE, 2112.

Wienhold, F.G., Fischer, H., Hoor, P. *et al.* (1998) TRISTAR–A tracer in situ TDLAS for atmospheric research, *Appl. Phys.* B, 67, 411–417.

Worden, H., Beer, R. & Rinsland, C.P. (1997) Airborne infrared spectroscopy of 1994 western wildfires, *J. Geophys. Res.*, 102, 1287–1299.

Yokelson, R.J., Goode, J.G., Ward, D.E. *et al.* (1999) Emissions of formaldehyde, acetic acid, methanol, and other trace gases from biomass fires in North Carolina measured by airborne Fourier transform infrared spectroscopy, *J. Geophys. Res.*, 104 (D23), 30, 109–130, 125.

Yokelson, R.J., Bertschi, I.T., Christian, T.J. *et al.* (2003) Trace gas measurements in nascent, aged, and cloud-processed smoke from African savanna fires by airborne Fourier transform infrared spectroscopy (AFTIR), *J. Geophys. Res.*, 108 (D13), 8478, doi: 10.1029/2002JD002322.

Zahniser, M.S., Nelson, D.D., McManus, J.B. & Kebabian, P.L. (1995) Measurement of trace gas fluxes using tunable diode laser spectroscopy, *Phil. Trans. R. Soc. Lond.* A., 351, 371–382.

Zhu, X. & Cassidy, D.T. (1995) Electronic subtracter for trace-gas detection with InGaAsP diode lasers, *Appl. Opt.* 34 (36), 8303–8308.

Chapter 3
UV-Visible Differential Optical Absorption Spectroscopy (DOAS)

John M.C. Plane and Alfonso Saiz-Lopez

3.1 Introduction

Differential optical absorption spectroscopy (DOAS) is the most widely used spectroscopic technique for making measurements of atmospheric composition. It is the basis of nearly all remote sensing measurements of chemical species in the troposphere and stratosphere, both from the surface and airborne platforms and from satellites. It is also employed in techniques operating in situ, such as cavity ring-down spectroscopy (CRDS). Satellite observations of the atmosphere are an extensive subject, which are dealt with briefly in Chapter 1. In this chapter we will describe the fundamentals of DOAS (Section 3.2), and then explore in some detail the various DOAS techniques which are used within the atmosphere. Note that this subject was reviewed in the mid-1990s (Platt, 1994; Plane & Smith, 1995), so we will here focus on progress made during 1994–2005.

The light source required for a DOAS measurement in the atmosphere can be artificial, for instance a high-pressure Xe lamp or a laser. If the light traverses the atmosphere in a well-defined optical path of usually several kilometres, and the source is observed directly after perhaps a single reflection, then this operating mode is known as long-path DOAS (LP-DOAS) or active DOAS. The light can also be injected into a multi-pass cell (usually in the White or Herriott configuration) where multiple reflections are used to obtain a sufficiently long optical path to produce measurable absorption. An extension of this scheme is cavity ring-down spectroscopy. All of these techniques are discussed in Section 3.3. Note that we will not cover two other well-established applications that employ differential spectroscopy, tunable diode laser absorption spectroscopy (TDLAS, covered in Chapter 2) and differential absorption lidar (DIAL).

As an alternative to an artificial light source, an extra-terrestrial light source (sun, moon or stars) can be employed for making the so-called passive DOAS measurements. In the case of sunlight, either direct observations of the sun or indirect observations of sunlight that has been scattered by air molecules (Rayleigh scattering) or particles (Mie scattering), or reflected by the earth's surface, can be used. Passive DOAS measurements determine the column abundance of a species, often over very long paths. In Section 3.4 we will discuss examples of passive DOAS measurements from a variety of platforms, and methods for retrieving spatial information from column abundance measurements.

Before beginning this review of the DOAS technique, it is worth considering in general terms the desirable characteristics of an analytical technique for making atmospheric measurements. First, the technique must have sufficient sensitivity to work at the low ambient concentration levels at which many chemically interesting species occur in the atmosphere. In this chapter we will mostly describe measurements of radical species, which are usually highly reactive and thus occur at mixing ratios of the order of a few parts per trillion (1 ppt is equivalent to a concentration of 2.4×10^7 molecules cm^{-3} at a pressure of 1 atm and temperature of 293 K). Therefore, sub-ppt detection limits are often desirable. Second, the instrument needs to measure a particular species with high specificity, in order to avoid interferences from other compounds present in the atmosphere. Third, being able to make measurements with high time resolution is necessary when studying species that are not spatially homogeneous, usually because they have short chemical lifetimes in the atmosphere. Finally, deploying an instrument that is largely automated and simple to use is a considerable advantage when making measurements in remote locations and over long periods. As we will see in this chapter, many of these characteristics apply to DOAS instruments.

3.2 Principles of differential optical absorption spectroscopy

3.2.1 *The Beer–Lambert law*

In absorption spectroscopy the Beer–Lambert law describes the exponential decrease in the intensity of light passing through a sample containing an absorbing species i:

$$I_{tr}(\lambda) = I_0(\lambda) \exp\left(- \int_0^l \sigma_i(\lambda, P, T)\, c_i \, dl \right) \tag{3.1}$$

where $I_{tr}(\lambda)$ is the light intensity, at wavelength λ, that is transmitted through the sample; $I_0(\lambda)$ is the intensity of light incident on the sample; l is the path length of the light through the sample containing species i with a concentration c_i and an absorption cross-section $\sigma_i(\lambda, P, T)$ at pressure P and temperature T. Hereafter, $\sigma_i(\lambda, P, T)$ will be denoted as $\sigma_i(\lambda)$. Note that $\sigma_i(\lambda)$ is defined with respect to base e; although this is now the more widespread convention, some authors define $\sigma_i(\lambda)$ to base 10. The natural logarithm of $I_0(\lambda)/I_{tr}(\lambda)$ then defines the optical density (OD), which is the sum of the optical densities of n contributing absorbers at wavelength λ:

$$\mathrm{OD}(\lambda) = \ln\left(\frac{I_0(\lambda)}{I_{tr}(\lambda)}\right) = \sum_{i=1}^{n} \mathrm{OD}_i(\lambda) = \sum_{i=1}^{n} \int_0^l \sigma_i(\lambda) c_i \, dl \tag{3.2}$$

The *average* concentration of species i over the light path of length l is then given by:

$$\overline{c_i} = \frac{\mathrm{OD}_i(\lambda)}{\sigma_i(\lambda) l} \tag{3.3}$$

In addition to molecular absorption, light is also extinguished in the atmosphere by Rayleigh and Mie scattering (Platt, 1994; Plane & Smith, 1995). The Rayleigh extinction coefficient describes elastic scattering of photons by air molecules:

$$\varepsilon_{\mathrm{R}}(\lambda) = \sigma_{\mathrm{R}}\, \lambda^{-4} c_{\mathrm{air}} \tag{3.4}$$

where σ_{R} is $4.4 \times 10^{-16}\,\mathrm{cm^2\,nm^4}$ for air, and c_{air} is the molecular concentration of air (2.4×10^{19} molecule $\mathrm{cm^{-3}}$ at 293 K and 1 atm pressure). Thus, the light intensity at a wavelength of 300 nm is reduced by 12% per kilometre of light path, compared with only 1% at 600 nm. The Mie extinction coefficient describes elastic scattering processes by aerosols:

$$\varepsilon_{\mathrm{M}}(\lambda) = \varepsilon_{\mathrm{M}}\, \lambda^{-n} \tag{3.5}$$

where ε_{M} and n (which varies between 1 and 4) depend on the aerosol size distribution and chemical composition. In the evaluation of the transmitted intensity, Rayleigh and Mie scattering can be treated analogously to molecular absorption, so that Equation 3.1 becomes:

$$I_{\mathrm{tr}}(\lambda) = I_0(\lambda) \exp\left(-\int_0^l \left(\sum_{i=1}^{n} \sigma_i(\lambda)\, c_i + \varepsilon_{\mathrm{R}}(\lambda) + \varepsilon_{\mathrm{M}}(\lambda) \right) \mathrm{d}l \right) \tag{3.6a}$$

Averaging over the optical pathlength l yields:

$$I_{\mathrm{tr}}(\lambda) = I_0(\lambda) \exp\left(-l \left(\sum_{i=1}^{n} \sigma_i(\lambda)\, \bar{c}_i + \bar{\varepsilon}_{\mathrm{R}}(\lambda) + \bar{\varepsilon}_{\mathrm{M}}(\lambda) \right) \right) \tag{3.6b}$$

3.2.2 *Differential absorption spectroscopy*

In the atmosphere the majority of molecular species are present continuously, and Rayleigh and Mie scattering always occur, so the unattenuated atmospheric spectrum (I_0) cannot be obtained free of any extinction. Hence, the Beer–Lambert law (Equation 3.6b) cannot be applied directly. Instead, differential absorption can be used *if the absorption spectrum of the species of interest contains marked structure within a fairly narrow wavelength range* (typically 10 nm or less). This structure will appear in the atmospheric absorption spectrum, and can be used as a 'fingerprint' of the molecular species. The differential absorption cross-section of the molecule, $\sigma_i'(\lambda)$, is defined by considering the absolute cross-section $\sigma_i(\lambda)$ as the sum of $\sigma_i'(\lambda)$, which varies rapidly with wavelength, and a slowly varying component $\sigma_i^s(\lambda)$ (Platt, 1994; Plane & Smith, 1995):

$$\sigma_i(\lambda) = \sigma_i'(\lambda) + \sigma_i^s(\lambda) \tag{3.7}$$

The partitioning of the absolute absorption cross-section into its differential and broad-band components is depicted in Figure 3.1, for the case of NO_2 over a wavelength

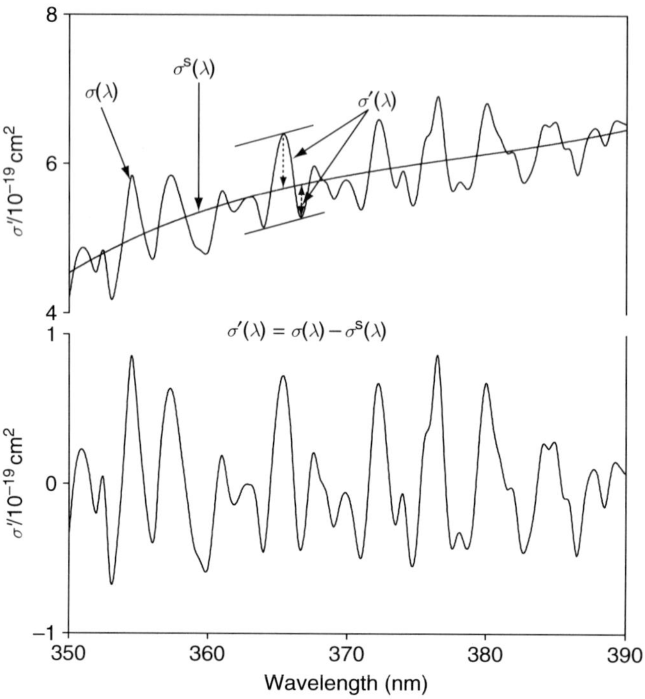

Figure 3.1 Absolute and differential absorption cross sections for NO_2 in the 350–390 nm region. In the top panel a 3rd degree polynomial (smooth line) is fitted to the absolute cross-section, $\sigma(\lambda)$, to account for the slowly varying trend of the spectrum $\sigma^s(\lambda)$. In the lower panel the differential absorption cross-section $\sigma'(\lambda)$, obtained after removing $\sigma^s(\lambda)$ is shown.

range commonly used by DOAS instruments. Note that the differential structure is much smaller than the absolute cross-section; nevertheless, there is sufficient structure for fitting to atmospheric absorption spectra. Substituting Equation 3.7 into 3.6b and including a factor $F_{\text{ins}}(\lambda)$ which incorporates instrument-related factors such as the wavelength dependence of the optical transmission in the spectrometer and the quantum efficiency of the detector elements yields:

$$I_{\text{tr}}(\lambda) = I_0(\lambda)\exp\left(-l\left(\sum_{i=1}^{n}\sigma_i'(\lambda)\,\overline{c}_i\right)\right)\exp\left(-l\left(\sum_{i=1}^{n}\sigma_i^s(\lambda)\,\overline{c}_i+\overline{\varepsilon}_{\text{R}}(\lambda)+\overline{\varepsilon}_{\text{R}}(\lambda)\right)\right)F_{ins}(\lambda) \tag{3.8}$$

The first exponential term in Equation 3.8 describes light extinction by narrow-band molecular absorption features, whereas the second exponential term and the instrument factor describe broad-band fluctuations. We can now define the measured intensity in the absence of differential absorption by $I_0'(\lambda)$:

$$I_{\text{tr}}(\lambda) = I_0'(\lambda)\exp\left(-l\left(\sum_{i=1}^{n}\sigma_i'(\lambda)\,\overline{c}_i\right)\right) \tag{3.9}$$

Hence, the differential optical density $\mathrm{OD}'(\lambda)$ is given by:

$$\mathrm{OD}'(\lambda) = \ln\left(\frac{I'_0(\lambda)}{I_{\mathrm{tr}}(\lambda)}\right) = l\sum_{i=1}^{n}\sigma'_i(\lambda)\bar{c}_i \tag{3.10}$$

By analogy with Equation 3.3, the concentration of a trace gas is now obtained from the differential absorption cross-section and differential optical density:

$$\bar{c}_i = \frac{OD'_i(\lambda)}{\sigma'_i(\lambda)l} \tag{3.11}$$

3.2.3 Wavelength regions and differential absorption cross-sections

In this section we discuss the criteria for selecting the wavelength region in which to observe a particular atmospheric species by DOAS. The first consideration is the wavelength dependence of light transmission through the atmosphere. At short wavelengths in the UV (<290 nm) the optical transmission of the atmosphere is significantly limited by the Hartley bands of O_3, molecular absorption by O_2 and Rayleigh scattering. These factors restrict DOAS measurements between about 220 and 290 nm to optical paths of less than about 1 km. In the infrared, optical transmission in the atmosphere is limited by the vibrational bands of molecules with strong absorption features such as H_2O, O_3 and CO_2. In addition, absorption coefficients in this spectral region tend to be much weaker when compared with those in the UV/visible (Okabe, 1978). Nevertheless, one advantage of working in the infrared is that individual ro-vibrational transitions can often be resolved because Lorentz-broadening is negligible at the range of pressures in the atmosphere (Okabe, 1978).

A second consideration when choosing a wavelength region is that the absorption spectrum of the molecule of interest contains significant absorption structures – for DOAS this is more important than the size of the absolute absorption cross-section. However, the absorption bands in the UV/visible region arise from the Lorentz-broadened ro-vibrational lines of electronic transitions (Okabe, 1978), so that the largest differential structures often occur at wavelengths where there is also significant absolute absorption. Figure 3.2 illustrates the differential absorption cross-sections in the UV/visible region for a number of important atmospheric species, over the wavelength intervals commonly used by DOAS instruments (Platt, 1994; Plane & Smith, 1995).

Another criterion to be considered is the spectral characteristics of the light source. For example, the NO_2 molecule exhibits a larger differential (and absolute) absorption cross-section in the 400–450 nm region compared with the 350–390 nm region. However, measuring NO_2 in the former spectral region is complicated by the pressure-broadened Xe lines in the spectra of high pressure Xe arc lamps usually employed in LP-DOAS instruments, so that the near-UV region illustrated in Figure 3.1 is usually preferred for measuring NO_2, especially for low concentrations in clean air (Platt, 1994; Plane & Smith, 1995).

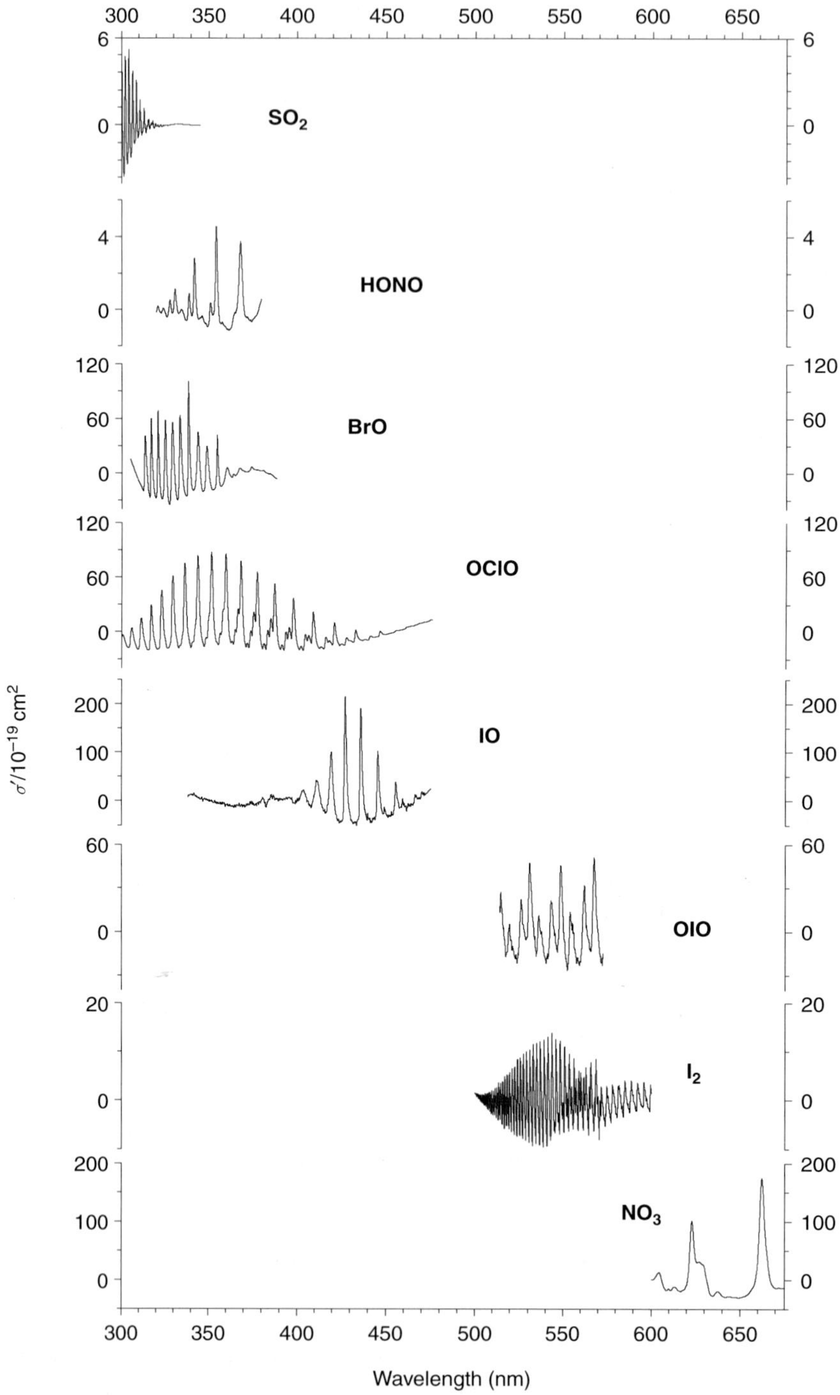

Figure 3.2 Examples of differential absorption cross-section spectra in the UV/visible region for a selection of important atmospheric species.

The molecular absorption cross-sections used as reference spectra when analysing atmospheric spectra are generally taken from the literature. Because $\sigma'(\lambda)$ is related directly to $\sigma(\lambda)$ (Equation 3.7), the concentrations derived from equation (3.11) are then traceable back to laboratory measurements of $\sigma(\lambda)$. This is particularly important in the case of absorption cross-sections for radical species (e.g. BrO, NO_3), whose absolute concentrations and hence cross-sections are often challenging to measure in the laboratory but where expert evaluations of several studies are available (Wahner *et al.*, 1988; Sander & Friedl, 1989; Yokelson *et al.*, 1994). The recommended literature cross-section, which will usually have been recorded at much higher resolution than required for atmospheric spectroscopy, must then be convolved with the instrument function F of the spectrometer employed in a particular DOAS instrument. That is, the resolution of the literature cross-section will be degraded to match that of the DOAS spectrometer. For instance, the DOAS reference cross-section, σ, can be expressed as:

$$\sigma(\lambda) = F\,(\sigma_{\mathrm{HR}}(\lambda)) \tag{3.12}$$

where σ_{HR} is the high resolution literature cross-section. The function F is typically a moving average of the high resolution spectrum around the central wavelength λ. The width and weighting of the average is usually obtained by measuring the wavelength profile of an atomic emission line (from a low pressure Hg/Ne lamp). Alternatively, reference spectra can be measured directly by placing appropriate gas cells into the light path of the DOAS instrument. This has the advantage that an instrument function is then not required for fitting to atmospheric spectra, and the wavelength calibration should exactly match atmospheric spectra taken with the same instrument. However, absolute calibration of the reference cross-section still requires de-convolving a high resolution spectrum from the literature.

A further consideration is that absorption cross-sections are quite often temperature dependent, and this needs to be accounted for when fitting reference spectra to atmospheric spectra. For example, the H_2O cross-section has been shown to vary linearly with temperature (Aliwell & Jones, 1996a):

$$\sigma_T(\lambda) = \sigma_{T_0}(\lambda) + (T - T_0)\frac{\partial\sigma(\lambda)}{\partial T} \tag{3.13}$$

where $\sigma_{T_0}(\lambda)$ refers to the cross-section at a reference temperature T_0. The differential in the second term of Equation 3.13 can then be fitted as a separate spectrum, with the factor $(T - T_0)$ either known or optimised in the fitting procedure.

3.2.4 Application to atmospheric measurements

There are a large number of species which play an important role in atmospheric chemistry and have absorption cross-sections in the near-UV/visible region suitable for detection by DOAS. Table 3.1 lists the molecules that have been observed by DOAS, including the wavelength regions employed and the detection limits achievable with a state-of-the-art instrument.

Table 3.1 Atmospheric species measured by DOAS in the UV/visible spectral range

Species	Wavelength interval (nm)	Detection limit[a] (ppt)
O_3	300–335	1900
NO	200–230	50[b]
NO_2	330–500	50
NO_3	623–662	0.4
HONO	330–380	30
NH_3	200–220	150[b]
SO_2	290–310	10
HCHO	260–360	50
CS_2	290–310	900
Benzene	230–280	200[b]
Toluene	260–280	250[b]
Naphthalene	310–320	100
Phenol	260–280	20[b]
p-Cresol	260–290	50[b]
OH	308	0.06[c]
ClO	260–320	5
OClO	300–450	0.8[d]
BrO	300–370	2
OBrO	450–550	1.5[d]
IO	415–450	1
OIO	535–575	4
I_2	535–575	9

[a] For a 5 km pathlength, assuming a minimum detectable optical density of 10^{-4}; ppt = parts per trillion (10^{-12}). [b] For a 1 km pathlength and an optical density of 10^{-3}, since absorption by O_3 and O_2 cause significant attenuation of light at wavelengths below 280 nm. [c] For a 2 km pathlength and a spectral resolution of 1.7×10^{-3} nm. [d] Stratospheric constituent, whose column abundance has been measured by ground-based, ballon-borne or satellite-borne DOAS instruments.

3.3 DOAS using artificial light sources

3.3.1 Broad-band light sources

This technique was developed at the end of the 1970s by Platt and Perner (Platt *et al.*, 1979, 1980a,b; Platt & Perner, 1980, 1983), and since then has been widely used for measuring many of the species in Table 3.1 (Hausmann & Platt, 1994; Platt, 1994; Brauers *et al.*, 1995; Plane & Smith, 1995; Smith *et al.*, 1995; Stutz & Platt, 1996; Tuckermann *et al.*, 1997; Alicke *et al.*, 1999; Hebestreit *et al.*, 1999; Allan *et al.*, 2001; Saiz-Lopez & Plane, 2004).

3.3.1.1 Experimental set-up

An LP-DOAS instrument comprises a transmitter, consisting of a high pressure Xe or incandescent lamp at the focal point of a parabolic or spherical mirror to provide a highly collimated light beam, and a receiver, consisting of a telescope coupled to a spectrometer, and a detector where the dispersed spectrum is recorded. In early DOAS instruments, the transmitter and the receiver were placed at opposite ends of the light path (Platt *et al.*, 1979), and a slotted-disk scanning device was employed as a detector in conjunction with a photomultiplier tube (Platt, 1994). Since the late 1980s, multiplexing detectors such as the photodiode array (PDA) and the charge coupled device (CCD) have replaced photomultiplier tubes. Another significant development in DOAS design has been to place the transmitter and the receiver in the same location, and fold the light path using a reflector (the total light path is then twice the distance between the DOAS and the reflector). In order to avoid the technical challenge of pointing a plane mirror with sufficient accuracy over an optical path of several kilometres, either retro-reflectors (which contain three mutually orthogonal plane mirrors) or corner-cubes have been used. Both types of reflectors utilise three internal reflections to return light back to its source with high accuracy (better than a few arc seconds is necessary if the reflector is more than 2 km from the source). These reflectors also have a large acceptance angle so that accurate pointing at the source is not necessary.

DOAS instruments using a folded light path have either placed the receiver alongside the transmitter (Plane & Nien, 1992; Hausmann & Platt, 1994) or combined them in a coaxial design in a single telescope (Axelsson *et al.*, 1990; Carslaw *et al.*, 1997a; Allan *et al.*, 1999, 2000a; Saiz-Lopez *et al.*, 2004). There are a number of important reasons for co-locating the transmitter and the receiver. First, only one location providing shelter and electrical power is needed, and the reflector can be placed in quite inaccessible terrain. Second, optimal processing of the atmospheric spectrum requires a spectrum of the lamp in the transmitter, unattenuated by transmission through the atmosphere (see p. 157). The lamp spectrum can be recorded on a routine basis by using an optical bypass to steer the light directly from the lamp to the spectrometer, if they are co-located. Third, by folding the light the optical path is doubled for the same path through the atmosphere. This reduces complications that might arise from chemically inhomogeneous air masses over very long distances. Fourth, there is a reduction in light losses due to divergence of the light beam because the reflected light converges back to its source. Also, light losses due to atmospheric scintillations occur on a much slower time scale than the time taken for the light to be transmitted and reflected. The reflected beam will therefore follow the same refracted path through the atmosphere and thus losses due to scintillations will be significantly less than for an unfolded path of double the length. Finally, the acceptance angle of a retro-reflector or corner-cube is typically about 20°, so that several DOAS instruments can use the same reflector without significant interference (Plane & Smith, 1995).

When designing a DOAS instrument there are a number of considerations, such as the type of lamp, the optimal length of the light path, the choice of spectral resolution and detector, and the desired degree of automation (Platt, 1994; Plane & Smith, 1995). The instrument illustrated in Figure 3.3 is a design that we have used in a variety of remote locations over the past decade (Carslaw *et al.*, 1997a,b; Allan *et al.*, 1999, 2000a,b, 2001; Saiz-Lopez & Plane, 2004; Saiz-Lopez *et al.*, 2004). It employs a coaxial design with a 32 cm diameter *f*/6 Newtonian telescope that houses both the transmitting and the receiving optics. The broadband light beam is powered by a 450 W Xe lamp, housed on the side of the telescope at the focal point of the spherical primary mirror. The

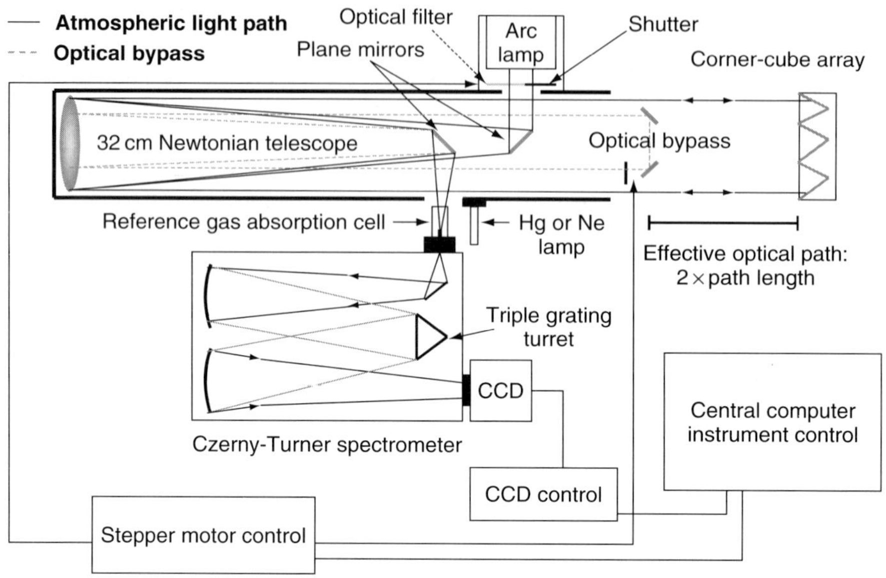

Figure 3.3 Schematic diagram of a coaxial LP-DOAS instrument where a single telescope houses both the transmitting and receiving optics, and the light beam is folded back by an array of corner cube reflectors.

outgoing beam is folded back to the telescope by an array of solid quartz corner-cubes (accuracy < 5 arc seconds). The received light is then focused onto a quartz optic fibre bundle. A mode mixer can be used to obtain a more uniform intensity distribution at the fibre exit into the spectrometer, independent of the light distribution at the entrance to the fibre bundle (Stutz & Platt, 1997). The light is then dispersed through a 0.5 m Czerny-Turner spectrometer, using either a 1200 or 600 groove mm^{-1} grating which produces spectral widths of about 35 or 70 nm, respectively, on a typical PDA or CCD detector. The resolution is then about 0.2–0.4 nm. CCD cameras have increasingly been employed as detectors in DOAS instruments. One reason is that CCD chips possess high quantum efficiencies (> 20%) in the near-UV, where a number of relevant atmospheric species have suitable absorption features (Table 3.1). High detection efficiency in the UV is important for making measurements with good signal-to-noise ratio, because the intensity of the transmitted light is substantially reduced by Rayleigh, Mie and molecular extinction, and the arc lamp intensity falls at wavelengths below 400 nm. CCDs tend to have large dark currents, so it is important to cool the devices (modern cameras are capable of reaching and maintaining temperatures down to −70°C, by combining a multi-stage Peltier thermoelectric cooler in an evacuated chamber). A final consideration is that étaloning can be created by interference on the surface-protective coating of some chips. We have found that front-illuminated CCDs largely eliminate this problem in contrast to back-illuminated chips.

The instrument shown in Figure 3.3 also contains a number of automated optical components. A stepper motor in the lamp housing allows the light to be shuttered off in order to record scattered light spectra from the atmosphere. An optical cut-on filter ($\lambda > 600$ nm) can also be inserted into the light beam when measuring the nitrate radical

(NO_3) at night (this prevents photolysis of the radical at shorter wavelengths in the light beam). A second stepper motor controls an optical bypass for recording unattenuated lamp spectra. Motorised translational stages are used to align the quartz optic fibre bundle with the image of the light beam received in the telescope. Finally, a motorised filter wheel inserts combinations of neutral density and cut-on filters into the light beam at the entrance of the spectrometer. These filters are used to attenuate the light entering the spectrometer to avoid saturating the CCD, and also to eliminate short wavelength light being diffracted in second order onto the detector.

3.3.1.2 Spectral de-convolution

Although different approaches have been used to analyse atmospheric spectra (Platt, 1994; Plane & Smith, 1995; Camy-Peyret *et al.*, 1996; Aliwell *et al.*, 2002), they are conceptually very similar. Analysis procedures mainly use either polynomials for the generation of the differential optical spectra and reference cross-sections or Fourier transforms with combinations of low- and high-pass filters. Here we will describe the spectral de-convolution procedure originally developed by Plane and Nien (1992).

Figure 3.4 illustrates the de-convolution of an atmospheric spectrum in the 637–677 nm wavelength region, in order to measure NO_3. Panel (a) shows a raw air spectrum (solid line) which is dominated by the ro-vibrational structure of an overtone band of H_2O (Rothman *et al.*, 2003). The raw spectrum is the sum of the transmitted intensity of the DOAS light beam and scattered light (usually only significant during the day). In the first stage of the de-convolution procedure, a processed atmospheric spectrum is generated by removing the scattered light contribution and any spectral features of the lamp source in the transmitter. This is achieved by collecting a set of three different spectra: an atmospheric + scattered light spectrum (A); a scattered light spectrum (S) obtained by shuttering off the Xe lamp; and a lamp + atmospheric + scattered light spectrum (L) obtained by opening an optical bypass which directs some of the Xe light directly into the receiving telescope. In order to reduce errors caused by fluctuations in the lamp and the scattered light intensity, and by changing atmospheric concentrations, the three spectra are collected in turn, with each spectrum being accumulated for 20 s over a period of typically 10 min. The processed spectrum is then given by:

$$I_{pr} = \frac{A - S}{L - A} \tag{3.14}$$

It is, therefore, the spectrum transmitted through the atmosphere, normalised by dividing by the unattenuated lamp spectrum. In principle, this procedure removes the Xe lamp spectral features contained in the transmitted spectrum and the pixel-to-pixel variability in light response across the surface of the CCD. Also, the three spectra contain the same dark current and electronic offset from the analog-to-digital converter in the detector, so these cancel by subtraction in the numerator and denominator of Equation 3.14.

The next step in the de-convolution procedure is to convert I_{pr} into a differential optical density (OD′) spectrum. This is achieved by Fourier transforming the spectrum, applying a low-pass filter, and then performing an inverse transform to obtain the broad-trend spectrum, I_0' (see Section 3.2.2). In panel (a) of Figure 3.4 the processed spectrum

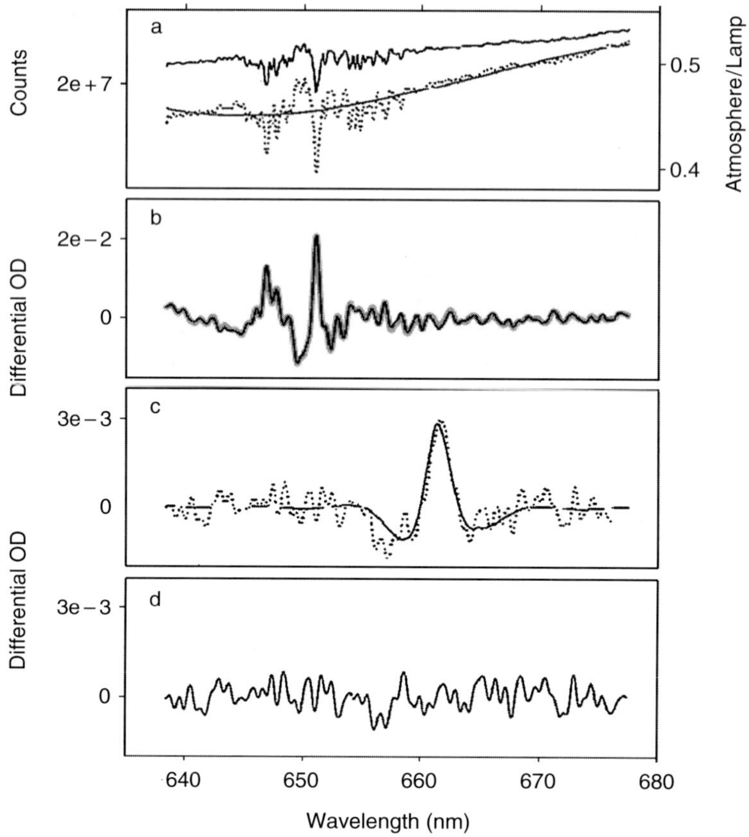

Figure 3.4 An atmospheric measurement of NO_3: (a) raw atmospheric spectrum (solid line, left-hand axis), and the processed atmospheric spectrum (grey line, right-hand axis) with a broad trend line obtained from a Fourier transform with a low-pass filter; (b) observed atmospheric optical density (black line) and fitted reference spectra (grey thick line); (c) atmospheric differential OD spectrum (grey line) after H_2O absorption has been fitted out but NO_3 contribution remains – the solid line is the fitted NO_3 reference differential cross-section once it has been degraded to the spectrometer resolution, corresponding to a mixing ratio of 11 ppt; (d) residual spectrum obtained after the known absorbers have been removed, indicating a minimum detectable mixing ratio for NO_3 of 0.4 ppt.

(grey line) is plotted together with I_0'. The OD′ spectrum is then given by the sum of the n individual absorber contributions at each wavelength (Equation 3.10), to which can be added a first-order polynomial that is optimised in the fitting procedure to take account of any residual off-sets (ideally, α and β should be zero):

$$\mathrm{OD}'(\lambda) = \ln\left(\frac{I_0'(\lambda)}{I_{\mathrm{pr}}(\lambda)}\right) = l\sum_{i=1}^{n}\sigma_i'(\lambda)\bar{c}_i + \alpha\lambda + \beta \tag{3.15}$$

Note that the reference cross-sections $\sigma_i'(\lambda)$ should be derived from absolute cross-sections with the same Fourier transform/low-pass filter routine that is applied to the processed atmospheric spectrum. Since $\mathrm{OD}'(\lambda)$ is measured over a large number of wavelengths in a multiplexing detector (typically 1024 with a CCD or PDA), but the unknowns in Equation 3.15 are the n molecular concentrations $\overline{c}_i$ and the parameters α and β (usually less than 10 unknowns in total), the solution of the Equation 3.15 is greatly over-determined. An optimal solution can be obtained from a least squares fitting routine that employs singular value decomposition (Press *et al.*, 1986). The analysis software should also permit shifting and stretching of the reference spectra with respect to wavelength, in order to correct for any differences in wavelength calibration between the DOAS spectrometer and the literature cross-sections. Note that this is less necessary if spectral calibrations are performed in the field using cells filled with reference gases.

The main absorbers in the 637–677 nm range are H_2O and NO_3. Once the total OD spectrum is obtained, both species need to be fitted out in the de-convolution procedure. Fitting a reference H_2O spectrum is complicated because some of the rotational lines in the vibrational overtone band around 651.5 nm are usually saturated at the column densities of H_2O found over several kilometre optical path in the lower troposphere. Hence, the absorption bands do not follow the Beer–Lambert law (Aliwell & Jones, 1996a). One solution to this problem is to take the high resolution H_2O reference spectrum from the HITRAN database (Rothman *et al.*, 2003), and compute a library of H_2O reference spectra, degraded to the spectrometer resolution, over a range of expected H_2O column densities and atmospheric temperatures. The fitting routine then finds the H_2O spectrum that best fits the observed atmospheric spectrum. Panel (b) of Figure 3.4 shows the atmospheric OD′ spectrum together with the overall fitted reference in the region used for NO_3 analysis. The reference spectrum for NO_3 is calculated from the reported cross-section (Yokelson *et al.*, 1994) of the $B^2E' - X^2A_2'$ absorption band at 662 nm. In panel (c) of Figure 3.4, the atmospheric OD′ spectrum is shown with the fitted H_2O reference spectrum subtracted out (grey line), together with the fitted NO_3 reference. Finally, the best fit is achieved when all known spectral features are removed, leaving the structureless residual spectrum shown in panel (d) of Figure 3.4(d).

In the particular case of NO_3, use can be made of the fact that the lifetime of the radical with respect to photodissociation is just a few seconds during the day (DeMore *et al.*, 1997), so a daytime spectrum will not contain any significant NO_3 absorption and can be used as a reference (Platt, 1991; Allan *et al.*, 2000b). The resulting division is analysed in the same manner as the processed spectrum, described above. Figure 3.5 is an example of the nocturnal time-profile of NO_3 measured by an LP-DOAS at Mace Head, Ireland, shown together with its rate of production (thick dotted line) calculated from the measured NO_2 and O_3 concentrations. Note the appearance of the radical after sunset, and the rapid disappearance due to photolysis, at sunrise.

3.3.1.3 Estimating errors and detection limits

The uncertainties in the retrieved column densities, and the detection limits of a DOAS instrument, are determined by a number of factors including the size and structure of the differential absorption cross-sections, the number of absorbers in the spectral region, the optical path length, and the intensity of the light signal transmitted through the atmosphere. Contributions to the systematic uncertainty include errors in the absolute absorption cross-sections of the reference spectra, and error in the length of the optical path, although with modern global positioning systems (GPS) this should not be significant.

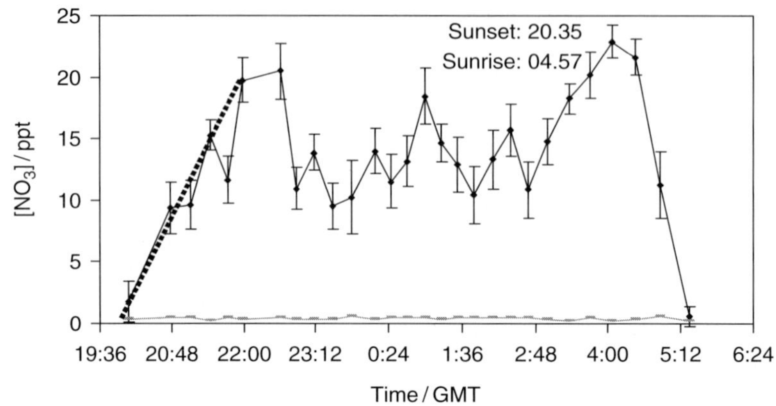

Figure 3.5 Example of NO_3 nocturnal behaviour as measured by LP-DOAS over a pathlength of 8.4 km at the remote coastal location of Mace Head, Ireland in 2002. The detection limit of the instrument is plotted as grey broken lines. The rate of production of NO_3 calculated from in situ measurements of NO_2 and O_3 is plotted after sunset (thick dotted line).

A measure of the goodness of fit in the de-convolution analysis is provided by a chi-square (χ^2) significance analysis:

$$\chi^2 = \frac{\sum_{\lambda_1}^{\lambda_2} \left(\mathrm{OD}'(\lambda) - l \sum_{i=1}^{n} \sigma_i'(\lambda)\bar{c}_i + \alpha\lambda + \beta) \right)^2}{m-1} \tag{3.16}$$

where m is the number of wavelength intervals between λ_1 and λ_2 (elements of the detector), and the other terms are defined in Equation 3.15. The smaller the χ^2 is, the better the fit of the reference spectra to the atmospheric spectrum, since less structure remains in the residual spectrum. The statistical standard errors of the retrieved concentrations are calculated directly from the singular value decomposition routine (Press *et al.*, 1986), and assume that the structures remaining in the residual are random. These errors are weighted by the χ^2 distribution and the $m - n$ degrees of freedom of the system – where n is the number of fitted parameters or unknowns, and m is the number of linear equations (i.e. number of detector elements). Hence, the least-squares fitting process determines a 1σ statistical error that normally underestimates the overall error, since the systematic errors discussed above also need to be included.

Often, consistent structures appear in a sequence of residual spectra. On rare occasions this might indicate the presence of molecular absorption features from a new species not included in the de-convolution routine (Equation 3.15). However, there are several more likely causes. First is incorrect matching of the spectral features between the reference spectra, often taken from the literature where they will have been recorded under different resolutions and conditions (pressure and temperature), and the atmospheric spectrum. This problem is more common when working in spectral regions where several reference spectra are simultaneously fitted in the de-convolution. Another cause of structures in the residual spectrum is incomplete removal of spectral features from the light source in the DOAS transmitter. The source is usually a Xe arc lamp, and the problem is that when using an optical bypass arrangement to obtain the lamp spectrum (e.g. see Figure 3.3), the image of the arc is larger than the entrance to an optic fibre or the spectrometer slits. There are large temperature gradients and hence variations in spectral output across a Xe arc, so the spectrum of the portion of the arc sampled by the bypass can be significantly different from the atmospheric spectrum, which is dominated by output from the hottest (and brightest) part of the arc. This will then lead to lamp structure remaining in the processed spectrum (Equation 3.14), and hence in the residual spectrum (Allan *et al.*, 2000a). If consistent structures are present in a sequence of residual spectra, these can be averaged to yield an 'instrumental reference' spectrum, which can then be fitted to the processed atmospheric spectrum together with the molecular reference spectra (Allan *et al.*, 2000a). This procedure often leads to a substantial reduction in the error and detection limits of trace species, without any significant effect on the retrieved column densities.

The minimum detectable OD can be estimated from the root mean square of the residual (Plane & Smith, 1995):

$$\mathrm{OD}'_{\min} = \sqrt{\frac{\chi^2}{m-1}} \tag{3.17}$$

The detection limit of an individual species can then be estimated from Equation 3.11. Note that the detection limit will depend on the atmospheric conditions affecting light transmission and scattered sunlight (i.e. precipitation, clouds and time of day). This is because χ^2 tends to vary inversely with the light intensity transmitted through the atmosphere to the detector, and χ^2 is roughly proportional to the level of scattered light. In principle, under conditions of good atmospheric visibility a minimum detectable OD of $\sim 1 \times 10^{-4}$ can be achieved (Platt, 1994; Plane & Smith, 1995).

3.3.1.4 Application to atmospheric measurements

The DOAS technique has been in use since the late 1970s, and has made a significant contribution to the understanding of atmospheric chemistry (Platt, 1994; Plane & Smith, 1995). Originally, Noxon and co-workers (Noxon, 1975; Noxon *et al.*, 1978, 1980) used an LP-DOAS instrument to measure NO_2 and NO_3 with the aim of complementing their stratospheric measurements (Section 3.4.1). However, it was Platt, Perner and co-workers (Perner & Platt, 1979; Platt *et al.*, 1979; Platt & Perner, 1980; Platt *et al.*, 1980a,b; Platt & Perner, 1983) who made many of the significant early advances. They first reported simultaneous observations of HCHO, O_3 and NO_2 in the boundary layer (Platt *et al.*, 1979), followed by measurements of HNO_2 and SO_2 (Perner & Platt, 1979; Platt & Perner, 1980). They also made the first direct measurement of NO_3 in the troposphere (Platt *et al.*, 1980b).

Since the last major reviews of the DOAS technique in the mid-1990s (Platt, 1994; Plane & Smith, 1995), a major intercomparison of different LP-DOAS designs and software retrieval algorithms has been published (Camy-Peyret *et al.*, 1996). In terms of atmospheric measurements, the most striking advances with DOAS have been pioneering observations of key halogen species in the lower troposphere. First, measurements of BrO were reported in the Arctic boundary layer during springtime (Hausmann & Platt, 1994);

the high levels observed (>20 ppt) were shown to be the cause of O_3 depletion events. In addition, BrO measurements have also been performed at mid-latitude locations such as the Dead Sea (Hebestreit *et al.*, 1999), a salt lake (Stutz *et al.*, 2002), and the mid-latitude coastal marine boundary layer (Saiz-Lopez *et al.*, 2004). ClO has also been detected by DOAS in a salt lake mid-latitude boundary layer (Stutz *et al.*, 2002).

A number of iodine species have been detected in the marine boundary layer by LP-DOAS, including IO (Alicke *et al.*, 1999; Allan *et al.*, 2000a), OIO (Allan *et al.*, 2001) and I_2 (Saiz-Lopez & Plane, 2004). Figure 3.6(a) shows a spectral fit confirming the presence of I_2: the grey line is the atmospheric OD spectrum after subtracting several absorbing species (e.g. H_2O, NO_2 and OIO, but not I_2), compared to a reference spectrum of I_2 (black line) (Saiz-Lopez & Plane, 2004). The molecule was detected at nighttime concentrations of up to 90 ppt together with OIO levels of 10 ppt (averaged over the 4.2 km pathlength, illustrated in Figure 3.6b). Figure 3.7 shows measurements of the IO radical off the north coast of Tenerife, Spain, which is probably more typical of open ocean conditions. The measurements (solid points) are compared with model predictions of IO from a constrained model that includes a full treatment of the processing of iodine-containing species in sea-salt aerosol (solid lines) (McFiggans *et al.*, 2000).

Finally, commercial DOAS systems have been also routinely used in air pollution–monitoring networks (e.g. Kourtidis *et al.*, 2000; Xie *et al.*, 2004). The commercial instruments usually consist of a transmitter and a receiver telescope attached to a computer-controlled spectrometer (Section 3.3.1.1). Because of the higher concentrations involved when studying urban pollution, optical pathlengths of typically a few hundred

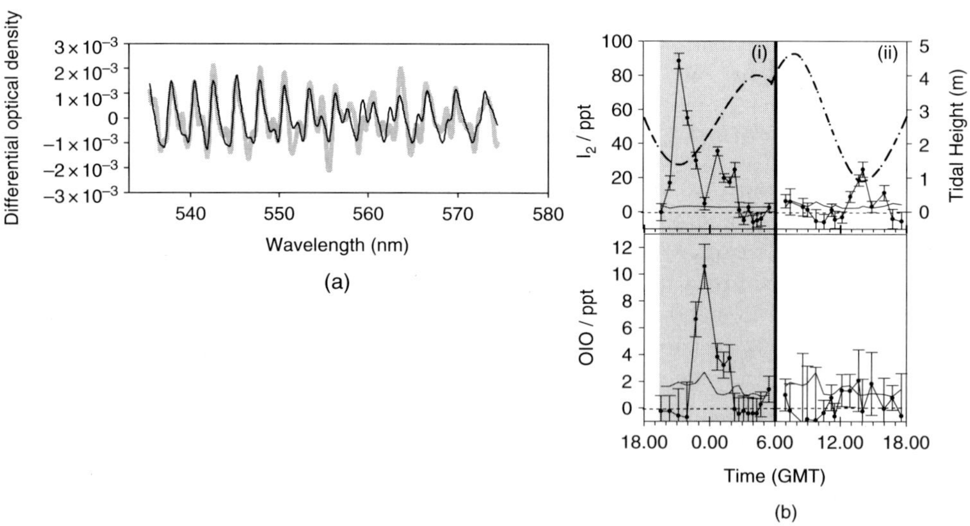

Figure 3.6 Spectral fit and concentration profiles of I_2. (a) The atmospheric OD spectrum (grey) is matched to an OD reference spectrum for I_2 (black) corresponding to a retrieved concentration of 88 ± 4.1 ppt. (b) The I_2 and OIO concentration profiles with the corresponding detection limits (thin black lines) are plotted against tidal height (thick broken lines). The white and grey backgrounds indicate day and nighttime periods, respectively. (Saiz-Lopez & Plane, 2004, reproduced with permission from the American Geophysical Union.)

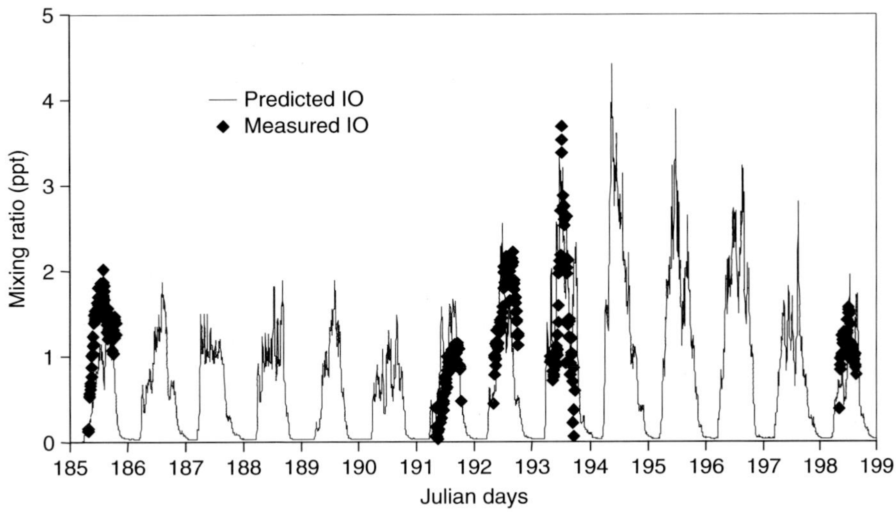

Figure 3.7 Time series measurements of IO made by LP-DOAS off Tenerife, Spain. The measured concentrations (solid points) are compared to model predictions of IO (solid lines) for a number of days during the field campaign. (McFiggans *et al.*, 2000, reproduced with permission from the American Geophysical Union.)

metres are used. Generally, DOAS monitoring networks emphasise measurements of urban pollutants such as SO_2, NO, NO_2, O_3 and aromatics.

3.3.2 Laser sources

3.3.2.1 OH radical measurements in the atmosphere

Hydroxyl radical has been detected in the atmosphere by optical absorption in the $A^2\Sigma^+ - X^2\Pi$ electronic transition around 308 nm (Perner *et al.*, 1987; Dorn *et al.*, 1988; Comes *et al.*, 1992; Mount, 1992). Because the rotational constant of OH is very large, rotational lines in the (0,0) vibrational band can be resolved even at atmospheric pressure. Hence, a laser can be used as the light source in a DOAS instrument, such as the example shown in Figure 3.8 (Hausmann *et al.*, 1997). A picosecond laser at 616 nm is frequency-doubled to 308 nm, and spectra are recorded with a high resolution Echelle spectrograph coupled to a PDA detector. This set-up provides simultaneous monitoring of a spectral width of 0.26 nm, which covers a number of OH absorption lines (Figure 3.9). An open multi-reflection or White cell (38.5 m long) is used to achieve an absorption path of more than 3 km. This is necessary because OH has a typical daytime peak concentration of only about 5×10^6 molecules cm^{-3}. The OD_{min} of the instrument is about 2×10^{-4}, so the detection limit of OH is about 1.5×10^6 molecules cm^{-3} (Hausmann *et al.*, 1997). There have been other laser systems used for OH detection, for example a pulsed XeCl excimer laser covering several OH lines (Mount, 1992) and a continuous wave narrow spectral linewidth laser which scans very rapidly over several OH rotational lines (Armerding *et al.*, 1996).

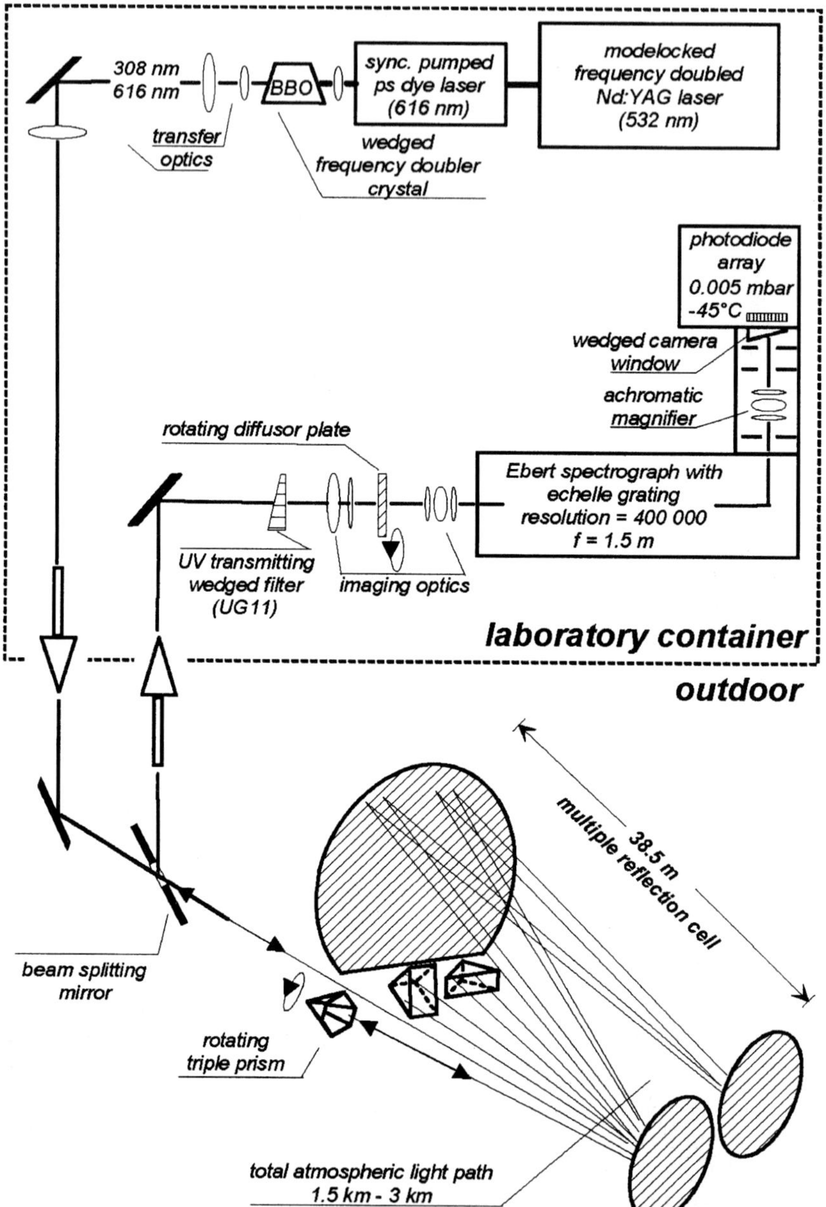

Figure 3.8 Schematic diagram of a long-path DOAS instrument employing a laser light source for tropospheric OH detection. The instrument consists of a high resolution Ebert spectrometer-photodiode array (PDA) detector system in combination with an open path multiple-reflection cell. (Hausmann *et al.*, 1997, reproduced with permission from the American Geophysical Union.)

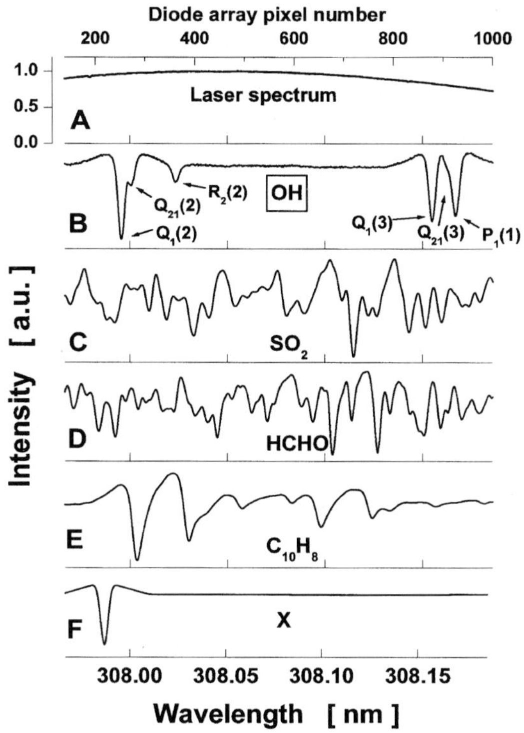

Figure 3.9 Wavelength interval used for OH tropospheric observations. The laser output spectrum, the absorption spectrum around 308 nm of the A-X(0,0) band of OH (with individual rotational lines identified), and the absorption spectra of other trace gases that potentially interfere in the LP-DOAS detection of OH are plotted in panels (A) to (F), respectively. Note that the absorption band depicted in (F) corresponds to an unidentified atmospheric component. (Dorn *et al.*, 1996, reproduced with permission from the American Geophysical Union.)

Note that in contrast to traditional LP-DOAS, where the retrieved concentration is integrated over several kilometres, the multi-pass arrangement provides a measurement integrated over a few tens of meters, which can be considered as in situ monitoring. Intercomparison of OH measurements by LP-DOAS and in situ laser-induced fluorescence (LIF) systems has shown very satisfactory agreement (Hofzumahaus *et al.*, 1998).

The spectra are analysed in a similar way to the broadband DOAS technique (Section 3.3.1.2), as illustrated in Figure 3.9. This shows the laser output spectrum (Panel (A)), and reference spectra for OH, SO_2, HCHO and naphthalene (Panels (B)–(E), respectively). Panel (F) records the absorption band of an unknown absorber. Very precise calibrations of the reference spectra are needed because of the very narrow wavelength interval used (307.96–308.19 nm) (Dorn *et al.*, 1996; Mount *et al.*, 1997; Brauers *et al.*, 2001). Figure 3.10(A) shows an atmospheric absorption spectrum at 308 nm, and the corresponding superimposed fitted reference spectra together with a ninth-order polynomial. The high-order polynomial is applied to correct the laser spectral baseline and the broadband component of the atmospheric spectrum. Figure 3.10(B) depicts the atmospheric spectrum after subtracting all the fitted reference spectra except for OH,

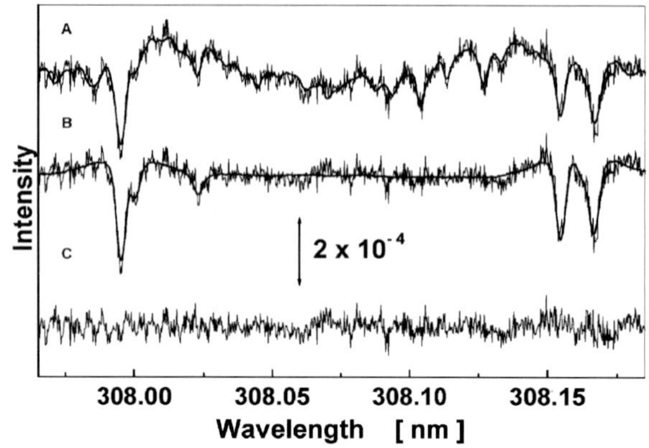

Figure 3.10 Spectral de-convolution of OH: (A) a fit (dark line) to an atmospheric absorption spectrum (grey line); (B) atmospheric spectrum with all species but OH fitted out (grey line), and the fitted OH reference spectrum (dark line); (C) the residual obtained after all species are fitted and subtracted from the atmospheric spectrum. (Dorn *et al.*, 1996, reprinted with permission from the American Geophysical Union.)

and Figure 3.10(C) shows the residual after subtracting all reference spectra. The OH absorption corresponds to an optical density of 2×10^{-4}, demonstrating the ability of the DOAS system to provide in situ measurements of OH radicals.

3.3.2.2 Cavity ring-down spectroscopy

Cavity ring-down spectroscopy (CRDS) was first developed in the late 1980s (O'Keefe & Deacon, 1988), and since then its use for laboratory and atmospheric spectroscopy has grown substantially (Paul & Saykally, 1997; Scherer *et al.*, 1997; Ball & Jones, 2003; Brown, 2003). The principle of the method is based on the gradual decrease of light intensity as a light pulse undergoes multiple reflections between two highly reflective mirrors in an optical cell. Figure 3.11 is an illustration of a typical CRDS set-up (Brown, 2003). The laser pulse is introduced into the CRD cavity through one of the mirrors. The intensity of light transmitted out of the cavity by leaking through the mirrors will decay exponentially with time (Paul & Saykally, 1997; Scherer *et al.*, 1997). This intensity is usually measured by placing a photomultiplier, with an appropriate optical filter, on the far side of the optical cavity from the laser. The cavity length L is the distance between the mirrors. The transit time between the mirrors should be large compared to the laser pulsewidth in order to discriminate discrete pulses or cavity modes inside the cell, thus ensuring the ring-down decay is exponential. If an absorbing species is now introduced into the cavity, then the decay in light intensity will be caused both by reflection losses and by molecular absorption, and as a result decay will be faster. The change in time decay can be used to measure the OD of the absorbing species (Ball & Jones, 2003; Brown, 2003).

The time decay depends ultimately on the mirror transmission $T_R (= 1 - R$, where R is the reflectivity of the mirrors (see Equation 3.19)). The time for the beam to traverse the cavity is given by L/c, where c is the speed of light. For a known absorption cross-section

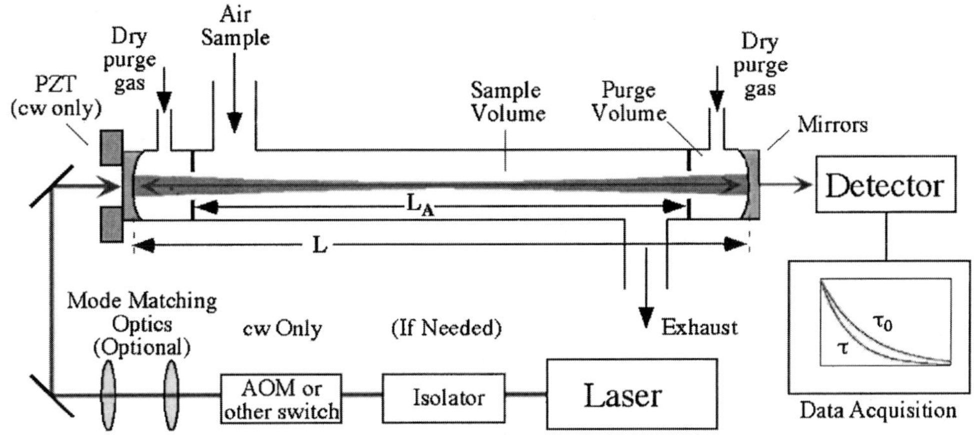

Figure 3.11 Schematic diagram of a cavity ringdown spectrometer. (Brown, 2003, reproduced with permission from American Chemical Society.)

(σ_i), the absorption coefficient of species i (A_i, in cm^{-1}) of the gas injected in the cavity cell can then be written as:

$$A_i = \frac{R_L}{c}\left(\frac{1}{\tau} - \frac{1}{\tau_0}\right) = \sigma_i c_i \quad (3.18)$$

where $1/\tau$ and $1/\tau_0$ are the first-order decay rates for the empty CRD cell and in the presence of the absorbing gas of concentration c_i, respectively. The factor R_L is the ratio of L to the length in the cell over which the absorbing species is present. For atmospheric applications, extinction due to Rayleigh and Mie scattering, and absorption by other gas-phase species, must be considered together with the losses by mirror transmission in the definition of τ_0 in the absence of the target gas (Brown, 2003):

$$\frac{1}{\tau_0} = c\left(\frac{T_R}{L} + \varepsilon_R + \frac{\varepsilon_M}{R_L} + \frac{\sum_i A_i}{R_L}\right) \quad (3.19)$$

where ε_R and ε_M are the wavelength-dependent extinction coefficients for Rayleigh and Mie scattering, and A_i are the absorption coefficients for the gas-phase species other than the target compound. Hence, monitoring the CRD time decay with the empty cavity and in the presence of the sample gas allows the measurement of the gas concentration following the Beer–Lambert law.

Cavity ring-down spectroscopy has been used increasingly since 1990s for kinetic and spectroscopic studies of a number of species of interest in atmospheric chemistry (Zhu & Johnston, 1995; Atkinson & Hudgens, 1997; Newman *et al.*, 1998, 1999; Schulz & Simpson, 1998; Wheeler *et al.*, 1998; Ashworth *et al.*, 2002; Mazurenka *et al.*, 2003; Nakano *et al.*, 2003). This proliferation of laboratory applications of CRDS has been due to the high sensitivity against minimum detectable fractional absorbance per pass ($\sim 10^{-6}$), and a good time resolution (10^{-6}–10^{-4} s) (Ball & Jones, 2003; Brown, 2003). In addition,

CRDS instruments have been used for field measurements of atmospheric species such as NO_3 and N_2O_5 (King *et al.*, 2000; Ball *et al.*, 2001; Brown *et al.*, 2001; Simpson, 2003). These nitrogen compounds are in thermal equilibrium ($NO_3 + NO_2 + M \leftrightarrow N_2O_5 + M$). In order to achieve simultaneous measurements of both species, the CRDS instrument contains two flow systems and ring-down cavities. The first flow system is kept at ambient temperature for the detection of ambient NO_3, whereas the second is heated up to ~80°C to thermally decompose N_2O_5, therefore yielding the sum of the NO_3 and N_2O_5 concentrations. An accurate measurement with CRDS requires a precise determination of the empty cavity ring-down time (τ_0). For instance, to measure the NO_3 radical τ_0 includes other absorbers such as water vapour, NO_2, O_3 and the aerosol contribution. A measurement of τ_0 can therefore be obtained by titrating with NO ($NO_3 + NO \rightarrow 2NO_2$), without affecting the concentrations of the other constituents (Brown *et al.*, 2003).

An extension of CRDS is to use a broadband laser source (typically a dye laser with a 20 nm bandwidth), and replace the photomultiplier with a spectrometer and a two-dimensional CCD with fast time-clocking to record spectra with a better than 1 μs time resolution (Ball *et al.*, 2001). Cavity ring-down absorption spectra are therefore obtained simultaneously over a range of wavelengths, so in principle this technique combines the advantages of CRDS with conventional DOAS. The technique has so far been employed to detect the NO_3 radical by monitoring the spectral region between 650 and 670 nm (Ball *et al.*, 2001).

Finally, it should be noted that CRDS and LP-DOAS are complementary techniques because they probe different spatial scales. The latter averages the trace gas concentration over typically several kilometres and thus provides some degree of remote sensing, whereas CRDS is an in situ measurement. The techniques have similar detection limits. Thus, combining the two techniques at a field site where there is likely to be significant heterogeneity because of local sources and sinks (e.g. in an urban environment or the marine coastal zone) provides a very useful way of studying fast chemistry and transport.

Another new locked cavity method which offers absorbance detection limits as low as 10^{-13} has been developed. The noise-immune cavity-enhanced Optical Heterodyne Modulation Spectrometry (NICE-OHMS) combines a high reflectivity cavity with significant noise reduction by modulation of the laser frequency (Ye *et al.*, 1998). Although, to date, the technique has not been applied to atmospheric measurements, its potential capabilities have been demonstrated by detecting absorbances of $\sim 3 \times 10^{-11}$ from C_2H_2, by using an ultra-stabilised Nd:YAG laser (Ma *et al.*, 1999).

3.4 DOAS using scattered sunlight

The earliest spectroscopic remote sensing measurement was the observation of O_3 in the atmosphere by Dobson and Harrison in 1926. In the mid-1970s, Noxon measured NO_2 in the upper troposphere and stratosphere by observing scattered sunlight at the zenith (Noxon, 1975, 1976). This pioneering work was followed by a wide variety of studies of many important atmospheric species (Noxon, 1981, 1983; McKenzie & Johnston, 1982; Solomon *et al.*, 1987a,b; Mount *et al.*, 1988; Pommereau & Goutail, 1988; Roscoe *et al.*, 1990; Smith & Solomon, 1990; Wahner *et al.*, 1990a,b; Perner *et al.*, 1991, 1994; Fiedler *et al.*, 1993; Sanders *et al.*, 1993; Arpag *et al.*, 1994; Pfeilsticker & Platt, 1994;

Pommereau & Piquard, 1994; Vanroozendael *et al.*, 1994; Aliwell & Jones, 1996b; Aliwell *et al.*, 1997; Eisinger *et al.*, 1997; Kreher *et al.*, 1997; Platt *et al.*, 1997; Richter *et al.*, 1999; Roscoe *et al.*, 1999; Wittrock *et al.*, 2000; Friess *et al.*, 2001, 2004; Allan *et al.*, 2002; Honninger & Platt, 2002; von Friedeburg *et al.*, 2002; Leser *et al.*, 2003). Note that DOAS measurements performed using lunar or stellar light are not the subject of this chapter, but have been discussed in previous reviews of the DOAS technique (Platt, 1994; Plane & Smith, 1995). This section contains a brief overview of the spectroscopic principles and the main applications of scattered-sunlight DOAS, or passive DOAS.

3.4.1 Spectroscopic principles

As the source of light is now outside the earth's atmosphere, the Beer–Lambert law (Equation 3.1) is modified (Solomon *et al.*, 1987a):

$$I_{tr}(\lambda) = I_0(\lambda)\exp(-\sigma(\lambda)\mathrm{SCD}) \tag{3.20}$$

where SCD is the slant column density of an atmospheric species, integrated along the path followed by solar photons through the atmosphere. If n species are present, the total OD is given by:

$$\mathrm{OD}(\lambda) = \ln\left(\frac{I_0(\lambda)}{I_{tr}(\lambda)}\right) = \sum_{i=1}^{n}\mathrm{OD}_i(\lambda) = \sum_{i=1}^{n}\sigma_i(\lambda)\mathrm{SCD}_i \tag{3.21}$$

Photons traversing the atmosphere are subject to Rayleigh and Mie scattering, and thus wavelength-dependent extinction coefficients are introduced in the analysis in the same manner as for the LP-DOAS technique (Section 3.2.1).

Passive DOAS spectra can therefore be analysed using the same analysis procedure as for LP-DOAS, but there are several additional considerations when using scattered solar radiation as the light source. The first is that the unattenuated $I_0(\lambda)$ spectrum cannot be measured (the same applies in LP-DOAS, but boundary layer spectra can at least be normalised by dividing by the spectrum of the lamp in the transmitter – Section 3.3.1.2). Instead, the $I_0(\lambda)$ spectrum is usually taken as the solar spectrum when the solar zenith angle (SZA) is at its minimum (i.e. at local noon), since the slant column through the atmosphere is then a minimum (Noxon, 1975, 1976; Noxon *et al.*, 1979). The objective is to remove the Fraunhofer lines present in the solar spectrum, since these structures correspond to up to 30% absorption, and are much larger than molecular absorption features. Taking the ratio of the spectra at high and low SZAs largely removes the Fraunhofer structure, as shown in Figure 3.12. The two atmospheric scattered sunlight spectra in Figure 3.12(a) are taken at 70° and 90° SZA, and the ratio compared to an O_3 reference spectrum in Figure 3.12(b). This demonstrates how the strong absorption bands of O_3 become visible in the 317–335 nm region after the solar Fraunhofer structure is removed.

Another factor to consider when retrieving atmospheric species from scattered sunlight spectra is the so-called 'Ring effect', named after Grainger and Ring (1962). This describes the partial 'filling-in' of the Fraunhofer structures in the solar spectrum as sunlight traverses through the atmosphere, resulting in a decrease of the ODs of Fraunhofer lines.

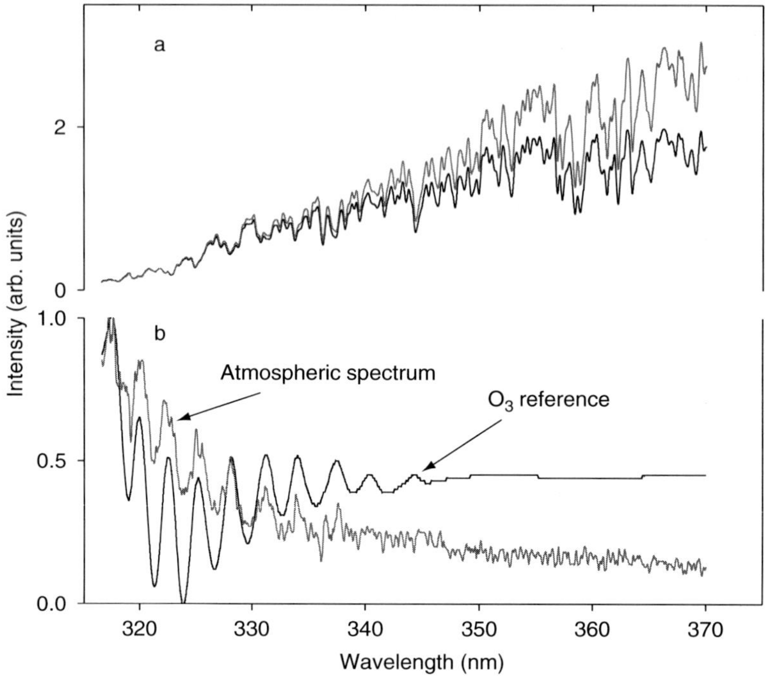

Figure 3.12 Scattered sunlight atmospheric spectra: (a) atmospheric spectra taken at 70° (grey line) and 90° (solid line) SZA, showing the dominant Fraunhofer structure (the signal intensity is shown on the left-hand axis); (b) the ratio of the atmospheric spectra in (a) (grey line) – the Fraunhofer lines largely disappear to reveal the strong O_3 absorption bands in the region 317–335 nm, which are compared to an O_3 reference spectrum (solid line). Note that both spectra have been normalised for comparison.

The Ring effect is most likely due to Raman scattering (Kattawar *et al.*, 1981; Fish & Jones, 1995; Vountas *et al.*, 1998; Sioris & Evans, 1999). The size of the Ring effect depends strongly on SZA, with the result that at larger SZA the observed Fraunhofer lines are weaker. Accurate removal of the Ring effect is necessary because it represents a change in OD that can be an order of magnitude larger than that from absorption by atmospheric trace gases. An artificially generated Ring spectrum is usually employed in the analysis of scattered sunlight spectra (Chance & Spurr, 1997). An additional complication is that cloudy conditions increase the optical path of scattered sunlight reaching the ground, which can greatly enhance the Ring effect (de Beek *et al.*, 2001).

Analysis of scattered sunlight spectra yields the SCD (Equation 3.21). However, the vertical column density (VCD) is a more useful quantity. The conversion from SCD to VCD is obtained by means of the so-called air mass factor (AMF) (Solomon *et al.*, 1987a; Perliski & Solomon, 1993):

$$\mathrm{VCD} = \frac{\mathrm{SCD}}{\mathrm{AMF}} \tag{3.22}$$

The AMF depends on the radiative transfer of the atmosphere and can be calculated from radiative transfer models (Marquard *et al.*, 2000; von Friedeburg *et al.*, 2002).

These models must include parameters such as wavelength-dependent Rayleigh and Mie scattering, vertical profiles of pressure, temperature and the species of interest, as well as the SZA and azimuth angle of the receiving telescope. For SZA $< 75°$, the AMF is well approximated by the simple geometrical relationship $1/\cos(\text{SZA})$.

3.4.2 Zenith-viewing geometry: Retrieval of NO_3 vertical profiles

Zenith-sky spectroscopy is based on the collection of scattered photons using a zenith-pointing telescope, from either a ground-based or airborne instrument. Although routine monitoring of O_3 and NO_2 in the stratosphere has been conducted for many years using zenith-pointing instruments (Noxon, 1975; Harrison, 1979; Noxon *et al.*, 1979; Syed & Harrison, 1980; Noxon, 1981; McKenzie & Johnston, 1982, 1983; Noxon, 1983; McKenzie & Johnston, 1984; Mount *et al.*, 1988; Pommereau & Goutail, 1988; Roscoe *et al.*, 1990; Goutail *et al.*, 1994; Gil *et al.*, 1996; Sarkissian *et al.*, 1997; VanRoozendael *et al.*, 1997; Vaughan *et al.*, 1997; Preston *et al.*, 1998; Roscoe *et al.*, 1999; Gil *et al.*, 2000; Liley *et al.*, 2000), here we will focus on ground-based DOAS measurements of NO_3. During the night, NO_3 has been monitored using direct observations of the moon as a light source (Noxon *et al.*, 1978, 1980; Noxon, 1983; Sanders *et al.*, 1987; Solomon *et al.*, 1989a,b; Wagner *et al.*, 2000). Since 1990s NO_3 columns have been measured for an hour or so before sunrise, by collecting scattered sunlight using zenith and off-axis DOAS when the SZA is less than about 95° (Weaver *et al.*, 1996; Aliwell & Jones, 1998; von Friedeburg *et al.*, 2002). In fact, NO_3 is photolysed so rapidly that the decrease in NO_3 column as the solar terminator sweeps downward before sunrise can be used to determine the vertical profile of the radical in the troposphere and lower stratosphere (Smith & Solomon, 1990; Smith *et al.*, 1993).

Figure 3.13(a) shows how NO_3 is sequentially removed by photolysis as the SZA decreases to 90° (sunrise). The VCDs of NO_3 are derived in a similar way to the NO_3 concentrations determined with an LP-DOAS instrument (Allan *et al.*, 2002) (Section 3.3.1.2), where each processed atmospheric spectrum is divided by a processed daytime reference spectrum. Note that the lifetime of NO_3 against photolysis is only a few seconds in full sunlight, so the persistence of NO_3 when SZA is less than 90° indicates a large quantity of N_2O_5 is present. This is in equilibrium with NO_3 ($NO_3 + NO_2 \leftrightarrow N_2O_5$), so the relatively slow thermal decomposition of N_2O_5 has the effect of delaying the disappearance of NO_3 after sunrise. This effect occurs in semi-polluted environments with NO_2 mixing ratios above 500 ppt (Coe *et al.*, 2002).

The generic optimal estimation method (OEM) developed by Rodgers (1976, 1990) for solving atmospheric data inversion problems has been successfully developed to retrieve vertical profile information of NO_3 from zenith-sky spectroscopic measurements of column abundance made through sunrise (Coe *et al.*, 2002). A forward model described by the expression $y = Kx$ predicts the column density time series, y, that would be observed from a vertical concentration profile, x, connected through a weighting function, K. The elements of the matrix K are the degree of photolysis of NO_3 as a function of height, SZA and time (Coe *et al.*, 2002). In fact, since x is unknown the forward model is inverted to yield the backward model $x = K^{-1}y$. The inversion process begins from an *a priori* guess of the NO_3 profile (x_0); the backward model is then used to calculate an

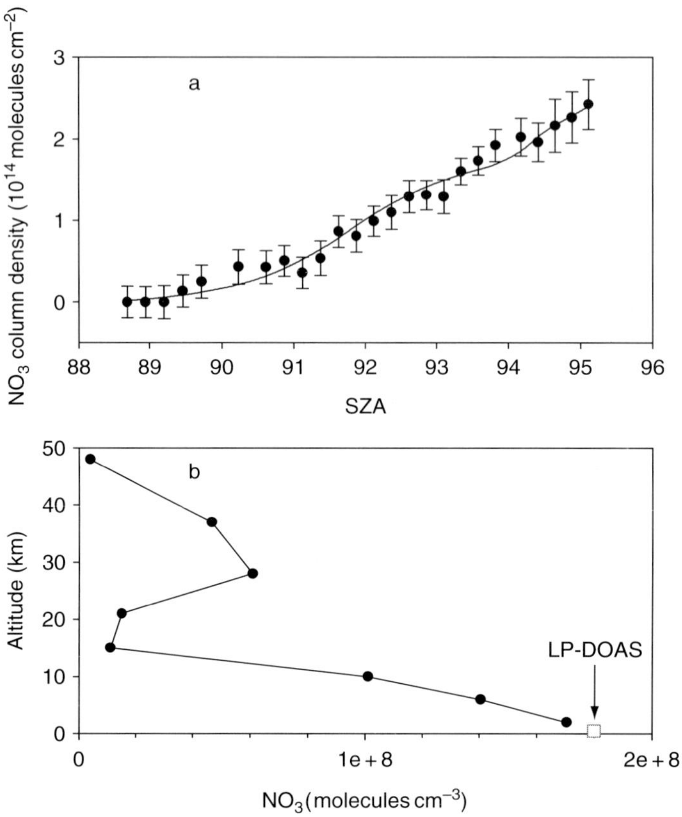

Figure 3.13 Vertical profiles of NO_3 obtained from zenith-sky measurements at Mace Head, west coast of Ireland on 2 August 2002: (a) measured NO_3 column density versus SZA; (b) NO_3 concentration profile up to 50 km retrieved from the data in (a), using an optimal estimation method (Coe *et al.* 2002). The solid line in (a) is calculated from the forward model (see text). The lowermost point of the profile in (b), assuming a 2 km layer height, is in good agreement with a measurement made by LP-DOAS in the boundary layer. (Data reproduced with kind permission of J.A. Shillito and H. Coe [UMIST].)

initial estimate of the column density time series, y_0. This is compared with the observed y, and the *a priori* profile is then refined iteratively to yield x, the optimised NO_3 profile. The technique yields significant profile information up to an altitude of about 40 km, if the signal-to-noise ratio is sufficiently high in spectra recorded with the very low light levels present when the terminator is in the upper stratosphere (i.e. SZA > 95°). The vertical resolution achievable is of the order of 3 km, depending on altitude.

Figure 3.13(b) shows the application of the OEM to the time series of NO_3 column measurements in Figure 3.13(a). In this case, the bulk of NO_3 is located below 5 km in the atmosphere, with a smaller peak between 25 and 30 km in the stratosphere. Also shown in Figure 3.13(b) is a measurement of NO_3 made in the same location with an LP-DOAS at the base of the boundary layer, showing that on this occasion NO_3 is apparently well mixed in the first 2 km. This situation contrasts with measurements made under very clean conditions in the Southern Ocean, where emission of dimethyl sulphide, which reacts rapidly with NO_3, creates a large gradient of NO_3 through the boundary layer (Allan *et al.*, 2002).

3.4.3 Multi-axis DOAS

The multi-axis DOAS (MAX-DOAS) technique involves recording scattered light spectra at several different zenith angles, either simultaneously or in rapid sequence, in order to infer the coarse height resolution of an atmospheric species (Honninger & Platt, 2002). First developed in 2002, it has been reviewed in 2004 (Honninger *et al.*, 2004a; Wittrock *et al.*, 2004). The technique consists of receiving scattered photons from multiple viewing directions above the horizon, allowing for the retrieval of better spatial information of the species of interest. The technique is especially powerful for monitoring atmospheric trace gases present in layers close to the earth's surface. Figure 3.14 shows the viewing geometry of a MAX-DOAS instrument where scattered sunlight is collected over a range of elevation angles (α). The zenith-scattered light describes a relatively longer path in the stratosphere than in the troposphere, compared to the scattered light collected by the lower elevation angle telescopes. Furthermore, when the SZA(θ) is large, trace gas absorption in the lowest layers of the atmosphere is enhanced because the photons traverse a longer absorption path through the lowermost atmospheric layer (the scattered sunlight also gets brighter at smaller θ, so that the signal-to-noise ratio is increased). Thus, absorption spectra collected at large θ with a telescope pointing at low α contain information mostly about lower tropospheric constituents, whereas spectra taken with a telescope at small (or zero) θ contain enhanced absorption by stratospheric constituents (Honninger & Platt 2002).

In order to achieve the collection of scattered sunlight at different viewing directions, different experimental set-ups can be deployed. First, observation at different elevation angles is obtained by sequentially moving one off-axis telescope, while a second

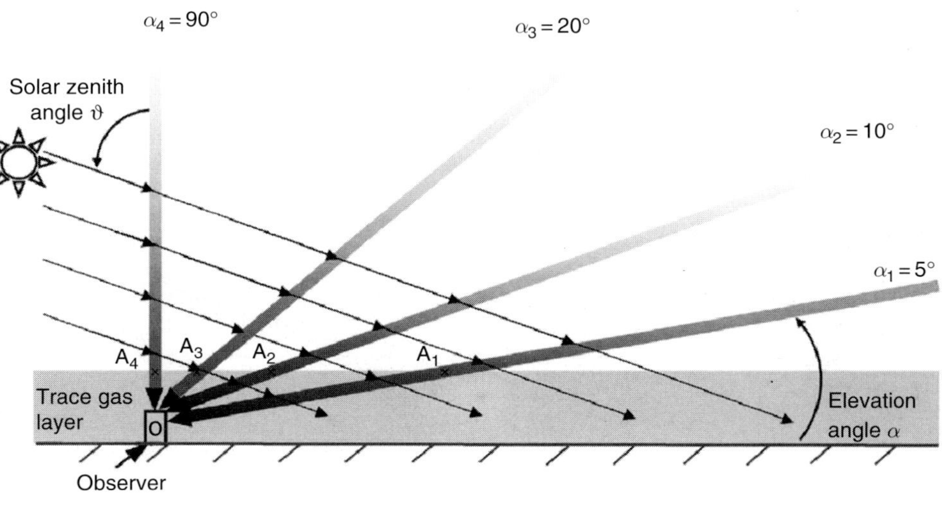

Figure 3.14 Schematic drawing of the viewing geometries of a ground-based MAX-DOAS instrument with four telescopes at elevation angles 5°, 10°, 20° and 90° above the horizon. The scattering points at a give altitude where solar rays intercept the field of view of the telescopes are shown as $A_1 - A_4$. The absorption path of the photons collected by the different telescopes decreases as a function of $1/\sin(\alpha)$. (Honninger and Platt, 2002, reproduced with permission from Elsevier.)

fixed telescope is sometimes employed pointing to the zenith. This approach has been employed in various studies since 2000 (Honninger & Platt, 2002; Bobrowski *et al.*, 2003; Leser *et al.*, 2003; Heckel *et al.*, 2004; Honninger *et al.*, 2004b,c). The set-up consists of one or two telescopes controlled by stepper motors coupled to a single spectrometer–detector system. The system is relatively simple to operate and therefore suitable for making measurements in remote locations. Alternatively, a number of static telescopes can be fixed at different elevation angles to allow simultaneous observations at more than two viewing geometries (Wagner *et al.*, 2002; Bobrowski *et al.*, 2003). The telescopes are then coupled to a single spectrometer by optic fibres, and dispersed onto a two-dimensional CCD (Section 3.3.1.1) where the spectra are separated as strips on different rows of the CCD. This offers the advantage of real-time observations at several elevation angles. The disadvantage of this approach is that the spectra may have slightly different line shapes, and the pixel-to-pixel variability across the CCD chip will not be removed when dividing the spectra since they are recorded on different portions of the chip. The former difficulty can be overcome to a degree by convoluting each spectrum with the corresponding line shape of the spectrum by which it is to be divided; the latter can be minimised by applying a pixel-to-pixel correction factor obtained by measuring diffuse (non-dispersed) light (e.g. from a tungsten lamp with a diffuser plate) across the CCD.

The MAX-DOAS technique yields the *SCD* of the species monitored at the different elevation angles. The VCD is then obtained by applying an AMF (Section 3.4.1). The AMF needs to be estimated from a radiative transfer model that determines the altitude of significant light scattering, for the SZA and elevation angle employed. For instance, for a stratospheric absorber there will be a strong dependence of the AMF on θ. In contrast, for an absorber in the boundary layer the AMF will have a small dependence on θ, but a strong dependence on α. Figure 3.15 illustrates an example where the BrO column abundances were strongly dependent on α, indicating that a significant fraction of BrO was in the lowest 2 km of the troposphere. This conclusion is explained by

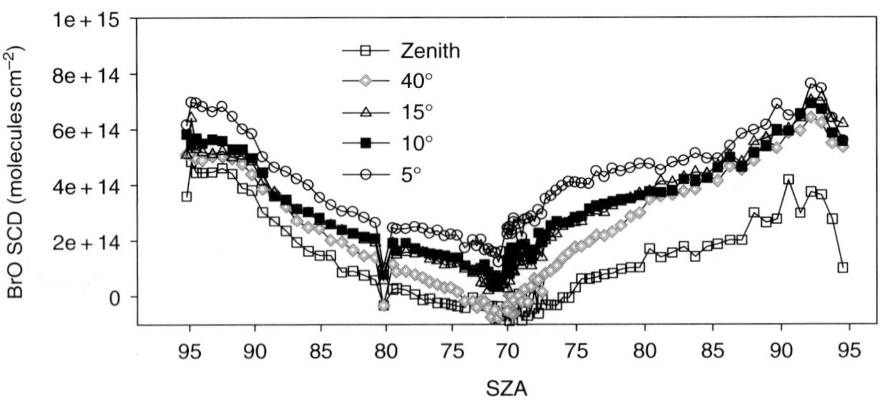

Figure 3.15 Example of the BrO slant column density recorded at five different elevation angles, as a function of the SZA. The strong dependence of the column abundance with elevation angle indicates that most of the BrO lies in the boundary layer. These spectra were recorded at Barrow, Alaska, on 25 March 2003 (Saiz-Lopez and Plane, unpublished data).

Figure 3.16, which compares the scattering geometries for an absorber present in the stratosphere and in the boundary layer (Honninger & Platt, 2002). For the case of single scattering occurring below the trace gas layer (stratospheric constituent), the *AMF* can be approximated as $AMF \approx 1/\cos\theta$. In contrast, for scattering above the trace gas layer (lower troposphere species) $AMF \approx 1/\sin\alpha$. Of course, the measured *SCD* contains contributions from both the stratosphere and the troposphere if the molecular species is present in both regions. However, for fixed θ the slope of a plot of *SCD* against $AMF\,(\approx 1/\sin\alpha)$ provides an estimate of the vertical column of the trace gas below the scattering altitude (i.e. in the lower troposphere), whereas the intercept is an estimate of the stratospheric component (Honninger & Platt, 2002).

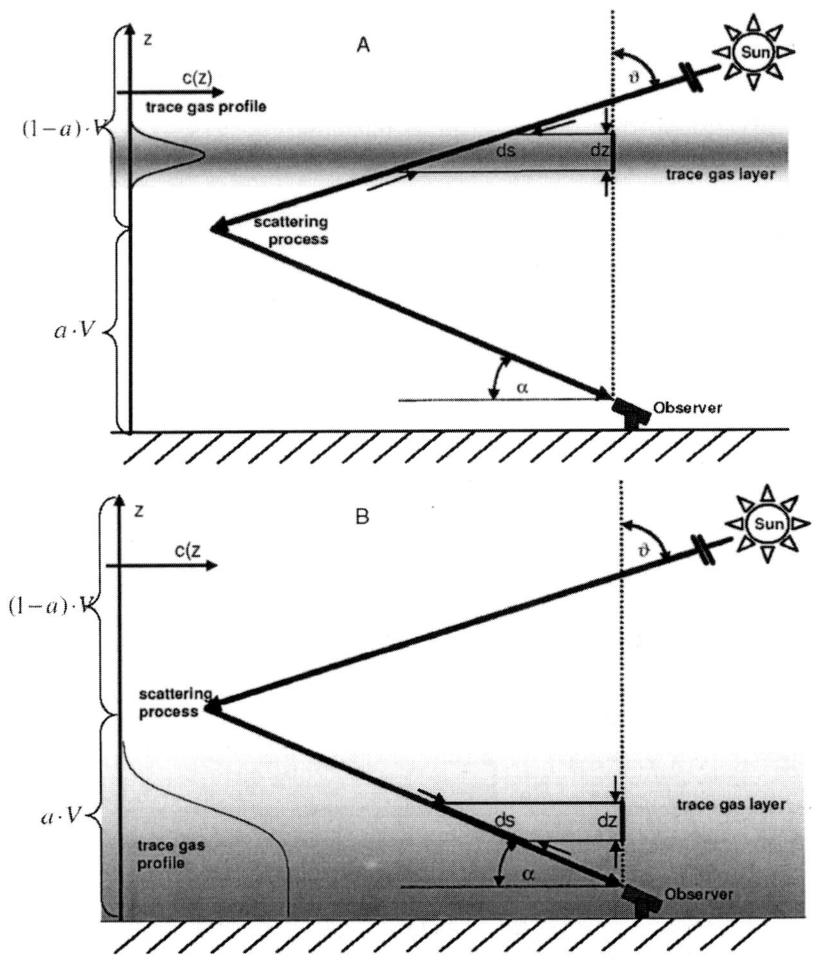

Figure 3.16 Illustration of an off-axis viewing geometry for single scattering, where the absorbing trace species occurs (A) in the stratosphere and (B) in the boundary layer. The observed SCD (along ds) is converted into VCD (along dz) by applying an AMF. (Honninger *et al.*, 2004a, reproduced with permission from the European Geosciences Union.)

3.4.3.1 Tomographic DOAS

The LP-DOAS technique measures the column abundance of an atmospheric species along an optical path, but does not provide information about the distribution in the atmosphere. However, the spatial distribution of a trace species can be obtained in two or even three dimensions by using a number of intersecting light paths and applying tomographic inversion (Hashmonay *et al.*, 1999; Price, 1999). This is known as tomographic DOAS (Tom-DOAS) (Pundt *et al.*, 2004; Laepple *et al.*, 2004). Instrumentally, it requires one or several artificial light sources in conjunction with a number of retro-reflectors. Additionally, a two-dimensional CCD together with an imaging spectrograph is needed to record simultaneously the spectra of light collected from a number of telescopes. Tom-DOAS can also be used to collect scattered sunlight by means of a number of mirrors that will collimate the light into different optic fibres in a similar way to MAX-DOAS.

Tom-DOAS has been applied to construct a two-dimensional map of NO_2 from a vehicle emission plume above a motorway (Pundt *et al.*, 2004). Figure 3.17 shows the experimental set-up of the Tom-DOAS above the motorway: two telescopes together with

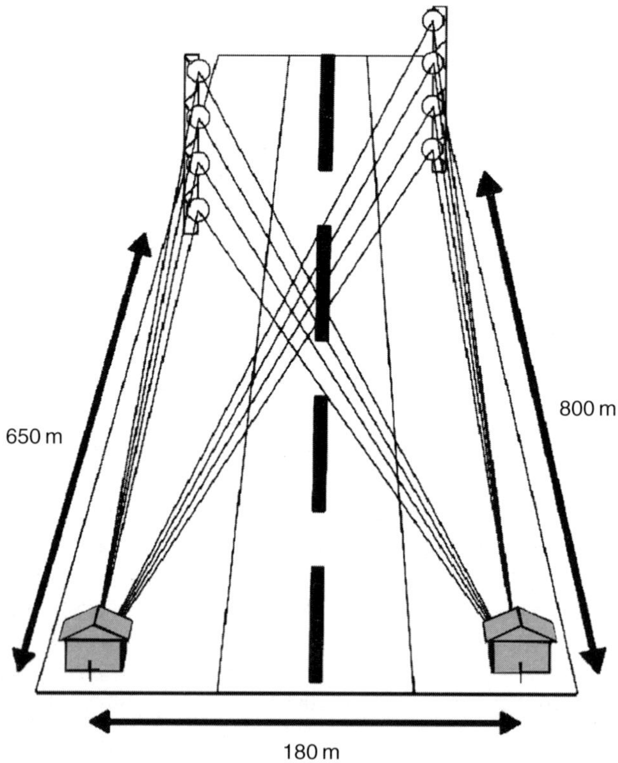

Figure 3.17 Experimental set-up of a Tomographic-DOAS to observe emissions around a motorway. The two DOAS telescopes housed at ground level on each side of the motorway can be steered to each of eight retroreflector arrays, positioned at different heights in two groups. This arrangement therefore provides a total of 16 light paths. (Reproduced with permission from Pundt *et al.*, 2004.)

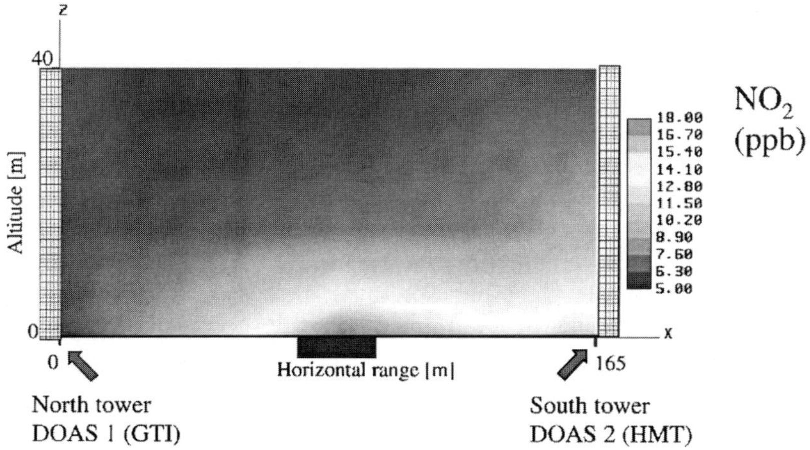

Figure 3.18 A contour plot of NO_2 as a function of altitude and orthogonal to the motorway (indicated by a solid black rectangle), measured by the Tom-DOAS system in Figure 3.14. GTI and HMT correspond to two different geographical locations from where the measurements are carried out. The reconstruction of the NO_2 concentration field was made using a simultaneous algebraic reconstruction technique. Note that the higher NO_2 concentrations are on the right-hand side, downwind from the motorway. (Reproduced with permission from Pundt *et al.*, 2004.)

eight retro-reflectors at different heights were used, providing 16 different light paths for evaluation. Figure 3.18 illustrates the two-dimensional NO_2 distribution determined using a simultaneous algebraic reconstruction technique (SART) (Pundt *et al.*, 2004). As expected, the higher values of NO_2 are observed downwind of the motorway.

3.4.3.2 Balloon-borne DOAS

Balloon-borne UV/visible spectrometers have been utilised to investigate vertical distributions of atmospheric species including O_3, O_4, BrO, NO_2, OClO, OBrO, IO, OIO, NO_3 and H_2O, by collecting direct sunlight or scattered sunlight at solar occultation (Pommereau & Goutail, 1988; Pommereau & Piquard, 1994; Ferlemann *et al.*, 1998; Harder *et al.*, 1998; Pundt *et al.*, 1998; Erle *et al.*, 2000; Ferlemann *et al.*, 2000; Fitzenberger *et al.*, 2000; Harder *et al.*, 2000; Pfeilsticker *et al.*, 2001; Lumpe *et al.*, 2002; Pundt *et al.*, 2002; Bosch *et al.*, 2003). In fact, this area is large enough to be reviewed in its own right, and only a brief description is given here. The principle of balloon-borne DOAS is similar to that of passive DOAS methods, in that scattered sunlight spectra are collected at different SZAs. The observations can be performed as the balloon-gondola ascends or by solar occultation once the balloon reaches its float altitude ($\sim$30 km). Solar occultation measurements begin as the sun moves down below the horizon. Pressure sensors and GPSs can track the position of the gondola precisely. During its ascent the instrument probes nearly the entire profile of the species of interest whereas while floating only the fraction of the profile above the balloon altitude is probed. This is because scattered light is collected in a near-horizontal direction with respect to the gondola. Hence, as the solar terminator moves upwards after sunset, scattering comes from higher (and further away) in the atmosphere. Generally, the float-occultation method provides stronger observable *SCDs* (Pommereau & Piquard, 1994).

The optical layout of a balloon-borne UV/visible spectrometer is similar to ground-based DOAS instruments. However, there are difficulties inherent in balloon flight. For instance, spectral shifts can be caused by rotation of the gondola, and therefore the entrance slit, with respect to the solar position. Some instruments employ a diffuser plate in the optical entrance to minimise this problem. Note that for tropospheric applications direct sunlight is often collected, without a diffuser, as the light describes a well-defined path in the troposphere (Fitzenberger *et al.*, 2000; Harder *et al.*, 2000). Pressure and temperature changes during flight can also cause spectral drift. This problem can be overcome by enclosing the entire spectrometer in a pressurised and thermostated unit. Additionally, the balloon needs to be maintained at a fixed azimuth angle, requiring an azimuthal suspension system on the gondola.

Finally, the spectral analysis by DOAS requires the solar features that dominate the absorption spectra to be removed (Section 3.4.1). The vertical distributions of species can be retrieved by inversion methods (Rodgers, 1976; Pommereau & Piquard, 1994), with a height resolution of about 1 km (Pundt *et al.*, 2002).

3.4.3.3 Airborne MAX-DOAS

Absorption spectroscopy measurements can also be performed from airborne platforms (Wahner *et al.*, 1989, 1990a,b; Schiller *et al.*, 1990; Glatthor *et al.*, 1998; McElroy *et al.*, 1999; Petritoli *et al.*, 2002, see also Chapter 2 for infrared applications). For instance, the column abundances of species such as NO_2, BrO and OClO have been monitored from airborne instruments.

The technique is based on the collection of UV/visible scattered sunlight, as a function of SZA, by a telescope–spectrometer system installed on an aircraft. The spectral analysis is performed using the scattered sunlight DOAS procedure (Sections 3.1.2 and 3.4.1). Numerous 'off-axis' scattered sunlight measurements with an airborne UV/visible spectrometer have been made (Petritoli *et al.*, 2002). Indeed, MAX-DOAS can also be applied from an airborne platform at different viewing directions above and below the aircraft (termed AMAX-DOAS) (Heue *et al.*, 2003; Wang *et al.*, 2003). Figure 3.19(a) compares the atmospheric observation geometries for an aircraft (assuming single scattering), compared to a ground-based zenith-sky instrument (Figure 3.19(b).) This shows the different paths taken by scattered light as it passes through trace gas layers in the troposphere and stratosphere. The upward-viewing geometry scans the atmosphere above the flight altitude whereas the downward-viewing telescopes collect scattered sunlight from above and below the aircraft. Following the MAX-DOAS approach, telescopes pointing to the altitude flight horizon (limb-viewing) will observe larger absorptions than those either looking towards the zenith or nadir. Depending on the scattering altitude two different modes of viewing can be employed. The nadir-viewing mode is used when the scattering occurs close to, or at, the earth's surface; limb-viewing is used when light is scattered above the surface. Therefore, depending on the flight altitude, the combination of spectra collected at different SZA by telescopes pointing at various elevation angles allows for discrimination between the contributions of the stratospheric and the tropospheric components to the

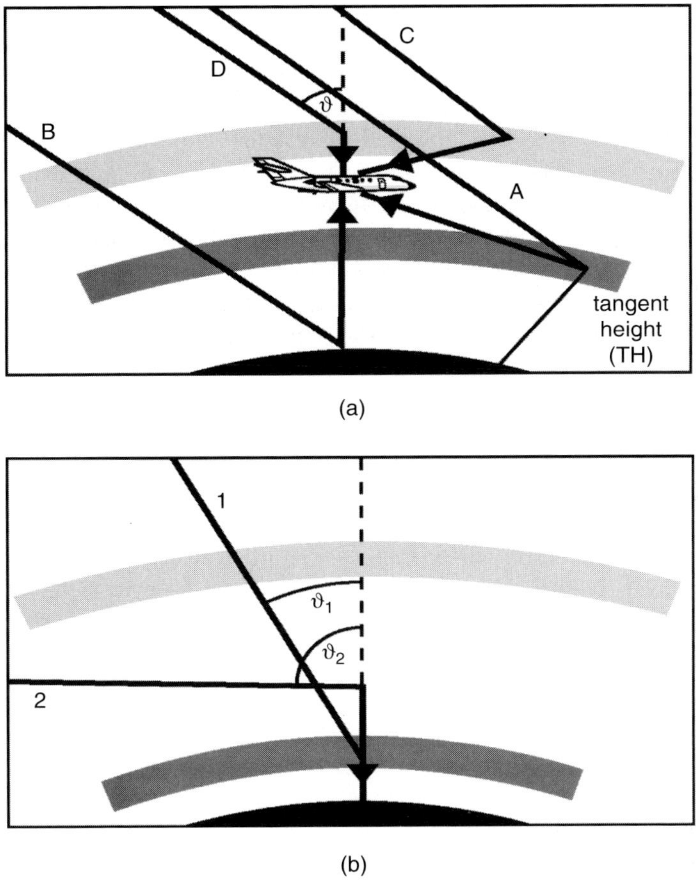

Figure 3.19 Different scattered sunlight paths seen from an airborne instrument (a), compared to those observed with a ground-based zenith-sky spectrometer (b). Note that the photon paths are shown for the simple case of single scattering. The upper and lower grey layers in panel (b) represent stratospheric and tropospheric components respectively. (Reproduced with permission from Prof. Ulrich Platt [University of Heidelberg].)

observed column density. The vertical profile information can then be derived by applying the corresponding *AMF* as in ground-based MAX-DOAS analysis.

3.5 Summary and conclusions

The aim of this chapter has been to describe the very significant advances that have been made in using UV/visible differential optical absorption spectroscopy to study the atmosphere, since the last major reviews of the field in 1994/95 (Platt, 1994; Plane & Smith, 1995). And, of course, we have excluded here a discussion of the remarkable progress that have been made in space-based DOAS observations of atmospheric constituents, which are covered briefly in Chapter 1. To summarise, the strengths inherent in the DOAS method are that any species with a highly structured spectrum in the UV/visible/near-IR region can be observed if its concentration is significant (Table 3.1), and the retrieved column density is directly traceable to an absolute laboratory spectrum. Since the late 1990s, several new species have been observed (uniquely by DOAS). These are IO, OIO and I_2. BrO and ClO have been observed for the first time in the lower troposphere. There have also been several major advances in DOAS techniques. These include the development of CRDS for making in situ measurements, and techniques such as MAX-DOAS and Tom-DOAS for obtaining spatial structure. Of particular note here has been the retrieval of vertical profiles of BrO and NO_3. In conclusion, the rate of progress in developing the theory and application of the DOAS method has been so rapid since 2000 that one can look forward with great anticipation to where the field will be in 2015.

Further reading

For more information on general atmospheric chemistry and spectroscopy, the following books are recommended:

Wayne, R.P. (2000) *Chemistry of Atmospheres*, Oxford University Press, Oxford.

Finlayson-Pitts, B.J. & Pitts, J.N. (2000) *Chemistry of the Upper and Lower Atmosphere*, Academic Press, San Diego.

Hollas, J.M. (1996) *Modern Spectroscopy*, John Wiley & Sons Ltd, New York.

References

Alicke, B., Hebestreit, K., Stutz, J. *et al.* (1999) Iodine oxide in the marine boundary layer, *Nature*, 397, 572–573.

Aliwell, S.R. & Jones, R.L. (1996a) Measurement of atmospheric NO_3. 1. Improved removal of water vapour absorption features in the analysis for NO_3. *Geophys. Res. Lett.*, 23, 2585–2588.

Aliwell, S.R. & Jones, R.L. (1996b) Measurement of atmospheric NO_3. 2. Diurnal variation of stratospheric NO_3 at midlatitude, *Geophys. Res. Lett.*, 23, 2589–2592.

Aliwell, S.R., Jones, R.L. & Fish, D.J. (1997) Mid-latitude observations of the seasonal variation of BrO. 1. Zenith-sky measurements, *Geophys. Res. Lett.*, 24, 1195–1198.

Aliwell, S.R. & Jones, R.L. (1998) Measurements of tropospheric NO_3 at midlatitude, *J. Geophys. Res.-Atmos.*, 103, 5719–5727.

Aliwell, S.R., Van Roozendael, M., Johnston, P.V. *et al.* (2002) Analysis for BrO in zenith-sky spectra: An intercomparison exercise for analysis improvement, *J. Geophys. Res.-Atmos.*, 107, art. no.-4199.

Allan, B.J., Carslaw, N., Coe, H. *et al.* (1999) Observations of the nitrate radical in the marine boundary layer, *J. Atmos. Chem.*, 33, 129–154.

Allan, B.J., McFiggans, G., Plane, J.M.C. *et al.* (2000a) Observations of iodine monoxide in the remote marine boundary layer, *J. Geophys. Res.-Atmos.*, 105, 14 363–14 369.

Allan, B.J., McFiggans, G., Plane, J.M.C. *et al.* (2000b) The nitrate radical in the remote marine boundary layer, *J. Geophys. Res.-Atmos.*, 105, 24 191–24 204.

Allan, B.J., Plane, J.M.C., Coe, H. *et al.* (2002) Observations of NO_3 concentration profiles in the troposphere, *J. Geophys. Res.-Atmos.*, 107, art. no.-4588.

Allan, B.J., Plane, J.M.C. & McFiggans, G. (2001) Observations of OIO in the remote marine boundary layer, *Geophys. Res. Lett.*, 28, 1945–1948.

Armerding, W., Spiekermann, M. & Walter, J. (1996) Multipass optical absorption spectroscopy: A fast-scanning laser spectrometer for the in situ determination of atmospheric trace-gas components, in particular OH, *Appl. Optics*, 35, 4206–4219.

Arpag, K.H., Johnston, P.V., Miller, H.L. *et al.* (1994) Observations of the Stratospheric BrO Column over Colorado, 40 Degrees N, *J. Geophys. Res.-Atmos.*, 99, 8175–8181.

Ashworth, S.H., Allan, B.J. & Plane, J.M.C. (2002) High resolution spectroscopy of the OIO radical: Implications for the ozone-depleting potential of iodine, *Geophys. Res. Lett.*, 29, art. no.-1456.

Atkinson, D.B. & Hudgens, J.W. (1997) Chemical kinetic studies using ultraviolet cavity ring-down spectroscopic detection: Self-reaction of ethyl and ethylperoxy radicals and the reaction $O_2 + C_2H_5 \rightarrow C_2H_5O_2$, *J. Phys. Chem. A*, 101, 3901–3909.

Axelsson, H., Galle, B., Gustzsson, K. *et al.* (1990) A transmitting/receiving telescope for DOAS-measurements using retroreflector technique, *Dig. Top. Meet. Opt. Remote Sens. Atmos., OSA*, 4, 641–644.

Ball, S.M., Povey, I.M., Norton, E.G. *et al.* (2001) Broadband cavity ringdown spectroscopy of the NO_3 radical, *Chem. Phys. Lett.*, 342, 113–120.

Ball, S.M. & Jones, R.L. (2003) Broad-band cavity ring-down spectroscopy, *Chem. Rev.*, 103, 5239–5262.

Bobrowski, N., Honninger, G., Galle, B. *et al.* (2003) Detection of bromine monoxide in a volcanic plume, *Nature*, 423, 273–276.

Bosch, H., Camy-Peyret, C., Chipperfield, M.P. *et al.* (2003) Upper limits of stratospheric IO and OIO inferred from center-to-limb-darkening-corrected balloon-borne solar occultation visible spectra: Implications for total gaseous iodine and stratospheric ozone, *J. Geophys. Res.-Atmos.*, 108, art. no.-4455.

Brauers, T., Hausmann, M., Bister, A. *et al.* (2001) OH radicals in the boundary layer of the Atlantic Ocean. 1. Measurements by long-path laser absorption spectroscopy, *J. Geophys. Res.-Atmos.*, 106, 7399–7414.

Brauers, T., Hausmann, M., Brandenburger, U. *et al.* (1995) Improvement of differential optical-absorption spectroscopy with a multichannel scanning technique, *Appl. Optics*, 34, 4472–4479.

Brown, S.S., Stark, H., Ciciora, S.J. *et al.* (2001) In-situ measurement of atmospheric NO_3 and N_2O_5 via cavity ring-down spectroscopy, *Geophys. Res. Lett.*, 28, 3227–3230.

Brown, S.S. (2003) Absorption spectroscopy in high-finesse cavities for atmospheric studies, *Chem. Rev.*, 103, 5219–5238.

Brown, S.S., Stark, H., Ryerson, T.B. *et al.* (2003) Nitrogen oxides in the nocturnal boundary layer: Simultaneous in situ measurements of NO_3, N_2O_5, NO_2, NO, and O_3, *J. Geophys. Res.-Atmos.*, 108, art. no.-4299.

Camy-Peyret, C., Bergqvist, B., Galle, B. *et al.* (1996) Intercomparison of instruments for tropospheric measurements using differential optical absorption spectroscopy, *J. Atmos. Chem.*, 23, 51–80.

Carslaw, N., Carpenter, L.J., Plane, J.M.C. *et al.* (1997a) Simultaneous observations of nitrate and peroxy radicals in the marine boundary layer, *J. Geophys. Res.-Atmos.*, 102, 18 917–18 933.

Carslaw, N., Plane, J.M.C., Coe, H. *et al.* (1997b) Observations of the nitrate radical in the free troposphere at Izana de Tenerife, *J. Geophys. Res.-Atmos.*, 102, 10 613–10 622.

Chance, K.V. & Spurr, R.J.D. (1997) Ring effect studies: Rayleigh scattering, including molecular parameters for rotational Raman scattering, and the Fraunhofer spectrum, *Appl. Optics*, 36, 5224–5230.

Coe, H., Allan, B.J. & Plane, J.M.C. (2002) Retrieval of vertical profiles of NO_3 from zenith sky measurements using an optimal estimation method. *J. Geophys. Res.-Atmos.*, 107, art. no.-4587.

Comes, F.J., Armerding, W., Grigonis, R. *et al.* (1992) Tropospheric OH – Local measurements and their interpretations. *Ber. Bunsen-Ges. Phys. Chem. Chem. Phys.*, 96, 284–286.

de Beek, R., Vountas, M., Rozanov, V.V. *et al.* (2001) The ring effect in the cloudy atmosphere. *Geophys. Res. Lett.*, 28, 721–724.

DeMore, W.B., Sander, S.P., Golden, D.M. *et al.* (1997) Chemical kinetics and photochemical data for use in stratospheric modelling. Eval. 11 Jet Propul. Lab., California.

Dobson, G.M.B. & Harrison, D.N. (1926) Measurements of the amount of ozone in the Earth's atmosphere and its relation with other geophysical conditions, *Proc. R. Soc. London*, 110, 660–693.

Dorn, H.P., Callies, J., Platt, U. *et al.* (1988) Measurement of tropospheric OH concentrations by laser long-path absorption spectroscopy, *Tellus*, 40B, 437–445.

Dorn, H.P., Brandenburger, U., Brauers, T. *et al.* (1996) In-situ detection of tropospheric OH radicals by folded long-path laser absorption. Results from the POPCORN field campaign in August 1994, *Geophys. Res. Lett.*, 23, 2537–2540.

Eisinger, M., Richter, A., Ladstatter-Weissenmayer, A. *et al.* (1997) DOAS zenith sky observations. 1. BrO measurements over Bremen (53 degrees N) 1993–1994, *J. Atmos. Chem.*, 26, 93–108.

Erle, F., Platt, U. & Pfeilsticker, K. (2000) Measurement of OBrO upper limits in the nighttime stratosphere, *Geophys. Res. Lett.*, 27, 2217–2220.

Ferlemann, F., Camy-Peyret, C., Fitzenberger, R. *et al.* (1998) Stratospheric BrO profiles measured at different latitudes and seasons: Instrument description, spectral analysis and profile retrieval. *Geophys. Res. Lett.*, 25, 3847–3850.

Ferlemann, F., Bauer, N., Fitzenberger, R. *et al.* (2000) Differential optical absorption spectroscopy instrument for stratospheric balloon borne trace-gas studies, *Appl. Optics*, 39, 2377–2386.

Fiedler, M., Frank, H., Gomer, T. *et al.* (1993) Ground based spectroscopic measurements of stratospheric NO_2 and OClO in Arctic Winter 1989/90, *Geophys. Res. Lett.*, 20, 963–966.

Fish, D.J. & Jones, R.L. (1995) Rotational raman-scattering and the ring effect in zenith-sky spectra, *Geophys. Res. Lett.*, 22, 811–814.

Fitzenberger, R., Bosch, H., Camy-Peyret, C. *et al.* (2000) First profile measurements of tropospheric BrO, *Geophys. Res. Lett.*, 27, 2921–2924.

Friess, U., Wagner, T., Pundt, I. *et al.* (2001) Spectroscopic measurements of tropospheric iodine oxide at Neumayer station, Antarctica, *Geophys. Res. Lett.*, 28, 1941–1944.

Friess, U., Hollwedel, J., Konig-Langlo, G. *et al.* (2004) Dynamics and chemistry of tropospheric bromine explosion events in the Antarctic coastal region, *J. Geophys. Res.-Atmos.*, 109, art. no.-D06305.

Gil, M., Puentedura, O., Yela, M. *et al.* (1996) OClO, NO_2 and O_3 total column observations over Iceland during the winter 1993/94, *Geophys. Res. Lett.*, 23, 3337–3340.

Gil, M., Puentedura, O., Yela, M. *et al.* (2000) Behaviour of NO_2 and O_3 columns during the eclipse of February 26, 1998, as measured by visible spectroscopy, *J. Geophys. Res.-Atmos.*, 105, 3583–3593.

Glatthor, N., Blom, C.E., von Clarmann, T. *et al.* (1998) Airborne remote sensing of NO_2 in the Arctic winter of 1994–1995 and comparison with a three-dimensional chemical transport model, *J. Geophys. Res.-Atmos.*, 103, 13 315–13 326.

Goutail, F., Pommereau, J.P., Sarkissian, A. *et al.* (1994) Total nitrogen dioxide at the Arctic polar circle since 1990, *Geophys. Res. Lett.*, 21, 1371–1374.

Grainger, J. & Ring, J. (1962) Anomalous Fraunhofer line profiles, *Nature*, 193, 762.

Harder, H., Camy-Peyret, C., Ferlemann, F. *et al.* (1998) Stratospheric BrO profiles measured at different latitudes and seasons: Atmospheric observations, *Geophys. Res. Lett.*, 25, 3843–3846.

Harder, H., Bosch, H., Camy-Peyret, C. *et al.* (2000) Comparison of measured and modeled stratospheric BrO: Implications for the total amount of stratospheric bromine, *Geophys. Res. Lett.*, 27, 3695–3698.

Harrison, A.W. (1979) Midsummer stratospheric NO_2 at latitude 45 Degrees S, *Can. J. Phys.*, 57, 1110–1117.

Hashmonay, R.A., Yost, M.G. & Wu, C.F. (1999) Computed tomography of air pollutants using radial scanning path-integrated optical remote sensing, *Atmos. Environ.*, 33, 267–274.

Hausmann, M. & Platt, U. (1994) Spectroscopic measurement of bromine oxide and ozone in the high Arctic during Polar Sunrise Experiment 1992, *J. Geophys. Res.-Atmos.*, 99, 25 399–25 413.

Hausmann, M., Brandenburger, U., Brauers, T. *et al.* (1997) Detection of tropospheric OH radicals by long-path differential-optical-absorption spectroscopy: Experimental setup, accuracy, and precision, *J. Geophys. Res.-Atmos.*, 102, 16 011–16 022.

Hebestreit, K., Stutz, J., Rosen, D. *et al.* (1999) DOAS measurements of tropospheric bromine oxide in mid-latitudes, *Science*, 283, 55–57.

Heckel, A., Richter, A., Tarsu, T. *et al.* (2004) MAX-DOAS measurements of formaldehyde in the Po-Valley, *Atmos. Chem. Phys. Diss.*, 4, 1151–1180.

Heue, K.-P., Bruns, M., Burrows, J.P. *et al.* (2003) Airborne multi axis DOAS measurements during the SCIAVALUES and FORMAT campaigns. *Geophys. Res. Abs.*, 5, 12 405.

Hofzumahaus, A., Aschmutat, U., Brandenburger, U. *et al.* (1998) Intercomparison of tropospheric OH measurements by different laser techniques during the POPCORN campaign 1994, *J. Atmos. Chem.*, 31, 227–246.

Honninger, G. & Platt, U. (2002) Observations of BrO and its vertical distribution during surface ozone depletion at Alert, *Atmos. Environ.*, 36, 2481–2489.

Honninger, G., von Friedeburg, C. & Platt, U. (2004a) Multi axis differential optical absorption spectroscopy (MAX-DOAS), *Atmos. Chem. Phys.*, 4, 231–254.

Honninger, G., Bobrowski, N., Palenque, E.R. *et al.* (2004b) Reactive bromine and sulphur emissions at Salar de Uyuni, Bolivia, *Geophys. Res. Lett.*, 31, art. no.-L04101.

Honninger, G., Leser, H., Sebastian, O. *et al.* (2004c) Ground-based measurements of halogen oxides at the Hudson Bay by active long-path DOAS and passive MAX-DOAS, *Geophys. Res. Lett.*, 31, art. no.-L04111.

Kattawar, G.W., Young, A.T. & Humphreys, T.J. (1981) Inelastic-scattering in planetary-atmospheres. 1. The ring effect, without Aerosols, *Astrophys. J.*, 243, 1049–1057.

King, M.D., Dick, E.M. & Simpson, W.R. (2000) A new method for the atmospheric detection of the nitrate radical (NO_3), *Atmos. Environ.*, 34, 685–688.

Kourtidis, K., Ziomas, I., Zerefos, C. *et al.* (2000) Benzene and toluene levels measured with a commercial DOAS system in Thessaloniki, Greece, *Atmos. Environ.*, 34, 1471–1480.

Kreher, K., Johnston, P.V., Wood, S.W. *et al.* (1997) Ground-based measurements of tropospheric and stratospheric BrO at Arrival Heights, Antarctica, *Geophys. Res. Lett.*, 24, 3021–3024.

Laepple, T., Knab, V., Mettendorf, K. *et al.* (2004) Longpath DOAS tomography on a motorway exhaust gas plume: Numerical studies and application to data from the BAB II campaign, *Atmos. Chem. Phys. Discuss.*, 4, 2435–2484.

Leser, H., Honninger, G. & Platt, U. (2003) MAX-DOAS measurements of BrO and NO_2 in the marine boundary layer, *Geophys. Res. Lett.*, 30, art. no.-1537.

Liley, J.B., Johnston, P.V., McKenzie, R.L. *et al.* (2000) Stratospheric NO_2 variations from a long time series at Lauder, New Zealand, *J. Geophys. Res.-Atmos.*, 105, 11 633–11 640.

Lumpe, J.D., Fromm, M., Hoppel, K. *et al.* (2002) Comparison of POAM III ozone measurements with correlative aircraft and balloon data during SOLVE, *J. Geophys. Res.-Atmos.*, 108, art. no.-8316.

Ma, L.S., Ye, J., Dube, P. *et al.* (1999) Ultrasensitive frequency-modulation spectroscopy enhanced by a high-finesse optical cavity: Theory, and application to overtone transitions of C_2H_2 and C_2HD, *J. Opt. Soc. Am. B*, 16, 2255–2268.

McFiggans, G., Plane, J.M.C., Allan, B.J. *et al.* (2000) A modelling study of iodine chemistry in the marine boundary layer, *J. Geophys. Res.-Atmos.*, 105, 14 371–14 385.

Marquard, L.C., Wagner, T. & Platt, U. (2000) Improved air mass factor concepts for scattered radiation differential optical absorption spectroscopy of atmospheric species, *J. Geophys. Res.-Atmos.*, 105, 1315–1327.

Mazurenka, M.I., Fawcett, B.L., Elks, J.M.F. *et al.* (2003) 410 nm diode laser cavity ring-down spectroscopy for trace detection of NO_2, *Chem. Phys. Lett.*, 367, 1–9.

McElroy, C.T., McLinden, C.A. & McConnell, J.C. (1999) Evidence for bromine monoxide in the free troposphere during the Arctic polar sunrise, *Nature*, 397, 338–341.

McKenzie, R.L. & Johnston, P.V. (1982) Seasonal variations in stratospheric NO_2 at 45 Degrees S, *Geophys. Res. Lett.*, 9, 1255–1258.

McKenzie, R.L. & Johnston, P.V. (1983) Stratospheric ozone observations simultaneous with NO_2 at 45 Degrees S, *Geophys. Res. Lett.*, 10, 337–340.

McKenzie, R.L. & Johnston, P.V. (1984) Springtime stratospheric NO_2 in Antarctica, *Geophys. Res. Lett.*, 11, 73–75.

Mount, G.H., Solomon, S., Sanders, R.W. *et al.* (1988) Observations of stratospheric NO_2 and O_3 at Thule, Greenland, *Science*, 242, 555–558.

Mount, G.H. (1992) The measurement of tropospheric OH by long path absorption. 1. Instrumentation, *J. Geophys. Res.-Atmos.*, 97, 2427–2444.

Mount, G.H., Brault, J.W., Johnston, P.V. *et al.* (1997) Measurement of tropospheric OH by long-path laser absorption at Fritz Peak Observatory, Colorado, during the OH Photochemistry Experiment, Fall 1993, *J. Geophys. Res.-Atmos.*, 102, 6393–6413.

Nakano, Y., Enami, S., Nakamichi, S. *et al.* (2003) Temperature and pressure dependence study of the reaction of IO radicals with dimethyl sulfide by cavity ring-down laser spectroscopy, *J. Phys. Chem. A*, 107, 6381–6387.

Newman, S.M., Howie, W.H., Lane, I.C. *et al.* (1998) Predissociation of the $A^2\Pi_{3/2}$ state of IO studied by cavity ring-down spectroscopy, *J. Chem. Soc.-Faraday Trans.*, 94, 2681–2688.

Newman, S.M., Lane, I.C., Orr-Ewing, A.J. *et al.* (1999) Integrated absorption intensity and Einstein coefficients for the O_2 $a^1\Delta_g$-$X^3\Sigma_g^-$ (0,0) transition: A comparison of cavity ringdown and high resolution Fourier transform spectroscopy with a long-path absorption cell, *J. Chem. Phys.*, 110, 10749–10757.

Noxon, J.F. (1975) Nitrogen dioxide in stratosphere and troposphere measured by ground-based absorption spectroscopy, *Science*, 189, 547–549.

Noxon, J.F. (1976) Atmospheric nitrogen fixation by lightning, *Geophys. Res. Lett.*, 3, 463–465.

Noxon, J.F., Norton, R.B. & Henderson, W.R. (1978) Observation of atmospheric NO_3, *Geophys. Res. Lett.*, 5, 675–678.

Noxon, J.F., Whipple, E.C. & Hyde, R.S. (1979) Stratospheric NO_2. 1. Observational method and behaviour at mid-latitude, *J. Geophys. Res.-Oceans and Atmos.*, 84, 5047–5065.

Noxon, J.F., Norton, R.B. & Marovich, E. (1980) NO_3 in the troposphere, *Geophys. Res. Lett.*, 7, 125–128.

Noxon, J.F. (1981) NOx in the mid-Pacific troposphere, *Geophys. Res. Lett.*, 8, 1223–1226.

Noxon, J.F. (1983) NO_3 and NO_2 in the mid-Pacific troposphere, *J. Geophys. Res.-Oceans and Atmos.*, 88, 1017–1021.

Okabe, H. (1978) *Photochemistry of Small Molecules*, Wiley-Interscience, New York.

O'Keefe, A. & Deacon, D.A.G. (1988) Cavity ringdown optical spectrometer for absorption – Measurements using pulsed laser sources, *Rev. Sci. Instrum.*, 59, 2544–2551.

Paul, J.B. & Saykally, R.J. (1997) Cavity ringdown laser absorption spectroscopy, *Anal. Chem.*, 69, A287–A292.

Perliski, L.M. & Solomon, S. (1993) On the Evaluation of air mass factors for atmospheric near-ultraviolet and visible absorption spectroscopy, *J. Geophys. Res.-Atmos.*, 98, 10 363–10 374.

Perner, D. & Platt, U. (1979) Detection of nitrous acid in the atmosphere by differential optical absorption, *Geophys. Res. Lett.*, 6, 917–920.

Perner, D., Platt, U., Trainer, M. *et al.* (1987) Measurements of tropospheric OH concentrations – A comparison of field data with model predictions, *J. Atmos. Chem.*, 5, 185–216.

Perner, D., Klupfel, T., Parchatka, U. *et al.* (1991) Ground-Based Uv-Vis Spectroscopy – Diurnal OClO profiles during January 1990 above Sondre-Stromfjord, Greenland, *Geophys. Res. Lett.*, 18, 787–790.

Perner, D., Roth, A. & Klupfel, T. (1994) Ground-based measurements of stratospheric OClO, NO_2, and O_3 at Sondre-Stromfjord in Winter 1991/92. *Geophys. Res. Lett.*, 21, 1367–1370.

Petritoli, A., Ravegnani, F., Giovanelli, G. *et al.* (2002) Off-axis measurements of atmospheric, trace gases by use of an airborne ultraviolet-visible spectrometer, *Appl. Optics*, 41, 5593–5599.

Pfeilsticker, K. & Platt, U. (1994) Airborne measurements during the Arctic Stratospheric Experiment – Observation of O_3 and NO_2, *Geophys. Res. Lett.*, 21, 1375–1378.

Pfeilsticker, K., Bosch, H., Camy-Peyret, C. *et al.* (2001) First atmospheric profile measurements of UV/visible O_4 absorption band intensities: Implications for the spectroscopy, and the formation enthalpy of the O_2-O_2 dimer, *Geophys. Res. Lett.*, 28, 4595–4598.

Plane, J.M.C. & Nien, C.F. (1992) Differential optical absorption spectrometer for measuring atmospheric trace gases, *Rev. Sci. Instrum.*, 63, 1867–1876.

Plane, J.M.C. & Smith, N. (1995) Atmospheric monitoring by differential optical absorption spectroscopy, in Clark, R.J.H. & Hester, R.E. (eds), *Spectrosc. Environ. Sci.*, John Wiley, London, pp. 223–262.

Platt, U., Perner, D. & Patz, H.W. (1979) Simultaneous measurement of atmospheric CH_2O, O_3, and NO_2 by differential optical absorption, *J. Geophys. Res.-Oceans and Atmos.*, 84, 6329–6335.

Platt, U. & Perner, D. (1980) Direct measurements of atmospheric CH_2O, HNO_2, O_3, NO_2, and SO_2 by differential optical absorption in the near UV, *J. Geophys. Res.-Oceans Atmos.*, 85, 7453–7458.

Platt, U., Perner, D., Harris, G.W. *et al.* (1980a) Observations of nitrous acid in an urban atmosphere by differential optical absorption, *Nature*, 285, 312–314.

Platt, U., Perner, D., Winer, A.M. *et al.* (1980b) Detection of NO_3 in the polluted troposphere by differential optical absorption, *Geophys. Res. Lett.*, 7, 89–92.

Platt, U. & Perner, D. (1983) in Killinger, D.K. & Mooradian, A. (eds), *Optical and remote sensing*, Springer-Verlag, Berlin, pp. 97–105.

Platt, U. (1991) Spectroscopic measurement of free radicals (OH, NO_3) in the atmosphere, *Fresenius J. Anal. Chem.*, 340, 633–637.

Platt, U., (1994) Differential optical absorption spectroscopy (DOAS), in Sigrist, M.W. (ed.), *Air monitoring by spectroscopy techniques*, John Wiley, London, pp. 27–83.

Platt, U., Marquard, L., Wagner, T. *et al.* (1997) Corrections for zenith scattered light DOAS, *Geophys. Res. Lett.*, 24, 1759–1762.

Pommereau, J.P. & Goutail, F. (1988) O_3 and NO_2 ground-based measurements by visible spectrometry during Arctic winter and spring 1988, *Geophys. Res. Lett.*, 15, 891–894.

Pommereau, J.P. & Piquard, J. (1994) Observations of the vertical distribution of stratospheric OClO, *Geophys. Res. Lett.*, 21, 1231–1234.

Press, W.H., Flannery, T.S. & Vetterlinng, W.T. (1986) Numerical recipes: The art of scientific computing, University Press, Cambridge.

Preston, K.E., Fish, D.J., Roscoe, H.K. *et al.* (1998) Accurate derivation of total and stratospheric vertical columns of NO_2 from ground-based zenith-sky measurements, *J. Atmos. Chem.*, 30, 163–172.

Price, P.N. (1999) Pollutant tomography using integrated concentration data from non-intersecting optical paths, *Atmos. Environ.*, 33, 275–280.

Pundt, I., Pommereau, J.P., Phillips, C. *et al.* (1998) Upper limit of iodine oxide in the lower stratosphere, *J. Atmos. Chem.*, 30, 173–185.

Pundt, I., Pommereau, J.P., Chipperfield, M.P. *et al.* (2002) Climatology of the stratospheric BrO vertical distribution by balloon-borne UV-visible spectrometry, *J. Geophys. Res.-Atmos.*, 107, art. no.-4806.

Pundt, I., Mettendorf, K.-U., Laepple, T. *et al.* (2005) Measurements of trace gas distributions by long-path DOAS-Tomography: 2D mapping of NO_2 during the motorway campaign BAB II, *Atmos. Environ.*, 39, 967–975.

Richter, A., Eisinger, M., Ladstatter-Weissenmayer, A. *et al.* (1999) DOAS Zenith sky observations: 2. Seasonal variation of BrO over Bremen (53 degrees N) 1994–1995, *J. Atmos. Chem.*, 32, 83–99.

Rodgers, C.D. (1976) Retrieval of atmospheric temperature and composition from remote measurements of thermal radiation, *Rev. Geophys.*, 14, 609–624.

Rodgers, C.D. (1990) Characterization and error analysis of profiles retrieved from remote sounding measurements, *J. Geophys. Res.-Atmos.*, 95, 5587–5595.

Roscoe, H.K., Kerridge, B.J., Pollitt, S. *et al.* (1990) Intercomparison of remote measurements of stratospheric NO and NO_2, *J. Atmos. Chem.*, 10, 111–144.

Roscoe, H.K., Johnston, P.V., Van Roozendael, M. *et al.* (1999) Slant column measurements of O_3 and NO_2 during the NDSC intercomparison of zenith-sky UV-visible spectrometers in June 1996, *J. Atmos. Chem.*, 32, 281–314.

Rothman, L.S., Barbe, A., Benner, D.C. *et al.* (2003) The HITRAN molecular spectroscopic database: Edition of 2000 including updates through 2001, *J. Quant. Spectrosc. Radiat. Transf.*, 82, 5–44.

Saiz-Lopez, A. & Plane, J.M.C. (2004) Novel iodine chemistry in the marine boundary layer, *Geophys. Res. Lett.*, 31, art. no.-L04112.

Saiz-Lopez, A., Plane, J.M.C. & Shillito, J.A. (2004) Bromine oxide in the mid-latitude marine boundary layer, *Geophys. Res. Lett.*, 31, art. no.-L03111.

Sander, S.P. & Friedl, R.R. (1989) Kinetics and product studies of the reaction ClO + BrO using flash photolysis ultraviolet absorption, *J. Phys. Chem.*, 93, 4764–4771.

Sanders, R.W., Solomon, S., Mount, G.H. *et al.* (1987) Visible spectroscopy at McMurdo Station, Antarctica. 3. Observations of NO_3, *J. Geophys. Res.-Atmos.*, 92, 8339–8342.

Sanders, R.W., Solomon, S., Smith, J.P. *et al.* (1993) Visible and near-ultraviolet spectroscopy at McMurdo Station, Antarctica. 9. Observations of OClO from April to October 1991, *J. Geophys. Res.-Atmos.*, 98, 7219–7228.

Sarkissian, A., Vaughan, G., Roscoe, H.K. *et al.* (1997) Accuracy of measurements of total ozone by a SAOZ ground-based zenith sky visible spectrometer, *J. Geophys. Res.-Atmos.*, 102, 1379–1390.

Scherer, J.J., Paul, J.B., O'Keefe, A. *et al.* (1997) Cavity ringdown laser absorption spectroscopy: History, development, and application to pulsed molecular beams. *Chem. Rev.*, 97, 25–51.

Schiller, C., Wahner, A., Platt, U. *et al.* (1990) Near UV atmospheric absorption–Measurements of column abundances during Airborne Arctic Stratospheric Expedition, January–February 1989. 2. OClO observations, *Geophys. Res. Lett.*, 17, 501–504.

Schulz, K.J. & Simpson, W.R. (1998) Frequency-matched cavity ring-down spectroscopy, *Chem. Phys. Lett.*, 297, 523–529.

Simpson, W.R. (2003) Continuous wave cavity ring-down spectroscopy applied to in situ detection of dinitrogen pentoxide (N_2O_5), *Rev. Sci. Instrum.*, 74, 3442–3452.

Sioris, C.E. & Evans, W.F.J. (1999) Filling in of Fraunhofer and gas-absorption lines in sky spectra as caused by rotational Raman scattering, *Appl. Optics*, 38, 2706–2713.

Smith, J.P. & Solomon, S. (1990) Atmospheric NO_3. 3. Sunrise disappearance and the stratospheric profile. *J. Geophys. Res.-Atmos.*, 95, 13 819–13 827.

Smith, J.P., Solomon, S., Sanders, R.W. *et al.* (1993) Atmospheric NO_3. 4. Vertical profiles at middle and polar latitudes at sunrise, *J. Geophys. Res.-Atmos.*, 98, 8983–8989.

Smith, N., Plane, J.M.C., Nien, C.F. *et al.* (1995) Nighttime radical chemistry in the San-Joaquin Valley, *Atmos. Environ.*, 29, 2887–2897.

Solomon, S., Schmeltekopf, A.L. & Sanders, R.W. (1987a) On the interpretation of zenith sky absorption measurements, *J. Geophys. Res.-Atmos.*, 92, 8311–8319.

Solomon, S., Mount, G.H., Sanders, R.W. *et al.* (1987b) Visible spectroscopy at McMurdo-Station, Antarctica. 2. observations of OClO, *J. Geophys. Res.-Atmos.*, 92, 8329–8338.

Solomon, S., Miller, H.L., Smith, J.P. *et al.* (1989a) Atmospheric NO_3. 1. Measurement technique and the annual cycle at 40-Degrees-N, *J. Geophys. Res.-Atmos.*, 94, 11 041–11 048.

Solomon, S., Sanders, R.W., Mount, G.H. *et al.* (1989b) Atmospheric NO_3. 2. Observations in polar regions, *J. Geophys. Res.-Atmos.*, 94, 16 423–16 427.
Stutz, J. & Platt, U. (1996) Numerical analysis and estimation of the statistical error of differential optical absorption spectroscopy measurements with least-squares methods, *Appl. Optics*, 35, 6041–6053.
Stutz, J. & Platt, U. (1997) Improving long-path differential optical absorption spectroscopy with a quartz-fiber mode mixer, *Appl. Optics*, 36, 1105–1115.
Stutz, J., Ackermann, R., Fast, J.D. *et al.* (2002) Atmospheric reactive chlorine and bromine at the Great Salt Lake, Utah, *Geophys. Res. Lett.*, 29, art. no.-1380.
Syed, M.Q. & Harrison, A.W. (1980) Ground based observations of stratospheric nitrogen dioxide, *Can. J. Phys.*, 58, 788–802.
Tuckermann, M., Ackermann, R., Golz, C. *et al.* (1997) DOAS-observation of halogen radical-catalysed arctic boundary layer ozone destruction during the ARCTOC-campaigns 1995 and 1996 in Ny-Alesund, Spitsbergen, *Tellus Ser. B-Chem. Phys. Meteorol.*, 49, 533–555.
Vanroozendael, M., Fayt, C., Bolsee, D. *et al.* (1994) Ground-based stratospheric NO_2 monitoring at Keflavik (Iceland) during EASOE, *Geophys. Res. Lett.*, 21, 1379–1382.
VanRoozendael, M., DeMaziere, M., Hermans, C. *et al.* (1997) Ground-based observations of stratospheric NO_2 at high and mid-latitudes in Europe after the Mount Pinatubo eruption, *J. Geophys. Res.-Atmos.*, 102, 19 171–19 176.
Vaughan, G., Roscoe, H.K., Bartlett, L.M. *et al.* (1997) An intercomparison of ground-based UV-visible sensors of ozone and NO_2, *J. Geophys. Res.-Atmos.*, 102, 1411–1422.
von Friedeburg, C., Wagner, T., Geyer, A. *et al.* (2002) Derivation of tropospheric NO_3 profiles using off-axis differential optical absorption spectroscopy measurements during sunrise and comparison with simulations, *J. Geophys. Res.-Atmos.*, 107, art. no.-4168.
Vountas, M., Rozanov, V.V. & Burrows, J.P. (1998) Ring effect: Impact of rotational Raman scattering on radiative transfer in earth's atmosphere, *J. Quant. Spectrosc. Radiat. Transf.*, 60, 943–961.
Wagner, T., Otten, C., Pfeilsticker, K. *et al.* (2000) DOAS moonlight observation of atmospheric NO_3 in the Arctic winter, *Geophys. Res. Lett.*, 27, 3441–3444.
Wagner, T., von Friedeburg, C., Wenig, M. *et al.* (2002) UV-visible observations of atmospheric O-4 absorptions using direct moonlight and zenith-scattered sunlight for clear-sky and cloudy sky conditions, *J. Geophys. Res.-Atmos.*, 107, art. no.-4424.
Wahner, A., Ravishankara, A.R., Sander, S.P. *et al.* (1988) Absorption cross-section of BrO between 312 and 385 nm at 298 and 223 K, *Chem. Phys. Lett.*, 152, 507–512.
Wahner, A., Jakoubek, R.O., Mount, G.H. *et al.* (1989) Remote sensing observations of daytime column NO_2 during the Airborne Antarctic Ozone Experiment, August 22 to October 2, 1987, *J. Geophys. Res.-Atmos.*, 94, 16 619–16 632.
Wahner, A., Callies, J., Dorn, H.P. *et al.* (1990a) Near UV atmospheric absorption measurements of column abundances during airborne Arctic Stratospheric Expedition, January–February 1989. 1. Technique and NO_2 observations, *Geophys. Res. Lett.*, 17, 497–500.
Wahner, A., Callies, J., Dorn, H.P. *et al.* (1990b) Near UV atmospheric absorption measurements of column abundances during Airborne Arctic Stratospheric Expedition, January–February 1989. 3. BrO observations, *Geophys. Res. Lett.*, 17, 517–520.
Wang, P., Bruns, M., Richter, A. *et al.* (2003) Validation of SCHIAMACHY with AMAX-DOAS measurements from the DLR Falcon, *Geophys. Res. Abs.*, 5, 9341.
Weaver, A., Solomon, S., Sanders, R.W. *et al.* (1996) Atmospheric NO_3. 5. Off-axis measurements at sunrise: Estimates of tropospheric NO_3 at 40 degrees N, *J. Geophys. Res.-Atmos.*, 101, 18 605–18 612.
Wheeler, M.D., Newman, S.M., Ishiwata, T. *et al.* (1998) Cavity ring-down spectroscopy of the $A^2\Pi_{3/2}$-$X^2\Pi_{3/2}$ transition of BrO, *Chem. Phys. Lett.*, 285, 346–351.

Wittrock, F., Muller, R., Richter, A. *et al.* (2000) Measurements of iodine monoxide (IO) above Spitsbergen, *Geophys. Res. Lett.*, 27, 1471–1474.
Wittrock, F., Oetjen, H., Richter, A. *et al.* (2004) MAX-DOAS measurements of atmospheric trace gases in Ny-Alesund, *Atmos. Chem. Phys. Diss.*, 3, 6109–6145.
Xie, P., Liu, W., Fu, Q. *et al.* (2004) Intercomparison of NOx, SO_2, O_3, and aromatic hydrocarbons measured by a commercial DOAS system and traditional point monitoring techniques, *Adv. Atmos. Sci.*, 21, 211–219.
Ye, J., Ma. L.S. & Hall, J.L. (1998) Ultrasensitive detections in atomic and molecular physics: Demonstrations in molecular overtone spectroscopy, *J. Opt. Soc. Am. B*, 15, 6–15.
Yokelson, R.J., Burkholder, J.B., Fox, R.W. *et al.* (1994) Temperature dependence of the NO_3 absorption-spectrum, *J. Phys. Chem.*, 98, 13 144–13 150.
Zhu, L. & Johnston, G. (1995) Kinetics and products of the reaction of the vinoxy radical with O_2, *J. Phys. Chem.*, 99, 15 114–15 119.

Chapter 4

Fluorescence Methods

Ezra C. Wood and Ronald C. Cohen

4.1 Introduction

Fluorescence is among the most sensitive analytical tools and consequently is applied widely in chemistry, physics, biology, and Earth and atmospheric sciences. The high sensitivity derives from the ability to configure fluorescence instruments so that they have high efficiency for detecting photons emitted by the molecule of interest and simultaneously a zero or near zero background signal. Fluorescence experiments have become more widely available and are applied in a broader range of applications as laser excitation sources, detectors and optical filters have become more reliable and less expensive. The decreases in size, weight and power consumption of laser sources are continuing to fuel the evolution of instrument design.

Fluorescence has been applied to measurements of atmospheric trace gases such as OH, NO, NO_2, Cl, Br, CO, PAHs and other molecules. Perhaps the most famous example is the anti-correlation of ClO with O_3 measured in the Antarctic vortex during the austral spring of 1987 (Anderson *et al.*, 1989), which strongly confirmed the role of chlorofluorocarbon (CFC) emissions in the chemistry of the ozone hole (Anderson *et al.*, 1991). Fluorescence measurements of OH, HO_2, ClO, BrO, NO and NO_2 were used to quantify the rate of stratospheric ozone removal in unprecedented detail in 1992 (Wennberg *et al.*, 1994b). These observations showed that, in contrast to previous modeling, the catalytic cycles involving the hydrogen oxides accounted for up to 50% of the total photochemical ozone loss at an altitude of 20 km.

In the troposphere, fluorescence measurements of OH have confirmed that our understanding of the sources and sinks of the Earth's major oxidant is approximately correct in a wide range of environs. Measurements of the OH lifetime have indicated the existence of unmeasured biogenic volatile organic compounds (VOCs) (Di Carlo *et al.*, 2004). A more quantitative description of OH and other HO_y compounds remains actively sought after.

Fluorescence has long been used to observe NO via its chemiluminescent reaction with ozone (Chapter 7). The application of this technique to the measurement of NO_2 has made indirect detection of a wide suite of nitrogen oxides possible, resolving the identity of the 'missing NO_y' and leading to new insights into the chemistry of organic nitrates (Day *et al.*, 2002, 2003).

In this chapter, we describe the fundamental principles that guide the development of instrumentation for detection of these species, emphasizing the common themes and areas where the application requires deviation from these commonalities. Useful scientific applications require instruments to be both highly sensitive and capable of demonstrating specificity, so that false signals from other species present in the atmosphere during a measurement are negligible or easily quantified. It is also crucial to have a calibration strategy that is accurate, precise over long time periods and transferable. In Sections 4.3–4.5 we review applications to specific molecules with attention to the above design criteria. Both direct detection methods and indirect strategies involving conversion of certain molecules (e.g. HO_2 or HNO_3) to those that can be detected by fluorescence techniques (e.g. OH or NO_2) are discussed. We conclude this chapter with an outlook on future progress.

4.2 Basic aspects of fluorescence measurements

4.2.1 Molecular and atomic photophysics

Fluorescence is the spontaneous emission of light resulting from the relaxation of an atom or molecule from an excited state to a lower state. Most atoms and small molecules absorb strongly in the ultraviolet (UV) region of the electromagnetic spectrum, corresponding to the energy gap between bound electronic states. These absorption spectra and subsequent fluorescence are unique and are generally characterized by a number of discrete peaks, which provide a convenient means by which measurements can be tailored to individual compounds. Although some large molecules exhibit fluorescence (e.g. naphthalene, synthetic dye molecules, tryptophan), in general, fluorescence is less efficient for larger compounds because quenching and internal conversion are competitive with emission.

Figure 4.1 depicts the major photophysical processes for a diatomic molecule such as OH. The solid arrow depicts excitation from the ground electronic state, $\tilde{X}$, to the first electronically excited state, $\tilde{A}$. The molecule can lose this electronic energy by fluorescence or by quenching. The wavy arrow depicts fluorescence and the corresponding transition to a vibrational level on the ground potential energy curve. The energy can also be 'quenched' by collisions with bath molecules (M), most commonly nitrogen and oxygen, as depicted by the dashed arrow. Quenching removes energy and converts it to thermal motion or internal excitation of a non-fluorescent species. In some cases only vibrational energy is removed and the molecule remains on potential energy surface $\tilde{A}$ (not depicted in Figure 4.1). Emission that follows vibrational quenching in the $\tilde{A}$ state or whose final destination is a vibrationally excited state of the $\tilde{X}$ state is red-shifted from the excitation wavelength.

In addition to quenching, which suppresses fluorescence, the electronically excited molecule may undergo the collisionless processes of internal conversion, intersystem crossing or dissociation. In internal conversion, the energy of the excited electronic state is converted to rotational and vibrational energy of the ground electronic state. An intersystem crossing occurs when the molecule switches to an isoenergetic state of different spin multiplicity. Finally, the molecule can dissociate if it enters an unbound

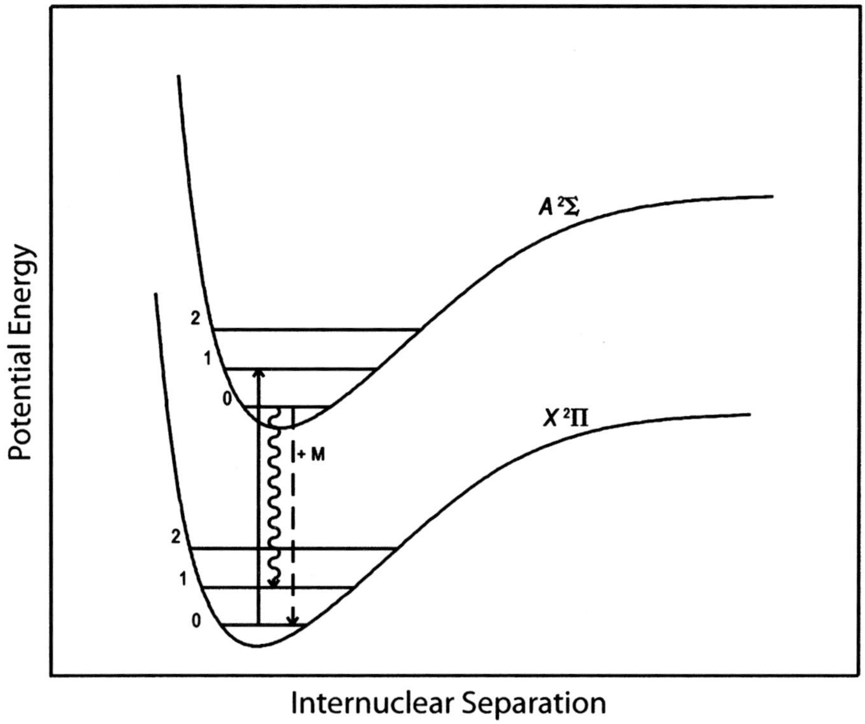

Figure 4.1 Schematic diagram of the potential energy curves of a diatomic molecule such as OH. Excitation, emission (fluorescence), and quenching are represented by the solid, wavy, and dashed arrows, respectively.

(repulsive) state either by direct excitation or after conversion from a bound excited state (predissociation). Following excitation of an ensemble of molecules, the population of the excited state will decay exponentially with a characteristic lifetime τ, defined as $\tau^{-1} = \tau_{\mathrm{R}}^{-1} + \tau_{\mathrm{Q}}^{-1} + \tau_{\mathrm{IC}}^{-1} + \tau_{\mathrm{ISC}}^{-1} + \tau_{\mathrm{DIS}}^{-1}$ where the terms on the right hand side are the rate constants for decay of the excited state due to emission, quenching, internal conversion, intersystem crossing and predissociation, respectively. The value of τ_{R} depends on the strength of the transition.

4.2.2 The fluorescence signal

The fluorescence signal S (photon counts/s) is governed by the equation

$$S = RQC \tag{4.1}$$

where R is the excitation rate, Q is the fluorescence quantum yield, and C is the collection efficiency of the instrument. The excitation rate is

$$R = E\left\{1 - \exp(-cl \int \Phi(\nu)\sigma(\nu, T, P)\mathrm{d}\nu)\right\} \tag{4.2}$$

or approximating $e^{-x} \approx (1-x)$,

$$R = Ecl\int \Phi(\nu)\sigma(\nu, T, P)\mathrm{d}\nu \tag{4.3}$$

where E is the excitation power in photons/s and the term inside the brackets of Equation (4.2) is the fraction of photons absorbed by the analyte of number density c over the path length l. The integral represents the overlap between the spectral profile of the excitation source $\Phi(\nu)$ and the absorption cross section of the molecule $\sigma(\nu)$, where ν is frequency. This equation assumes the fraction of molecules in the excited state to be small. Use of the full Einstein equations for stimulated absorption, stimulated emission, and spontaneous emission is required if populations approach 50%. Under such conditions the transition is said to be *saturated* and the fluorescence is independent of input power. Since most fluorescence instruments are limited by noise, which is linearly proportional to the power of the light source, saturation is not advantageous.

The fluorescence quantum yield, Q, is the fraction of excited molecules that fluoresce, that is (photons emitted)/(photons absorbed). For molecules that can be treated as two-level systems in which the only means of relaxation from the excited state to the ground state are radiative or collisionally induced, the fluorescence quantum yield is given by

$$Q = \frac{k_{\mathrm{rad}}}{k_{\mathrm{rad}} + \sum_i k_{\mathrm{q}i}[\mathrm{M}_i]} \tag{4.4}$$

where k_{rad} is the radiative rate constant of the excited species $(1/\tau_{\mathrm{R}})$ and $k_{\mathrm{q}i}$ are the fluorescence quenching rate constants due to collisional deactivation by bath molecules M_i, mainly nitrogen, oxygen, water, argon and carbon dioxide. Table 4.1 summarizes these relevant parameters for compounds commonly measured by fluorescence.

Often the quenching rate $\Sigma k_{\mathrm{q}i}[\mathrm{M}_i]$ is much greater than k_{rad}. Under these circumstances the $1/[\mathrm{M}]$ term in the fluorescence quantum yield combines with the c term in the excitation rate to make the overall signal rate proportional to $c/[\mathrm{M}]$, which is the mixing ratio. This allows the detection cell to be held at lower than ambient pressure without a penalty in the fluorescence signal, as the negative effect of the decreased number density is nearly equally offset by an increase in the fluorescence quantum yield due to decreased quenching. For example, the (0–0) transition of OH at 308 nm has a radiative lifetime of 676 ns (Bailey *et al.*, 1997), and $\Sigma k_{\mathrm{q}i}[\mathrm{M}_i]$ is ten times this radiative rate at 7 torr. For NO_2 and NO_3, which have excited state lifetimes of ~100 μs, the quenching rate is faster than the rate of radiative decay at pressures well below 1 torr.

The collection efficiency, C, is given by

$$C = \Omega\int T(\nu)\eta(\nu)\varepsilon(\nu)\mathrm{d}\nu \int_{t_1}^{t_2} \exp\left(\frac{-t}{\tau}\right)\mathrm{d}t \tag{4.5}$$

where Ω represents the fraction of the solid angle of the fluorescence intercepted by the collection lens and focused onto the photocathode; the first integral represents the transmission through the collection optics $T(\nu)$, the quantum efficiency of the PMT $\eta(\nu)$ and the emission spectrum of the molecule $\varepsilon(\nu)$; and the second integral is the fraction of the fluorescence which occurs in the detection time gate of width $t_2 - t_1$ for an excited state lifetime τ.

Table 4.1 Relevant spectroscopic parameters of a few compounds measured by fluorescence

Species	Absorption cross section σ (base e) (cm^2)	Excited state lifetime τ_0	Quenching rate constants k_q (cm^3 $molecule^{-1}$ s^{-1})	Additional compounds measurable
Cl	3.9×10^{-13} (118.0 nm) (1)	2.6 ns		ClO, $ClONO_2$, ClOOCl
OH	1.4×10^{-15} (282 nm) 7.1×10^{-15} (308 nm) (2)	676 ns (3)	$\tilde{A}^2\Sigma^+(\nu' = 0)$: 3.3×10^{-11} (N_2) 1.4×10^{-10} (O_2) (3)	HO_2, H_2O
NO	1.9×10^{-17} cm^2 (4) (single-photon)	205 ns (5)	$\tilde{A}^2\Sigma(\nu' = 0)$: 3.7×10^{-11} (N_2) 1.5×10^{-10} (O_2) (6)	NO_2, NO_y
NO_2	5.9×10^{-19} 408 nm 1.4×10^{-20} 640 nm (7)	65–130 μs (8)	6×10^{-11} (N_2) 5.4×10^{-11} (O_2) $\lambda_{excited} = 532$ nm	Peroxy nitrates, alkyl and hydroxy-alkyl nitrates, HNO_3
NO_3	2.2×10^{-17} 662 nm 1.5×10^{-17} 623 nm (9)	biexponential, 3–30 μs fast, 340 μs slow (10, 11)	$B^2E(\nu' = 0000)$: 1.7×10^{-11} (N_2) 2.1×10^{-11} (O_2) (11)	N_2O_5
HCHO	1.6×10^{-20} (354 nm) (12)	50 ns–4 μs (13, 14)	5.2×10^{-13} (N_2) 2.7×10^{-12} (O_2) $\lambda_{excite} = 353.2$ nm (15)	
SO_2	1.58×10^{-17} $\lambda_{excite} = 203.56$ nm (16)	13–600 μs (17, 18)	1.45×10^{-11} (He) $\lambda_{excite} = 266$ nm (19)	

Sources: (1) Anderson *et al.*, 1980; (2) McGee & McIlrath, 1984; (3) Bailey *et al.*, 1997; (4) Bradshaw *et al.*, 1982; (5) Luque & Crosley, 2000; (6) Nee *et al.*, 2004; (7) Voigt *et al.*, 2002; (8) Patten *et al.*, 1990; (9) Yokelson *et al.*, 1994; (10) Ishiwata *et al.*, 1983; (11) Nelson *et al.*, 1983; (12) Cantrell *et al.*, 1990; (13) Moore & Weisshaar, 1983; (14) Miller & Lee, 1978; (15) Shibuya & Lee, 1978; (16) Martinez & Joens, 1992; (17) Holtermann *et al.*, 1980; (18) Brus & McDonald, 1974; (19) Zhang *et al.*, 1998.

Although detailed knowledge of each of these components of the signal is essential to predict the performance of a fluorescence measurement (Bloss *et al.*, 2003, Holland *et al.*, 1995, Stevens *et al.*, 1994, Wennberg *et al.*, 1994a, Thornton *et al.*, 2000), in practice most of these quantities do not need to be explicitly evaluated. Rather, laser-induced fluorescence (LIF) instruments are calibrated using a known concentration of the sampled compound. Calibration is also essential because of potential sampling losses.

4.2.3 The background

The background (or zero) in a fluorescence instrument typically consists of

$$B = B_{chamber} + B_{P\text{-}scatter} + B_{solar} + B_{PMT} + B_{interference} \tag{4.6}$$

where $B_{chamber}$ is optical scatter from surfaces within the optical cell; $B_{P\text{-}scatter}$ is the sum of pressure-dependent Raman, Rayleigh and Mie scattering; B_{solar} is solar-induced; B_{PMT} is from the dark current of the PMT; and $B_{\text{interference}}$ represents any potentially time-varying signal, for example fluorescence or scattering by other atmospheric gases or aerosol.

General strategies for reducing the background of a fluorescence measurement include use of geometric baffles to block scattered laser light, operation at low pressure to reduce $B_{\text{P-scatter}}$ and to enable time-gated detection, and use of optical filters.

The background in fluorescence instruments is commonly measured by one or more of the following methods: introducing zero air to the instrument, selectively scrubbing the sampled air of the targeted compound, or tuning the excitation frequency off-resonance with the chosen absorption. As B inevitably varies for reasons that are inscrutable, it is advantageous to make $S \gg B$. Indeed the success of many fluorescence instruments stems from the near-zero backgrounds achieved.

4.2.4 The signal-to-noise ratio and detection limit

The mixing ratio χ is related to the fluorescence signal S by

$$\chi = \frac{S}{\alpha} \tag{4.7}$$

where α is the instrument's calibration constant in counts/s/ppb, S is experimentally determined by subtracting the background count rate B from the observed gross signal $S+B$. The associated uncertainty, σ_S, with this subtraction is given by:

$$\sigma_S = \sqrt{\sigma_{S+B}^2 + \sigma_B^2} \tag{4.8}$$

where σ_{S+B} and σ_B are the uncertainties of the total count rate and background count rates, respectively. The noise (σ_X) associated with a photon count rate X over an interval of time t is equal to $(Xt)^{1/2}$. Thus if equal amounts of time are spent measuring B and S, σ_S is equal to $(St + 2Bt)^{1/2}$ and the signal-to-noise ratio, SNR, can be expressed as

$$\text{SNR} = \frac{St}{\sigma_S} = \frac{St}{\sqrt{St + 2Bt}} = \frac{S\sqrt{t}}{\sqrt{S + 2B}} \tag{4.9}$$

Note that if the background and signal are proportional to excitation power, then the SNR will be proportional to the square root of the excitation power and to the square root of the averaging time.

At the detection limit, $S \ll 2B$. Combining Equations 4.7 and 4.9 gives the minimum detectable mixing ratio, χ_{min}:

$$\chi_{\text{min}} = \frac{\text{SNR}}{\alpha}\sqrt{\frac{2B}{t}} \tag{4.10}$$

If the uncertainty in B is negligible because it is not varying or only varying slowly with time, then the time spent measuring B can be reduced to a negligible fraction of the total measurement time for an additional factor of $\sqrt{2}$ in the SNR. This is often the case for instruments where B is proportional to laser power or determined by thermally induced dark counts. Although longer averaging times can eliminate shot noise, pure statistical noise rarely sets the ultimate detection limit on instrument performance. Factors such as Mie scattering from a time-varying atmospheric aerosol concentration, fluorescence of other gases, instability of the laser power, laser wavelength or laser alignment often determine the fundamental limit that cannot be improved upon by signal averaging.

4.2.5 General design of fluorescence instruments

At the heart of a fluorescence instrument is the optical cell, in which three elements intersect: gas flow, excitation and detection. Typical designs orient these three quantities along orthogonal axes so that the fluorescence is observed against a dark background. Figure 4.2 shows a typical detection cell. Laser light enters and exits a cell through Brewster-angle windows and passes through a series of narrow aperture baffles to reduce scattered laser light. Fluorescence emanating from the center of the cell is collected and collimated by a lens, passes through an optical filter and is then focused onto the photocathode of a PMT by a second lens.

Lamps have been successfully used for numerous fluorescence measurements, including $O(^3P)$ (Anderson, 1975; Kita *et al.*, 1996), halogens (Anderson *et al.*, 1980), sulphur dioxide (Luke, 1997), carbon monoxide (Holloway *et al.*, 2000), and OH measurements (Anderson, 1976). However, the use of lasers offers many advantages:

1) Lasers generally offer more intense beams of light.
2) Pulsed lasers allow for time-gated detection methods which greatly reduce background noise.
3) Scattered light can be minimized by aligning the electric field vector of the laser light parallel to the detection axis, as all types of scattering (Raman, Rayleigh, etc.) are primarily emitted in the plane perpendicular to the electric field vector.
4) Laser light can be relatively-easily multi-passed through an optical cell.
5) The tunability and narrow linewidths of some lasers allow for spectral modulation as a test of interferences or for measurement of backgrounds.

Time gating is most effective when the detection cell is held at a pressure low enough that the excited state lifetime is much longer than the duration of the laser pulse. Delayed detection of fluorescence is then nearly free of laser-induced background. The application of low-pressure time-gated fluorescence, techniques which had been frequently used in the laboratory, for atmospheric measurements was pioneered by Thomas Hard, Robert O'Brien and co-workers at Portland State University and they named it FAGE: fluorescence assay by gas expansion (Hard *et al.*, 1984). In FAGE, sampled air is expanded through a critical orifice into a low-pressure detection cell. In addition to the reduction in background attained by temporal rejection of the laser-induced background, FAGE offers the following advantages:

1) Pressure-dependent sources of background noise (Rayleigh, Raman and Mie scattering) are reduced.
2) The decrease in pressure broadening allows for spectral modulation of the fluorescence signal using the rotational structure of the analyte's absorption spectrum.
3) The reduced gas number density minimizes chemical and photolytic interferences.

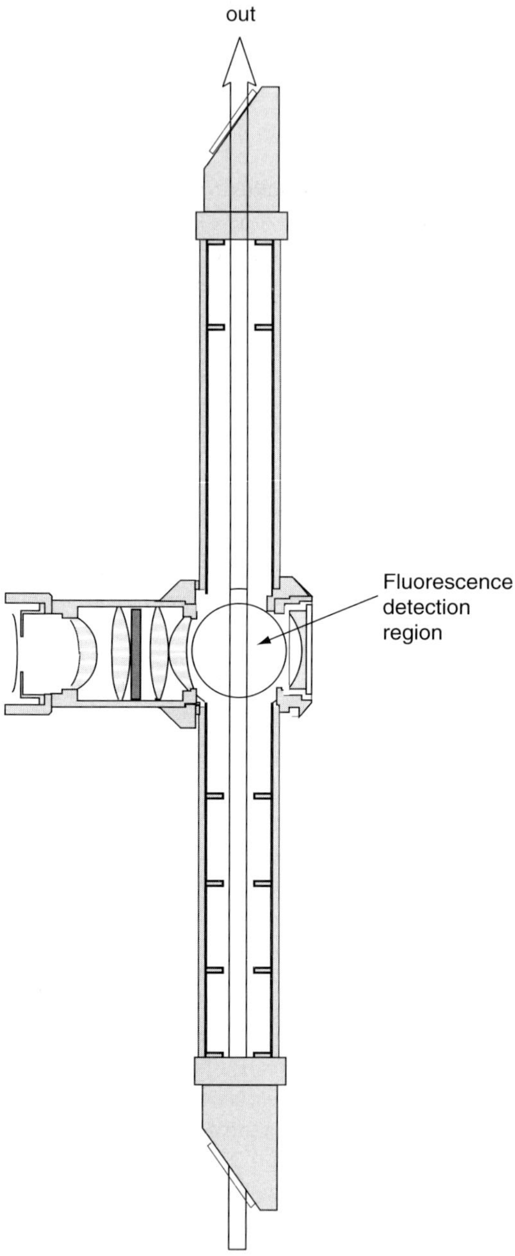

Figure 4.2 A single-pass fluorescence cell. Laser light enters and exits the cell through Brewster-angle windows and passes through a series of baffles. A fraction of the isotropically emitted fluorescence from the detection region is collected and collimated by a set of lenses, passed through a spectral filter, and is then focused onto the photocathode of a PMT by a second set of lenses. Gas flow is perpendicular to the page. (From Kanaya *et al.*, 2001, with kind permission of Springer Science and Business Media.)

The disadvantages of the low-pressure technique are the need for vacuum pumps and potential surface losses on the pressure-reducing critical orifice of the inlet. There is also a fall-off in the calibration constant at pressures low enough that the reduction in quenching rate no longer offsets the reduced number density.

Figure 4.3 illustrates the time gates used for detection of NO_2 at a cell pressure of 2 torr. The fluorescence lifetime of NO_2 at this pressure is approximately 270 ns. The laser-induced noise is almost entirely coincident with the 40 ns transit time of the laser pulse in the detection cell. Thus an 800 ns wide detection gate, delayed by 55 ns, reduces the background by 95% while collecting ~75% of the fluorescence. With a repetition rate of 8 kHz, this gives an experimental duty cycle of 0.64%, reducing any continuous noise sources (e.g. B_{solar}, B_{PMT}) by a factor of 156. To synchronize the excitation pulse with the detection electronics, a beamsplitter diverts a small fraction of the laser beam and directs it to a fast photodiode. The signal from this photodiode serves as the trigger for the photon counting electronics.

A single-pass design was used for the instrument shown in Figure 4.2; however, many fluorescence instruments use a multi-pass design utilizing highly reflective (>99.9%) mirrors. The obvious advantage of multi-passing is the increased excitation rate produced by the elevated photon flux. The price of this enhancement is usually an increased background resulting from light scattered by the mirrors themselves. This scattered light is often red-shifted, diffuse and emanates from a large area, making it difficult to completely eliminate using geometric baffles. The most common multi-pass design is a White cell. Herriott cells (Herriott *et al.*, 1964) are also used (Cleary *et al.*, 2002; Wood *et al.*, 2003), though these are less common due to the increased likelihood of saturation

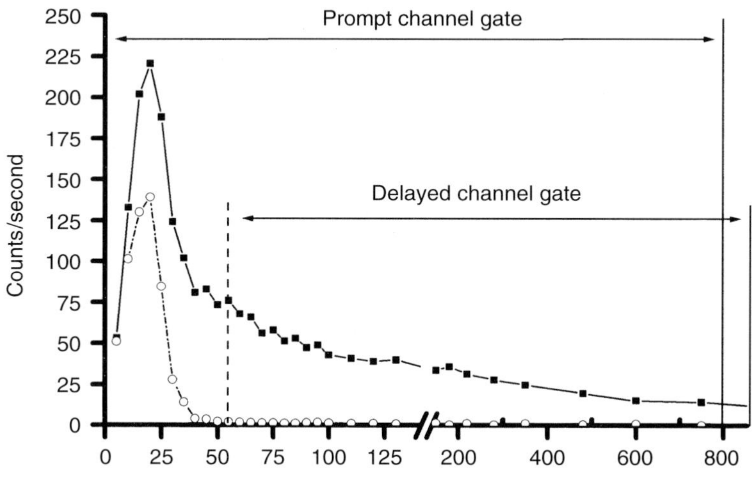

Figure 4.3 Reduction of the background by time gated detection. Total signal for 50 ppb NO_2 (squares) and 0 ppb NO_2 (circles), i.e., background signal, at 2 torr are plotted versus delay from when the laser pulse enters the detection cell. One channel of the counter catches all signal prompt with the laser pulse and has an 800 ns wide gate. The other channel has a 55 ns delayed, 800 ns wide gate to eliminate scatter prompt with the laser pulse. The area under the fluorescence curve remains large (~75%) in the delayed gate, while the background noise has essentially decayed to zero. (From Thornton *et al.*, 2000, reprinted with permission from American Chemical Society.)

of the electronic transition as each pass of the laser light is focused into the centre of the cell. A spherical mirror, placed at a distance of its radius of curvature below the centre of the optical cell, is often used to double the solid angle collected. As illustrated in Figure 4.2, baffles are used to shield the detector optics from scattered laser light. Additional baffles are often placed on the outer edge of the collection lens and retro-reflector. These baffles and the interior of detection cells are often painted with a low fluorescence optical black paint.

Optical filters are used to transmit the desired fluorescence in a certain spectral region while rejecting all other light. Fluorescence can be resonant with the excitation wavelength, as for OH detection at 308 nm; red-shifted from the excitation wavelength, as for NO_2 detection; or blue-shifted, as for two-photon LIF detection of NO. Bandpass filters, long-pass filters, short-pass filters or some combination of these three are used. The photon flux of 100 mW of 532 nm light is 2.7×10^{17} photons/s. With this light source, the Rayleigh-scattered photon flux from nitrogen and oxygen in a typical multi-pass optical cell at 2 torr is about 10^{10} photons/s, and the magnitude of chamber scatter is usually comparable. Typically about 1% of the scatter is imaged onto a photocathode with a quantum efficiency of 10%. Thus attenuation of order 10^{-7} is desirable to completely reject this scattering.

The main types of optical filters used are coloured glass filters, liquid chemical solution filters and interference filters. The first two are based on absorption, while interference filters have dielectric coatings which reject light by reflection. Coloured glass filters have sharp cut-on wavelengths and high transmissions in their spectral pass. Increased attenuation can be attained by use of multiple absorbing filters. The major drawback of absorbing filters is that the filters themselves fluoresce (Fong & Brune, 1997; Pfeiffer & Porter, 1964). Liquid filters (Anderson, 1976; Bradshaw *et al.*, 1985; Fong & Brune, 1997) contain a concentrated chemical solution with a distinct absorption spectrum contained between two solid windows. Interference filters exhibit high attenuation and high transmission and when coated with high grade quartz usually do not fluoresce; however, use of multiple filters does not lead to increased attenuation. Additionally, the cut-on wavelength of interference filters depends on the angle of incidence, requiring geometric baffling for rejection of scattered light that has sources in a different physical location than the desired molecular or atomic fluorescence. Detection of resonance fluorescence precludes such filtering (besides the use of bandpass filters) and requires efficient temporal gating as previously described, in order to reject laser scatter and Rayleigh scattering.

The overview above provides an introduction to the design criteria for measurement of atmospheric trace gases by fluorescence. In the following sections we discuss specific examples. Issues related to sampling which were given limited attention above but which are usually as challenging as detection will be discussed in more detail.

4.3 The hydroxyl radical

The hydroxyl radical, OH, regulates the concentration of the majority of trace gases, including ozone (Wennberg *et al.*, 1994b), and defines the oxidative capacity of the atmosphere (Levy, 1972; Logan *et al.*, 1981). Measurement of OH is difficult: peak concentrations in the troposphere and stratosphere are in the order of 10^6–10^7 molecules/cm^3. Heterogeneous reactions are extremely efficient, complicating sampling. Measurement of OH is also challenging because of photolytic interferences. OH is produced in the atmosphere primarily by the reaction sequence:

$$O_3 + h\nu \xrightarrow{\lambda<340\,\text{nm}} O_2 + O(^1D) \qquad (4.11)$$

$$O(^1D) + H_2O \rightarrow 2OH \qquad (4.12)$$

An intense light source whose intended purpose is excitation of OH can also photolyze ozone to produce $O(^1D)$. The false OH concentration is proportional to the square of the excitation power and linear in ozone and water vapour number density. Early attempts (Wang & Davis, 1974; Wang *et al.*, 1975, 1976; Davis *et al.*, 1976) to measure tropospheric OH by LIF detected only laser-generated OH.

The $(\tilde{A}^2\Sigma^+ \leftarrow X^2\Pi)$ electronic transition of OH is intense and has structured absorption and fluorescence. The $(\nu' = 0) \leftarrow (\nu'' = 0)$ transition at 308 nm and the $(\nu' = 1) \leftarrow (\nu'' = 0)$ vibronic transition at 282 nm are Doppler-limited and have peak absorption cross sections of $7.13 \times 10^{-15}\,\text{cm}^2$ and $1.43 \times 10^{-15}\,\text{cm}^2$, respectively (McGee & McIlrath, 1984), and are used for all measurements of atmospheric OH.

4.3.1 *Instruments for stratospheric and upper tropospheric OH measurements*

The first measurements of atmospheric OH using fluorescence were made in 1971 with a rocket-borne spectrometer in the upper stratosphere and lower mesosphere (Anderson, 1971a,b). These measurements were based on solar-induced fluorescence of OH at twilight conditions and gave an estimate of the OH column density in the mesosphere.

In 1975 Anderson measured stratospheric OH using an instrument that descended through the stratosphere while suspended from a stabilized parachute (Anderson, 1976). Ambient air was sampled as it flowed through the interior of the instrument during descents from altitudes as high as 43 km. OH was excited in the $(0 \leftarrow 0)$ transition at 308 nm and fluorescence was detected at the same wavelength (resonance fluorescence). The UV light source was a low-pressure microwave discharge of helium mixed with a small amount of water vapour housed in a quartz tube lamp. A spherical mirror on one side and a quartz lens on the other were used to produce a collimated beam of photons corresponding to OH emission lines in the range 306.4–312.0 nm. The lamp flux was monitored continuously during flight. Fluorescence from excited OH was detected by a PMT after passing through a narrow-bandpass interference filter centered at 309 nm, a polarization plate oriented so as to reduce lamp-induced Rayleigh scatter, and a quartz lens. The instrument background was determined by titration of ambient OH with HCl injected shortly before the detection point. Resonance fluorescence measurements of atomic oxygen, $O(^3P)$, were observed using similar instruments (Anderson, 1975; Takegawa *et al.*, 2001).

Using this instrument, HO_2 was observed by conversion to OH (Anderson *et al.*, 1981):

$$HO_2 + NO \rightarrow OH + NO_2 \tag{4.13}$$

The conversion efficiency of reaction 4.13 is less than unity due to the three-body reaction between OH and NO to form nitrous acid (HONO):

$$OH + NO + M \rightarrow HONO + M \tag{4.14}$$

The efficiency of the conversion was quantified by using two or more NO flow rates for the conversion. The fluorescence observed while the NO is injected stems from both OH and HO_2, and so OH measurements (without NO flow) were subtracted to extract HO_2 concentrations.

Despite the success of this instrument it was desirable to have an instrument that would have higher sensitivity and be capable of more effective tests for interferences. Elimination of the chemical zero was also desirable. Stimpfle and Anderson (1988) developed a dye laser pumped by a pulsed copper vapour laser designed to meet these objectives. The $Q_1(1)$ line at 281.91 nm or the $Q_1(2)$ line at 282.07 nm was pumped, and red-shifted fluorescence in the $(0 \leftarrow 0)$ band at 308 nm was detected. This laser system avoided the problem of laser-generated OH and maintained high sensitivity by using a high repetition rate (17 kHz) and a low pulse energy. The laser system was housed in a gondola and the UV beam was directed into the detection pod where the light was focused into a multi-pass White cell. The fluorescence was collected by an array of 450 fibre optic cables and relayed back into the gondola where the fluorescence was delivered to a spectrometer with a bi-alkali PMT fitted at the exit slit. Nine of the fibre optic cables were directed to a separate PMT with a 282 nm interference filter in order to measure the Rayleigh scatter from the optical cell. Observation of this Rayleigh scatter was used to monitor changes in the sensitivity of the instrument. At a given pressure, any change in the calibration constant of the instrument (e.g. change in laser alignment or power) will manifest as an equal change in the observed Rayleigh scatter. Thus, observation of this scatter can serve as an in-flight proxy calibration and the quantity that must be measured in the laboratory is the ratio of the instrument's response to OH and Rayleigh scatter. This is a design element common to most of the Anderson group's instruments. The idea is to find a transfer standard for the instrument calibration that can be accurately and precisely measured in both the laboratory and under field conditions where thermal changes, vibrations or other factors result in sub-optimal optical alignment.

The optical collection efficiency of the laser-equipped balloon instrument was subsequently improved by installing the PMT directly on the detection pod (Stimpfle *et al.*, 1990). The spectrometer was replaced by two filters: a chemical solution filter (potassium hydrogen phthalate dissolved in methyl alcohol) followed by an interference filter (4 nm full width at half maximum, FWHM, bandpass centred at 310 nm).

Because OH is highly reactive and removed with near-unit efficiency upon collision with surfaces, sampling requires that the detection volume be unperturbed. This requirement, which argues for a short, wide, straight sampling tube, is balanced against the requirement for baffling against solar scatter, which favours the exact opposite – a long, narrow, bent tube. In the balloon instruments, the sample chamber had a 15 cm diameter, and a steady descent velocity of ~5 m/s was maintained through radio control of the balloon. Interaction with the instrument wall was monitored by a thermistor array after the gas passed through the detection cell. Elevated temperatures near the centre were a diagnostic for air that had touched the instrument walls and then moved into the detection volume. Baffles prevented excessive solar scatter in the instrument. The low-altitude limit of operation was between 23 and 29 km due to either excessive Rayleigh scatter or for, the laser-equipped instruments, thermal instability of the laser. The 1σ uncertainties in the OH and HO_2 measurements were 18 and 19%, respectively.

The logistical demands of high altitude ballooning limited these instruments to infrequent flights of short duration, and consequently in-flight diagnostics were limited. Also, the feasibility of installing a comprehensive chemical and dynamical payload on a balloon was low, limiting the potential scientific return. Thus an instrument that could be mounted on the premier stratospheric chemistry research platform, NASA's ER-2, was highly desirable (information on the ER-2 aircraft is available at http://www.dfrc.nasa.gov/Newsroom/FactSheets/FS-046-DFRC.html). In May 1993, OH and HO_2 were measured in the lower stratosphere aboard the ER-2 aircraft during the Stratospheric Photochemistry, Aerosols and Dynamics Expedition (SPADE) (Wennberg *et al.*, 1994a). A diagram of this instrument is shown in Figure 4.4. This instrument was much smaller than its balloon-borne predecessor and incorporated an array of diagnostics that were not possible to implement during the short duration of a balloon flight. The laser system was a compact ($<0.2\,m^3$, <40 kg) dye laser pumped by two diode-pumped, 6 kHz repetition rate 450 mW Nd:YLF lasers. The conversion efficiency of the dye laser fundamental into the UV was ~10%, with a final 282 nm UV beam of linewidth $0.1\,cm^{-1}$.

The instrument was mounted on the nose of the ER-2 aircraft where a sampling system was devised that allowed controlled flow speeds deep into the instrument without contact

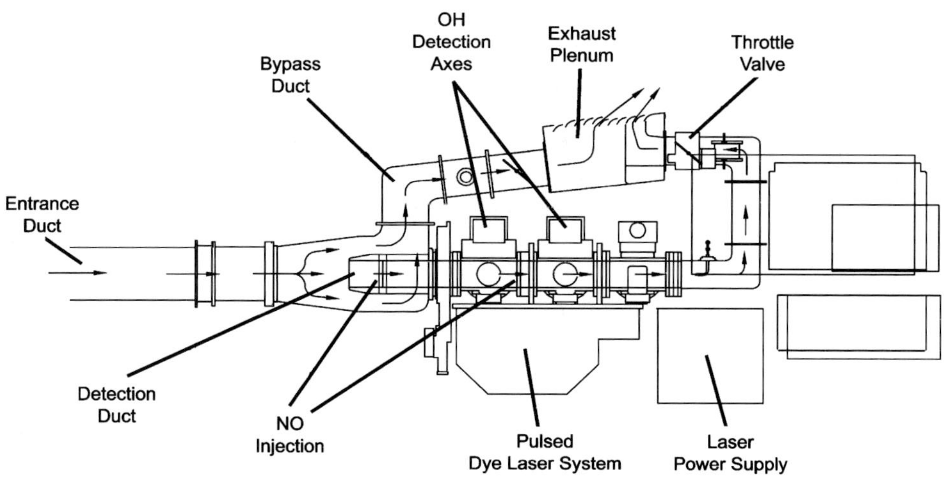

Figure 4.4 The HO_x instrument mounted in the nose of the NASA ER-2 aircraft. Most of the flow bypasses the inner duct. A laminar subsection of the flow enters the inner detection duct and passes through three consecutive OH detection axes. HO_2 is converted to OH by reaction with NO injected at one of two positions. Water vapour is detected in the third axis (see Section 4.3.3) (From Wennberg *et al.*, 1995, with permission of American Meteorological Society.)

with the instrument walls. The air flow was slowed from the aircraft velocity of 200 to 100 m/s in a large bypass duct. A sub-sample of air in this duct entered the detection duct where flow speeds were regulated between 10 and 100 m/s. Observations showed that turbulence in the flow at speeds below 40 m/s resulted in loss of OH. The gas passed by three detection cells used for measurement of OH, HO_2, and H_2O (Section 4.3.4). NO was injected after the first cell to allow for detection of HO_2 in the second cell. An NO injector was also positioned before the first cell as a diagnostic for the HO_2-to-OH conversion efficiency.

The aircraft platform flew dozens of times in the first deployment of this HO_x instrument and hundreds since. Diagnostics showed [OH] was linear in laser power, confirming non-quadratic behaviour, and showing that retrieved [OH] was independent of laser pulse energy. Also [OH] was observed to be zero at night, further confirming the absence of laser-generated OH. Finally, titration of ambient OH with perfluoropropene (C_3F_6) was also used to confirm the absence of laser-generated OH. Perfluoropropene is the compound of choice for chemical modulation as it contains no hydrogen atoms that could possibly re-form OH while titrating ambient OH (Dubey *et al.*, 1996). Introduction of perfluoropropene to the inlet only removes ambient OH. Laser-generated OH molecules are detected in nanoseconds after their creation by the same laser pulse, and their reaction with perfluoropropene is too slow for removal prior to detection.

The background of OH LIF instruments is routinely measured by tuning the laser on and off the chosen rovibronic absorption feature in the OH absorption spectrum. An OH 'reference' chamber, in which high concentrations of OH are produced, is used to lock the laser frequency. Ideally all components of the background are invariant over this small shift in excitation wavelength. Most OH LIF instruments continuously employ spectral modulation during operation and occasionally use chemical modulation to test for interferences.

In principle, the calibration of the balloon and aircraft-borne instruments could be calculated from detailed knowledge of the transmission of the optical system and the detailed temperature-dependent state-to-state quenching parameters (Smith & Crosley, 1990; Wennberg *et al.*, 1994a). In practice, the optical transmission is not known well enough and a direct calibration is necessary. Such calculations do, however, provide a useful check on the signal and magnitude of the observed pressure and temperature dependence. The balloon-borne and aircraft-borne instruments were calibrated using a known quantity of OH that was produced by reacting an NO_2 reference standard with excess H atoms:

$$H + NO_2 \rightarrow OH + NO \tag{4.15}$$

These calibrations were performed over a range of temperatures and pressures and using N_2 and $N_2/O_2/Ar$ mixtures (Anderson, 1976; Stimpfle and Anderson, 1988; Stimpfle *et al.*, 1990; Wennberg *et al.*, 1994a).

For the ER-2 instrument, the calibration was checked against absorption at very high mixing ratios of OH to test the assumption that the OH delivered to the detection region could be calculated from the concentration of the NO_2 standard. These two methods agreed to within 10%. The ER-2 instrument utilized the N_2 Stokes Raman scattering at 302 nm as its transfer standard. This was more reliable than the Rayleigh transfer standard used in the balloon instrument because the laser scatter from the detection chamber is much lower at red-shifted wavelengths than at the Rayleigh lines.

The uncertainty in the ER-2 OH measurements (Perkins *et al.*, 2001) was $\pm 13\%(1\sigma)$, which is a combination of the uncertainty of the laboratory calibration and the uncertainty in the inference of the sensitivity from in-flight diagnostics. The precision was 0.03 ppt in a two-second integration. For HO_2, additional uncertainty in the chemical conversion leads to a slightly greater uncertainty of $\pm 15\%$, with a precision of 0.15 ppt in a two-second integration. Typical detection limits were 0.05 ppt.

4.3.2 Instruments for lower tropospheric OH measurements

The central role of OH in tropospheric chemistry was recognized in the late 1960s and early 1970s (Weinstock, 1969). In the troposphere, OH removes VOCs such as methane and isoprene by initiating oxidation. In the presence of NO_x and sunlight, the oxidation results in catalytic production of ozone. Secondary aerosols may also be produced. In the lower troposphere, ozone is an irritant and the principal component of photochemical smog. In the upper troposphere, ozone is a potent greenhouse gas.

The high water vapour concentrations in the troposphere have required instruments developed for tropospheric OH measurement to use resonant excitation and detection at 308 nm. This is superior to excitation at 282 nm because the product of the ozone absorption cross section and the photolysis quantum yield of $O(^1D)$ is reduced by a factor of ~30 at 308 nm relative to 282 nm. Early instruments for tropospheric OH detection (Davis *et al.*, 1976; Wang and Davis, 1974; Wang *et al.*, 1975, 1976) failed due to laser-generated OH. Hard and co-workers (1984) reduced laser-generated OH using FAGE, as described in Section 2.2.5, and their early success combining FAGE with excitation at 308 nm using high repetition–rate lasers has been adopted as a model by most research groups using LIF to observe tropospheric OH (Creasey *et al.*, 1997; Faloona *et al.*, 2004; Hard *et al.*, 1995; Holland *et al.*, 1995; Kanaya *et al.*, 2001).

All of the current generation instruments operate at low pressure, pump optically with 1–20 mW of 1–10 kHz tunable light at 308 nm, and detect resonance fluorescence at 308 nm. Most use similar methods for calibration and background measurement and have similar inlet and optical cell designs. Some of the reviews that have described these instruments and scientific results in detail are Crosley, 1995, and Heard and Pilling, 2003.

Inlets consist of a critical orifice (~1 mm diameter) centred on either a flat plate or a cone. Inlet losses are typically minimized by use of unreactive coatings such as halocarbon wax and by judicious design of the inlet (Faloona *et al.*, 2004; Stevens *et al.*, 1994). NO is injected immediately downstream of the orifice, with the reaction time determined by the volume of the inlet and by the instrument's pumping speed. Most OH instruments utilize a single-pass optical cell and detect fluorescence with a PMT. The Pennsylvania State University design (Faloona *et al.*, 2004) is the exception and uses a multi-pass White Cell with a microchannel plate detector. This instrument has the best sensitivity reported to date with a detection limit of 1.4×10^5 molecules/cm^3 in a 30-second integration, SNR of 2, and an uncertainty of $\pm 16\%$ (1σ) for both OH and HO_2. Most other instruments have reported detection limits in the range of 2.8×10^5 to 10^6 molecules/cm^3 in the same averaging times and SNR values (Heard & Pilling, 2003).

Tropospheric OH instruments are usually calibrated by photolysis of water in the presence of O_2 to produce OH and HO_2 in a 1:1 ratio. The 184.9 nm line of a mercury lamp is used to photolyze water:

$$H_2O + h\nu \xrightarrow{\lambda=184.9\,\text{nm}} H + OH \tag{4.16}$$

$$H + O_2 + M \rightarrow HO_2 + M \tag{4.17}$$

The OH production rate is given by

$$\frac{d[OH]}{dt} = [H_2O]\sigma_{H_2O}\varphi_{OH}F \tag{4.18}$$

where σ_{H_2O} is the absorption cross section of water at 184.9 nm, φ_{OH} is the photolysis quantum yield of water to form OH, and F is the photon flux of the mercury lamp. Integration yields

$$[OH]_t = [H_2O]\sigma_{H_2O}\varphi_{OH}Ft \tag{4.19}$$

The absolute water vapour concentration is measured by use of chilled hygrometers or by IR spectroscopy, F is measured with an absolutely calibrated photomultiplier tube, and the residence time is determined using calibrated flow measurements. Alternately, the product Ft, which is difficult to determine accurately, can be measured by an 'O_2 chemical actinometer' (Bloss *et al.*, 2004; Holland *et al.*, 1995; Kanaya *et al.*, 2001; Schultz *et al.*, 1995). The same mercury lamp photolyzes oxygen:

$$O_2 \xrightarrow{184.9\,\text{nm}} O + O \tag{4.20}$$

$$2(O + O_2 + M \rightarrow O_3 + M) \tag{4.21}$$

creating ozone, which is measured with a commercial ozone analyzer and compared to the calculated $[O_3]$ produced:

$$[O_3] = [O_2]\sigma_{O_2}\phi_{O_3}Ft \tag{4.22}$$

$$\text{or } Ft = \frac{[O_3]}{[O_2]\sigma_{O_2}\phi_{O_3}}$$

where σ_{O_2} is the absorption cross section of molecular oxygen and Φ is the quantum yield for ozone production. Actinometry can also be based on photolysis of N_2O instead of ozone (Faloona *et al.*, 2004). The $O(^1D)$ photolysis product reacts with N_2O to form 2NO molecules, which can be detected by chemiluminescence (see Chapter 7).

Alternate calibration methods include monitoring the decay of a hydrocarbon by gas chromatography in an irradiated, continuously stirred reactor in which OH is produced by photolysis of water vapour (Hard *et al.*, 1984), production of a steady state OH concentration by alkene ozonolysis (Hard *et al.*, 2002), and absorption of super-ambient OH concentrations (Kanaya *et al.*, 2001). Typically, these calibration strategies are used at a single value (often close to zero) of the water vapour concentration. Ambient measurements require accounting for water vapour quenching in the sample because it quenches OH 22 times faster than nitrogen (Bailey *et al.*, 1999). Additionally, losses of radicals due to clustering, which is very dependent upon water vapour concentation, have been observed in some instruments (Heard & Pilling, 2003), and thus calibrations are ideally performed at ambient water vapour concentrations.

HO_2 is detected after chemical conversion to OH by reaction with NO. In tropospheric HO_x instruments, NO is added in the low pressure region downstream of the pressure-reducing orifice. HO_2-to-OH conversion efficiencies range from 30 to 95% (Creasey *et al.*, 2003; Kanaya *et al.*, 2001; Stevens *et al.*, 1994). In addition to the formation of HONO (Reaction 4.14), a further complication encountered in the troposphere is conversion of organic peroxy radicals (RO_2) to OH by the reaction scheme

$$RO_2 + NO \rightarrow RO + NO_2 \qquad (4.23)$$

$$RO + O_2 \rightarrow R'CHO + HO_2 \qquad (4.24)$$

$$HO_2 + NO \rightarrow OH + NO_2 \qquad (4.13)$$

where R′CHO represents an aldehyde or ketone. Conversion rates of 5% for CH_3O_2 and $C_2H_5O_2$ have been measured (Kanaya *et al.*, 2001; Holland *et al.*, 2003). Ren *et al.* (2004) report negligible interference from up to 100 ppt of methane-, ethane-, propane-, and n-butane-derived RO_2 species.

The accuracy of LIF HO_x measurements has also been assessed by comparison to measurements using different techniques and therefore presumably subject to different interferences, sampling artefacts, and calibration errors. Intercomparisons of OH measurements have been reviewed in detail by Heard and Pilling (2003). For example, OH measurements by LIF and by folded-path differential optical absorption spectrometry (DOAS) were obtained in rural north-eastern Germany in an experiment during the summer of 1994. A linear fit to the data gives DOAS measurement of $1.09 \times \text{LIF} + (0.28 \times 10^6)$ molecules cm^{-3} with an R^2 correlation coefficient of 0.8 (Hofzumahaus *et al.*, 1998). Most of the scatter is accounted for by the imprecision of the measurements. A comparison study of Penn State's GTHOS (Ground-based Tropospheric Hydrogen Oxides Sensor) LIF detector with NCAR's perCIMS (peroxy radical chemical ionization mass spectrometer) detector in HO_2 mode was conducted at a rural site in Pennsylvania, USA, in 2002 (Ren *et al.*, 2003a). Measurements of ambient HO_2 and both instruments' calibration sources were compared. The comparison showed the calibration sources were identical to within 2%. The accuracy of perCIMS in HO_2 mode was estimated to be 41% at the 95% confidence interval, with 15% of the measured HO_2 in perCIMS stemming from CH_3O_2. The accuracy of GTHOS was estimated to be 32% at the 95% confidence level. Figure 4.5 shows a time series of ambient HO_2 measurements. A linear regression of the comparison data, with an integration period of 140 seconds, was described by PerCIMS $HO_2 = 0.96 \times \text{GTHOS} + 0.6$ ppt, with an R^2 value of 0.85.

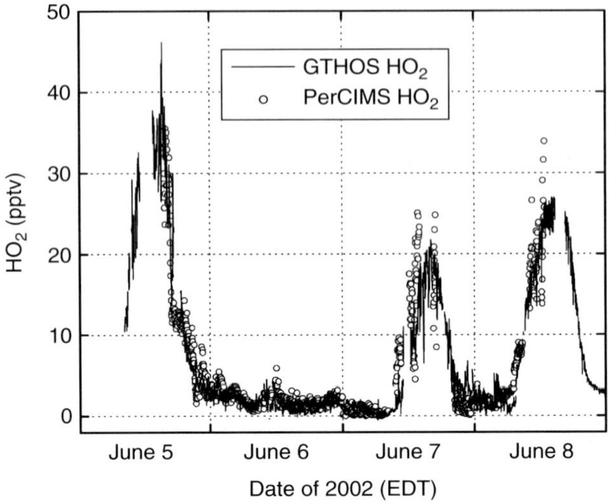

Figure 4.5 Time series of ambient HO_2 measurements by LIF (GTHOS, solid line) and perCIMS (circles). (From Ren *et al.*, 2003a.)

4.3.3 OH fluorescence as an indirect detection strategy

In addition to HO_2, other species can be measured indirectly by OH fluorescence. Water is measured by the stratospheric hygrometer described by Weinstock *et al.* (1994). Photolysis of water vapour by Lyman α radiation at 121.6 nm produces electronically excited OH:

$$H_2O + h\nu(121.6\,\text{nm}) \rightarrow OH(A^2\Sigma^+) + H \qquad (4.25)$$

A fraction of the OH produced is highly rotationally excited in the $\nu' = 1$ level of the $(A^2\Sigma^+)$ electronic state. Following vibrational relaxation, emission from highly excited rotational levels of the $\tilde{A}^2\Sigma^+(\nu' = 0)$ state is detected at 305–325 nm in an optical cell similar to that of the stratospheric HO_x instrument aboard the ER-2.

Ambient detection of nitrous acid (HONO) by photo-fragmentation–LIF should also be feasible and prototypes have been described (Liao *et al.*, 2003; Rodgers and Davis, 1989).

4.3.4 Naphthalene

Martinez *et al.* (2004) discovered that naphthalene could be measured by their HO_x detector thanks to a fortuitous overlap between the absorption spectra of naphthalene and OH. Naphthalene has a sharp absorption peak at 308.0026 nm, slightly to the red of the $Q_1(2)$ rotational line of OH at 307.9951 nm. The naphthalene absorption is negligible to the blue of the $Q_1(2)$ line which was used as the off-resonance wavelength for spectral modulation of OH. Since the Penn State group alternated their offline measurement to high and low frequencies of the OH peak, they were able to produce accurate OH and naphthalene observations after the interference was identified as a useful signal.

4.3.5 Measurement of the OH lifetime

Most of the techniques described in this chapter are used to measure the concentration of a particular atmospheric species. However, the concentration is just one of many parameters that are useful for testing our understanding of atmospheric chemistry. A few LIF instruments have been constructed which measure the total reactivity of OH (k_{total}) or equivalently its lifetime (τ_{OH}) defined as:

$$k_{total} = \tau_{OH}^{-1} = \sum_i k_i C_i = k_{CO}[CO] + k_{NO_2}[NO_2] + k_{CH_4}[CH_4] + \cdots \qquad (4.26)$$

where $k_i C_i$ is the product of an individual reactant concentration and its second order rate constant with OH. The advantage of measuring the total loss rate is that it is not possible to measure every compound with which OH reacts; indeed the identity of every compound that reacts with OH is not known. The two methods of measuring the total reactivity of OH combine LIF detection of OH with standard techniques borrowed from laboratory kinetics.

Kovacs and Brune (2001) demonstrated measurement of the OH lifetime in 2001. Their method, TOHLM (Total OH Loss Measurement), also used by Heard and co-workers at Leeds (Johnson, 2004), is based on the decay of OH concentrations when mixed with ambient air in a flow tube as depicted in Figure 4.6. Approximately 500 ppt of OH and HO_2 in approximately a 1:1 ratio, produced by photolysis of water vapour by a mercury lamp, is released into a 5 cm diameter, glass flow tube by a moveable injector. Ambient air flows into the flow tube and the relative OH concentration is measured downstream by LIF. As the total flow rate and reaction distance are known, the reaction time between the injected OH and the ambient air is known for any injector tube position. Plotting OH concentration versus reaction time yields decay profiles that are fit to single exponentials to extract the first order rate constant. Injection of a known and high concentration of a compound with a well-quantified reaction rate with OH, such as carbon monoxide, is used to 'calibrate' the flow conditions, and injection of zero air is used to quantify wall losses of OH which must be subtracted from the total measured loss. The measurement is simplified by the fact that the OH concentration does not have to be quantified absolutely. Rather, only the relative decay of the OH concentration over time is measured.

At a few sites, the total OH reactivity measured by TOHLM was compared with the total OH reactivity calculated by summing the product of the individually measured OH sinks (e.g. CO, CH_4, NO_2, VOCs) with their respective reaction rate coefficients with OH (i.e. by explicit evaluation of the right hand side of Equation 4.26). In Nashville during the SOS '99 study, observed OH reactivities were on average 40% higher than the calculated reactivities (Kovacs *et al.*, 2003), whereas during the summer of 2001 in the PMTACS-NY field study in New York City, the calculated and observed OH reactivities agreed to within 10% (Ren *et al.*, 2003b). In a forest in northern Michigan, the difference between the observed and the calculated reactivities values was found to be exponentially dependent on temperature and to closely resemble the temperature profile of terpene emissions, suggesting the existence of significant concentrations of unmeasured biogenic VOCs (Di Carlo *et al.*, 2004).

In an alternative strategy described by Calpini and co-workers (Calpini *et al.*, 1999; Jeanneret *et al.*, 2001) and successfully implemented by Sadanaga and co-workers (2004b),

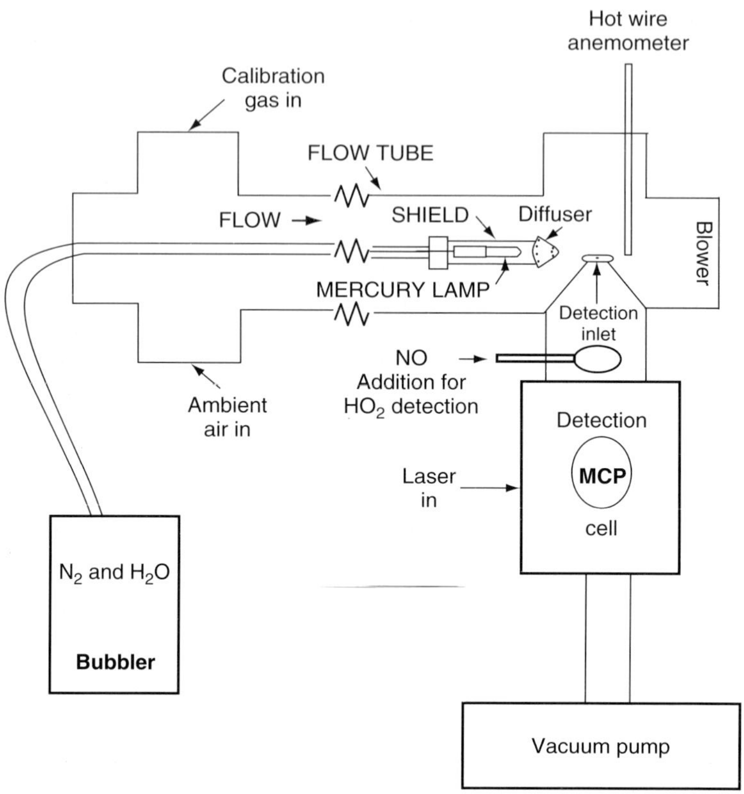

Figure 4.6 The OH lifetime instrument of Kovacs and Brune. Ambient air enters the flow tube from the left and mixes with high concentrations of OH released at a series of distances from the LIF inlet. (From Kovacs and Brune, 2001, with kind permission of Kluwer Academic Publishers.)

OH is produced by flash photolysis of ambient ozone using the fourth harmonic of an Nd:YAG laser at 266 nm (80 mJ/pulse, 10 Hz) followed by reaction of the $O(^1D)$ photolysis product with water vapour. The subsequent decay of the high OH concentration produced (up to 10^{11} molecules/cm^3) is measured by LIF in real time, thus indicating the OH reactivity.

4.4 The nitrogen oxides

The nitrogen oxide radicals ($NO_x \equiv NO + NO_2$) are central to the production of ozone in the troposphere and the destruction of ozone in the stratosphere. Accurate and precise measurements of the atmospheric abundance of NO_x and their oxidation products – peroxy nitrates, alkyl nitrates and nitric acid – are thus important to understanding and modeling air quality (Liang *et al.*, 1998; Russell & Dennis, 2000; Thornton *et al.*, 2002), acid rain and nitrogen deposition (Galloway, 2001; Munger *et al.*, 1998), greenhouse forcing (Mickley *et al.*, 1999), the carbon cycle (Lerdau *et al.*, 2000; Holland *et al.*, 1997), and stratospheric ozone (Wennberg *et al.*, 1994b). Measurements of NO_x are also important to combustion science (Docquier & Candel, 2002) and are used in some strategies for detection of explosives (Steinfeld & Wormhoudt, 1998). Detection of NO_2 is difficult because high concentrations of other NO_y species represent interferences for many techniques. Additionally, concentrations vary by 1–2 orders of magnitude at a single location and by several orders of magnitude between the combustion sources and the remote atmosphere, where mixing ratios are as small as 1–2 ppt. Till the time of writing, no technique has become the singular method of choice. Instruments based on IR absorption (Chapter 2), visible absorption (Chapter 3), LIF (this chapter) and chemiluminescence (Chapter 7) are all in active use.

Although NO is routinely measured accurately by chemiluminescence in scientific and a wide variety of other settings, high quality measurements of NO_2 by chemiluminescence are relatively rare and have been limited to scientific instruments. The most sensitive, in situ instruments for observing NO_2 are photofragmentation – chemiluminescence and direct LIF. The latter is emerging as competitive with the former in both cost and sensitivity. In many monitoring networks, surface catalysis is used to convert NO_2 to NO prior to chemiluminescence detection. These measurements virtually always overestimate NO_2 because of the conversion of nitric acid, alkyl nitrates and peroxy-nitrates into NO_2. As a result, the data from such networks is ignored by the scientific community, although it serves some useful purposes for development of public policy.

4.4.1 Photophysics of NO_2

Figure 4.7 depicts the absorption spectrum of NO_2. It is highly congested due to the presence of numerous overlapping ro-vibronic transitions involving the X^2A_1, A^2B_2, B^2B_1, and the C^2A_2 electronic states and as a result of large differences in the geometries of the ground and the excited states. The rotational structure is embedded in a broad continuum, which peaks at 413 nm. Photodissociation to $NO + O(^3P)$ with near-unity quantum yield occurs for excitation wavelengths shorter than 398 nm. At short times

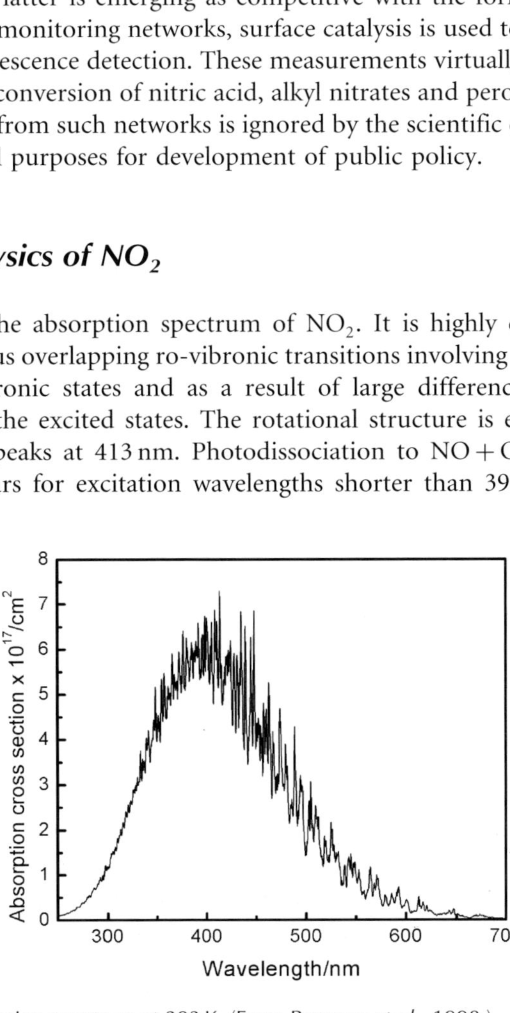

Figure 4.7 NO_2 absorption spectrum at 293 K. (From Burrows *et al.*, 1998.)

($<1\,\mu s$) following excitation, the emission spectrum of NO_2 exhibits vibrational structure at pressures below ~ 10 m torr (Donnelly *et al.*, 1979). At longer times or higher pressures, the emission spectrum resembles a broad peak centred in the near-IR due to efficient stepwise vibrational relaxation of the excited electronic state (Clough & Thrush, 1967, Mazely *et al.*, 1994, Patten *et al.*, 1990). The excited state lifetime of NO_2, $\sim 100\,\mu s$ is much longer than expected based on its integrated absorption cross section and is a strong function of excitation wavelength, pressure, and spectral region observed during measurement (Donnelly & Kaufman, 1978, Patten *et al.*, 1990, Sackett & Yardley, 1972). This long radiative lifetime is a result of strong interaction between the excited electronic states and upper vibrational levels of the ground electronic state as first described by Douglas (1966). Quenching of electronically excited NO_2 is rapid; the air-average value with 532 nm excitation is $6 \times 10^{-11}\,cm^3\,molecule^{-1}\,s^{-1}$ (Donnelly *et al.*, 1979).

4.4.2 Single-photon NO_2 instrument design

In contrast to the design of tropospheric OH LIF instruments, which have for the most part converged, the design and operation of NO_2 LIF instruments are widely varied and evolving rapidly. There is no standard excitation wavelength or optical cell design. Most instruments operate at reduced pressure and use pulsed lasers with time-gated fluorescence detection. Early attempts at detection of tropospheric NO_2 via LIF were conducted at atmospheric pressure but suffered from an aerosol interference (Tucker *et al.*, 1975). Interest was not renewed until work by Fong and Brune (1997), Perkins *et al.* (2001), and Thornton *et al.* (2000) in the late 1990s.

The choice of excitation wavelength is a trade-off between specificity and sensitivity. Specificity is maximized in the red end of the spectrum because of the high ratio of on-resonance to off-resonance absorption, while sensitivity is maximized in the blue end where the absolute absorption cross section is highest. Due to the greater ease in producing tunable light and the specificity afforded at longer wavelengths, excitation at all wavelengths is competitive. Current NO_2 LIF instruments use excitation wavelengths ranging from the blue to the red region of the spectrum. Emission from excited NO_2 peaks in the near-IR, dictating the type of PMT to use. Multi-alkali or cooled GaAs photocathodes, with quantum efficiencies of $\sim$8–15% in the range of 800–900 nm are typically used.

Most NO_2 LIF instruments use a dual-wavelength technique to demonstrate specificity to NO_2. This spectral modulation is similar to the background measurement technique used by OH LIF instruments except that for NO_2 there is still appreciable fluorescence at the off-resonance frequency due to the continuum in the absorption spectrum. This technique has been applied in both the stratosphere (Perkins *et al.*, 2001) and the troposphere (Cleary *et al.*, 2002; Fong & Brune, 1997; Matsumoto and Kajii, 2003; Matsumi *et al.*, 2001; Matsumoto *et al.*, 2001; Thornton *et al.*, 2000). In the instrument described in detail by Thornton *et al.* (2000), the output from a narrow linewidth dye laser pumped by a 3 W, 8 kHz Nd:YAG laser excites a pair of overlapping rotational lines in the (430) $A^2B_2 \leftarrow (000)X^2A_1$ vibronic band of NO_2 at $17086.5\,cm^{-1}$ (585.257 nm). The laser frequency is repeatedly tuned on- and off-resonance with this peak, spectrally modulating the fluorescence signal. The NO_2 concentration is proportional to the difference between the 'online' and 'offline' signal.

The detection limit of the instrument described by Thornton *et al.* is 6 ppt for a 1-minute integration, with SNR = 2. The precision is shot-noise limited and at 1.5 ppb, NO_2 equals 16% (1σ) in 1 second. If combined with the supersonic jet technology described by Cleary *et al.* (Section 4.4.3), an enhancement of ~ 20 in the signal rate could be expected.

Figure 4.8 depicts NO_2 spectral modulation. The bottom trace is the fluorescence excitation spectrum of ambient NO_2. The top trace is the transmission through a 2.5 cm reference cell containing 20 torr of pure NO_2. After scanning the laser frequency in order to locate the position of the online and offline peaks, the laser frequency alternates between the two indicated frequencies. Use of spectral modulation conveys a spectroscopic fingerprint of NO_2 in the fluorescence signal itself, greatly increasing the specificity of the instrument. Any potential interference would only increase both the offline and the online signals but would not affect the difference between the two, as it is unlikely that any interference would respond to such a small change in excitation frequency. This leaves 'false NO_2' as the most likely potential interference, for example that produced by the reaction of NO and O_3 or thermal decomposition of HO_2NO_2 or N_2O_5 in the sampling lines. Such interferences can be quantified by calculation and by laboratory experiment, and are usually insignificant.

Dye lasers (Fong & Brune, 1997; Perkins *et al.*, 2001; Thornton *et al.*, 2000), tunable diode lasers (Cleary *et al.*, 2002), and optical parametric oscillators (Matsumi *et al.*, 2001) have all been successfully used for NO_2 detection. The linewidth of dye lasers

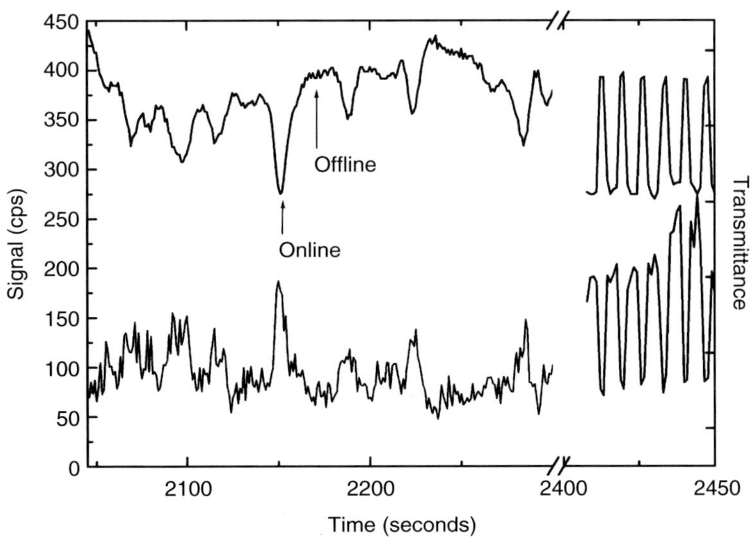

Figure 4.8 Spectral modulation of NO_2. From time 2050 seconds to 2300 seconds, the laser frequency is scanned. The bottom trace is fluorescence from ambient NO_2 and the top trace is the transmission through a 2.5 cm reference cell containing 20 torr NO_2. At time 2400 seconds, the laser frequency alternates between the online and offline positions. The increase in fluorescence at the right of the figure is from an increase in the ambient NO_2 mixing ratio.

with external etalons and of tunable diode lasers with external cavities are all sufficiently narrow so as to excite a particular rotational feature in the NO_2 absorption spectrum. The dye laser used by Thornton *et al.* has a linewidth of $0.06\,cm^{-1}$. The linewidth of the optical parametric oscillator used by Matsumi *et al.* is $21\,cm^{-1}$, and so it is tuned on and off the base and peak of an entire vibronic band, enveloping numerous rotational lines. The separation of their chosen online and offline frequencies is $100\,cm^{-1}$, while the separation chosen by Fong and Brune, for example, is less than $0.1\,cm^{-1}$.

Long-pass filters with a cut-on wavelength of 700 nm are typically used. Interference filters are most common, though liquid-solution filters (Fong & Brune, 1997) and red coloured glass filters (Matsumoto *et al.*, 2001; Matsumoto & Kajii, 2003) have all been used. The low detection limit realized by Thornton *et al.* is in large part due to the superior filter set used, which reduces the background to 1.5 counts/s at 100 mW of laser power.

An alternative approach to LIF detection of NO_2 is to use a single wavelength and demonstrate specificity to NO_2 by non-spectral methods. The advantage of such an approach is that simple, high power laser systems can be used instead of the relatively complicated narrow-linewidth laser systems described above. George and O'Brien demonstrated measurement of NO_2 in the laboratory using the 532 nm line from a 1.4 W Nd:YAG laser at 30 Hz (George & O'Brien, 1991). To measure the instrument's background, ambient air was sampled after flowing through $FeSO_4$, which reduces NO_2 to NO. Specificity of the instrument to NO_2 was implied in that a false signal would have to be chemically modulated by $FeSO_4$, absorb light at 532 nm and re-emit red-shifted fluorescence or scattering in a time gate similar to the excited lifetime of NO_2 and at wavelengths transmitted by the detection optics. However, aerosol effects were identified as a potential interferences as the $FeSO_4$ pack could act as a particulate filter and create a 'false' zero.

Matsumoto *et al.* used a single wavelength approach to tropospheric measurements in first a multi-pass alignment (Matsumoto *et al.*, 2001) and later in a higher power, single-pass alignment (Matsumoto and Kajii, 2003). This latter version uses a frequency-doubled, 6.5 W Nd:YVO_4 laser at 532 nm at a repetition rate of 10 kHz and detects NO_2 fluorescence at wavelengths longer than 640 nm. The cell pressure is held at 2 torr. Matsumoto *et al.* demonstrated that an aerosol interference exists and they showed that it is non-linear in the laser pulse energy. The background for this instrument is measured using an annular diffusion scrubber which consists of a 9 mm inner diameter cylindrical glass tube coated with a mixture of titanium dioxide (TiO_2) and hydroxyapatite ($Ca_{10}(PO_4)_6(OH)_2$). Ideally all NO_2 adsorbs onto the surface of the tube while aerosol passes through. During NO_2 measurement, ambient air is sampled through a similar, uncoated glass tube. The detection limit of this single-pass instrument is 3.6 ppt for a 1-minute average and SNR of 2. A similar instrument was used in conjunction with a chemical amplifier for detection of peroxy radicals (Sadanaga *et al.*, 2004a).

4.4.3 Supersonic expansions

One method to increase the sensitivity of an NO_2 LIF instrument is to sample air through an orifice small enough and with sufficient pumping so as to produce a supersonic expansion. As the gas expands adiabatically on the low pressure side, it cools and the population of states with low rotational quantum number J is greatly increased. This greatly simplifies the absorption spectrum as the integrated absorption cross section of one peak will 'inherit' the intensity of many other transitions, which will in turn decrease greatly. Cleary *et al.* took advantage of this absorption enhancement to build an NO_2 LIF instrument of comparable sensitivity relative to its contemporaries but of considerably smaller size and complexity (Cleary *et al.*, 2002). By sampling ambient air through a 350 μm pinhole and with a pumping speed of 30 L/s, an air sample of rotational temperature between 25 and 50 K at 200 m torr is produced. This expanding jet extends to a Mach disk 14 mm past the pinhole, at which point the air is rapidly thermalized by collisions with the surrounding room-temperature gas. This expansion is intersected by 56 passes of 640.2 nm light from a continuous-wave (cw) tunable diode laser. The absorption cross section of NO_2 at this wavelength at room temperature is $1.3 \times 10^{-20}\,cm^2$, but is magnified by a factor of 30 by this supersonic expansion to $3.9 \times 10^{-19}\,cm^2$. The usual spectral modulation of dual-wavelength LIF instruments is highly effective in this case, as the off-resonance laser frequency excites an area of very low absorption as indicated in Figure 4.9. The detection limit of this instrument is 145 ppt in a 1-minute integration, and SNR = 2.

4.4.4 Calibration of LIF NO_2 instruments

Calibration of NO_2 LIF instruments is straightforward because NO_2 is stable in dry, compressed air tanks. Without exception, all field-tested NO_2 LIF instruments are calibrated through dynamic dilution of NO_2 with zero air. The error in the resulting calibration constant is principally driven by the uncertainty in the calibration source mixing ratio and the subsequent passing efficiency of NO_2 through regulators, flow

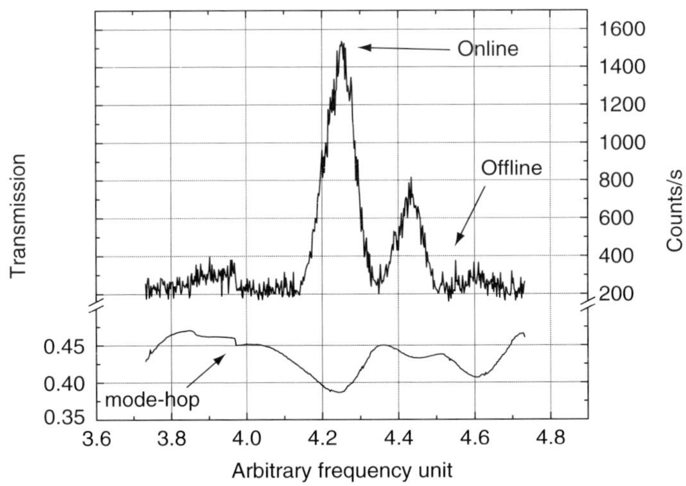

Figure 4.9 Fluorescence of 82 ppb of NO_2 in zero air and corresponding reference cell transmission spectrum of NO_2 in the region of excitation of 15 620 cm^{-1}. The frequencies used for on-line and off-line measurements are indicated with arrows. A diode laser cavity mode hop is also indicated with an arrow. (From Cleary *et al.*, 2002, with permission of Optical Society of America.)

controllers and other tubing. Experiments at the National Institute of Standards and Technology (NIST) have shown agreement to within 1.0% (Bertram *et al.*, 2005) between NO_2 permeation tube standards (Fried & Hodgeson, 1982) (a gravimetric standard exhibiting a combined standard uncertainty of 0.2% at 1 ppm) and the NO_2 cylinder standards commonly used for field calibrations, in this case Scott Specialty Gas, 10 ppm (Bertram *et al.*, 2005). In our laboratory, NO_2 standards (2–50 ppmv) have been compared to each other regularly over a three-year period. Most tanks were stable to within 1.0% although a few deteriorated rapidly. Experiments show that NO_y is conserved in the tanks and that the decrease in NO_2 is balanced by an increase in HNO_3 as has been reported by Fried *et al.*, (Bertram *et al.*, 2005; Fried *et al.*, 1988).

4.4.5 Intercomparisons

LIF measurements of NO_2 have been compared to photofragmentation–chemiluminescence (PF-CL). The LIF instrument described by Thornton *et al.* (2000) was compared to a PF-CL instrument in summer 1999 during the SOS 99 campaign in Nashville, Tennessee, and in summer 2000 during the Texas Air Quality Study (TexAQS 2000) in Houston, Texas (Thornton *et al.*, 2003). The least-squares regression for 1-minute data over the entire SOS 99 campaign showed agreement to within 1% with good correlation ($R^2 = 0.98$), as depicted in Figure 4.10. Adjustment of the PF-CL data to account for a discontinuous shift of 12% found halfway through the campaign yielded higher correlation ($R^2 > 0.99$) and agreement still within the uncertainties of the two instruments (5%, 1σ). Comparison of data from the Texas campaign showed similar agreement. In

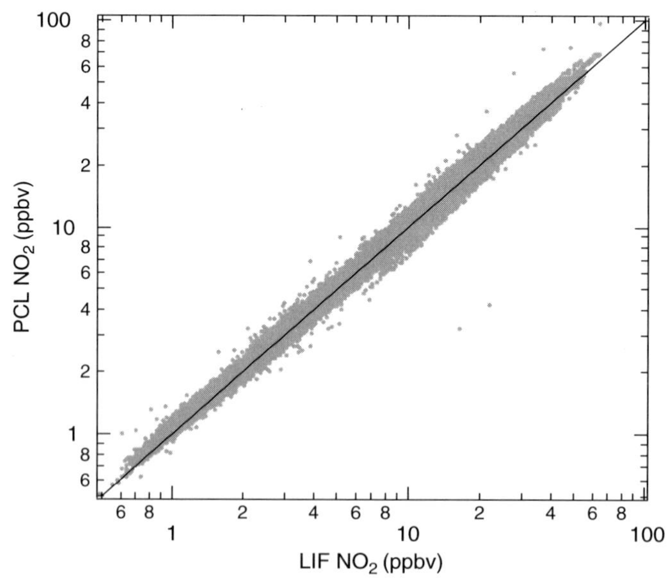

Figure 4.10 Comparison of LIF NO_2 measurements to photofragmentation-chemiluminescence (PF-CL) NO_2 measurements in Nashville, TN. Note the logarithmic scale. (From Thornton *et al.*, 2003.)

both polluted locations, NO_2 mixing ratios varied from slightly under 1 ppb to well over 50 ppb.

The most likely potential interference for LIF detection of NO_2 is thermal or heterogeneous conversion of NO_2-containing compounds such as PAN, N_2O_5, and HO_2NO_2 to yield NO_2 in the sample lines. Relatively inert, perfluoro-alkoxy (PFA) teflon tubing is used to minimize heterogeneous reactions, and thermal decomposition interferences were calculated to be less than 1% in the LIF instrument. Measurement of NO_2 by PF-CL is equally susceptible to these interferences as well as conversion of any NO-containing compound to NO by the photolysis lamp (Ryerson *et al.*, 2000). The most likely photolytic interference is from HONO, which, calculations indicated, would present at most a 2% interference depending on the [NO]:[HONO] ratio. Both instruments were also affected by oxidation of NO to NO_2 by O_3 in the sample lines. The magnitude of this interference was calculated to be always less than 0.55% of ambient NO_2.

Comparisons of the LIF-based NO_2 instruments to PF-CL instruments have also been described by Matsumoto *et al.* (2001) and by Matsumi *et al.* (2001). The comparison of all data points ($[NO_2] = 0$–6 ppb) described by Matsumoto *et al.* in the marine boundary layer in Okinawa Island, Japan, showed agreement to within 2% with an R^2 of 0.99. However, at NO_2 mixing ratios less than 500 ppt, the disagreement increased to 17% with an R^2 of 0.57. The correlation of comparisons at low values of $[NO_2]$ was limited by the precision of both instruments. Additionally, the long residence time in the PF-CL system allowed for significant formation of NO_2 by reaction of NO with O_3 (up to ~16% of ambient NO_2). Matsumi *et al.* demonstrated agreement to within 10% with high correlation ($R^2 > 0.96$) for NO_2 measurements in suburban Nagoya, Japan, where NO_2 mixing ratios were usually less than 500 ppt.

These comparisons have confirmed that NO_2-specific measurements can be made with total uncertainty of 10% or less over a wide range of concentrations and measurement environments.

4.4.6 Thermal dissociation: LIF

The sum of oxidized nitrogen species, 'NO_y', is customarily measured by reducing all compounds to NO using a heated gold or molybdenum converter and chemiluminescence detection of NO (Williams *et al.*, 1998). Day *et al.* demonstrated an alternative technique, thermal dissociation – laser induced fluorescence (TDLIF) (Day *et al.*, 2002), which preserves some information on the measured air's chemical speciation. The basic premise of the TDLIF technique is that when heated, entire classes of nitrogen oxides (e.g., peroxy nitrates, HNO_3, etc.) will thermally dissociate to yield NO_2 and a companion radical:

$$RONO_2 + \text{heat} \rightarrow RO + NO_2 \qquad (4.27)$$

where RO represents $CH_3C(O)O_2$ for peroxy acetyl nitrate, HO for nitric acid, CH_3O for methyl nitrate, etc. The NO_2 product is then measured by LIF as described in Section 4.4.3. Different groups of compounds can be measured separately by taking advantage of the differences in the bond dissociation energy of their RO–NO_2 bond. As an air sample is heated starting at ambient temperature, compounds with weak RO–NO_2 bonds such as

PAN, MPAN, N_2O_5, HO_2NO_2, and peroxy propyl nitrate (PPN) will dissociate first. At higher temperatures, non-peroxy organic nitrates consisting of alkyl nitrates and hydroxy alkyl nitrates will eventually dissociate, and finally at the hottest temperatures nitric acid will dissociate into OH and NO_2. A temperature scan of the atmosphere is shown in Figure 4.11.

In practice, air is sampled first through a short section of warm PFA tubing and then into two or more quartz tubes wrapped in resistively heated nichrome wire. The initial section of tubing is designed so as to minimize HNO_3 adsorption (i.e. use of warm PFA tubing) (Neuman *et al.*, 1999; Ryerson *et al.*, 1999). Each inlet is connected to its own dedicated NO_2 detection cell. One inlet is set at ambient temperature and measures only NO_2. Additional inlets are set at the characteristic decomposition temperature of a group of compounds, so that each inlet yields a measurement of the sum of both ambient NO_2 and NO_2 produced by thermal dissociation of the more highly oxidized compounds. The residence times in each heated section range from 20 to 50 ms. It is found that peroxy nitrates are fully dissociated at 180°C, alkyl nitrates are fully dissociated at 350°C, and nitric acid is fully dissociated at 550°C. Consider an inlet pair with one set at 180°C and another set at 350°C. The lower temperature inlet's NO_2 detection cell will measure an NO_2 concentration equal to the sum of ambient NO_2 and the concentration of ambient peroxy nitrates. The higher temperature inlet's detection cell will measure an NO_2 concentration equal to that of the lower inlet and the concentration of alkyl nitrates. The difference in measured NO_2 concentrations yields the alkyl nitrate concentration. The hottest inlet temperature, at which nitric acid decomposes, yields a measurement similar to NO_y with the exception that non-NO_2 containing compounds (e.g. NO, HONO, RONO) are not

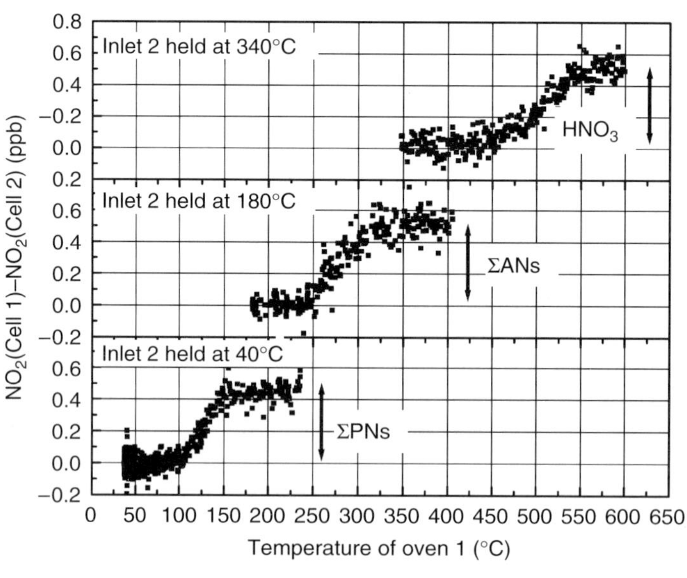

Figure 4.11 Temperature scan of ambient air taken at Big Hill, CA. In each of these three plots, inlet 2 is held at the specified temperature while the temperature of inlet 1 is scanned. The difference in NO_2 measured by each inlet's dedicated NO_2 LIF detector yields a measurement of a class of NO_y compounds – in this case ~0.4 ppb peroxy nitrates, 0.5 ppb alkyl nitrates, and 0.5 ppb nitric acid.

measured. After the heating section, the sampled gas flows through a cooling section for ~180 ms before passing through a pressure-reducing pinhole. From there, the gas is rapidly transported through PFA tubing to the NO_2 detection cells.

The advantage of TDLIF over a standard NO_y instrument is that information on chemical speciation is preserved, while the advantage of TDLIF over measurements of individual compounds (e.g. 2-hyroxy propyl nitrate, PAN, methyl nitrate, etc.) is that theoretically all NO_2-containing compounds are measured by one instrument.

In several atmospheric field campaigns since the latter half of 1980s, the sum of individually measured NO_y species (NO, NO_2, HNO_3, PAN, etc.) has frequently fallen short of the total NO_y as measured by catalytic reduction to NO and chemiluminescence. Based on measurements of speciated NO_y by TDLIF at three locations, Day *et al.* proposed the identification of this so-called 'missing NO_y' as total alkyl nitrates (ΣANs) (Day *et al.*, 2003). At the University of California – Blodgett Research Forest Station in the foothills of the Sierra Nevada, at Granite Bay, California (a suburban site), and at La Porte, Texas, ΣANs comprised a much higher fraction of NO_y and NO_z ($NO_z \equiv NO_y - NO_x$) than previously reported, from 8 to 50% depending on site, time of day, and season. These are much higher values than previously measured in other studies, which were typically an order of magnitude lower. In all prior analyses of the NO_y chemical budget, only a few types of alkyl nitrates (e.g. C_2–C_5 alkyl nitrates) were individually measured (e.g. by gas chromatography and/or Mass Spectrometry, Chapters 8 and 5). ΣANs, as measured by TDLIF, consists of all alkyl nitrates – straight-chain alkyl nitrates, hydroxy-alkyl nitrates, and nitrates of biogenic origin, most prominently isoprene nitrates. Thus the ability of TDLIF to measure all alkyl nitrates as a lump sum provides a more accurate measure of ΣANs/NO_y and most likely has explained the 'missing NO_y' problem.

4.4.7 Two-photon LIF detection of NO

Detection of nitric oxide (NO) via LIF has primarily relied on two-photon excitation as first presented by Bradshaw and Davis in 1982 (Bradshaw & Davis, 1982; Sandholm *et al.*, 1997, 1990). Figure 4.12 depicts the excitation and detection scheme used. NO is sequentially excited at atmospheric pressure from the $X^2\Pi$ ground state to the $\tilde{A}^2\Sigma$ state and finally to the $D^2\Sigma$ state by pulsed light of wavelengths 226 nm and 1.1 μm, respectively. The final collected signal is emission at 187 nm from the $D^2\Sigma$ state to the ground $X^2\Pi$ state, blue-shifted from all background noise sources. Laser scatter, Rayleigh scatter, Stokes Raman scattering, and all other noise sources are resonant with or red-shifted from the two excitation frequencies.

In early instruments (Bradshaw *et al.*, 1985; Sandholm *et al.*, 1990), the 226 nm light was produced by frequency mixing 1064 μm light from the Nd:YAG laser with 287 nm light produced by frequency-doubling the 574 nm output of a 10 Hz Nd:YAG-pumped dye laser. The IR light for the second excitation was either the fixed 1064 μm fundamental output of the same Nd:YAG laser or tunable IR light produced by injecting 580 nm light from a second dye laser into a H_2-filled Raman shifter. In a later version, both the IR and UV probe light were produced by optical parametric oscillators pumped by the third harmonic of a 10 Hz Nd:YAG laser at 355 nm. In all cases, the two synchronized pumping beams are combined before entering the detection cell. The cylindrical fluorescence chamber is outfitted with up to four sets of detection optics and PMTs.

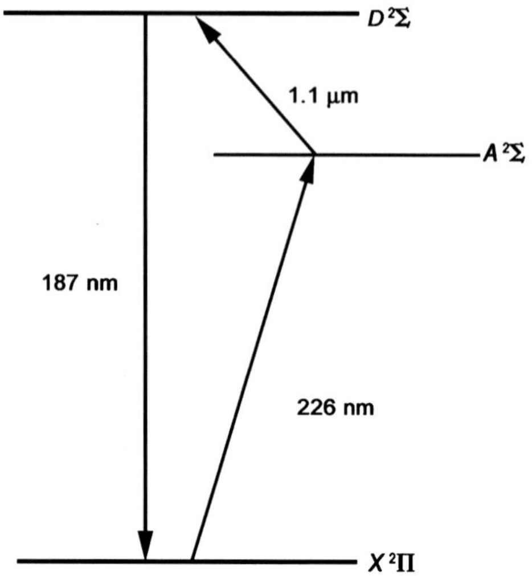

Figure 4.12 Energy level diagram for two-photon LIF detection of NO. The final emission of 187 nm light is blue-shifted from both pump wavelengths.

Early versions of this instrument used liquid-solution filters to discriminate against noise at 226 nm in combination with extreme solar blind CuI PMTs (Bradshaw *et al.*, 1985; Sandholm *et al.*, 1990). The instrument described by Sandholm *et al.* (1997) uses a series of dielectrically coated mirrors which provide great enough rejection at 226 nm ($T < 10^{-13}$) to allow use of a CsTe PMT, which although not solar blind has a much higher quantum efficiency than CuI PMTs (16% vs 0.1% near 190 nm). Instrumental backgrounds are measured either by blocking the second probe beam at 1064 nm or by spectral modulation of *both* probe lasers.

The detection limit reported for the instrument used during the NASA GTE/CITE 2 program (Sandholm *et al.*, 1990) was 3.5 ppt in a 2-minute integration, and SNR = 2 with a precision of 4.5 ppt at NO mixing ratios of 35 ppt. The 1997 version (Sandholm *et al.*, 1997) quoted a detection limit of under 1 ppt for a 100-second integration, and SNR = 2.

The main advantages of the two-photon technique are the high sensitivity attained, the specificity afforded by use of two separate ro-vibronically resolved electronic transitions, and the ability to operate at ambient pressure. The obvious disadvantage is the complexity and size of an instrument with multiple non-trivial lasers.

4.4.8 Single-photon LIF detection of NO

In 2003, Bloss *et al.* demonstrated single-photon LIF detection of NO in the atmosphere using an all solid-state laser system (Bloss *et al.*, 2003). Following excitation at 226 nm, emission from the $\tilde{A}^2\Sigma$ state to the $X^2\Pi$ state between 240 and 390 nm was observed using time-gated detection. The excitation source was a frequency quadrupled Nd:YAG-pumped Ti:Sapphire laser. With only 0.5 mW of 226 nm laser power, the instrument was capable of detecting atmospheric NO in the ppb range with good agreement with a commercial chemiluminescence detector. Further improvements, namely use of an improved non-linear crystal and improved optics, should produce a detection limit of 0.07 ppt for a 60-second signal integration.

4.4.9 Photofragmentation two-photon LIF detection of NO_2

NO_2 can be detected indirectly by two-photon NO LIF after conversion to NO (Bradshaw *et al.*, 1999; Sandholm *et al.*, 1990). Either a XeF excimer laser at 353 nm or the 3rd harmonic of a Nd:YAG laser at 355 nm can be used to photolyze NO_2 to NO and $O(^3P)$. As this produces a measurement of $NO_x(NO + NO_2)$, concurrent measurements of ambient NO must be subtracted to obtain NO_2 measurements. The advantages of using a laser as a photolysis source over the more commonly used xenon arc lamps are that the monochromaticity of the light reduces photolytic interferences, and the high degree of collimation and high intensity of the laser allow for use of long path length photolysis cells which provide more efficient photolysis of NO_2 and shorter residences times (<1 s), reducing interferences by thermal decomposition of thermally labile species such as HO_2NO_2, PAN and N_2O_5. The photolysis pulse precedes the overlaying excitation pulses by 30–50 μs to allow for complete thermalization of the vibrationally excited NO photolysis product. Rapid photolysis of NO_2 is necessary to avoid the interferences presented by surface-catalyzed decomposition of NO_y-species as experienced by early versions of this and other indirect NO_2 instruments (Bradshaw *et al.*, 1999). The detection limit of this instrument to NO_2 during the NASA GTE/CITE 2 program (Sandholm *et al.*, 1990) was 10 ppt for a 6-minute integration, and SNR = 2 with a 1σ precision of 6.5 ppt at NO_2 mixing ratios of 33 ppt.

4.4.10 NO_3 and N_2O_5

In situ measurements of NO_3 and N_2O_5 are desirable for a full understanding of nocturnal chemistry. NO_3 oxidises VOCs, and reacts with NO_2 to form N_2O_5 which can hydrolyze heterogeneously to form HNO_3, thus removing NO_x from the atmosphere. Due to the rapid photolysis and reaction with NO of NO_3 and the fast equilibrium between NO_3 and N_2O_5, daytime concentrations of both species are low.

The $\tilde{B}^2E' \leftarrow \tilde{X}^2A_2'$ transition of NO_3 spans a series of vibronic peaks in the red region of the spectrum, with a peak absorption cross section of 1.7×10^{-17} cm^2 at 662 nm, the (0000 ← 0000) transition. Photolysis of NO_3 to either $NO + O_2$ or $NO_2 + O$ competes with fluorescence for excitation wavelengths less than 640 nm (Johnston *et al.*, 1996). Fluorescence is negligible for excitation wavelengths less than 595 nm. The decay of electronically excited NO_3 is bi-exponential, with a short component of 3–30 μs and a long component of 340 μs (Carter *et al.*, 1996; Nelson *et al.*, 1983), much longer than expected based on its intense absorption. This is attributed to perturbation of the B^2E' excited state by high lying vibrational states of the $X^2\,A_2'$ ground state, similar to NO_2. NO_3 fluoresces in a series of vibronic bands in the red and near-IR regions of the spectrum.

The efficacy of LIF for detection of atmospheric NO_3 was first demonstrated by Wood *et al.* (2003). Their laboratory prototype instrument excited NO_3 at 662 nm in a multi-pass Herriott Cell with a 36 mW cw diode laser, and detected fluorescence between 700 and 750 nm. The sensitivity of this instrument was 137 counts/s/ppb at 393 K which they extrapolated to 176 counts/s/ppb at 298 K. Combined with a background of 1340 counts/s, this gave a detection limit of 76 ppt for a 60-second integration, and SNR = 2. The background consisted of approximately half laser scatter and half Stokes Raman scattering of O_2. Calibration of the instrument was performed by thermolysis of N_2O_5 followed by quantitative conversion of NO_3 to NO_2 by reaction with NO and subsequent LIF detection of NO_2 using a tunable diode laser at 638 nm.

The observed sensitivity described above is too low for detection of NO_3 as ambient concentrations are rarely this high. However, since in moderately polluted environments N_2O_5 can exist in significantly higher concentrations than NO_3, it is adequate for detection of N_2O_5 after thermal dissociation to NO_2 and NO_3. Direct measurement of N_2O_5 by LIF is difficult because its electronic transitions are weak.

Wood and co-workers (Wood *et al.*, 2005) presented measurements of atmospheric N_2O_5 near the San Francisco Bay Area using a field version of their prototype instrument. This instrument used a high power (300 mW) cw laser in a single pass configuration, and sampled ambient air through a section of heated PFA teflon tubing before reducing the pressure through a 0.8 mm orifice. The background was measured by periodic addition of a small flow of 100 ppm NO to the sampled air.

Matsumoto *et al.* (2003) presented measurements of N_2O_5 in a suburban environment near Tokyo using a similar instrument. This instrument excited NO_3 from thermolysis of ambient N_2O_5 at the (1000) ← (0000) transition at 623 nm using a commercial dye laser pumped by a frequency-doubled, pulsed Nd:YVO_4 laser at 6.5 W. Ambient air passed through a 40 cm length of 1/2″ PFA tube heated to 90℃ and then entered a 0.4 mm diameter critical orifice. The sensitivity of this instrument was 90 counts/s/ppb with a background of 30 counts/s, which gave a detection limit of 22 ppt with a 60-second average, SNR = 2. This instrument's internal detection axis was calibrated using a method similar to that described by Wood *et al.*, with heterogeneous losses of NO_3 estimated at 25%.

4.5 Detection of halogen compounds

Molina and Rowland (1974) suggested a link between anthropogenic CFCs and gas-phase catalytic destruction of ozone in the stratosphere by chlorine radicals. A similar catalytic cycle involving bromine was proposed by Wofsy *et al.* (1975). In situ measurements of ClO and BrO by Jim Anderson and co-workers using resonance fluorescence techniques were crucial to establishing that these hypotheses were correct and to understanding stratospheric chemistry in detail.

Low concentrations of bare halogen atoms preclude their routine detection. ClO and BrO weakly absorb in the UV, but the excited states predissociate on a time scale short compared to their radiative lifetimes and consequently the fluorescence yields are low. However, ClO and BrO can be measured indirectly after conversion to the bare halogen atom by reaction with nitric oxide. Early instruments were installed on balloon-borne payloads (Anderson *et al.*, 1977; Brune & Anderson, 1986) and balloon-borne observations continue to provide useful insights (Avallone *et al.*, 1993; Toohey *et al.*, 1993). Later measurements were executed aboard the ER-2 aircraft. The design of these instruments has been reviewed by Brune and Stimpfle (1993). The capacity to measure chlorine nitrate ($ClONO_2$) and the chlorine monoxide dimer (ClOOCl) has been developed using thermal dissociation and ClO detection (Stimpfle *et al.*, 1999, 2004).

4.5.1 Spectroscopy and calibration

The $^2D_{5/2} \leftarrow {}^2P_{3/2}$ transition at 118.9 nm, which coincides with a local minimum in the O_2 absorption spectrum, is used to excite atomic chlorine using a resonance lamp in all measurement platforms (Anderson *et al.*, 1977, 1980; Avallone *et al.*, 1993; Brune & Anderson, 1986; Brune *et al.*, 1989b; Pierson *et al.*, 1999; Vogel *et al.*, 2003). The light source consists of a sealed, low pressure (2–8 torr He) 2.45 GHz microwave plasma discharge operated at 120 W to which a trace amount of molecular chlorine is added by heating $PtCl_2$. The lamps require high ratios of 118.9 nm to Lyman alpha radiation at 121.6 nm to be effective light sources because the impurity Lyman alpha light contributes noise but not signal. Barium is used to react away impurities, especially those containing H atoms. The output of this lamp passes through a cell with MgF_2 windows containing 260 torr of O_2 which absorbs all emission lines in the three major multiplets of atomic chlorine except for the 118.9 nm line. Fluorescence is detected with a cesium iodide or potassium bromide photocathode PMT. Measurement of Br is similar to that of Cl. The light source is a Br resonance lamp, which excites Br using eight atomic emission lines of bromine between 115.9 and 126.1 nm after attenuation by a gaseous oxygen filter.

ClO and BrO are measured as Cl and Br after reaction with NO:

$$XO + NO \rightarrow X + NO_2 \qquad (4.28)$$

where X represents Cl or Br. Competing with Reaction 4.28 is the association reaction between NO and X:

$$X + NO + M \rightarrow XNO + M \qquad (4.29)$$

Reaction 4.29 is six times faster for Cl than for Br. A range of NO flow rates (e.g. 8 to 50 sccm (standard cubic centimeters per minute) of NO for ClO) is used to quantify the extent to which Reactions 4.28 and 4.29 go to completion. Reaction 4.29 is usually negligible.

All instruments were calibrated before and after each flight by installing the detection pod into a laboratory flow system in which a known Cl or Br number density was produced over a range of pressures. For Cl, three different calibration methods were used – a chemical method, a photometric method, and an absorption method.

In the chemical method, a known Cl number density is produced by the following two reactions:

$$Cl + O_3 \rightarrow ClO + O_2 \qquad (4.30)$$

$$NO + ClO \rightarrow NO_2 + Cl \qquad (4.31)$$

For Reaction 4.30, Cl is produced by microwave discharge of Cl_2 and mixed with excess ozone of known concentration. The concentration of the ClO product is inferred from the decrease in ozone concentration and is quantitatively converted to Cl by addition of NO.

The photometric method is based on explicit evaluation of each of the components of equation 4.1 (Section 4.1). In the absorption method, a flow of Cl is simultaneously probed by the flight instrument and measured by an external absorption apparatus at

134.7 nm. The calibration constants produced by these three independent methods agreed to within their stated uncertainties. The absorption technique proved the most versatile and convenient method for routine calibrations. An absorption technique was similarly used to calibrate the Br detection axes at 145.8 nm.

4.5.2 Balloon-borne instruments

Early balloon-borne instruments measured Cl, ClO, and BrO in air that flowed through the interior of an aerodynamically shaped pod (or *nacelle*) as it descended from altitudes as high as 44 km (Anderson *et al.*, 1977, 1980; Brune & Anderson, 1986), similar to balloon-borne OH measurements. After ascending to float altitudes while attached to a balloon, instruments were detached and descended on a stabilized parachute through the stratosphere at speeds of 40–100 m/s. The first instruments measured Cl and ClO on separate identical detection pods mounted on a common support frame (Anderson *et al.*, 1977, 1980) so that both species could be measured simultaneously. The background of the Cl instrument was measured by titrating ambient Cl with ethane injected below the detection point. Later versions (Brune & Anderson, 1986) measured only ClO and BrO on the two pods.

4.5.3 Aircraft measurements

Instruments that measured ClO and BrO were installed on the wing pod of the ER-2 aircraft for missions flying into the polar vortex (Brune *et al.*, 1989a,b). The instrument's inlet design was the basis for the design described for OH measurements (Section 4.3.2). A double-duct design was used to deliver a laminar flow of unperturbed air to three consecutive detection axes. NO was added prior to the first detection axis. ClO was measured using the first two axes while BrO was measured with the third. Sample flow conditions were monitored by measurement of the gas temperature using a thermistor bead array downstream of the detection axes.

All axes were calibrated on the ground before and after flights. After calibration of the Cl axes, the bromine axis was then calibrated to a known Br concentration produced by the reaction

$$Cl + Br_2 \rightarrow BrCl + Br \tag{4.32}$$

where the concentration of the limiting reagent Cl is measured by the calibrated Cl detection axes. Both axes monitored Rayleigh scattering as an in-flight proxy calibration.

4.5.4 $ClONO_2$ and ClOOCl

Later developments made possible the measurement of chlorine nitrate ($ClONO_2$) (Stimpfle *et al.*, 1999) and the chlorine monoxide dimer (ClOOCl) (Stimpfle *et al.*, 2004) aboard the ER-2 aircraft. Both compounds are measured by thermal dissociation followed by detection of ClO:

$$ClONO_2 + M \xrightarrow{\Delta} ClO + NO_2 + M \qquad \Delta H = 113\,\text{kJ/mol} \qquad (4.33)$$

$$ClOOCl + M \xrightarrow{\Delta} ClO + ClO + M \qquad \Delta H = 71\,\text{kJ/mol} \qquad (4.34)$$

The air flow is heated by a grid of resistively heated silicon strips. Due to the difference in bond energies, $ClONO_2$ and ClOOCl dissociate fully at different temperatures. A parallel pair of flows is utilized, each with two consecutive detection axes. One flow measures ClO in the first detection axis and either $ClONO_2$ or ClOOCl in the second depending on heater temperature. The other flow measures BrO in the first axis and uses the second detection axis for diagnostics of the dissociation temperature of the mirror image $ClONO_2$/ClOOCl axis.

The detection limit for ClO is 3 ppt in a 35-second integration with an uncertainty of 17% (1σ) (Stimpfle *et al.*, 2004). $ClONO_2$ and ClOOCl are measured with a detection limit of 10 ppt in the same integration period, with an uncertainty of 21% (1σ) (Stimpfle *et al.*, 2004). At an air density of 2.6×10^{18} molecules cm^{-3}, the detection limit for BrO aboard the ER-2 was 16 ppt in a 1-minute integration, and SNR = 2, with an uncertainty of 18% (1σ) (Brune *et al.*, 1989a).

4.6 Future directions

The application of fluorescence to measurements of atmospheric trace gases is expanding rapidly. New laser technologies based on diodes or frequency mixing are greatly extending the range of excitation wavelengths that can be produced and hence molecules that can be detected in a portable package. For example, the use of blue laser diodes by the data storage industry, currently in its infancy, will likely fuel new developments in NO_2 detection (or after frequency-doubling, NO detection). Reductions in the size and increases in the autonomy of fluorescence instruments will allow for more widespread application of the technique, enabling advances in atmospheric sciences. In particular, it will become (1) simpler to assemble the comprehensive suites of chemical measurements that are now critical to advancing mechanistic atmospheric chemistry research, and (2) become possible to construct large scale networks with hundreds of sensors that permit observations of spatial and long-term trends in key indicators of the health of the atmosphere.

Further reading

Atkins, P.W., de Paula J. (2002) *Physical Chemistry*, 7th edition, Oxford University Press, New York.

McClenny, W.A., Williams, E.J., Cohen, R.C. & Stutz, J. (2002) Preparing to measure the effects of the NO_x SIP call: Methods for ambient monitoring of NO, NO_2, NO_y and individual NO_z species, *J. Air Waste Manag.*, 52(5), 542–562.

Brune, W.H. & Richard M. Stimpfle (1993) In Situ Measurements of Stratospheric Reactive Trace Gases, Chapter 5 of *Measurement Challenges in Atmospheric Chemistry*, Leonard Newman (ed.), Advances in Chemistry series 232.

Anderson, J.G. (1995) Polar Processes in Ozone Depletion, Chapter 17 of Progess and Problems in Atmospheric Chemistry, Editor John R. Barker, Advanced Series in Physical Chemistry, vol. 3, World Scientific, 1995.

References

Anderson, J.G. (1971a) *J. Geophys. Res.*, 76, 7820–7824.
Anderson, J.G. (1971b) *J. Geophys. Res.*, 76, 4634–4652.
Anderson, J.G. (1975) *Geophys. Res. Lett.*, 2, 231–234.
Anderson, J.G. (1976) *Geophys. Res. Lett.*, 3, 165–168.
Anderson, J.G., Brune, W.H. & Proffitt, M.H. (1989) *J. Geophys. Res. Atmos.*, 94, 11 465–11 479.
Anderson, J.G., Grassl, H.J., Shetter, R.E. & Margitan, J.J. (1980) *J. Geophys. Res.*, 85, 2869–2887.
Anderson, J.G., Grassl, H.J., Shetter, R.E. & Margitan, J.J. (1981) *Geophys. Res. Lett.*, 8, 289–292.
Anderson, J.G., Margitan, J.J. & Stedman, D.H. (1977) *Science*, 198, 501–503.
Anderson, J.G., Grassl, H.J., Shetter, R.E., Margitan, J.J. (1980) Stratospheric free chlorine measured by balloon-borne in situ resonance fluorescence. *J. Geophys. Res.*, 85(C5), 2869–2887.
Anderson, J.G., Toohey, D.W. & Brune, W.H. (1991) *Science*, 251, 39–46.
Avallone, L.M., Toohey, D.W., Brune, W.H., Salawitch, R.J., Dessler, A.E. & Anderson, J.G. (1993) *Geophys. Res. Lett.*, 20, 1795–1798.
Bailey, A.E., Heard, D.E., Henderson, D.A. & Paul, P.H. (1999) *Chem. Phys. Lett.*, 302, 132–138.
Bailey, A.E., Heard, D.E., Paul, P.H., Pilling, M.J. (1997) Collisional quenching of OH ($A\,^2\Sigma^+, \nu' = 0$) by N_2, O_2 and CO_2 between 204 and 294 K. Implications for atmospheric measurements of OH by laser-induced fluorescence. *J. Chem. Soc.-Faraday Trans.*, 93(16), 2915–2920.
Bertram, T.H., Cohen, R.C., Thorn III, W.J. & Chu, P.M. (2005) *J. Air Waste Manag. Assoc.*, 55(10).
Bloss, W.J., Gravestock, T.J., Heard, D.E., Ingham, T., Johnson, G.P. & Lee, J.D. (2003) *J. Environ. Monit.*, 5, 21–28.
Bloss, W.J., Lee, J.D., Bloss, C., Heard, D.E., Pilling, M.J., Wirtz, K., Martin-Reviejo, M. & Siese, M. (2004) *Atmos. Chem. Phys.*, 4, 571–583.
Bradshaw, J., Davis, D., Crawford, J., Chen, G., Shetter, R., Muller, M., Gregory, G., Sachse, G., Blake, D., Heikes, B., Singh, H., Mastromarino, J. & Sandholm, S. (1999) *Geophys. Res. Lett.*, 26, 471–474.
Bradshaw, J. & Davis, D.D. (1982) *Opt. Lett.*, 7, 224–226.
Bradshaw, J.D., Rodgers, M.O. & Davis, D.D. (1982) Single photon laser-induced fluorescence detection of NO and SO_2 for atmospheric conditions of composition and pressure. *Appl. Opt.*, 21(14), 2493–2500.
Bradshaw, J.D., Rodgers, M.O., Sandholm, S.T., Kesheng, S. & Davis, D.D. (1985) *J. Geophys. Res.*, 90, 12 861–12 873.
Brune, W.H. & Anderson, J.G. (1986) *Geophys. Res. Lett.*, 13, 1391–1394.
Brune, W.H., Anderson, J.G. & Chan, K.R. (1989a) *J. Geophys. Res.-Atmos.*, 94, 16 639–16 647.
Brune, W.H., Anderson, J.G. & Chan, K.R. (1989b) *J. Geophys. Res.-Atmos.*, 94, 16 649–16 663.
Brune, W.H. & Stimpfle, R.M. (1993) *Advances in Chemistry Series*, 133–184.
Brus, L.E. & McDonald, J.R. (1974) Time-resolved fluorescence kinetics and B-1 1(Delta-1 G) vibronic structure in tunable ultraviolet laser-excited SO_2 vapor. *J. Chem. Phys.*, 61(1), 97–105.
Burrows, J.P., Dehn, A., Deters, B., Himmelmann, S., Richter, A., Voigt, S. & Orphal, J. (1998) Atmospheric remote-sensing reference data from GOME: Part I. Temperature-dependent

absorption cross-sections of NO_2 in the 231–794 nm range, *J. Quant. Spectrosc. Radiat. Transf.*, 60, 1025–1031.

Calpini, B., Jeanneret, F., Bourqui, M., Clappier, A., Vajtai, R. & van den Bergh, H. (1999) *Analusis*, 27, 328–336.

Cantrell, C.A., Davidson, J.A., McDaniel, A.H., Shetter, R.E. & Calvert, J.G. (1990) Temperature-dependent formaldehyde cross sections in the near-ultraviolet spectral region. *J. Phys. Chem.*, 94(10), 3902–3908.

Carter, R.T., Schmidt, K.F., Bitto, H. & Huber, J.R. (1996) *Chem. Phys. Lett.*, 257, 297–302.

Cleary, P.A., Wooldridge, P.J. & Cohen, R.C. (2002) Laser-induced fluorescence detection of atmospheric NO_2 with a commercial diode laser and a supersonic expansion, *Appl. Opt.*, 41, 6950–6956.

Clough, P.N. & Thrush, B.A. (1967) *Trans. Faraday Soc.*, 63, 915–925.

Creasey, D.J., Evans, G.E., Heard, D.E. & Lee, J.D. (2003) *J. Geophys. Res.-Atmos.*, 108.

Creasey, D.J., HalfordMaw, P.A., Heard, D.E., Pilling, M.J. & Whitaker, B.J. (1997) *J. Chem. Soc.-Faraday Trans.*, 93, 2907–2913.

Crosley, D.R. (1995) *J. Atmos. Sci.*, 52, 3299–3314.

Davis, D.D., Heaps, W. & McGee, T. (1976) *Geophys. Res. Lett.*, 3, 331–333.

Day, D.A., Dillon, M.B., Wooldridge, P.J., Thornton, J.A., Rosen, R.S., Wood, E.C. & Cohen, R.C. (2003) *J. Geophys. Res.-Atmos.*, 108.

Day, D.A., Wooldridge, P.J., Dillon, M.B., Thornton, J.A. & Cohen, R.C. (2002) *J. Geophys. Res.-Atmos.*, 107.

Di Carlo, P., Brune, W.H., Martinez, M., Harder, H., Lesher, R., Ren, X., Thornberry, T., Carroll, M.A., Young, V., Shepson, P.B., Riemer, D., Apel, E. & Campbell, C. (2004) *Science*, 304, 722–725.

Docquier, N. & Candel, S. (2002) *Prog. Energy Combustion Sci.*, 28, 107–150.

Donnelly, V.M. & Kaufman, F. (1978) *J. Chem. Phys.*, 69, 1456–1460.

Donnelly, V.M., Keil, D.G. & Kaufman, F. (1979) *J. Chem. Phys.*, 71, 659–673.

Douglas, A.E. (1966) *J. Chem. Phys.*, 45, 1007–1015.

Dubey, M.K., Hanisco, T.F., Wennberg, P.O. & Anderson, J.G. (1996) *Geophys. Res. Lett.*, 23, 3215–3218.

Faloona, I.C., Tan, D., Lesher, R.L., Hazen, N.L., Frame, C.L., Simpas, J.B., Harder, H., Martinez, M., Di Carlo, P., Ren, X.R. & Brune, W.H. (2004) *J. Atmos. Chem.*, 47, 139–167.

Fong, C. & Brune, W.H. (1997) *Rev. Sci. Instrum.*, 68, 4253–4262.

Fried, A. & Hodgeson, J. (1982) *Anal. Chem.*, 54, 278–282.

Fried, A., Sams, R., Dorko, W., Elkins, J.W. & Cai, Z.T. (1988) *Anal. Chem.*, 60, 394–403.

Galloway, J.N. (2001) *Water Air and Soil Pollut.*, 130, 17–24.

George, L.A. & O'Brien, R.J. (1991) *J. Atmos. Chem.*, 12, 195–209.

Hard, T.M., George, L.A. & O'Brien, R.J. (1995) *J. Atmos. Sci.*, 52, 3354–3372.

Hard, T.M., George, L.A. & O'Brien, R.J. (2002) *Environ. Sci. Technol.*, 36, 1783–1790.

Hard, T.M., O'Brien, R.J., Chan, C.Y. & Mehrabzadeh, A.A. (1984) *Environ. Sci. Technol.*, 18, 768–777.

Heard, D.E. & Pilling, M.J. (2003) *Chem. Rev.*, 103, 5163–5198.

Herriott, D., Kompfner, R. & Kogelnik, H. (1964) *Appl. Opt.*, 3, 523–526.

Hofzumahaus, A., Aschmutat, U., Brandenburger, U., Brauers, T., Dorn, H.P., Hausmann, M., Hessling, M., Holland, F., Plass-Duelmer, C. & Ehhalt, D.H. (1998) *J. Atmos. Chem.*, 31, 227–246.

Holland, E.A., Braswell, B.H., Lamarque, J.F., Townsend, A., Sulzman, J., Muller, J.F., Dentener, F., Brasseur, G., Levy II, H., Penner, J.E. & Roelofs, G.J. (1997) *J. Geophys. Res.*, 102, 15 849–15 866.

Holland, F., Hessling, M. & Hofzumahaus, A. (1995) *J. Atmos. Sci.*, 52, 3393–3401.

Holland, F., Hofzumahaus, A., Schafer, A., Kraus, A. & Patz, H. (2003) *J. Geophys. Res.*, 108, doi:10.1029/2001JD001393.

Holloway, J.S., Jakoubek, R.O., Parrish, D.D., Gerbig, C., Volz-Thomas, A., Schmitgen, S., Fried, A., Wert, B., Henry, B. & Drummond, J.R. (2000) *J. Geophys. Res.-Atmos.*, 105, 24 251–24 261.

Holtermann, D.L., Lee, E.K.C. & Nanes, R. (1980) Single rovibronic level lifetimes of the $^1A_2$ state of SO_2 excited in the 3043 A ("*E*") band – Rotationally resolved fluorescence emission-spectrum. *Chem. Phys. Lett.*, 75(1), 91–93.

Ishiwata, T., Fujiwara, I., Naruge, Y., Obi, K. & Tanaka, I. (1983) Study of NO_3 by laser-induced fluorescence. *J. Phys. Chem.*, 87(8), 1349–1352.

Jeanneret, F., Kirchner, F., Clappier, A., van den Bergh, H. & Calpini, B. (2001) *J. Geophys. Res.*, 106, 3083–3093.

Johnson, G.P. (2004) Ph.D. Thesis, University of Leeds, Leeds.

Johnston, H.S., Davis, H.F. & Lee, Y.T. (1996) *J. Phys. Chem.*, 100, 4713–4723.

Kanaya, Y., Sadanaga, Y., Hirokawa, J., Kajii, Y. & Akimoto, H. (2001) *J. Atmos. Chem.*, 38, 73–110.

Kita, K., Imamura, T., Iwagami, N., Morrow, W.H. & Ogawa, T. (1996) *Ann. Geophysicae-Atmos. Hydrospheres Space Sci.*, 14, 227–237.

Kovacs, T.A. & Brune, W.H. (2001) Total OH loss rate measurement, *J. Atmos. Chem.*, 39, 105–122.

Kovacs, T.A., Brune, W.H., Harder, H., Martinez, M., Simpas, J.B., Frost, G.J., Williams, E., Jobson, T., Stroud, C., Young, V., Fried, A. & Wert, B. (2003) *J. Environ. Monit.*, 5, 68–74.

Lerdau, M.T., Munger, L.J. & Jacob, D.J. (2000) *Science*, 289, 2291, 2293.

Levy, H. (1972) *Planetary Space Sci.*, 20, 919–935.

Liang, J.Y., Horowitz, L.W., Jacob, D.J., Wang, Y.H., Fiore, A.M., Logan, J.A., Gardner, G.M. and Munger, J.W. (1998) *J. Geophys. Res.-Atmos.*, 103, 13 435–13 450.

Liao, W., Hecobian, A., Mastromarino, J. & Tan, D. (2003) American Geophysical Union, Fall Meeting.

Logan, J.A., Prather, M.J., Wofsy, S.C. & McElroy, M.B. (1981) *J. Geophys. Res.-Oceans Atmos.*, 86, 7210–7254.

Luke, W.T. (1997) *J. Geophys. Res.-Atmos.*, 102, 16 255–16 265.

Luque, J. & Crosley, D.R. (2000) Radiative and predissociative rates for NO A $^2\Sigma^+ \nu' = 0$–5 and D $^2\Sigma^+ \nu' = 0$–3. *J. Chem. Phys.*, 112(21), 9411–9416.

Martinez, R.D. & Joens, J.A. (1992) SO_2 Absorption cross-section measurements from 197 nm to 240 nm. *Geophys. Res. Lett.*, 19(3), 277–279.

Martinez, M., Harder, H., Ren, X., Lesher, R.L. & Brune, W.H. (2004) *Atmos. Chem. Phys.*, 4, 563–569.

Matsumi, Y., Murakami, S., Kono, M., Takahashi, K., Koike, M. & Kondo, Y. (2001) *Anal. Chem.*, 73, 5485–5493.

Matsumoto, J., Hirokawa, J., Akimoto, H. & Kajii, Y. (2001) *Atmos. Environ.*, 35, 2803–2814.

Matsumoto, J., Imai, H., Kosugi, N. & Kajii, Y. (2005) In situ measurement of N_2O_5 in the urban atmosphere by thermal decomposition/laser-induced fluorescence technique. *Atmos. Environ.*, 39, 6802–6811.

Matsumoto, J. & Kajii, Y. (2003) *Atmos. Environ.*, 37, 4847–4851.

Mazely, T.L., Friedl, R.R. & Sander, S.P. (1994) *J. Chem. Phys.*, 100, 8040–8046.

McGee, T.J., McIlrath, T.J. (1984) Absolute OH absorption cross-sections (for Lidar measurements). *J. Quant. Spectrosc. Radiat. Transf.*, 32(2), 179–184.

Mickley, L.J., Murti, P.P., Jacob, D.J., Logan, J.A., Koch, D.M. & Rind, D. (1999) *J. Geophys. Res.*, 104, 30 153–30 172.

Miller, R.G. & Lee, E.K.C. (1978) Single vibronic level photochemistry of formaldehydes in A $^1a_2$ state – Radiative and nonradiative processes in H_2CO, HDCO, and D_2CO. *J. Chem. Phys.*, 68(10), 4448–4464.

Molina, M.J. & Rowland, F.S. (1974) *Nature*, 249, 810–812.

Moore, C.B. & Weisshaar, J.C. (1983) Formaldehyde photochemistry. *Ann. Rev. Phys. Chem.*, 34, 525–555.

Munger, J.W., Song-Miao, F., Bakwin, P.S., Goulden, M.L., Goldstein, A.H., Colman, A.S. & Wofsy, S.C. (1998) *J. Geophys. Res.*, 103, 8355–8368.

Nee, J.B., Juan, C.Y., Hsu, J.Y., Yang, J.C. & Chen, W.J. (2004) The electronic quenching rates of NO(A $^2\Sigma^+, \nu' = 0$–2). *J. Chem. Phys.*, 300(1–3), 85–92.

Nelson, H.H., Pasternack, L. & McDonald, J.R. (1983) Excited state dynamics of NO_3. *J. Chem. Phys.*, 79(9), 4279–4284.

Neuman, J.A., Huey, L.G., Ryerson, T.B. & Fahey, D.W. (1999) *Environ. Sci. Technol.*, 33, 1133–1136.

Patten, K.O., Burley, J.D. & Johnston, H.S. (1990) Radiative lifetimes of nitrogen-dioxide for excitation wavelengths from 400 nm to 750 nm. *J. Phys. Chem.*, 94(20), 7960–7969.

Perkins, K.K., Hanisco, T.F., Cohen, R.C., Koch, L.C., Stimpfle, R.M., Voss, P.B., Bonne, G.P., Lanzendorf, E.J., Anderson, J.G., Wennberg, P.O., Gao, R.S., Del Negro, L.A., Salawitch, R.J., McElroy, C.T., Hintsa, E.J., Loewenstein, M. & Bui, T.P. (2001) *J. Phys. Chem. – A*, 105, 1521–1534.

Pfeiffer, L.N. & Porter, J.F. (1964) *Appl. Opt.*, 3, 317.

Pierson, J.M., McKinney, K.A., Toohey, D.W., Margitan, J., Schmidt, U., Engel, A. & Newman, P.A. (1999) *J. Atmos. Chem.*, 32, 61–81.

Ren, X., Edwards, G.D., Cantrell, C.A., Lesher, R.L., Metcalf, A.R., Shirley, T. & Brune, W.H. (2003a) Intercomparison of peroxy radical measurements at a rural site using laser-induced fluorescence and Peroxy Radical Chemical Ionization Mass Spectrometer (PerCIMS) techniques, *J. Geophys. Res.*, 108, 4605.

Ren, X.R., Harder, H., Martinez, M., Lesher, R.L., Oliger, A., Simpas, J.B., Brune, W.H., Schwab, J.J., Demerjian, K.L., He, Y., Zhou, X.L. & Gao, H.G. (2003b) *Atmos. Environ.*, 37, 3639–3651.

Ren, X.R., Harder, H., Martinez, M., Faloona, I.C., Tan, D., Lesher, R.L., Di Carlo, P., Simpas, J.B. & Brune, W.H. (2004) *J. Atmos. Chem.*, 47, 169–190.

Rodgers, M.O. & Davis, D.D. (1989) *Environ. Sci. Technol.*, 23, 1106–1112.

Russell, A. & Dennis, R. (2000) *Atmos. Environ.*, 34, 2283–2324.

Ryerson, T.B., Huey, L.G., Knapp, K., Neuman, J.A., Parrish, D.D., Sueper, D.T. & Fehsenfeld, F.C. (1999) *J. Geophys. Res.-Atmos.*, 104, 5483–5492.

Ryerson, T.B., Williams, E.J. & Fehsenfeld, F.C. (2000) *J. Geophys. Res.-Atmos.*, 105, 26 447–26 461.

Sackett, P.B. & Yardley, J.T. (1972) *J. Chem. Phys.*, 57, 152–163.

Sadanaga, Y., Matsumoto, J., Sakurai, K., Isozaki, R., Kato, S., Nomaguchi, T., Bandow, H. & Kajii, Y. (2004a) *Rev. Sci. Instrum.*, 75, 864–872.

Sadanaga, Y., Yoshino, A., Watanabe, K., Yoshino, A., Wakazono, Y., Kanaya, Y. & Kajii, Y. (2004b) *Rev. Sci. Instrum.*, 75, 2648–2655.

Sandholm, S., Smyth, S., Bai, R. & Bradshaw, J. (1997) *J. Geophys. Res.*, 102, 28 651–28 661.

Sandholm, S.T., Bradshaw, J.D., Dorris, K.S., Rodgers, M.O. & Davis, D.D. (1990) *J. Geophys. Res.*, 95, 10 155–10 161.

Schultz, M., Heitlinger, M., Mihelcic, D. & Volz-Thomas, A. (1995) *J. Geophys. Res.*, 100, 18 811–18 816.

Shibuya, K. & Lee, E.K.C. (1978) Vibrational and electronic-energy transfers from a single vibronic level of $H_2CO(a, {}^1a_2)$, 4^1 level. *J. Chem. Phys.*, 69(2), 758–766.

Smith, P.S. & Crosley, D.R. (1990) *J. Geophys. Res.*, 95, 16 427–16 422.

Steinfeld, J.I. & Wormhoudt, J. (1998) *Ann. Rev. Phys. Chem.*, 49, 203–232.

Stevens, P.S., Mather, J.H. & Brune, W.H. (1994) *J. Geophys. Res.*, 99, 3543–3557.

Stimpfle, R.M. & Anderson, J.G. (1988) *Geophys. Res. Lett.*, 15, 1503–1506.

Stimpfle, R.M., Cohen, R.C., Bonne, G.P., Voss, P.B., Perkins, K.K., Koch, L.C., Anderson, J.G., Salawitch, R.J., Lloyd, S.A., Gao, R.S., Del Negro, L.A., Keim, E.R. & Bui, T.P. (1999) *J. Geophys. Res.*, 104, 26 705–26 714.

Stimpfle, R.M., Wennberg, P.O., Lapson, L.B. & Anderson, J.G. (1990) *Geophys. Res. Lett.*, 17, 1905–1908.

Stimpfle, R.M., Wilmouth, D.M., Salawitch, R.J. & Anderson, J.G. (2004) *J. Geophys. Res.*, 109, JD003811.

Takegawa, N., Kita, K., Kondo, Y., Matsumi, Y., Parrish, D.D., Holloway, J.S., Koike, M., Miyazaki, Y., Toriyama, N., Kawakami, S. & Ogawa, T. (2001) *J. Geophys. Res.-Atmos.*, 106, 24 237–24 244.

Thornton, J.A., Wooldridge, P.J. & Cohen, R.C. (2000) Atmospheric NO_2: In situ laser-induced fluorescence detection at parts per trillion mixing ratios, *Anal. Chem.*, 72, 528–539.

Thornton, J.A., Wooldridge, P.J., Cohen, R.C., Martinez, M., Harder, H., Brune, W.H., Williams, E.J., Roberts, J.M., Fehsenfeld, F.C., Hall, S.R., Shetter, R.E., Wert, B.P. & Fried, A. (2002) *J. Geophys. Res.-Atmos.*, 107, 4146.

Thornton, J.A., Wooldridge, P.J., Cohen, R.C., Williams, E.J., Hereid, D., Fehsenfeld, F.C., Stutz, J. & Alicke, B. (2003) Comparisons of in situ and long path measurements of NO_2 in urban plumes, *J. Geophys. Res.-Atmos.*, 108, 4496.

Toohey, D.W., Avallone, L.M., Allen, N.T., Demusz, J.N., Hazen, J.N., Hazen, N.L. & Anderson, J.G. (1993) *Geophys. Res. Lett.*, 20, 1791–1794.

Tucker, A.W., Birnbaum, M. & Fincher, C.L. (1975) *Appl. Opt.*, 14, 1418–1422.

Vogel, B., Muller, R., Deshler, T., Grooss, J.U., Karhu, J., McKenna, D.S., Muller, M., Toohey, D., Toon, G.C. & Stroh, F. (2003) *J. Geophys. Res.-Atmos.*, 108.

Voigt, S., Orphal, J., Burrows, J.P. (2002) The temperature and pressure dependence of the absorption cross-sections of NO_2 in the 250–800 nm region measured by Fourier-transform spectroscopy. *J. Photochem. Photobiol. a-Chemistry*, 149(1–3), 1–7.

Wang, C.C., Davis, L.I., Wu, C.H. & Japar, S. (1976) *Appl. Phys. Lett.*, 28, 14–16.

Wang, C.C., Davis, L.I., Wu, C.H., Japar, S., Niki, H. & Weinstock, B. (1975) *Science*, 189, 797–800.

Wang, C.C. & Davis, L.I.J. (1974) *Phys. Rev. Lett.*, 32, 349–352.

Weinstock, B. (1969) *Science*, 166, 224–225.

Weinstock, E.M., Hintsa, E.J., Dessler, A.E., Oliver, J.F., Hazen, N.L., Demusz, J.N., Allen, N.T., Lapson, L.B. & Anderson, J.G. (1994) *Rev. Sci. Instrum.*, 65, 3544–3554.

Wennberg, P.O., Cohen, R.C., Hazen, N.L., Lapson, L.B., Allen, N.T., Hanisco, T.F., Oliver, J.F., Lanham, N.W., Demusz, J.N. & Anderson, J.G. (1994a) *Rev. Sci. Instrum.*, 65, 1858–1876.

Wennberg, P.O., Cohen, R.C., Stimpfle, R.M., Koplow, J.P., Anderson, J.G., Salawitch, R.J., Fahey, D.W., Woodbridge, E.L., Keim, E.R., Gao, R.S., Webster, C.R., May, R.D., Toohey, D.W., Avallone, L.M., Proffitt, M.H., Loewenstein, M., Podolske, J.R., Chan, K.R. & Wofsy, S.C. (1994b) *Science*, 266, 398–404.

Wennberg, P.O., Hanisco, T.F., Cohen, R.C., Stimpfle, R.M., Lapson, L.B. & Anderson, J.G. (1995) In-situ measurements of OH and HO_2 in the upper troposphere and stratosphere, *J. Atmos. Sci.*, 52, 3413–3420

Williams, E.J., Baumann, K., Roberts, J.M., Bertman, S.B., Norton, R.B., Fehsenfeld, F.C., Springston, S.R., Nunnermacker, L.J., Newman, L., Olszyna, K., Meagher, J., Hartsell, B., Edgerton, E., Pearson, J.R. & Rodgers, M.O. (1998) *J. Geophys. Res.-Atmos.*, 103, 22 261–22 280.

Wofsy, S.C., McElroy, M.B. & Yung, Y.L. (1975) *Geophys. Res. Lett.*, 2, 215–218.

Wood, E.C., Wooldridge, P.J., Freese, J.H., Albrecht, T. & Cohen, R.C. (2003) *Environ. Sci. Technol.*, 37, 5732–5738.

Wood, E.C., Bertram, T.H., Wooldridge, P.J. & Cohen, R.C. (2005) Measurements of N_2O_5; NO_2 and O_3 east of the San Francisco Bay, *Atmospheric Chemistry and Physics*, 5, 483–491.

Yokelson, R.J., Burkholder, J.B., Fox, R.W., Talukdar, R.K. & Ravishankara, A.R. (1994) Temperature dependence of the NO_3 absorption spectrum. *J. Phys. Chem.*, 98(50), 13 144–13 150.

Zhang, Q., Chen, C.X., Yu, S.Q., Ma, X.X. (1998) Kinetic studies on state-state coupling and collisional quenching of excited sulfur dioxide. *Int. J. Chem. Kinet.*, 30(11), 831–837.

Chapter 5
Mass Spectrometric Methods for Atmospheric Trace Gases

Jonathan Williams

5.1 Introduction

Mass spectrometry is the most powerful analytical technique available to atmospheric scientists. Being applicable to all elements, it can provide qualitative and quantitative information on organic and inorganic species in complex mixtures such as air. The linearity accuracy and detection limits of mass spectrometry are unequalled in analytical science, permitting the molecular structures of complex species up to 10^8 atomic mass units (dalton) to be determined and attomole (10^{-18}) levels to be detected. The technique is highly versatile and its application in physics, chemistry and biology has led to key discoveries in all fields. Today, the air we breathe is routinely analysed by mass spectrometry to determine ambient concentrations of a wide range of compounds, radicals and isotopes. Such instrumentation can be operated 'in-situ' on the ground, on ships, from aircraft, or even mounted on balloons or rockets. Furthermore mass spectrometry is also used in the laboratory to investigate key atmospheric reactions or processes under controlled conditions, as well as to analyse air compressed in containers or trapped in ice cores. This chapter introduces the fundamentals of trace gas mass spectrometry (Section 5.2), explains how a mass spectrometer functions (Section 5.3) and describes instrumentation that is currently being used to explore atmospheric composition (Section 5.4). Examples are given of some of the results in atmospheric research and finally future developments in the field are predicted (Section 5.5).

5.1.1 Generating a mass spectrum

Generally speaking we use a mass spectrometer to obtain a 'mass spectrum' of the air we wish to analyse. A mass spectrum provides information on the molecular composition and abundance in the sample as a function of mass. In the case of ambient air it can tell us the concentrations of important trace gases such as acetone, hydroxyl radicals or nitric acid as well as the major constituents such as oxygen. We may even use the mass spectrum to identify unknown species in the air. To obtain this very useful mass spectrum, we must be able to sort the air molecules in some way by mass, prior to detection and ideally quantification. Intuitively we may think of exploiting the force of gravity to differentiate molecular masses. However, this force is not strong or tractable enough to effectively mass filter the sample. Manipulation or filtering of molecules is only possible if the species are ionised

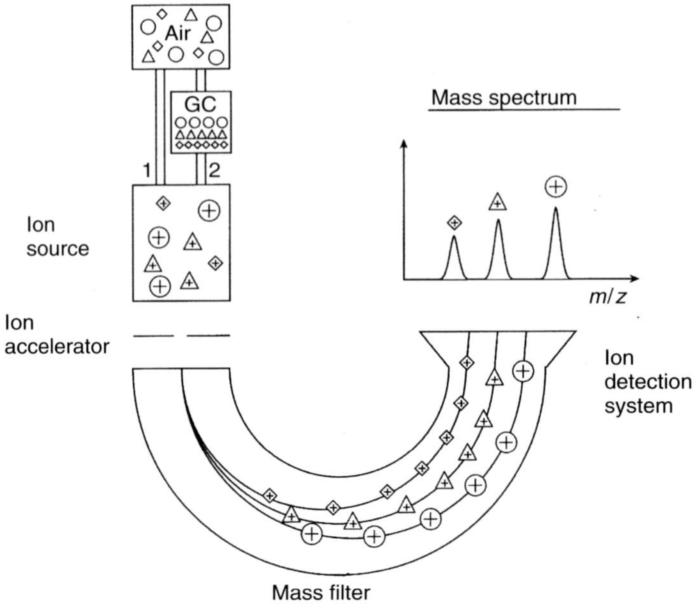

Figure 5.1 Schematic diagram of a mass spectrometer for air analysis.

or charged, since filtering is achieved by the application of electric and magnetic fields. The resulting force acting on the species is then proportional to the charge of the species. By carefully applying these forces to the ionised sample we may segregate the sample according to the mass-to-charge ratio (m/z). Figure 5.1 shows a schematic diagram of a mass spectrometer used to measure ambient air. The first step is the introduction of the air into the mass spectrometer. This may be done directly, i.e. inlet 1, or via some form of pre-separation method such as gas chromatography, inlet 2 (see Chapter 8). Following the introduction of the sample to the mass spectrometer, the first step in analysis is ionisation. A range of direct and indirect ionisation methods have been developed for mass spectrometry and those pertinent to atmospheric trace gas analysis are described in Section 5.3.1. The ions are then accelerated into a region where electric and magnetic fields are applied to filter ions by masses allowing them to be separately detected. Several mass filtering techniques are discussed in Section 5.3.2. After an ion passes through the mass filter, it enters the detection system, triggering a response that is recorded on a data system. For quantification, the response of an unknown sample is compared to that of a standard of known concentration. A summary of the most widespread detector types is given in Section 5.3.3.

5.1.2 Historical development of the mass spectrometer for air measurements

A brief history of the development of mass spectrometry is shown in Table 5.1. The first experiments of J.J. Thompson in 1912 using electric fields to deflect charged particles onto

Table 5.1 Timeline of developments in mass spectrometry for trace gas analysis in air

1803	Dalton formulates atomic theory – all atoms of the same element have the same mass.
1800s	The development of pumps makes it possible to achieve vacuums better than 10^{-3} torr. Vacuum is a critical aspect in mass spectrometry, see Section 5.2.1.
1897	J.J. Thomson discovers electron *m/z* ratio with combined electric and magnetic fields. Receives the Nobel Prize in 1906.
1898	W. Wien analyses anode rays by magnetic deflection and establishes their positive charge.
1912	The first mass spectra of O_2, N_2 and CO_2 are recorded by J.J. Thomson.
1919	F.W. Aston develops the first mass spectrometer with velocity focussing, and with it determines mass defects in 1923.
1930s	Mattauch–Herzog type magnetic sector instrument developed at the University of Vienna.
1940s	Magnetic sector-type instruments used extensively. O.C. Nier used such an instrument in the Second World War to isolate plutonium, aiding construction of the first atomic bomb.
1953	The quadrupole analyser and the ion trap are described by W. Paul and H.S. Steinwedel. Paul received the Nobel Prize in 1989.
1956	Mass spectrometers are first coupled with gas chromatographs by McLafferty and R.S. Gohlke.
1958	Ions produced in different types of flames measured with a mass spectrometer. (Knewstubb & Sudgen, 1958).
1963	Ions are first detected in the atmosphere.
1966	M.S.B. Munson and F.H. Field develop the principle of chemical ionisation.
1968	First commercial quadrupole instrument is launched by Finnigan.
1975	First commercial GC-MS systems become available with capillary columns.
1977–78	Positive and negative ions are detected for the first time in the stratosphere by Arnold and co-workers (Arnold *et al.*, 1977; Arnold & Henschen, 1978).
1980	Sulphuric acid measured for the first time in stratosphere (Arnold & Fabian, 1980).
1984	Acetone detected in upper troposphere/lower stratosphere (Hauck & Arnold, 1984).
1984	First continuous flow isotope ratio mass spectrometer measurements (Barrie *et al.*, 1984).
1985	Antarctic ozone hole discovered by J. Farman.
1987	Isotope mass spectrometer establishes 160 000 year temperature recorded in ice core (Jouzel *et al.*, 1987).
1989	Nitric acid measurements by atmospheric pressure ionisation mass spectrometer in the Antarctic vortex and contribute to understanding of ozone hole chemistry (Crutzen & Arnold, 1986).
1998	Proton transfer reaction mass spectrometer launched commercially by Ionicon and first airborne measurements made over the Tropical rainforest (Crutzen *et al.*, 2000).
2003	Fast GC-MS systems developed for aircraft measuring ca 50 species, at picomol levels within 2 minutes (Apel *et al.*, 2003).
2004	Mass spectrometers and air samplers for laboratory analysis by MS are flown monthly in the upper troposphere and lower stratosphere on commercial airliners, http://www.caribic-atmospheric.com/.

photographic plates led to the discovery of the electron (Thomson, 1913). Substantial background material relating to this discovery is available through the London Science Museum Website http://www.sciencemuseum.org.uk/on-line/electron/section2/index.asp. Thomson's student F.W. Aston, in 1919, refined the instrument, which led to the discovery of isotopes in 1923 (Aston, 1942). Both scientists received Nobel prizes for their work. Over the following 30 years the technique was rapidly developed and used in diverse fields such as organic chemistry and physics. The first commercially available mass spectrometer with magnetic sector mass filter was delivered in 1942 for the purpose of analysing synthetically prepared organic compounds. Further developments quickly followed, particularly in mass filters and detectors. The time of flight mass spectrometer (see Section 5.3.2.2), quadrupole analyzer (Section 5.3.2.1) and the ion trap (Section 5.3.2.4) were developed in the 1950s, and with the advent of the first computers systems became more automated. The successful coupling of gas chromatography and mass spectrometry improved the detection limits considerably, and as more commercial systems became available in the 1960s, the technique proliferated throughout the field of science.

Mass spectrometers were first applied to atmospheric research in the 1960s. This work was stimulated by the discovery that radio waves are reflected in the ionosphere and a desire to understand the atmospheric distribution of ions. We now know that ions are produced naturally in the atmosphere and their concentration controls the electrical properties of the atmosphere as well as influencing important processes such as aerosol production. Cosmic rays comprising extraterrestrial protons and alpha particles impinging on the earth have sufficiently high energy to ionise N_2 (15.58 eV) and O_2 (12.07 eV). Being charged, the resulting ions are affected by the earth's magnetic field, analogously to those in the mass spectrometer described above. The radiative emission of these particles is the source of the Aurora seen at high latitudes. Over the first 100 km of the atmosphere the ionisation density is relatively constant, around 10^3 ions cm^{-3}. Since the neutral gas density varies by a factor of 4×10^6 in this range, the mixing ratio of ions to neutrals decreases substantially from high altitude to low altitude. At the surface there is only one ion for circa 10^{16} neutral molecules and the only significant source is from radon (Viggiano & Arnold, 1995). Interestingly, for the reasons given above, the use of mass spectrometry in the atmosphere began not at the ground but at high altitude (Narcisi & Bailey, 1965). This was partly due to the interest in the ionosphere but mostly because the high ion-to-neutrals ratio made it technically easier to make these measurements in this region despite the inherent difficulty in getting the instruments to 60–100 km above the Earth (see Section 5.2.1). Indeed it was only much later that the first measurements of atmospheric ions were made from aircraft in the troposphere (Heitmann & Arnold, 1983), and then at ground level (Perkins & Eisele, 1984). Only positive atmospheric ions were measured until 1971. In 1971, negative atmospheric ions were first measured using rocket and balloon borne instruments (Arnold *et al.*, 1971). This group discovered that mass spectrometric measurements of naturally occurring ions in the middle to lower atmosphere also provided information on the neutral trace gas composition (e.g. acetone). It was a natural step to extend the measurement apparatus to include an internal ion source, leading to the development of the atmospheric pressure ionisation chemical ionisation mass spectrometer (API-CIMS) technique (see Section 5.4.1), based on the principle of chemical ionisation first demonstrated by Munson and Field in 1966. Through the 1980s these techniques permitted the measurement of important trace gases in the upper troposphere and stratosphere. Airborne nitric acid measurements over the Antarctic significantly contributed to the understanding of ozone hole chemistry. In the laboratory, the improved sensitivity of continuous flow GC-MS allowed isotope ratios of many important species to be determined (Barrie *et al.*, 1984). In the 1990s chemical ionisation mass spectrometry was developed for the measurement of important radical species such as OH in the lower atmosphere. Despite its low concentrations and reactivity, this species is now routinely measured at Hohenpeissenberg, Germany (see Figure 5.2). The capillary GC-MS has become the most widespread MS technique in atmospheric research, being particularly suited for measurement of organohalogens, CFCs, HCFCs and alkanes. In 2000, the proton transfer reaction mass spectrometry technique became commercially available and has proved popular for measurements of selected oxygenates on short time scales (e.g. methanol). The number of atmospheric species measured by mass spectrometry continues to increase, while detection limits, sampling time, size and weight of instrumentation decreases. Within a century, mass spectrometry has developed from a single laboratory experiment to become the most powerful and prolific technique in atmospheric trace gas research at the time of writing, being applied from the tropics to the poles and from the ground to the top of the atmosphere. Several examples of the application of mass spectrometry in atmospheric field studies are given in Figure 5.2.

Figure 5.2 Examples of where trace gas mass spectrometry is used in atmospheric science.
(1) Ground-based APICIMS at Hohenpeissenberg for the routine measurement of OH. http://www.dwd.de/en/FundE/Observator/MOHP/MOHP.html.
(2) Balloon launch of mass spectrometer to stratosphere from Kiruna, Sweden (Arnold & Henschen, 1978).
(3) The German research ship *Meteor* carrying PTR-MS and GC-MS measurements (Williams *et al.*, 2004).
(4) Aircraft flying PTR-MS over the Amazonian rainforest (Crutzen *et al.*, 2000).
(5) Commercial passenger aircraft carrying mass spectrometers and taking air samples on a regular basis for long-term monitoring of the atmosphere (http://www.caribic-atmospheric.com/).

5.2 Fundamentals

5.2.1 The importance of vacuum

In Section 5.1 we have described mass spectrometry as involving the selective manipulation of ions to a detector by electric and magnetic fields. Accurate deflection of ions can only be achieved at low pressures as otherwise frequent collisions with neutral molecules would destroy the order in the trajectories and possibly produce unwanted ion–molecule reactions. Thus vacuum is essential for mass spectrometric analysis. From kinetic theory the mean free path (L in m) of an ion at room temperature and pressure can be calculated by:

$$L = \frac{kT}{\sqrt{2} \times p \times \sigma} \tag{5.1}$$

Where k is the Boltzman constant ($1.381 \times 10^{-23}\,\mathrm{JK^{-1}}$), T is the temperature (in K), p is the pressure (in Pa) and σ is the collision cross section (typically $50 \times 10^{-20}\,\mathrm{m^2}$). For mass spectrometry a mean free path of circa 1 m is required, generally somewhat longer than the mass filter. It follows that the maximum pressure should be 6.6×10^{-3} Pa (66 nbar). Furthermore, where high voltages are used in ion sources or mass filters, the pressure must be further reduced to prevent arcing.

For the reasons given above, the invention of reliable and effective pumping methods was key to the development of mass spectrometry, see Table 5.1. This is normally done in two steps. First, a mechanical pump is used to reduce the pressure from ambient pressure 1 bar (10^5 Pa) to around a millibar (10^2 Pa). Nowadays clean diaphragm pumps are preferred in place of the older rotary oil sealed pumps. To achieve high vacuum (better than a millibar) an alternative pumping principle to that of compression must be used. For this a diffusion pump is the simplest form; however, in modern instruments these are being superseded by turbo molecular pumps. The later consist of several stages of fast rotating blades (typically 60 000 rpm) that drive molecules to a mechanical pump. Since the molecular pumps cannot operate at near ambient pressure they must be used in combination with a mechanical pre-pump. When used to analyse ambient air in the field, care must be taken to ensure that pumps are not shaken or vibrated excessively. This is particularly true for the fast-rotating turbo pumps. When operated on aircraft or ships, pumps are often mounted on shock-absorbing mounts or the entire mass spectrometer is switched off before landing to prevent damage.

5.2.2 The atomic mass scale

Atomic and molecular weights used in mass spectrometry are generally expressed in terms of atomic mass units (amu) or daltons. This unit is based on a relative scale in which the reference is the carbon isotope ^{12}C, which has been assigned the mass of exactly 12 amu. Therefore, by definition, 1 amu is 1/12 of the mass of one neutral ^{12}C atom.

$$\begin{aligned} 1\,\mathrm{amu} &= 1\ \mathrm{dalton} = 1/12(12\,\mathrm{g}\ ^{12}\mathrm{C/mol}\ ^{12}\mathrm{C}/6.0221 \times 10^{23}\ \mathrm{atoms}\ ^{12}\mathrm{C/mol}\ ^{12}\mathrm{C}) \\ &= 1.66054 \times 10^{-24}\,\mathrm{g/atom}\ ^{12}\mathrm{C} = 1.66054 \times 10^{-27}\,\mathrm{kg/atom}\ ^{12}\mathrm{C} \end{aligned}$$

The atomic weight of an isotope such as ^{35}Cl is related to the reference ^{12}C by comparing the two masses. Since the chlorine isotope is 2.914707 times heavier than that of the carbon isotope the mass of the chlorine isotope is:

$$^{35}Cl = 12.0000 \text{ dalton} \times 2.19407 = 34.9688 \text{ dalton}$$

As 1 mol of ^{12}C weighs 12.0000 g, the atomic weight of ^{35}Cl is 34.9688 g/mol.

One result of this relative scale is that the masses of most ions in a mass spectrum fall near but not at integer values. Acetone (propanone), for example, has an exact mass of 58.0418.

$$\underset{C}{3 \times 12.0000} + \underset{H}{(6 \times 1.00783)} + \underset{O}{(1 \times 15.99491)}$$

Table 5.2 gives the exact masses of a selection of isotopes of common elements and selected atmospheric trace gas species.

Note that for stoichiometric calculations, chemists use the average mass calculated using the atomic weight, which is an average of the isotopes of each element of the molecule. The average mass of methane would therefore be $12.01115 + 4 \times 1.00797 = 16.0434$ daltons (see Table 5.2). However, in mass spectrometry, the monoisotopic mass generally is used. This mass is calculated by using the mass of the most abundant isotope for each constituent element. Methane is therefore $12.0000 + 4 \times 1.00783 = 16.03132$ daltons. In mass spectrometry we may need to distinguish between ions with almost identical masses. The extent to which they can be differentiated depends on the resolution of the mass

Table 5.2 Exact masses of selected elements and molecules

Element	Atomic mass/daltons
^{1}H	1.00783
^{2}H	2.01410
^{12}C	12.00000
^{13}C	13.00335
^{14}N	14.00307
^{15}N	15.00011
^{16}O	15.99491
^{18}O	17.99916
^{35}Cl	34.96590
^{37}Cl	36.96590
C_5H_8	68.06264
C_4H_5N	67.04222 (Low resolution needed)
CH_3COCH_3	58.0418
HC(O)CH(O)	58.00548 (Medium resolution needed)
$^{12}CH_3D$	17.03758
$^{13}CH_4$	17.03466 (High resolution needed)

spectrometer. In Table 5.2 the masses are given with four or five figures to the right of the decimal point. High resolution mass spectrometers have this precision.

As has been noted above, ions are required to perform mass spectrometry. How these ions interact with applied electric and magnetic fields depends on the mass and the charge of the ions. It is important to note that a mass spectrometer determines the mass-to-charge ratio (m/z) of these ions, see Figure 5.1. For example, singly charged methane $^{12}C_1H_4^+$, has $m/z = 16.03132$ whereas $^{13}CH_4^{2+}$ has $m/z = 8.518$. Where z is the total charge of the ion (Q) divided by the charge of one electron (e):

$$z = Q/e \quad e = 1.6 \times 10^{-19} \text{ Coulomb} \tag{5.2}$$

If mass is given in daltons and charge is given in electron charges, m/z is given in thomson (Th). In some cases the unitless terms mass number and charge number are given, in which case m/z is also unitless. Most ions encountered in mass spectrometry have a charge corresponding to the loss of one electron. For this reason the term 'mass-to-charge ratio' is often shortened to mass.

5.2.3 Resolving power and resolution

The resolving power of a mass spectrometer is a measure of its ability to separate two ions of any defined mass difference. The theoretical resolving power necessary to separate two ions differing in their m/z is defined as $M/\Delta M$, where M is the mean of the two masses and ΔM is the difference between the two masses. In order to separate an ion of the major biogenic emission product isoprene (m/z 68) from an ion of the biomass burning product pyrrole (m/z 67) a theoretical resolution of about 68 is required. This is considered to be low resolving power. A medium resolving power of circa 1600 is required to resolve the atmospherically important molecule acetone (m/z 58.04186) from the shorter-lived whole mass isomer glyoxal (m/z 58.00547) (see Table 5.2). Note that the isomers acetone and propanal have exactly the same mass and therefore cannot be resolved. Methane is an important trace gas in the atmosphere and valuable information about its sources and sinks can be obtained by measuring its isotopes. This is complicated as the methane isotopes $^{13}CH_4$ and $^{12}CH_3D$ require a mass spectrometer with higher resolving power, in this case ca. 5800. The original mass spectrometer built by Thomson had a resolving power of about 13 which was improved tenfold by his student Aston in 1933, see Table 5.1. Commercial mass spectrometers are available with resolving powers from a few hundred to 500 000.

In Figure 5.3, two overlapping mass peaks M_1 and M_2 in a mass spectrum are depicted. The resolving powers calculated above do not specify the acceptable degree of separation between the ions with slightly different masses. This is given by the resolution R, which is defined as the smallest mass difference for which two masses (M_1 and M_2) are resolved. By general agreement the peaks M_1 and M_2 are said to be resolved if the height of the valley between the peaks (h) is equal or less than 10% of the peak height (H). The resolution R therefore defines the smallest mass difference ΔM of peak width for which the two peaks M_1 and M_2 are resolved to this extent. Many quadrupole instruments operate at constant

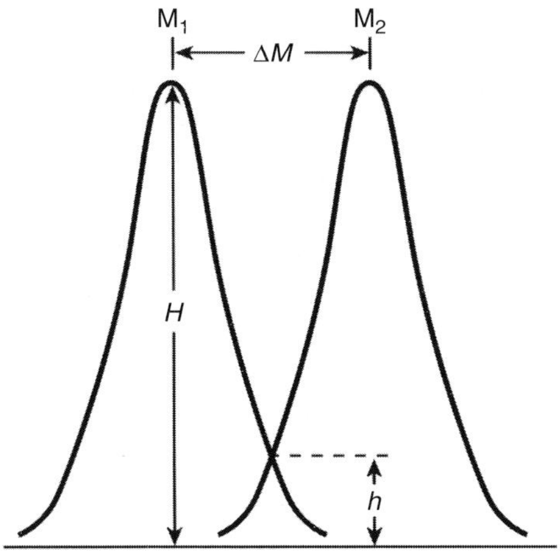

Figure 5.3 Parameters defining peak resolution.

resolution. This means that the ability to separate mass 100 and mass 1000 is the same. Note that if the resolution ΔM is 1 amu at $m/z = 100$, then the resolving power is 100. However if ΔM is 1 amu at $m/z = 1000$, then the resolution is still one but the resolving power is 1000. In practice, the resolution may be set by introducing compounds at a suitable mixture into the mass spectrometer to generate peaks M_1 and M_2. Adjustment of the various operating parameters of the mass spectrometer is performed until the peaks have the desired resolution, i.e., 10% valley. Since sometimes it is difficult to get two mass spectral peaks of equal height adjacent to one another, an alternative method of calculating ΔM is to use the full width at half maximum (FWHM) of a single m/z value peak. Resolving power calculated using the FWHM method gives values for R that are about twice that determined by the 10% valley method. By determining the resolving power by using the full width at 5% full maximum, an R value close to the 10% valley method can be derived. However, this is very difficult in practice because of background signals.

5.3 Key elements of a mass spectrometer

5.3.1 Ionisation

The first step in the mass spectrometric analysis of an air sample is the ionisation of the sample. The method used to ionise a substance profoundly influences the appearance of its mass spectrum. For trace gas species in ambient air, two methods of ionisation are currently prevalent. These are termed 'electron ionisation' and 'chemical ionisation' and discussed in Sections 5.3.1.1 and 5.3.1.2 respectively.

5.3.1.1 Electron ionisation

Electrons produced from a heated filament in an evacuated chamber can be accelerated towards an anode by applying a voltage (V). The energy of these electrons will be eV, where e is the charge of the electron. Each electron is associated with a wave whose wavelength λ,

$$\lambda = \frac{h}{mv} \tag{5.3}$$

where m is its mass, v is its velocity and h is Planck's constant. For kinetic energies of 70 eV the wavelength is 1.4×10^{-10} m. When this wavelength is close to the bond lengths, the electron waves and the electric field of the molecule mutually distort one another. The distorted electron wave can be considered to be many waves with different frequencies and if one of the waves has the correct frequency (or energy) to interact with molecular electrons (i.e. the energy corresponding to a transition), an energy transfer can occur. The interaction of an electron with a molecule is rather non-specific. One possibility is that the molecule may be electronically excited, in similar fashion to ultraviolet spectroscopy (see Chapter 3). This can be represented by

$$\mathrm{M + e \rightarrow M^{*} + e^{-}} \tag{5.4}$$

Alternatively a molecular electron may be ejected from the molecule to leave a radical cation. This is the best outcome for subsequent mass spectrometry since it produces an ion that can be manipulated in electric and magnetic fields.

$$\mathrm{M + e^{-} \rightarrow M^{+}\cdot + 2e^{-}} \tag{5.5}$$

A further possibility is direct capture of the electron by the molecule to give a stable radical anion. However, the probability of this is low. Although less than one per cent of the sample molecules are converted to positive ions, electron capture as depicted in Equation (5.6) is probably one hundred times less probable.

$$\mathrm{M + e^{-} \rightarrow M^{-}\cdot} \tag{5.6}$$

The energy of the ionising electrons may be varied by varying the applied voltage. By convention, standard mass spectra are obtained at 70 eV since the maximum ion yield is obtained near this energy. Furthermore, at this level small changes in energy do not greatly affect the mass spectrum, helping the reproducibility of the analysis. Mass spectra can be obtained at any electron energy above the ionisation energy of the molecule (typically between 6 and 12 eV). Indeed it is often desirable to measure the mass spectrum at energy levels less than 70 eV, despite the reduction in ion abundance, since the amount of molecular fragmentation following ionisation can be reduced and therefore identification facilitated. At higher potentials the electron wavelengths become too small and molecules become effectively transparent. Extensive mass spectral libraries have been compiled for substances ionised by electron ionisation at 70 eV. These can be matched against unknown spectra from a sample to identify components.

5.3.1.2 Chemical ionisation

In chemical ionisation, ions are produced through the collision of the molecule to be analysed with primary ions present in the source. The indirect method of chemical ionisation (CI) is a complementary technique to the direct method of electron ionisation (EI). While the high energies involved in EI commonly lead to fragmentation of the molecular ion, chemical ionisation produces ions with little excess energy. Less fragmentation facilitates identification of molecular species. Ion sources for chemical ionisation are very similar to those for EI but they typically operate at high pressures (e.g. 0.1 mbar compared to 10^{-6} mbar for EI). In other words, in EI the gas sample is reduced in pressure before the ionisation directly before the mass filter, while in CI the ionisation generally occurs at higher pressure but the resulting ions must be directed via ion lenses to the low pressure region containing the mass filter. In CI, both positive ions (e.g. H_3O^+ ions) and negative (e.g. SF_6^- ions) can be used. Reactant gas ions are created by leaking a reactant gas into an ion source where they interact with an electron beam or high energy particles (e.g. emitted by radioactive sources) to produce primary ions. The pressure in the ionisation region is adjusted so that the concentration ratio of reagent to sample is 10^3–10^4. Because of this large concentration difference the electron interacts predominantly with the reagent. The reagent ions then react with the sample molecules of interest to produce ions. This form of chemical ionisation has two advantages over electron ionisation described in Section 5.3.1.1. First, electron ionisation of a molecule can give rise to a multitude of fragments, and for a complex mixture like ambient air, the resulting mass spectrum can be indecipherable. Secondly by judiciously choosing the reagent ion the trace gases of interest in the air can be detected preferentially over the hugely more abundant species such as nitrogen and oxygen. Typical reagents are methane, water and ammonia. Water is used for measurement of species such as acetone by API-CIMS and PTR-MS (Sections 5.4.1 and 5.4.2). Methane has been used in negative ion chemical ionisation in combination with gas chromatography, for the analysis of alkyl nitrates and organohalogens in air (Section 5.4.4).

5.3.2 Mass filters

The mass filter is the heart of the mass spectrometer. It is here that the ions that have been produced in the ion source are separated according to their m/z. The ability to separate ions is characterised by the resolution (see Section 5.2.3). From the geometrical configuration of the mass filters described below a theoretical resolution can be calculated. However, in all cases the actual resolution is dependent on many other factors specific to the device in question (e.g. constancy of the electric fields etc.) and should be empirically assessed. The mass filters described below are all used in atmospheric research. In this section the principle of each is described and in Sections 5.4 and 5.5 various applications of these filters are described.

5.3.2.1 Quadrupole

The quadrupole mass filter is a device that uses the stability of the trajectories in oscillating electric fields to separate ions according to their m/z, see Figure 5.4. An ideal quadrupole consists of four symmetrically arranged hyperbolic shaped surfaces of infinite length. In reality, quadrupole mass filters consist of four parallel cylindrical rods. Typically these are 10–20 mm in radius and 20 cm long. Opposing rods are electrically connected and a potential comprising a DC voltage (U) and an RF component ($V \cdot \cos \omega t$) is applied. For example when a positive ion enters the space between the rods, the electric field will act to draw it to a negative rod. If the rod potential changes sign before the ion discharges on the rod, the ion will change direction. Since there is no field applied in the z axis the ions continue with the same speed as they enter the quadrupole. The DC and RF fields can be varied so that an ion of a specific m/z can pass through the rods while others do not.

The equations of motion of an ion in a quadrupole can be represented by the Mathieu differential equations:

$$\frac{d^2x}{dt^2} + (a + 2q\cos \omega t) = 0 \tag{5.7}$$

$$\frac{d^2y}{dt^2} - (a + 2q\cos \omega t)y = 0 \tag{5.8}$$

$$\frac{d^2z}{dt^2} = 0 \tag{5.9}$$

where,

$$a = \frac{8eU}{mr_0^2\omega^2} \tag{5.10}$$

$$q = \frac{4eV}{mr_0^2\omega^2} \tag{5.11}$$

$$\omega t = 2\zeta \tag{5.12}$$

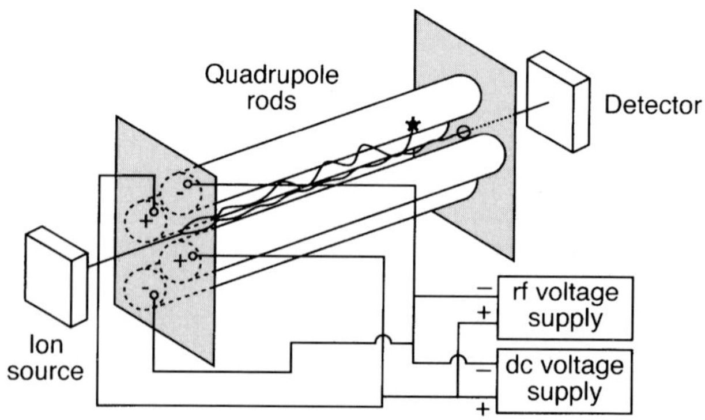

Figure 5.4 A schematic diagram of a quadrupole mass filter.

The solutions to these differential equations lie in two classes, corresponding to either 'stable' or 'unstable' trajectories. When the rod potentials (U, V and/or ω) are varied in a suitable manner then an ion of selected m/z (and only this ratio) may pass through the quadrupole to the detector. By adjusting these potentials with time to satisfy a small mass and then sequentially the larger masses, a mass spectrum can be generated. Figure 5.5 shows a so-called stability diagram for a quadrupole. Within the area defined by the shark's fin triangle are values of a and q generating stable solutions. The operating line of the quadrupole is defined by the ratio $a/q = 2U/V$. By increasing this ratio, the region of stable solutions can be so restricted that only ions of a specific mass pass through. In this way the resolution of the mass filter can be varied, i.e. higher the ratio the higher the resolution. However, when too high a ratio is chosen, the operating line is no longer secant of the stable area and no ions will pass at all. For a real quadrupole, the efficiency of transmission of ions depends on the exact locations and directions of the ions when entering the rod system. Ions entering the rod system inclined to the z-axis may possibly not pass the filter despite having the correct m/z. In fact the transmission in the mass filter (number of ions entering/number of ions exiting) is approximately inversely proportional to the resolution. Both transmission and resolution can influence the accuracy and precision of a measurement and a compromise must be made between the two for optimum performance.

Quadrupole mass filters are often operated at unit mass resolution (i.e. 1 amu, low resolution), in contrast to the magnetic mass filters described in Section 5.3.2.3. However, quadrupoles are relatively cheap and easy to operate. This is because the necessary electronics (precise RF generators) have become standard technology instruments.

5.3.2.2 *Time of flight*

When ions of different masses are produced at the same place with the same energy, and then accelerated by a potential V over a distance d, they will arrive at different times. The time of flight mass filter relies on an accurate measurement of these times to distinguish ions of different masses. In reality, ions are usually continually produced but expelled in bundles from an ion source by the transient application of a potential to the source

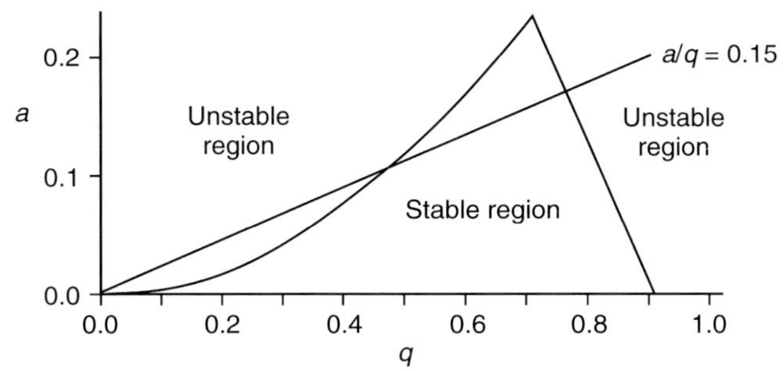

Figure 5.5 The stability diagram for a quadrupole.

lenses. In a magnetic field-free linear separation system, an ion with mass m and total charge $q = ze$, has a kinetic energy KE defined by Equation (5.13).

$$\text{K.E.} = \frac{mv^2}{2} = zeV \tag{5.13}$$

Since $t = v/d$, we can rearrange this equation to give

$$t^2 = \frac{m}{z} \times \left(\frac{d^2}{2Ve}\right) \tag{5.14}$$

Equation 5.14 shows that ions with lower masses will therefore reach the detector first. By accurately measuring the time t and keeping V and d constant m/z can be derived. An advantage of this technique is that although the time differences are extremely small, time can be very accurately measured. The principle is depicted in Figure 5.6.

Time of flight mass filters have very high transmission efficiencies, which leads to very high sensitivities. All the ions formed are directed to the detector and so, in principle, all ions are analysed. This is in contrast to the quadrupole mass filter, described in Section 3.2.1, which measures masses sequentially. An important drawback of time of flight instruments is difficultly in achieving good resolution. This stems from the physical limits of creating ions, of the same energy, instantaneously, in the same place. These are the following: the length of time in the ion formation pulse is finite, creating a time distribution; there is a minimum size of the ion-generating volume creating a spatial distribution; and the ions produced do in reality have a kinetic energy distribution. All these factors contribute to uncertainty in the measurement of the flight time and hence to lower resolution. To counter this, TOF-MS practitioners lengthen the flight tube and hence the flight time, thus increasing the resolution. A long flight tube would be typically 1–2 m in length. This is a considerable restriction when operating this instrument on small jet aircraft where space is a precious commodity. One major advantage of current TOF systems is that they may be operated at high measurement frequencies. Commercially available systems may be operated at 500 Hz (in contrast to typical quadrupole data acquisition frequency of 10 Hz), making them ideally suited to comprehensive chromatography applications, see Chapter 8.

5.3.2.3 Magnetic sector

The mass separating effect of a magnetic field comes from the fact that ions of equal energy but different m/z describe different radial trajectories when passing through a

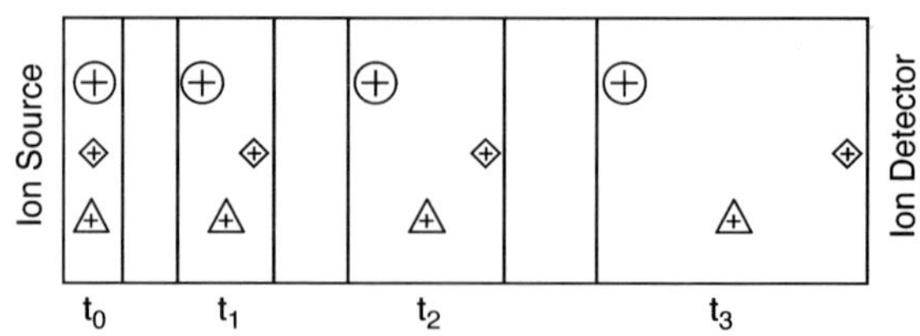

Figure 5.6 A representation of the principle of time of flight mass spectrometry (Jochum, 1988 with permission from Expert Verlag GmbH).

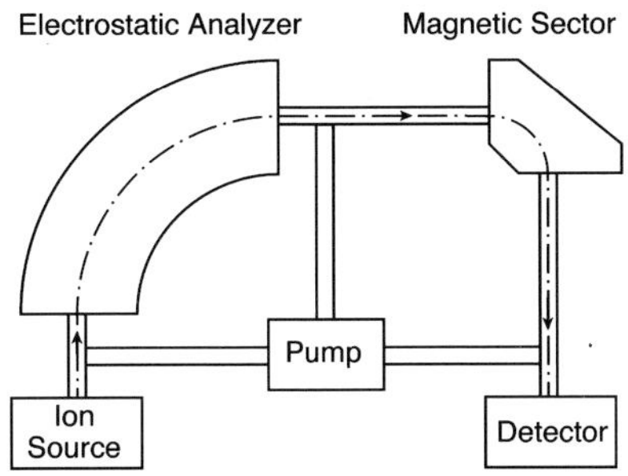

Figure 5.7 A schematic of the Nier-Johnson type magnetic sector mass spectrometer design.

magnetic field. A schematic diagram of a magnetic sector mass filter is given in Figure 5.7. In a homogeneous magnetic field of strength B, ions with a velocity v experience a force F, in a direction perpendicular to the field lines and to the direction of motion of the ions. In order to traverse the circular path, the magnetic or Lorentz force FL and the centripetal force Fc must be equal. The radius of the resulting trajectory can be calculated by equating the Lorentz force $\mathrm{FL} = evB$ with the centripetal force $\mathrm{Fc} = mv^2/r$. Substituting the relationship $eU = mv^2/2$ for the ion energy (where U is the acceleration potential), we obtain:

$$r = \frac{mv}{eB} = \frac{1}{B}\frac{2mU}{e} \tag{5.15}$$

Rearranging we can obtain an expression for m/z.

$$\frac{m}{z} = \frac{B^2 r^2 e}{2V} \tag{5.16}$$

From Equation (5.16) it can be seen that mass spectra can be obtained by varying one of three variables (B, V, or r) while holding the other two constant. Classical magnets were not well suited to the fast scanning necessary for coupling with GC due to magnetic heating by Foucauld currents induced by rapidly changing the magnetic field. Most modern mass spectrometers contain an electromagnet in which ions are sorted by holding V and r constant while varying the current in the magnet and hence the field strength B.

In practice, most magnetic sector instruments are built to deflect ions through circular paths of 60, 90 or 180°. When a collection of ions with the same m/z but with small diverging directional distribution is acted upon by the magnetic field, a converging directional distribution is produced as the ions leave the field (directional focussing). For this reason the simplest magnetic sector mass filters, are called single focussing spectrometers. It is important for mass spectrometry with magnetic sectors that the incident ions have the same energy as far as possible. A large span of energies leads to a low resolution. Therefore to obtain the high resolution required for isotope measurements an electric and a magnetic field are combined so that the energy dispersing effect of the magnet is compensated by the electric field. By this so-called double focussing method, directional and energy focussing is achieved. Figure 5.7 shows a schematic diagram of the so-called 'Nier-Johnson geometry' design.

A wide variety of double focussing mass spectrometers are available commercially. Resolutions of up to 10^5 can be obtained and this is necessary when analysing isotopic ratios in ambient air.

5.3.2.4 Ion trap

An ion trap is a device in which gaseous ions can be formed and confined for extended periods by electric and/or magnetic fields. In a sense, the ion trap may be thought of as a 'three dimensional' quadrupole (see Section 3.2.1.). The electric fields in a standard quadrupole assembly constrain ions into stable two-dimensional motion between the quadrupole rods. By constructing an arrangement of two end-cap electrodes, one on either side of a third annular electrode and all of hyperbolic cross section, ions injected into the volume between the three electrodes can be 'trapped' into a complex circular motion by applying suitable DC and RF electric fields. Ions with an appropriate m/z circulate in a stable orbit within the cavity. When the radio frequency is increased, the orbits of the heavier ions become stabilised, while those of the lighter ions become destabilised and eventually collide with the electrode. The ions repel one another in the trap and as a result their trajectories expand as a function of time. To avoid ion losses through this expansion, a small pressure of helium gas (typically 0.13 Pa) is maintained in the trap. This removes excess energy from the ions by collision.

Like a quadrupole, the ion trap is usually a low resolution mass spectrometer, being usually able to resolve unit mass (1 amu). A considerable advantage of the ion trap mass filter is that it can be used to perform MS/MS experiments. This involves trapping ions of a characteristic m/z as described above and then collisionally fragmenting them further before detection. In this way two different species having the same m/z can be differentiated provided they produce distinct fragments. Usually the precursor ions are given more kinetic energy through the use of an extra RF voltage applied to the end caps so that total collision energy can be increased and the selected ions induced to fragment. This is of great help in atmospheric studies where many different species may lie on the same mass (e.g. propanal, acetone, glyoxal all have mass 58 at unit mass resolution, see Table 5.2) but when fragmented can be expected to be different.

5.3.3 Detectors

Following the successful application of the mass filters described above we obtain a current of ions with a well-defined m/z. These ion currents lie typically between 10^{-8} and 10^{-16} A. To measure these currents a number of systems are available. The three most commonly in use in atmospheric science are described below.

5.3.3.1 The Faraday cup

The Faraday cup collector is the simplest and most inexpensive of detectors available. The cup is aligned so that ions exiting the mass filter strike the collector electrode. The charge carried by the ions that hit the cup is earthed via a high ohmic resistance and the potential difference across this resistance is measured with a high impedance amplifier.

The electrode is surrounded by a cage that prevents the escape of reflected ions and ejected secondary electrons. The collector electrode is inclined with respect to the entering ions so that particles striking or leaving the electrode are reflected away from the entrance to the cup. The response of this detector is independent of the mass, the energy and the chemical nature of the ion. The main disadvantage of the Faraday cup is the amplifier required, since this limits the speed at which a spectrum can be obtained. Moreover, as the cup detector uses no internal amplification (see Sections 5.3.3.2 and 5.3.3.3) it is less sensitive. The Faraday cup is widely used in systems requiring high precision, as the charge of the cup is not dependent on the mass and energy of the detected ion. It is necessary to use such detectors for isotopic analyses, see Section 5.4.3.

5.3.3.2 The Secondary electron multiplier (SEM)

Where extremely high sensitivity is required to detect ions, a secondary electron amplifier is used. The secondary electron multiplier comprises a conversion dynode, a series of further dynodes and a collector. Every mass-filtered ion that hits the conversion dynode releases a burst of electrons, which are accelerated to the next dynode. In turn, each electron hitting the second dynode release a similar burst of secondary electrons. With each successive dynode the electron current is amplified with the result that the current measured at the final cup is several orders of magnitude greater than the incident ion current. A 20-stage SEM can produce a current gain of ca 10^7. A combined schematic diagram and picture of an SEM is shown in Figure 5.8. The SEM pictured is one used in the proton transfer mass spectrometer described in Section 5.4.2. To better judge the size of this detector a 1 Euro coin is included in the photograph.

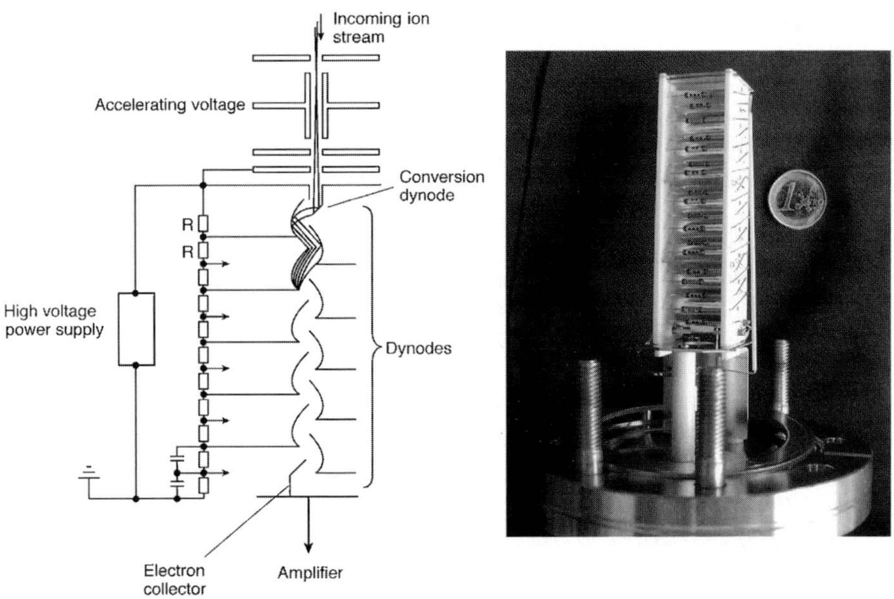

Figure 5.8 Schematic diagram and photograph of a secondary electron multiplier.

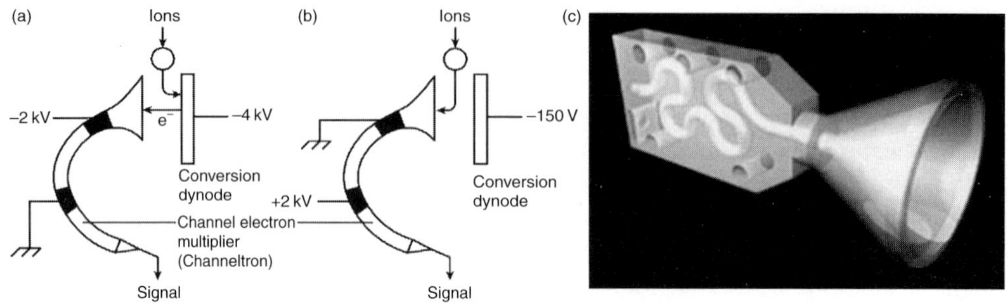

Figure 5.9 A schematic diagram of a channeletron when used for (a) negative and (b) positive ion detection and (c) a cross section picture of a commercial version (Picture courtesy of Sjuts Optotechnik GmbH).

In general, SEM detectors are robust and reliable, providing high current gains in nanosecond response times. The detector response is, however, dependent on the mass, the charge, the speed and the molecular structure. This necessitates empirical assessment of the ion calibration factors for quantification work. For magnetic sector instruments, the detector may be set directly behind the mass filter exit slit as ions reaching the detector usually have enough energy to eject electrons from the first dynode. When used with quadrupoles the ions must be accelerated prior to the first impact. The lifetime of these detectors is 1–2 years because of surface contamination of the Cu/Be surface from the incident ions. They are not as precise as Faraday cups but the higher sensitivity allows for rapid scanning.

5.3.3.3 The Channeletron

The channeletron (or channel electron multiplier) is a variant of the secondary electron multiplier described above, but it has a single continuous dynode rather than several in series. It consists of a glass or ceramic tube with semiconducting inner surfaces. This detector is used in the API-CIMS instrument described in Section 5.4.1. When used for positive ions, see Figure 5.9a, a conversion dynode at high negative potential (3–4 kV) is used to release electrons when the ion impacts. The resulting electrons are led into the channeletron where they cause an electron avalanche, sufficient to be detected by a preamplifier. In the case of negative ions, Figure 5.9b, the ions themselves are led into the channeletron and detected. Current gains of 10^5 are typical for such detectors. A cross section of a commercially available channeletron is shown in Figure 5.9c.

5.4 Mass spectrometry used in atmospheric chemistry

5.4.1 Atmospheric pressure ionisation-chemical ionisation mass spectrometer (API-CIMS)

As the name suggests, in API-CIMS primary ions are produced in an ion source operating under approximately atmospheric pressure. This instrument has also been termed as ion molecule reaction mass spectrometer (IMRMS). In such an instrument, positive or negative reagent ions interact with the trace gas molecules of interest and produce new ions by chemical ionisation, as described in Section 5.3.2. For these reactions to be fast and effective, the ionisation has to occur at high pressures, and for atmospheric scientists it is practical to choose approximately atmospheric pressure.

Figure 5.10 shows a schematic diagram of a mass spectrometer that has been used to measure SO_2 (when in negative ion mode) and acetone (when in positive ion mode) on various field campaigns on the ground and from on-board aircraft. Ions are produced by corona discharge between a needle and the aperture lens housed within the ion source, which is kept at constant temperature (80°C) and pressure (280 hPa). The polarity of the needle is chosen to have the same polarity as the ions to be detected since this improves the transmission of ions through the system. Further details of such chemical ionisation systems can be found in Huang *et al.* (1990). In this case the pressure in the ion source is chosen to be slightly lower than the lowest pressure encountered in flight (at maximum altitude of the aircraft, in this case 8.5 km–350 hPa).

The ion source is separated from the decluster chamber by a small 0.25 mm orifice. As the decluster chamber is at lower pressure 0.5–1.0 hPa, a small flow (200 ml) of air is drawn through the source region into the decluster chamber. When ionisation of ambient air is done at these pressures, significant clustering of water molecules can occur on the ions of interest. Since the number of water clusters can vary greatly, the ion signal can be spread over many masses (e.g. $X^+(H_2O)_n$), complicating the mass spectra in similar fashion to fragmentation. In the decluster chamber the mean free path of the gas particles is typically 100 μm, which is enough for them to be accelerated in the electric field present in the decluster chamber. The energy thus imparted is sufficient to induce declustering

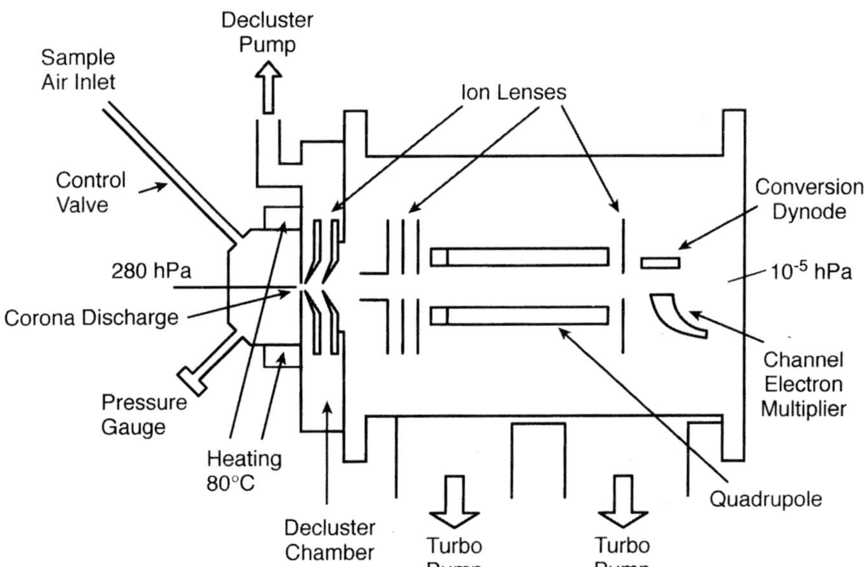

Figure 5.10 A schematic diagram of an atmospheric pressure chemical ion mass spectrometry. (Jost, 2002)

of the ions upon collision with another gas molecule. The desired ion-molecule reactions can occur from when the ions are first produced until the ions are drawn into the high vacuum of the spectrometer, where the pressure is too low for ion-molecule reactions to be effective. A number of ion lenses are used to improve the transmission of the ions from the declustер chamber to the channel electron detector. The ion lenses always have the same polarity as the ions to be detected, and the lens potential must be optimised. If it is too low the transmission will be poor, while too high causes the ions to enter the quadrupole too fast and poorly resolved mass spectra result.

Let us now examine the underlying principle of operation. Having produced a positive or negative reagent ion $A^{\pm}$, we wish to react it with a neutral molecule of interest, X. This can occur in a time dt, between ion source and high vacuum.

The rate of reaction can therefore be described,

$$\frac{d[A^{\pm}]}{dt} = -k[A^{\pm}][X] \qquad (5.17)$$

where k is the rate coefficient and the terms in square brackets are concentrations. Assuming that only a small fraction of the trace gas X reacts then it follows that [X] is approximately constant. Therefore we may integrate to derive the expression,

$$[X] = \frac{1}{k} \ln \frac{[A^{\pm}]_{t=0}}{[A^{\pm}]_{t=T}} \qquad (5.18)$$

By taking into account that the number of ions does not change in-ion molecule reactions we may further transform Equation (5.18) to include $B^{\pm}$, the product ions, as shown below

$$[X] = \frac{1}{kT} \ln \frac{[A^{\pm}]_{t=T} + [B^{\pm}]_{t=T}}{[A^{\pm}]_{t=T}} = \frac{1}{kT} \ln \frac{(1 + [B^{\pm}]_{t=T})}{[A^{\pm}]_{t=T}} \qquad (5.19)$$

If the reaction time T is sufficiently short, the concentration of reactant ions $A^{\pm}$ remains much higher than the concentration of the product ions $B^{\pm}$, and the approximation $\ln(1+x) \approx x$ when $|x| << 1$ can be applied, simplifying this equation to

$$[X] = \frac{1}{kT} \frac{[B^{\pm}]_{t=T}}{[A^{\pm}]_{t=T}} \qquad (5.20)$$

The concentration ratio of the reagent ion $[A^{\pm}]$ to the product ion $[B^{\pm}]$ can be determined from the relative count rates of the mass spectrometer. Therefore the concentration of gas X may be determined if we know the reaction time t, the rate coefficient k and the count rate ratio of $B^{\pm}$ and $A^{\pm}$. The same principle is used in the proton transfer mass spectrometer described in Section 5.4.2. The reaction time T can be calculated or measured although the measurement is complicated. The rate of an ion–molecule reaction is very fast and the rate coefficient k typically has a value of 10^{-9} cm^3 molecule^{-1} s^{-1}.

In the case of the atmospheric pressure ionisation mass spectrometer described here, multiple reagent ions can be formed (e.g. in positive mode $H_3O^+(H_2O)_n$). The distribution of these ions can change markedly with humidity. Each may react with X with different rates, which are known only with ca 30% accuracy. Additionally, back reactions and continuing reactions of the product ions can also be sources of error. Clearly, careful calibration is essential to determine the proportionality coefficient between the X, the molecule of interest and the product/reagent ion ratio under a variety of temperature and humidity conditions. The measurement principle of API-CIMS is in many ways similar to the PTR-MS technique. The accuracy, precision and errors of atmospheric measurements made with these techniques are discussed in Section 5.4.2.

The API-CIMS instrument described here has been successfully deployed on several ground and airborne campaigns. Over the west coast of Africa it has been used to track the emission and subsequent photochemical production of acetone (using H_3O^+ as a reagent ion) and sulphur dioxide (using the CO_3^- ion) from the large-scale biomass burning in the dry season (Jost *et al.*, 2003). Furthermore, the instrument was employed in the laboratory to investigate the direct emission of compounds (e.g. acetonitrile, CH_3CN) when selected African biomass materials were burned under controlled conditions, see Figure 5.11.

If the global distribution of biomass is known, then laboratory emission rates determined in this way may be used to derive global emission strengths. Furthermore the emissions rates may be included in global models in order to estimate the atmospheric effect of these global emissions on species such as ozone.

By adopting different reagent ions, other gases of atmospheric interest may be selectively detected. The SF_6^- reagent ion has been used in laboratory or field studies to detect molecules of atmospheric interest such as HNO_3, NO_2, O_3, HOCl, CF_2O and SO_2 (Huey *et al.*, 1995) and in 2004 hydrochloric acid (HCl) was measured from aircraft in the upper troposphere/lower stratosphere with a 5 pptv detection limit and an accuracy of 25% (Marcy *et al.*, 2004).

5.4.2 The proton transfer reaction mass spectrometer

The proton transfer reaction mass spectrometer (PTR-MS) is a specific example of the general chemical ionisation scheme described above, being developed by Lindinger and co-workers at the University of Innsbruck (Lindinger *et al.*, 1998). The underlying principle is the same as described in Section 5.4.1 but in this case, the reagent ion is H_3O^+. Complications resulting from water clusters present in API-CIMS are reduced by judicious

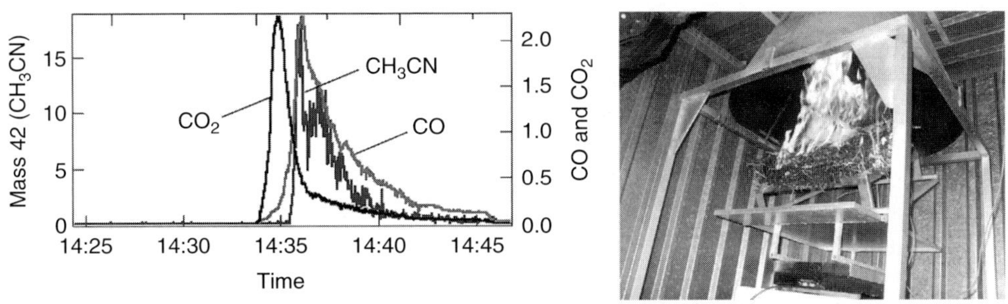

Figure 5.11 Example of a laboratory application of API-CIMS – Biomass burning analysis.

Table 5.3 Proton affinities

Compound	Formula	Proton affinity (kJ mol^{-1})
Nitrogen	N_2	464.6
Oxygen, Ozone	O_2, O_3	396.3, 595.9
Noble gases	Ar, Ne, He	346.3, 174.4, 148.5
Carbon mon/di oxide	CO, CO_2	562.8, 515.8
Methane	CH_4	544
Water	H_2O	*660.0*
Methanol	CH_3OH	725.5
Acetonitrile	CH_3CN	779.2
Acetone	CH_3COCH_3	782.1
Dimethylsulphide	CH_3SCH_3	801.2
Isoprene	$CH_2C(CH_3)CHCH_2$	797.6

Source: Data from Hunter and Lias (1998).

choice of pressure and energy parameters, as detailed below. The key step is the transfer of a proton from the reagent ion (H_3O^+) to the molecule of interest (X). Through proton transfer reactions it is possible to measure all compounds with a proton affinity larger than that of H_2O. Fortunately for atmospheric scientists, it is 'blind' to the major air constituents N_2, O_2, Ar, and CO_2, as can be seen from Table 5.3. However, it can detect many important atmospheric species such as acetone and isoprene. The ionisation occurs at relatively low energies so that protonation usually does not cause the molecule to fragment, or if it does, by ejecting only one H_2O molecule. This considerably simplifies the mass spectra. The proton affinity or gas-phase basicity is defined as the negative molar Gibbs energy, $-\Delta G$, of the hypothetical protonation reaction,

$$H^+ + X \rightarrow XH^+ \tag{5.21}$$

A selection of these proton affinities is given here in Table 5.3. A comprehensive list can be found at the NIST website – http://webbook.nist.gov/chemistry/. All the species in Table 5.3 with proton affinities greater than water have been measured by PTR-MS.

A schematic diagram of the PTR-MS apparatus is shown in Figure 5.12. The ion source consists of a hollow cathode and an earthed anode, through which water vapour flows at about 8 cm^3 min^{-1}. When held at a pressure of about 2 mbar (200 Pa), a cathode voltage of 600 V is sufficient to cause an electric discharge between cathode and anode. The result is an intense source of H_3O^+ ions: count rates of typically 1×10^6 counts s^{-1} being detected by the mass spectrometer. An alternative source of H_3O^+ ions has also been developed using alpha particles emitted from a strip of Am^{241} to ionise water vapour (Hanson *et al.*, 2003). Connected to the ion source is the reaction chamber or 'drift tube', where the proton transfer reactions occur between H_3O^+ and the compounds of interest in ambient air. The reactor consists of a series of stainless steel rings separated by thin isolating teflon rings. The steel rings are connected by resistors so that a voltage of up to 600 V can be applied over the entire set of rings to obtain a homogenous electric field. Ambient air is pumped through the drift tube which is maintained at 2 mbar (200 Pa) pressure. The residence time of air in the drift tube is about 1 second. During this time

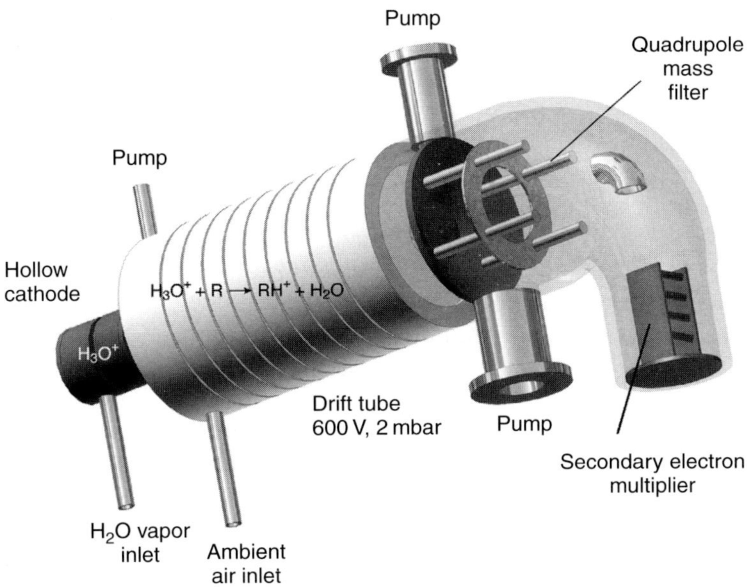

Figure 5.12 A schematic diagram of the PTR-MS.

H_3O^+ ions are accelerated through the drift tube containing ambient air and ionising reactions may occur. As we have seen in Section 5.4.1, H_3O^+ ions can form clusters with water of the type $H_3O^+(H_2O)_n$. The distribution of the clusters is highly dependent on the sample humidity. More humid conditions promote larger clusters. Their presence complicates the mass spectra and for some species it can lead to a strong humidity dependence in the calibration (see Section 5.4.1). This is because some trace gas species react with H_3O^+ but not with $H_3O^+(H_2O)_n$ ions, an effect that has been shown to be important for non-polar compounds such as benzene (Warneke *et al.*, 2001). The drift tube electric field is applied to increase the average kinetic energy of the ions to prevent clustering so that in PTR-MS, H_3O^+ is the predominant reagent ion. The disadvantage of the electric field is that it effectively limits the reaction time and pressure in the reaction chamber, making the PTR-MS technique less sensitive than API-CIMS. From the drift tube the ions are directed to a quadrupole mass filter held at 10^{-5} bar by a turbo molecular pump.

Quantification is based on the difference in counts between ambient air and air that has been scrubbed of organics by passing it through a heated trap containing a platinised wool catalyst. The interpolated background signal before and after an ambient measurement is then subtracted from the ambient signal to derive a mass signal corrected for any background contamination originating from the mass spectrometer. As the effectiveness of the mass filter and detector is mass dependent the ion counts must be corrected for this transmission difference. The transmission as a function of mass can readily be determined by sequentially injecting high concentrations of species covering the entire mass range. At sufficiently high concentrations all the reagent ions (e.g. H_3O^+) are converted to product ions (e.g. $CH_3COCH_3^+$). The ratio of the initial reagent ion counts per second (typically 1×10^6) to the acetone counts indicates the relative transmission of the two ions. Normally the acetone ions, mass 59, are transmitted around 2.4 times better than the H_3O^+ and this must be accounted for in the quantification. Since the transmission of the ions measured can change markedly as the detector ages it is advisable to ascertain the transmission regularly when operating such instrumentation in the field.

In measurement mode, the resulting signal is the number of counts detected at each mass per second, which can be converted to a mixing ratio if the rate constant for the protonation reaction and the reaction time are known (see Section 5.4.1). All exothermic proton transfer reactions proceed at, or close to, the collisional rate, and where specific reactions rates have not been measured they can be calculated (Su & Chesnavich, 1982). In practice the PTR-MS is calibrated in the laboratory against commercially available gas standards but derived rate coefficients are checked for consistency with theoretical ones. In addition to the uncertainty in the gas standards there is statistical error associated with the count rates that must be considered in the overall error assessment. This noise signal (NS) is determined by Equation (5.21) (Hayward *et al.*, 2002), where mean signals are in counts per second and dwell time is the time spent counting at a given mass.

$$\mathrm{NS} = \frac{\text{mean signal (cps)}}{\sqrt{(\text{mean signal (cps)} \times \text{dwell time (s)}}} \qquad (5.22)$$

The statistical error of PTR-MS data with a 2-second integration time is typically ±10% at volume mixing ratios around 5 nmol/mol and ±30% at 1 nmol/mol. The errors in the absolute values are between 10 and 30%, being largely determined by uncertainty in the gas-phase standard. Optimisation of the PTR-MS technique for airborne measurement resulted in detection limits for most species between 10 and 100 pmol/mol, which are determined from the background (air through catalyst) values at each mass. The accuracy, precision and detection limit of this method are currently worse than gas chromatographic methods (see Section 5.3.4). However, the measurement frequency, and hence data coverage, is considerably better. The high frequency of measurement is ideally suited to fast-moving platforms such as jet aircraft that travel at typical speeds of $150\,\mathrm{ms}^{-1}$. The PTR-MS has been operated on the ground (Salisbury *et al.*, 2003), on planes (Crutzen *et al.*, 2000) and on ships (Williams *et al.*, 2004). In the study conducted on ships two PTR-MS systems were used to measure species such as DMS, methanol, acetone and acetonitrile in the air and the surface seawater simultaneously.

The main disadvantage of the technique for atmospheric research is that several compounds can contribute to the signal at a given mass. For example, mass 59 corresponds to protonated acetone, propanal and glyoxal, all of which may be present in ambient air (Williams *et al.*, 2001). In some cases it is possible, by examining the potential contributing compounds or the variation of the mass signal, to deduce a most likely candidate for a given mass. However, in many cases the concentration determined for a given mass cannot be assigned to a single compound unambiguously. Making the PTR-MS technique more compound-specific is a key goal in the future, see Section 5.5. Some groups have replaced the commonly used quadrupole detector with a TOF MS (Blake *et al.*, 2004). While the PTR-TOF system has demonstrated improved mass resolution over a quadrupole, there are considerable drawbacks. First, improved resolution will never resolve some of the key overlapping species in the atmosphere (e.g. acetone and propanal) since they have identical masses. Secondly the PTR-TOF system is necessarily much larger and heavier than the quadrupole, and the sensitivity at this stage in the development is some 2 orders of magnitude less. At the time of writing, it would seem that a PTR-ion trap would be a prudent line of instrument development.

An example of a vertical profile of acetone, methanol and acetonitrile mixing ratios up to 13 km is shown in Figure 5.13. These data were obtained on board a German research aircraft flying over the Mediterranean. It is clear that there is a high degree of structure in the profiles. By studying the back trajectories at different altitudes it was possible to determine the origins of the airmasses as: 8–13 km from Southeast Asia; 4–8 km from the Atlantic Ocean region; and 0–4 km from Eastern Europe. The chemical signatures of these airmasses are also distinct and reflect the sources in the aforementioned regions. From previous studies we know that acetonitrile is a good marker for biomass burning, methanol is primarily released from vegetation and that acetone can be produced from both anthropogenic and biogenic sources. Thus we can see that in July, biomass burning is important in Eastern Europe and Southeast Asia but not in the airmasses from the West. By careful analysis of such datasets we may make estimates of the global sources for these compunds. These data are described in detail by Lelieveld *et al.* (2002). The clear vertical structure in this plot is obtained because of the key advantage of the PTR-MS (and API-CIMS) techniques, namely high measurement frequency.

5.4.3 The use of CIMS for radical measurement (OH)

The hydroxyl radical is the most important oxidant in the lower atmosphere, and therefore the primary removal mechanism for most trace gas species in ambient air. As a result there has been considerable interest in measuring the concentration of this radical. However, as OH is found at very low concentrations (ca. 1×10^6 molecules cm^{-3}) and the radical is inherently very reactive (chemical lifetime 0.1–1 seconds), so its measurement represents

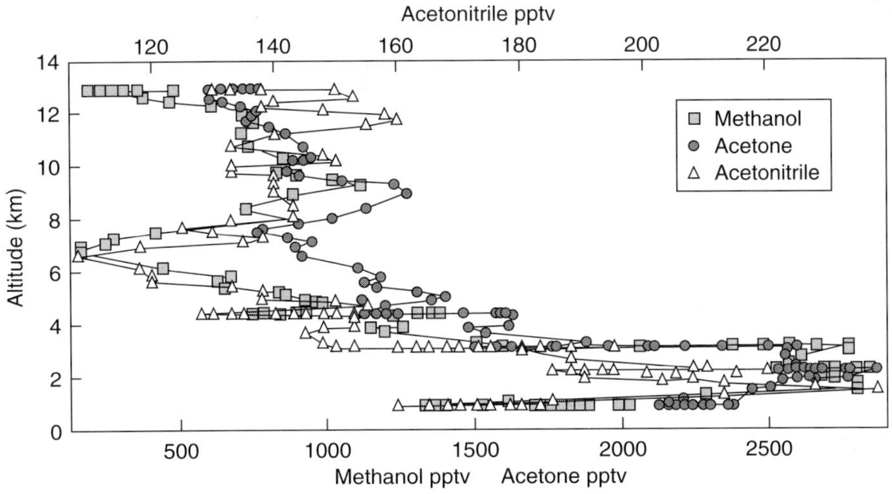

Figure 5.13 An example of PTR-MS data collected from an aircraft.

a considerable challenge. Three methods have emerged as capable of making routine atmospheric measurements of OH, two optical methods (DOAS and LIF described in Chapters 3 and 4, respectively) and one mass spectrometric. The mass spectrometric method described here has an OH sensitivity similar to LIF but is about 10 times more sensitive than DOAS. In the mass spectrometric method, the determination of OH is indirect as the OH radicals are rapidly titrated by $^{34}SO_2$ prior to measurement. This titration of OH is achieved via the reactions:

$$OH + {}^{34}SO_2 + M \rightarrow H^{34}SO_3 + M \tag{5.23}$$

$$H^{34}SO_3 + O_2 \rightarrow {}^{34}SO_3 + HO_2 \tag{5.24}$$

$$^{34}SO_3 + H_2O + M \rightarrow H_2{}^{34}SO_4 + M \tag{5.25}$$

A typical experimental arrangement is shown in Figure 5.14. Air is drawn into the instrument at a rate of about $10\,L\,min^{-1}$. Sufficient levels of $^{34}SO_2$ are introduced into the inlet to ensure that the OH is almost completely reacted within a few milliseconds. This prevents significant loss of the OH to the walls of the inlet. $H_2{}^{34}SO_4$ is not present at significant concentrations in the atmosphere and can be accurately measured by negative ion chemical ionisation mass spectrometry. A number of important features distinguish this radical measurement from the API-CIMS system described in Section 5.4.1, although in both cases the ions are filtered with a quadrupole (Section 5.3.2.1). For the ionisation it was found that direct ionisation of the air sample was unsuitable for OH measurements as it generated a high background signal and interferences by other ion–molecule reactions and other radicals or unstable species also generated in the air stream. Chemical ionisation is achieved by NO_3^- reagent ions which are generated separately from the sample air and

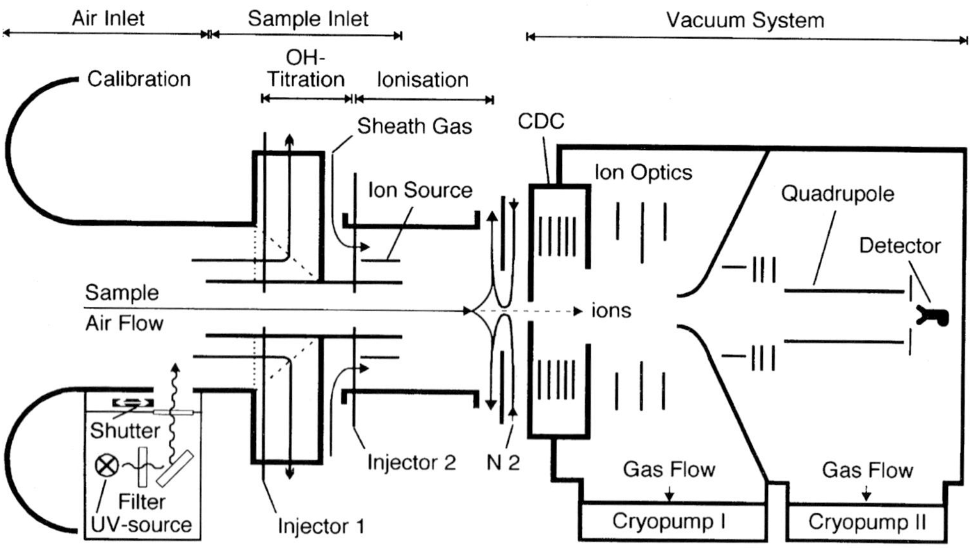

Figure 5.14 A schematic diagram of a mass spectrometer to measure OH. (Berresheim *et al.*, 2000 with permission from Elsevier.)

then drawn through the sample by an electric field. To generate these ions, a radioactive emitter (^{241}Am) is placed in a sheath flow of purified air enriched with HNO_3 vapour and high purity propane. The chemical ionisation then proceeds via the reaction:

$$H_2{}^{34}SO_4 + NO_3^- \rightarrow H^{34}SO_4^- + HNO_3 \tag{5.26}$$

Although the majority of the NO_3^- reactant ions are clustered with neutral HNO_3 and/or H_2O molecules, they may be nominally considered to be NO_3^- as all clusters have been found to react at very similar rates. As described in Section 5.4.1, by measuring the reagent ion NO_3^- at $m/z = 63$ and the product $H^{34}SO_4^-$ at $m/z = 99$ a concentration of $H_2{}^{34}SO_4$ (and hence OH) can be deduced, if the rate coefficient and the reaction time are known. Nonetheless, because of uncertainties in the rate coefficient and possible inlet effects, direct calibration is also performed by producing a known concentration of OH in the sample air by online photolysis of atmospheric water vapour whose concentration in turn is accurately determined by dewpoint hygrometry (see Chapter 4 Section 5.4.3.2). A further complication of this measurement is that under certain atmospheric conditions significant quantities of OH can be generated (or recycled) within the system through the reactions of NO or ozone with HO_2 and RO_2. To minimise this potential positive measurement artefact, excess propane is added downstream from the $^{34}SO_2$ titration step to remove OH that might be generated from HO_2. For background signal measurements, propane is added at the beginning of the titration zone to remove ambient OH.

OH measurement time resolution with this system is typically 30 seconds. The detection limit is about 2×10^5 molecules cm^{-3} based on 5-minute signal integration, while precision and accuracy are estimated to be 48 and 54% respectively. Such a system has been deployed on the ground for long-term measurements of OH at the German Weather Service's (DWD) Meteorological Observatory, Hohenpeissenberg (Berresheim *et al.*, 2000). A regularly updated evaluation of the OH monitoring data can be seen on the website http://www.dwd.de/en/FundE/Observator/MOHP/. A similar system has also been adapted for use on aircraft and flown in several field measurement campaigns (e.g. Eisele *et al.*, 2001).

A novel extension to this method has been described by Hanke *et al.* (2001). This has enabled the on-line measurement of peroxy radical measurements, namely HO_2 and the sum of the organic peroxy radicals, RO_2. This method relies on the amplifying chemical conversion of peroxy radicals via the chain reaction with NO and SO_2, to make gaseous sulphuric acid, which is detected by CIMS. The ratio of the OH reaction rates with SO_2 and NO is equal to the ratio of the H_2SO_4 and the total peroxy radical concentration ($HO_2 + RO_2$). By reducing the O_2/N_2 ratio, the H_2SO_4-forming reactions of RO_2 can be so disfavoured that measured H_2SO_4 is almost entirely formed by HO_2. In this way it has proven possible to measure HO_2 and the sum of HO_2 and RO_2 radicals in the atmosphere, with a time resolution of 1 minute and a detection limit of 0.5 pptv. These radicals were accurately measured at mixing ratios between 1 and 30 pmol/mol (pptv). An alternative but related method for measuring $RO_2 + HO_2$ has been developed which does not use the amplification, and instead relies on the extreme sensitivity of the CIMS system to determine OH (Edwards *et al.*, 2003). In this case the peroxy radicals are converted with NO to OH, in a pre-reactor held at constant low pressure. This instrument has been employed to measure the evolution of peroxy radicals in the northern hemisphere winter–spring transition (Cantrell *et al.*, 2003).

5.4.4 Coupled GC-MS capabilities

The combination of a gas chromatograph and a mass spectrometer has become one of the most powerful and widely used tools for the analysis of complex mixtures such as air. In this application, spectra are collected for separated compounds as they exit from a chromatographic column. By monitoring specific ions the mass spectrometer can detect trace gas species more specifically and sensitively than many conventional GC detectors (see Chapter 8). A variety of GC-MS quadrupole systems are commercially available which can be reliably automated and deployed for long-term measurements. Furthermore, the MS can be operated in either EI or CI mode, offering a broad range of detectable species. While the GC-MS can detect many species specifically and sensitively, analysis times of typically 1 hour limit the temporal data coverage. This is in contrast with the CIMS methods described above that are capable of measuring on-line albeit with less sensitivity and specificity. Since 2000, considerable effort has been invested to make the GC-MS measurements more rapidly. Some current examples of these applications are given below.

A global network of GC-MS systems (Agilent 5973) has been set up by Prinn *et al.* (2000) in order to monitor the long-term changes of trace gas species. The sites have been selected in remote locations to assess changes in the background concentrations of chemically and radiatively important trace gas species. The sites include Cape Grim, Tasmania (41°S, 145°E), Cape Matatula, American Samoa (14°S, 171°E), Ragged Point, Barbados (13°N, 59°W), Mace Head, Ireland (53°N, 10°W), and Trinidad Head, California (41°N, 124°W).

Of particular interest are the man-made hydrochloroflourocarbons (HCFCs) 134a, 141b and 142b that have been introduced as replacements for the stratospheric ozone depleting chlorofluorocarbons (e.g. CFC 11 and 12). At least four measurements are made per day with approximately 1% precision and the calibrated data are available on the network homepage http://cdiac.ornl.gov/ndps/alegage.html. Lifetimes of HCFCs are significantly shorter than that of CFCs because they are capable of being scavenged by OH in the troposphere. For this reason, HCFCs do not cause as much ozone depletion per molecule on a 100-year timescale as compared to the CFCs (Montzka & Fraser, 2002). However the mole fractions of these species are steadily increasing in both hemispheres. If the release rate of such species is known, and global measurements are available, then it is theoretically possible to determine the global average OH. In this way GC-MS measurements can contribute to the long-term assessment of the global oxidising capacity.

The high sensitivity of the GC-MS can be exploited to measure trace gas species at extremely low mixing ratios. For example, the ambient concentrations of halon 1202 (CF_2Br_2) can be reliably measured in ambient air even though it is present at only 0.045 pmol/mol (pptv) (Fraser *et al.*, 1999). These measurements were made using a magnetic sector mass spectrometer with an EI source. The presence of previously unknown but potentially important gas SF_5CF_3 was discovered using the same instrument (Sturges *et al.*, 2000). These compounds were first observed at mixing ratios of around 0.005 pmol/mol (pptv). Assuming this as a detection limit we may estimate that if 200 tonnes of such material was emitted anywhere in the world it would be detectable by this instrument, provided the troposphere is well mixed. Therefore even when relatively small emissions of long-lived species occur in the atmosphere they can be detected by GC-MS

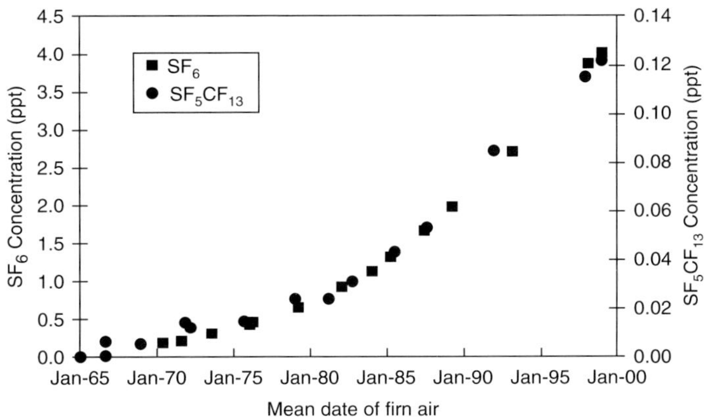

Figure 5.15 An example of GC-MS data collected from firn air. (Courtesy of Dr W. Sturges, UEA.)

all over the world. Figure 5.15 shows the atmospheric mixing ratios of SF_6 and SF_5CF_3 since January 1965. The air for this analysis was pumped up from where it was naturally trapped in consolidated deep snow, in eastern Antarctica. Air from 1965 was determined to be 100 m below the snow surface, where after freezing in the ice matrix it has remained isolated from mixing with present-day air. SF_5CF_3 has a current growth rate of about 0.008 pptv per year (ca. 6% per year). The similarity of the growth curves for SF_5CF_3 and SF_6 has the researchers to speculate that the former is a breakdown product of the latter in high-voltage equipment.

The examples given above all involve the application of GC-MS with EI ion sources. The commercially available negative ion chemical ionisation source has also been applied to atmospheric measurements. The results are very promising and worth mentioning here. By feeding a small flow of methane into the ionisation chamber, thermalised electrons can be generated that can interact with organohalogen and organonitrate species, to produce measurable ions at $m/z = 35$ (Cl), 79 (Br), 127 (I) or 46 (NO_2^-).

$$CH_4 + e\ (70\,eV) \rightarrow CH_4^+ + 2e\ (thermal) \quad (5.27)$$

$$RONO_2 + e\ (thermal) \rightarrow RO + NO_2^- \quad (5.28)$$

$$RX + e\ (thermal) \rightarrow R + X^-(m/z = {}^{35}Cl, {}^{79}Br, {}^{127}I) \quad (5.29)$$

The results show simple spectra with little co-elution and low background signals. The sensitivity is comparable to that of the electron capture detector (0.01 pmol/mol (pptv) in a 1000 ml sample) and the combination of retention time and ion measurement means compounds can be identified easily and with a high degree of certainty. Such measurements can be of great interest over the open ocean where emissions of organohalogen and alkyl nitrates have been identified.

In the ground-based examples given above, the GC-MS analysis times are typically about 1 hour. Considerable progress has been made in reducing the analysis time of GC-MS systems used in atmospheric analysis. By miniaturising the gas chromatograph and cryogenic preconcentrator, a 2–5 minute duty cycle has shown to be possible for measurements of the C_2 to C_4 carbonyl compounds and methanol (Apel *et al.*, 2003). This instrument has been flown on a jet aircraft and shown to be able to detect low pmol/mol levels off the target analytes. While this is not as fast as the systems described in Section 5.4.1 and 5.4.2, the GC-MS system is generally more sensitive, offers a broader range of potentially detectable compounds and unlikely to be affected by interfering compounds at the same mass.

5.4.5 The isotope ratio mass spectrometer

The isotope ratio of a gas phase compound is generally defined as the ratio of a less abundant isotope to the most abundant isotope. The 'natural' abundance of the ^{13}C to ^{12}C is 1.1%. However, there are subtle but significant variations in the ratios and these can be determined using high precision isotope ratio mass spectrometry (IRMS). From a knowledge of the isotopic composition for a trace gas species in the atmosphere, we may learn much about its sources and chemistry. Gases emitted from certain sources have characteristic isotopic ratios. For example it has been found that methane emitted from rice paddies contains 2% less carbon 13 (^{13}C) than atmospheric methane. Moreover, removal of methane from the atmosphere by OH, O^1D and Cl causes the isotope fraction to change, and so the isotope ratio can also indicate the relative importance of the removal mechanisms (Brenninkmeijer *et al.*, 2003). This is information that is not accessible by measuring the concentration of the trace gases alone. Indeed, our understanding of the global cycles of CH_4 and CO_2 stems partly from isotopic analysis, and our knowledge of past climate change is based on isotopic analysis of H_2O in ice cores or carbonates in sediments. Isotopic measurements of stratospheric carbon dioxide have shown how such measurements can be used to examine carbon exchanges between the atmosphere and biosphere on interannual to glacial–interglacial time scales (Boering *et al.*, 2004). It is interesting to note that one of the oldest atmospheric measurements, the determination of oxygen abundance, is now most accurately performed with an isotope mass spectrometer able to measure subtle variations in the approximately 20% oxygen content. The same techniques are now being applied to volatile organic carbon compounds which are present at much lower (nmol/mol-pmol/mol or ppbv-pptv) atmospheric concentrations (Goldstein & Shaw, 2003).

In the case of C^{13}/C^{12}, the largest deviation from the standard for naturally occurring molecules is approximately 10%, but more usually deviations are observable in the fourth significant digit. These small but important changes are normally expressed as the per mil (‰) difference to an agreed standard. To measure these effects requires an analytical set-up of extremely high precision, with standard deviations in the range 4–6 significant figures. The precision of trace gas concentrations determined by the GC-quadrupole systems described in Section 5.4.3 are typically between 1 and 5%. This is clearly not sufficient for the isotopic analyses, which are performed exclusively on magnetic sector instruments with electron impact sources and multiple collectors as detectors to ensure the requisite stability and hence precision. A schematic diagram of suitable apparatus is shown in Figure 5.16. Normally the source is operated at 70 eV and ionisation occurs at high efficiency. Typically out of 1000 neutrals entering the source, one ion reaches the detector. In IRMS, mass resolution is sacrificed to optimise precision. Hence an electric sector that can degrade the later is uncommon. A uniform magnetic field is required to produce multiple well-defined beams emerging towards the detectors. This is typically provided by a permanent magnet of around 0.75 tesla. A series of fixed Faraday cups (multicollector system) are used as the detector system with one cup being allocated for each ion (see Figure 5.16 and Section 5.3.3). Although Faraday cups are less sensitive than electron multipliers, they are superior for IRMS because: the high count rates needed for high precision measurements are in their normal operating range whereas high ion currents would damage an electron multiplier; the Faraday cup is able to detect very large ion currents without gain adjustment; and the cup is stable and has a long working lifetime (ca. 10 years).

J.J. Thompson and F.W. Aston were the first to establish the nominal terrestrial isotopic abundances of the elements. Between 1940 and 1980 the straightforward method of dual inlet analysis was used to determine isotopic ratios in many fields. For CO_2 analysis this method involved a pre-separated and combusted sample, and a reference gas sample each being attached to an IRMS inlet. The sample and reference are then alternately admitted into the IRMS, allowed to stabilise and the response (e.g. for m/z 44, 45, 46) recorded at a plateau level. The measurement cycle is then repeated many times. This method was superseded in the 1980s by the continuous flow (CF) method (Barrie *et al.*, 1984). This approach involves separating the individual compounds by gas chromatography, combusting or reducing the eluting compounds to CO_2, N_2 or H_2O, and continuously analysing each peak with the IRMS. When analysing air for trace gases at low concentrations such as methane or volatile organic hydrocarbons, the sample is first cryogenically concentrated (see Chapter 8), before separation by gas chromatography. Ambient air contains approximately 1.8 μmol/mol (ppmv) of methane and it is necessary to concentrate 25 mL to measure the C^{13}/C^{12}. For other organic gases at lower ambient volume mixing ratios (<100 pptv) a correspondingly higher volume is required (10–20 L). The organic gases exiting the GC enter a combustion chamber where analytes such as methane or other hydrocarbons are combusted to CO_2 with 99.99% efficiency in a NiO filled chamber heated to around 1000°C. Following combustion the CO_2 peaks corresponding to the separated and combusted trace gases are introduced to the mass spectrometer entrained in a stream of helium carrier gas. For VOCs, depending on the size of the peaks the precision varies from 0.2 to 3.6 per mil (‰, 10^{-4}) (Rudolph *et al.*, 1997).

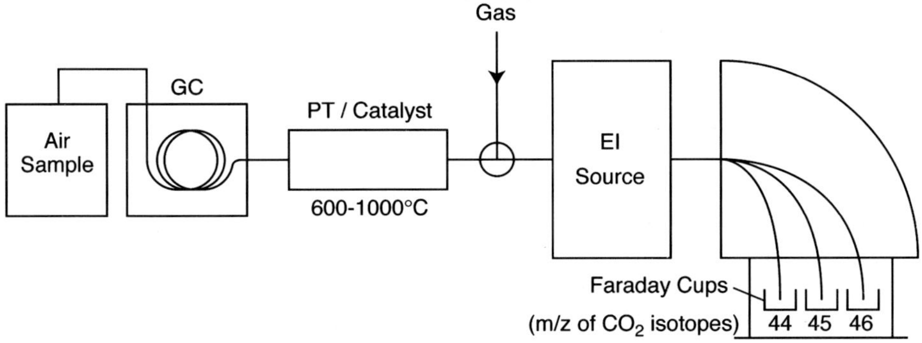

Figure 5.16 A schematic diagram of a continuous flow isotope ratio mass spectrometer.

5.5 Future developments

The importance of mass spectrometry in atmospheric research will almost certainly grow in the future. Continuing technological improvements are expected to reduce the size, power, analysis times and cost of MS systems. Increased efforts are being made to develop miniature mass spectrometers which are hand portable but retain the performance of laboratory based instruments. For atmospheric research this would considerably simplify the investigation of sources in the field and installation on balloon or aircraft platforms where weight and space are critical. It is expected that as vacuum technology is improved and pumps become smaller considerable progress can be made. This is because in many cases it is not the mass analyser but the associated electronics and vacuum technology that limit the miniaturisation (Badman & Cooks, 2000). Considerable funds have been made available to develop this field with the aim of providing specific, portable detectors capable of detecting toxic gases rapidly. Atmospheric research stands to gain from these rapid developments, particular in the analysis of organic trace species which play a key role in global air chemistry (Williams, 2004).

Presently both GC-MS and CIMS type instrumentation are used to measure trace gases in the atmosphere. Extensive validation though intercomparison has taken place between GC-MS and CIMS methods (Kato *et al.*, 2004 and references therein); between PTR-MS and API-CIMS methods (Sprung *et al.*, 2001), and between two differently configured PTR-MS systems (de Gouw *et al.*, 2004). All the intercomprisons to date have reported generally satisfactory results. The GC-MS system offers the atmospheric researcher high selectivity and sensitivity but long analysis times. In contrast, CIMS systems provide better spatial and temporal coverage through faster measurements, but are currently less sensitive and selective. It can be expected that the capabilities of both systems will converge in the coming years. By replacing the quadrupole used in the CIMS with an ion trap, much higher selectivity (but not yet sensitivity) has been shown to be possible with a PTR-MS type system on the ground (Prazeller *et al.*, 2003). The more sensitive API-CIMS system has also been coupled to an ion trap and even aircraft measurements conducted successfully (Kiendler & Arnold, 2003). An alternative approach using a triple quadrupole to sequentially select ions, fragment ions and then detect the fragmented ions has also been shown to be viable (Reiner *et al.*, 1998). For CIMS systems it seems that further ion schemes will be employed to detect new ranges of molecules. For example, it was shown that peroxy acetyl nitrate (PAN) type compounds can be detected by a CIMS type method using a I^- reagent ion (Slusher *et al.*, 2004). GC-MS systems will continue to improve the frequency of measurement and the number of quantifiable compounds. New narrow bore columns and miniature GC ovens with low thermal mass enable faster chromatography to be achieved. Fast in situ measurements by GC will become more common. In both Europe and the United States new high altitude, long range aircraft will shortly become available to researchers. As space will be limited, competition between these techniques can be expected. It will be interesting to see which proves more successful.

The field of isotope ratio mass spectrometry is expected to develop rapidly in the coming years. Advances in the analysis techniques for compound specific measurements of carbon isotopes in small samples and the availability of reliable commercial mass spectrometers open a new field to atmospheric researchers. Considerable progress in the isotopic characterisation of sources and photochemical processes can be expected. The aforementioned carbon isotope studies will be further expanded through the simultaneous examination of N, Cl, Br or O isotopes.

Generally it can be expected that in the future mass spectrometers will become more accessible to the atmospheric researcher, replacing less sensitive and selective systems currently in use. Mass spectrometric determinations of trace gases in the atmosphere will become more routine, both on the ground though global instrument networks, and in the air on commercial and research aircraft. Experience gained in this way will aid the investigation of extra terrestrial atmospheres by mass spectrometry for trace gas species, which has already begun (Fenselau & Caprioli, 2003).

Acknowledgements

Information and helpful advice is gratefully acknowledged from my colleagues Prof. F. Arnold (Max Planck Institute – Heidelberg), Dr C.A.M. Brenninkmeijer and Dr J. Crowley (Max Planck Institute – Mainz), Dr H. Berresheim (Deutsche Wetter Dienst), Dr W. Sturges (University of East Anglia), T. Klüpfel and G. Eerdekens (ORSUM Group – Max Planck Institute – Mainz). Mrs Feyerherd from the Max Planck Institute graphics department is thanked for her enormous help and patience in the preparation of figures.

Further reading

de Hoffmann, E. & Stroobant, V. (2002) *Mass Spectrometry: Principles and Applications*, Second edition, Wiley, ISBN 0–471–48566–7.

Budde, W.L. (2001) *Analytical Mass Spectrometry: Strategies for Environmental and Related Applications*, Oxford University Press, ISBN 0–8412–3664–X.

Moore, J.H., Davis, C.C., Coplan, M.A. & Greer, S. (2002) *Building Scientific Apparatus: A Practical Guide to Design and Construction*, Third edition Westfield Press, ISBN 0813340063.

Viggiano, A.A. & Hutton, D.E. (1999) Airborne mass spectrometers: Four decades of atmospheric and space research at the air force laboratory, *J. Mass Spectrom.*, 34, 1107–1129.

General internet mass spectrometry resource, http://www.i-mass.com/

Atmospheric Chemistry and Physics, http://www.copernicus.org/EGU/acp/acp.html. Frequently contains atmospheric measurements of trace gases by mass spectrometers.

References

Apel, E.C., Hills, A.J., Lueb, R., Zindel, S., Eisele, S., Riemer, D.D. (2003) A fast-GC/MS system to measure C-2 to C-4 carbonyls and methanol aboard aircraft, *J. Geophys. Res. Atmos.*, 108, D20, 8794.

Arnold F., Kessel, J., Krankowsky, D., Wieder, H. & Zahringer, J. (1971) Negative ions in the lower ionosphere: A mass spectrometric measurement, *J. Atmos. Terr. Phys.*, 33, 1169.

Arnold F., Krankowsky, D. & Marien, K.H. (1977) First mass spectrometric measurements of positive ions in the stratosphere, *Nature*, 267 (5606), 30–32.

Arnold, F. & Henschen, G. (1978) First mass analysis of stratospheric negative ions, *Nature*, 275 (5680), 521–522.

Arnold F. & Fabian, R. (1980) First measurements of gas phase sulphuric acid in the stratosphere, *Nature*, 283, 55.

Aston, F.W. (1942) Mass spectra and Isotopes, Second edition, Edward Arnold and Co., London, 1942.

Badman, E.R. & Cooks, R.G. (2000) Miniature mass analysers, *J. Mass Spectrom.*, 35, 659–671.

Barrie, A., Bricout, J. & Koziet, J. (1984) Gas-chromatography – Stable isotope ratio analysis at natural abundance levels, *Biomed. Mass Spectrom.*, 11, 583.

Berresheim, H., Elste, T., Plass-Dülmer, C., Eisele, F.L. & Tanner, D.J. (2000) Chemical ionization mass spectrometer for long term measurements of OH and H_2SO_4, *Int. J. Mass Spec.*, 202, 91–109.

Blake, R.S., White, C., Hughes, C.O., Ellis, A.M. & Monks, P.S. (2004) Demonstration of proton tranfer reaction time of flight mass spectrometry for real time analysis of trace volatile organic compounds, *Anal. Chem.*, 76, 3841–3845.

Boering, K.A., Jackson, T., Hoag, K.J., Cole A.S., Perri, M.J., Thiemens, M. & Atlas, E. (2004) Observations of the anomalous oxygen isotopic composition of carbon dioxide in the lower stratosphere and the flux of the anomaly to the troposphere, *Geophys. Res. Lett.*, 31, 3, Art. No. L03109.

Brenninkmeijer, C.A.M., Janssen, C., Kaiser, J., Röckmann, T., Rhee, T.S. & Assonov, S.S. (2003) Isotope effects in the chemistry of atmospheric trace compounds, *Chem. Rev.*, 103, 5125–5161.

Cantrell, C.A., Edwards, G.D., Stephens, S., Mauldin, L., Kosciuch, E., Zondlo, M. & Eisele, F. (2003) Peroxy radical observations using chemical ionization mass spectrometry during TOPSE, *J. Geophys. Res.*, 108, D6, Art. No. 8371.

Crutzen P.J. & Arnold, F. (1986) Nitric acid cloud formation in the cold Antarctic stratosphere – A major cause for the springtime ozone hole, *Nature*, 324, 6098, 746–749.

Crutzen, P.J., Williams, J., Pöschl, U. *et al.* (2000) High spatial and temporal resolution measurements of primary organics and their oxidation products over the tropical forests of Surinam, *Atmos. Environ.*, 34, 1161–1165.

de Gouw, J., Warneke, C., Holzinger, R., Klüpfel, T. & Williams, J. (2004) Inter-comparison between airborne measurements of methanol, acetonitrile and acetone using two differently configured PTR-MS instruments, *Int. J. Mass Spec.*, 239, 129–137.

Edwards, G.D., Cantrell, C.A., Stephens, S., Hill, B., Goyea, O., Shetter, R.E., Mauldin, R.L., Kosciuch, E., Tanner, D.J. & Eisele, F.L. (2003) Chemical ionization mass spectrometer instrument for the measurement of tropospheric HO2 and RO2, *Anal. Chem.*, 75, 20, 5317–5327.

Eisele, F.L., Mauldin, R.L., Tanner, D.J. *et al.* (2001) Relationship between OH measurements on two different NASA aircraft during PEM Tropics B, *J. Geophys. Res. Atmos.*, 106, 32 683.

Fenselau, C. & Caprioli, R. (2003) Mass spectrometry in the exploration of Mars – Introduction, *J. Mass Spectrom.*, 38, 1.

Fraser, P.J., Oram, D.E., Reeves, C.E., Penkett, S.A. & McCulloch, A. (1999) Southern hemispheric halon trends (1978–1998) and global halon emissions, *J. Geophys. Res. Atmos.* 1999, 104, D13, 15 985–15 999.

Goldstein, A.H. & Shaw, S.L. (2003) Isotopes of volatile organic compounds: An emerging approach for studying atmospheric budgets and chemistry, *Chem. Rev.*, 103, 5025–5048.

Hanke, M., Uecker, J., Reiner, T. & Arnold, F. (2001) Atmospheric peroxy radicals: ROXMAS, a new mass-spectrometric methodology for speciated measurements of HO2 and ΣRO2 and first results, *Int. J. Mass Spec.*, 213, 91–99.

Hanson, D.R., Greenberg, J, Henry, B.E. & Kosciuch, E. (2003) Proton transfer reaction mass spectroscopy at high drift tube pressure, *Int. J. Mass Spectrom.*, 223, 507–518.

Hauck, G. & Arnold F. (1984) Improved positive-ion composition measurements in the upper troposphere and lower stratosphere and the detection of acetone, *Nature*, 311(5986) 547–550.

Hayward, S., Hewitt, C.N., Sartin J.H. & Owen, S.M. (2002) Performance characteristics and applications of a proton transfer reaction mass spectrometer for measuring volatile organic compounds in ambient air, *Environ. Sci. Tech.*, 36, 1554–1560.

Heitmann, H. & Arnold, F. (1983) Composition of tropospheric ions, *Nature*, 306, 747.

Huang, E.C., Wachs, T., Conboy, J.J. & Henion, J.D. (1990) Atmospheric pressure ionisation mass spectrometry – Detection for the separation sciences, *Anal. Chem.*, 62, 713–725.

Huey, L.G., Hanson, D.R. & Howard, C.J. (1995) Reactions of SF6- and I- with atmospheric trace gases, *J. Phys. Chem.*, 99, 5001–5008.

Hunter, E.P. & Lias S.G. (1998) Evaluated gas phase basicities and proton affinities of molecules: An update, *J. Phys. Chem. Ref.* Data, 27, 413–656.

Jochum, K.P. (1988) Massenspektrometrie, in *Physikalische Analytik*, Chapter 6, 198, 206, 207.

Jost, C. (2002) Applications of atmospheric pressure chemical ionization mass spectrometry: Emissions from biomass burning and distribution of SO_2 over Northern Europe, Doctoral thesis, 42.

Jost, C., Trentmann, J., Sprung, D., Andreae, M.O., McQuaid, J.B. & Barjat, H. (2003) Trace gas chemistry in a young biomass burning plume over Namibia: Observations and model simulations, *J. Geophys. Res. Atmos.*, 108, D13, 10.1029/2002JD002431.

Jouzel, J., Lorius, C., Petit, J.R., Genthon, C., Barkov, N.I., Kotlyakov, V.M. & Petrov, V.M. (1987) Vostok ice core: A continuous isotope temperature record over the last climatic cycle (160 000 years), *Nature*, 329, 403–408.

Kato, S., Miyakawa, Y., Kaneko, T. & Kajii, Y. (2004) Urban air measurements using PTR-MS in Tokyo area and comparison with GC-FID measurements, *Int. J. Mass Spec.*, 235, 103–110.

Kiendler, A. & Arnold, F. (2003) Detection of gaseous oxygenated hydrocarbons in upper troposphere and lower stratospheric aircraft borne experiments, *Int. J. Mass Spec.*, 223–224, 733–741.

Knewstubb, P.F. & Sudgen, T.M. (1958) Mass spectrometric observations of ions in flames, *Nature*, 181, 474–475.

Lelieveld, J., Berresheim, H., Borrmann, S., Crutzen, P.J., Dentener, F.J., Fischer, H., Feichter, J., Flatau, P.J., Heland, J., Holzinger, R., Korrmann, R., Lawrence, M.G., Levin, Z., Markowicz, K.M., Mihalopoulos, N., Minikin, A., Ramanathan, V., de Reus, M., Roelofs, G.J., Scheeren, H.A., Sciare, J., Schlager, H., Schultz, M., Siegmund, P., Steil, B., Stephanou, E.G., Stier, P., Traub, M., Warneke, C., Williams, J. & Ziereis, H. (2002) Global air pollution crossroads over the Mediterranean, *Science*, 298(5594), 794–799.

Lindinger, W., Hansel, A. & Jordan, A. (1998) On-line monitoring of volatile organic compounds at ppt levels by means of Proton Transfer Reaction Mass Spectrometery (PTR-MS). Medical applications, food control and envirnonmental research. *Int J. Mass Spectrom. Ion Processes*, 173, 191–241.

Marcy, T.P., Fahey, D.W., Gao, R.S. *et al.* (2004) Quantifying stratospheric ozone in the upper troposphere with in situ measurements of HCl, *Science*, 304, 261–265.

Montzka, S.A. & Fraser P.J. (2002) (Chapter 1: Controlled substances and other source gases. Scientific Assessment of Ozone depletion. C.A. Ennis, World Meteorological Organisation, Report number 47.

Munson, M.S.B. & F.H. Field (1966) Chemical ionization mass spectrometry. 1. General introduction, *J. Am. Chem. Soc.*, 88, 2621–2630.

Narcisi, R.S. & Bailey A.D. (1965) Mass spectrometric measurements of positive ions at altitudes of 64 to 112 km, *J. Geophys. Res. Atmos.*, 70, 3687.

Perkins, M.D. & Eisele, F.L. (1984) First mass spectrometric measurements of atmospheric ions at ground level, *J. Geophys. Res. Atmos*, 89, 9649.

Prazeller, P., Palmer, P.T., Boscaini, E., Jobson, T. & Alexander, M. (2003) Proton transfer reaction ion trap mass spectrometer, *Rapid Communications in Mass Spectrometry*, 17(14), 1593–1599.

Prinn, R.G., R.F. Weiss, R.F., Fraser, P.J. *et al.* (2000) A history of chemically and radiatively important gases in air deduced from ALE/GAGE/AGAGE, *J. Geophys. Res. Atmos.*, 115, 17 751–17 792.

Reiner, T., Mohler, O. & Arnold F. (1998) Improved atmospheric trace gas measurements with an aircraft-based tandem mass spectrometer: Ion identification by mass-selected fragmentation studies, *J. Geophys. Res. Atmos.*, 103, D23, 31 309–31 320.

Rudolph, J., Lowe, D.C., Martin, R.J. & Clarkson, T.S. (1997) A novel method for compound specific determination of delta 13C in volatile organic compounds at pptv levels in ambient air, *Geophys. Res Lett.*, 24, 659.

Salisbury, G., Williams, J., Holzinger, R., Gros, V., Mihaloponlos, N., Vrekoussis, M., Sarda-Estere, R., Berresheim, H., von Kuhlmann, R., Lawrence, M., Lelieveld, J. (2003) Ground-based PTR-MS measurements of reactive organic compounds during the MINOS campaign in Crete, July–August 2001, *Atmos. Chem. Phys.*, 3, 925–940, 3 July.

Slusher, D.L, Huey, G.L., Tanner, D., Flocke, F. & Roberts, J. (2004) A thermal dissociation-chemical ionization mass spectrometry (TD-CIMS) technique for the simultaneous measurement of peroxyacyl nitrates and dinitrogen pentoxide. *J. Geophys. Res. Atmos.*, 109(D19), Art. No. D19315, 14 October.

Sprung, D., Jost, C., Reiner, T., Hansel, A. & Wisthaler, A. (2001) Acetone and acetonitrile in the tropical Indian Ocean boundary layer and free troposphere: Aircraft-based intercomparison of AP-CIMS and PTR-MS measurements. *J. Geophys. Res.*, 106 (D22), 28 511–28 527.

Sturges, W.T., Wallington, T.J., Hurley, M.D. *et al.* (2000) A potent greenhouse gas identified in the atmosphere: SF5CF3, *Science*, 289 (5479), 611–613.

Su, T. & Chesnavich, W.J. (1982) Parameterization of the ion polar molecule collision rate constant by trajectory calculations, *J. Chem. Phys.*, 76, 5183.

Thomson, J.J. (1913) Rays of positive electricity and their application to chemical analysis, Longmans Green, London, 1913.

Viggiano, A.A. & Arnold, F. (1995) I Chemistry and composition of the atmosphere, in *Handbook of Atmospheric Electrodynamics*, Hans Holland (ed.), Vol. 1, Chapter 1, CRC Press, ISBN 0–8493–2520–X

Warneke, C., van der Veen, C., Luxembourg, S. *et al.* (2001) Measurements of benzene and toluene in ambient air using proton-transfer-reaction mass spectrometry: Calibration, humidity dependence, and field intercomparison, *Int. J. Mass Spec.*, 207(3), 167–182.

Williams, J., Pöschl, U., Crutzen, P.J. *et al.* (2001) An atmospheric chemistry interpretation of mass scans obtained from a proton transfer mass spectrometer flown over the tropical rainforest of Surinam, *J. Atmos. Chem.*, 38, 133–166.

Williams, J., Holzinger, R., Gros, V. *et al.* (2004) Measurements of organic species in air and seawater from the tropical Atlantic, *Geophys. Res. Lett.*, 31(23), Art. No. L23S06, 5 October.

Williams, J. (2004) Organic trace gases in the atmosphere: An overview. *Environ. Chem.*, 1, 125–136 doi., 10.1071E N040507.

Chapter 6

Mass Spectrometric Methods for Aerosol Composition Measurements

Hugh Coe and James D. Allan

6.1 Introduction to atmospheric aerosols

Aerosols are ubiquitous and are an important feature of the earth's atmosphere that are produced and modified by many anthropogenic and natural processes. There are many ways that aerosols impact directly upon the quality of life in polluted environments, for example through unsightly smogs and their effect on human respiratory and cardiovascular health (Dockery *et al.*, 1993; Pope *et al.*, 2002; Schwartz, 1994; White & Roberts, 1977). On a wider scale, atmospheric aerosols are also known to play a major part in many global issues, such as the hydrological cycle, climate forcing and the general chemistry of the atmosphere (Penner *et al.*, 2001; Ramanathan *et al.*, 2001; Ravishankara, 1997). In addition, aerosol particles control the formation of polar stratospheric clouds (PSCs), which are responsible for converting halogen species locked in longer-lived reservoirs into more labile forms that have the ability to catalytically remove ozone in the springtime lower polar stratosphere in both hemispheres.

There are many identified sources for atmospheric particulate matter. The transport sector is known to be a major source, through the direct emission of particles into the atmosphere in the form of exhaust emissions (Colvile *et al.*, 2001; Mayer, 1999). Other combustion sources such as industry, forest fires and coal burning are also known to be important. Particles may also be mechanically suspended into the atmosphere; mechanisms include the action of wind on loose soil or dust, volcanic eruptions and sea spray.

Another major source of the mass of particulate matter in the atmosphere is the conversion of gas-phase chemical species to the condensed phase. These include not only lower volatility gas-phase oxidation products such as sulphuric acid and partly oxidised organic chemicals, but also semivolatile species that will partition into the particle phase under suitable conditions, such as ammonia, nitric acid and water. Under certain conditions, particles are known to spontaneously form from their gas-phase precursors if the vapour pressures become high enough. These are observed in the atmosphere in the form of discrete, localised bursts of very small particles that quickly grow to larger sizes through coagulation or further condensation (Kulmala, 2003; Kulmala *et al.*, 2004).

Atmospheric aerosol particles exist over a wide range of shapes and sizes. Those that are created mechanically, such as sea salt or dust particles, tend to be of around a micron upwards in diameter, a size definition normally referred to as the coarse fraction.

These particles, while relatively low in number, tend to make up the majority of the mass of particulate matter in the atmosphere. They also tend to be the most important when considering the optical properties of the atmosphere, due to their size. The most persistent fraction of particles in the atmosphere is that of around 0.1–1 μm in diameter. This fraction is often referred to as the accumulation mode and consists mainly of matter that has condensed from the gas phase through various processes. These particles are generally the most important when considering atmospheric chemistry, as they tend to collectively posses the largest amount of chemically active surface area. They also tend to be the particles that most readily form the nuclei for cloud droplets. There are also particles of smaller sizes in the atmosphere, and these are often believed to be very important when considering human health, as they can penetrate deeper into the lungs when inhaled (Harrison & Yin, 2000; Seaton *et al.*, 1995). These exist in high numbers, especially in urban environments, but do not represent a high fraction of the total particulate mass.

Despite significant research to study aerosol processes and effects, much of our understanding has consistently been limited by the small size of aerosols, making their study intrinsically difficult. To understand many of the important properties of aerosol particles it is necessary to know both their composition and size. For example, these properties dictate (a) the toxicity of the particles and their transmission through the human respiratory system; (b) the hygroscopicity and refractive index of the particles and hence their radiative properties; (c) their ability to act as cloud condensation nuclei and so affect cloud albedo; and (d) their transport in the atmosphere and subsequent deposition rate of pollutants. For many years, the study of this field utilised filter or impaction technology and subsequent offline analysis. Progress in the physical characterisation of aerosol particles significantly improved during the 1970s and 1980s but online chemical analysis of aerosol particles has only taken place in the 1990s or so and much progress is still being made. This chapter provides a brief overview of traditional techniques and physical measurements of aerosol particles, but concentrates largely on the developments in online mass spectrometry of aerosol particles since 2000.

6.2 Physical aerosol measurements and sizing definitions

The simplest measure of the amount of particulate matter in an air sample is the gravimetric mass; particles are collected on a filter or impactor substrate over a period and are subsequently weighed. As larger particles are not thought to be significant to human health, these are often removed using impactors or cyclones before sampling. This work has formed the basis for official environmental monitoring and compliance standards such as PM_{10} and $PM_{2.5}$, where the total mass of particles with aerodynamic diameters less than 10 and 2.5 μm respectively is measured. The technology has become available to measure the mass of particulate matter in a sample online and with a high time resolution. Examples include the beta attenuation monitor (e.g. Met One Instruments Inc. BAM-1020, Grants Pass, OR, USA), which works by continuously passing beta radiation through a collected filter sample and measuring the reduction in measured radiation on the other side, or the tapered element oscillating microbalance (e.g. Thermo Electron Corporation Series 1400a TEOM, East Greenbush, NY, USA), which collects the sample on a filter mounted on the end of an oscillating element and records the changes to the resonant frequency (Patashnick & Rupprecht, 1991).

The most common method of counting the volumetric number concentration of particles present in a sample is the condensation particle counter (CPC, e.g. GRIMM Aerosol Technik Model 5.403, Airnring, Germany; TSI Inc. Model 3025a, Shoreview, MN, USA). These use a thermal gradient to supersaturate the air with respect to a working fluid (usually butan-1-ol, as this condenses readily on both polar and non-polar surfaces, although some recent instruments such as the TSI model 3785 use water) and grow particles to micron sizes, before counting their number using light scattering. The smallest size of particles observable with a commercial instrument of this kind is 3 nm.

The size of aerosol particles in the atmosphere is an important property to measure. This dictates many properties such as their mechanical behaviour (e.g. where in the human respiratory system it will be likely to deposit), optical properties, surface areas and curvatures. The latter two properties are important in how the particle will interact with the gas phase, and so knowing the sizes and numbers of particles is vital when considering condensation, heterogeneous chemistry and water uptake.

The technology to accurately measure the size of particles is similarly well established and many commercial instruments are available for these purposes. However, at this stage it is important to define what is meant by the 'size' of an aerosol particle. If all particles were spherical, it would be sufficient to measure particle size in terms of their geometric diameter but in the atmosphere this is not the case; electron microscopy studies have consistently shown that ambient particles can exist in a wide variety of other morphologies, such as amorphous, crystalline, fractal, agglomerate and aggregate, depending on their sources and compositions (e.g. Dye *et al.*, 2000; Li *et al.*, 2003; McMurry *et al.*, 1996).

The simplest empirical measure of a particle's size is the 'volume equivalent' diameter or D_v. This is defined as the diameter of a sphere with the same volume as the subject particle's condensed-phase constituents. However, there is no straightforward and universally reliable way of probing this quantity directly. Aerosol sizing instrumentation uses a variety of methods to determine the size of particles and these report many different metrics, which must be considered as being physically different properties of the subject particles (DeCarlo *et al.*, 2004).

Particles are frequently counted and sized optically in situ and in real time using instruments referred to as optical particle counters or OPCs (e.g. GRIMM Model 1.108; Met One Model 9012; Droplet Measurement Technologies Cloud Droplet Probe, Boulder, CO, USA). A particle is illuminated with a light source, either monochromatic (normally from a laser) or polychromatic (from an incandescent source), and the scattered light is measured at a fixed angle. The 'optical' diameter or D_o can be derived from the peak scattering intensity. This measurement can be based on the responses to aerosols generated using manufactured spheres of known sizes but can also be predicted from first principles using Mie theory (Mie, 1908).

As a quantity of a given particle, D_o depends upon its physical diameter and refractive index relative to the material used for calibration. The advantages of this technique are that it is fast, relatively cheap and easy to implement. Depending upon the light source used and the scatting angle monitored, it can be used to study particles of around 0.1 μm or larger. However, the method is not generally very well suited for the study of irregularly shaped particles, as these can give unpredictable scattering intensities.

Aerodynamic particle sizers, or APSs (e.g. TSI Model 3321), work by accelerating the sampled aerosol through a nozzle into a partly evacuated chamber, where they pass through two laser fringes. As well as being able to derive the optical diameter from the scattering intensity, the particle velocities can be calculated based on the difference in detection times by the two lasers. This velocity can then be used to calculate another property of the particles, the aerodynamic diameter or D_a. This is an important property to measure, as it influences particle behaviours in many real-world scenarios, such as dry deposition. Indeed, most environmental monitoring standards are based on selecting particles according to their aerodynamic diameter (e.g. PM_{10}, $PM_{2.5}$), as it dictates where in the human respiratory system a particle is likely to become deposited and potentially harm health.

The aerodynamic behaviour of a particle is influenced by both its inertia and the drag forces it experiences, so it is dependent upon the particle's physical size, shape and density. Therefore, the D_a of a particle is defined as the diameter of a sphere of unit density ($1\,g\,cm^{-3}$) that would attain the same velocity in the instrument. The APS is again limited by the wavelength of the light used but is also limited by the fact that the speeds of the smaller particles are very similar to that of the gas, so the lower limit of this instrument is around $0.5\,\mu m$ (Chen *et al.*, 1985). As with the OPCs, the measurements are in real time so the time resolution of the data generated is very high.

Smaller particles cannot be reliably detected optically but those of around $1\,\mu m$ and smaller can be charged and then filtered by electrostatic classifiers, such as differential mobility analysers or DMAs (Flagan, 1998, 2004; Knutson & Whitby, 1975; Winklmayr *et al.*, 1991). These use carefully controlled gas flows and variable electrical fields to select charged aerosol particles according to the speed at which they migrate through a region of flowing, particle-free air. Terminal velocity is reached when the electrostatic forces they experience are balanced by their drag forces and is dictated by the electric field strength and the charge, size and shape of the particle. A different size definition, the electro-mobility diameter (D_m), is defined as the diameter of a sphere charged to $-1e$ that would have the same velocity of the subject particle under the same conditions. Note that this quantity is completely independent of the particle's density. Also note that the effect of a particle's shape on D_m is qualitatively the opposite to its effect on D_a; an irregularly-shaped particle will have an increased D_m but a decreased D_a relative to a spherical particle of the same volume. It is important to remember details such as these when comparing data from different instruments.

By stepping or scanning the DMA voltage and taking multiple measurements of the number concentration in the output using a CPC, a complete size-distributed number concentration can be built up. This forms the basis of instruments such as the scanning mobility particle sizer (e.g. TSI model 3034) (Wang & Flagan, 1990). The sizing limits of these instruments are dictated by the geometry and flow rates of the DMA and the capabilities of the CPC used.

Prior to size selection, the aerosol particles must be charged. A bipolar neutraliser (e.g. TSI model 3077) is usually used for this purpose. A radioactive source (such as ^{85}Kr, ^{90}Sr, ^{210}Po or ^{241}Am) charges the gas molecules present in the aerosol and the particles with positive or negative charges are neutralised by successive collisions with the oppositely charged gas molecules. When equilibrium is reached, most particles are neutralised, while a fraction will have a small positive or negative charge of one elementary charge. As a certain fraction inevitably become charged with two or more elementary charges, an amount of the particles of a given D_m will be of larger physical sizes than those intended. However, as this charging behaviour is predictable, it can be taken account of at the analysis stage through weighted subtractions of the appropriate higher size bins (Adachi *et al.*, 1985; Fuchs, 1963; Liu and Pui, 1974; Wiedensohler, 1988).

A disadvantage of coupling a DMA to a CPC is that the scanning procedure normally takes minutes, which makes it undesirable for the study of rapidly changing systems. To address this problem, fast electrostatic particle spectrometers such as the TSI model 3090 and Cambustion Ltd. model DMS500 (Cambridge, UK) have been developed to electrostatically count and size particles in real time. These use an array of electrometers within the classifier chamber to count the charged particles of multiple mobilities simultaneously, allowing complete size distributions to be delivered at a rate of up to 10 Hz. The overall sensitivity of this design is lower, making it best suited to environments with high particle numbers, such as when sampling emissions at their source. Unipolar diffusion chargers are normally used to charge the aerosol particles with these instruments, which use ionised gas molecules of a single polarity produced from a radioactive source or corona discharge to charge the particles. These yield higher numbers of charged particles but as they work under complex, non-equilibrium conditions, can add extra ambiguities to the sizing and counting abilities of the instrument (Biskos *et al.*, 2004; Tammet *et al.*, 2002).

6.3 Particle composition

In addition to the number, size and mass, the structure and compositions of particles in an aerosol sample are very important quantitative factors to consider in atmospheric science. These can be a fundamental factor in determining the important properties of particles, such as how they interact with incident light, what physiological damage they may cause and whether they are likely to form suitable nuclei for cloud droplets. There are many instruments that are designed to probe these changes in behaviour directly, such as integrating nephelometers (e.g. TSI model 3563) and cloud condensation nucleus (CCN) counters (e.g. Droplet Measurement Technologies CCN). However, in order to better understand the underlying processes it is desirable to explicitly know the chemical compositions of the aerosol particles.

It is also desirable to know how the component chemical species are distributed among the particles, or their mixing state, as this is again important in how they will interact with the gas phase. Two particulate constituents are said to be internally mixed if they coexist in the individual particles in a population, whereas if they are externally mixed, the two components are present in two distinct particle populations mixed within the aerosol. In the real world, more complex mixing states are often observed.

6.3.1 Offline methods

The chemical analysis of aerosols is not as straightforward as the counting and sizing, due to the small amount of mass present. The most direct method is to pass a sample flow through a filter and analyse the accumulated material in the laboratory using standard analytical procedures. A common method of analysing the collected material is to dissolve the material in water and analyse the solution using ion chromatography (IC). This well-established analytical technique quantitatively measures the amounts of aqueous inorganic ions present. However, other aqueous-phase analysis techniques are possible; for example, proton nuclear magnetic resonance (HNMR), which is useful for the identification of functional groups in the water soluble organic fraction (Decesari *et al.*, 2000). Analysis using other solvents and extraction techniques can also be performed.

Other techniques include evolved gas analysis (EGA), where the material is heated and the resultant gasses analysed using any one of a number of gas-phase techniques. This is particularly useful for analysing insoluble organic components and the evolved gasses are frequently analysed by gas chromatography coupled with mass spectrometry, another common analytical procedure, which is described in detail in Chapter 8. While this is useful for identifying specific compounds, much of the matter within particles typically remains unresolved. The thermo-optical organic carbon (OC)/elemental carbon (EC) analyser (Sunset Laboratories, Forest Grove, OR) (Birch & Cary, 1996) is another common example of EGA; the sample is first heated in an inert atmosphere (e.g. helium) and the amount of evolved carbon measured to give a measure of the organic carbon (OC), after which it is exposed to oxygen and the subsequent amount of gas-phase carbon that is evolved gives a measure of the amount of elemental carbon present (EC). The advantage of this method is that while it does not speciate the organic matter present, it does account for all of the mass.

There are also common methods that can analyse the sample in place, without transferring it to the mobile phase, such as X-ray fluorescence, which is useful for the detection of specific elements, metals in particular, and Fourier transform infrared spectroscopy, which can be used to identify organic functional groups and structure.

The material used for the filter (the substrate) can be tailored to the intended analytical technique, and types in common use include cellulose (paper), PTFE and quartz fibre. A degree of size-selection is possible with this technique. For instance, upstream cyclones or impactors can be used to remove larger particles and filters with larger pores can be chosen to so that smaller particles are not collected. However, it is not generally possible to obtain any detailed size-resolved information with these methods.

An alternative method of sample collection is the cascade impactor (e.g. MSP Corp. MOUDI, Shoreview, MN, USA). The principle behind an impactor is to direct the sample flow at high speed towards the substrate, mounted on a solid surface. The rapid change in direction as the air passes around the surface causes particles over a certain D_a (dictated by the flow rate, pressure and impactor geometry) to leave the gas streamlines and impact on the substrate. By arranging several impactor stages in series with successively decreasing cutpoints, particles can be separated and collected according to their sizes. Because the substrates do not need to be porous, a wider choice of materials (e.g. aluminium) is available. The Dekati Measurements Ltd. Electrical Low Pressure Impactor (Tampere, Finland) is a variation on this technique, where the particles are charged prior to collection. By measuring the current being passed to each impactor stage using electrometers, a real-time aerodynamic size distribution can be generated during sample collection.

A problem with impactors in general is that they rely on the particles sticking to the substrates when they impact; they are frequently known to 'bounce' or otherwise become

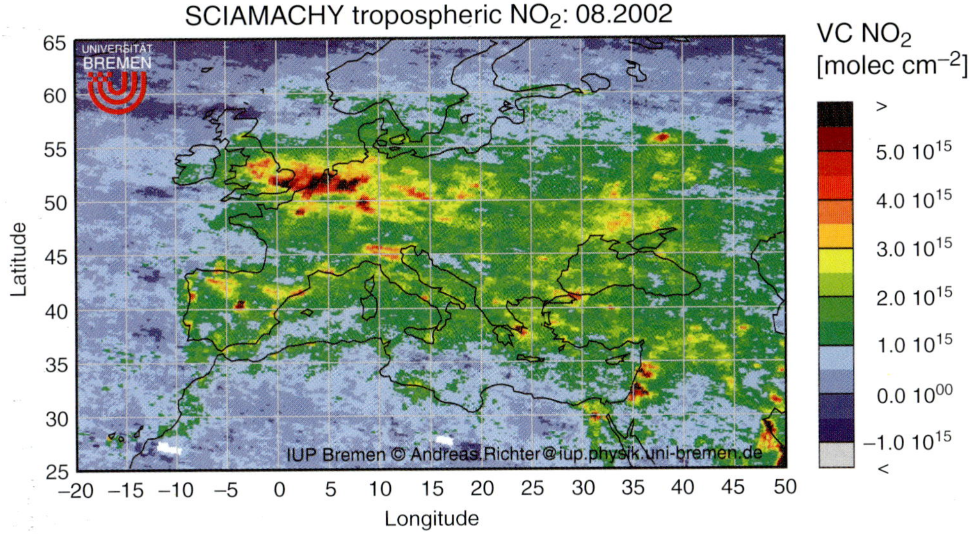

Plate 1 Tropospheric vertical column for NO_2 over Europe measured by the SCIAMACHY instrument aboard the ENVISAT satellite. The pixel size is 30 × 60 km, and the retrieval used the reference sector method. (Reproduced with permission from Richter *et al.*, 2004, 2005, University of Bremen.) (See Figure 1.8.)

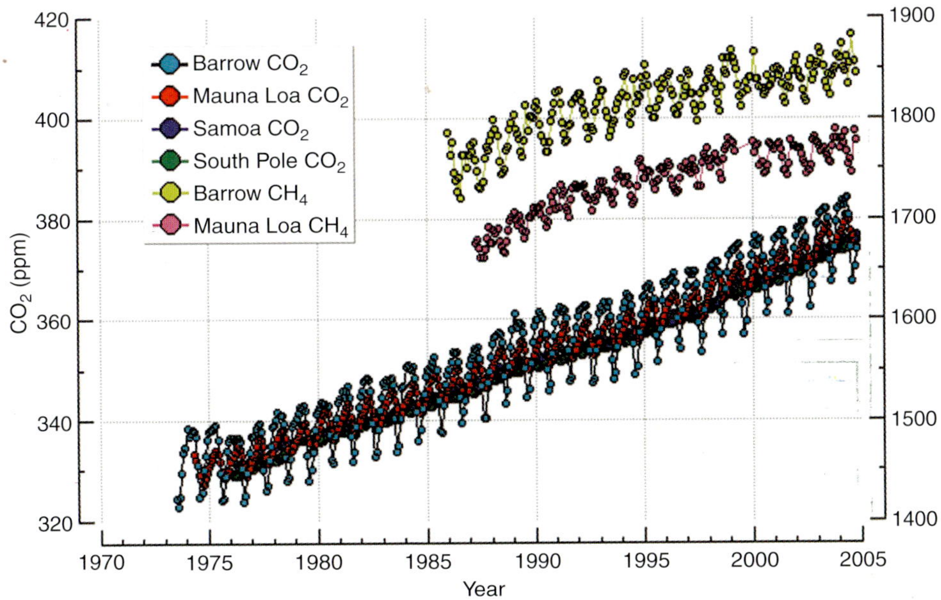

Plate 2 Concentrations of the greenhouse gases methane and CO_2 measured by the NOAA CMDL monitoring network since 1973. Methane is measured using a gas chromatograph with a flame ionisation detector and CO_2 is measured using a non-dispersive infrared (NDIR) analyser. (Data courtesy of Pieter Tans, NOAA.) (See Figure 1.10.)

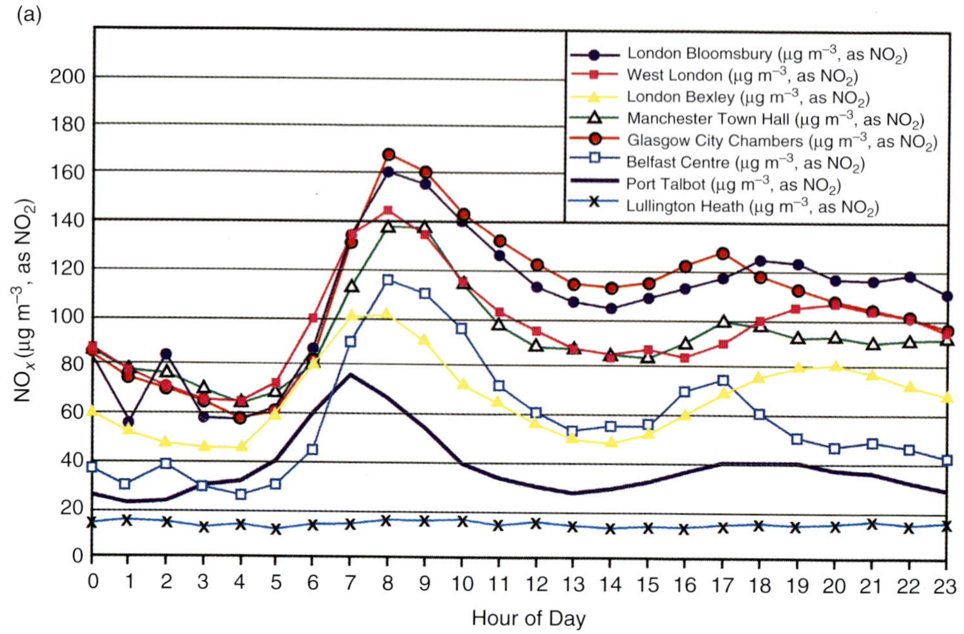

Plate 3a The diurnal cycle of NO_x at eight sites in the UK, showing the morning rush hour maximum due to vehicle traffic and a secondary maximum in the evening. (Figure 6.14 of AQEG, "Nitrogen Dioxide in the United Kingdom", 2004, printed with permission from Crown.) (See Figure 7.10a.)

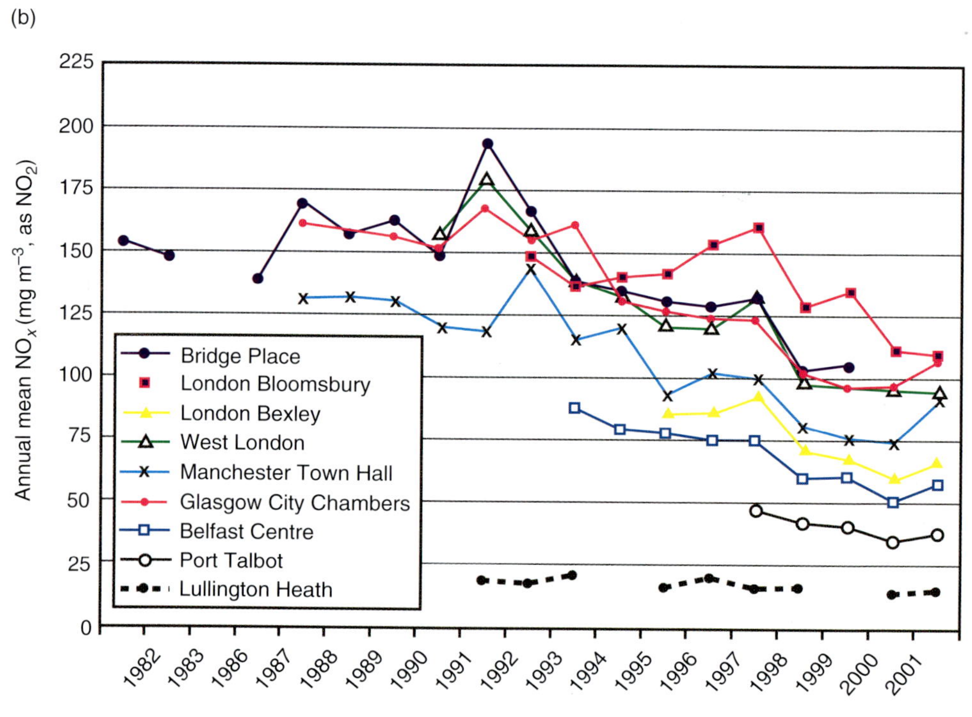

Plate 3b The decadal trend in NO_x at nine sites in the UK, showing the gradual decline due to the tightening of vehicle emission standards. (Figure 6.5 of AQEG, "Nitrogen Dioxide in the United Kingdom", 2004, printed with permission from Crown.) (See Figure 7.10b.)

(c)

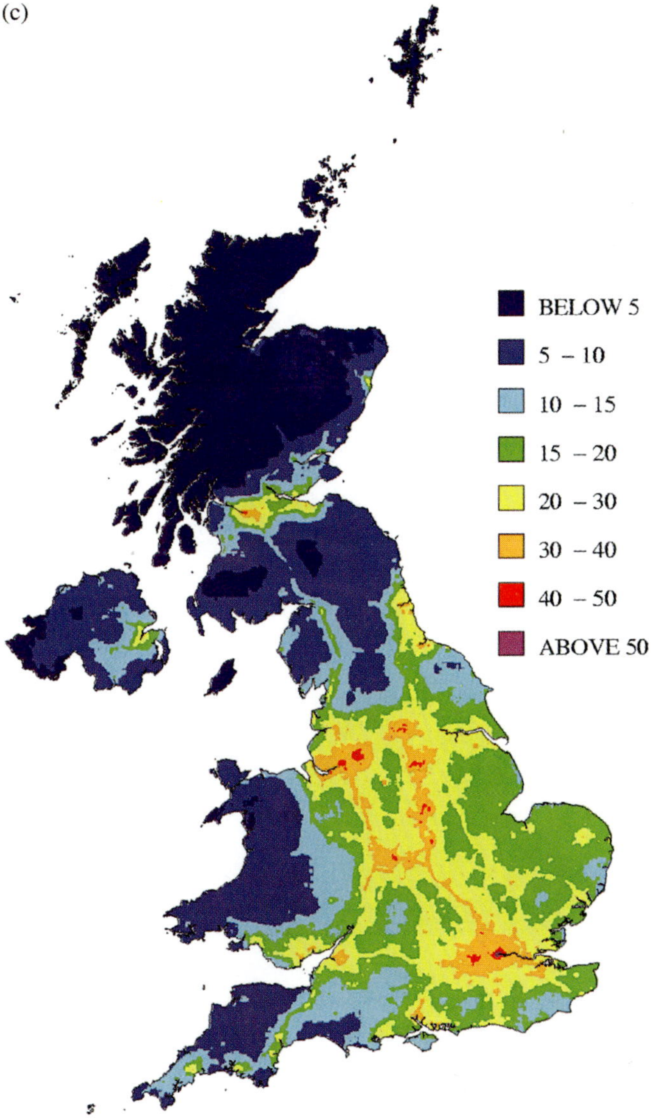

Plate 3c The geographical distribution of NO_2 in the UK, derived from monitoring-network data interpreted with an empirical model. (Figure 3.1 of Stedman *et al.*, 2002, printed with permission from AEA Technology.) (See Figure 7.10c.)

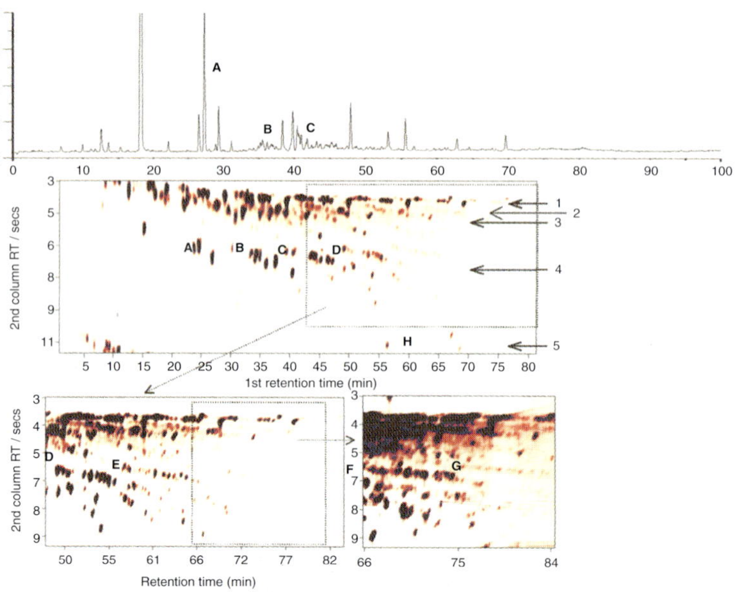

Plate 4 Comparison of single column (upper) and GC × GC separations (lower) of a Leeds urban air sample. Areas of the full chromatogram are successively extracted at higher gain to illustrate increasing isomeric complexity at higher boiling points. GC × GC chromatograms are annotated with start of individual C_x isomer band (running right to left) where A = C_2, B = C_3, C = C_4, D = C_5, E = C_6, F = C_7, G = C_8, H = naphthalene. Chemical banding assignments, 1; aliphatics, 2; olefinics, 3; oxygenated, 4; mono aromatics, 5; polyaromatics. (From Hamilton & Lewis, 2003, reproduced with permission from Elsevier.) (See Figure 8.10.)

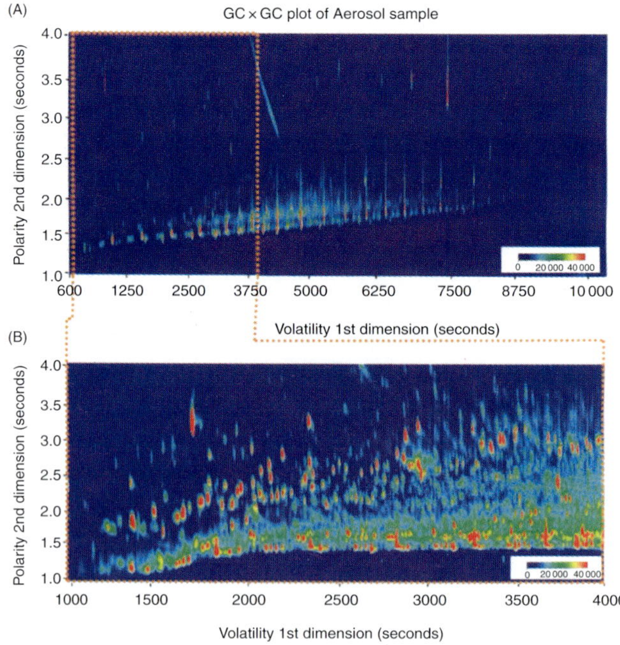

Plate 5 Two-dimensional GC × GC-TOFMS total ion current (tic)-plot of an aerosol sample in 2D-contour plot: (A) showing the full chromatogram of the analysed aerosol with (B) the extraction of the selected section for data analysis. (From Welthagen *et al.*, 2003, reproduced with permission from Elsevier.) (See Figure 8.11.)

dislodged after impaction and become re-entrained in the air flow, meaning they may be collected on the wrong stage or not at all. To mitigate this problem, the substrates are often pre-treated or coated before sample collection to promote adhesion. However, these methods are only permitted when the treatment or coating will not perturb the aspect of particle chemistry of interest or interfere with the analytical procedure being used.

While these bulk sampling methods can employ many powerful and well-established analytical techniques to determine the chemical nature of the particles, they carry many intrinsic limitations. First, a measurable amount of particulate material has to be collected, and the detection limits of laboratory chemical analysis instrumentation usually require sampling times from several hours to days, so temporal resolution is generally poor. Secondly, because of the intervening time between sample collection and analysis, volatile components of the aerosol may evaporate and be lost or chemically unstable compounds may react in the interim (Chow, 1995; Zhang & McMurry, 1987, 1992). Semi-volatile chemicals may also interchange with the gas phase during sampling or before analysis and there is the potential for contamination during handling. Thirdly, while cascade impactors provide some size-resolved information, the resolution is generally poor and particle bounce may result in erroneous size information. Finally, because the aerosol is handled in bulk, no information on the extent of internal or external chemical mixing is retained.

A common remedy for the problem of gas-phase interchanges during bulk sampling is to place a suitable denuder (tubing lined with a reagent) upstream of the collector to absorb any gas phase chemicals that would otherwise contaminate the substrate. A second denuder or chemically treated filter can also be placed downstream and subsequently analysed to measure the amount of particle-phase material that has evaporated from the substrate during sampling (e.g. Mader *et al.*, 2003).

6.3.2 Online bulk sampling and analysis methods

Some of the problems associated with bulk sampling such as gas-phase partitioning after collection and contamination during handling can be avoided by performing the analysis in situ by an automated instrument, for example the Thermo Electron Series 8400N Ambient Particulate Nitrate Monitor. This collects a sample over time through impaction and then, through heating, converts the nitrate fraction of a sample to NO_x, which it then measures to calculate the particulate nitrate mass concentration (Stolzenburg & Hering, 2000). While this offers an increased time resolution (minutes are possible), it is limited to one aspect of particle chemistry, although instruments employing variations of this principle can be used in parallel to study other chemical species (e.g. sulphate and OC/EC).

The particle into liquid sampler (PILS) (Orsini *et al.*, 2003; Weber *et al.*, 2001) and the steam jet aerosol collector (SJAC) (Khlystov *et al.*, 1995) are examples of automated, in situ methods that use a more generalised approach. Particles are grown to large sizes by introducing them into a supersaturated water environment, and then impacted on a surface, over which water is flowed continuously. The sample water is collected periodically and analysed, frequently using IC, to quantify the amounts of various chemicals present in the aerosol. This allows high time resolution data (minutes) on many ionic components of the aerosol to be collected simultaneously. The collection technique is also not limited to IC; with a suitable detector, it can be used to quantify fractions such as water-soluble organic carbon as well.

Another novel in situ method is the thermal desorption aerosol GC-MS/FID (TAG) (Williams *et al.*, 2004). This collects particles in bulk on a cooled surface through impaction over a period of time. After collection, the surface is isolated and then heated, vaporising the components. The organic fraction is then transferred to a gas chromatograph for analysis, using either a flame ionisation detector for total mass measurements or a mass spectrometer for compound identification. This method is similar to the online methods involving thermal decomposition, discussed later. The analysis produced by this instrument is in many ways analogous to the results obtained by offline GC analysis; while many specific compounds can be positively identified and quantified, the majority of the organic matter in particles remains unresolved. These methods are similar to those employed for the gas-phase speciation of organic compounds, as discussed in Chapter 8.

These instruments are automated and therefore less labour-intensive and so are ideally suited for continuous monitoring operations. However, while these online instruments are very powerful and reliable, they have inherent limitations in that size-resolved data are not delivered, and as is common to all forms of bulk sampling instrumentation, cannot provide any information regarding mixing states.

6.4 Online aerosol mass spectrometry

The field of online aerosol mass spectrometry has emerged to provide a real-time method for the chemical analysis of aerosols. These techniques overcome the problems associated with bulk collection and offline analysis by analysing the composition in situ on a particle-by-particle basis. The basic principle of an aerosol mass spectrometer is to introduce airborne particles into the instrument, vaporise and ionise the material and then analyse the ions produced using mass spectrometry. There have been many different designs of aerosol mass spectrometer over the years, most developed in-house at research institutions around the world and tailored to specific tasks, usually related to either ambient sampling or specific laboratory studies. However, one of the more useful features of these instruments is their versatility; in most cases, the individual instruments have proved useful in multiple applications.

At the time of writing, two models of aerosol mass spectrometer are available commercially: the TSI model 3800 Aerosol Time Of Flight Mass Spectrometer (ATOFMS), an instrument based on the pioneering work of Prather *et al.* (1994) and developed further by Gard *et al.* (1997), and the Aerodyne Research Inc. Aerosol Mass Spectrometer (AMS; Billerica, MA, USA), a design introduced by Jayne *et al.* (2000). Both of these instruments are designed to be portable and suitable for both field and laboratory studies, but the data produced are fundamentally different. The ATOFMS, being a laser desorption/ionisation and time of flight-based instrument, delivers quantitative size and largely qualitative composition information on individual particles, while the AMS, being based around thermal desorption, electron impact ionisation and quadrupole mass spectrometry, gives quantitative data on both the size and composition of the entire aerosol ensemble, but gives only limited data on specific particles and cannot study refractory components.

The technical aspects of these and many other examples of similar instruments will be discussed in this section and the specific advantages of the various approaches explored. Instruments developed at the beginning of twenty-first century that will be referred to by name include the Aerosol Composition Mass Spectrometer (ACMS), Atmospheric Pressure Chemical Ionisation Mass Spectrometer (APCI-MS), the Chapel Hill instrument, the Laser Mass Analyzer for Particles in the Airborne State (LAMPAS), the Particle Analysis by Laser Mass Spectrometry (PALMS), the Particle Blaster, the Rapid Single particle Mass Spectrometry (RSMS), the Surface Ionisation Particle Beam Mass Spectrometer (SI-PBMS), the Thermal Desorption Chemical Ionisation Mass Spectrometer (TD-CIMS) and the Thermal Desorption Particle Beam Mass Spectrometer (TDPBMS). Table 6.1 lists these instruments, their measurement technologies and selected publications. The list is not exhaustive, but represents a wide spectrum of the novel techniques used. Examples of further implementations of these techniques can be found elsewhere in the literature. (e.g. Hearn & Smith, 2004; Hunt & Petrucci, 2002; Öktem *et al.*, 2004).

6.4.1 Mass spectrometer types

The principle of a mass spectrometer is to separate and count ions according to their mass-to-charge ratios (m/z). A detailed discussion is given in Chapter 5, and so the operation of a mass spectrometer is considered only briefly here. Quantitatively, m is taken as the mass of the ion relative to the standard atomic mass (defined as one twelfth of the rest mass of a ^{12}C atom, or 1.6606×10^{-27} kg) and z is the charge relative to e, the elementary charge (1.602×10^{-19} C). The charge of most ions detected is normally ± 1 e, although multiple charging is possible, depending upon the composition of the ion and the ionisation technique used. In this chapter, m/z is treated as being dimensionless, although atomic mass units (amu or u), Daltons (Da) and Thompsons (Th) are used as units for the same quantity elsewhere in the literature.

The most basic design of a mass spectrometer is the magnetic sector mass spectrometer, which accelerates and focuses ions using electric fields and then bends their paths with transverse magnetic fields. As the ions are accelerated and deflected by electric and magnetic fields of specific strengths, their velocities and deflected trajectories are therefore dependent on their m/z. Therefore, by using a fixed detector (such as an electrometer), the ions are filtered according to their m/z prior to counting. This type of mass spectrometer can be scanned by varying the electric or magnetic field strengths. While this method is generally not favoured in current designs of aerosol mass spectrometers due to its bulk, it is capable of very high resolutions, so is often used in other laboratory applications where mass measurements of fractions of amu are needed, such as when identifying specific elements. An exception is the RCMS, which has used the compact design introduced by Nier & Schlutter (1985).

Quadrupole mass spectrometers work by again accelerating the ions using electric fields but this time, the ions are selected by passing them between four parallel rods. A voltage is applied between the two sets of opposing rods, which consists of AC and DC components. The ions adopt oscillating trajectories as they travel the length of the rods (see for example Chapter 5, Figure 5.4), the magnitude of the oscillations dependent on their m/z, the AC

Table 6.1 Instruments to measure aerosol composition using mass spectrometry

Instrument	Vaporisation and ionisation	Mass spectrometer	Sizing	Selected literature
ACMS	Thermal-EI	Quadrupole or magnetic sector	None	Schreiner *et al.*, 2002
AMS	Thermal-EI	Quadrupole	Polydisperse aerodynamic	Jayne *et al.*, 2000; Jimenez *et al.*, 2003a
APCI-MS	Thermal-CI	Ion Trap	None	Hoffmann *et al.*, 2002
ATOFMS (TSI 3800)	Single laser *	Bipolar reflectron (originally monopolar linear) TOF	Polydisperse opto-aerodynamic	Gard *et al.*, 1997; Prather *et al.*, 1994
Chapel Hill Instrument	Dual laser or thermal-laser	Linear TOF	Polydisperse opto-aerodynamic	Sykes *et al.*, 2002; Woods *et al.*, 2001
LAMPAS/ LAMPAS 2	Single laser	Bipolar (originally monopolar) linear TOF	Polydisperse (originally monodisperse) opto-aerodynamic	Hinz *et al.*, 1994; Trimborn *et al.*, 2000
PALMS/ WB-57F PALMS	Single laser	Reflectron (originally PSPF) TOF	Polydisperse optical, later with opto-aerodynamic	Murphy and Thomson, 1995; Thomson *et al.*, 2000
Particle Blaster	Single laser	Reflectron TOF	None	Reents *et al.*, 1995; Reents and Ge, 2000
RSMS/RSMS II/RSMS III	Single laser	Linear (originally reflectron) TOF, later bipolar	Monodisperse aerodynamic (originally none)	Carson *et al.*, 1995; 1997a; Lake *et al.*, 2003
SI-PBMS	Thermal	Quadrupole	None	Svane *et al.*, 2004
TD-CIMS	Thermal-CI	Triple quadrupole	None	Voisin *et al.*, 2003
TDPBMS	Programmable thermal-EI	Quadrupole	None	Tobias and Ziemann, 1999; Tobias *et al.*, 2000

* Two-step laser desorption and ionisation was first demonstrated on an instrument of this type (Morrical *et al.*, 1998), although it has not been widely used since.

and DC voltages applied and the frequency of the AC voltage. Normally, the frequency is kept fixed during operation, as it is easier to vary the magnitudes of the voltages. The AC voltage acts as a high-pass m/z filter, as ions with too low an m/z adopt trajectories with oscillations larger than the spacing of the rods, causing them to strike the rods, become neutralised and not be detected. In a similar manner, the DC voltage acts as a low-pass

m/z filter by further perturbing the oscillations of high m/z ions. Therefore, if the two voltages are selected correctly, a quadrupole can act as a filter for all ions but those of the desired m/z.

After the rods, the ions are detected using devices such as electron multipliers or conversion dynodes, sometimes after extraction using additional electric fields. While essentially performing the same filtering function as a magnetic sector mass spectrometer, quadrupoles are much more compact, robust and generally easier to implement, making them ideal for field instrumentation. Their accuracy is also not affected by the initial velocities of the ions prior to selection. However, a disadvantage of selecting the ions this way is that the resolving power of quadrupoles is limited by the amount of time the ions spend in the rod region, which is typically less than a microsecond. Because of this, the m/z resolution of quadrupole mass spectrometers is normally around unity.

A third common type of mass spectrometry is the time-of-flight (TOF) method. The basic principle is to again accelerate the ions over a specific electric potential, but the time taken to travel a set distance to the detector (e.g. microchannel or microsphere plates) is measured, from which the velocity and therefore the m/z can be calculated (see Figure 5.6 in Chapter 5). This technique requires accurate timing electronics, rapid data collection and the precise gating of the start of the time of flight, meaning the ions must be delivered in discrete pulses. It does have a major advantage in that ions of many m/z can be detected simultaneously, which is not possible with the previous two methods, and does not have a maximum m/z imposed upon it. The TOF mass spectrometer is capable of sub-unity m/z resolution, but is limited by several factors, including the duration of the ion pulse, the spatial distribution of the ions before extraction and the distribution of velocities of the ions prior to acceleration. There are ways of mitigating these losses in resolution, post source pulse focusing (PSPF) being one such method (Kinsel & Johnston, 1989). This involves applying an additional voltage pulse to part of the drift section after the ionisation event, which causes any lagging ions to catch up with those of the same m/z. While this produces a satisfactory improvement in the resolution, it is intrinsically limited to a particular portion of the mass spectrum. The wider the section it is applied to, the less pronounced the improvement. Many implementations make use of a reflectron instead, as shown in Figure 6.1, which specifically reduces the broadening due to uncertainties in the ion velocities in the drift region. Electric fields are used to reverse the direction of travel of the ions halfway through their flight. As ions with a velocity higher than intended for their m/z penetrate deeper into the reflectron, the reflection time is increased, so the otherwise shortening of their time of flight is largely

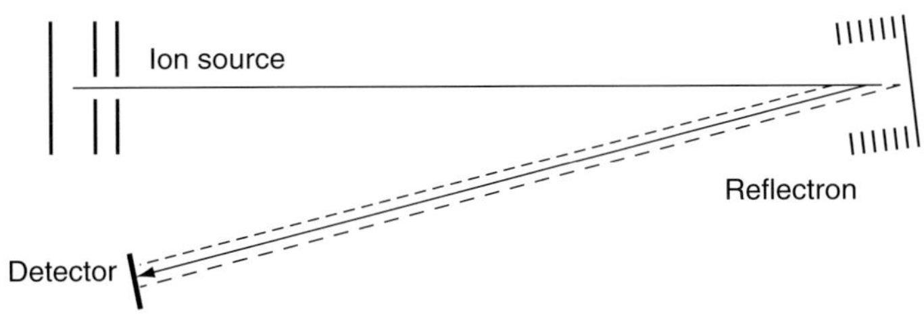

Figure 6.1 The basic principle of a reflectron time-of-flight mass spectrometer. Trajectories of ions with higher and lower velocities than desired are shown with dashed lines.

cancelled out, providing the electric fields are tuned correctly (Weickhardt *et al.*, 1996). An advantage of this over PSPF is that providing it is correctly tuned, the increase in resolution benefits the ions of all m/z.

Ion trap mass spectrometers use a system of two end-cap electrodes either side of a ring electrode, which is effectively the rod geometry of a quadrupole wrapped around on itself. By applying the AC and DC voltages between the caps and the ring in the same manner as in a quadrupole, ions of a range of m/z's are indefinitely held in stable three-dimensional orbits between the electrodes. The voltages can be initially set to capture and hold all ions and then subsequently ramped to selectively release ions according to their m/z for detection. Like the TOF mass spectrometer, this is capable of delivering a complete mass spectrum from a single ionisation event, but is not reliant upon having a pulsed ion source. Also, it is capable of higher resolutions than quadrupole mass spectrometers due to the increased resolving time. However, because it must be scanned, it takes much longer than a TOF mass spectrometer to deliver a complete mass spectrum, so is unsuitable for rapid data collection.

A method sometimes used to further investigate the origin of specific peaks within a mass spectrum is tandem mass spectrometry or MS-MS. This can be performed using a triple quadrupole mass spectrometer, which is in effect three quadrupoles in series. The first quadrupole selects an m/z of interest, while the second does not act as a filter but as a region where the ions are further fragmented by techniques such as collision induced dissociation (CID). The third quadrupole then performs a scan to give a spectrum of the secondary fragments, which is of particular use in identifying organic chemicals. By disabling the initial selection and secondary fragmentation, this configuration is also capable of performing the same task as a standard quadrupole, so the peaks of interest in a sample can be identified. An MS-MS is also possible with ion trap mass spectrometers; the basic method is to release all ions except for those of the m/z of interest from the trap before returning it to a state where it will hold all ions. The remaining ions are then fragmented using CID before the mass spectrometer is scanned as normal.

6.4.2 Inlet technologies

Mass spectrometers must be operated at high or ultra-high vacuum, because random collisions with gas molecules can prevent ions from being detected. Therefore, a key feature of aerosol mass spectrometry involves removing as much of the gas-phase material from the aerosol sample as possible while retaining the particulate fraction for analysis. The exceptions are the thermal desorption-chemical ionisation methods (e.g. APCI-MS, TD-CIMS) because while the mass spectrometer is operated at high vacuum, the collection, desorption and ionisation can take place at atmospheric or near atmospheric pressure. The most common method of removing gas from the sample is to form a collimated particle beam from the aerosol at the inlet and skim off the majority of the gas-phase material using a series of apertures and differential pumping. In most early designs, the particle beam was formed by accelerating the particles through a capillary tube or nozzle into the vacuum of the instrument. Due to their inertia, the particulate fraction of the aerosols forms a less divergent jet than the gas fraction at the point of expansion. The particles pass through a series of channels or apertures, allowing the excess gas to be removed by the pumping system. There is still some divergence in the particle beams, with the amount of divergence dependent on the aerodynamic size of the particles. Particles must hit a physical target or pass through laser focal points in order to be detected and so there is a tolerance in the solid angle in which particles can travel and be detected. This in turn leads not only to a reduction in sensitivity, but also to a bias towards particles of a given aerodynamic size. These can both be mitigated by reducing the distance between the nozzle and the detection point.

Since mid-1990s, the implementation of aerodynamic lenses prior to a nozzle expansion has sought to address these problems (Liu *et al.*, 1995a,b; Petrucci *et al.*, 2000; Zhang *et al.*, 2002, 2004), as shown in Figure 6.2. Particles are drawn through a series of concentric apertures with successively decreasing diameters, which causes the gas streamlines to rapidly compress and expand as the aerosol flows through the lens system. These successive compressions cause the particle streamlines to converge on the axis of the lens, so that when they are accelerated through the nozzle, the radial components of their velocities are small and the streamlines take the form of a tightly collimated beam. As the lens must be operated at a low pressure to reduce defocusing due to Brownian motion, the aerosol must normally pass through a critical orifice before entering the lens. The main advantage of aerodynamic lenses is that particles over a wide range of sizes are focused into a narrow and very parallel beam, thereby helping to eliminate any size-dependent detection biases an instrument may have over a given size range and improving quantitative capabilities. The tighter beam also increases the number of particles detected, which in turn improves the instrument sensitivity and overall data quality, particularly for the smaller particles (Kane *et al.*, 2001; Su *et al.*, 2004). The TDPBMS, ACMS, AMS, SI-PBMS and Chapel Hill instrument (see Table 6.1) are among the instruments that use this technology and TSI have manufactured an aerodynamic lens (model 3801-030 AFL) as an optional accessory for the ATOFMS. In the case of the ACMS, the lens design was modified to allow sampling at high altitudes (Schreiner *et al.*, 1998).

Normally, a particular lens implementation can effectively focus a specific range of particle sizes; for instance, the 3801-030 is specified for particles between 30 and 300 nm in aerodynamic diameter. Smaller particles do not posses enough inertia to be focused

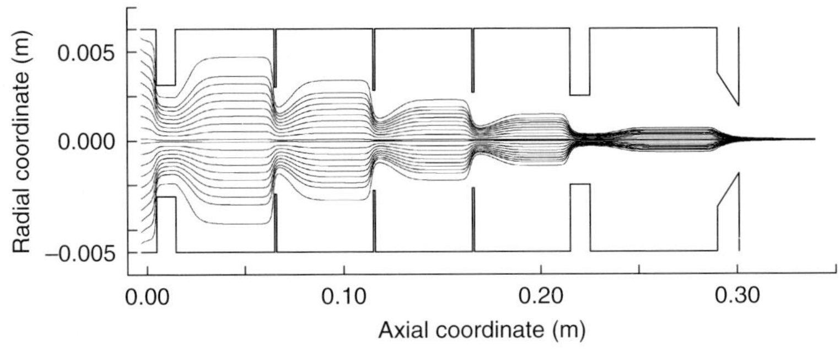

Figure 6.2 Computed 100 nm particle trajectories within an aerodynamic lens system. Several aerosol trajectories enter from the left and become focused down the axes of the lenses during the successive gas expansions and contractions. The right side is where the aerosol enters the first vacuum chamber of the instrument. (Jayne *et al.*, 2000, reprinted with permission from 2000 American Association for Aerosol Research, published by Taylor and Francis.)

properly, so their trajectories tend to follow the gas streamlines, and they are also the particles most affected by Brownian motion, which adds an inherently random component to their velocities. Larger particles can overshoot the axis during focusing at a particular lens stage, in effect becoming defocused, or more simply they can impact on the interior surfaces of the lens system during the rapid changes of gas velocity. These factors result in decreased focusing efficiencies for particles outside these diameter ranges. As aerodynamic lenses are normally best suited for submicron particles, a more traditional nozzle inlet is often preferable when studying coarse particles. An unavoidable limitation of aerodynamic inlets is encountered when sampling certain types of non-spherical particles, as these are subject to lift forces perpendicular to the direction of acceleration, in effect causing extra divergence.

6.4.3 Particle sizing methods

As well as the chemical nature of aerosol particles, aerosol mass spectrometers are often capable of providing information on the size of the individual particles sampled, allowing the direct linkage of chemical compositions with particle size. The ability to do this is invaluable when deriving mixing states, applying compositions to particle populations measured by sizing instruments, or indeed applying accurate size and time information to chemical composition data from offline aerosol analysis techniques. A method favoured in some of the earlier instruments was to measure the total response in the mass spectrometer, in effect measuring the total amount of constituents and therefore the volume of the aerosol (e.g. Allen & Gould, 1981; Myers & Fite, 1975), but as this is only reliable for aerosols of known composition, it is not suitable for ambient measurements. For this reason, it is desirable for an aerosol mass spectrometer to employ an independent sizing method.

6.4.3.1 Sizing by monodisperse sampling

Certain designs of aerosol mass spectrometer are intrinsically incapable of sizing the particles on their own, due to the handling of material in bulk during analysis (e.g. TDPBMS, TD-CIMS) or a lack of triggering (e.g. RSMS-II). However, they are still capable of delivering size-resolved particle data if the sampled aerosol particles are selected according to size before analysis, which can be performed by coupling the instrument in series with a DMA. By stepping the DMA and taking repeated measurements with the aerosol mass spectrometer, size-resolved (in D_m space) data can be derived. Corrections for multiply charged particles can be made in a similar manner as an SMPS. An alternative method, used by the RSMS II and III, is to exploit the size biases that nozzle inlets incur. The particles are selected aerodynamically by carefully designing the instrument's inlet to have an optimum aerodynamic diameter for delivery to the instrument, with the particles of other sizes being removed by the differential pumping system. This optimum can then be controlled by varying the inlet pressure (Mallina *et al.*, 2000; Phares *et al.*, 2002). The problem with monodisperse sizing is that multiple data captures are needed to generate a complete distribution, leading to long scan times. While this is not a problem in laboratory scenarios where there are no shortages of particles, and systems are stable or repeatable, this sizing method is not ideal for ambient sampling, especially in relatively clean environments or during aircraft measurements.

6.4.3.2 *Optical sizing*

Optical sizing by scattering intensity measurements is in principle compatible with aerosol mass spectrometry because it is passive, i.e. the particles are unaffected by the measurement, so an in situ probe can be used to measure the particles prior to desorption. Also, many designs of laser-based instruments require the particles to be detected in order to trigger the desorption laser (see Section 6.4.4), so in this manner a detection laser can serve a dual purpose. This method has been used in the PALMS instrument in particular. However, it is intrinsically limited by the wavelength and intensity of the light used and subject to all the extra complications associated with shape and composition, and variations in scattering intensity with angle that are associated with all optical sizing methods.

6.4.3.3 *Aerodynamic sizing*

At the time of writing, most designs of aerosol mass spectrometer use aerodynamic sizing to deliver size-resolved information of the particles in a similar manner as an APS. The basic principle is to measure the velocity of the particles after the nozzle acceleration, which can be linked to their aerodynamic diameter. Salt *et al.* (1996) showed this to be a much more reliable sizing method than single-angle optical scattering, due to the elimination of the complex issues associated with optical sizing mentioned above. A version of selective, monodisperse aerodynamic sizing was used in the original LAMPAS instrument. This featured a spatial separation on the particle beam path between the optical detection point and the focal point of the desorption laser. By adjusting the time delay between detection and triggering of the desorption laser, only particles of a particular velocity are detected. A polydisperse aerodynamic method first introduced by Prather *et al.* (1994) is used in more recent instruments such as the ATOFMS, Chapel Hill Instrument and LAMPAS 2. With these, the velocity is measured by using two optical scattering detectors at different stages of the particle beam and the velocity calculated accordingly. As well as providing sizing information, fast electronics derive the appropriate triggering delay for the desorption laser on the fly, so that analysis can be performed on particles of all sizes with an optimum hit rate. In exactly the same manner as OPCs and APSs, limitations are imposed on this sizing method by the wavelength and intensity of the light used in the detectors, so sizes of particles of around 200 nm or less in optical diameter cannot be delivered.

The AMS adopts a slightly different approach to aerodynamic sizing. During normal operation, the instrument alternates between two modes (Jimenez *et al.*, 2003a). In the first mode, particles of all sizes are sampled as the mass spectrometer scans to deliver complete average mass spectra of the entire particle ensemble sampled. The second mode uses a chopper wheel (a rotating disc with radial slits cut into it) to deliver pulses of particles into the instrument (at a rate of around 100–130 Hz) while the mass spectrometer stays fixed at one of a selection of specific m/z's. A TOF region separates the particles temporally according to their velocities before they are detected. As flash vaporisation and quadrupole mass spectrometry operate on very short timescales for non-refractory components (of the order of μs), it can measure the particles' times of flight (of the order of ms) by comparing the detection times with the slit positions as measured by an optical sensor. This method can detect individual particles as pulses in the mass spectrometer signal or, by averaging the measured signals over successive chopper periods, can derive an ion count rate as a function of particle TOF, which can in turn be transformed into particle diameter (Allan *et al.*, 2003a). This method has the advantage of not being limited by the optical size of the particles, therefore the lower sizing limit of the instrument is only limited by the inlet transmission. However, it is intrinsically limited to continuous ion source methods, so it is not compatible with laser-based instruments. It also cannot deliver size information as accurately as the opto-aerodynamic method due to the finite width of the chopper slits (typically 1–2% of the chopper circumference). Finally, it assumes that the particle vaporisation and ion generation and detection times are short; particles that vaporise slowly or those that generate very high mass ions can be sized inaccurately. Another limitation of aerodynamic sizing in general is the need to introduce a particle TOF region to the instrument between the inlet and the detection region. The larger the region, the more accurate the sizing, but this increases the instrument's susceptibility to particle beam divergence, as described in the inlet Section 6.4.2 above, thereby reducing its sensitivity and possibly increasing any relative biases towards particle of certain sizes.

The differences between methods must be taken into consideration when comparing size-resolved aerosol mass spectrometry data with those of another particle sizing instrument. In particular, the aerodynamic diameters delivered by aerosol mass spectrometers must be treated with some caution due to the low pressures that these instruments are operated at. Traditional aerodynamic sizing methods, for example impactor or cyclone inlets, cascade impactors and aerodynamic particle sizers, typically size at near atmospheric pressure, where the size of the particle is comparable to or much greater than the mean free path of the gas molecules (known as the transition and continuum regimes respectively). Providing the Reynolds' numbers remain low, the particle mechanics are normally adequately described by Stokes' formula, where the drag force exerted on a particle is proportional to its diameter (Fuchs, 1964). This leads to the aerodynamic diameter (D_a), as measured by these instruments, to have a square root dependence on the particle density when relating it to the volume-equivalent diameter (D_v). The relationship is given by (DeCarlo *et al.*, 2004; Kelly & McMurry, 1992; McMurry *et al.*, 2002):

$$D_a\sqrt{C(D_a)} = D_v\sqrt{\frac{1}{\chi}\frac{\rho_p}{\rho_0}}\sqrt{C(D_v)} \tag{6.1}$$

where ρ_p is the particle density (in g cm^{-3}), ρ_0 the unity density (1 g cm^{-3}) and χ the dynamic shape factor of the particle. The latter quantity is unity for perfect spheres and increases as the particle becomes more irregular and therefore more subject to drag forces (Brockmann & Rader, 1990). The ratio of the relative particle density to the dynamic shape factor is often referred to as the particle 'effective density' or ρ_{eff} for convenience. $C(D)$ is a slip correction function, applied to take account of the kinetic discontinuity at the particle's surface (Cunningham, 1910). For large particles, continuum fluid dynamics applies and this quantity is close to unity, but as the particle sizes begin to approach the mean free path of the gas molecules (67.3 nm for air at 23°C and 730 torr), it increases and can become significant (Allen & Raabe, 1982).

Because the aerodynamic acceleration in aerosol mass spectrometers usually takes place in the form of a supersonic expansion into high vacuum, the particles can be sized under conditions where the mean free path of the gas molecules is much greater than the size of the particles (known as the free molecular regime). This is especially true of systems using aerodynamic lenses or those operated at high-altitude, where the pressure upstream of the nozzle is also low. Under these conditions, the drag force exerted on a particle is proportional to its cross-sectional area rather than its diameter (Fuchs, 1964) and therefore the aerodynamic diameter measured is a fundamentally different property to the 'classical' aerodynamic diameter measured at higher pressures. Jimenez *et al.* (2003b,c) introduced this as the 'vacuum aerodynamic diameter' or D_{va} of a particle. The empirical definition is essentially the same as the classical version, being the diameter of a sphere of unit density that would behave the same way under the same conditions, but it is related to the particle's physical properties as follows (DeCarlo *et al.*, 2004):

$$D_{va} = \frac{1}{\chi}\frac{\rho_p}{\rho_0}D_v \tag{6.2}$$

Because of this, assumptions about particle shape and density must be made when comparing size-resolved aerosol mass spectrometry data with those of other instruments, even when using other aerodynamic methods. For instance, with heavier-than-unity particles typically found in the atmosphere, a vacuum aerodynamic instrument will report slightly larger diameters than a classical aerodynamic instrument. These differences can be used to a scientist's advantage, as the coupling of two sizing methods in series can be used to derive properties such as the effective density of the particle (e.g. Jimenez *et al.*, 2003b; McFiggans *et al.*, 2004; Murphy *et al.*, 2004).

Another limitation of this sizing technique is the effect the sudden change of pressure has on the particles, which is again especially pronounced for systems that use aerodynamic lenses, as there is a significant residence time between the critical orifice and the nozzle. For example, aerosol particles are often sampled as aqueous droplets but an amount of the water fraction of these may evaporate during aerodynamic focusing and sizing. As the precise amount of water lost is difficult to quantify or predict, this often causes problems when trying to report the 'true' sizes of particles (Buzorius *et al.*, 2002). For this reason, it is often pertinent to control the humidity of the aerosols sampled.

6.4.4 *Ablation and ionisation techniques*

One of the principle features of any aerosol mass spectrometry technique is the conversion of some or all of the condensed phase particulate matter into gas-phase ions. The methods employed by the various instrument designs can be separated into three main categories: thermal methods, where the particle constituents are vaporised thermally and ionised using heat, electrons or chemical ions; single-step laser methods, where a single laser pulse is used to simultaneously vaporise and ionise the particulate matter; and two-step laser methods, where a laser pulse performs the ionisation but the vaporisation is performed thermally or using a second, weaker laser. The technical details of the different methods, along with their strengths and weaknesses, examples of instruments that use them and how the data they generate can be used are presented in the following sub-sections.

6.4.4.1 Thermal methods

The earliest designs of aerosol mass spectrometers vaporised and ionised the particulate matter in a single step by impacting the particle beam upon a (normally metallic) surface heated to a very high temperature, known as the vaporiser (e.g. Davis, 1973; Myers & Fite, 1975). This technique, known as surface ionisation, is used in some modern instruments such as the SI-PBMS. Its advantage is that it yields large numbers of ions and is very useful for the quantitative study of specific inorganic components of aerosol particles, in particular the alkali elements (Davidsson *et al.*, 2002; Jäglid *et al.*, 1996). This makes the instrument very useful for the study of sea-salt based particles, in particular. However, the strong biases towards specific elements, based on their electronic structure and the work function of the vaporiser material, makes surface ionisation unsuitable for general-purpose aerosol instrumentation.

These issues can be overcome by separating the vaporisation and ionisation stages of the analysis; the impaction surface is heated to a lower temperature so that the particles are vaporised in the form of a neutral gas and are subsequently ionised independently. Early examples of instruments that could do this include CAART (Chemical Analysis of aerosols in Real Time) (Allen & Gould, 1981), and PAMS (Particle Analysis by Mass Spectrometry) (Sinha *et al.*, 1982). The CAART design is shown in Figure 6.3, as this employed the main basic principles of decoupled thermal vaporisation. The variations on these principles that have been employed since will be discussed below.

Particles enter the instrument through a nozzle, where they are collimated into a beam before passing through two differentially pumped chambers to reduce the gas-phase pressure to less than 10^{-6} torr. Once in the detection region, they enter a vaporiser-ioniser cell, which serves to vaporise the particles, contain the neutral gas molecules until a reasonable fraction of them can be ionised, and direct the ions to the mass spectrometer. The vaporisation surface consists of a resistively heated tantalum wire in a coiled cone shape, as shown in Figure 6.4, which presents a continuous surface to the particle beam but is 50% open to the side. The wire is heated to around 1500 K and the particles are flash vaporised upon impaction. The ionisation region consists of a cylindrical cage 2 cm in length by 1 cm in diameter. Electrons are produced by a thoriated tungsten filament and are accelerated into the cage by the potential difference (70 V by default), creating a cloud of high-energy electrons in the centre. The energy transferred to the vapour particles through random electron impacts (EI) is enough to cause ionisation, creating positive ions that are extracted and focused by electric fields and analysed by the quadrupole mass spectrometer. Normally, 70 eV is sufficient to break covalent bonds within the molecules, so the ions detected from a single chemical species are frequently taken from a selection of ionised fragments of the parent molecules. However, providing that the electrons have the correct energy, these fragmentation patterns are very repeatable, even between instruments (McLafferty & Turecek, 1993).

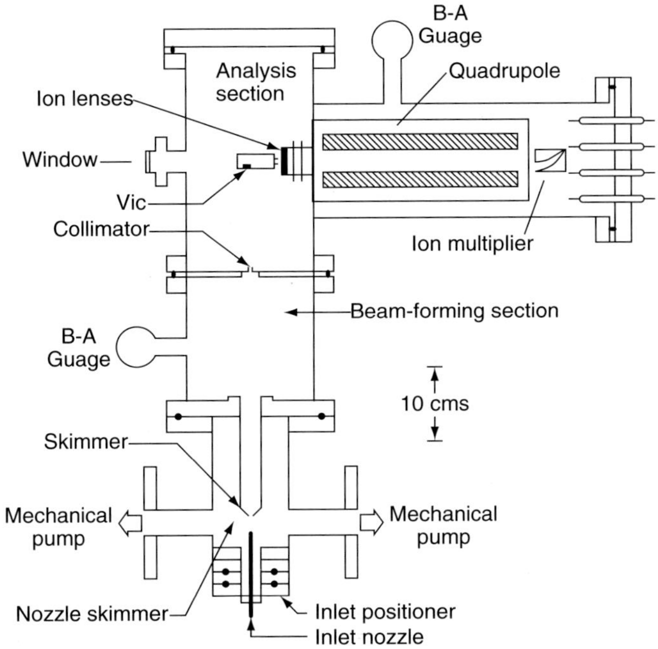

Figure 6.3 Schematic of the CAART. This instrument employed the basic principles of decoupled thermal vaporisation and ionisation used by all modern thermal aerosol mass spectrometers. The B-A gauges are beta attenuation pressure gauges. The 'VIC' is the vaporiser-ioniser cell referred to in Figure 6.4. (Allen & Gould, 1981, reprinted with permission from American Institute of Physics.)

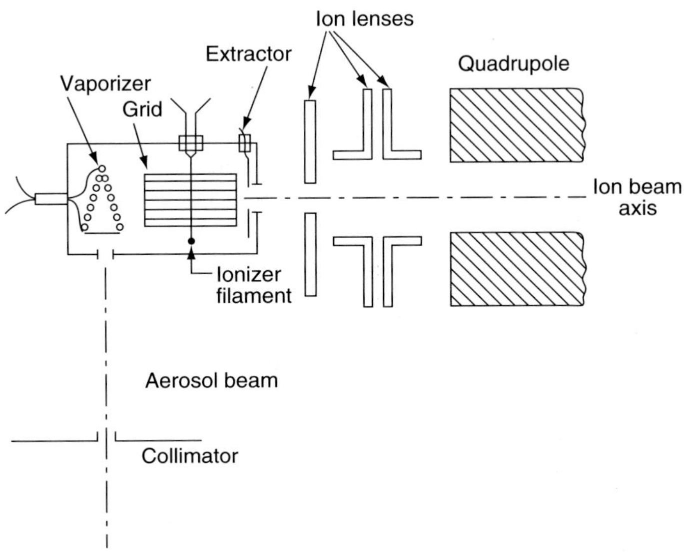

Figure 6.4 The vaporiser-ioniser cell and ion optics of the CAART. (Allen & Gould, 1981, reprinted with permission from American Institute of Physics.)

Most decoupled thermal methods employ variations on this basic principle, whereby particles are collected on a surface, vaporised using heat and ionised separately using a continuous ion source before analysis. In instruments such as the AMS, the vaporiser is contained within the ionisation region itself, to maximise the fraction of gas-phase molecules that are ionised, thereby increasing the sensitivity. However, care must be taken when using this configuration to ensure that the presence of the metallic surface does not interfere with the electrical potential of the ionisation region or the field geometries of the mass spectrometer. The filament also radiatively heats the vaporisation surface, imposing a minimum vaporisation temperature and making the configuration unsuitable for instruments such as the TDPBMS (see below).

There are two general methods of performing the vaporisation itself: one is to keep the surface at a constant temperature, allowing the particles to vaporise as they hit the surface, a technique used by the surface ionisation methods, CAART and the AMS amongst others. As this is normally performed in ultra high vacuum with temperatures of hundreds of °C, the particles flash vaporise almost instantly. To prevent particles leaving the surface before vaporisation is complete, in the AMS the conductively heated molybdenum vaporiser was coated with layers of mesh. This was later replaced with an inverted cone vaporiser made from porous tungsten. A second method is to collect the particles on a cooler surface in bulk and subsequently heat it during a separate analysis stage. While this does interrupt sampling, thereby reducing the time resolution of the instrument, it does allow a large amount of material to be collected for the analysis runs. Examples of this technique include the TDPBMS when operated in temperature-programmed mode, which impacts the particles upon a cryogenically cooled surface (a conductively heated, v-shaped molybdenum foil) during sampling. By carefully ramping the surface temperature at the analysis stage and scanning the mass spectrometer continuously, it permits the study of composition as a function of component volatility. Another example is the TD-CIMS. As this collects the particles at atmospheric pressure, it cannot form a particle beam and use impaction. Instead, the particles pass through a bi- or unipolar charger and precipitate electrostatically onto the collection surface, which is in the form of a nichrome filament. After collection, the filament is injected into a separate analysis chamber for vaporisation and ionisation.

An advantage of 70 eV EI is that it is powerful enough to ionise almost all organic and inorganic molecules, which is useful when trying to achieve mass closure of secondary aerosol species. As the use of 70 eV electrons for ionisation is a very well established convention in mass spectrometry, the mass spectral data produced (such as in the AMS and TDPBMS) are directly comparable with those obtained using other laboratory instruments. The established signatures of fragmentation of many different specific chemicals are held on databases/atlases, compiled, for example, by NIST (2003), making the interpretation of data much more straightforward. When analysing ambient particles, common inorganic species such as nitrate and sulphate manifest themselves at specific m/z ratios in the mass spectrum, making it easy to identify and quantify them (Allan *et al.*, 2003a). Organic species tend to generate much more complex patterns and are therefore harder to identify, especially when multiple species are present.

Thermal vaporisation is not limited to standard quadrupole mass spectrometry and EI ionisation. For instance, the APCI-MS and TD-CIMS use chemical ionisation, the former in conjunction with an ion trap and the latter with a triple quadrupole.

Chemical ionisation (CI) is a much 'gentler' method that charges the gas-phase molecules by reaction with ions generated from a radioactive source or corona discharge (e.g. N_2^+, O_2^+, H_3O^+, O_2^- and CO_3^-). For more details of CI, see Chapter 5, Sections 3.2.2 and 3.2.4. The resultant ions produced are chemically different from the parent analyte molecules, a common result being the addition or removal of a proton, but as there is no molecular fragmentation, it is highly suited to the study of the organic fraction in particular. Chemical ionisation is a very selective form of ionisation that can be tailored to specific chemical fractions by controlling the initial ions produced. Another advantage of chemical ionisation is that the particle collection, vaporisation and ionisation stages can be performed at higher pressures. The neutral gas molecules can be removed by differential pumping as the ions are accelerated into the mass spectrometer, rather than at the aerosol inlet before vaporisation. This means perturbations to the particles through exposure to high vacuum can be eliminated, as too can the losses of smaller particles associated with particle beam formation, allowing the online study of the chemistry of ambient ultrafine particles as small as 6 nm with the TD-CIMS (Smith *et al.*, 2004).

The main advantage of the decoupled thermal methods so far discussed is the ability to be quantitative, as it can be assumed that all (or a defined fraction) of the available particulate matter is collected and vaporised. As the ionisation occurs in the gas phase, the components do not interact once vaporised, allowing the measurements of individual chemical fractions to be made independently of one another (Bley, 1988), in turn meaning that accurate mass concentrations can be derived for many chemical species in a sample simultaneously. With the AMS and TDPMS, their absolute quantitative capabilities are dependent upon the particles reliably hitting the vaporisers, so they are in principle susceptible to the biases associated with beam divergence. Their use of aerodynamic lenses mitigates this over large ranges of sizes. Thermal vaporisation is not without its issues and limitations. The act of heating imparts extra energy to the subject molecules, which can change the fragmentation behaviour during ionisation when compared with purely gas-phase mass spectrometry. In the more extreme cases, the heat may fragment the molecules as they are vaporised, meaning the neutral gas-phase species produced are not chemically the same as the parent condensed-phase components of the particles. This phenomenon is most pronounced for complex organic molecules. A specific example is the behaviour of di- and polycarboxylic acids in the AMS, a common group of chemical species in atmospheric aerosols. Rather than vaporising in their molecular forms, they tend to pyrolyse on the vaporiser surface and produce large numbers of simpler molecules such as CO_2, CO and H_2O instead. However, these molecules can still be used as quantitative markers for the parent organic species.

Another intrinsic limitation is that these instruments are blind to the refractory fraction of the particles. While they can very effectively study components such as ammonium, nitrate, sulphate and organics, the common components with much higher melting points such as elemental carbon and crustal material cannot be vaporised and are therefore not detected. Finally, the systems that use quadrupole or magnetic sector mass spectrometers are limited to studying one m/z at any given time, which, coupled with the fact that ions are generated on a continuous basis, means that complete composition data cannot be derived for individual particles; they can only either generate averaged mass spectra for an aerosol sample or information on a single m/z for a given particle. This makes probing mixing states inherently difficult if not impossible.

Electron impact and CI are generally not well suited for conventional TOF mass spectrometry, which requires a pulsed ion source. Coggiola *et al.* (2000) presented a thermal vaporisation-EI aerosol mass spectrometer that held the ions produced in a potential well before being ejected by a short (≈100 ns) voltage pulse that provided the start of the TOF in the mass spectrometer (Grix *et al.*, 1989). The voltage pulses were gated by an optical particle detector located in the particle beam. Unfortunately, this design suffered from the problem that the number of ions coexisting in the ion source region was intrinsically space-charge limited to around 10^5. This meant that when a particle's composition was dominated by one particular component (e.g. water), the other ion species were detected with drastically reduced sensitivity.

Future thermal instrument developments could benefit from Orthogonal extraction Reflectron Time of Flight (ORTOF) mass spectrometers (Chen *et al.*, 1999), a new type of mass spectrometer designed for continuous ion sources. Pulses of ions are extracted at right angles from a focused ion beam, itself extracted from the ionisation region, which are analysed using a reflectron TOF configuration. As the spatial and velocity distributions of the ions in the axis of extraction are intrinsically constrained, very high m/z resolutions are possible, permitting a short TOF region, which in turn permits a high pulse rate and near-continuous data collection. At the time of writing, Aerodyne Research Inc. is offering an ORTOF mass spectrometer as an interchangeable alternative to the standard quadrupole in the AMS. Initial results are promising, effectively giving size-resolved data for all m/z channels simultaneously (Drewnick *et al.*, 2005).

6.4.4.2 Single-step laser methods

Laser desorption and ionisation (LDI) is a different approach that uses a high-powered, focused laser to deliver a large amount of energy to the particle while airborne to simultaneously vaporise and ionise the components (Johnston, 2000; Noble & Prather, 1998). The earliest examples of this technique being used to study mobile particles include those presented by Sinha *et al.* (1982) and McKeown *et al.* (1991). The original PALMS instrument, as described by Murphy & Thomson (1995), is shown as an example in Figure 6.5, as this has all the common elements of most current LDI designs, although many new innovations have occurred since and will be discussed below. The aerosols enter the instrument through a capillary tube and pass through differentially pumped chambers before entering a high vacuum region, where they pass through the focal point of the desorption laser. A KrF eximer laser was used, which delivers UV light at a wavelength of 248 nm. As the particles tend to be fast moving after introduction into the vacuum of the instrument, high amounts of energy must be transferred in very short spaces of time to hit and vaporise them successfully. The most effective way of providing enough laser energy is through the Q-switching of the desorption laser, which delivers a short, efficient and powerful pulse orthogonal to the particle beam. To trigger these laser pulses, an additional, lower-powered but continuous laser beam (He:Ne laser at 633 nm) in conjunction with an avalanche photodiode was used along the same beam path to detect the particles as they entered the instrument. The scattering events as seen by the photodiode were used to trigger the Q-switch of the desorption laser. The pulse

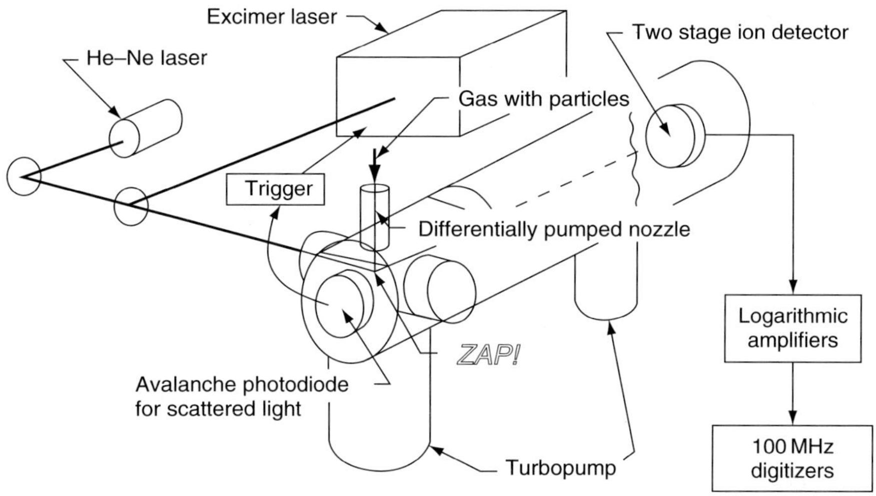

Figure 6.5 Basic schematic of the original PALMS instrument, which incorporates most of the fundamental principles of laser-based aerosol mass spectrometry. (Murphy & Thomson, 1995, reprinted with permission from American Association of Aerosol Research, published by Elsevier Science Inc.)

energy delivered by the desorption laser was sufficient to both vaporise and ionise the components of the particles. The use of a pulsed ion source has the added advantage of being able to act as the start gate in a TOF mass spectrometer, which is used to analyse ions of all m/z produced from the individual particles simultaneously. The mass spectrometer could be operated in either positive or negative ion mode. The PSPF method was also used, acting on the ions of an m/z between 24 and 350, and gave the instrument an $m/\Delta m$ FWHM resolution of >200.

There have been many variations on this principle over the years, employing many different alternative technologies. For instance, other designs have used Ar^+ (514 or 488 nm) (e.g. Salt *et al.*, 1996), He:Cd (442 nm) (e.g. Carson *et al.*, 1995), frequency doubled Nd:YAG or Nd:YVO_4 (532 nm) (e.g. Thomson *et al.*, 2000) or diode lasers (e.g. Woods *et al.*, 2001) in the particle detection systems, and photomultipliers rather than photodiodes are now used in most implementations to detect the scattered light. In addition to the power consumption, cost and portability, the choice of laser technology also affects the detection capabilities, as was discussed in the sizing Section 6.4.3.2, together with aerodynamic sizing of the particles using two detection lasers. Particles smaller than the lower limit of the detection and triggering system can be studied by firing the desorption laser arbitrarily, without triggering. This results in a lack of direct particle sizing information, but a DMA placed in series or a size-selective inlet as described in the sizing section can be used to generate size-resolved data (e.g. Ge *et al.*, 1998; Kane *et al.*, 2001). As there is no triggering, the hit rate is significantly reduced but this is mitigated in the RSMS II, where the desorption laser beam and the particle beam are counter-propagating, as opposed to the orthogonal alignment used in other instruments, thereby increasing the coincident volume of the laser and particle beams.

To reliably be detected, a particle must accurately pass through the focal points of all the lasers and absorb a sufficient amount of the desorption laser energy to generate ions. This will not occur for all of the particles, as the particle beam is slightly divergent. Allen *et al.* (2000) measured detection efficiencies ranging from 1 hit in 10^2 to 10^6 particles when sampling ambient particles with the ATOFMS. The scattering and absorbance of the respective laser photons are dependent upon the composition and structure of the particle and the laser wavelengths used. These factors can lead to several size and composition biases in the detection efficiency (Carson *et al.*, 1997b; Kane & Johnston, 2000; Silva & Prather, 2000). While present in all differentially pumped instrument designs, the particle beam focusing biases can be more pronounced with LDI than with the thermal methods, as the laser focal points are generally smaller than the thermal vaporiser surfaces. Again, the use of aerodynamic lenses at the beginning of the twenty-first century has helped to mitigate these over certain size ranges (Su *et al.*, 2004).

Many laser designs have been used for the desorption, such as ArF (193 nm) (e.g. Kane *et al.*, 2001), XeCl (308 nm) (e.g. Yang *et al.*, 1996), CO_2 (10.6 μm) (e.g. Prather *et al.*, 1994), N_2 (337 nm) (e.g. Trimborn *et al.*, 2000) and frequency tripled or quadrupled Nd:YAG (355 or 266 nm) (e.g. Gard *et al.*, 1997). The choice of laser used in a particular implementation is influenced by several factors. These include the pulse power and duration capabilities, power consumption, cost, portability and the ionisation threshold (minimum power flux required to generate ions). However, an individual instrument is not normally limited to a specific laser and many groups have experimented with multiple types. The ionisation threshold is dictated by both the wavelength of the laser light used and the composition of the particles under analysis, as the absorption bands and ionisation energies vary greatly between particle constituents. Generally speaking, less laser power is needed to achieve ionisation for shorter wavelengths, as fewer photons per molecule are needed (Thomson & Murphy, 1993). As the amount of energy absorbed by the particle increases, a greater fraction of the constituents become desorbed and/or ionised, and a greater amount of molecular fragmentation occurs. These phenomena continue to increase until a second threshold is reached, whereby a plasma is formed. The key technical parameter is the total amount of energy delivered per unit area during a laser pulse, known as the fluence.

A potential problem with using TOF mass spectrometers with LDI is that the high-powered vaporisation and ionisation process produces ions with a wide distribution of initial velocities, which if unchecked, would lead to loss of resolution. The PSPF method has been used to remedy this (e.g. Hinz *et al.*, 1996; Murphy & Thomson, 1995), but the favoured method for many of the instrument designs since mid-1990s is the use of reflectrons, and $m/\Delta m$ resolutions of the order of hundreds are typically reported (Gard *et al.*, 1997; Reents & Ge, 2000). The ionisation process also routinely produces many ions of both positive and negative charges, so another common technique is to use two mass spectrometers of opposite polarities simultaneously, so two complete mass spectra are delivered per particle detected. This technique was introduced by Hinz *et al.* (1996), applied to the original LAMPAS, and has been used in other instruments since, such as the LAMPAS 2, ATOFMS and RSMS III. This increases the amount of composition information gained for individual particles, vastly improving the value of the data. Laser desorption and ionisation is not limited to TOF mass spectrometry, although it is the most common method. Yang *et al.* (1996) and Parker *et al.* (2000) presented laser-based instruments that employ ion trap mass spectrometers. One advantage of this configuration is that it permits MS-MS, see Section 6.4.1. For instance, detection and identification of specific chemical species such as bis(2-ethylhexyl)-phosphoric acid and elements such as uranium in particles have been demonstrated using this method (Gieray *et al.*, 1998; Reilly *et al.*, 1997).

Unlike electron or chemical ionisation methods, laser ionisation events are discrete, intrinsically limited to individual particles and can produce complete mass spectra when coupled to TOF-MS, so in this sense, these instruments can truly be described as single particle mass spectrometers. When sampling ambient data, their real strength is their ability to deliver composition and size information for specific particles, allowing the explicit determination of mixing states with a very high time resolution. As the laser pulses deliver much more energy to the particles than thermal methods, they are able to vaporise all of the components of the particles and are not limited to the non-refractory fraction. For this reason, they are also ideal for the study of trace components in particles and are incredibly useful in assigning sources to particles based on these. On the other hand, providing quantitative chemical information with LDI is inherently difficult. The many complexities associated with the combined desorption and ionisation process result in the limitation that for most LDI instruments, the chemical composition data derived from sampling ambient particles are mainly qualitative in nature. If enough laser energy to cross the plasma formation threshold can be delivered, the Particle Blaster can deliver quantitative data on the elementary compositions of particles, as all the atoms present are converted to positive ions (Reents & Ge, 2000; Reents & Schabel, 2001). A similar quantitative method was also demonstrated by Mahadevan *et al.* (2002). However, as the constituents are completely atomised in these processes, no molecular information is retained.

At lower fluences, the constituent molecules remain more intact, allowing some chemical information to be obtained, but at the same time, the precise ionisation process can be highly variable in both the ions it produces and their abundances. A significant component of this variability is due to uncertainties in the amount of laser energy delivered to the particles and work has indicated that the reproducibility may be significantly improved if the beam homogeneity and shot-to-shot power variability of the desorption laser are better constrained (Wenzel & Prather, 2004). However, individual chemical components can interact with each other during the combined desorption and ionisation process, unevenly distributing energy and charge between the chemical components (Neubauer *et al.*, 1998; Reilly *et al.*, 2000). These phenomena are known as 'matrix effects' and the results are often difficult to predict, as they are known to be non-linear and dependent on both the explicit compositions of the particles and the precise amounts of energy absorbed. The study of complex molecules is in turn hampered by the highly variable fragmentation behaviours that result from this.

When looking at the contributions to the mass spectra from inorganic ions, it has been found that their relative contributions, if not their absolute magnitudes, can be quantified, providing that the chemical matrices are fairly consistent (Gross *et al.*, 2000; Middlebrook *et al.*, 2003). In principle, quantitative inversions based on this behaviour can be applied to ambient data, providing that an *a priori* particle matrix can be assumed for a given population of particles and the mass spectra obtained can be linked with either laboratory standards or composition data from other collocated instruments in the field (Bhave *et al.*, 2002; Dall'Osto *et al.*, 2004; Fergenson *et al.*, 2001). Work to produce a reliable inversion method based on these techniques is ongoing.

The interactions between constituents can be used to the instrument's advantage. For instance, the technique of matrix-assisted LDI (MALDI) deliberately uses these matrix effects to generate more predictable ionisation behaviours and can be applied to aerosol mass spectrometers (Mansoori *et al.*, 1996). It involves mixing or coating the material under analysis with a specific substance designed to absorb the laser light and transfer the energy during the desorption and ionisation in a more controlled manner. While this is a powerful technique in the laboratory or for identifying very specific components (e.g. Arnold & Reilly, 1998), it is not ideal for general ambient sampling as it involves perturbing the aerosol composition. Another technique where interactions between constituents is taken advantage of is in situ chemical ionisation, where the desorption laser is operated at a lower fluence to generate specific ions (mainly Na^+ and K^+) from the particle. These ions combine with the neutral desorbed molecules and like other forms of chemical ionisation, do not result in molecular fragmentation, making it useful for the identification of organics. Lazar *et al.* (2000) were able to demonstrate the capability of this technique to study surface coatings of tributyl phosphate on soil particles, but as this is a very specific technique, it is not suitable for general-purpose work.

6.4.4.3 Two-step laser methods

The steps of desorption and ionisation do not necessarily need to be performed together in laser-based instruments. Instead, a pulse from a lower energy CO_2 or eximer laser can be used to vaporise the particle before a second ionisation laser ionises the resultant vapours immediately afterwards (Cabalo *et al.*, 2000; Morrical *et al.*, 1998; Zelenyuk *et al.*, 1999). As the ionisation is performed in the gas phase, most of the problems associated with matrix effects are avoided. Also, as less energy is required to perform the ionisation, the power delivered by the ionisation laser can be reduced so the molecules are subject to less fragmentation. However, the trade-off with this general technique is that the instrument's capability for studying inorganics is reduced, especially the refractory fraction. By controlling the desorption laser's power, it is also possible to selectively study the outer layers of the particle (e.g. Lazar *et al.*, 1999; Woods *et al.*, 2002). Molecular fragmentation associated with the single-step processes is reduced, but not eliminated entirely, and because the energy of individual photons is too small to cause ionisation, the molecules must be bombarded with laser light intense enough to induce multiple photon absorption, leading to uncertainties in the amount of energy delivered to individual molecules. To address this, the Chapel Hill instrument uses the 9th harmonic of an Nd:YAG laser (118 nm, in the vacuum UV) to perform the ionisation in a two laser based system. As only single photons of this energy (10.5 eV) are required to ionise most molecules, the uncertainty in the energy delivered and the amount of molecular fragmentation that occurs are both reduced substantially. Sykes *et al.* (2002) presented a modification to the Chapel Hill instrument whereby the laser vaporisation stage could be replaced with a heated surface in a similar manner to the thermal desorption techniques, but the ionisation was still performed with the VUV laser. This allows the temperature of the vaporisation to be controlled, allowing study into the volatility of the chemical constituents. The two CW detection lasers were still used to detect the presence of the particle, as these controlled the triggering of the ionisation laser. The instrument has been shown to be very powerful in both the identification and quantification of organic species within particles but its relatively low sensitivity makes the implementations best suited for laboratory studies, at least so far.

An alternative ionisation method that has been demonstrated recently is photoelectron resonance capture ionization (PERCI) (LaFranchi & Petrucci, 2004; LaFranchi *et al.*, 2004), which uses the low-energy electrons produced from a metallic surface when illuminated with a pulsed laser. These electrons transfer onto the gas-phase vapour products of the particles after laser or thermal desorption, forming negative ions. As the energy of the electrons is small and controllable, this too results in reduced fragmentation of the organic molecules. It is also possible to use laser desorption coupled with continuous ionisation techniques. Reilly *et al.* (2004) presented a method where particles are collected in bulk on a surface before being vaporised either with a pulsed laser or alternatively, heating the surface in a similar manner to the thermal methods. The ionisation can be performed by electron impact, chemical ionisation or glow discharge, and an ion trap is used for analysis. This technique could potentially benefit from the quantitative capability that decoupled desorption/ionisation offers with the sensitivity needed for ambient sampling. However, as the particles are handled in bulk, there is no sizing or single particle capability.

6.5 Analysis of ambient data

Many of these instruments described above have been used to sample ambient aerosols for a variety of different purposes at a multitude of locations. These include ground-based sites in urban, rural and remote environments, and mobile platforms such as on-road vans, ships and aircraft. These have presented extra challenges at the analysis stage, as the aerosol climatology can be very complex and often highly unpredictable, given that aerosol mass spectrometry delivers data on a much more detailed level than was previously possible. As each different instrument design produces fundamentally different data, many novel approaches towards data analysis have been developed.

As stated before, the real strength of the thermal methods is the ability to deliver quantitative composition data. The AMS in particular is capable of delivering simultaneous mass concentrations, in $\mu g\, m^{-3}$, for multiple chemical components by applying a matrix-based inversion to the mass spectra and calibrations of ammonium nitrate standards (Allan *et al.*, 2004a). Typical data products include chemical loadings of nitrate, sulphate, ammonium and organic matter. When combined with the sizing data described previously, size-resolved $dM/d\log D$ distributions (also in $\mu g\, m^{-3}$) can also be generated for the individual species (Allan *et al.*, 2003a). Processed data of these types have the advantage of being directly comparable to those produced by other aerosol chemistry instrumentation such as bulk samplers and in situ analysers (e.g. Allan *et al.*, 2004b; Drewnick *et al.*, 2003; Topping *et al.*, 2004) and the products of numerical modelling tools (e.g. Tang *et al.*, 2004), facilitating overall data synthesis and interpretation.

While the APCI-MS and TD-CIMS can use chemical ionisation and tandem mass spectrometry to deliver rich data on the organic chemicals present in a sample, EI and single quadrupole analysis is a limited technique when trying to identify specific organic species, as fragmentation produces many different peaks in the mass spectrum. Atmospheric particles contain a myriad of different organic compounds (Jacobson *et al.*, 2000; Saxena & Hildemann, 1996), so in most real-world environments, the ensemble organic data are impossible to completely deconvolve. The analysis of ambient organic data in the AMS is often limited to the study of generalised classes of organic species such as hydrocarbon like and more oxidised types, based on markers within the mass spectra (Allan *et al.*, 2003b; Canagaratna *et al.*, 2004; Zhang *et al.*, 2005). The temperature-programmed mode of operation of the TDPMS is useful in addressing this, as the mass spectra are delivered as a function of volatility, separating the components.

Instruments using LDI tend to produce very large data volumes during ambient sampling, as individual mass spectra for as many as hundreds of thousands of particles can be captured during a field campaign. The need to reduce this data to a form that is both usable and scientifically valid has prompted the utilisation of some innovative new techniques and their ongoing development is seen by many as one of the main challenges currently facing the single-particle mass spectrometry community (Sipin *et al.*, 2003).

Most methods are based around grouping particles according to common features in their mass spectra and linking these groups to particle compositions and sources. Providing that the particle sampling and detection biases can accurately be taken into account (Wenzel *et al.*, 2003), these analyses can be used to deliver quantitative data on the size and number concentrations (instrument capability permitting) of speciated particle populations with a high time resolution. Due to the inherent amount of variability in the spectra captured, the criteria for grouping the particles must be carefully chosen. Many methods have been used over the years, each with their own advantages and disadvantages, and these are very much in a state of continual development.

A common generalised method is to compare all the particle spectra after data collection has been completed, grouping them according to their numerical similarity and identifying the common features within the groups to gain information about their compositions. Examples of such approaches include principle component analysis (Hinz *et al.*, 1996), fuzzy cluster analysis (Hinz *et al.*, 1999), multivariate patch analysis (Parker *et al.*, 2000) and hierarchical cluster analysis (Murphy *et al.*, 2003). The grouping produced by these methods can be further inspected when making scientific conclusions. While these techniques are very powerful, they are computationally intensive and can only be applied to complete datasets, making near-real-time analysis very difficult.

An alternative approach used by some groups is to analyse spectra sequentially using pattern matching algorithms such as the ART-2A (Adaptive Resonance Theory) neural network (Bhave *et al.*, 2001; Carpenter *et al.*, 1991; Phares *et al.*, 2001; Song *et al.*, 1999) and ADAMS (Algorithm for Discriminant Analysis of Mass Spectra) (Tan *et al.*, 2002). These can group particle spectra according to similarities with previously seen spectra and can adapt to unexpected results and outliers by creating new groups on the fly or progressively altering the reference spectra. While useful for offline analysis, these techniques have the added advantage of permitting real-time analysis. The scientific validity of the clustering produced during an analysis run is highly dependent on the matching and learning parameters of the algorithm and any previous training it has had, so these must be controlled carefully.

6.6 Applications of aerosol mass spectrometric techniques to atmospheric aerosol sampling

The purpose of this section is not to provide a thorough review of every field application of aerosol mass spectrometers since mid-1990s, but to give a series of examples of different applications to demonstrate the wide applicability of the techniques.

Until the mid-1990s aerosol mass spectrometers were laboratory-based developmental instruments. However, since that time their use in sampling the ambient atmosphere has increased substantially and many current large field campaigns contain one or more types of such instruments. Many of the early field applications not only demonstrated proof of concept of aerosol mass spectrometric techniques as valuable tools, but supplied new insights into aerosol composition and mixing state. During the first Aerosol Characterisation Experiment (ACE-1) in the southern ocean, Murphy and co-workers used PALMS to investigate the marine boundary layer aerosol under pristine conditions (Murphy *et al.*, 1998a,b). Sample data are shown in Figure 6.6. The study showed that almost all of the submicron aerosols detected contained sea salt, nearly all the observed sulphate was mixed with the sea-salt particles, the chloride-to-bromide ratio was approximately that of seawater, and iodine was enhanced to approximately the concentration of bromine. The iodine appeared to be correlated with organic particles and the chloride and bromide were anticorrelated with sulphate (Murphy *et al.*, 1997). Silva *et al.*, (1999) used both positive and negative ion spectra to differentiate particles formed during biomass burning from those of other combustion sources. This is possible, since potassium produces strong positive ion signals and these provide a key marker for biomass burning as a source.

Since these earlier studies, aerosol mass spectrometric techniques have rapidly developed and are now used widely in field experiments. There are several studies of long-range transport of aerosols that link their mixed composition to source and particle history. For example, Guazzotti *et al.* (2003) measured single particle mass spectra with the ATOFMS instrument during INDOEX (INDian Ocean EXperiment), from a research ship in the Indian Ocean, and compared sources of particulate emanating from the Indian subcontinent and from the Arabian Peninsula. Clear differences in the mass spectra were discernable between these two source regions, with the Indian subcontinent being heavily influenced by biomass burning from household fires whereas fossil fuel combustion dominated in the Arabian Sea area. In the former case 74% of the observed aerosols were shown to be of biomass burning origin, whereas in the latter case 68% of the measured aerosols were of fossil fuel combustion origin. Source identification and processing was also studied by Dall'Osto *et al.* (2004), who studied single particle mass spectra at Mace Head, a coastal Atlantic Ocean site on the west coast of Ireland, during the 2002 NAMBLEX (North Atlantic Marine Boundary Layer Experiment) field campaign, using a commercial ATOFMS. Particles of several different sources were measured during the course of a two-week measuring period (Figure 6.7) and Saharan dust influences were identified with particles displaying strong aluminium and silicon signatures, whereas more locally produced dust was rich in calcium. Sea salt was often measured but it was shown to be processed to varying degrees, displaying a spectrum similar to pure sea salt when locally produced but demonstrated displacement of the chloride by nitrate in more aged particles, as nitric acid uptake has displaced the less volatile chloride from the particles.

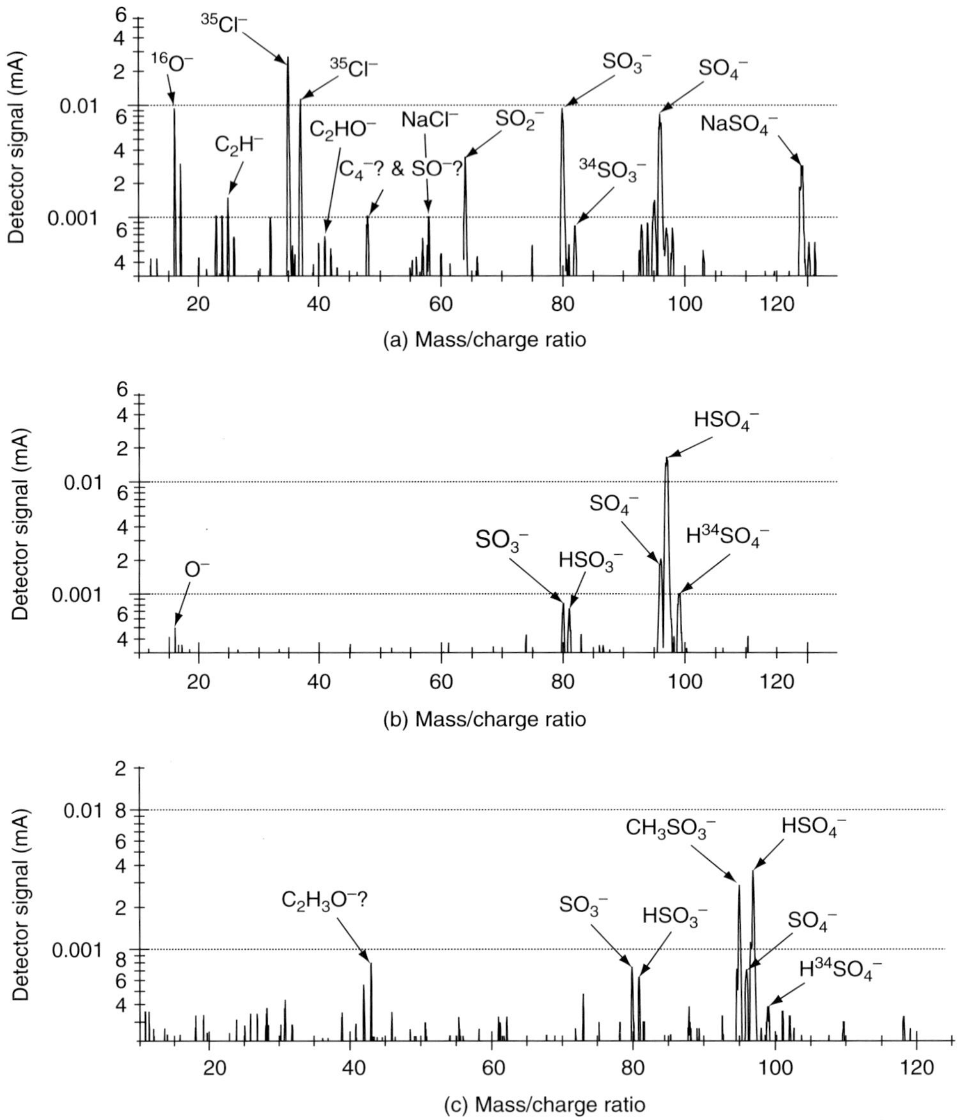

Figure 6.6 Sample mass spectra from the PALMS instrument taken during ACE-1 at Cape Grim. The three negative ion spectra show that (a) sodium sulphate particles, which were the most abundant type of sulphate-containing mass spectra, can be distinguished from (b) either ammonium sulphate or sulphuric acid particles, and (c) a probable mixture of methane sulphonic acid (MSA) and ammonium sulphate or sulphuric acid. The latter particles were very rare. The assignments were made by comparison with laboratory data (Murphy *et al.*, 1998a).

Urban measurements have also been made extensively using aerosol mass spectrometry. Motor vehicle and diesel truck emissions were studied on a single particle basis by Gross *et al.* (2000) during a tunnel study using the ATOFMS. Three clear classes of compounds were observed: those that were composed of polycyclic aromatic hydrocarbons (PAHs);

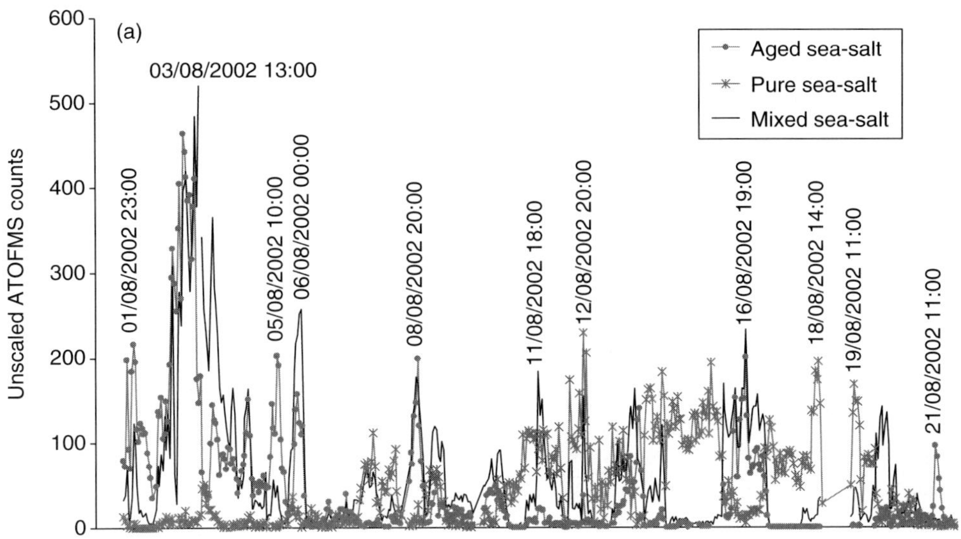

Figure 6.7 Demonstration that the ATOFMS can be used to assess the extent of ageing of sea salt particles in the marine boundary layer. Mass spectra from sea salt were further grouped according to the ratio of the nitrate-to-chloride ion ratios, indicating the level of displacement of chloride by the nitrate ion (Dall'Osto *et al.*, 2004).

carbon particles, or soot; and particles containing inorganic compounds that produce strong ion signals of aluminium, calcium, iron and barium. The study was able to separate the number fraction of these groups and showed that the PAH number fraction is approximately 60% with the other two groups approximately 10% each. The work shown in Figure 6.8, measured during a study in Baltimore, by Lake *et al.* (2003) demonstrated that the RSMS-III instrument was capable of obtaining submicron particle mass spectra as a function particle size for particles greater than 45 nm aerodynamic diameter. In the urban environment, this ability allows the instrument to probe the highly dynamic ultra-fine particles produced in large numbers by motor vehicle emissions. The authors demonstrated that carbon-containing particles were mostly present in the smallest sizes measured and correlated with the rush hour traffic flow peak.

The aerodyne aerosol mass spectrometer (AMS) is ideally suited to measuring organic aerosols in the urban environment and their interactions with inorganic species. Sample data are shown in Figures 6.9–6.11. This is because it can sample small sized aerosol, it targets non-refractory aerosol components and it produces quantitative mass loading information. The field application of the instrument was first demonstrated by Jimenez *et al.* (2003a) and the behaviour of urban aerosol has been investigated by Allan *et al.* (2003b). Comparative studies in an urban environment, as displayed in Figure 6.9, have shown that the instrument is indeed quantitative for common, submicron aerosol components provided that several efficiencies are determined (Drewnick *et al.*, 2003). The modification of the particles as the urban pollution is advected from a city to the surrounding region can also be observed (Alfarra *et al.*, 2004; Boudries *et al.*, 2004). The observations show a separate small mode of particles in urban areas composed almost entirely of organic species whose mass spectra are similar to those observed from engine

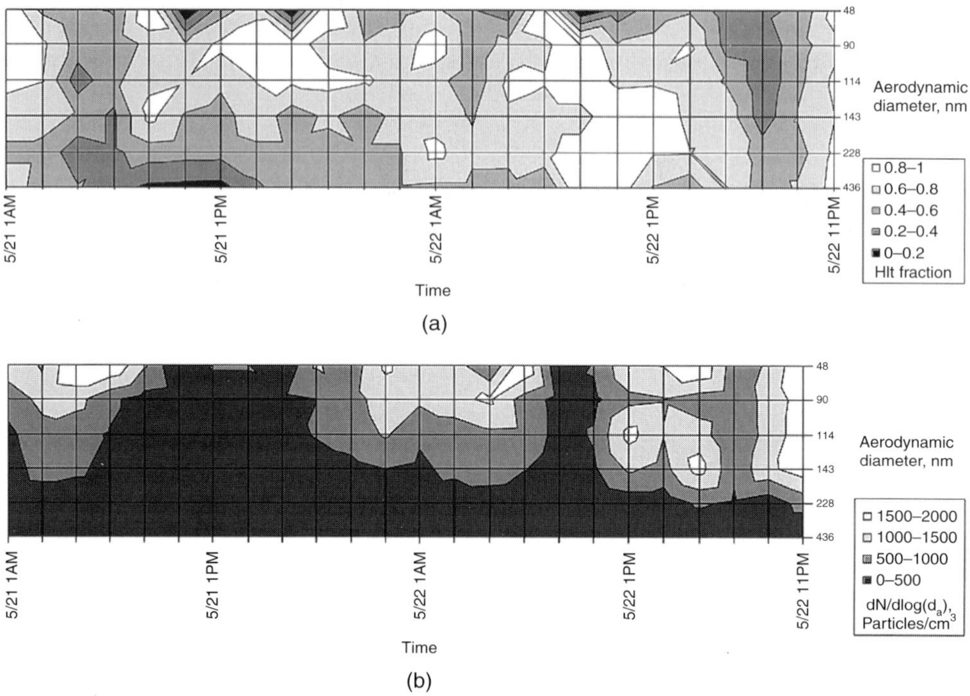

Figure 6.8 The top panel shows a contour plot of the fraction of carbon containing particles in the measured particle population (identified by large signals at m/z 28) as a function of their aerodynamic diameter. The bottom figure shows the number distribution of the same particle population as a function of aerodynamic diameter and time. It is clear that the carbon containing particles are mostly below 100 nm and peak in the early morning rush hour, due to a combination of increased traffic flow and the suppression of mixing caused by the presence of a nocturnal inversion (Lake *et al.*, 2003).

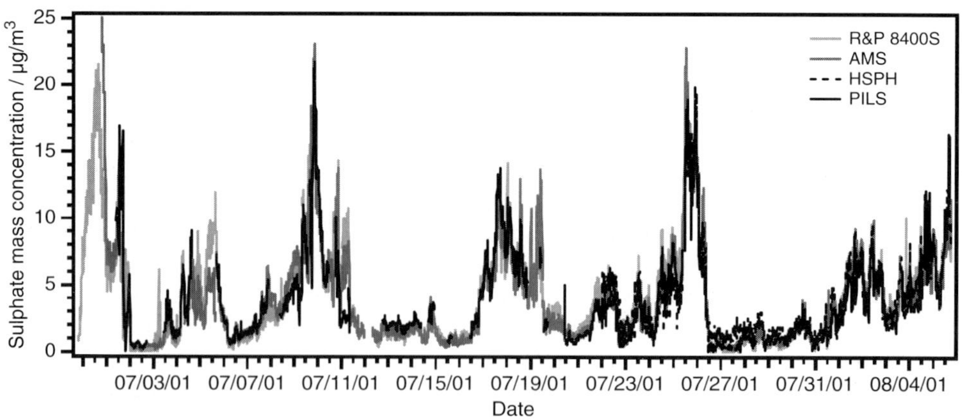

Figure 6.9 A comparison of submicron sulphate mass concentration between the AMS, a Particle Into Liquid Sampler (PILS) (Weber *et al.*, 2001), a Rupprecht and Patashnick 8400S and a continuous sulphate monitor developed at the Harvard School of Public Health (HSPH).

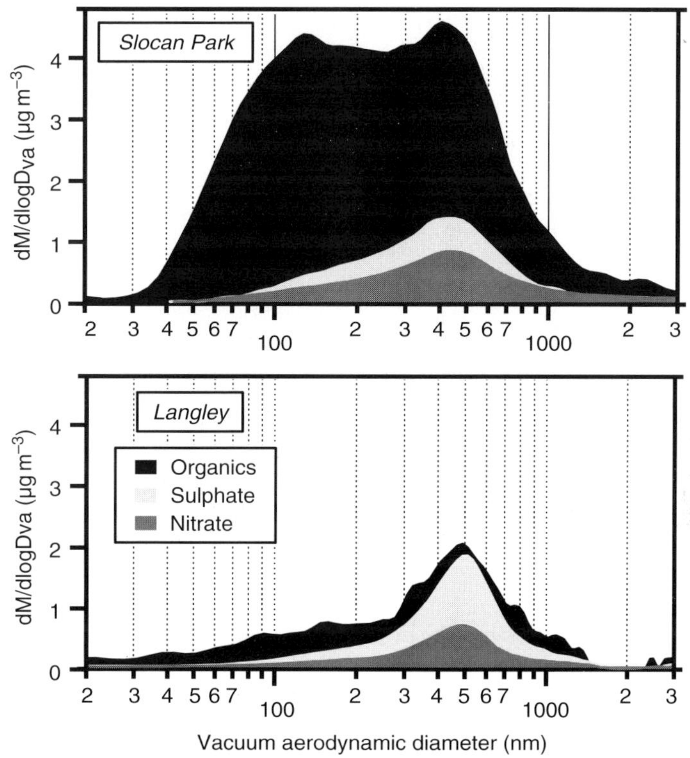

Figure 6.10 The AMS produces mass loadings as a function of particle vacuum aerodynamic diameter for organics, sulphate and nitrate fractions. The upper panel shows an average distribution for the city of Vancouver, and the lower panel shows a similar period at Langley, a rural location impacted by aged air from the Seattle area. The data were collected during the PACIFIC 2001 air quality study in Canada during 2001 (Alfarra *et al.*, 2004).

oil particles. The accumulation mode appears to be a mixture of ammonium sulphate and organic material with a variable amount of nitrate dependent on the source profile of the region. The organic fraction in the larger mode has a very different mass spectral signature and shows large peaks at m/z of 18 and 44 (H_2O^+ and CO_2^+), which have only been observed to be dominant in test particles when sampling di- and polycarboxylic acids. The AMS has also been used in several US air quality studies (e.g. Middlebrook *et al.*, 2003). In background air, the particle composition has been modified. For example, during ACE-ASIA it was found that the submicron particulate mass was dominated by accumulation mode acidified sulphate and organic material, whose mass spectral fingerprint was again dominated by the $CO_2{}^+$ fragment. Nitrate is very much reduced and is possibly organic in nature and the components appear to be internally mixed (Topping *et al.*, 2004). Comparisons between the AMS and offline techniques or bulk online submicron species mass loading have shown good agreement between nitrate, sulphate and organic fractions in many situations (e.g. Drewnick *et al.*, 2003).

Ice nucleation is an extremely important phenomenon in the atmosphere, as liquid droplets will not freeze without the presence of a seed until temperatures as cold as

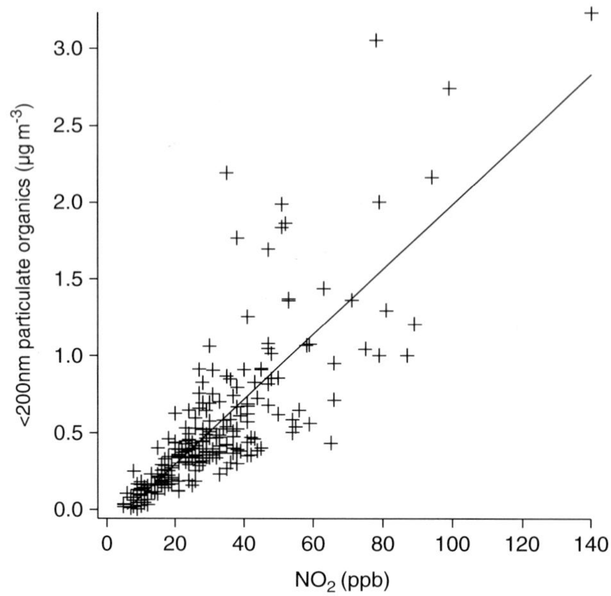

Figure 6.11 A regression analysis of NO_x and the organic mass loading in particles less than 200 nm aerodynamic diameter measured by an AMS in central Manchester, UK. There is a high degree of correlation demonstrating that the source of small organic particles is dominated by traffic. These data were used to calculate an emission rate for primary organic mass from motor vehicles in Manchester of 0.45 tonnes $km^{-2}year^{-1}$ based on an emission ratio from the National Atmospheric Emissions Inventory. This compares with the PM_{10} emission rate from transport of 1.38 tonnes $km^{-2}year^{-1}$ (Allan *et al.*, 2003b).

−40°C. The presence of ice nuclei increases the probability of freezing at temperatures as high as −5 to 10°C. However, the number of ice nuclei in the atmosphere is poorly known and their composition is highly uncertain. As ice nuclei often control the ice crystal concentration in glaciated clouds, dictating the radiative properties of the cloud and its precipitation efficiency, the properties of ice nuclei are extremely important. Single particle mass spectrometric studies are beginning to make a real contribution in this area. Cziczo *et al.* (2004) have used the PALMS instrument to investigate ice nuclei both in the laboratory and in the field and have shown that particles containing organic material are found preferentially in the droplet phase not the ice phase, indicating that organics may inhibit ice formation. If that is the case then this has a significant impact on the way cold clouds will behave towards pollution aerosol.

Aerosol mass spectrometers have now been flown on aircraft to address several very important questions. A version of the PALMS instrument was fitted to a WB-57F high altitude research aircraft by Thomson *et al.* (2000) and is shown in Figure 6.12. The instrument was mounted into the nose cone to avoid sampling artefacts when entraining aerosol into the fuselage of a rapidly moving aircraft, with its inlet facing forward of the aircraft in undisturbed flow. The instrument design had to be capable of withstanding operation in an environment without ambient pressure control on a platform capable of reaching altitudes of 19 km and so probe the stratosphere. As shown in Figure 6.13,

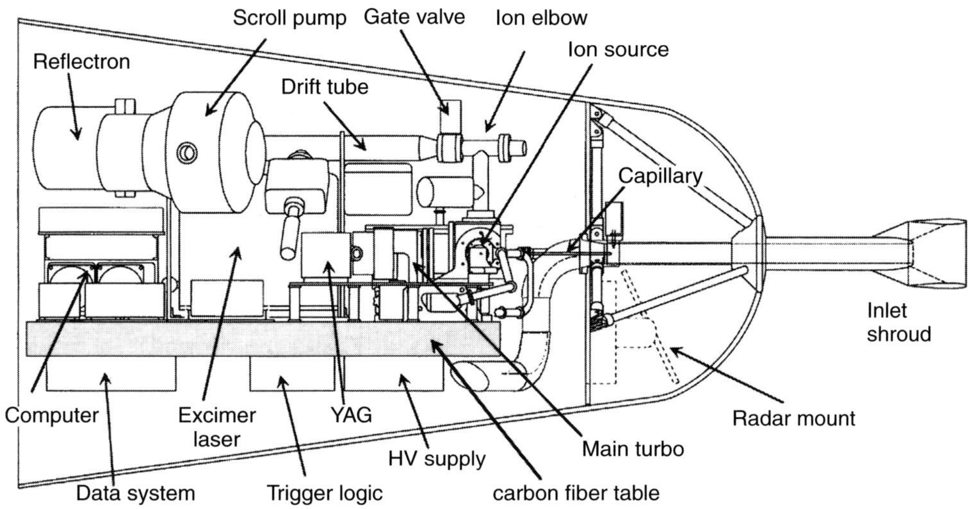

Figure 6.12 A schematic of the PALMS instrument fitted into the nose cone of the WB-57F aircraft. (Thomson *et al.*, 2000).

Murphy *et al.* (1998b) used this instrument to characterise stratospheric aerosol and showed that in the upper troposphere very often there are more organic containing particles than those containing sulphate. In the lower stratosphere they showed that most of the aerosols were made of sulphate–water mixtures. Signatures of meteoritic material were seen in the lower stratospheric aerosol, which contained over 45 different elements, demonstrating the complexity of atmospheric aerosol.

Aerosol mass spectrometry has also been used to probe polar stratospheric clouds. Schreiner *et al.* (2002) developed a balloon borne mass spectrometer (the ACMS) to measure the composition of PSCs in the arctic stratosphere. They showed that within the polar vortex high backscatter ratios, indicating the presence of PSCs, correlated with high particle water and nitric acid signals, the ratios of which indicated the presence of drops of ternary solution rather than solid hydrate formation.

At the beginning of the twenty-first century an Aerodyne AMS system was deployed on an aircraft by Bahreini *et al.* (2003). The instrument was installed on a Twin Otter aircraft and made a series of low- to mid-level flights in the troposphere to investigate the structure of highly polluted layers advected from southeast Asia during the Aerosol Characterisation Experiment – Asia (ACE-Asia). The data often showed several distinct and highly polluted layers over the lower 4 km of the atmosphere that were rich in both sulphate and organic particulate material. The loadings of these components exceeded $10\,\mu g\,m^{-3}$ on some occasions and arose in one broad mode with a central mode vacuum aerodynamic diameter of around 400–500 nm. These data and those from subsequent missions are being obtained at sufficiently high time resolution that it is now possible to obtain radiative closure estimates along the flight track. This level of detail has never before been possible and opens up routes to help address the current challenges facing the scientific community to reduce the uncertainty surrounding the impact of aerosols and clouds on climate.

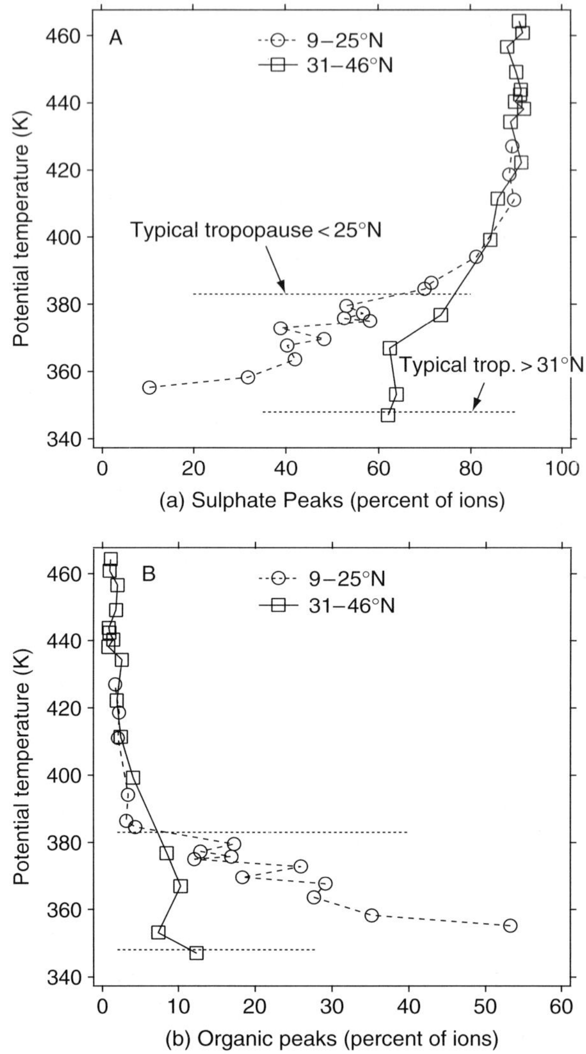

Figure 6.13 Average profiles showing the percentage of the total ion current measured from a single particle that can be demonstrated to be sulphate (A) and organic (B). The data are shown in potential temperature coordinates and show that the organic containing particles are very significant in the upper troposphere, especially in the tropics, whereas lower stratospheric aerosols are principally composed of sulphate (Murphy *et al.*, 1998b).

6.7 Summary

The technology of aerosol mass spectrometry has developed extremely quickly over the last few years and since the mid-1990s has moved from a laboratory demonstration of feasibility to a wide range of field instruments capable of adding significant and unique information to large experiments. The instruments are not only being targeted at novel problems such as polar stratospheric cloud composition, stratospheric aerosols and ice nucleation but are also being used extensively in regional and local source apportionment studies. The quantitative instruments are being used to investigate aerosol processing and mass transfer rates, in particular the organic fraction. The single particle instruments, often utilising laser ablation/ionisation, offer single particle detection capability and can observe refractory material. However, though they can quantify size, mass quantification remains elusive. The advances made at the time of writing applying a two step desorption-ionisation process using a dual laser approach may help to improve the situation significantly in this area; however, at present this has not been demonstrated routinely. In contrast, the thermal ablation systems, such as the AMS have been developed to quantitatively measure non-refractory material as a function of particle size but at present these instruments cannot identify mass loadings and mixing state of single particles, though by combining such instruments with TOF mass spectrometer technology this should be feasible in the future. These instruments are also capable of utilising other forms of ionisation methods that are less harsh than electron impact ionisation used currently. Implementation of techniques such as photo- or chemi-ionisation will lead to less fragmentation and clearer identification of individual compounds. However, the thermal ablation methods will largely be restricted to non-refractory material, though in certain circumstances, surface ionisation can be useful for determining information of certain ions, in particular the alkali metals.

No instrument so far developed can claim to deliver explicit, complete, quantitative physiochemical data for all atmospheric aerosol particles and aerosol mass spectrometers are no exception to this. Because of this, the science of analysing aerosol data is most effective when the results from many instruments are properly combined within the context and goals of the study being performed. Bulk sampling and analysis still usually provides the most authoritative data on overall composition, due to the breadth of well-established analytical tools available, and the dedicated sizing instruments are still considered the most reliable for generating number and size data. However, as described before, the aerosol mass spectrometry methods offer many valuable insights that were hitherto impossible, so by including their greatly enhanced size and time resolutions, benefits of in situ analysis and abilities to extensively study individual particles, a vastly more accurate and detailed picture of aerosol climatology can be obtained in a given scenario, if the data are used correctly.

Further reading

Seinfeld, J.H. & S.N. Pandis (1998) *Atmospheric Chemistry and Physics: From Air Pollution to Climate Change*, John Wiley & Sons, New York, Chichester.

Suess, D.T. and K.A. Prather (1999) Mass spectrometry of aerosols, *Chem. Rev.*, 99 (10), 3007–3035.

McMurry, P.H. (2000) A review of atmospheric aerosol measurements, *Atmos. Environ.*, 34 (12–14), 1959–1999.

Noble, C.A. and K.A. Prather (2000) Real-time single particle mass spectrometry: A historical review of a quarter century of the chemical analysis of aerosols, *Mass Spectrom. Rev.*, 19 (4), 248–274.

Baron, P.A. & K. Willeke (2001) *Aerosol Measurement: Principles, Techniques, and Applications*, 2nd Edition, Wiley-Interscience, New York, Chichester.

Aerosol Science & Technology, 33 (1–2) (Special Issue), Taylor & Francis, 2000.
Jimenez, J.L. *Aerosol Mass Spectrometry Web Page*, http://cires.colorado.edu/jimenez/ams.html

References

Adachi, M., Kousaka, Y. & Okuyama, K., (1985) Unipolar and bipolar diffusion charging of ultrafine aerosol-particles, *J. Aerosol Sci.*, 16 (2), 109–123.

Alfarra, M.R., Coe, H., Allan, J.D., Bower, K.N., Boundries, H., Canagaratna, M.R., Jimenez, J.L., Jayne, J.T., Garforth, A., Li, S. & Worsnop, D.R. (2004) Characterization of urban and rural organic particulate in the Lower Fraser Valley using two aerodyne aerosol mass spectrometers, *Atmos. Environ.*, 38 (34), 5745–5758.

Allan, J.D., Jimenez, J.L., Williams, P.I., Alfarra, M.R., Bower, K.N., Jayne, J.T., Coe, H. & Worsnop, D.R. (2003a) Quantitative sampling using an Aerodyne aerosol mass spectrometer – 1. Techniques of data interpretation and error analysis, *J. Geophys. Res.-Atmos.*, 108 (D3), 4090.

Allan, J.D., Alfarra, M.R., Bower, K.N., Williams, P.I., Gallagher, M.W., Jimenez, J.L., McDonald, A.G., Nemitz, E., Canagaratna, M.R., Jayne, J.T., Coe, H. & Worsnop, D.R. (2003b) Quantitative sampling using an Aerodyne aerosol mass spectrometer – 2. Measurements of fine particulate chemical composition in two U.K. cities, *J. Geophys. Res.-Atmos.*, 108 (D3), 4091.

Allan, J.D., Coe, H., Bower, K.N., Alfarra, M.R., Delia, A.E., Jimenez, J.L., Middlebrook, A.M., Drewnick, F., Onasch, T.B., Canagaratna, M.R., Jayne, J.T. & Worsnop, D.R. (2004a) A generalised method for the extraction of chemically resolved mass spectra from Aerodyne aerosol mass spectrometer data, *J. Aerosol Sci.*, 35 (7), 909–922.

Allan, J.D., Bower, K.N., Coe, H., Boudries, H., Jayne, J.T., Canagaratna, M.R., Millet, D.B., Goldstein, A.H., Quinn, P.K., Weber, R.J. & Worsnop, D.R. (2004b) Submicron aerosol composition at Trinidad Head, California, during ITCT 2K2: Its relationship with gas phase volatile organic carbon and assessment of instrument performance, *J. Geophys. Res.-Atmos.*, 109 (D23), D23S24.

Allen, J. & Gould, R.K. (1981) Mass-spectrometric analyzer for individual aerosol-particles, *Rev. Sci. Instrum.*, 52 (6), 804–809.

Allen, J.O., Fergenson, D.P., Gard, E.E., Hughes, L.S., Morrical, B.D., Kleeman, M.J., Gross, D.S., Galli, M.E., Prather, K.A. & Cass, G.R. (2000) Particle detection efficiencies of aerosol time of flight mass spectrometers under ambient sampling conditions, *Environ. Sci. Technol.*, 34 (1), 211–217.

Allen, M.D. & Raabe, O.G. (1982) Re-evaluation of Millikan's oil drop data for the motion of small particles in air, *J. Aerosol Sci.*, 13 (6), 537–547.

Arnold, R.J. & Reilly, J.P. (1998) Fingerprint matching of E-coli strains with matrix-assisted laser desorption ionization time-of-flight mass spectrometry of whole cells using a modified correlation approach, *Rapid Commun. Mass Spectrom.*, 12 (10), 630–636.

Bahreini, R., Jimenez, J.L., Wang, J., Flagan, R.C., Seinfeld, J.H., Jayne, J.T. & Worsnop, D.R. (2003) Aircraft-based aerosol size and composition measurements during ACE-Asia using an aerodyne aerosol mass spectrometer, *J. Geophys. Res. Lett.*, 108(D23), 8645.

Bhave, P.V., Fergenson, D.P., Prather, K.A. & Cass, G.R. (2001) Source apportionment of fine particulate matter by clustering single-particle data: Tests of receptor model accuracy, *Environ. Sci. Technol.*, 35 (10), 2060–2072.

Bhave, P.V., Allen, J.O., Morrical, B.D., Fergenson, D.P., Cass, G.R. & Prather, K.A. (2002) A field-based approach for determining ATOFMS instrument sensitivities to ammonium and nitrate, *Environ. Sci. Technol.*, 36 (22), 4868–4879.

Birch, M.E. & Cary, R.A. (1996) Elemental carbon-based method for monitoring occupational exposures to particulate diesel exhaust, *Aerosol Sci. Technol.*, 25 (3), 221–241.

Biskos, G., Mastorakos, E. & Collings, N. (2004) Monte-Carlo simulation of unipolar diffusion charging for spherical and non-spherical particles, *J. Aerosol Sci.*, 35 (6), 707–730.

Bley, W.G. (1988) Quantitative measurements with quadrupole mass spectrometers – Important specifications for reliable measurements, *Vacuum*, 38 (2), 103–109.

Boudries, H., Canagaratna, M.R., Jayne, J.T., Alfarra, M.R., Allan, J.D., Coe, H., Pryor, S.C., Jimenez, J.L., Brook, J.R., Li, S. & Worsnop, D.R. (2004) Chemical and physical processes controlling the distribution of aerosols in the Lower Fraser Valley, Canada, during the Pacific 2001 field campaign, *Atmos. Environ.*, 38 (34), 5759–5774.

Brockmann, J.E. & Rader, D.J. (1990) APS Response to nonspherical particles and experimental-determination of dynamic shape factor, *Aerosol Sci. Technol.*, 13 (2), 162–172.

Buzorius, G., Zelenyuk, A., Brechtel, F. & Imre, D. (2002) Simultaneous determination of individual ambient particle size, hygroscopicity and composition, *Geophys. Res. Lett.*, 29 (20), 1974.

Cabalo, J., Zelenyuk, A., Baer, T. & Miller, R.E. (2000) Two-color laser induced evaporation dynamics of liquid aerosols probed by time-of-flight mass spectrometry, *Aerosol Sci. Technol.*, 33 (1–2), 3–19.

Canagaratna, M.R., Jayne, J.T., Ghertner, D.A., Herndon, S., Shi, Q., Jimenez, J.L., Silva, P.J., Williams, P., Lanni, T., Drewnick, F., Demerjian, K.L., Kolb, C.E. & Worsnop, D.R. (2004) Chase studies of particulate emissions from in-use New York City vehicles, *Aerosol Sci. Technol.*, 38 (6), 555–573.

Carpenter, G.A., Grossberg, S. & Rosen, D.B. (1991) Art 2-A – An adaptive resonance algorithm for rapid category learning and recognition, *Neural Netw.*, 4 (4), 493–504.

Carson, P.G., Neubauer, K.R., Johnston, M.V. & Wexler, A.S. (1995) Online chemical-analysis of aerosols by rapid single-particle mass-spectrometry, *J. Aerosol Sci.*, 26 (4), 535–545.

Carson, P.G., Johnston, M.V. & Wexler, A.S. (1997a) Laser desorption/ionization of ultrafine aerosol particles, *Rapid Commun. Mass Spectrom.*, 11 (9), 993–996.

Carson, P.G., Johnston, M.V. & Wexler, A.S. (1997b) Real-time monitoring of the surface and total composition of aerosol particles, *Aerosol Sci. Technol.*, 26 (4), 291–300.

Chen, B.T., Cheng, Y.S. & Yeh, H.C. (1985) Performance of a TSI aerodynamic particle sizer, *Aerosol Sci. Technol.*, 4 (1), 89–97.

Chen, Y.H., Gonin, M., Fuhrer, K., Dodonov, A., Su, C.S. & Wollnik, H. (1999) Orthogonal electron impact source for a time-of-flight mass spectrometer with high mass resolving power, *Int. J. Mass Spectrom.*, 185–187, 221–226.

Chow, J.C. (1995) Measurement methods to determine compliance with ambient air-quality standards for suspended particles, *J. Air Waste Manag. Assoc.*, 45 (5), 320–382.

Coggiola, M.J., Shi, Z. & Young, S.E. (2000) Airborne deployment of an instrument for the real-time analysis of single aerosol particles, *Aerosol Sci. Technol.*, 33 (1–2), 20–29.

Colvile, R.N., Hutchinson, E.J., Mindell, J.S. & Warren, R.F. (2001) The transport sector as a source of air pollution, *Atmos. Environ.*, 35 (9), 1537–1565.

Cunningham, E. (1910) On the velocity of steady fall of spherical particles through fluid medium, *Proceedings of the Royal Society of London. Series A-Containing Papers of a Mathematical and Physical Character*, 83 (563), 357–365.

Cziczo, D.J., DeMott, P.J., Brooks, S.D., Prenni, A.J., Thomson, D.S., Baungardner, D., Wilson, J.C., Freideniweis, S.M. & Murphy, D.M. (2004) Observations of organic species and atmospheric ice formation, *Geophys. Res. Lett.*, 31 (12), 42 116.

Dall'Osto, M., Beddows, D.C.S., Kinnersley, R.P., Harrison, R.M., Donovan, R.J. & Heal, M.R. (2004) Characterization of individual airborne particles by using aerosol time-of-flight mass spectrometry at Mace Head, Ireland, *J. Geophys. Res.-Atmos.*, 109 (D21), D21302.

Davidsson, K.O., Engvall, K., Hagström, M., Korsgren, J.G., Lönn, B. & Pettersson, J.B.C. (2002) A surface ionization instrument for on-line measurements of alkali metal components in combustion: Instrument description and applications, *Energy & Fuels*, 16 (6), 1369–1377.

Davis, W.D. (1973) Surface ionization mass spectroscopy of airborne particulates, *J. Vacuum Sci. Technol.*, 10 (1), 278.

DeCarlo, P.F., Slowik, J.G., Worsnop, D.R., Davidovits, P. & Jimenez, J.L. (2004) Particle morphology and density characterization by combined mobility and aerodynamic diameter measurements. Part 1: Theory, *Aerosol Sci. Technol.*, 38 (12), 1185–1205.

Decesari, S., Facchini, M.C., Fuzzi, S. & Tagliavini, E. (2000) Characterization of water-soluble organic compounds in atmospheric aerosol: A new approach, *J. Geophys. Res.-Atmos.*, 105 (D1), 1481–1489.

Dockery, D.W., Pope, C.A., Xu, X.P., Spengler, J.D., Ware, J.H., Fay, M.E., Ferris, B.G. & Speizer, F.E. (1993) An association between air-pollution and mortality in 6 United States cities, *N. Engl. J. Med.*, 329 (24), 1753–1759.

Drewnick, F., Schwab, J.J., Hogrefe, O., Peters, S., Husain, L., Diamond, D., Weber, R. & Demerjian, K.L. (2003) Intercomparison and evaluation of four semi-continuous PM2.5 sulfate instruments, *Atmos. Environ.*, 37 (24), 3335–3350.

Drewnick, F., Hings, S.S., DeCarlo, P., Jayne, J.T., Gonin, M., Fuhrer, K., Weimer, S., Jimenez, J.L., Demerjian, K.L., Borrmann, S. and Worsnop, D.R. (2005) A new time-of-flight aerosol mass spectrometer (TOF-AMS) instrument description and first field deployment, *Aerosol Sci. Technol.*, 39 (7), 637–658.

Dye, A.L., Rhead, M.M. & Trier, C.J. (2000) The quantitative morphology of roadside and background urban aerosol in Plymouth, UK, *Atmos. Environ.*, 34 (19), 3139–3148.

Fergenson, D.P., Song, X.H., Ramadan, Z., Allen, J.O., Hughes, L.S., Cass, G.R., Hopke, P.K. & Prather, K.A. (2001) Quantification of ATOFMS data by multivariate methods, *Anal. Chem.*, 73 (15), 3535–3541.

Flagan, R. (2004) Opposed migration aerosol classifier (OMAC), *Aerosol Sci. Technol.*, 38 (9), 890–899.

Flagan, R.C. (1998) History of electrical aerosol measurements, *Aerosol Sci. Technol.*, 28 (4), 301–380.

Fuchs, N. (1963) On the stationary charge distribution on aerosol particles in a bipolar ionic atmosphere, *Geofisica pura e applicata*, 56, 185–193.

Fuchs, N.A. (1964) Steady rectilinear motion of aerosol particles, in *The Mechanics of Aerosols: Revised and Enlarged Edition*, pp. 21–69, Pergamon Press, Oxford; London; Edinburgh; New York; Paris; Frankfurt.

Gard, E., Mayer, J.E., Morrical, B.D., Dienes, T., Fergenson, D.P. & Prather, K.A. (1997) Real-time analysis of individual atmospheric aerosol particles: Design and performance of a portable ATOFMS, *Anal. Chem.*, 69 (20), 4083–4091.

Ge, Z.Z., Wexler, A.S. & Johnston, M.V. (1998) Laser desorption/ionization of single ultrafine multicomponent aerosols, *Environ. Sci. Technol.*, 32 (20), 3218–3223.

Gieray, R.A., Reilly, P.T.A., Yang, M., Whitten, W.B. & Ramsey, J.M. (1998) Tandem mass spectrometry of uranium and uranium oxides in airborne particulates, *Anal. Chem.*, 70 (1), 117–120.

Grix, R., Grüner, U., Li, G., Stroh, H. & Wollnik, H. (1989) An electron-impact storage ion-source for time-of-flight mass spectrometers, *Int. J. Mass Spectrom. Ion Processes*, 93 (3), 323–330.

Gross, D.S., Galli, M.E., Silva, P.J. & Prather, K.A. (2000) Relative sensitivity factors for alkali metal and ammonium cations in single particle aerosol time-of-flight mass spectra, *Anal. Chem.*, 72 (2), 416–422.

Guazzotti, S.A., Suess, D.T., Coffee, K.R., Quinn, P.K., Bates, T.S., Wisthalu, A., Hansel, A., Ball, W.P., Dickerson, R.R., Neususs, C., Crutzen, P.J. & Prathee, K.A. (2003) Characterization of carbonaceons aerosols outflow from India and Arabia: Biomass/biofuel burning and fossil fuel combustion, *J. Geophys. Res. Atmos.*, 108 (D15), 4485.

Harrison, R.M. & Yin, J.X. (2000) Particulate matter in the atmosphere: Which particle properties are important for its effects on health?, *Sci. Total Environ.*, 249 (1–3), 85–101.

Hearn, J.D. & Smith, G.D. (2004) A chemical ionization mass spectrometry method for the online analysis of organic aerosols, *Anal. Chem.*, 76 (10), 2820–2826.

Hinz, K.P., Kaufmann, R. & Spengler, B. (1994) Laser-induced mass analysis of single particles in the airborne state, *Anal. Chem.*, 66 (13), 2071–2076.

Hinz, K.P., Kaufmann, R. & Spengler, B. (1996) Simultaneous detection of positive and negative ions from single airborne particles by real-time laser mass spectrometry, *Aerosol Sci. Technol.*, 24 (4), 233–242.

Hinz, K.P., Greweling, M., Drews, F. & Spengler, B. (1999) Data processing in on-line laser mass spectrometry of inorganic, organic, or biological airborne particles, *J. Am. Soc. Mass Spectrom.*, 10 (7), 648–660.

Hoffmann, T., Bandur, R., Hoffmann, S. & Warscheid, B. (2002) On-line characterization of gaseous and particulate organic analytes using atmospheric pressure chemical ionization mass spectrometry, *Spectrochimica Acta Part B-Atomic Spectroscopy*, 57 (10), 1635–1647.

Hunt, A.L. & Petrucci, G.A. (2002) Analysis of ultrafine and organic particles by aerosol mass spectrometry, *Trac-Trends Anal. Chem.*, 21 (2), 74–81.

Jacobson, M.C., Hansson, H.C., Noone, K.J. & Charlson, R.J. (2000) Organic atmospheric aerosols: Review and state of the science, *Rev. Geophys.*, 38 (2), 267–294.

Jäglid, U., Olsson, J.G. & Pettersson, J.B.C. (1996) Detection of sodium and potassium salt particles using surface ionization at atmospheric pressure, *J. Aerosol Sci.*, 27 (6), 967–977.

Jayne, J.T., Leard, D.C., Zhang, X.F., Davidovits, P., Smith, K.A., Kolb, C.E. & Worsnop, D.R. (2000) Development of an aerosol mass spectrometer for size and composition analysis of submicron particles, *Aerosol Sci. Technol.*, 33 (1–2), 49–70.

Jimenez, J.L., Jayne, J.T., Shi, Q., Kolb, C.E., Worsnop, D.R., Yourshaw, I., Seinfeld, J.H., Flagan, R.C., Zhang, X., Smith, K.A., Morris, J.W. & Davidovits, P. (2003a) Ambient aerosol sampling using the aerodyne aerosol mass spectrometer, *J. Geophys. Res.-Atmos.*, 108 (D7), 8425.

Jimenez, J.L., Bahreini, R., Cocker, D.R., Zhuang, H., Varutbangkul, V., Flagan, R.C., Seinfeld, J.H., O'Dowd, C.D. & Hoffmann, T. (2003b) New particle formation from photooxidation of diiodomethane (CH2I2), *J. Geophys. Res.-Atmos.*, 108 (D10), 4318.

Jimenez, J.L., Bahreini, R., Cocker, D.R., Zhuang, H., Varutbangkul, V., Flagan, R.C., Seinfeld, J.H., O'Dowd, C.D. & Hoffmann, T. (2003c) Correction to "New particle formation from photooxidation of diiodomethane (CH2I2)", *J. Geophys. Res.-Atmos.*, 108 (D23), 4733.

Johnston, M.V. (2000) Sampling and analysis of individual particles by aerosol mass spectrometry, *J. Mass Spectrom.*, 35 (5), 585–595.

Kane, D.B. & Johnston, M.V. (2000) Size and composition biases on the detection of individual ultrafine particles by aerosol mass spectrometry, *Environ. Sci. Technol.*, 34 (23), 4887–4893.

Kane, D.B., Oktem, B. & Johnston, M.V. (2001) Nanoparticle detection by aerosol mass spectrometry, *Aerosol Sci. Technol.*, 34 (6), 520–527.

Kelly, W.P. & McMurry, P.H. (1992) Measurement of particle density by inertial classification of differential mobility analyzer generated monodisperse aerosols, *Aerosol Sci. Technol.*, 17 (3), 199–212.

Khlystov, A., Wyers, G.P. & Slanina, J. (1995) The steam-jet aerosol collector, *Atmos. Environ.*, 29 (17), 2229–2234.

Kinsel, G.R. & Johnston, M.V. (1989) Post source pulse focusing – A simple method to achieve improved resolution in a time-of-flight mass-spectrometer, *Int. J. Mass Spectrom. Ion Processes*, 91 (2), 157–176.

Knutson, E.O. & Whitby, K.T. (1975) Aerosol classifcation by electric mobility: Apparatus, theory, and application, *J. Aerosol Sci.*, 6 (6), 443–451.

Kulmala, M. (2003) How particles nucleate and grow, *Science*, 302 (5647), 1000–1001.

Kulmala, M., Vehkamaki, H., Petajda, T., Dal Maso, M., Lauri, A., Kerminen, V.M., Birmili, W. & McMurry, P.H. (2004) Formation and growth rates of ultrafine atmospheric particles: A review of observations, *J. Aerosol Sci.*, 35 (2), 143–176.

LaFranchi, B.W. & Petrucci, G.A. (2004) Photoelectron resonance capture ionization (PERCI): A novel technique for the soft-ionization of organic compounds, *J. Am. Soc. Mass Spectrom.*, 15 (3), 424–430.

LaFranchi, B.W., Zahardis, J. & Petrucci, G.A. (2004) Photoelectron resonance capture ionization mass spectrometry: A soft ionization source for mass spectrometry of particle-phase organic compounds, *Rapid Commun. Mass Spectrom.*, 18 (21), 2517–2521.

Lake, D.A., Tolocka, M.P., Johnston, M.V. & Wexler, A.S. (2003) Mass spectrometry of individual particles between 50 and 750 nm in diameter at the Baltimore supersite, *Environ. Sci. Technol.*, 37 (15), 3268–3274.

Lazar, A., Reilly, P.T.A., Whitten, W.B. & Ramsey, J.M. (1999) Real-time surface analysis of individual airborne environmental particles, *Environ. Sci. Technol.*, 33 (22), 3993–4001.

Lazar, A.C., Reilly, P.T.A., Whitten, W.B. & Ramsey, J.M. (2000) Laser desorption/in situ chemical ionization aerosol mass spectrometry for monitoring tributyl phosphate on the surface of environmental particles, *Anal. Chem.*, 72 (9), 2142–2147.

Li, J., Anderson, J.R. & Buseck, P.R. (2003) TEM study of aerosol particles from clean and polluted marine boundary layers over the North Atlantic, *J. Geophys. Res.-Atmos.*, 108 (D6), 4189.

Liu, B.Y.H. & Pui, D.Y.H. (1974) Electrical neutralization of aerosols *J. Aerosol Sci.*, 5 (5), 465–472.

Liu, P., Ziemann, P.J., Kittelson, D.B. & McMurry, P.H. (1995a) Generating particle beams of controlled dimensions and divergence: I. Theory of particle motion in aerodynamic lenses and nozzle expansions, *Aerosol Sci. Technol.*, 22 (3), 293–313.

Liu, P., Ziemann, P.J., Kittelson, D.B. & McMurry, P.H. (1995b) Generating particle beams of controlled dimensions and divergence: II. Experimental evaluation of particle motion in aerodynamic lenses and nozzle expansions, *Aerosol Sci. Technol.*, 22 (3), 314–324.

Mader, B.T., Schauer, J.J., Seinfeld, J.H., Flagan, R.C., Yu, J.Z., Yang, H., Lim, H.J., Turpin, B.J., Deminter, J.T., Heidemann, G., Bae, M.S., Quinn, P., Bate, T., Eatough, D.J., Huebert, B.J., Bertram, T. & Howell, S. (2003) Sampling methods used for the collection of particle-phase organic and elemental carbon during ACE-Asia, *Atmos. Environ.*, 37 (11), 1435–1449.

Mahadevan, R., Lee, D., Sakurai, H. & Zachariah, M.R. (2002) Measurement of condensed-phase reaction kinetics in the aerosol phase using single particle mass spectrometry, *J. Phys. Chem. A*, 106 (46), 11 083–11 092.

Mallina, R.V., Wexler, A.S., Rhoads, K.P. & Johnston, M.V. (2000) High speed particle beam generation: A dynamic focusing mechanism for selecting ultrafine particles, *Aerosol Sci. Technol.*, 33 (1–2), 87–104.

Mansoori, B.A., Johnston, M.V. & Wexler, A.S. (1996) Matrix-assisted laser desorption/ionization of size- and composition selected aerosol particles, *Anal. Chem.*, 68 (20), 3595–3601.

Mayer, H. (1999) Air pollution in cities, *Atmos. Environ.*, 33 (24–25), 4029–4037.

McFiggans, G.B., Coe, H., Burgess, R., Allan, J.D., Cubison, M., Alfarra, M.R., Saunders, R., Saiz-Lopez, A., Plane, J.M.C., Wevill, D., Carpenter, L., Rickard, A.R. & Monks, P.S. (2004) Direct evidence for coastal iodine particles from Laminaria macroalgae – Linkage to emissions of molecular iodine, *Atmos. Chem. Phys.*, 4 (3), 701–713.

McKeown, P.J., Johnston, M.V. & Murphy, D.M. (1991) Online single-particle analysis by laser desorption mass-spectrometry, *Anal. Chem.*, 63 (18), 2069–2073.

McLafferty, F.W. & Turecek, F. (1993) *Interpretation of Mass Spectra*, 4th edition, University Science Books, Mill Valley, CA.

McMurry, P.H., Litchy, M., Huang, P.F., Cai, X.P., Turpin, B.J., Dick, W.D. & Hanson, A. (1996) Elemental composition and morphology of individual particles separated by size and hygroscopicity with the TDMA, *Atmos. Environ.*, 30 (1), 101–108.

McMurry, P.H., Wang, X., Park, K. & Ehara, K. (2002) The relationship between mass and mobility for atmospheric particles: A new technique for measuring particle density, *Aerosol Sci. Technol.*, 36 (2), 227–238.

Middlebrook, A.M., Murphy, D.M., Lee, S.H., Thomson, D.S., Prather, K.A., Wenzel, R.J., Liu, D.Y., Phares, D.J., Rhoads, K.P., Wexler, A.S., Johnston, M.V., Jimenez, J.L., Jayne, J.T., Worsnop, D.R., Yourshaw, I., Seinfeld, J.H. & Flagan, R.C. (2003) A comparison of particle mass spectrometers during the 1999 Atlanta Supersite Project, *J. Geophys. Res.-Atmos.*, 108 (D7), 8424.

Mie, G. (1908) Beitraege zur Optik trueber Medien, speziell kolloidaler Metalosungen, *Annales de Physik*, 25, 377–445.

Morrical, B.D., Fergenson, D.P. & Prather, K.A. (1998) Coupling two-step laser desorption/ionization with aerosol time-of-flight mass spectrometry for the analysis of individual organic particles, *J. Am. Soc. Mass Spectrom.*, 9 (10), 1068–1073.

Murphy, D.M. & Thomson, D.S. (1995) Laser ionization mass-spectroscopy of single aerosol-particles, *Aerosol Sci. Technol.*, 22 (3), 237–249.

Murphy, D.M., Thomson, D.S. & Middlebrook, A.M. (1997) Bromine, iodine, and chlorine in single aerosol particles at Cape Grim, *Geophys. Res. Lett.*, 24 (24), 3197–3200.

Murphy, D.M., Thomson, D.S., Middlebrook, A.M. & Schein, M.E. (1998a) In situ single-particle characterization at Cape Grim, *J. Geophys. Res. Lett.*, 103 (D13), 16 485–16 491.

Murphy, D.M., Thomson, D.S. & Mahoney, T.M.J. (1998b) In situ measurements of organics, meteoritic material, mercury, and other elements in aerosols at 5 to 19 kilometers, *Science*, 282 (5394), 1664–1669.

Murphy, D.M., Middlebrook, A.M., & Warshawsky, M. (2003) Cluster analysis of data from the Particle Analysis by Laser Mass Spectrometry (PALMS) instrument, *Aerosol Sci. Technol.*, 37 (4), 382–391.

Murphy, D.M., Cziczo, D.J., Hudson, P.K., Schein, M.E. & Thomson, D.S. (2004) Particle density inferred from simultaneous optical and aerodynamic diameters sorted by composition, *J. Aerosol Sci.*, 35 (1), 135–139.

Myers, R.L. & Fite, W.L. (1975) Electrical detection of airborne particulates using surface ionization techniques, *Environ. Sci. Technol.*, 9 (4), 334–336.

Neubauer, K.R., Johnston, M.V. & Wexler, A.S. (1998) Humidity effects on the mass spectra of single aerosol particles, *Atmos. Environ.*, 32 (14–15), 2521–2529.

Nier, A.O. & Schlutter, D.J. (1985) High-performance double-focusing mass-spectrometer, *Rev. Sci. Instrum.*, 56 (2), 214–219.

NIST Mass Spec Data Center (Director S.E. Stein) (2003) Mass Spectra, in *NIST Chemistry WebBook, NIST Standard Reference Database Number 69*, P.J. Linstrom & W.G. Mallard (eds), National Institute of Standards and Technology, Gaithersburg, MD.

Noble, C. & Prather, K. (1998) Air pollution: The role of particles, *Physics World*, 11 (1), 39–43.

Öktem, B., Tolocka, M.P. & Johnston, M.V. (2004) On-line analysis of organic components in fine and ultrafine particles by photoionization aerosol mass spectrometry, *Anal. Chem.*, 76 (2), 253–261.

Orsini, D.A., Ma, Y.L., Sullivan, A., Sierau, B., Baumann, K. & Weber, R.J. (2003) Refinements to the particle-into-liquid sampler (PILS) for ground and airborne measurements of water soluble aerosol composition, *Atmos. Environ.*, 37 (9–10), 1243–1259.

Parker, E.P., Trahan, M.W., Wagner, J.S., Rosenthal, S.E., Whitten, W.B., Gieray, R.A., Reilly, P.T.A., Lazar, A.C. & Ramsey, J.M. (2000) Detection and classification of individual airborne microparticles using laser ablation mass spectroscopy and multivariate analysis, *Field Anal. Chem. Technol.*, 4 (1), 31–42.

Patashnick, H. & Rupprecht, E.G. (1991) Continuous PM-10 measurements using the tapered element oscillating microbalance, *J. Air Waste Manag. Assoc.*, 41 (8), 1079–1083.

Penner, J.E., Andreae, M., Annegarn, H., Barrie, L., Feichter, J., Hegg, D., Jayaraman, A., Leaitch, R., Murphy, D., Nganga, J. & Pitari, G. (2001) Aerosols, their direct and indirect effects, in *Climate Change 2001: The Scientific Basis. Contribution of Working Group I to the Third Assessment Report of the Intergovernmental Panel on Climate Change*, J.T. Houghton, Y. Ding, D.J. Griggs, M. Noguer, P.J. van der Linden, X. Dai, K. Maskell & C.A. Johnson (eds), pp. 289–348, Cambridge University Press, Cambridge, New York.

Petrucci, G.A., Farnsworth, P.B., Cavalli, P. & Omenetto, N. (2000) A differentially pumped particle inlet for sampling of atmospheric aerosols into a time-of-flight mass spectrometer: Optical characterization of the particle beam, *Aerosol Sci. Technol.*, 33 (1–2), 105–121.

Phares, D.J., Rhoads, K.P., Wexler, A.S., Kane, D.B. & Johnston, M.V. (2001) Application of the ART-2a algorithm to laser ablation aerosol mass spectrometry of particle standards, *Anal. Chem.*, 73 (10), 2338–2344.

Phares, D.J., Rhoads, K.P. & Wexler, A.S. (2002) Performance of a single ultrafine particle mass spectrometer, *Aerosol Sci. Technol.*, 36 (5), 583–592.

Pope, C.A., Burnett, R.T., Thun, M.J., Calle, E.E., Krewski, D., Ito, K. & Thurston, G.D. (2002) Lung cancer, cardiopulmonary mortality, and long-term exposure to fine particulate air pollution, *J. Am. Med. Assoc.*, 287 (9), 1132–1141.

Prather, K.A., Nordmeyer, T. & Salt, K. (1994) Real-time characterization of individual aerosol-particles using time-of-flight mass-spectrometry, *Anal. Chem.*, 66 (9), 1403–1407.

Ramanathan, V., Crutzen, P.J., Kiehl, J.T. & Rosenfeld, D. (2001) Atmosphere – Aerosols, climate, and the hydrological cycle, *Science*, 294 (5549), 2119–2124.

Ravishankara, A.R. (1997) Heterogeneous and multiphase chemistry in the troposphere, *Science*, 276 (5315), 1058–1065.

Reents, W.D. & Ge, Z.Z. (2000) Simultaneous elemental composition and size distributions of submicron particles in real time using laser atomization/ionization mass spectrometry, *Aerosol Sci. Technol.*, 33 (1–2), 122–134.

Reents, W.D. & Schabel, M.J. (2001) Measurement of individual particle atomic composition by aerosol mass spectrometry, *Anal. Chem.*, 73 (22), 5403–5414.

Reents, W.D., Downey, S.W., Emerson, A.B., Mujsce, A.M., Muller, A.J., Siconolfi, D.J., Sinclair, J.D. & Swanson, A.G. (1995) Single-particle characterization by time-of-flight mass-spectrometry, *Aerosol Sci. Technol.*, 23 (3), 263–270.

Reilly, P.T.A., Gieray, R.A., Yang, M., Whitten, W.B. & Ramsey, J.M. (1997) Tandem mass spectrometry of individual airborne microparticles, *Anal. Chem.*, 69 (1), 36–39.

Reilly, P.T.A., Lazar, A.C., Gieray, R.A., Whitten, W.B. & Ramsey, J.M. (2000) The elucidation of charge-transfer-induced matrix effects in environmental aerosols via real-time aerosol mass spectral analysis of individual airborne particles, *Aerosol Sci. Technol.*, 33 (1–2), 135–152.

Reilly, P.T.A., Harris, W.A., Wright, K.C., Whitten, W.B. & Ramsey, J.M. (2004) External to the trap vaporization and ionization for real-time quantitative particle analysis, in *23rd Annual AAAR Conference*, p. 5 (1B2), Hyatt Regency Atlanta, Atlanta, GA.

Salt, K., Noble, C.A. & Prather, K.A. (1996) Aerodynamic particle sizing versus light scattering intensity measurement as methods for real time particle sizing coupled with time-of-flight mass spectrometry, *Anal. Chem.*, 68 (1), 230–234.

Saxena, P. & Hildemann, L.M. (1996) Water-soluble organics in atmospheric particles: A critical review of the literature and application of thermodynamics to identify candidate compounds, *J. Atmos. Chem.*, 24 (1), 57–109.

Schreiner, J., Voigt, C., Mauersberger, K., McMurry, P. & Ziemann, P. (1998) Aerodynamic lens system for producing particle beams at stratospheric pressures, *Aerosol Sci. Technol.*, 29 (1), 50–56.

Schreiner, J., Voigt, C., Zink, P., Kohlmann, A., Knopf, D., Weisser, C., Budz, P. & Mauersberger, K. (2002) A mass spectrometer system for analysis of polar stratospheric aerosols, *Rev. Sci. Instrum.*, 73 (2), 446–452.

Schwartz, J. (1994) Air-pollution and daily mortality – A review and meta analysis, *Environ. Res.*, 64 (1), 36–52.

Seaton, A., Macnee, W., Donaldson, K. & Godden, D. (1995) Particulate air-pollution and acute health-effects, *Lancet*, 345 (8943), 176–178.

Silva, P.J. & Prather, K.A. (1997) On-line characterization of individual particles from automobile emissions, *Environ. Sci. Technol.*, 31 (11), 3074–3080.

Silva, P., Liu, D.Y., Noble, C.A. & Prather, K.A. (1999) Size and chemical characterization of individual particles resulting from liomass burning of local southern California species, *Environ. Sci. Technol.*, 33 (18), 3068–3076.

Silva, P.J. & Prather, K.A. (2000) Interpretation of mass spectra from organic compounds in aerosol time-of-flight mass spectrometry, *Anal. Chem.*, 72 (15), 3553–3562.

Sinha, M.P., Giffin, C.E., Norri, D.D., Estes, T.J., Vilker, V.L. & Friedlander, S.K. (1982) Particle analysis by mass-spectrometry, *J. Colloid Interface Sci.*, 87 (1), 140–153.

Sipin, M.F., Guazzotti, S.A. & Prather, K.A. (2003) Recent advances and some remaining challenges in analytical chemistry of the atmosphere, *Anal. Chem.*, 75 (12), 2929–2940.

Smith, J.N., Moore, K.F., McMurry, P.H. & Eisele, F.L. (2004) Atmospheric measurements of sub-20 nm diameter particle chemical composition by thermal desorption chemical ionization mass spectrometry, *Aerosol Sci. Technol.*, 38 (2), 100–110.

Song, X.H., Hopke, P.K., Fergenson, D.P. & Prather, K.A. (1999) Classification of single particles analyzed by ATOFMS using an artificial neural network, ART-2A, *Anal. Chem.*, 71 (4), 860–865.

Stolzenburg, M.R. & Hering, S.V. (2000) Method for the automated measurement of fine particle nitrate in the atmosphere, *Environ. Sci. Technol.*, 34 (5), 907–914.

Su, Y.X., Sipin, M.F., Furutani, H. & Prather, K.A. (2004) Development and characterization of an aerosol time-of-flight mass spectrometer with increased detection efficiency, *Anal. Chem.*, 76 (3), 712–719.

Svane, M., Hagstrom, M. & Pettersson, J.B.C. (2004) Chemical analysis of individual alkali-containing aerosol particles: Design and performance of a surface ionization particle beam mass spectrometer, *Aerosol Sci. Technol.*, 38 (7), 655–663.

Sykes, D.C., Woods, E., Smith, G.D., Baer T., & Miller, R.E. (2002) Thermal vaporization-vacuum ultraviolet laser ionization time-of-flight mass spectrometry of single aerosol particles, *Anal. Chem.*, 74 (9), 2048–2052.

Tammet, H., Mirme, A. & Tamm, E. (2002) Electrical aerosol spectrometer of Tartu University, *Atmos. Res.*, 62 (3–4), 315–324.

Tan, P.V., Malpica, O., Evans, G.J., Owega, S. & Fila, M.S. (2002) Chemically-assigned classification of aerosol mass spectra, *J. Am. Soc. Mass Spectrom.*, 13 (7), 826–838.

Tang, Y.H., Carmichael, G.R., Horowitz, L.W., Uno, I., Woo, J.H., Streets, D.G., Dabdub, D., Kurata, G., Sandu, A., Allan, J., Atlas, E., Flocke, F., Huey, L.G., Jakoubek, R.O., Millet, D.B., Quinn, P.K., Roberts, J.M., Worsnop, D.R., Goldstein, A., Donnelly, S., Schauffler, S., Stroud, V., Johnson, K., Avery, M.A., Singh, H.B. & Apel, E.C. (2004) Multiscale simulations of tropospheric chemistry in the eastern Pacific and on the US West Coast during spring 2002, *J. Geophys. Res.-Atmos.*, 109 (D23), D23S11.

Thomson, D.S. & Murphy, D.M. (1993) Laser-induced ion formation thresholds of aerosol-particles in a vacuum, *Appl. Opt.*, 32 (33), 6818–6826.

Thomson, D.S., Schein, M.E. & Murphy, D.M. (2000) Particle analysis by laser mass spectrometry WB-57F instrument overview, *Aerosol Sci. Technol.*, 33 (1–2), 153–169.

Tobias, H.J. & Ziemann, P.J. (1999) Compound identification in organic aerosols using temperature-programmed thermal desorption particle beam mass spectrometry, *Anal. Chem.*, 71 (16), 3428–3435.

Tobias, H.J., Kooiman, P.M., Docherty, K.S. & Ziemann, P.J. (2000) Real-time chemical analysis of organic aerosols using a thermal desorption particle beam mass spectrometer, *Aerosol Sci. Technol.*, 33 (1–2), 170–190.

Topping, D.O., Coe, H., McFiggans, G.B., Burgess, R., Allan, J.D., Alfarra, M.R., Bower, K.N., Choularton, T.W., Decesari, S. & Facchini, M.C. (2004) Aerosol chemical characteristics from sampling conducted on the island of Jeju, Korea during ACE Asia, *Atmos. Environ.*, 38 (14), 2111–2123.

Trimborn, A., Hinz, K.P. & Spengler, B. (2000) Online analysis of atmospheric particles with a transportable laser mass spectrometer, *Aerosol Sci. Technol.*, 33 (1–2), 191–201.

Voisin, D., Smith, J.N., Sakurai, H., McMurry, P.H. & Eisele, F.L. (2003) Thermal desorption chemical ionization mass spectrometer for ultrafine particle chemical composition, *Aerosol Sci. Technol.*, 37 (6), 471–475.

Wang, S.C. & Flagan, R.C. (1990) Scanning electrical mobility spectrometer, *Aerosol Sci. Technol.*, 13 (2), 230–240.

Weber, R.J., Orsini, D., Daun, Y., Lee, Y.N., Klotz, P.J. & Brechtel, F. (2001) A particle-into-liquid collector for rapid measurement of aerosol bulk chemical composition, *Aerosol Sci. Technol.*, 35 (3), 718–727.

Weickhardt, C., Moritz, F. & Grotemeyer, J. (1996) Time-of-flight mass spectrometry: State-of-the-art in chemical analysis and molecular science, *Mass Spectrom. Rev.*, 15 (3), 139–162.

Wenzel, R.J., Liu, D.Y., Edgerton, E.S. & Prather, K.A. (2003) Aerosol time-of-flight mass spectrometry during the Atlanta Supersite Experiment: 2. Scaling procedures, *J. Geophys. Res.-Atmos.*, 108 (D7), 8427.

Wenzel, R.J. & Prather, K.A. (2004) Improvements in ion signal reproducibility obtained using a homogeneous laser beam for on-line laser desorption/ionization of single particles, *Rapid Commun. Mass Spectrom.*, 18 (13), 1525–1533.

White, W.H. & Roberts, P.T. (1977) On the nature and origins of visibility-reducing aerosols in the Los Angeles air basin, *Atmos. Environ.*, 11 (9), 803–812.

Wiedensohler, A. (1988) An Approximation of the bipolar charge-distribution for particles in the sub-micron size range, *J. Aerosol Sci.*, 19 (3), 387–389.

Williams, B.J., Goldstein, A.H., Kreisberg, N.M. & Hering, S.V. (2004) Speciated organic compositions of atmospheric aerosols: A new, in-situ instrument, in *23rd Annual AAAR Conference*, p. 113 (5D4), Hyatt Regency Atlanta, Atlanta, GA.

Winklmayr, W., Reischl, G.P., Lindner, A.O. & Berner, A. (1991) A new electromobility spectrometer for the measurement of aerosol size distributions in the size range from 1 to 1000 Nm, *J. Aerosol Sci.*, 22 (3), 289–296.

Woods, E., Smith, G.D., Dessiaterik, Y., Baer, T. & Miller, R.E. (2001) Quantitative detection of aromatic compounds in single aerosol particle mass spectrometry, *Anal. Chem.*, 73 (10), 2317–2322.

Woods, E., Smith, G.D., Miller, R.E. & Baer, T. (2002) Depth profiling of heterogeneously mixed aerosol particles using single-particle mass spectrometry, *Anal. Chem.*, 74 (7), 1642–1649.

Yang, M., Reilly, P.T.A., Boraas, K.B., Whitten, W.B. & Ramsey, J.M. (1996) Real-time chemical analysis of aerosol particles using an ion trap mass spectrometer, *Rapid Commun. Mass Spectrom.*, 10 (3), 347–351.

Zelenyuk, A., Cabalo, J., Baer, T. & Miller, R.E. (1999) Mass spectrometry of liquid aniline aerosol particles by IR/UV laser irradiation, *Anal. Chem.*, 71 (9), 1802–1808.

Zhang, X.Q. & McMurry, P.H. (1987) Theoretical-analysis of evaporative losses from impactor and filter deposits, *Atmos. Environ.*, 21 (8), 1779–1789.

Zhang, X.Q. & McMurry, P.H. (1992) Evaporative losses of fine particulate nitrates during sampling, *Atmos. Environ. Part A – Gen. Top.*, 26 (18), 3305–3312.

Zhang, X.F., Smith, K.A., Worsnop, D.R., Jimenez, J., Jayne, J.T. & Kolb, C.E. (2002) A numerical characterization of particle beam collimation by an aerodynamic lens-nozzle system: Part I. An individual lens or nozzle, *Aerosol Sci. Technol.*, 36 (5), 617–631.

Zhang, X.F., Smith, K.A., Worsnop, D.R., Jimenez, J.L., Jayne, J.T., Kolb, C.E., Morris, J. & Davidovits, P. (2004) Numerical characterization of particle beam collimation: Part II – Integrated aerodynamic-lens-nozzle system, *Aerosol Sci. Technol.*, 38 (6), 619–638.

Zhang, Q., Alfarra, M.R., Worsnop, D.R., Allan, J.D., Coe, H., Canagaratna, M.R. and Jimenez, J.L. (2005) Deconvolution and quantification of hydrocarbon-like and oxygenated organic aerosols based on aerosol mass spectrometry, *Environ. Sci. Technol.*, 39 (13), 4938–4952.

Chapter 7
Chemical Methods: Chemiluminescence, Chemical Amplification, Electrochemistry, and Derivatization

Andrew J. Weinheimer

7.1 Introduction

7.1.1 Scope

This chapter covers techniques that require chemical conversion of the target species for a measurement to take place. Foremost among these is chemiluminescence, which is the phenomenon of light production by a chemical reaction. Whether the fluorescent species is one of the initial reaction products or one produced after energy transfer to another species, this phenomenon enables the measurement of many species of atmospheric interest. It is a fundamental process in many of the measurement techniques described in this chapter. In some cases the chemiluminescence mechanism is understood in detail, in others it is not.

The chapter will cover the measurement of O_3 by its chemiluminescent reaction with a number of dyes, either in a liquid form or adsorbed onto solid substrates. These are inexpensive, low-power, light-weight techniques that require careful and frequent calibration for quantitative results. When the experimental situation allows, use is made of heavier, more stable, and more accurate instruments based on the gas-phase chemiluminescent reaction of O_3 with either ethene or NO. Conversely nitric oxide is normally detected by its chemiluminescent reaction with O_3, and this technique allows other nitrogen compounds to be converted to NO for subsequent measurement, provided the conversion is good – a paramount concern for this class of measurements.

Like O_3, NO_2 can be measured by its chemiluminescent reaction with a dye, either as liquid or adsorbed onto solid substrates, and this technique is used not only for measuring NO_2 itself but also for the measurement of peroxy radicals by measuring the amplified signal resulting from the conversion of a single peroxy radical to a larger number of NO_2 molecules by a chemical amplifier technique. In a related technique, a peroxy radical is converted to a larger number of H_2SO_4 molecules, which are detected using the CIMS technique (Chapter 5).

Also included are several techniques involving the 'scrubbing' of soluble gases into solution for detection by liquid-based techniques, generally after conversion into other species and sometimes involving chemiluminescence or fluorescence detection as the final step in the measurement process. We also include an electrochemical technique for O_3 upon which much of our knowledge of atmospheric ozone is based.

The order of coverage within the chapter goes somewhat chronologically, but with some grouping according to classes of species.

7.1.2 *Measurement highlights*

Some measurements resulting from this class of instruments will be highlighted in the course of describing the respective instruments. These measurements span the spectrum from the routine monitoring of pollutants such as NO_x on the ground in urban environments to specialized high-altitude aircraft studies of topics such as stratospheric ozone depletion. These instruments have been essential to providing an accurate picture of pollution in the urban environment as well as providing insight into the processes that control ozone throughout the atmosphere, whether it relates to how anthropogenic emissions affect stratospheric O_3 or how natural processes such as lightning can foster O_3 production in the troposphere. The light weight of chemiluminescence detectors has allowed them to be carried to great heights by balloons and rockets to sample regions of the atmosphere not accessible to aircraft sampling. The use of electrochemical sondes for measuring altitude profiles of O_3 from balloons has enabled the determination of temporal trends and spatial distributions of atmospheric O_3. The fast response of chemiluminescence detectors allows for eddy flux measurements of pollutants such as O_3 as well as natural emissions such as isoprene from trees. The fast response also makes possible the detection of fine-scale filamentary structures and thereby sheds light on atmospheric mixing processes. It also allows the detection of fine-scale pollution plumes from power plants and aircraft exhaust, thereby enabling our understanding of the processes that occur there. The high sensitivity of chemiluminescence O_3 detectors enables meaningful measurements in near-surface O_3 depletion events where a low detection limit is required. This class of instruments also includes measurements of species central to the radical chemistry of the atmosphere, both through the chemical amplifier measurements of peroxy radicals themselves, HO_2 and RO_2, and the liquid-based techniques used for the measurement of reactive intermediates or radical reservoir species, such as H_2O_2, CH_3OOH, HCHO, and HONO, whose measurement and modeling are essential to the understanding of the radical chemistry in the atmosphere.

7.2 O_3 via heterogeneous chemiluminescence

The chemiluminescent reactions of O_3 with a variety of dyes have been used in instruments that are generally portable and lightweight, and so have found application on balloon, rocket, and aircraft platforms. Their rapid response and high sensitivity have also made them suitable for flux measurements via eddy correlation. Some have even been developed commercially. This class of instruments was initiated by the work of Bernanose & René (1959) and Bowman & Alexander (1966), who made laboratory observations of the strong chemiluminescence observed when a large number of organic compounds were exposed to O_3 and made the suggestion that O_3 could be thus measured. For these reactions, the mechanism for light production is generally poorly known, but it has been suggested to involve the production of an excited product that transfers energy to the dye which then fluoresces, as the chemiluminescence wavelengths are generally near the respective fluorescent wavelengths.

7.2.1 Early years (luminol and rhodamine-B)

The first practical chemiluminescence instrument for atmospheric measurements was developed by Regener (1960), who adsorbed luminol onto silica gel to produce a solid-phase reagent for use in a balloon instrument that could go to high altitudes (>30 km) where a liquid-based instrument would not work due to the low pressures and temperatures. A chemiluminescent disk was viewed by a photomultiplier tube across a small gap, through which air was drawn for sampling, generating a signal proportional to the partial pressure of O_3. This type of arrangement is illustrated in Figure 7.1 taken from a later work (Stevens & O'Keeffe, 1970), which illustrates common features of this class of devices. Air is pumped or otherwise forced through a small gap (~1 cm) between a chemiluminescent disk, or pool of liquid, and the window of a photomultiplier tube (PMT) which views the chemiluminescence occurring at the surface of the reagent, or also within the volume, in the case of a liquid. The small sample volumes facilitate rapid response.

Citing an effect of moisture on the response of the luminol instrument, Regener (1964) later adsorbed rhodamine-B on silica gel. Since moisture was found to gradually degrade the response, a moisture trap was inserted upstream of the detection cell. In addition, light also reduces the sensitivity, and O_3 consumes the reagent, so periodic calibrations are needed to monitor the gradual degradation of the detector. The instrument appeared to be specific to O_3, and no significant interferences from NO_2 and SO_2 were found. Such instruments were flown on hundreds of balloons in a sounding network as well as on a jet aircraft. The balloon instrument weighed 1.0 kg and consumed very little power (the luminol version used 0.4 W).

Hodgeson *et al.* (1970), working from one of Regener's prototypes, developed an improved version of the rhodamine-B instrument and described its operation in greater detail than done in prior publications. Hodgeson *et al.* (1970) found aspects of the prototype to be cumbersome, including the complexity of the recipe for preparation of the chemiluminescent surface. They worked on improvements and published results of a number of tests on the new version. They employed a small cylindrical reaction chamber closely coupled to a two-inch photomultiplier tube, with the disk less than 13 mm from the red-sensitive photocathode (the rhodamine-B emission peaks near 580 nm). This small volume made possible a time constant for the detection of step changes in O_3 of only a few seconds.

Hodgeson *et al.* (1970) determined that the most serious problem with the Regener technique was the effect of water vapor on the rhodamine-B silica gel surface, which they suggested was likely due to H_2O adsorption and quenching of the excited state prior to fluorescence. They substantially eliminated this effect by adding a hydrophobic agent, silicone resin, to the silica gel surface. The sensitivity was then independent of water vapor, as demonstrated in a test at 80% relative humidity, where the resin-treated surface

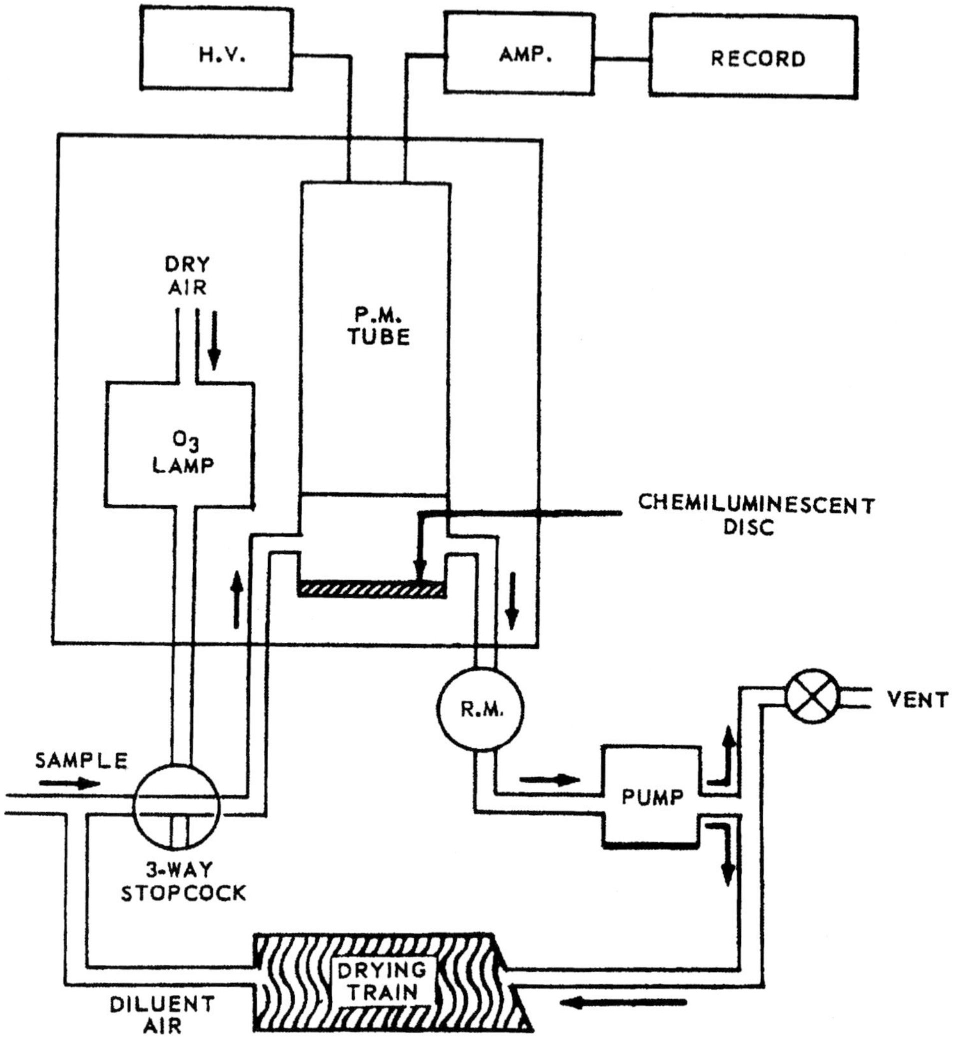

Figure 7.1 Diagram of a typical heterogeneous O_3 chemiluminescence instrument. Sample air is pumped across a chemiluminescent disk which is viewed by a photomultiplier tube across a small gap (~1–2 cm). In this case the sample air is diluted before detection, and provision is made for O_3 generation for calibration purposes. (Figure 1 of Stevens & O'Keeffe, 1970, printed with permission from American Chemical Society.)

maintained a constant sensitivity, while an untreated surface suffered a 50% decrease over 30 min.

Potential interferents were tested. No interference was found for NO_2, SO_2, PAN, H_2O_2, peracetic acid, H_2S, propane, NH_3, NO, and Cl_2. In tests with variable flow rates, it was found that once the flow was above a minimum rate, there was no dependence on flow rate. At lower flows, a significant fraction of the O_3 is apparently consumed and so the response is reduced, while at higher flows the consumed fraction is negligible, so the response is independent of flow.

The instrument had a detection limit of better than 1 ppbv and was linear up to 400 ppbv. Its primary limitation was the gradual decay of sensitivity over time, so periodic calibration was required. The calibration source was an Hg penray lamp operated under controlled conditions, its output having been calibrated against a primary standard.

Hilsenrath and co-workers (Hilsenrath *et al.*, 1969; Hilsenrath & Kirschner, 1980) deployed a rhodamine-B detector on rocket-borne sondes to measure some 30 profiles of ozone concentration, starting in the late 1960s. Flow through the instrument is driven by the flow into an evacuated ballast volume that fills as the pressure equalizes during descent on a parachute. With flows and pressures not as high as in other applications, the rocket sonde operates in a low-flow regime where the signal is proportional to the flow rate as well as the ozone concentration, so it is necessary to derive the flow from measurements of pressure and temperature inside the ballast volume. With careful sensor preparation and calibration processes, Hilsenrath and co-workers were able to achieve a calibration stability of 5% or better, leading to measurements over the wide altitude range of 20–65 km with an absolute accuracy of 12% and a precision of 6%. An altitude profile measured with this instrument is shown as the solid line in Figure 7.2. The measurements compare extremely well both with a rocket-borne optical technique (height-dependence of downwelling UV flux) and with two in situ electrochemical sondes (Section 7.3), except that the chemiluminescence technique shows a high bias at the lowest altitudes (<20 km), likely due to inaccuracies in the flow rate measurement and a residual glow from the chemiluminescent disk.

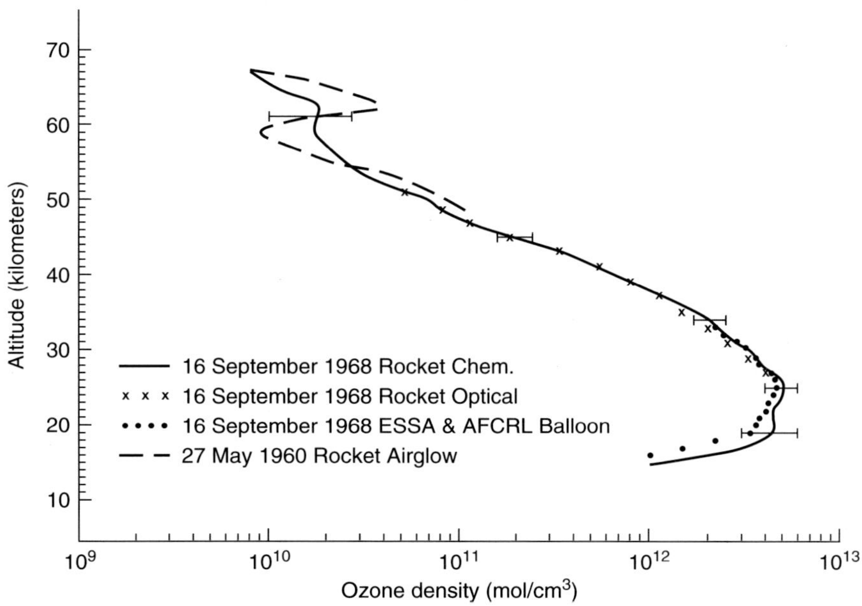

Figure 7.2 O_3 profiles over the altitude range 15–70 km measured using different techniques: (a) rhodamine-B chemiluminescent dye, (b) altitude profile of downwelling UV flux, (c) electrochemical sondes, (d) rocket airglow. Techniques (a) and (c) are described in the text. (Figure 10 of Hilsenrath *et al.*, 1969, printed with permission from American Geophysical Union.)

7.2.2 Since mid-1980s (eosin Y, coumarin-47, chromotropic acid)

Since the mid-1980s additional dyes have been used for the chemiluminescence detection of O_3, and two have found application in commercial instruments. The first of these is eosin-Y, used as a solution in ethylene glycol. Ray *et al.* (1986) built a detection cell in which a red-sensitive photomultiplier tube views a fiber pad across which the dye solution flows continuously at 1 mL/min, as it is recirculated by a pump. Air is drawn through the small gap between the PMT and the wetted fiber pad at flows of 1–9 L/min. Hodgeson *et al.* (1970) had earlier tested eosin-Y adsorbed on silica gel and found high sensitivity, but their surface showed a much more rapid long-term decay than for rhodamine-B, making it unsuitable for long-term use. Hodgeson *et al.* (1970) report that the chemiluminescence peaks near 560 nm.

Like rhodamine-B, eosin-Y detection is very fast and sensitive. At an air flow of 1 L/min the response frequency is 1.5 Hz. Reduction of cell volume and use of an elevated air flow of 9 L/min gave a 7 Hz response (Ray *et al.*, 1986). In addition to slightly better sensitivity, eosin-Y's main advantage over rhodamine-B is that it does not suffer a water vapor dependence. Drummond *et al.* (1991) describe a commercial, battery-powered eosin-Y instrument with a response time of less than 0.5 s and a detection limit ($S/N = 3$) of 1 ppbv for a 4 s measurement. Ray *et al.* (1986) found no interference from NO, NO_2, or PAN, while Drummond *et al.* (1991) found none from NO, NO_2, HNO_3, H_2O_2, NH_3, or HCHO. As is typical with these instruments, calibration is achieved by periodically supplying a known amount of O_3 to the inlet from a stable source calibrated with another instrument.

A second commercial instrument employs coumarin-47 adsorbed on silica gel, as described by Schurath, Güsten, and co-workers (Schurath *et al.*, 1991; Güsten *et al.*, 1992; Güsten & Heinrich, 1996). An advantage of this dye is that its chemiluminescence peaks in the 400–500 nm range, where blue-sensitive PMTs are available, and these are less expensive than those required for the red-emitting dyes (rhodamine-B, eosin-Y). The coumarin-47 instrument is also very fast and sensitive. The 90% response time is significantly better than 0.1 s, with a 100 L/min flow, and the detection limit ($S/N = 3$) is better than 50 pptv. A disadvantage is that its sensitivity increases with ambient H_2O vapor, and it is necessary to measure the H_2O partial pressure and apply a laboratory-derived sensitivity-enhancement correction factor. Attempts to eliminate the H_2O dependence by the use of hydrophobic or hygroscopic compounds, as done by Hodgeson *et al.* (1970) for rhodamine-B, were unsuccessful. The O_3 sensitivity has a 22 s response time (90%) to changes in water vapor, so it does not interfere with the measurement of rapid O_3 changes. There is a similar effect due to SO_2. It was found that the silica gel disks freshly impregnated with coumarin-47 needed to be sensitized by exposure to ozone (100 nbar for 2 hr). Once this was done, the disks had a useful lifetime of about 50 hr, during which the sensitivity is constant, but after which it decays rapidly. In a typical application, the instrument needs calibration against an absolute instrument every half hour (Güsten & Heinrich, 1996). No interferences were found for NO_x, H_2O_2, or PAN.

The shortcomings of the coumarin-47 instrument are balanced by its high sensitivity and fast response, and by its ease of deployment because of its low weight (1.5 kg) and low power consumption. It has found use in long-term ground-based measurements of the vertical deposition flux of O_3 using eddy correlation (Güsten & Heinrich, 1996), as well as use on a light aircraft to make measurements in complex, mountainous terrain, where it would not be possible to fly larger, faster research aircraft (McKendry *et al.* 1998). In addition, the instrument has been used to study small-scale, filamentary features of the ozone field generated by wave breaking in the lower stratosphere. For this it was flown on a high-altitude aircraft (Egrett) together with a slow response (10–20 s) ozone UV absorption instrument against which the much faster chemiluminescence instrument could be continuously calibrated (Bradshaw *et al.*, 2002).

Takayanagi *et al.* (2003) have developed a liquid-phase detector using chromotropic acid. A novel feature of this technique is the need to photo-activate the chromotropic acid by a few-second exposure to UV light just before it flows into view of the PMT. This results in a stable, sensitive, and fast detector with a detection limit of 40 pptv and 10–90% and 90–10% rise/fall times of 130 and 80 ms. These are comparable to the speed of the coumarin-47 technique but with flows of only 1.5 L/min as opposed to 100 L/min. As with the other techniques, no interferences from NO_2 and H_2O_2 were found, and, as with other liquid techniques, it was not affected by relative humidity in the way that the dry sensors are.

7.2.3 Summary

These heterogeneous techniques are appealing due to their high specificity for ozone, low cost, low weight, and low power consumption, while having high sensitivity and fast response. The fast response makes them useful for aircraft use and eddy correlation measurements of O_3 fluxes. The dry techniques suffer from sensitivity changes due to ambient water vapor, and all suffer from drift in the sensitivity, primarily as a long-term decline in sensitivity as the reagent is consumed. Thus all require relatively frequent calibration. Compared with the homogeneous chemiluminescence techniques (Section 7.5), these heterogeneous techniques avoid the use of compressed, hazardous gases (nitric oxide and ethene) used as reagents for the chemiluminescence reaction. However, this is traded for the need to take great care in the preparation and/or activation of the solid or liquid-phase chemiluminescent surfaces. In sum, the heterogeneous techniques find their use where weight and power are at a premium, such as from balloons, aircraft, and rockets, and where speed is of utmost importance, as in measurements of fluxes and in measurements from aircraft, but great care in calibration must be exercised.

7.3 O_3 via electrochemistry

O_3 profiles up to 35 km altitude have been routinely measured by a network of several dozen stations spread over the globe, with records going back to the 1960s and 1970s for many of the stations. Two types of sondes have been used for the bulk of these measurements and both are based on the reaction of dissolved potassium iodide (KI) with O_3 in air as it is bubbled through the solution during balloon ascent:

$$2KI + O_3 + H_2O \rightarrow I_2 + O_2 + 2KOH$$

The Brewer-Mast cell (Brewer & Milford, 1960) consists of a single electrochemical cell with a silver anode and a platinum cathode. A polarizing voltage of 0.42 V is applied. The free iodine generated in the reaction is converted back to iodide ions at the cathode:

$$I_2 + 2e^- \rightarrow 2I^-$$

At the anode two electrons are released to the circuit via ionization of two silver atoms:

$$2Ag \rightarrow 2Ag^+ + 2e^-$$

In this manner each O_3 molecule bubbled through the solution gives rise to the flow of two electrons in the circuit. From knowledge of the pump-driven flow (220 cc/min) through the bubbler, the concentration of O_3 in air is derived.

The electrochemical concentration cell (ECC) consists of two chambers, an anode chamber and a cathode chamber of differing KI concentrations, connected by an ion bridge, which provides an ion pathway for the current and prevents mixing of the two solutions. A schematic diagram appears in Figure 7.3 (Komhyr *et al.*, 1995). As for the Brewer-Mast cell, a current is measured which is proportional to the ozone concentration.

As part of the data reduction procedures, a correction factor for loss of pumping volume at reduced pressure is applied, a zero-ozone background current is subtracted, and the profile is scaled to match a measurement of the total ozone column, making an allowance for ozone above the maximum height of the sounding (WMO, 1998). Comparison studies have shown that between the tropopause and ~28 km, different sondes show consistent results. Precisions are less than 3% and systematic biases are

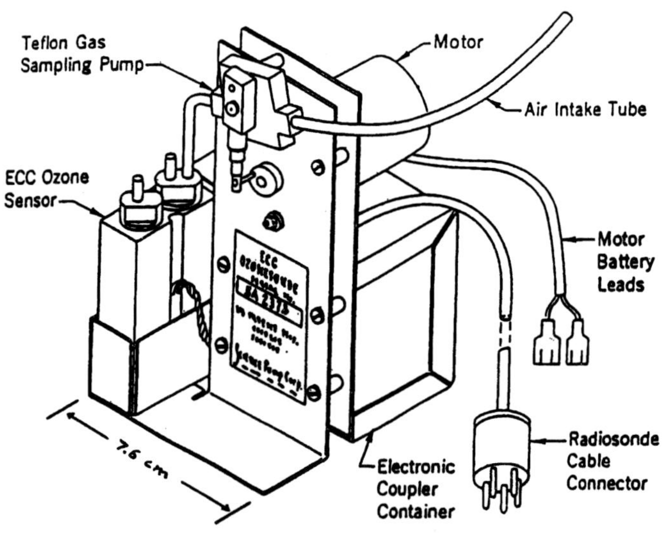

Figure 7.3 A schematic diagram of an electrochemical concentration cell (ECC) used for the routine measurement of O_3 profiles by ozonesondes. (Figure 1 of Komhyr *et al.*, 1995, printed with permission from American Geophysical Union.)

smaller than 5%. For the troposphere, with its small ozone amounts, errors are larger, with systematic differences of 10–15%, with the Brewer-Mast sonde being less precise than ECC sondes. The ECC sonde has a precision better than 5–10% and a positive bias of 3%. However, the differences are not simply due to the sonde type, but different stations using the same type of sonde can show systematic differences due to different preparation and correction procedures. The network would benefit from greater standardization.

The extensive data sets generated from the sounding stations have enabled analyses of trends and distributions of O_3 in the troposphere, which is not well sampled by satellite instruments. Logan (1999) determined, for example, that the broad summertime maximum in O_3 at northern middle and high latitudes extends all the way up to the tropopause, with median concentrations at their peak of 125–200 ppbv in June–July, while being a factor of 2 smaller in the winter. Also, the effects of biomass burning can be observed in the secondary O_3 peak in the South Atlantic in December–January due to the transport of O_3 and its precursors from the Northern Hemisphere. A similar biomass burning feature is observed at Samoa in the middle and upper troposphere in October.

7.4 Nitrogen compounds (NO, NO_2, NO_y) via chemiluminescence with O_3

7.4.1 The $NO + O_3$ reaction

Reactive nitrogen is generally introduced into the atmosphere as NO, which readily (few minutes) establishes a steady state with NO_2 during the daytime. Further oxidation gives rise to other members of the reactive nitrogen family (NO_y). Active nitrogen ($NO_x = NO + NO_2$) is important for its role in tropospheric O_3 formation and its role in catalytic O_3 destruction in the stratosphere. Fundamental to the principle of measurement of reactive nitrogen compounds is the chemiluminescent reaction of NO and O_3, and contrary to the case of the reactions used in the measurement of O_3 by heterogeneous chemiluminescence, the mechanism of this gas-phase reaction has been thoroughly studied. Clyne *et al.* (1964) determined its mechanism to involve the production of excited NO_2 in some fraction of the reactions (7.1), followed by fluorescence (7.3) of those excited NO_2 molecules that are not collisionally quenched (7.4):

$$NO + O_3 \rightarrow NO_2^* + O_2 \qquad (7.1)$$

$$NO + O_3 \rightarrow NO_2 + O_2 \qquad (7.2)$$

$$NO_2^* \rightarrow NO_2 + h\nu \qquad (7.3)$$

$$NO_2^* + M \rightarrow NO_2 + M \qquad (7.4)$$

Also, the spectrum of the radiation was found to be electronic in origin, with a short wavelength limit of 590 nm and extending to at least 1100 nm. The temperature dependence of the chemiluminescence is almost entirely due to the temperature dependence of the NO_2^* production reaction, and it is favored by an increase in temperature. Clough and Thrush (1967) extended this work by measuring the intensity distribution of the chemiluminescence over the range 600–2600 nm, mapping out the peak near 1200 nm, as well as the rate constants for the two branches (7.1) and (7.2). They also determined the temperature dependence of the quantum yield for the chemiluminescent branch (7.1), and while there is some discrepancy about the magnitude of this quantum yield, this yield increases with increasing temperature. Clough and Thrush (1967) found a yield of 7% at 300 K, while Schurath *et al.* (1991) found 20%; both found an increase with temperature. Following this initial work, some half dozen determinations of the rate of the overall reaction have been made and form the basis of the currently recommended values ($A = 3 \times 10^{-12}\,cm^3$ molecule$^{-1}\,s^{-1}$, $E/R = 1500\,K$, Sander *et al.*, 2000). This detailed knowledge of the fundamental properties of the chemiluminescent reaction of NO and O_3 is an aid in the design of instruments for the measurement of reactive nitrogen and O_3, as illustrated in the next section.

7.4.2 The measurement of nitric oxide (NO)

Fontijn *et al.* (1970), Stuhl and Niki (1970), and Stedman *et al.* (1972) recognized that the chemiluminescent reaction of NO and O_3 could be used to measure either NO or O_3 in ambient air. Fontijn *et al.* (1970) demonstrated the feasibility of the method for NO measurements in polluted air by finding a linear response over the range of 4 ppbv to 100 ppmv with no interferences from NO_2, CO_2, CO, C_2H_4, NH_3, SO_2, and H_2O. Stuhl and Niki (1970) and Stedman *et al.* (1972) built a prototype instrument that was used for measurements in a polluted urban environment. It had a detection limit of 1.5 ppbv ($S/N = 3$). They found modest interferences from metal carbonyls and C_2H_4, but these are greatly diminished by the use of a filter to cut off light with $\lambda < 648\,nm$, where the chemiluminescence due to these species primarily occurs, and which is below the wavelength range of the bulk of the NO_2 emission. Red filters are common features of later designs as well, so that other potential interferences are also minimized.

Ridley and Howlett (1974) built an automated instrument for deployment in the stratosphere, motivated by the need to assess the potential impact of supersonic aircraft on stratospheric ozone. A cylindrical reaction vessel was built and viewed axially by a PMT. Provided that there is an adequate flow of reagent O_3 to maintain a high enough concentration for near-complete conversion of NO to NO_2 (dependent on mass flow and pressure of the reaction vessel), and provided that collisional quenching (7.4) dominates chemiluminescence (7.1), they derived that the chemiluminescence intensity is given by

$$I = \frac{k_1 k_3}{(k_1 + k_2) \sum_i k_{4i} [M]_i} F\,[NO]$$

where k_1, k_2, k_3, and k_4 are the rate constants for the above reactions, F is the volume flow rate into the reaction vessel, and [NO] is the concentration of NO to be measured (Ridley *et al.*, 1972; Ridley & Howlett, 1974). This can be further simplified to demonstrate the fundamental dependence of instrumental sensitivity (S) on flow and pressure:

$$S = \frac{C F_0 \mu}{P}$$

where F_0 is the mass flow rate of sample air, P is the reaction vessel pressure, μ is the NO mixing ratio, and C is a proportionality factor (see also Ridley, 1978). This shows that sensitivity is enhanced by increasing the sample flow, within the constraints of practicality imposed by the need to deploy the instrument on an aircraft or balloon, and also by minimizing the pressure so that quenching is minimized and the excited NO_2 molecules are allowed to fluoresce. Based on these design principles, Ridley & Howlett (1974) were able to achieve a sensitivity improvement to 30 pptv for a 1 s detection limit. Chemiluminescence instruments generally operate in a regime where the O_3 concentration is high enough that virtually all of the NO reacts with O_3 within view of the PMT. Of those that react, the fraction that become excited (NO_2^*) is in the range 5–20% (depending on temperature and laboratory results adopted; see above), and of these about one in a thousand (at most) radiate prior to collisional quenching. Thus the signal is proportional to total flow and inversely proportional to the pressure due to the dominance of the quenching process over fluorescence.

Optimization of a chemiluminescence instrument is also improved by modest elevation of the reaction vessel temperature (to 30–40°C) to increase the rate of overall reaction ((7.1) and (7.2)) and also to favor the chemiluminescence branch (7.1). However, there is no point in increasing the temperature too far as it results in too much heat transfer to the red-sensitive PMT, which must be kept cool (e.g. using dry ice) to minimize background dark counts within the PMT itself. Moreover background counts from the reaction vessel also increase at elevated temperatures. Temperature- and pressure-control of the reaction vessel, when possible, can also contribute to the stability of instrument sensitivity. Another critical factor is the reflective property of the reaction vessel, and gold-plating is often employed.

Ridley and Grahek (1990) describe a widely-used reaction vessel design (Figure 7.4). It is machined from stainless steel and then highly polished and gold plated to maximize the reflectivity and the number of photons collected from the chemiluminescent reaction. It has a volume of 230 cm^3 (L = 4.5″, D = 2″) and is designed for operation with a sample flow of 1000 sccm at a pressure of 5–10 torr. The reagent flow is 100–200 sccm of 3–4% O_3 in O_2, which is produced by flowing high-purity O_2 through a silent discharge. The sample and reagent gases are each distributed in an annular ring at the photomultiplier end of the reaction vessel, and they are mixed in an annular fashion as they enter the vessel just in front of the PMT, in order to maximize the amount of light collected. With its modest flow requirements and its small size, it is suitable for deployment on aircraft where space savings are critical. Yet the instrument sensitivity (7 counts per second per pptv) and detection limit (1–2 pptv at 10 s) are still quite good for most purposes.

Chemiluminescence instruments are zeroed by switching the ozone flow to a volume just upstream of the reaction vessel to allow the $NO + O_3$ reaction and attendant chemiluminescence to occur upstream of the reaction vessel and out of view of the PMT. This allows measurement of the background signal due to the PMT dark current plus a signal due to presence of O_3 in the reaction vessel, likely due to wall luminescence (Ridley & Howlett, 1974; Kley & McFarland, 1980). This background signal is subtracted from the measure mode signal to give the signal proportional to ambient NO. A measurement artifact is commonly observed when sampling synthetic air with no NO, a gas for which there should be no difference between the zero and measure mode signals. However, there generally is a small difference, and its origin is not well understood, but since it

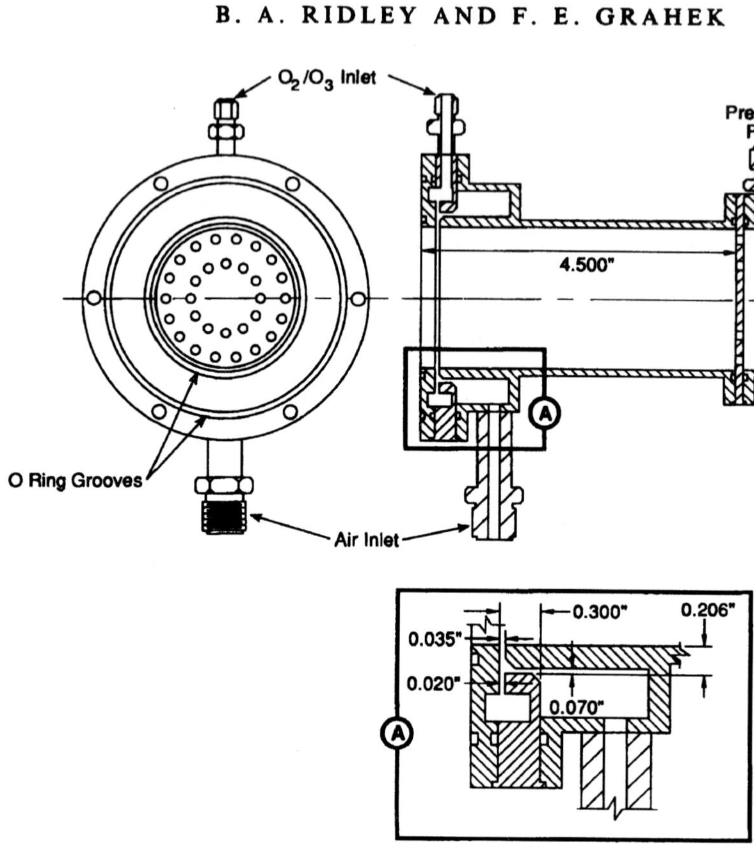

Figure 7.4 Mechanical design of an NO reaction vessel used in many research-grade chemiluminescence instruments. (Figure 1 of Ridley & Grahek, 1990, printed with permission from American Meteorological Society.)

is related to cleanliness of the reaction vessel, it may be related to wall luminescence (Kley & McFarland, 1980). At any rate, this artifact is routinely measured and applied as a correction to the measure-minus-zero signal. Since this artifact signal may be of the order of 10 pptv or less, it is not significant in polluted environments, but in clean, remote regions the magnitude and uncertainty of the artifact signal may be a limiting factor. As NO is stable at ppmv levels in a mixture with N_2 in compressed gas cylinders, calibration is generally achieved by the periodic addition of a known mixing ratio to the sample inlet.

A complicating issue is the variation of quenching efficiency with the identity of the third body in reaction (7.4). Matthews *et al.* (1977) drew attention to this for studies of combustion systems where a large variety and quantity of combustion products were present. As a practical matter for most atmospheric measurements, it is water, and its variability, that is most critical, as it is abundant in the atmosphere and a much better quencher than dry air. As Matthews *et al.* (1977) demonstrate, the quenching efficiency, and therefore the instrument sensitivity, can vary with the concentration of water and other species in the reaction vessel. Hence calibrations should be performed via NO addition on top of ambient air, rather than synthetic air. Also, for measurements in humid air, it points to the need for a correction to the instrument sensitivity based on simultaneous measurements of H_2O vapor. Interestingly, Carroll *et al.* (1985) found it beneficial to add a constant flow of water, at 2–3% of the total flow, directly to the reaction vessel. Even though the increased quenching due to H_2O decreases instrument sensitivity by about 15%, overall performance is improved because the background signal is also reduced and stabilized, the artifact is reduced, and the start-up settling time is shortened. Ridley *et al.* (1994) describe the proper way to correct for sensitivity to ambient H_2O vapor. It essentially requires adjusting real-time sensitivity calibrations back to zero ambient H_2O, employing ambient H_2O vapor measurements combined with a relative quenching coefficient for H_2O, while taking into account the dilution of the sample flow by the ozonizer flow and the added H_2O flow into the reaction vessel. The sensitivity at the time of measurement can then be similarly determined using the ambient H_2O vapor measurements.

Chemiluminescence instruments for nitrogen species are widely used for atmospheric measurements. They have been deployed for research studies on aircraft of many types, as well as balloons and rockets, covering remote regions of the globe and also extending through most of the stratosphere. In addition, they are used for routine monitoring of pollution at ground sites, generally using commercially available instruments. An exciting early result from these instruments was the observation of extremely low mixing ratios, only a few pptv, in the boundary layer of the remote Pacific (McFarland *et al.*, 1979). While initially difficult to reconcile with models, these low values were later confirmed by aircraft measurements (e.g. Ridley *et al.*, 1987). As an example of more recent research, Figure 7.5 illustrates the installation of an automated NO-NO_y instrument in the lower fuselage of a high-altitude aircraft (NASA WB-57), and Figure 7.6 shows NO measurements from this deployment which demonstrate that Florida thunderstorms inject prodigious amounts of lightning-produced NO into the upper troposphere, where lightning can be a dominant source of NO_x which can affect O_3 levels in the upper troposphere well downstream of thunderstorm regions (Ridley *et al.*, 2004).

7.4.3 The measurement of nitrogen dioxide (NO_2)

The ready detection of NO via chemiluminescence in turn enables the detection of other reactive nitrogen species which can, *with specificity*, be converted to NO, and while a number of instruments have been built and a number of papers have been written, it is the aspect of specificity which has generated the most concern. It appears that Sigsby *et al.* (1973) were the first to describe such an instrument in the literature. It employed a 6-ft length of 1/8-inch stainless steel tubing heated to 750–900°C as an NO_2-to-NO converter, and by switching the converter in and out of the sample flow path, signals for (a) NO plus NO_2 and (b) just NO could alternately be obtained. The difference would be the signal for NO_2 but for the fact that NH_3 was also converted to NO by this system. By the addition of a scrubber for NH_3, Sigsby *et al.* (1973) were able to correct for this, but compounding the number of differences required of signals which were not actually simultaneous added uncertainty to the measurement. Also, it was not

Figure 7.5 Photograph of an automated NO-NO_y instrument mounted in a lower-fuselage pallet (1 of 4) on the NASA WB-57F. The instrument is hoisted into position so that the pallet forms the lower fuselage skin. The heated sample inlet extends below the pallet. (Photo courtesy of David J. Knapp.)

clear to what extent other potential interferents were specifically tested. On the other hand, Winer *et al.* (1974) did a systematic study of several potential interferents to the determination of NO_2 using molybdenum and carbon converters, namely PAN, ethyl nitrate, ethyl nitrite, nitroethane, and HNO_3. They found that PAN was converted just as efficiently as NO_2, as were ethyl nitrate and ethyl nitrite. HNO_3 was converted less efficiently but to a significant extent. Thus the measurements from such a converter are difficult to interpret, but they concluded that they more nearly yield a measure of total oxides of nitrogen (NO_y), rather than NO plus NO_2, but of course if HNO_3 were the dominant oxide of nitrogen, it would not be an adequate measurement since it is not converted with unit efficiency. Matthews *et al.* (1977) performed a similar study for a stainless steel converter and with similar results, including significant conversion of HCN and incomplete conversion of NH_3.

A far more specific technique, a photolytic one rather than a surface catalytic one, was proposed by Kley and coworkers (Kley & McFarland, 1980; Kley *et al.*, 1981), who shone a xenon arc lamp on a Pyrex photolysis cell in which sampled air had a residence time of the order of 2–3 s. There was enough UV power and residence time to achieve conversion of 60% of the NO_2 to NO. Via calibration with a known flow of NO_2, this conversion efficiency can be determined, and this, combined with a measurement of NO, allows NO_2 to be determined. The NO signal is subtracted from the signal via the illuminated

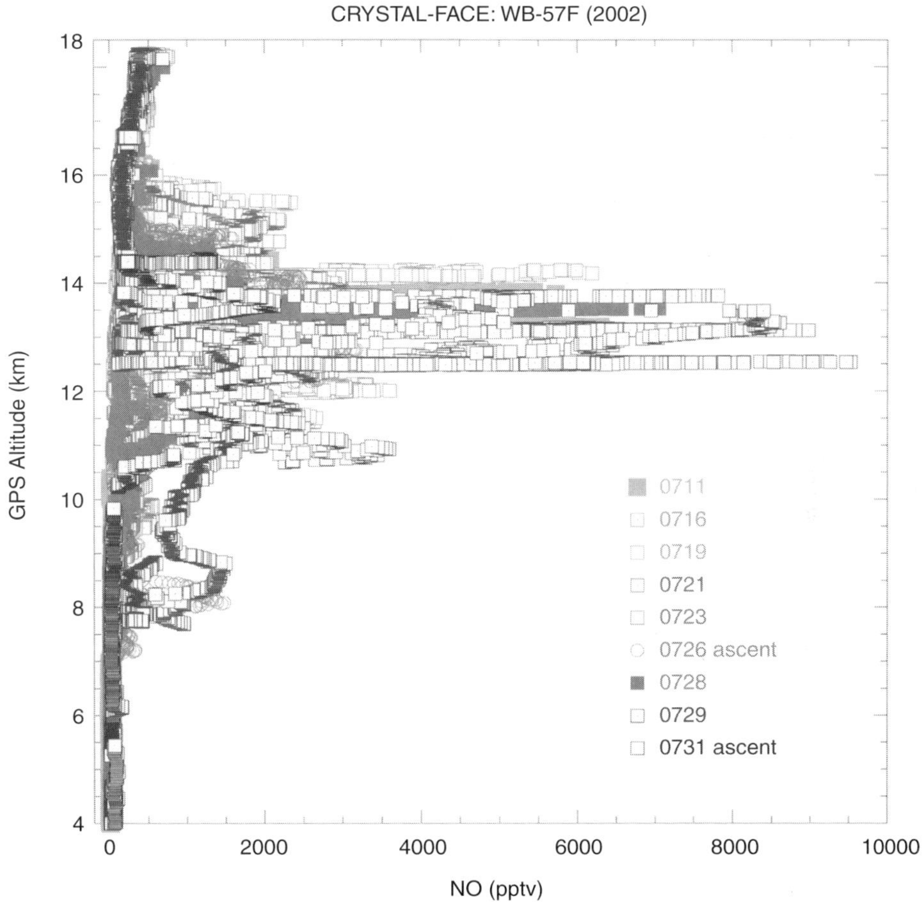

Figure 7.6 Chemiluminescence measurements of NO, largely produced by lightning, at near-tropical latitudes over and near southern Florida, illustrating thunderstorm outflow up to the local tropopause, but with peaks values a couple of km below. NO mixing ratios are 10–50 times background amounts, showing that lightning can be major source of O_3-producing NO_x to this region. (Figure 12c of Ridley *et al.*, 2004, printed with permission from American Geophysical Union.)

photolysis cell, and the difference is divided by the conversion efficiency to yield NO_2. The known flow of NO_2 is generated either from an NO_2 permeation device or using a known flow of NO converted to NO_2 via near-complete reaction with O_3.

The photolytic technique has potential for much greater specificity than does a technique using a hot catalytic surface. Many of the potential interferents for an NO_2 measurement, such as HO_2NO_2, N_2O_5, $ClONO_2$, and PAN, are thermally unstable and prone to convert to NO_2 when exposed to heat. On the other hand, the absorption spectra of these species as well as other interferents (e.g. HONO and NO_3) show much weaker absorption in the 300–400 nm range compared with NO_2 (Figure 7.7). Specificity is further enhanced by the use of a 320 nm cutoff filter (Kley & McFarland, 1980), because of the spectral differences evident in Figure 7.7.

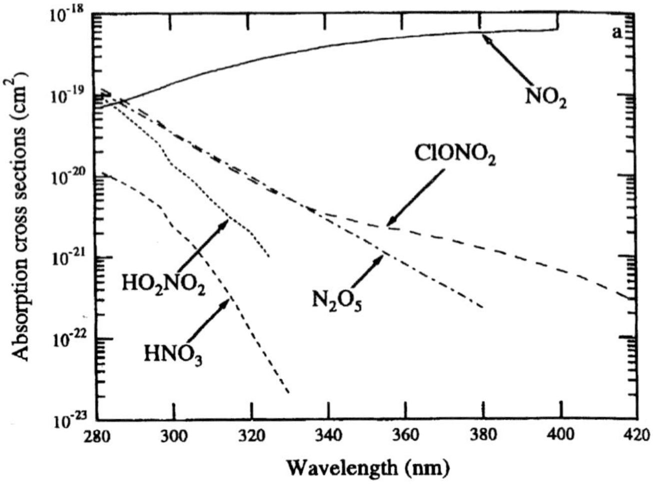

Figure 7.7 UV absorption spectra of NO_2 and several potential interferent compounds, illustrating the discrimination that can be obtained via photolytic conversion at wavelengths in the 350–400 nm range. (Figure 1 of Gao *et al.*, 1994, printed with permission from American Geophysical Union.)

A complication with the technique is an artifact signal that occurs, even when no NO_2 is present in the sample flow. In such a case, the signal should be the same whether the light is shining on the photolysis cell or not. However, Kley and McFarland (1980) found such a signal to be present and that its intensity was related to the cleanliness of the cell, suggesting that this artifact is caused by wall contaminants such as HNO_3 adsorbed from the gas phase or as nitrate in aerosol form. As with the artifact in the NO measurement, this NO_2 artifact may routinely be determined using synthetic air, and a correction applied. If it is too large, then the cell requires cleaning.

Bollinger *et al.* (1984) used the photolytic technique to obtain measurements of NO_2 with a detection limit of 10 pptv for a 1 s integration time. Combined with NO measurements with a detection limit of 2 pptv at 1 s, very low mixing ratios of below 20 pptv were measured at a surface site in the Colorado mountains. The high sensitivity of these instruments allowed the determination of what at the time seemed to be exceptionally low mixing ratios of NO_x for a continental surface site, but it was possible to explain the low values in terms of the synoptic meteorology which brought air to the high-altitude site unperturbed by ground-level sources.

In view of the number of NO_2 conversion techniques that have been used for field measurements, it has been necessary to conduct side-by-side comparisons of instruments to insure identical conditions. In a ground-based study, Fehsenfeld *et al.* (1987) compared the photolytic technique to catalytic conversion on a solid ferrous sulfate surface (Dickerson *et al.*, 1984) and found the catalytic technique to be subject to significant interferences from n-propyl nitrate (NPN) and PAN based on tests with known amounts of those species. Ridley *et al.* (1988a) made a comparison of the same two techniques but aboard a NASA Convair 990 aircraft and with a similar conclusion. Since the airborne ferrous sulfate converter gave rise to levels about three times higher than the photolytic converter, it was concluded that the ferrous sulfate converter was likely subject to an interferent, possibly PAN. In a companion paper, Ridley *et al.* (1988b) presented a modified version of photolysis/chemiluminescence instrument. A 45° ultraviolet mirror was used to reflect only the useful radiation from the arc lamp into the photolysis cell, preventing the unwanted visible and infrared light from illuminating the cell. The use of the ultraviolet mirror combined with a lens to keep most of the light from shining on the cell walls reduced the NO_2 artifact signal described above. In addition, the mass flow rate through the cell and its pressure were held constant, giving rise to a constant conversion efficiency and a stable calibration. The paper also describes a sampling issue that can lead to erroneous NO_2 values if not properly calibrated and corrected for, and that is due to the conversion of NO to NO_2 by reaction with O_3, both homogeneous and heterogeneous, in the sampling line and photolysis cell. The apparent sensitivity will vary with ambient O_3, but factors are derived to correct for this (Ridley *et al.*, 1988b). The 1-min detection limit at 0–2 km (5–7 km) was 7 pptv (15 pptv) for a 1 min average. This allowed the airborne measurement of NO_x in regions of a tropopause fold and demonstrated that stratospheric NO_x could be introduced over a wide vertical range in the troposphere and not just at the tropopause, as assumed in many models at the time.

Given the proven benefits of using photolytic converters rather than heated catalysts, further developments in photolysis/chemiluminescence measurements of NO_2 have come from improved light sources. Gao *et al.* (1994) used a metal halogen lamp in a photolytic converter deployed on the NASA ER-2. This lamp is much more efficient in generating useful UV light than the xenon arc lamp, while the heat load is still significant but manageable on the high-flying ER-2, where there is a ready supply of cold air. The initial instrument had a conversion efficiency of only about 0.37, and it was adequate to resolve the NO_2 in the exhaust plume of the ER-2 (Figure 7.8). Eventual improvements in the instrument centered on a change in the combination of filters used (del Negro, 1999)

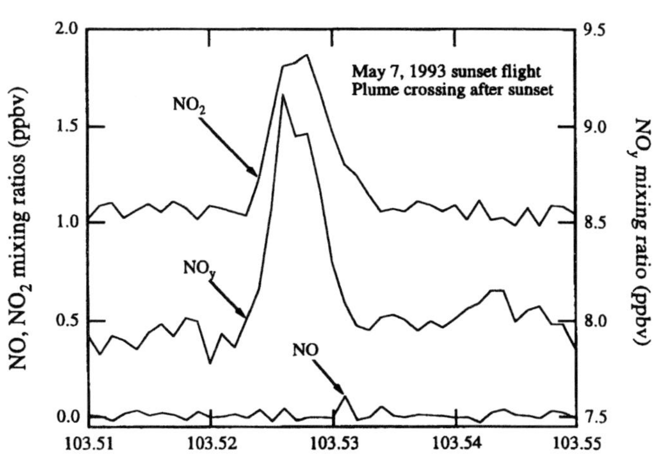

Figure 7.8 Measurements obtained with NO, NO_2, and NO_y chemiluminescence instruments aboard the NASA ER-2 as it samples its own exhaust. The good time response (1-sec points are plotted) of both the chemiluminescent detector itself and the photolytic converter enable good resolution of the plume. The sampling occurred after local sunset so that all of the NO has been converted to NO_2. (Figure 14 of Gao *et al.*, 1994, printed with permission from American Geophysical Union.)

and resulted in conversion efficiencies of about 0.50, with an instrumental uncertainty range of 10–30%. These NO_2 measurements were used in combination with NO, O_3, ClO, and HO_2 measurements to make a comparison with calculated NO_2 values in the sunlit lower stratosphere, with the finding of a 19% difference, which was well within the combined uncertainties of 50–70%.

Around 2000 Ryerson *et al.* (2000) built an NO_2 converter using a high-pressure Hg arc lamp. This design was driven by the need for high time resolution aircraft measurements, which in turn requires a high NO_2-to-NO conversion fraction during a relatively short time in the photolysis cell. The Hg arc lamp makes this possible while at the same time reducing the heat load, also a significant consideration. An ellipsoidal reflector is used to direct the light toward the photolysis cell through a set of optical filters designed to minimize sample heating and attendant interferences from thermally labile species, as well as to restrict the light to longer wavelengths. This results in the loss of 35% of the useful lamp output, but this is compensated for by increased specificity due to reduced photolysis of potential interferent species (HNO_3, HO_2NO_2, N_2O_5; see Figure 7.7). Interferences from PAN and HNO_3 were measured in the laboratory and found to be negligible. The long pass filter (350 nm) was found to significantly reduce the UV artifact described earlier. This combined with the apparently high water solubility of the contaminant species serves to confirm earlier suggestions that HNO_3 contamination of the cell walls plays a role in the artifact signal. The cell pressure is maintained at 250 torr (typically flown on a mid-altitude P3 aircraft) for a cell residence time of 0.8 s and 1/e response time of 0.65 s. The conversion factor of NO_2 to NO is 0.7. Achieving such high conversion at relatively short residence times has enabled high resolution aircraft measurements to be made in atmospheric features with small spatial scales, such as plumes from power plants and petrochemical facilities (Ryerson *et al.*, 2003).

7.4.4 The measurement of total reactive nitrogen (NO_y)

The lack of converter specificity for NO_2 or any other particular species is turned to an advantage in the measurement of total reactive nitrogen, or NO_y, which includes NO, NO_2, NO_3, N_2O_5, HONO, HNO_3, HO_2NO_2, $ClONO_2$, PANs, and other organic nitrates. For an NO_y measurement it is desired that all reactive nitrogen species be converted to NO, but not other nitrogen-containing species (e.g. NH_3, N_2O, HCN). As a practical matter, it is required that near-unity conversion be achieved for all NO_y species contributing significantly to the total NO_y amount, and that there be only minimal conversion of potential interferents, as determined by the product of the conversion efficiency and abundance for a particular species.

Several early studies used a variety of converters that were originally intended for use as NO_2-to-NO converters, but which suffered interferences from the conversion of other nitrogen-containing species as well. So, these converters were pursued as means to measure HNO_3 and other nitrogen species. Joseph & Spicer (1978) used a dual channel instrument, each channel having a molybdenum converter in line, and placed a nylon filter in line for one of them to scrub HNO_3, so that the difference signal would yield the HNO_3 mixing ratio. Kelly and Stedman (1979) used the same principle to measure HNO_3, but instead with a converter consisting of a quartz tube at 350°C packed with Pyrex beads. Helas & Warneck

(1981) used the difference in signals from molybdenum and ferrous sulfate converters to infer a measure of what they termed 'excess NO_x', or non-NO_x NO_y compounds, which they recognized might include N_2O_5, HO_2NO_2, $CH_3O_2NO_2$, and PAN, although the conversion efficiencies for these species were not demonstrated. The molybdenum converter found application in commercial instruments, as flown by Dickerson (1984) to measure total reactive nitrogen in the free troposphere, where convection was demonstrated to have a significant impact on reactive nitrogen mixing ratios. One difficulty observed with the molybdenum converter is a 'memory effect' in which, after exposure to elevated levels of reactive nitrogen, the background signal remains high, introducing uncertainty into the measurement. Controlled, stepped heating has been used to thermally dissociate PANs, organic nitrates, and HNO_3 to NO_2 for detection by LIF (Chapter 4).

Bollinger *et al.* (1983) and Fahey *et al.* (1985) introduced the use of gold catalysts for the measurement of NO_y via the reduction of NO_y species to NO by CO. Bollinger *et al.* (1983) employed a gold-coated quartz tube and tested the conversion efficiency of NO_2, HNO_3, and n-propyl nitrate (NPN), selected as being representative of organic and inorganic reactive nitrogen compounds. Conversion was measured as a function of converter temperature and CO mixing ratio. Under appropriate conditions, complete conversion was found for all three species. Fahey *et al.* (1985) extended this work by testing the conversion of a thin-walled solid gold tube, which was found to be more reliable than the coated quartz tube used earlier. Nickel and stainless steel converters were also tested, but were not as good. NO_y species chosen for tests of conversion efficiency and linearity were NO_2, HNO_3, N_2O_5, and PAN. For a converter at 300°C, the conversion efficiencies of all species exceeded 90%, and the conversion was linear for mixing ratios in the range 0.1–50 ppbv. Temperatures higher than 300°C were not used in order to minimize potential interferences, such as from N_2O.

In addition, the non-NO_y compounds NH_3, HCN, N_2O, CH_4, and various chlorine and sulfur compounds were tested as interferents. NH_3 and HCN were found to be the principal interferents, at least in dry, synthetic air. Conversions of 2–8% and 2–20% were found for NH_3 and HCN, respectively. The addition of 0.5–2.5% H_2O was found to reduce these conversion efficiencies to about 2%. Using laboratory air at 20% relative humidity resulted in conversion efficiencies of less than 1%. In contrast with this, later work by Kliner *et al.* (1997) found near-complete conversion of HCN. On the other hand, Weinheimer *et al.* (1998) measured HCN conversion efficiencies for three different converters sampling ambient air from a DC-8 aircraft, with results for two of the converters being consistent with the results of Fahey *et al.* (1985), and one, not consistent. Conversion in humidified ambient air for two converters was small, at about 5%. However, the third converter showed 30% conversion. Taken all together, the results from the various studies show that outwardly identical converters can experience different interferences from HCN, perhaps due to their own particular histories (air sampled, cleaning history, temperature history). However, these effects are minimized by ambient humidity, and irrespective of humidity, conversion is smaller in ambient air than in synthetic air. This result points to the value of testing each converter for its own specific interference from HCN in order to verify that ambient levels of NH_3 and HCN are generally small enough that interferences will usually be negligible in field measurements. This also points to the general value in testing converters for the conversion efficiency of several component species, as well as interferents. Fahey *et al.* (1986) used the then new gold-converter technique in conjunction with measurements of many of the component NO_y species at a ground site in the Colorado mountains to show that PAN was often the dominant NO_y species. They also found that correlations of NO_y with O_3 gave evidence for the production of O_3 in polluted air masses and showed that some NO_y species are subject to surface loss.

The molybdenum and gold converters have been compared in side-by-side tests, with mixed results. Fehsenfeld *et al.* (1987) found that the two gave similar estimates of NO_y in ambient air over a wide range of mixing ratios, ranging from 0.4 ppbv in clean continental air to about 100 ppbv in very polluted conditions. In this test, the molybdenum converter did not suffer from memory effects noted above. However, it did suffer significant degradation in conversion efficiency after extended operation in polluted conditions. In a second ground-based test, Williams *et al.* (1998) report generally more favorable results. Seven converters were compared, all followed by chemiluminescence detection of NO, but some with gold catalysts using either CO or H_2 as the reducing agent and some using molybdenum converters. No significant difference between gold and molybdenum converters was found for polluted conditions experienced, where NO_y was in the range 2–100 ppbv. This limits these favorable conclusions to these high mixing ratio conditions. The authors note that differences could become significant at lower mixing ratios, and especially for conditions when NO_x is a smaller fraction of NO_y.

Another important condition for measurement reliability is the transmission of HNO_3 through inlets and sample lines, as HNO_3 can be a dominant component of NO_y, and it is also very sticky. That is, it is prone to adsorb onto surfaces, either reversibly or irreversibly. Neuman *et al.* (1999) used a sensitive, fast-response HNO_3 CIMS instrument to test various surfaces for their ability to pass HNO_3. They found that teflon above 10°C is very good for this purpose, but that stainless steel, glass, fused silica, aluminum, nylon, silica-steel, and silane-coated glass all take up a large fraction of the HNO_3. HNO_3 is reversibly taken up by teflon at low temperatures, but if it is kept moderately warm (>10°C) it is a suitable material for inlets and sample lines. Inlets for NO_y are typically composed of teflon and are often heated to somewhat higher temperatures of 30–70°C to insure good transmission (e.g. Ridley *et al.*, 1994).

NO_y measurements from the NASA ER-2 aircraft have proven to be a valuable component of extensive field missions aimed at understanding polar ozone loss processes. Fahey *et al.* (1990) found that correlation of NO_y and N_2O, which derives from N_2O being the source of stratospheric NO_y, could be used to infer the degree of denitrification in the winter polar stratosphere, a process critical to ozone chemistry, as the removal of reactive nitrogen from an air mass allows chlorine species to catalytically destroy ozone. This nitrogen loss process was later observed by Fahey *et al.* (2001) in the form of large HNO_3-containing particles that were precipitating in the winter arctic stratosphere, causing denitrification and hence making ozone loss more likely (Figure 7.9).

A new and interesting use of an NO_y instrument is unattended measurements aboard civil in-service commercial aircraft. The MOZAIC (Measurements of Ozone and Water Vapour aboard Airbus in-service Aircraft) project includes a small NO_y instrument installed on long-haul aircraft to provide regular surveys of the poorly sampled tropopause region. These are difficult constraints under which to operate a research-grade instrument, and especially with the long inlet line required, but with the large number of flights taken, and with proper attention to quality control, a large quantity of valuable data are being generated (Volz-Thomas *et al.*, 2004).

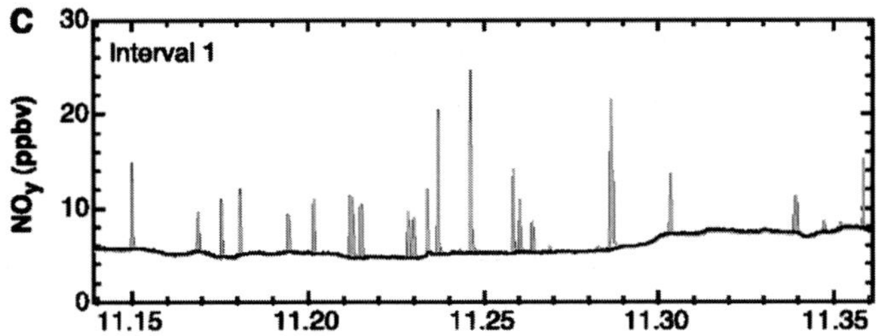

Figure 7.9 Simultaneous measurements from two Au-converter/chemiluminescence NO_y channels aboard the NASA ER-2 in the Arctic stratosphere. The darker trace is for air sampled through an aft-facing inlet and so is primarily gas-phase NO_y (mostly HNO_3 in this case) as particles are inertially separated. The lighter trace is for air and particles sampled through a forward-facing inlet, and the two traces closely overlap, except at the "spikes" which are due to the presence of HNO_3-containing particles, likely nitric acid trihydrate, which evaporate and release a burst of HNO_3. By means of sedimentation, these particles irreversibly remove HNO_3 from the air mass and this denitrification allows chlorine to remain active for a longer time in destroying ozone. (Figure 2c of Fahey *et al.*, 2001, reprinted with permission from AAAS.)

7.4.5 Routine NO_x monitoring

As NO_x is a major urban pollutant, central to photochemical ozone production, its routine monitoring has become an important aspect of pollution control in urban environments. In the UK, a network of over 200 automated sites employs chemiluminescence NO_x instruments, most with molybdenum converters, to allow an assessment of ambient NO_x levels for the purpose of determining whether NO_2 concentration objectives, imposed by government agencies, have been met (AQEG, 2004). These objectives are not to exceed 200 $\mu g\, m^{-3}$ over a one-hour average, nor to exceed 40 $\mu g\, m^{-3}$ as an annual average (divide these values by 2 to get the rough equivalent in ppbv for standard temperature and pressure). These data also allow the observation of patterns, cycles, trends, and correlations with measurements of other species that can be tested against our understanding of the basic processes of pollution chemistry as embodied in numerical models. In particular, it enables the connection of emission rates to ambient concentrations, so that predictions of future concentrations may be inferred from projections of future emissions.

Figure 7.10a shows the diurnal variation in NO_x at various sites in the monitoring network (AQEG, 2004). That road traffic is the largest source is reflected in the diurnal cycle with a maximum during the morning rush hour and secondary maximum in the evening, except for the most remote site, which shows little in the way of a diurnal cycle. The monitors are also valuable for the determination of longer-term, multi-year trends. Emissions in the UK have declined by 37% from 1990 to 2000, and this is reflected in the decadal trend in NO_x mixing ratios shown in Figure 7.10b (AQEG, 2004). Emissions are expected to decrease another 25% by 2010 due to tighter vehicle emission standards, so it is important to continue to monitor ambient levels to see whether their decline is commensurate with this. The monitoring network may also be used to construct the large-scale spatial distribution of NO_x and correlate the measured concentrations to emission estimates. Figure 7.10c shows the distribution within the UK of annual-average NO_x concentrations derived from the network dataset interpreted through use of an empirical model (Stedman *et al.*, 2002). As the density of sites throughout the monitoring network

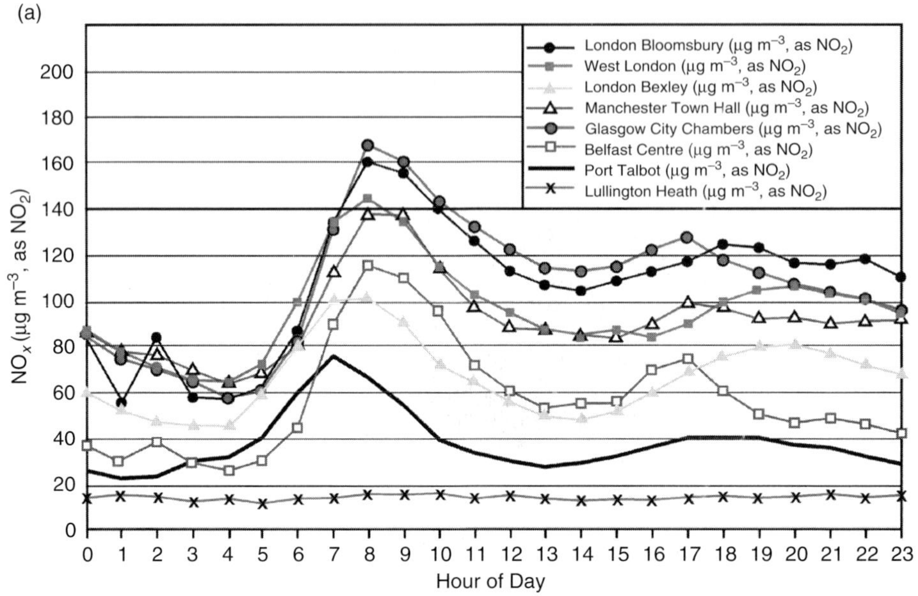

Figure 7.10a The diurnal cycle of NO_x at eight sites in the UK, showing the morning rush hour maximum due to vehicle traffic and a secondary maximum in the evening. (Figure 6.14 of AQEG, "Nitrogen Dioxide in the United Kingdom", 2004, printed with permission from Crown.) (Reproduced in colour as Plate 3a after page 270.)

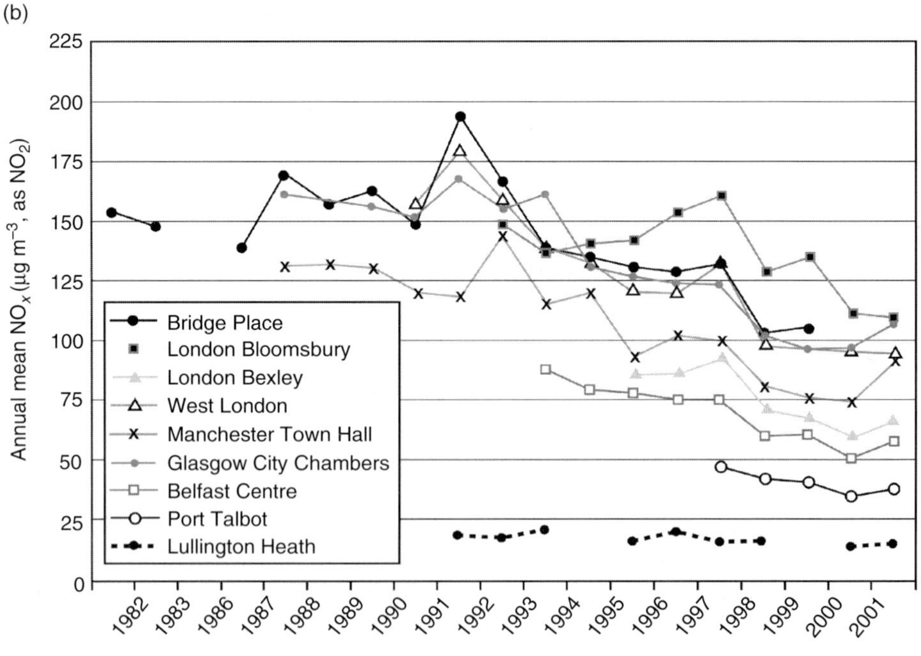

Figure 7.10b The decadal trend in NO_x at nine sites in the UK, showing the gradual decline due to the tightening of vehicle emission standards. (Figure 6.5 of AQEG, "Nitrogen Dioxide in the United Kingdom", 2004, printed with permission from Crown.) (Reproduced in colour as Plate 3b after page 270.)

(c)

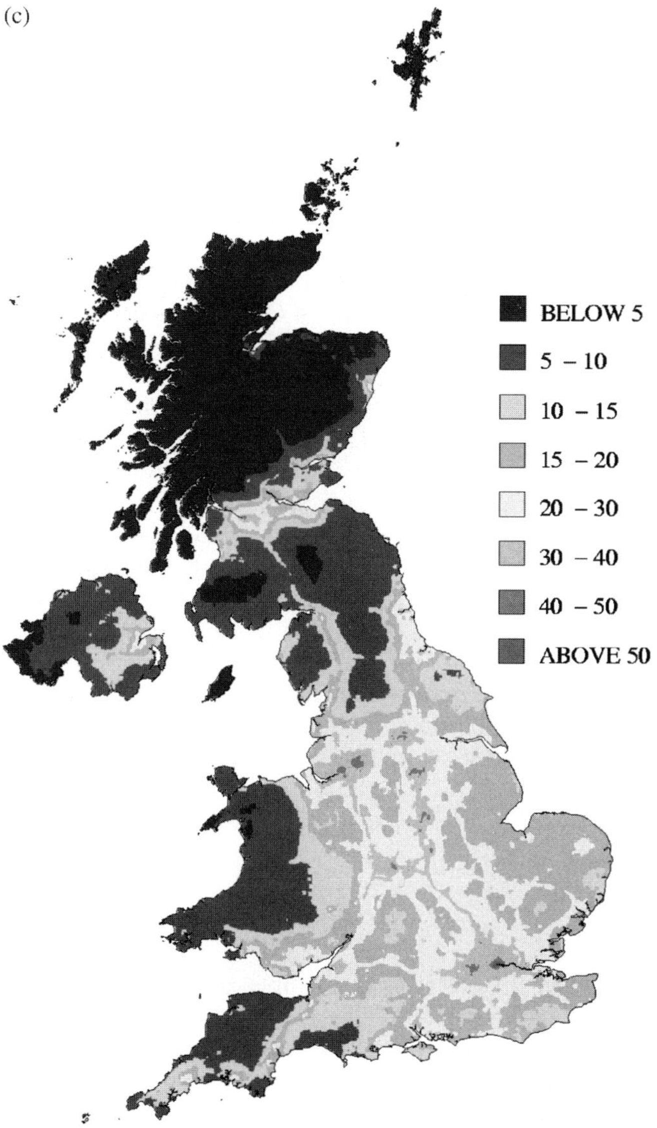

Figure 7.10c The geographical distribution of NO_2 in the UK, derived from monitoring-network data interpreted with an empirical model. (Figure 3.1 of Stedman *et al.*, 2002, printed with permission from AEA Technology.) (Reproduced in colour as Plate 3c after page 270.)

is not always high, and in order to elucidate the dependence of measured concentrations on NO_x sources, it is useful to interpolate the network measurements and to model empirically the contributions of local and more distant sources to the NO_x measured at a particular site. In turn, the empirical model is validated by comparison of its results with measurements at specific sites. Maps such as that shown in Figure 7.10c can be used to formulate statistics on the regions in which concentration objectives for NO_x are exceeded, as well as the fraction of the population thus affected. Furthermore, it serves to illustrate the impact of urban areas on more rural regions.

7.5 Ozone via homogeneous chemiluminescence

7.5.1 O_3 + Ethene

The first gas-phase chemiluminescence instrument was for the measurement of O_3 and was based on the chemiluminescent reaction of O_3 with ethene. Nederbragt *et al.* (1965) mixed flows of ozone-containing air and ethene directly in front of a PMT. The detection limit was 30 ppbv, so it was not useful for routine atmospheric monitoring but was useful for monitoring the elevated levels (1000s of ppbv) that are produced by beams of electrons near accelerators. Warren and Babcock (1970) made further improvements, especially with regard to portability, but, as the application was still for accelerators, they were satisfied with a detection limit too high for atmospheric applications.

Commercial versions based on this principle were produced and used for atmospheric monitoring, and some were modified for use in research applications, including aircraft deployment (Routhier *et al.*, 1980; Gregory *et al.*, 1983; Kondo *et al.*, 1987). The emission spectrum for the O_3–ethene reaction peaks near 430 nm (Finlayson *et al.*, 1972), and the products include excited OH and HCHO. The instruments consist of inlet flow system, an ethene flow system, and a reaction chamber viewed by a PMT. The modifications to the commercial instruments for aircraft use generally include flow control and pressure regulation, so that instrument sensitivity does not change with altitude or cabin pressure. Gregory *et al.* (1983, 1984) report a detection limit of 2 ppbv, with a 90% response in about 2 s, and an accuracy of 5% or 5 ppbv, whichever is larger.

7.5.2 O_3 + NO

Just as O_3 can be added in excess to the sample air stream for the measurement of NO, so can NO be added in excess for the measurement of O_3. Stedman *et al.* (1972) built a prototype instrument based on this principle. A typical reaction vessel was a cylindrical tube, 4″ long and 2″ in diameter, viewed from one end by a PMT through a red filter (as for NO instruments). Conical reaction vessels were also tested. A typical sample flow was 20 cc min^{-1}, with a reagent flow of 3 cc min^{-1} of pure NO. The response time was less than 1 s, excluding inlet sampling. Calibration was achieved in multiple ways using a photolytic O_3 source. One is via NO titration, which adds a measured flow of a known concentration of NO to the sample stream before it enters the instrument to allow reaction prior to that in the reaction vessel. As NO is increased the O_3 signal decreases. An extrapolation to zero O_3 gives the known amount of NO corresponding to the O_3 level being used. Another method used an infrared absorption measurement of O_3 from the photolytic source. A linear response was found for the wide range of 50 ppbv–50 ppmv. Several species were tested for interference at the 100 ppmv level: SO_2, NO_2, Cl_2, H_2O, C_2H_2, C_2H_4, and C_3H_6. None gave detectable chemiluminescence. The instrument was tested in the field where it showed excellent response to atmospheric variability. The chemiluminescence technique using NO is much more sensitive than that using ethene due to the high rate of reaction of O_3 with NO.

The high sensitivity and fast response of the chemiluminescence detector using NO has led to its use in making eddy-correlation flux measurements of O_3 to the ground, as surface deposition is a major sink for O_3. Pearson (1990) describes an instrument designed to make such measurements from an aircraft, where frequency response requirements are much more demanding than for an instrument fixed to the ground. A filtered frequency response of 12 Hz was achieved. An earlier version of the instrument (Pearson & Stedman, 1980) was improved upon by doubling the sample flow to 3.7 L min^{-1} and by increasing the diameter of the photocathode of the red-sensitive PMT to 51 mm. The reagent flow is 2.8 $cm^3\,s^{-1}$ of 5–10% NO. The cylindrical reaction vessel has an ID of 4.7 cm and a length of 8 cm and is operated at 8–10 mb. The inlet flow is regulated by a stainless steel frit in an aft-facing inlet to prevent cloud and rain water from entering. The reaction vessel is maintained at 39°C in order to keep the sensitivity constant in what may be expected to be a warm aircraft flying in the boundary layer. The PMT is operated in the analog mode and is low-pass filtered at 10–12 Hz. An operational consideration when flying the poisonous NO reagent gas is the safety requirement for a containment vessel to house the compressed gas bottle with a line plumbed to the outside of the aircraft, so that in the event of a leak, the NO is vented outside the aircraft.

Lenschow *et al.* (1981) used an earlier version of this instrument to analyze the relative contributions of chemistry and transport to the ozone budget in the boundary layer over range and crop land and to show that the time rate of change of the O_3 concentration was several times larger than contributions from horizontal advection and the divergence of the vertical eddy flux, thus demonstrating the dominance of local chemistry over transport in determining the local time rate of change of the O_3 concentration.

Ridley *et al.* (1992) describe a high-sensitivity chemiluminescence instrument that is designed to have small size and weight, suitable for use as part of a multi-instrument package on a small aircraft and still with good frequency response. Photon counting is employed to give a sensitivity of 2000 counts s^{-1} per ppbv of O_3 and a detection limit well below 0.1 ppbv for a 2 s integration. The overall measurement uncertainty at higher mixing ratios is estimated to be 2 ppbv plus 5%. The reaction vessel is conically shaped, with a volume of 17 cm^3 (eight times smaller than that of Pearson, 1990), and is made of highly polished stainless steel coated with gold. It operates at a pressure of about 10 torr, with a small sample flow of 180 sccm that is combined with a reagent flow of 1.5 sccm of pure NO. The instrument is calibrated prior to flight using a standard UV absorption instrument. Ridley *et al.* (1992) offer several interesting points about the chemiluminescence measurement of O_3 compared to that for NO. First, tropospheric O_3 is often several orders of magnitude larger than that of NO, so high signal-to-noise ratio is expected. Second, the reagent NO is much easier to provide (from a bottle) than is the reagent O_3 for the NO measurement (silent discharge). Third, since the reagent NO is pure compared to 3–5% for reagent O_3, smaller reaction vessel volumes are possible as the reaction occurs more quickly and less residence time is required. And fourth, the limiting background signal in an NO instrument due to reagent O_3 is not present in the O_3 instrument. All these factors combine to make possible instruments which are very fast and sensitive. The high sensitivity and low noise of this instrument were put to test in low altitude (~100 m) aircraft measurements in the Arctic, where ozone depletion events occurred in which O_3 mixing ratios were measured to be as low as 0.03 ppbv due to processes not yet fully understood but likely involving bromine chemistry.

7.6 NO_2 via luminol

Luminol has found application in the measurement of NO_2 via chemiluminescence near 425 nm. As luminol was described earlier to be used for the chemiluminescence measurement of O_3, an obvious concern is the interference from O_3 to the NO_2 determination. Despite this potential limitation, the luminol technique initially offered some hope of avoiding the significant interferences encountered in the NO_2 techniques described earlier (Section 7.4.3) that coupled heated catalytic converters with NO chemiluminescence detectors. Although photolytic converters for NO_2 detection do not suffer serious interference problems and are generally used instead of the catalytic converters, the luminol technique still enjoys application in commercial NO_2 instruments, as well as in some peroxy radical instruments (Section 7.9.1). It has the advantages of small size, low power consumption, and high sensitivity.

Maeda *et al.* (1980) converted a gas-phase reaction vessel so that it could hold a pool of liquid luminol solution with a constant surface area, and they varied the luminol concentration to optimize the response to NO_2. KOH was added to the solution, and its concentration was optimized at 0.05 M. KOH was determined to facilitate the dissolution of NO_2 in the luminol solution, and the OH^- may participate in the chemiluminescent reaction. Potential interferents were tested. No interference was found from NH_3, organic nitrate, organic nitrite, NO, and hydrocarbons. A negative interference was found with CO_2, but this was minimized by increasing the pH with additional KOH (1.0 M). O_3 is a very strong interferent with a chemiluminescence signal 30 times that for NO_2, but that is remedied by the addition of sodium sulfite to the luminol solution, which also removes an SO_2 interference. The detection limit for NO_2 for this prototype was 50 pptv.

Wendel *et al.* (1983) improved on the earlier design. Rather than use a pool of luminol, they used a wetted strip of cellulose fiber filter paper so that the detector would be less sensitive to movement and orientation, and so that the PMT would not view the reservoir of liquid. This also improved the time response by eliminating the signal that comes from the liquid volume at some time lag after the NO_2 encounters the surface. Other modifications were a cotton trap to remove O_3 in the inlet and the use of NaOH instead of KOH. Also, the measurement of NO was achieved by its conversion (60–70%) to NO_2 by passing the sample over CrO_3 adsorbed on silica gel (in contrast with the more common conversion of NO_2 to NO followed by NO detection). Also, Na_2SO_3 (0.01 M) was used to reduce the sensitivity to O_3 and at the same time enhance the sensitivity to NO_2. This also reduces the induction period (the time period after changing the solution during which the sensitivity to NO_2 increases) to one hour rather than days. It was also found that the addition of methanol to the solution further increases the sensitivity and specificity for NO_2. No interference was found from 20 ppbv each of HNO_3, NH_3, and HCN. At best, the NO_2 response is 30 times that for O_3. PAN was found to respond like NO_2 (indeed a gas chromatograph in conjunction with a luminol detector was later used to measure PAN and NO_2; Burkhardt *et al.*, 1988). The detection limit for field measurements was reported to be 30 pptv ($S/N = 2$), with a 2 Hz response to 20 ppb changes. However, Ray *et al.* (1986) subsequently report for a later version of this instrument a detection limit of 1 pptv and a 3 Hz response.

Schiff *et al.* (1986) describe a commercial instrument based on luminol (Luminox LMA-3, Unisearch/Scintrex). In contrast to the complex solution described in the Wendel *et al.* (1983) paper, comparatively little information is available in the open literature about this commercial instrument. However, it was evaluated by Fehsenfeld *et al.* (1990) in a comparison with a photolysis/chemiluminescence instrument (Section 7.4.3) and a tunable diode laser absorption spectrometer (TDLAS, Chapter 2). For high levels of NO_2 above 2 ppbv, similar results were found for all three instruments. On the other hand, below 2 ppbv, interferences from O_3 and PAN were evident. However, they were consistent enough that they could be corrected for down to 0.3 ppbv, provided there are simultaneous measurements of O_3 and PAN. An O_3 scrubber placed on the inlet to remove this interference worked to do that, but it did not remove PAN and it removed about 50% of the NO_2 as well. Kelly *et al.* (1990) also assessed the Luminox instrument for airborne measurements and determined several corrections that needed to be applied for low mixing ratios. Schmidt *et al.* (1995) determined that recommended O_3 corrections did not work, in contrast to the Fehsenfeld *et al.* (1990) study, but they, too, found that the proprietary scrubber had the unacceptable trait of removing NO_2 as well. However, they created a scrubber using natural rubber that removed most of the O_3 without significantly affecting NO_2.

In summary, the luminol detector is sensitive, lightweight, fast, and has low power consumption. It suffers from strong interferences from O_3 and PAN, which, in the past, have been corrected for with varying degrees of success. Also of concern are calibration instabilities, nonlinearities, and zero offsets, but Kelly *et al.* (1990) found that with care, errors as low as 15% could be attained.

7.7 Peroxides, HCHO, and HONO via liquid techniques

7.7.1 H_2O_2 via dissolution and chemiluminescence

Peroxides are products of atmospheric photochemistry that serve as reservoirs of odd-hydrogen radicals and as indicators of prior photochemistry in an air mass. One of the early techniques for the measurement of H_2O_2 employed its collection from the gas phase for detection in the liquid phase via the chemiluminescent reaction of H_2O_2 with the ubiquitous luminol reagent. Kok *et al.* (1978a) designed two methods for 'scrubbing' H_2O_2 from the gas phase, one of them being a coil through which both the sample air and distilled water flowed, extracting ambient H_2O_2 into an aqueous solution. Once in solution, it is mixed with a luminol solution and a solution of copper ions directly in front of a blue-sensitive PMT to detect the 450 nm emission. The chemiluminescence mechanism is not fully known, but it involves the decomposition of H_2O_2 by the metal catalyst, thus giving good potential for specificity. It was found to have no significant interference from benzene, hexane, PAN, NO_2, NO, SO_2, O_3, or particulate carbon. The system was linear over the 1–100 ppbv range, and the detection limit was about 1 ppbv, adequate for studies in polluted regions but not for the clean atmosphere. In a study in the Los Angeles area (Kok *et al.*, 1978b), it was found that during moderate smog episodes, when O_3 reached 150–200 ppbv, H_2O_2 rose to the range of 10–30 ppbv.

7.7.2 Peroxides via dissolution, derivatization, and fluorimetry

A more sensitive, liquid-based technique was developed by Lazrus *et al.* (1986). After the peroxides (H_2O_2 and organic peroxides) are in solution, the peroxidase enzyme is used to catalyze the reaction of peroxides with p-hydroxyphenylacetic acid (POPHA) to form a fluorescent dimer which is delivered to a fluorimeter, where the dimer is excited at 320 nm and its emission detected at 400 nm, thereby giving a measurement of total peroxides, $H_2O_2 + ROOH$. A second channel is treated with the enzyme catalase to destroy H_2O_2 before the POPHA reaction, so as to provide a measurement of ROOH only. This is thus described as a dual enzyme technique.

As in the prior instrument, sample air and liquid solution travel together through a coil, 'stripping' the peroxides from the gas phase and into the stripping solution (5×10^{-3} M potassium acid phthalate adjusted to pH = 6 with NaOH). The sample air travels through the coil at a high flow of 2 L/min, giving an air-to-solution ratio in the coil of 4800, causing the solution to form a thin film on the wall of the coil to maximize the transfer of the peroxides to the solution. After exiting the coil, the liquid is gravitationally separated from the air for analysis, the first step of which is the addition of a conditioning reagent, which eliminates an interference from SO_2. The conditioning reagent contains the H_2O_2-destroying catalase for the ROOH-only channel but not for the $H_2O_2 + ROOH$ channel. Each of the two streams is then mixed with a fluorescence reagent (containing POPHA) before delivery to a fluorimeter cell. The purity of the POPHA and H_2O_2 is of primary concern for minimizing noise and background signals. Baseline noise is about 10–33 pptv under field conditions, and the 10–90% rise time is 30 s. No interferences were found from a large number of species.

A modified version of this technique was employed by Heikes (1992) for measurements at Mauna Loa Observatory. The detection limits (three times the standard deviation of blanks measured on zero air) were 30 pptv for both H_2O_2 and ROOH. Average mixing ratios were 1050 pptv H_2O_2 and 140 pptv ROOH in free tropospheric air. Values were lower than models predicted, possibly due to inadequate treatment of removal process in the models.

A significant enhancement in this technique is the application of high-pressure liquid chromatography (HPLC, Chapter 8) to the peroxides in solution prior to the formation of the dimer. This allows the peroxides to be separated from one another and measured separately. Heikes and co-workers have deployed such an instrument on the NASA DC-8, NASA P3-B, and the NCAR C130 in several field campaigns, as well as on the ground (Lee *et al.*, 1995; Snow *et al.*, 2003). With a sample resolution of a few minutes, they have achieved detection limits for H_2O_2, hydroxymethyl hydroperoxide, CH_3OOH, peroxyacetic acid, and ethyl hydroperoxide of 5, 7, 13, 72, and 84 pptv, respectively (Lee *et al.*, 1995). By examination of data from five NASA missions in the Pacific covering different seasons over a decade, it was found that the distribution of peroxides was determined by the seasonal cycles of transport and photochemical activity (O'Sullivan *et al.*, 2004).

Morgan and Jackson (2002) also describe an HPLC instrument for ground-based use with a detection limit of 20 pptv. To collect the peroxides in solution, they use a nebulization reflux concentrator to collect the peroxides on a mist of droplets prior to separation by HPLC (Figure 7.11). The high surface area of the cloud of droplets facilitates collection of the gases. H_2O_2 and CH_3OOH were generally measurable, while hydroxymethyl hydroperoxide, peroxyacetic acid, and ethyl hydroperoxide were occasionally detectable but not measurable. In a study at Mace Head, Ireland, highest H_2O_2 levels were encountered when clean maritime air flowed over the sampling site, as opposed to continental air where NO_x species removed the peroxide precursors. A correlation of H_2O_2 with the tidal cycle was observed, with high H_2O_2 observed during low tide, coincident with particle nucleation events, and at a time when the newly exposed surfaces were likely a source of chemical species to the boundary layer.

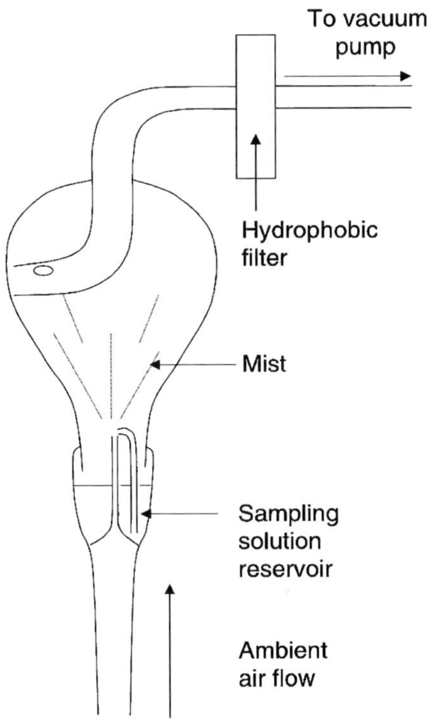

Figure 7.11 A nebulizer reflux concentrator used to collect soluble gas–phase species – peroxides in this case – into solution for subsequent detection by liquid-based analytical techniques. (Figure 1 of Morgan & Jackson, 2002, printed with permission from American Geophysical Union.)

7.7.3 HCHO via dissolution, derivatization, and fluorimetry

Formaldehyde (HCHO) is an important intermediate in atmospheric photochemistry. It is formed in the oxidation of hydrocarbons; in turn it is a source of radicals. It is also directly emitted in urban areas. It is measured by optical techniques described earlier (DOAS, TDLAS, Chapter 2) and also by techniques similar to those for peroxides.

As for peroxides, there is an enzymatic/fluorimetric technique for the measurement of HCHO (Lazrus *et al.*, 1988). A similar scrubber is used, and the aqueous HCHO is oxidized by nicotinamide adenine dinucleotide (NAD^+) to produce the reduced form (NADH) in a reaction that is catalyzed by formaldehyde dehydrogenase (FDH). The NADH is measured by fluorimetry, with excitation at 340 nm and emission at 460 nm. Heikes (1992) used an improved version of the technique at Mauna Loa. The detection limit (three times the standard deviation of blanks measured on zero air) was 60 pptv for 5 min averages. In free tropospheric air, the average mixing ratio was 100 pptv, generally lower than model predictions, while it reached a maximum of 450 pptv during a 2-day photochemical haze event. The fact that clean air can have levels near the detection limit is a persistent challenge for HCHO instruments.

In another fluorimetric technique, the dissolved HCHO reacts with a dione and ammonium acetate to ultimately form a fluorescent dihydropyridine derivative in a Hantzsch reaction (Dong & Dasgupta, 1987; Kormann *et al.*, 2003). For example, Dong and Dasgupta (1987) react HCHO with 2,4-pentanedione to ultimately form 3,5-diacetyl-1,4-dihydrolutidine (DDL), and the DDL is detected by fluorescence at 510 nm after excitation at 400 nm. Kormann *et al.* (2003) achieved a detection limit of 42 pptv (1 s) for a 3 min integration for a commercial instrument flown on the DLR Falcon. The instrument was adapted for airborne use by means of a constant-pressure inlet, and it was calibrated using liquid-phase standard solutions in the field and with a gas-phase permeation device in the laboratory. In the marine boundary layer over the Mediterranean, measured mixing ratios of 1300 pptv agreed relatively well with model results. At higher levels, however, significant discrepancies were found, possibly associated with unknown source compounds associated with biomass burning.

Cardenas *et al.* (2000) conducted a comparison of a Hantzsch-fluorometric instrument with DOAS and TDLAS instruments. This fluorimetric instrument had a detection limit of 85 ± 49 pptv($S/N = 2$) for 1 min data and was calibrated with a gas-phase permeation device that enabled the determination that inlet losses were only about 3%. At higher levels of HCHO, the fluorimeter exhibited good agreement with both techniques. As detection limits were approached (400 pptv/20 min for DOAS; 270 pptv/10 min for TDLAS), however, agreement was not so good, and this points to a significant uncertainty in HCHO measurements at background levels and the need for improved sensitivity.

7.7.4 HCHO via dissolution, derivatization, and HPLC

After scrubbing from the gas phase into solution, another technique for HCHO employs derivatization with 2,4-dinitrophenylhydrazine (DNPH), followed by HPLC separation with UV-visible detection of hydrazones to enable the measurement of HCHO and other aldehydes. While initially developed for use on the ground, Lee *et al.* (1996) report a version of this technique adapted to airborne use on the Canadian NRC Twin Otter by separating the system into (a) a sampling module based on a coil scrubber that is flown and (b) an HPLC-based analysis system that is used on the ground within about 2 hr of the end of the flight. It was later deployed on the NOAA WP-3 and has been compared with a TDLAS aboard the same aircraft (Fried *et al.*, 2002). On the WP-3 the 5 min precisions for the derivatization technique were in the range 40–80 pptv with systematic uncertainties of 10–20%, and the TDLAS errors were comparable. On average, the measurements agreed to within about 100 pptv over the range of ambient levels of 0–800 pptv and do not display a systematic difference. However, significant differences (of random sign) were frequently observed. The two instruments found similar altitude trends with relatively high levels (257 pptv) at the higher altitudes sampled (4–8 km) off the NE coast of the US, indicative of vertical transport of HCHO or its precursors to these altitudes.

7.7.5 HONO via liquid techniques

Since it is readily photolyzed, HONO can be an important source of OH radicals at sunrise, before daytime photochemistry becomes efficient. The atmospheric formation of HONO is not well understood, but surface sources and aerosol sources are likely. Harrison *et al.* (1996) describe two liquid-based techniques for the measurement of HONO. In the first, two vertically oriented annular denuders (3 mm gap) in series sample an air flow at 10 L min^{-1} for a period of time, after which each is extracted into 20 mL of deionized water and stored for later analysis. Ion chromatography (Chapter 8) is used to analyze for NO_2^- (from the ambient HONO). The first denuder collects all of the ambient HONO, so the signal from the second one represents a background due to NO_2 and other interferents, so its value (typically only 2–5% of that in the primary) is subtracted to determine the amount of ambient HONO collected in the sample. This is converted to an ambient concentration from a record of the total volume of air sampled, which is determined with a calibrated dry gas meter.

A second method is a continuous analyzer which starts by scrubbing HONO into a solution of dilute sodium carbonate which is pumped continuously through the collection coil and then mixed with a solution of ascorbic acid in sulfuric acid. This reduces the dissolved NO_2^- to NO, which is released to the gas phase and is measured by a standard, commercial chemiluminescence analyzer (Section 7.4.2). The scrubber removes all of the HONO but also a small fraction of the ambient NO_2 and PAN, both of which contribute a background signal. To measure this background, the already sampled air is passed through a second scrubber so that the effects of ambient NO_2 and PAN (the HONO is gone) can be measured. The second channel of the NO analyzer is used to measure the NO from this scrubber. Interference tests using a mixture of 240 ppbv PAN and 225 ppbv NO_2 yield an interference corresponding to only 1.9% of PAN and 0.4% of NO_2. The instrument generates continuous data, but it is effectively smoothed to a 2–5 min averaging time.

These techniques have been used to elucidate interesting aspects of the land surface exchange of HONO over grass and sugar beet surfaces. It is concluded that there is a surface reaction between NO_2 and H_2O which produces HONO, and when NO_2 is greater than about 10 ppbv this heterogeneous source causes a net upward flux of HONO from the land surface, but when NO_2 is less than about 10 ppbv, there is a net downward flux (Harrison *et al.*, 1996). HONO production on surfaces also raises a caution for sampling into instruments. Zhou *et al.* (2002) found evidence for the production of HONO on the surface of a sunlit glass sampling manifold, presumably derived from adsorbed HNO_3 and H_2O.

Some newer instruments employ photometric detection in the liquid phase after derivatization into an azo dye. For example, Heland *et al.* (2001) describe a long-path absorption photometer (LOPAP) which has a continuous stripping coil where HONO is converted into a diazonium salt, followed by an azo dye unit where the dye is formed, and then followed by a teflon tube which acts as a long-path absorption cell. Due to the multiple internal reflections a long path length is attained, and a grating spectrometer with a diode array measures the absorption spectrum. As for the continuous technique described just above (detecting gaseous NO_2), a second channel in series with the primary is used to eliminate potential interferences by providing a background for subtraction. It has been found to agree with the IC technique, described above, in a two-point comparison in the laboratory. Additionally, LOPAP compared reasonably well when tested in a large outdoor smog chamber against a DOAS device (Chapter 3), which is less likely to suffer interferences. Deviations were within error limits, but the new instrument was systematically higher by $(13 \pm 5)\%$. The value of LOPAP is its low cost and small size, while still achieving a detection limit of 3 pptv for a 4 min integration.

7.8 Isoprene via O_3 chemiluminescence

The largest known flux of a reactive, biogenic hydrocarbon to the atmosphere is that of isoprene. It even influences atmospheric chemistry in urban areas. Hills and Zimmerman (1990) describe an instrument for the measurement of isoprene based on its chemiluminescent reaction with O_3. The instrument is fast and so has been used for eddy covariance measurements of isoprene fluxes (Guenther & Hills, 1998). The 95% response time to a step change is about 0.5 s, so the system is capable of measuring isoprene fluctuations of 1 Hz and less. Specificity is a concern, and there is potentially an interference from NO (Section 7.4.2), but this is discriminated against spectrally, as NO_2^* emits over the range 600–2600 nm with a peak at 1200 nm, while the isoprene–O_3 reaction produces light in the 500 nm region, with a peak at 490 nm from excited HCHO and at 550 nm from glyoxal. Within the context of flux measurements over a forest canopy, the interference primarily comes from other organic compounds that are emitted, such as monoterpenes, propene, ethene, and methyl butenol, or from the deposition of methacrolein and methyl vinyl ketone. Using a knowledge of typical North American emission fluxes, it is estimated that the total interference would be about +5% for emission fluxes and −3% for deposition fluxes.

Field measurements at a mixed hardwood site, where isoprene emissions are expected to be substantial, found daytime fluxes in the range of 1–14 $mg\,C\,m^{-2}\,h^{-1}$. Measured values follow model predictions of the diurnal cycle, which is driven by photosynthetically active radiation and temperature (Figure 7.12), and are generally in accord with results from other, more labor-intensive techniques.

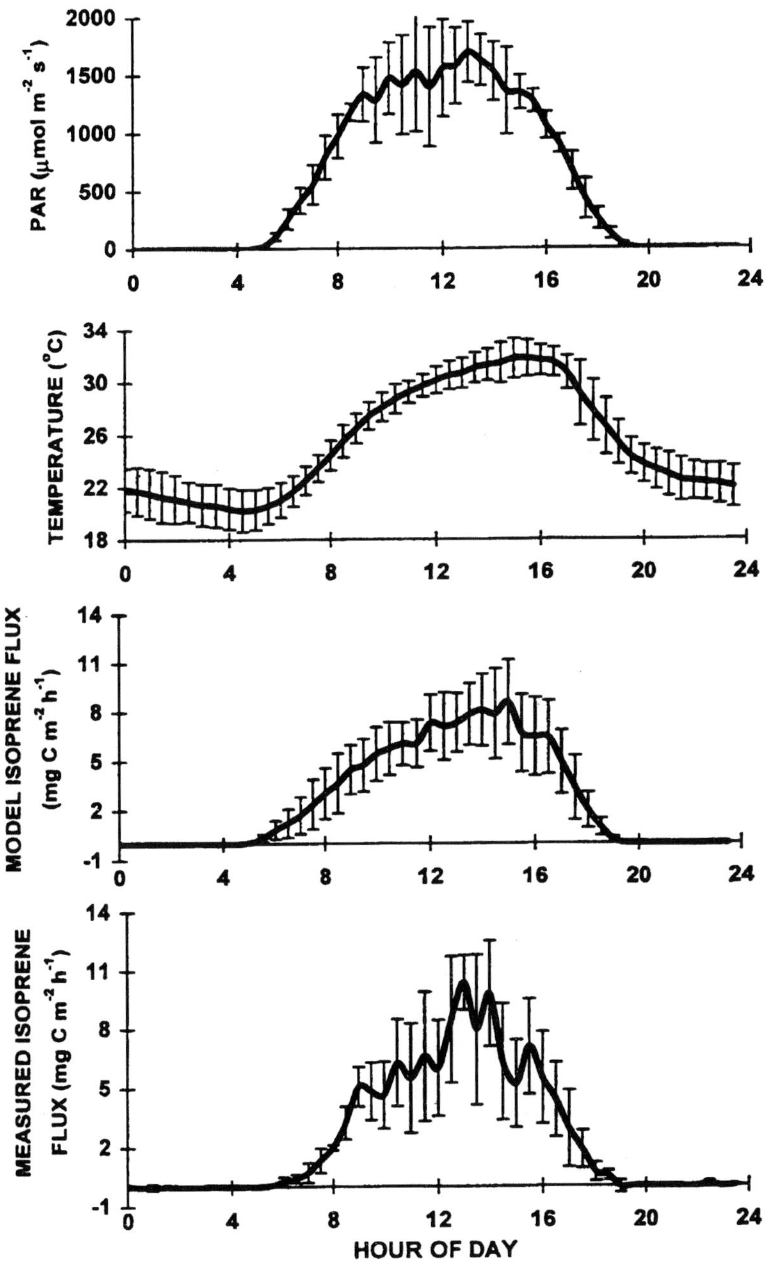

Figure 7.12 The flux of isoprene (bottom panel) is measured above the canopy of a mixed hardwood site in North Carolina using a fast-response chemiluminescence detector. The measured diurnal cycle, driven by photosynthetically active radiation and temperature, agrees well with the model. (Figure 8 of Guenther & Hills, 1998, printed with permission from American Geophysical Union.)

7.9 Peroxy radicals via chemical amplification

Peroxy radicals (HO_2 and RO_2) play an essential role in atmospheric O_3 formation through their ability to transfer an O atom to NO to form NO_2, which in turn can be photolyzed to liberate an O atom which combines with an O_2 molecule to from O_3. For the measurement of peroxy radicals, two methods of chemical conversion, with amplification, have been employed. Common to the two methods, for HO_2 detection, is the initial conversion of HO_2 to OH via reaction with NO, with subsequent reactions to recycle most of the product OH back to HO_2, which in turn reacts with the abundant NO to yield even more HO_2, thereby giving amplification. The methods differ in the reactions used to recycle OH back to HO_2 as well as in the species chosen for detection of the amplified signal, but amplification makes it possible for both methods to detect a species more abundant than the original HO_2. As described below, the original method employs CO in the recycling of OH, with NO_2 as the detected species. Techniques at the time of writing employ SO_2 in the recycling of OH, with H_2SO_4 as the detected species.

7.9.1 Using reagent NO and CO, followed by detection of NO_2

The technique for the measurement of peroxy radicals by chemical amplification (PERCA) was pioneered by Cantrell and Stedman (1982) and Cantrell *et al.* (1984), with later developments by Hastie *et al.* (1991) and others. The reaction mechanism starts with the conversion of HO_2 to NO_2 for detection via the luminol technique described earlier (Section 7.5):

$$HO_2 + NO \rightarrow OH + NO_2 \quad (7.5)$$

$$OH + CO \rightarrow H + CO_2 \quad (7.6)$$

$$H + O_2 + M \rightarrow HO_2 + M \quad (7.7)$$

The OH produced in reaction (7.5) is recycled back to HO_2 by the reactions (7.6, 7.7) with CO and O_2 to further increase the level of NO_2. The sensitivity of the technique is limited by radical loss processes such as reactions on the walls and radical–radical reactions which terminate the chain:

$$OH + NO + M \rightarrow HONO + M \quad (7.8)$$

$$HO_2 + NO_2 + M \rightarrow HO_2NO_2 + M \quad (7.9)$$

$$HO_2 + HO_2 \rightarrow H_2O_2 + O_2 \quad (7.10)$$

The number of NO_2 molecules produced per initial, ambient HO_2 molecule is dependent on the concentration of reagent NO as well as the allowed reaction time, but for typical operational conditions, it is of order 100, or often greater. This large amplification factor, or chain length, results in an elevated level of NO_2 that can readily be detected. The overall chain reaction oxidizes NO and CO by HO_2 and OH, respectively, and it can be initiated by either HO_2 or OH. Thus an instrument responds to ambient OH as well, so its use as an HO_2 measurement relies on the ambient OH abundance typically being about a factor of 100 smaller than that of HO_2.

Organic peroxy radicals are also detected by this technique and with comparable, albeit somewhat smaller, sensitivities (or chain lengths):

$$RO_2 + NO \rightarrow RO + NO_2 \tag{7.11a}$$

$$RO + O_2 \rightarrow R''COR' + HO_2 \tag{7.12}$$

The effective chain length is smaller than for HO_2 due to competing reactions which lead to the production of nitrogen compounds and terminate the chain:

$$RO_2 + NO + M \rightarrow RONO_2 + M \tag{7.11b}$$

$$RO + NO + M \rightarrow RONO + M \tag{7.13}$$

Also the alkoxy radicals can undergo decomposition or isomerization processes which may in turn lead to HO_2 formation in varying degrees. Because the chain length varies from one peroxy radical species to the next, this poses a potential difficulty for the interpretation of the measurement as one of the strict sum of HO_2 and RO_2 concentrations. This was addressed in Cantrell *et al.* (1993) by modeling the set of possible reactions for a representative set of organic peroxy radicals to obtain the various chain lengths. These were combined with estimated abundances to determine the effective chain length for a typical mix of peroxy radicals (HO_2 plus organics). The effective chain length was modeled to be about 91% of the HO_2 chain length. Because the chain lengths of the (presumably) most abundant organic peroxy radicals generally approach that of HO_2, in most situations it is sufficient to use an effective chain length. This result is generally supported by the work of Ashbourn *et al.* (1998), who not only performed similar modeling but measured the response of a PERCA to various organic peroxy radicals relative to the response for HO_2. Furthermore, this study highlighted the importance of variable heterogeneous losses to inlet surfaces as another source of variation in response among radicals, as well as a potentially even more significant uncertainty in the measurement of peroxy radicals unless great care is taken in the design of the inlet.

Figure 7.13 is a schematic diagram of a chemical amplifier instrument (Cantrell *et al.*, 1993). It consists of an inlet section (e.g. glass-lined stainless steel), where the conversion chemistry occurs, coupled to a luminol-based NO_2 detector (Section 7.5). The typical inlet flow is 1 lpm with reagent mixing ratios of 3 ppmv NO and 10% CO. Air is pumped through the inlet and then to the detector, with provision for the addition of reagent gases at the upstream end of the inlet. An NO_2 source is also shown as a means of calibrating the NO_2 detector.

To calibrate a PERCA instrument, two things are required: (1) an absolute calibration of the NO_2 detector to quantify the amplified NO_2 signal, and (2) a knowledge of the effective chain length in order to reduce the amplified NO_2 mixing ratio back to the ambient peroxy radical mixing ratio. However, there is a significant background signal present that must be removed prior to application of this sensitivity calibration. Cantrell and coworkers have achieved this through modulation of the signal by replacing the

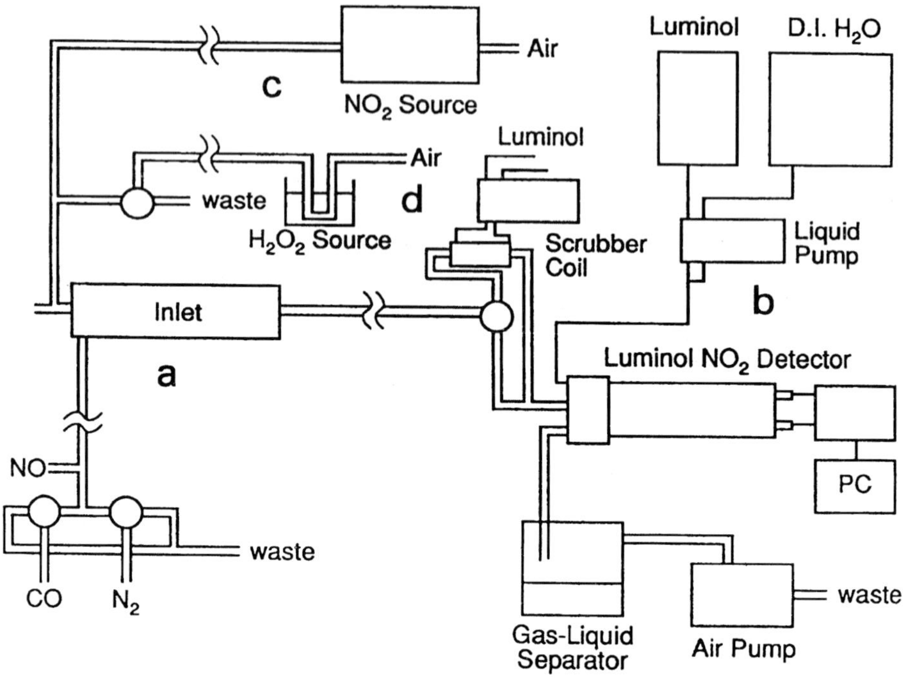

Figure 7.13 Schematic diagram of an instrument for the measurement of peroxy radicals by chemical amplification (PERCA), based on the luminol detection of NO_2. (Figure 8 of Cantrell *et al.*, 1993, printed with permission from American Geophysical Union.)

CO flow with an N_2 flow (Figure 7.13). This turns off the chain reaction and allows the luminol detector to see the large background signal due to ambient NO_2 as well as the NO_2 produced from ambient O_3 by the large levels of reagent NO in the reactor. The amplitude of the modulation is then the amplified radical signal. A limitation of the technique is that the amplitude of the modulation can be small compared to the background (e.g. ~10% in Figure 3 of Cantrell *et al.*, 1984). Thus any variability in the background due to variations in ambient NO_2 and O_3 can lead to errors in the measurement.

Another difficulty with the technique is interferences from peroxyacetyl nitrate ($CH_3C(O)O_2NO_2$, PAN) and peroxynitric acid (HO_2NO_2, PNA). These species dissociate thermally and produce, respectively, peroxyacetyl radicals ($CH_3C(O)O_2$, PA) and HO_2:

$$CH_3C(O)O_2NO_2 \rightarrow CH_3C(O)O_2 + NO_2 \quad (7.14)$$

$$HO_2NO_2 \rightarrow HO_2 + NO_2 \quad (7.15)$$

The NO_2 so produced will simply contribute to the background and is removed by the modulation technique. However, the HO_2 and PA radicals will lead to amplified signals and are true interferents. Hastie *et al.* (1991) determined that the magnitude of these interferences could be large, and they addressed this by a modification of the modulation technique which reduces the interferences by a factor of ~25. Instead of replacing the CO flow by N_2, modulation was accomplished by employing an alternative CO addition point 200–300 ms downstream from the NO addition point used to implement the amplifier chemistry. This allows enough time for the radical chemistry to run its course, terminating in the formation of HONO, via reaction (7.8), before CO is introduced to enable the recycling of OH back to HO_2 and also before there is enough time for a significant fraction of the thermally labile species to dissociate. Thus the ambient peroxy radicals are not amplified in this background mode, but in the subsequent few seconds of reaction time, PAN and PNA decompose and the amplified signals from the resultant PA and HO_2 radicals contribute to the background signal, and so are removed by the modulation. Another advantage of this approach is that the nitrogen flow is no longer needed, and the luminol detector does not suffer shifts due to changes in gas composition, which can make an errant contribution to the radical (modulation) signal if not carefully accounted for (as in Cantrell *et al.*, 1993). Hu and Stedman (1994) minimize interferences from PAN and PNA by using a smaller reaction chamber and thereby a shorter reaction time.

Once the amplitude of the modulation signal is obtained it must be converted to an absolute amount of NO_2, using the absolute calibration of the luminol-based NO_2 detector. Cantrell *et al.* (1993) found a number of complicating factors that need to be addressed. One is that the detector sensitivity changes due to the reagent gases. There is a quadratic response to NO_2 with the high NO that is present, and this is minimized by adjusting the luminol solution. Also, there is a dependence on CO level, as well as a signal from any metal carbonyls that originate in the CO tank. In any event, these dependencies necessitate calibration of the detector at more than one level of NO_2 and in the presence of the different gas flows required of the modulation scheme employed in any particular instrument, and this is done.

Once the amplitude of the modulation signal is expressed as an absolute mixing ratio of NO_2, this must be converted to the mixing ratio of ambient radicals by dividing by the amplification factor, or chain length. The determination of the chain length has been approached in a number of ways. Early calculations of Cantrell and co-workers indicated chain lengths of over 1000, yet experimentally determined chain lengths were typically smaller by an order of magnitude. Hastie *et al.* (1991) pointed out that the loss of radicals to the walls could be a dominant, and underestimated, loss process, consistent with chain lengths more typically of order 100. Additionally, they proposed a calibration procedure using PA radicals from the thermal decomposition of PAN, with the PAN amount being determined via gas chromatography (Chapter 8) coupled with a molybdenum NO_y chemiluminescence instrument (Section 7.4.4). Their modeling of measured chain lengths, using fitted values for the wall loss coefficient led to the conclusion that more radicals are lost to the walls than via either of the terminating reactions (7.8) and (7.9). Cantrell *et al.* (1993) employed the thermal decomposition of H_2O_2 to produce an arbitrary amount of radicals for calibration purposes. After measuring the magnitudes of the signals above background, with and without the radical chemistry turned on, the ratio of the two gives the amplification factor directly.

Another calibration method that has been used widely with chemical amplifiers (Section 7.9.2) is the 185 nm photolysis of H_2O in the presence of O_2 to yield equal amounts of OH and HO_2 (Schultz *et al.*, 1995). O_2 is also photolyzed at 185 nm, and the concentration of O_3 so produced can be measured and used in combination with the known absorption cross section of O_2 to determine the actinic flux at 185 nm and therefore the amount of radicals produced, given the cross section for H_2O. Schultz *et al.* (1995) note such a system is versatile since CO can be added to convert all of the OH to HO_2, or that a hydrocarbon can be used instead of CO, so that an instrument can be calibrated for specific RO_2 radicals that derive from the hydrocarbon. An early difficulty with the technique centered on an uncertainty in the H_2O cross section, but later measurements have converged on a newly accepted value, about 30% higher than the prior one (Cantrell *et al.*, 1997; Hofzumahaus *et al.*, 1997; Creasey *et al.*, 2000). The O_2 cross section can also be problematic as it exhibits a lot of structure in the vicinity of 185 nm, and the actual spectrum from the Hg lamps used may vary with its operating conditions (current, temperature, age) and with the optical depth between the lamp and the volume photolyzed (Cantrell *et al.*, 1997; Lanzendorf *et al.*, 1997). However, if care is taken to determine the actinic flux with conditions that match those in effect for the HO_2 calibration, then this method is reliable (Creasey *et al.*, 2000). Alternatively, rather than accomplishing actinometry via O_2 absorption, N_2O absorption may be employed coupled with a chemiluminescence instrument to measure the NO so produced (Edwards *et al.*, 2003). An additional photolytic calibration technique employs the photolysis of CH_3I at 254 nm in the presence of O_2 to produce CH_3O_2 (Clemitshaw *et al.*, 1997), with the actinometry accomplished by monitoring the decay of CH_3I with an absorption measurement at 265 nm.

A significant dependence of chain length on H_2O has been discovered and investigated. Mihele and co-workers (Mihele & Hastie, 1998; Mihele *et al.*, 1999) found a reduction in chain length of a factor of 2 at a relative humidity of 40%, and they attributed this to increased uptake of radicals on the walls with an increase in relative humidity. Mihele and Hastie (2000) determined that it could be significantly improved by heating the walls of the reactor and reducing the residence time there. Also, they derived a correction to the chain length that is a function of relative humidity. On the other hand, Reichert *et al.* (2003) determined that the H_2O effect is not all due to walls, but that it also involves water complexes. No matter the cause it would seem that the older PERCA measurements are in need of correction for this effect.

A variation on PERCA is the use of laser-induced fluorescence (LIF, Chapter 4) to measure NO_2 rather than luminol. This avoids problems of specificity associated with the luminol technique (especially O_3), and there is potential for improved time response as well. However, the technique is still susceptible to uncertainties due to background variability, but this should be smaller, as O_3 does not contribute. Sadanaga *et al.* (2004) describe such an instrument which uses the same amplification chemistry as standard PERCAs, as well as substituting N_2 for CO for background determination. It also uses H_2O photolysis at 185 nm coupled with O_3 actinometry to provide an HO_2 calibration. For a 1 min integration time the detection limit ($S/N = 2$) of the instrument is in the range of 2–4 pptv, depending on ambient relative humidity (and assuming 30 ppbv O_3 and 20 ppbv NO_2).

Instruments using PERCA are typically deployed on the ground (e.g. Monks *et al.*, 1996; Clemitshaw *et al.*, 1997; Handisides *et al.*, 2003). For a study at the Mauna Loa Observatory, Cantrell *et al.* (1996b) measured peroxy radicals for four one-month intensives. For a typical midday RO_2 mixing ratio of 25 pptv the accuracy was $\pm$(8–16) pptv for a 1 min average. These data were used to compare with steady state model calculations based on a large suite of trace gas measurements and other parameters to test for model agreement with the measured diurnal cycle. On a number of days the agreement is quite good. However, on other days the measured values are lower than those calculated suggesting, as one possibility at least, that peroxy radicals are lost to aerosol surfaces. Green *et al.* (2003) report the first use of the PERCA on an aircraft (C130). The performance was good (40% uncertainty) provided O_3 was constant, but O_3 variability could be compounded by aircraft speed, causing large variations in the background signal, resulting in unusable data. Such variability is a common feature of aircraft sampling, but it may be possible to compensate for it using a dual-inlet approach, with one channel dedicated to simply measuring the background (Cantrell *et al.*, 1996a), so that the background is actually measured simultaneous with the peroxy radical signal.

7.9.2 Using reagent NO and SO_2, followed by CIMS detection of H_2SO_4

A new method of chemical amplification and detection was implemented by Reiner *et al.* (1997) using sulfur chemistry:

$$HO_2 + NO \rightarrow OH + NO_2 \qquad (7.5)$$

$$OH + SO_2 + M \rightarrow HOSO_2 + M \qquad (7.16)$$

$$HOSO_2 + O_2 \rightarrow SO_3 + HO_2 \qquad (7.17)$$

$$SO_3 + 2H_2O \rightarrow H_2SO_4 + H_2O \qquad (7.18)$$

The initial step in this scheme is the same as in the original, but now the OH is recycled back to HO_2 via reaction with SO_2 (7.16) rather than CO. The SO_2 is ultimately oxidized to H_2SO_4 (7.18), which is the detected species. A distinct advantage of this scheme is the much smaller atmospheric background for H_2SO_4 compared to the NO_2 detected in the PERCA scheme. Because the background of the detected species is smaller, less amplification is required, and the reaction time can be shorter, lessening interferences and loss processes from, as before, chain terminators like reaction (7.8) and losses to the walls. Also, since the background is smaller, less measurement uncertainty is introduced by its variability. The technique also does not suffer the strong H_2O dependence of the PERCA. The H_2SO_4 produced by this amplification scheme is detected by chemical ionization mass spectrometry (CIMS), a technique already employed in the measurement of atmospheric H_2SO_4 (Chapter 5). The instrument acronym is ROxMAS (RO_x mass spectrometer) or PerCIMS (peroxy CIMS).

Organic peroxy radicals are also detected in this instrument. For methyl peroxy radical:

$$CH_3O_2 + NO \rightarrow CH_3O + NO_2 \qquad (7.19)$$

$$CH_3O + O_2 \rightarrow HCHO + HO_2 \qquad (7.20)$$

The HO_2 so produced (7.20) can then participate in the scheme (7.5, 7.16–7.18). However, since reactions other than (7.20) are possible for CH_3O, the efficiency of conversion of CH_3O_2 to H_2SO_4 is somewhat less than that for HO_2. This was measured in the laboratory and was found to be in the range of 80–100%, depending on reagent gas concentrations (Reiner *et al.*, 1999). So, the incomplete conversion of this particular RO_2 species is only modest and not a dominant source of error, so incomplete conversion of organic peroxy radicals is not a serious problem in the clean atmosphere where CH_3O_2 will be the dominant one. However, in more polluted environments, other organic peroxy radicals are present whose conversion is not guaranteed, so there may be increased measurement uncertainty.

Reiner *et al.* (1998, 1999) developed an instrument for deployment on an aircraft (Figure 7.14). Ambient air is forced through the flow reactor by the motion of the aircraft, so it operates near ambient pressure. The reagent gases, NO and SO_2, are added at the upstream end of the reactor at port RG1 (shown in Figure 7.14). These convert the peroxy radicals to H_2SO_4 by the time the sample reaches the inlet to the ion-molecule reaction (IMR) section which operates at a reduced pressure near 100 mb. Here the CIMS technique is employed to detect H_2SO_4 (Chapter 5). The amplification factor (chain length) and the ultimate H_2SO_4 concentration are determined by the concentrations of the reagent gases, but for typical conditions, this factor is about 7, and the lifetime for HO_2 against reaction with NO is small, at less than 0.1 s, helping to minimize the effects of unwanted loss processes. A background signal is obtained by switching the reagent SO_2 from RG1 to RG2 (Figure 7.14), while still adding NO at RG1. This allows the OH + NO termination reaction to occur, but since SO_2 is still added to the system, interferent peroxy radicals (e.g. from PAN or HO_2NO_2) formed downstream of RG2 will be converted to H_2SO_4 and will contribute to the background determination.

Calibration is achieved using the UV photolysis of H_2O at 185 nm, as described in Section 7.9.1, with care taken to use the effective absorption cross sections characteristic of the particular system employed. The effects of wall losses were carefully measured on the ground, and the airborne measurements of peroxy radicals were corrected for these

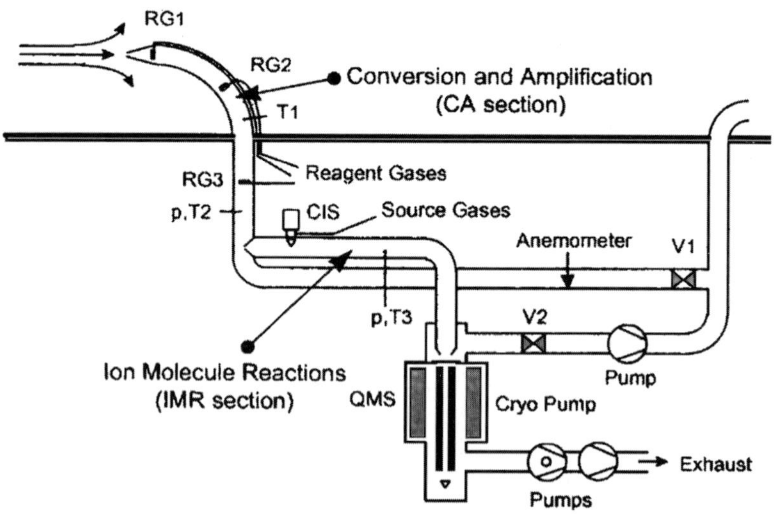

Figure 7.14 Schematic diagram of an airborne instrument for the measurement of peroxy radicals by chemical amplification, based on the CIMS detection of H_2SO_4 (RO_xMAS). (Figure 1 of Reiner *et al.*, 1998, printed with permission from American Geophysical Union.)

effects. The accuracy of the airborne measurements is ±(40–50%), with the dominant contribution coming from the 35% uncertainty in the output of the calibration source. The precision of the measurements under stable in-flight conditions is about 15–20%.

Airborne peroxy radical measurements were made over Germany with values of $HO_2 + RO_2$ typically in the range of 10–50 pptv over the 4–8 km altitude range. Additional measurements of O_3, NO, H_2O, and the ozone photolysis frequency provided enough constraints to allow a comparison of the measurements with a simple steady state model. For a particular flight under clear-sky conditions, good agreement was obtained, indicating that the production of peroxy radicals was dominated in that case by the production of OH via O_3 photolysis followed by O-atom reaction with H_2O, and that radical loss was dominated by self and permutation reactions among peroxy radicals.

This instrument can also be used to discriminate between HO_2 and RO_2 by exploiting the dependence on O_2 of the conversion efficiency of RO_2 species to HO_2 (7.20), and the manner in which that conversion to HO_2 competes with addition reactions of RO_2 with NO and of RO with NO and SO_2. By switching the buffer gas used to dilute the sample flow between O_2 and N_2, the O_2 mixing ratio can be reduced to about 2%, thereby suppressing the conversion of RO to HO_2 in (7.20). This gives rise to an approximate HO_2-only mode, whose signal can be subtracted from the usual $HO_2 + RO_2$ mode to yield a measure of RO_2, with careful consideration of conversion efficiencies for the two modes. Figure 7.15 shows measurements made on a mountaintop in Italy, illustrating the diurnal cycles of HO_2, RO_2, and their sum, with midday values of 30 pptv for the sum, and with comparable contributions from HO_2 and RO_2 (Hanke *et al.*, 2002). Also shown is the production rate for OH that is calculated from measurements of O_3, H_2O, and the O_3 photolysis rate. The observed strong correlation between the OH production rate and the peroxy radical mixing ratios, even for some short time scale features, generally results from the fast photochemical equilibrium established among this set of radicals and serves to confirm aspects of our understanding of this photochemistry.

Using the same amplification chemistry as Reiner and co-workers, with NO and SO_2 as reagent gases, Cantrell and coworkers (Edwards *et al.*, 2003) developed a similar instrument (PerCIMS), also detecting the resultant H_2SO_4 using CIMS. The instrument also has a mode which discriminates against RO_2 species, for an HO_2-only mode, but it is achieved differently from ROxMAS. The radical loss reaction

$$RO + NO + M \rightarrow RONO + M \qquad (7.21)$$

is made to compete with the HO_2-producing reaction (7.20), not by reducing O_2, but by increasing the NO mixing ratio by about two orders of magnitude from the order of 10 ppmv up to 1000 ppmv or more. Extensive laboratory tests of the discrimination against RO_2 species were conducted by measuring the response in the HO_x mode (high NO) to a large variety of RO_2 radicals generated from these precursors: methane, propane, butane, isobutane, ethene, propene, *cis*-butene, *trans*-butene, isoprene, acetone, benzene, and toluene. Conversion efficiencies in the HO_x mode were generally about 10–15%, and by modeling the conversion chemistry it was concluded that $RO + SO_2$ chemistry is quite significant in bringing the conversion up from the 2–5% range to the 10–15% range. The

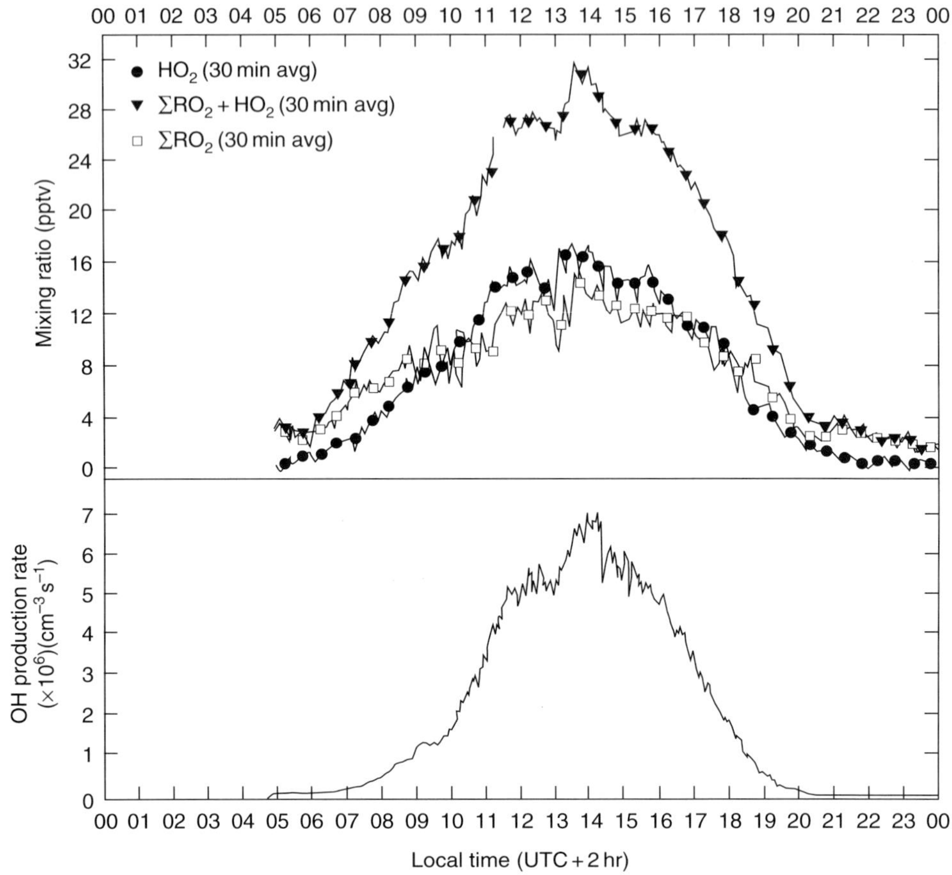

Figure 7.15 RO_xMAS measurements of the diurnal cycle of HO_2, RO_2, and their sum (upper panel) on a mountaintop in Italy, along with the calculated OH production rate (lower panel). (Figure 3 of Hanke *et al.*, 2002, printed with permission from Elsevier.)

conversion efficiencies were also measured in HO_xRO_x mode and were mostly in the range of 75–90%. These laboratory tests demonstrate that the PerCIMS instrument should yield HO_xRO_x values which are 92–100% of the true amounts. Moreover, by modeling the RO_2 partitioning, a correction may be applied for these modest biases. The accuracy for HO_xRO_x is ±35% (at 95% confidence), and for HO_2 it is 41%. The typical precision is 10%, and the typical detection limit is $1 \times 10^7\,\text{cm}^{-3}$ for a 15 s average (~1 pptv at ~500 mb). PerCIMS is calibrated using H_2O photolysis with N_2O actinometry (described previously).

The PerCIMS measurement of $HO_2 + RO_2$ on a P-3B flying alongside a DC-8 carrying an LIF instrument for HO_2 showed generally favorable agreement, with the sum of peroxy radicals on average being about 2.5 times that of HO_2 alone, as might be expected for these conditions, and with a degree of scatter generally consistent with estimated errors. The PerCIMS instrument has been deployed on aircraft in field campaigns with large suites of complementary instruments which, via modeling, allow meaningful tests of our understanding of the radical chemistry of the atmosphere (Cantrell *et al.*, 2003). For example, the species that determine the radical production rate are measured, and the relations among the calculated production rate, the measured $HO_2 + RO_2$ amount, and NO_x have been evaluated and found to be generally consistent with our understanding. Also, peroxy radical measurements, when combined with measurements of these other species, enable the calculation of the time rate of change of O_3 in an air mass. Interestingly it has also been observed that radical concentrations are reduced in the presence of clouds (Cantrell *et al.*, 2003), indicating the potential for a significant effect of clouds on oxidant chemistry.

7.10 Summary and future directions

For several decades these 'chemical method' techniques have provided valuable measurements of a wide variety of species, and many of them will continue to do so for the foreseeable future. However, there are some areas where shortcomings can be identified, shortcomings that limit advances in our understanding or ability to monitor. An area where some of these instruments fail to tell the whole story is with respect to speciation within groups of species. For example, the organic peroxy radical measurement by chemical amplification provides a lumped measure of these species, whereas speciated measurement of the different RO_2 would be useful and could be related to measurements of their source hydrocarbons in the air mass. In the same vein, speciated measurement of organic peroxides would also be useful. The HPLC techniques described in this chapter succeed in some cases, but it seems that mixing ratios are often below detection limits. The same is true for HCHO measurements, which still have large uncertainties in clean, background air. Likewise, with respect to speciation, the measurement of NO_y, while valuable, could gainfully be replaced by speciated measurements of the many NO_y component species. Progress in this area is being made with use of the CIMS technique already for HNO_3, N_2O_5, and speciated PANs (peroxy acyl nitrates). To some extent this renders the NO_y measurement not as useful as it once was; however, it is still a valuable tracer, especially for those campaigns in which an insufficient number of component species are measured. Improvements in the photolytic/chemiluminescence detection of NO_2 may be on the horizon with the advent of smaller, lighter, less heat-producing light sources such as LEDs emitting in the 400 nm region (Buhr, 2004). On the other hand, LIF techniques (Chapter 4) can be an attractive alternative to photolytic/chemiluminescence detection when NO is a large fraction of NO_x (as in the upper troposphere), because under such conditions the latter technique will always be limited by the need to calculate NO_2 as a difference between two relatively large numbers. As for measurement platforms, the use of in-service aircraft, as by the MOZAIC program, could be used to assemble expansive climatologies of a variety of species and provide a type of specialized (not-so-routine) monitoring of the important UT/LS region. All aircraft data, and especially those from a monitoring program such as MOZAIC, are valuable for validating the ever-growing suite of satellite measurements, which provide tremendous information on the broad scale, but which still require in situ aircraft data to aid in their interpretation.

Further reading

Chemiluminescence in Analytical Chemistry, edited by Ana M. García-Campaña, Willy R.G. Baeyens, Marcel Dekker, Inc., 2001.

Handbook of Instrumental Techniques for Analytical Chemistry, edited by Frank A. Settle, Prentice-Hall, 1997.

R.E. Sievers (1995) *Selective Detectors: Environmental, Industrial, and Biomedical Applications*, Wiley.

K.C. Clemitshaw (2004) A review of instrumentation and measurement techniques for ground-based and airborne field studies of gas-phase tropospheric chemistry, *Critical Reviews in Environmental Science and Technology*, 34, 1–108.

C.A. Cantrell (2003) Observation for chemistry (in situ), chemiluminescent techniques, in *Encyclopedia of Atmospheric Sciences*, J.R. Holton, J.A. Curry, J.A. Pyle (eds), Academic Press, Vol. 4, pp. 1454–1460.

References

Air Quality Expert Group (AQEG), M. Pilling, chair (2004) Nitrogen dioxide in the United Kingdom, http://www.defra.gov.uk/environment/airquality/aqeg.

Ashbourn, S.F.M., Jenkin, M.E., & Clemitshaw, K.C. (1998) Laboratory studies of the response of a peroxy radical chemical amplifier to HO_2 and a series of organic peroxy radicals, *J. Atmos. Chem.*, 29(3), 233–266.

Bernanose, A.J., & René, M.G. (1959) Oxyluminescence of a few fluorescent compounds of ozone, *Advan. Chem. Ser.*, 21, 7–12.

Bollinger, M.J., Sievers, R.E., Fahey, D.W., & Fehsenfeld, F.C. (1983) Conversion of nitrogen dioxide, nitric acid, and n-propyl nitrate to nitric oxide by gold-catalyzed reduction with carbon monoxide, *Anal. Chem.*, 55, 1980–1986.

Bollinger, M.J., Hahn, C.J., Parrish, D.D., Murphy, P.C., Albritton, D.L., & Fehsenfeld, F.C. (1984), NO_x measurements in clean continental air and analysis of the contributing meteorology, *J. Geophys. Res.*, 89, 9623–9631.

Bowman, R.L., & Alexander, N. (1966) Ozone-induced chemiluminescence of organic compounds, *Science*, 154, 1454–1456.

Bradshaw, N.G., Vaughan, G., Busen, R., Garcelon, S., Jones, R., Gardiner, T., & Hacker, J. (2002) Tracer filamentation generated by small-scale Rossby wave breaking in the lower stratosphere, *J. Geophys. Res.*, 107(D23), 4689, doi:10.1029/2002JD002086.

Brewer, A.W., & Milford, J.R. (1960) The Oxford-Kew Ozone Sonde, *Proc. Roy. Soc.* London, Ser. A, 470–495.

Buhr, M. (2004) Measurement of NO_2 in ambient air using a solid-state photolytic converter, *Proceedings of the Symposium on Air Quality Measurement Methods and Technology 2004*, Air and Waste Management Association, Research Triangle Park, NC, April 20–24.

Burkhardt, M.R., Maniga, N.I., & Stedman, D.H. (1988) Gas chromatographic method for measuring nitrogen dioxide and peroxyacetyl nitrate in air without compressed gas cylinders, *Anal. Chem.*, 60, 816–819.

Cantrell, C.A., & Stedman, D.H. (1982) A possible technique for the measurement of atmospheric peroxy radicals, *Geophys. Res. Lett.*, 9, 846–849.

Cantrell, C.A., Stedman, D.H., & Wendel, G.J. (1984) Measurement of atmospheric peroxy radicals by chemical amplification, *Anal. Chem.*, 56, 1496–1502.

Cantrell, C.A., Shetter, R.E., Lind, J.A., McDaniel, A.H., Calvert, J.G., Parrish, D.D., Fehsenfeld, F.C., Buhr, M.P., & Trainer, M. (1993) An improved chemical amplifier technique for peroxy radical measurements, *J. Geophys. Res.*, 98, 2897–2909.

Cantrell, C.A., Shetter, R.E., & Calvert, J.G. (1996a), Dual-inlet chemical amplifier for atmospheric peroxy radical measurements, *Anal. Chem.*, 68, 4194–4199.

Cantrell, C.A., Shetter, R.E., Gilpin, T.M., Calvert, J.F., Eisele, F.L., & Tanner, D.J. (1996b) Peroxy radical concentrations measured and calculated from trace gas measurements in the Mauna Loa Observatory Experiment 2, *J. Geophys. Res.*, 101, 14 653–14 664.

Cantrell, C.A., Zimmer, A., & Tyndall, G.S. (1997) Absorption cross sections for water vapor from 183 to 193 nm, *Geophys. Res. Lett.*, 24(17), 2195–2198.

Cantrell, C.A., *et al.* (2003) Peroxy radical behavior during the Transport and Chemical Evolution over the Pacific (TRACE-P) as measured aboard the NASA P-3B aircraft, *J. Geophys. Res.*, 108(D20), 8797, doi:10.1029/2002JD003674.

Cardenas, L.M., *et al.* (2000) Intercomparison of formaldehyde measurements in clean and polluted environments, *J. Atmos. Chem.*, 37, 53–80.

Carroll, M.A., McFarland, M., Ridley, B.A., & Albritton, D.L. (1985) Ground-based nitric oxide measurements at Wallops Island, Virginia, *J. Geophys. Res.*, 90(D7), 12 853–12 860.

Clemitshaw, K.C., Carpenter, L.J., Penkett, S.A., & Jenkin, M.E. (1997) A calibrated peroxy radical chemical amplifier for ground-based tropospheric measurements, *J. Geophys. Res.*, 102, 25 405–25 416.

Clough, P.N., & Thrush, B.A. (1967) Mechanism of chemiluminescent reaction between nitric oxide and ozone, *Trans. Faraday Soc.*, 63, 915–925.

Clyne, M.A.A., Thrush, B.A., & Wayne, R.P. (1964) Kinetics of the chemiluminescent reaction between nitric oxide and ozone, *Trans. Faraday Soc.*, 60, 359–370.

Creasey, D.J., Heard, D.E., & Lee, J.D. (2000) Absorption cross-section measurements of water vapour and oxygen at 185 nm. Implications for the calibration of field instruments to measure OH, HO_2 and RO_2 radicals, *Geophys. Res. Lett.*, 27(11), 1651–1654.

Del Negro, L.A., *et al.* (1999) Comparison of modeled and observed values of NO_2 and J_{NO_2} during the Photochemistry of Ozone Loss in the Arctic Region in Summer (POLARIS) mission, *J. Geophys. Res.*, 104(D21), 26 687–26 703.

Dickerson, R.R., Delany, A.C., & Wartburg, A.F. (1984) Further modification of a commercial NO_x detector for high sensitivity, *Rev. Sci. Instrum.*, 55, 1995–1998.

Dong, S., & Dasgupta, P.K. (1987) Fast fluorometric flow injection analysis of formaldehyde in atmospheric water, *Environ. Sci. Technol*, 21, 581–588.

Drummond, J.W., Topham, L.A., Mackay, G.I., & Schiff, H.I (1991) Use of chemiluminescence techniques in portable, lightweight, highly sensitive instruments for measuring NO_2, NO_x, and O_3, in Schiff, H.I. (ed.), SPIE, Vol. 1433, *Measurement of Atmospheric Gases*, 224–231.

Edwards, G.D., Cantrell, C.A., Stephens, S., Hill, B., Goyea, O., Shetter, R.E., Mauldin, R.L., Kosciuch, E., Tanner, D.J., & Eisele, F.L. (2003) Chemical ionization mass spectrometer instrument for the measurement of tropospheric HO_2 and RO_2, *Anal. Chem.*, 75, 5317–5327.

Fahey, D.W., Eubank, C.S., Hübler, G., & Fehsenfeld, F.C. (1985) Evaluation of a catalytic reduction technique for the measurement of total reactive odd-nitrogen NO_y in the atmosphere, *J. Atmos. Chem.*, 3, 435–468.

Fahey, D.W., *et al.* (1986) Reactive nitrogen species in the troposphere: Measurements of NO, NO_2, HNO_3, particulate nitrate, peroxyacetyl nitrate (PAN), O_3, and total reactive odd nitrogen (NO_y) at Niwot Ridge, Colorado, *J. Geophys. Res.*, 91, 9781–9793.

Fahey, D.W., *et al.* (1990) A diagnostic for denitrification in the winter polar stratosphere, *Nature*, 345, 698–702.

Fahey, D.W., *et al.* (2001) The detection of large HNO_3-containing particles in the winter Arctic stratosphere, *Science*, 291, 1026–1031.

Fehsenfeld, F.C., *et al.* (1987) A ground-based intercomparison of NO, NO_x, and NO_y measurement techniques, *J. Geophys. Res.*, 92, 14 710–14 722.

Fehsenfeld, F.C., *et al.* (1990) Intercomparison of NO_2 measurement techniques, *J. Geophys. Res.*, 3579–3597.

Finlayson, B.J., Pitts, J.N., Jr., & Akimoto, H. (1972) Production of vibrationally excited OH in chemiluminescent ozone-olefin reactions, *Ch. Phys. Lett.*, 12(3), 495–498.

Fontijn A., Sabadell, A.J., & Ronco, R.J. (1970) Homogeneous chemiluminescent measurement of nitric oxide with ozone, *Anal. Chem.*, 42, 575–579.

Fried, A., Lee, Y.-N., Frost, G., Wert, B., Henry, B., Drummond, J.R., Hübler, G., & Jobson, T. (2002) Airborne CH_2O measurements over the North Atlantic during the 1997 NARE campaign: Instrument comparisons and distributions, *J. Geophys. Res.*, 107(D4), 4039, doi:10.1029/2000JD000260, 2002.

Gao, R.S., Keim, E.R., Woodbridge, E.L., Ciciora, S.J., Proffitt, M.H., Thompson, T.L., McLaughlin, R.J., & Fahey, D.W. (1994) New photolysis system for NO_2 measurements in the lower stratosphere, *J. Geophys. Res.*, 99(D10), 20 673–20 681.

Gregory, G.L., Hudgins, C.H., & Edahl, R.A. (1983) Laboratory evaluation of an airborne ozone instrument that compensates for altitude/sensitivity effects, *Environ. Sci. Technol.*, 17, 100–103.

Gregory, G.L., Beck, S.M., & Williams, J.A. (1984) Measurements of free tropospheric ozone: An aircraft survey from 44° North to 46° South latitude, *J. Geophys. Res.*, 89, 9642–9648.

Green, T.J., Reeves, C.E., Brough, N., Edwards, G.D., Monks, P.S., & Penkett, S.A. (2003) Airborne measurements of peroxy radicals using the PERCA technique, *J. Environ. Monit.*, 5, 75–83.

Guenther, A.B., & Hills, A.J. (1998) Eddy covariance measurement of isoprene fluxes, *J. Geophys. Res.*, 103(D11), 13 145–13 152.

Güsten, H., Heinrich, G., Schmidt, R.W.H., & Schurath, U. (1992) A novel ozone sensor for direct eddy flux measurements, *J. Atmos. Chem.*, 14, 73–84.

Güsten, H., & Heinrich, G. (1996) On-line measurements of ozone surface fluxes: Part I. Methodology and instrumentation, *Atmos. Environ.*, 30(6), 897–909.

Handisides, G.M., Plass-Dulmer, C., Gilge, S., Bingemer, H., & Berresheim, H. (2003) Hohenpeissenberg Photochemical Experiment (HOPE 2000): Measurements and photostationary state calculations of OH and peroxy radicals, *Atmos. Chem. Phys.*, 3, 1565–1588.

Hanke, M., Uecker, J., Reiner, T., & Arnold, F. (2002) Atmospheric peroxy radicals: ROXMAS, a new mass-spectrometric methodology for speciated measurements of HO_2 and ΣRO_2 and first results, *Int. J. Mass Spectrometry*, 213, 91–99.

Harrison, R.M., Peak, J.D., & Collins, G.M. (1996) Tropospheric cycle of nitrous acid, *J. Geophys. Res.*, 101(D9), 14 429–14 439.

Hastie, D.R., Weissenmayer, M., Burrows, J.P., & Harris, G.W. (1991) Calibrated chemical amplifier for atmospheric RO_x measurements, *Anal. Chem.*, 63, 2048–2057.

Heikes, B.G. (1992) Formaldehyde and hydroperoxides at Mauna Loa Observatory, *J. Geophys. Res.*, 97, 18 001–18 013.

Helas, G., & Warneck, P. (1981) Background NO_x mixing ratios in air masses over the North Atlantic Ocean, *J. Geophys. Res.*, 86, 7283–7290.

Heland, J., Kleffmann, J., Kurtenbach, R., & Wiesen, P. (2001) A new instrument to measure gaseous nitrous acid (HONO) in the atmosphere, *Environ. Sci. Technol.*, 35, 3207–3212.

Hills, A.J., & Zimmerman, P.R. (1990) Isoprene measurement by ozone-induced chemiluminescence, *Anal. Chem.*, 62, 1055–1060.

Hilsenrath, E., Seiden, L., & Goodman, P. (1969) An ozone measurement in the mesosphere and stratosphere by means of a rocket sonde, *J. Geophys. Res.*, 74, 6873–6880.

Hilsenrath, E., & Kirschner, P.T. (1980) Recent assessment of the performance and accuracy of a chemiluminescent rocket sonde for upper atmospheric ozone measurements, *Rev. Sci. Instrum.*, 51, 1381–1389.

Hofzumahaus, A., *et al.* (1997) Reply to Lanzendorf *et al.*, *Geophys. Res. Lett.*, 24(23), 3039–3040.

Hodgeson, J.A., Krost, K.J., O'Keeffe, A.E., & Stevens, R.K. (1970) Chemiluminescent measurement of atmospheric ozone, *Anal. Chem.*, 42, 1795–1802.

Hu, J., & Stedman, D.H. (1994) Free radical detector for tropospheric measurement using chemical amplification, *Anal. Chem.*, 66, 3384–3393.

Joseph, D.W., & Spicer, C.W. (1978) Chemiluminescence method for atmospheric monitoring of nitric acid and nitrogen oxides, *Anal. Chem.*, 50, 1400–1403.

Kelly, T.J., & Stedman, D.H. (1979) Measurements of H_2O_2 and HNO_3 in rural air, *Geophys. Res. Lett.*, 375–378.

Kelly, T.J., Spicer, C.W., & Ward, G.F. (1990) An assessment of the luminol chemiluminescence technique for measurement of NO_2 in ambient air, *Atmos. Environ.*, 24A, 2397–2403.

Kley, D., & McFarland, M. (1980) Chemiluminescence detector for NO and NO_2, *Atmos. Technol.*, 12, 63–69.

Kley, D., Drummond, J.W., McFarland, M., & Liu, S.C. (1981) Tropospheric profiles of NO_x, *J. Geophys. Res.*, 86, 3153–3161.

Kliner, D.A.V., Daube, B.C., Burley, J.D., & Wofsy, S.C. (1997), Laboratory investigation of the catalytic reduction technique for measurement of atmospheric NO_y, *J. Geophys. Res.*, 102(D9), 10 759–109 776.

Kok, G.L., Holler, T.P., Lopez, M.B., Nachtrieb, H.A., & Yuan, M. (1978a), Chemiluminescent method for determination of hydrogen peroxide in the ambient atmosphere, *Environ. Sci. Technol.*, 12, 1072–1076.

Kok, G.L., Darnall, K.R., Winer, A.M., Pitts, J.N., & Gay, B.W. (1978b) Ambient air measurements of hydrogen peroxide in the California South Coast Air Basin, *Environ. Sci. Technol.*, 12, 1077–1080.

Komhyr, W.D., Barnes, R.A., Brothers, G.B., Lathrop, J.A., & Opperman, D.P. (1995) Electrochemical concentration cell ozonesonde performance evaluation during STOIC 1989, *J. Geophys. Res.*, 100(D5), 9231–9244.

Kondo, Y., Kojima, H., Toriyama, N., Morita, Y., Takagi, M., & Matthews, W.A. (1987) Chemiluminescent ozone instrument for aircraft observation, *J. Meteorol. Soc. Japan*, 65, 795–802.

Kormann, R., Fischer H., de Reus, M., Lawrence M., Brühl, C., von Kuhlmann, R., Holzinger, R., Williams, J., Lelieveld, J., Warneke, C., de Gouw, J., Heland, J., Ziereis, H., & Schlager, H. (2003) Formaldehyde over the eastern Mediterranean during MINOS: Comparison of airborne in-situ measurements with 3D-model results, *Atmos. Chem. Phys.*, 3, 851–861.

Lanzendorf, E.J., Hanisco, T.F., Donahue, N.M., & Wennberg, P.O. (1997) Comment on "The measurement of tropospheric OH radicals by laser-induced fluorescence spectroscopy during the POPCORN field campaign" by Hofzumahaus *et al.* and "Intercomparison of tropospheric OH radical measurements by multiple folded long-path laser absorption and laser induced fluorescence" by Brauers *et al.*, *Geophys. Res. Lett.*, 24(23), 3037–3038.

Lazrus, A.L., Kok, G.L., Lind, J.A., Gitlin, S.N., Heikes, B.G., & Shetter, R.E. (1986) Automated fluorometric method for hydrogen peroxide in air, *Anal. Chem.*, 58, 594–597.

Lazrus, A.L., Fong, K.L., & Lind, J.A. (1988) Automated fluorometric determination of formaldehyde in air, *Anal. Chem.*, 60, 1074–1078.

Lee, M., Noone, B.C., O'Sullivan, D., & Heikes, B.G. (1995) Method for the collection and HPLC analysis of hydrogen peroxide and C1 and C2 hydroperoxides in the atmosphere, *J. Atmos. Oceanic Technol.*, 12, 1060–1070.

Lee, Y.-N., Zhou, X., Leaitch, W.R., & Banic, C.M. (1996) An aircraft measurement technique for formaldehyde and soluble carbonyl compounds, *J. Geophys. Res.*, 101(D22), 29 075–29 080.

Lenschow, D.H., Pearson, R., Jr., & Stankov, B.B. (1981) Estimating the ozone budget in the boundary layer by use of aircraft measurements of ozone eddy flux and mean concentration, *J. Geophys. Res.*, 86(C8), 7291–7297.

Logan, J.A. (1999) An analysis of ozonesonde data for the troposphere: Recommendations for testing 3-D models and development of a gridded climatology for tropospheric ozone, *J. Geophys. Res.*, 104(D13), 16 115–16 149.

Maeda, Y., Aoki, K., & Munemori, M. (1980) Chemiluminescence method for the determination of nitrogen dioxide, *Anal. Chem.*, 52, 307–311.

Matthews, R.D., Sawyer, R.F., & Schefer, R.W. (1977) Interferences in chemiluminescent measurement of NO and NO_2 emissions from combustion systems, *Environ. Sci. Technol.*, 12, 1092–1096.

McFarland, M., Kley, D., Drummond, J.W., Schmeltekopf, A.L., & Winkler, R.H. (1979) Nitric oxide measurements in the equatorial Pacific region, *Geophys. Res. Lett.*, 6, 605–608.

McKendry, I.G., Steyn, D.G., O'Kane, S., Zawar-Reza, P., & Heuff, D. (1998) Lower tropospheric ozone measurements by light aircraft equipped with chemiluminescent sonde, *J. Atmos. Oceanic Technol.*, 15, 136–143.

Mihele, C.M. & Hastie, D.R. (1998) The sensitivity of the radical amplifier to ambient water vapor, *Geophys. Res. Lett.*, 25(11), 1911–1913.

Mihele, C.M., Mozurkewich, M., & Hastie, D.R. (1999) Radical loss in a chain reaction of CO and NO in the presence of water: Implications for the radical amplifier and atmospheric chemistry, *Int. J. Chem. Kinet.*, 31, 145–152.

Mihele, C.M. & Hastie, D.R. (2000) Optimized operation and calibration procedures for radical amplifier-type detectors, *J. Atmos. Technol.*, 17, 788–794.

Monks, P.S., Carpenter, L.J., Penkett, S.A., & Ayers, G.P. (1996) Night-time peroxy radical chemistry in the remote marine boundary layer over the Southern ocean, *Geophys. Res. Lett.*, 23(5), 535–538.

Morgan, R.B., & Jackson, A.V. (2002) Measurements of gas-phase hydrogen peroxide and methyl hydroperoxide in the coastal environment during the PARFORCE project, *J. Geophys. Res.*, 107(D19), 8109, doi:10.1029/2002JD000257.

Nederbragt, G.W., Van der Horst, A., & Van Duijn, J. (1965) Rapid ozone determination near an accelerator, *Nature*, 206, 87.

Neuman, J.A., Huey, L.G., Ryerson, T.B., & Fahey, D.W. (1999) Study of inlet materials for sampling atmospheric nitric acid, *Environ. Sci. Technol.*, 33, 1133–1136.

O'Sullivan, D.W., *et al.* (2004) Long-term and seasonal variations in the levels of hydrogen peroxide, methylhydroperoxide, and selected compounds over the Pacific Ocean, *J. Geophys. Res.*, 109, D15S13, doi:10.1029/2003JD003689.

Pearson, R. & Stedman, D. (1980) Instrumentation for fast-response ozone measurements from aircraft, *Atmos. Technol.*, 12, 51–55.

Pearson, R. (1990) Measuring ambient ozone with high sensitivity and bandwidth, *Rev. Sci. Instrum.*, 61, 907–916.

Ray, J.D., Stedman, D.H., & Wendel, G.J. (1986) Fast chemiluminescent method for measurement of ambient ozone, *Anal. Chem.*, 58, 598–600.

Regener, V.H. (1960) On a sensitive method for the recording of atmospheric ozone, *J. Geophys. Res.*, 65, 3975–3977.

Regener, V.H. (1964) Measurement of atmospheric ozone with the chemiluminescent method, *J. Geophys. Res.*, 69, 3795–3800.

Reichert, L., Hernández, M.D.A., Ströbener, D., Burkert, J., & Burrows, J.P. (2003) Investigation of the effect of water complexes in the determination of peroxy radical ambient concentrations: Implications for the atmosphere, *J. Geophys. Res.*, 108(D1), 4017, doi:10.1029/2002JD002152.

Reiner, T., Hanke, M., & Arnold, F. (1997) Atmospheric peroxy radical measurements by ion molecule reaction-mass spectrometry: A novel analytical method using amplifying chemical conversion to sulfuric acid, *J. Geophys. Res.*, 102, 1311–1326.

Reiner, T., Hanke, M., & Arnold, F. (1998) Aircraft-borne measurements of peroxy radicals in the middle troposphere, *Geophys. Res. Lett.*, 25, 47–50.

Reiner, T., Hanke, M., Arnold, F., Ziereis, H., Schlager, H., & Junkermann, W. (1999) Aircraft-borne measurements of peroxy radicals by chemical conversion/ion molecule reaction mass spectrometry: Calibration, diagnostics, and results, *J. Geophys. Res.*, 104, 18 647–18 659

Ridley, B.A., Schiff, H.I., & Welge, K.H. (1972) Measurement of NO in the stratosphere by NO/O_3 chemiluminescence, Report to the National Oceanic and Atmospheric Administration, Contract No. N22-15-72N. (Available through National Technical Information Service, Springfield, Va.)

Ridley, B.A. & Howlett, L.C. (1974) An instrument for nitric oxide measurements in the stratosphere, *Rev. Sci. Instrum.*, 45, 742–746.

Ridley, B.A. (1978) Measurements of minor constituents in the stratosphere by chemiluminescence, *Atmos. Technol.*, 9, 27–34.

Ridley, B.A., Carroll, M.A., & Gregory, G.L. (1987) Measurements of nitric oxide in the boundary layer and free troposphere over the Pacific Ocean, *J. Geophys. Res.*, 92(D2), 2025–2047.

Ridley, B.A., Carroll, M.A., Torres, A.L., Condon, E.P., Sachse, G.W., Hill, G.F., & Gregory, G.L. (1988a) An intercomparison of results from ferrous sulphate and photolytic converter techniques for measurements of NO_x made during the NASA GTE/CITE 1 aircraft program, *J. Geophys. Res.*, 93, 15 803–15 811.

Ridley, B.A, Carroll, M.A., Gregory, G.L., & Sachse, G.W. (1988b) NO and NO_2 in the troposphere: Technique and measurements in regions of a folded tropopause, *J. Geophys. Res.*, 93, 15 813–15 830.

Ridley, B.A. & Grahek, F.E. (1990) A small, low flow, high-sensitivity reaction vessel for NO chemiluminescence detectors, *J. Atmos. Oceanic Technol.*, 7, 307–311.

Ridley, B.A., Grahek, F.E., & Walega, J.G. (1992) A small, high-sensitivity, medium-response ozone detector suitable for measurements from light aircraft, *J. Atmos. Oceanic Technol.*, 9, 142–148.

Ridley, B.A., Walega, J.G., Dye, J.E., & Grahek, F.E. (1994) Distributions of NO, NO_x, NO_y, and O_3 to 12 km altitude during the summer monsoon over New Mexico, *J. Geophys. Res.*, 99(D12), 25 519–25 534.

Ridley, B. *et al.* (2004) Florida thunderstorms: A faucet of reactive nitrogen to the upper troposphere, *J. Geophys. Res.*, 109, D17305, doi:10.1029/2004JD004769.

Routhier, F., Dennett, R., Davis, D.D., Wartburg, A., Haagenson, P., & Delany, A.C. (1980) *Free Tropospheric and Boundary-layer Airborne Measurements of Ozone over the Latitude Range of 58 S to 70 N*, 7307–7321.

Ryerson, T.B., Williams, E.J., & Fehsenfeld, F.C. (2000), An efficient photolysis system for fast-response NO_2 measurements, *J. Geophys. Res.*, 105(D21), 26 447–26 461.

Ryerson, T.B. *et al.* (2003) Effect of petrochemical industrial emissions of reactive alkenes and NO_x on tropospheric ozone formation in Houston, TX, *J. Geophys. Res.*, 108(D8), 4249, doi:10.1029/2002JD003070.

Sadanaga, Y., Matsumoto, J., Sakurai, K., Isozaki, R., Kato, S., Nomaguchi, T., Bandow, H., & Kajii, Y. (2004) Development of a measurement system of peroxy radicals using a chemical amplification/laser-induced fluorescence technique, *Rev. Sci. Instrum.*, 75, 864–872.

Sander, S.P., *et al.* (2000) Chemical kinetics and photochemical data for use in stratospheric modeling, Supplement to evaluation 12: Update of key reactions, Evaluation No. 13, NASA JPL Pub. 00-003.

Schiff, H.I., Mackay, G.I., Castledine, C., Harris, G.W., & Tran, Q. (1986) Atmospheric measurements of nitrogen dioxide with a sensitive luminol instrument, *Water, Air and Soil Pollution*, 30, 105–114.

Schmidt, R.W.H., Kames, J., Kanter, H.J., Schurath, U., & Slemr, F. (1995) A selective ozone scrubber for application in ambient nitrogen dioxide measurements using the commercial Luminox (LMA-3, Scintrex/Unisearch, Inc.), *Atmos. Environ.*, 29, 947–950.

Schultz, M., Heitlinger, M., Mihelcic, D., & Volz-Thomas, A. (1995) Calibration source for peroxy radicals with built-in actinometry using H_2O and O_2 photolysis at 185 nm, *J. Geophys. Res.*, 100, 18 811–18 816.

Schurath, U., Speuser, W., & Schmidt, R. (1991) Principle and application of a fast sensor for atmospheric ozone, *Fresenius J. Anal. Chem.*, 544–547.

Sigsby, J.E., Black, F.M., Bellar, T.A., & Klosterman, D.L. (1973) Chemiluminescent method for analysis of nitrogen compounds in mobile source emissions (NO, NO_2, and NH_3), *Environ. Sci. Technol.*, 7, 51–54.

Snow, J.A., Heikes, B.G., Merrill, J.T., Wimmers, A.J., Moody, J.L., & Cantrell, C.A. (2003) Winter-spring evolution and variability of HO_x reservoir species, hydrogen peroxide, and methyl hydroperoxide, in the northern middle to high latitudes, *J. Geophys. Res.*, 108(D4), 8362, doi:10.1029/2002JD002172.

Stedman, D.H., Daby, E.E., Stuhl, F., & Niki, H. (1972) Analysis of ozone and nitric oxide by a chemiluminescent method in laboratory and atmospheric studies of photochemical smog, *J. Air Pollution Control Assoc.*, 22, 260–263.

Stedman, J.R., Bush, T.J., & Vincent, K.J. (2002) UK air quality modelling for annual reporting 2001 on ambient air quality assessment under Council Directives 96/62/EC and 1999/30/EC. AEA Technology, National Environmental Technology Centre. Report AEAT/ENV/R/1221.

Stevens, R.K. & O'Keeffe, A.E. (1970) Modern aspects of air pollution monitoring, *Anal. Chem.*, 42, 143A–148A.

Stuhl, F. & Niki, H. (1970) An optical detection method for NO in the range of 10^{-2} to 10^3 ppm by the chemiluminescent reaction of NO with O_3, *Scientific Research Staff Report*, Ford Motor Company.

Takayanagi, T., Su, X.-L., Dasgupta, P.K., Martinelango, K., Li, G., Al-Horr, R.S., & Shaw, R.W. (2003) Chemiluminometric measurement of atmospheric ozone with photoactivated chromotropic acid, *Anal. Chem.*, 75, 5916–5925.

Volz-Thomas, A., Berg, M., Heil, T., Houben, N., Lerner, A., Petrick, W., Raak, D., & Pätz, H.-W. (2004) Measurements of total odd nitrogen (NO_y) aboard MOZAIC in-service aircraft: Instrument design, operation and performance, *Atmos. Chem. Phys. Discuss.*, 4, 6149–6183.

Warren, G.J. & Babcock, G. (1970) Portable ethylene chemiluminescence ozone monitor, *Rev. Sci. Inst.*, 280–282.

Weinheimer, A.J., Campos, T.L., & Ridley, B.A. (1998) The in-flight sensitivity of gold-tube NO_y converters to HCN, *Geophys. Res. Lett.*, 25, 3943–3946.

Wendel, G.J., Stedman, D.H., & Cantrell, C.A. (1983) Luminol-based nitrogen dioxide detector, *Anal. Chem.*, 55, 937–940.

Williams, E.J., *et al* . (1998) Intercomparison of ground-based NO_y measurement techniques, *J. Geophys. Res.*, 103(D17), 22 261–22 280.

Winer, A.M., Peters, J.W., Smith, J.P., & Pitts, J.N. (1974) Response of commercial chemiluminescent NO-NO_2 analyzers to other nitrogen-containing compounds, *Env. Sci. Technol.*, 8, 1181–1121.

World Meteorological Organization (WMO) (1998) Assessment of trends in the vertical distribution of ozone, SPARC Report No. 1, Global Ozone Res. and Monit. Proj., WMO Rep. 37.

Zhou, X., He, Y., Huang, G., Thornberry, T.D., Carroll, M.A., & Bertman, S.B. (2002) Photochemical production of nitrous acid on glass sample manifold surface, *Geophys. Res. Lett.*, 29(14), 1681, doi:10.1029/2002GL015080.

Chapter 8
Chromatographic Methods

Jacqueline F. Hamilton and Alastair C. Lewis

8.1 Introduction

The measurement of biogenic and anthropogenic organic species in the atmosphere is a key area of atmospheric research. Nearest to the Earth's surface, within the tropospheric boundary layer, determining the concentration of reactive hydrocarbon species is important because of their ability to react rapidly with OH radicals (formed from sunlight) in the presence of NO_x to form ozone and photochemical smog. The oxygenated and nitrated reaction products can then react further or partition to the particulate phase due to their relatively lower vapour pressures when compared to the parent hydrocarbon. The use of gas chromatography to determine the organic content of the atmosphere was a key factor in understanding the reasons behind the large-scale smog events in Los Angeles in the 1970s.

At the other extreme, in the stratosphere, remote measurements of very long–lived halogenated species are required because of their critical impact as precursors to ozone destruction. In addition, other stable molecules such as methane, present throughout the atmosphere, have importance as greenhouse gases and influence global climate and temperature change.

The wide range of both short- and long-lived species that are of interest in atmospheric science, coupled to extremely low concentrations and a requirement often for in situ automated analysis, has led to the development of many novel chromatographic techniques and methodologies. Whilst some atmospheric species such as organic acids, peroxides, and aldehydes are measured using high performance liquid chromatography (HPLC), the majority of organic species are analysed using capillary gas chromatography (GC). The diversity of compounds that are of interest has resulted in almost every kind of analytical detector finding a role within atmospheric analysis by chromatography. Many organic compounds present in the atmosphere are bound to particles or are in aerosol form, for example polycyclic aromatic hydrocarbons (PAH), polychlorinated biphenyls (PCBs), and dioxins. A vast number of methodologies for the analysis of these species exist, although GC with mass spectrometry is the core technique for many.

Determining the exact chemical makeup of such multi-component mixtures is often beyond the specificity of a single analytical property (e.g. absorption of ultraviolet (UV) radiation or ionisation followed by separation by mass). Chromatography is one of the most common methods of separating complex mixtures into their constituent components. Chromatography traces its origins back to the early twentieth century when botanist Mikhail Tswett used a liquid – solid adsorption separation to resolve many coloured components within chlorophyll. This original application gave the technique its name from the Greek for 'colour', or *Chroma*, and 'writing', *graphein*. From the mid-twentieth century onwards, chromatography expanded rapidly in many directions: 1940, the development of partition chromatography (for which A.J.P. Martin and R.L.M. Synge were awarded the Nobel Prize); 1952, gas chromatography; 1970, bonded particles for liquid chromatography; and in 1981, the pioneering of electrically driven separations, or electrophoresis (Bartle & Myers, 2002). At the heart of all chromatographic techniques, however, is a common set of principles. The mixture of components is solvated in a fluid (referred to as the *mobile phase*) used to transport the analytes through the separation system. The flow of mobile phase passes over or through an immiscible bed of material or *stationary phase*. Interactions occur between analytes and the stationary phase, and this slows the rate at which they pass through the bed. Each analyte may interact (e.g. by adsorption) to a different extent with the stationary phase, resulting in the physical separation of the mixture into its component parts. Figure 8.1 shows a schematic view of the chromatographic process as it occurs in packed column elution chromatography.

Chromatography covers a wide range of methods and techniques, so it is useful to classify these into a number of key subsets. Perhaps the most basic description that can be made is of the physical state of the mobile phase: gas, liquid or supercritical fluid. A second classification describes the way in which the mobile phase interacts with the stationary phase. Where the stationary phase is either packed into or coated onto the walls of a length of tube, it is referred to as *column chromatography*. The physical dimensions can vary considerably: gas chromatography uses columns as long as 100 m

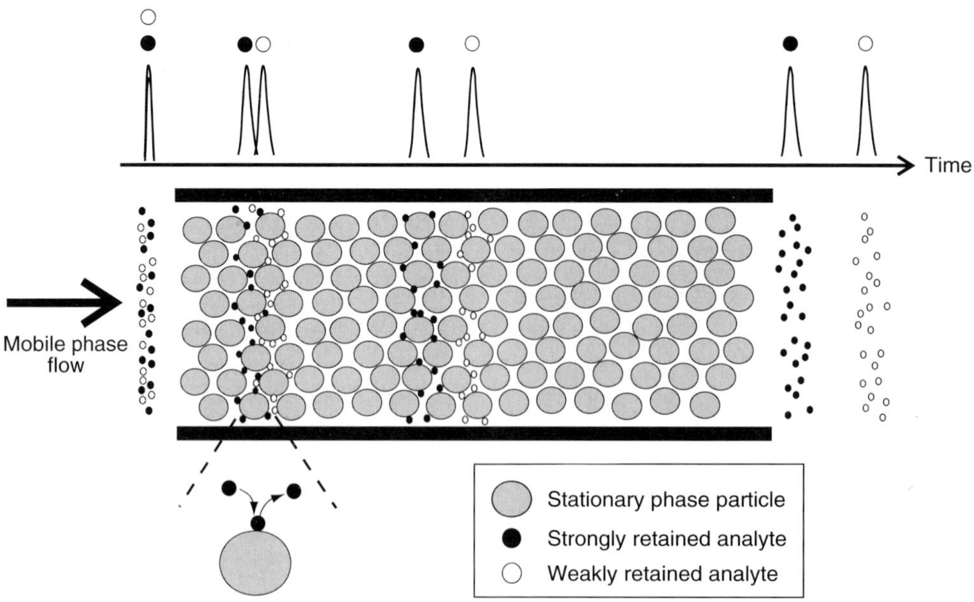

Figure 8.1 Chromatographic separation of compounds in a packed column.

and as narrow as 50 μm whereas liquid chromatography takes place on columns of relatively modest length but up to several centimetres wide. By applying positive pressure at the start of the column, the mobile phase can be driven through the column with a parabolic flow profile. Packed column chromatography involves the flow of mobile phase through a column filled with a finely powdered stationary phase, in the order of μm in diameter, such as molecular sieve or silica. In capillary column chromatography, the stationary phase, usually a highly viscous liquid, is coated onto the walls of the column, and the mobile phase flows over this thin film. *Planar chromatography* (often referred to as thin layer chromatography), where the stationary phase is coated onto a plate and samples introduced as spots, finds widespread use in the pharmaceutical and biochemical industries but is generally unsuitable as separation technique in atmospheric analysis. A description of the various chromatographic techniques available is given in Table 8.1, along with typical atmospheric examples of their use. For a more detailed description of the fundamentals and applications of chromatography, see the material suggested in 'Further reading'.

8.2 Gas chromatography

In gas chromatography the mobile phase, a highly diffuse gas, is used to solvate and transport components within a mixture through or over a stationary phase, with separation occurring due to differences in the rates of migration. The mobile phase gas must have a high diffusion coefficient to ensure maximum numbers of gas–stationary phase interactions. The most common gases used in gas chromatography are nitrogen, hydrogen, and helium. Maximum separation efficiency is achieved using the lighter gases and as such nitrogen does not find many applications in atmospheric separations. The highest separation efficiencies are achieved using hydrogen gas over a wider gas velocity range than helium. However, the carrier gas of choice is usually helium, due to the safety implications and in-column hydrogenation of some unstable species associated with hydrogen. High purity gases are used and it is normal to further purify the carrier gas with the use of a combination of oxygen, moisture, and hydrocarbon traps.

The time a compound takes to be eluted from the column is known as the *retention time* and is used in the identification of analytes through standard injections. In a correctly optimised chromatographic separation, the analyte band will reach the detector in a narrow plug with a degree of band broadening to produce a Gaussian-shaped peak. The height or area under the peak is representative of the sample concentration giving a quantitative measurement. The overall resolving power of a chromatographic column can be given in terms of peak capacity. This is the maximum number of component peaks that can be theoretically resolved on a given column. In gas chromatography, peak capacities are generally considerably larger than in liquid chromatography, with typical values being 200 and 20, respectively. To completely resolve highly complex atmospheric samples would require extremely high peak capacities and selective detectors, extraction techniques and derivatisation are often used to simplify the separation required.

The range of techniques that are in use is so broad that a complete review of analytical methods is impractical. Many individual methods, however, have common components or key procedural steps and these will be discussed. A general outline of a typical atmospheric

Table 8.1 Classification of chromatographic techniques and examples of their use in atmospheric chemistry

Description		Mobile phase	Stationary phase	Equilibrium	Analyte properties	Atmospheric examples	Typical detectors
Liquid chromatography (LC)	Liquid–solid	Liquid-organic	Solid particles (e.g. silica)	Adsorption/ desorption	Non-polar	PAH, organic fractionation	UV, FL
	Liquid-bonded phase	Liquid-aqueous	Organic species bonded or stabilised on a solid particle surface (e.g. C_{18})	Partition between liquid and bonded surface	All types	Many applications, e.g. PAHs, NO_2-PAH, peroxides, PCBs, carbonyls and derivatives	UV, FL, IR, MS, ELS
	Liquid–liquid	Liquid	Liquid supported on solid particles	Partition between immiscible liquid phases	Non-ionic polar		UV, FL, IR, MS, ELS
	Size exclusion/gel permeation or filtration	Liquid	Porous particles	Retardation due to penetration of particle pores	All types	Large molecules in aerosols; polymers, proteins, biomolecules	UV, FL, IR, MS, ELS, LIF
	Ion exchange	Liquid	Charged ion exchange resins	Ion exchange	Ionic	Organic acids, inorganic ions	MS, FL, UV
	Affinity	Liquid	Binding ligand on solid support particle	Molecule-specific binding (e.g. protein/enzyme)	Non-ionic polar	None to date	MS, UV
Capillary electrochromatography (CEC)	Liquid–solid	Liquid (electro-osmotic flow)	Organic species bonded or stabilised on a solid particle surface (e.g. C_{18})	Partition between liquid and bonded surface	Non-polar, non-ionic polar and ionic	None to date	UV, MS, LIF

Gas(GC)	Gas–Liquid	Gas	Liquid coated onto solid particles in a packed column	Partition between phases	Non-polar and non-ionic polar	'BP-1' type pure volatility separations	FID, MS, ECD
	Gas-bonded phase	Gas	Stabilised stationary phase on solid surface (e.g. particles or capillary walls)	Partition between phases	Non-polar and non-ionic polar	Typical semivolatile VOC separations, aromatics, CFCs, halocarbons, PAN, terpenes, hydrocarbons, carbonyls	FID, MS, ECD, FPD, PID
	Gas–solid	Gas	Solid support, either particles or on capillary walls	Adsorption/ desorption	Non-polar and volatile	Light non-methane hydrocarbons, small carbonyls,	FID, MS
	Size exclusion	Gas	Porous particles or supported on capillary walls	Penetration of porous media	Non-polar and permanent gases	Permanent gases, CO, CH_4,	TCD, FID, HID
Supercritical fluid (SFC)	SF-bonded phase	SF	Stabilised stationary phase on solid surface (e.g. particles or capillary walls)	Partition between phases	Non-polar and non-ionic polar	PAH, substituted aromatics, terpenes, thermally labile compounds.	UV, FID, MS

Abbreviations: UV – Ultraviolet, FL – Fluorescence, MS – Mass spectrometry (generally quadrupole or ion trap but also TOF GC and LC and magnetic sector for GC), IR – Infrared, ELS – Evaporative Light Scattering, LIF – Laser induced fluorescence, FID – Flame ionization detector, ECD – Electron capture detector, FPD – Flame photometric detector, PID – photoionisation detector, HID – Helium ionization detector, TCD – Thermal conductivity detector.

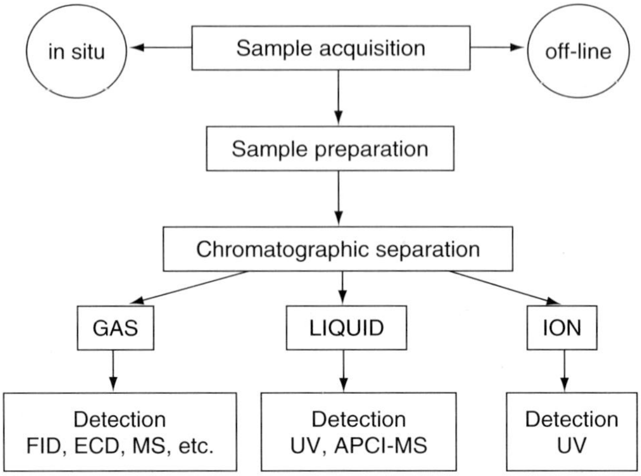

Figure 8.2 Flow diagram of a typical chromatographic separation regime. FID = flame ionisation detector, ECD = electron capture detector, MS = mass spectrometer, APCI = atmospheric pressure chemical ionisation.

determination can be broken down into the following: sample acquisition, preparation, separation, and detection, as shown in Figure 8.2, with the first two stages of acquisition and preparation often proving the most challenging. A number of chromatograms obtained from atmospheric analysis are also presented.

8.2.1 Sample acquisition

The initial step of sample acquisition is a far from trivial task for atmospheric measurements, and defines a major grouping of techniques into either in situ or post acquisition analysis. The ability to store an atmospheric sample is a critical factor as to whether analysis may be performed back in the laboratory or on-site immediately following acquisition. Stable species such as methane, CO and CFCs may be successfully stored in sample vessels for many weeks without affecting sample integrity, and widespread measurements of these species have been performed. Atmospheric degradation products such as peroxyacetyl nitrate (PAN, $CH_3C(O)OONO_2$) are so unstable, however, that analysis must be performed immediately. For many other important reactive species such as alkenes, monoterpenes and dimethyl sulphide (DMS), the storage of samples has been shown to lead to a degree of analyte loss due to reaction with co-sampled pollutants such as ozone and oxides of nitrogen (Sin *et al.*, 2001). To overcome these problems of reaction during storage, many forms of scrubber (e.g. potassium iodide and glycerol to remove ozone) have been tested to remove such species without affecting sample integrity (Vairavamurthy *et al.*, 1992).

Much atmospheric sampling is performed using stainless steel canisters of 1–10 L in volume (often referred to as whole air samples), filled either by vacuum release or by pressurising the sample using stainless steel bellows or Teflon diaphragm pumps. To retain sample integrity, there must be minimal interaction with the canister walls, and coating methods such as electropolishing, silica or Teflon coating are currently in use. The preparation and cleaning of canisters require careful attention, and high vacuums are often applied along with elevated temperatures for periods of hours or even days. A similar method to canister samples is the use of collapsible Teflon or Tedlar sample bags. These may be filled by a pump and returned to a laboratory for analysis using almost identical procedures to that of canister samples. Whilst generally of lower unit cost, they often produce a greater degree of sample artefact and analyte losses.

Many lower volatility species are unsuitable for collection using canister methods because of problems associated with analyte condensation onto the walls of the container. For this reason a second method of sample acquisition is widely used based on the use of a solid phase adsorbent (which is essentially a very high capacity packed column) as an analyte trap. The adsorbent used in the trap may be chosen to introduce an element of selectivity to the trapping mechanism, although in practice a trap-all approach is commonly used. Recycling of a single adsorbent allows for instrument automation, which is not possible with canister methods, and such recycled traps now form the basis of many national monitoring programs in the urban environment, as well as automated instruments for research in clean air environments (http://www.epa.gov/ttn/amtic/airtoxpg.html).

A huge range of adsorbent supports are commercially available, ranging from high surface area ($>1000\,m^3\,g^{-1}$) carbon material (both charcoals and graphitised) with strong retention characteristics, to lower surface area ($<50\,m^3\,g^{-1}$) polymerics such as Tenax™ TA. Whilst being relatively low cost compared to sample canisters, care is often required in the cleaning and preparation of sample tubes. Samples may be introduced to the adsorbent tubes either dynamically over short periods of time (typically minutes) or via diffusional sampling over longer periods (typically several days). Carbon-based adsorbents are suitable for a wide range of species ranging from volatile hydrocarbons and chlorofluorocarbons (CFCs) to organic nitrates. Polymeric materials are used mainly for the concentration of lower volatility species such as aromatics and monoterpenes, although compounds as large as 2- and 3-ring polycyclic aromatic species may also be successfully trapped and thermally or solvent desorbed. Sulphur species trapping is often performed via chemisorption onto gold wool traps that provide a stable matrix for the sample to be stored for reasonable periods of time (Berresheim *et al.*, 1998).

The complete retention of all target analytes on the adsorbent bed must be carefully evaluated, and calculation of retention volume (often referred to as breakthrough volume) is essential. For sampling very volatile species (e.g. ethane or ethene) the retention volume for an adsorbent system is often the critical factor in determining system sensitivity. Trapping at sub-ambient temperatures is commonly employed to increase the maximum sample volume, often using liquid nitrogen or carbon dioxide to cool the adsorbent trap. For operation of any instrument in a field location, cryogen usage may be a problem and modern instruments often employ Peltier or thermoelectric heat pumps to cool adsorbent traps.

The analysis of low molecular weight, oxygenated compounds in the atmosphere can be difficult, with irreversible retention of polar compounds onto absorbents and losses to canister walls. A selective and quantitative method of sampling oxygenated compounds is the use of derivatisation during sampling to form an analyte more suited to chromatographic separations. Samples can be collected onto solid-phase microextraction fibres

(SPME) or solid sorbent tubes coated with a derivatising agent (Aschmann *et al.*, 2003; Moortgat *et al.*, 2002). Extraction of derivatised analytes with organic solvent and subsequent liquid injection into either a GC or HPLC system is the most common analysis method of carbonyls in the atmosphere. Tubes can also be manufactured in the same dimensions as the GC-inlet allowing direct thermal desorption of derivatised analytes into the column (Ho & Yu, 2002). Detection limits vary but are typically in the low- to sub-ppbv levels with sampling times in the order of hours (Ho & Yu, 2004).

Derivatisation reagents can be selected on the basis of their functional group reactivity. The most common derivatisation reagents are dinitro-phenylhydrazine (DNPH) (Druzik *et al.*, 1990) and O-(2,3,4,5-pentafluorobenzyl)-hydroxylamine (PFBHA) (Cecinato *et al.*, 2001), which react with carbonyl functionalities. 3,5-bis(trifluoromethyl)-phenylhydrazine (TFMPH) has been used to measure formaldehyde concentrations (Marsella *et al.*, 2000). A development since 2000 is online in-inlet derivatisation of mono- and di-carboxylic acids using trimethysilyl (TMS) (Docherty & Ziemann, 2001). Derivatisation is also used extensively with HPLC analysis and will be discussed later.

Particulate phase material may be collected for organic analysis using a number of techniques, typically drawing air through quartz or Teflon filter paper using a large capacity pump. The selectivity of the sampling procedure can be modified by altering the size fraction of aerosol collected, either into a small size fraction or into a 'smaller than' fraction such as PM10 or PM2.5. Since the sample preparation of particulate-bound PAH is convoluted, large amounts of material must be collected for suitable sensitivity – often running to thousands of cubic meters of air drawn though a single filter. Semi-volatile species may desorb back to the gas phase from the particle matrix during sampling, and an adsorbent for gas phase material is often placed downstream of the filter medium. This is generally a weak polymeric medium such as Tenax™ or XAD resin. Given the very large volumes of air often drawn through the filter and adsorbent medium, artefact formation is a serious concern with methods such as this, and the true degree of partitioning of species between gas and particle phases is often very difficult to establish.

8.2.2 Sample preparation and injection

8.2.2.1 Removal of water

The inevitable presence of water in atmospheric samples and its removal prior to gas chromatography separation is a complex area. In certain circumstances its presence may be both beneficial, for example with canister samples where it occupies the active sites on the vessel walls, and detrimental, notably through affecting either the detector or reproducibility of the separation. Alumina porous layer open tubular (PLOT) columns are particularly affected by moisture in the sample, and large changes in stationary phase affinity occur when water is introduced. Detectors such as mass spectrometers are also extremely sensitive to the water introduced with the sample, and high background noise may result. Even the robust flame ionisation detector can be affected by injection of water, where the flame may be extinguished when water elutes from the column.

Many forms of selective water removal exist, the simplest of which is the use of condensation traps or stripping coils. Losses of light molecular species are insignificant although condensation of higher boiling organic material may arise at very low temperatures. Inorganic adsorbents are also commonly used, notably potassium carbonate and magnesium perchlorate. Adsorbents such as these, however, have limited capacity and often require frequent regeneration or replacement. A combination of initial condensation and second stage adsorbent scrubber often provides sufficient capacity to dry a sample stream of air for many hours or days. Continuous drying may be achieved using permeation membranes such as Nafion®. These forms of dryers operate by generating a steep concentration gradient across a membrane permeable only to highly polar material such as water. A counter-current of dry gas is passed around the outside of the membrane as a sheath gas and carries moisture away to waste. However, this type of drier is unsuitable for samples where the quantification of polar materials is required.

8.2.2.2 Direct loop injection

Atmospheric species present at high concentration require the least amount of sample preparation, often only the removal of excess water from the sample. The analysis of CH_4, CO, H_2, N_2O, and some abundant CFCs is performed simply by filling a known volume injection loop (between 0.1 and 5 mL in volume) at fixed temperature and pressure, followed by direct injection to the analytical column. Backflushing is often then performed to remove remaining components from the analytical column. A further example of direct loop injection is in the analysis of PAN where it is isolated using a direct loop injection to a cooled isothermal pre-column, followed by a short analytical capillary column coupled to an electron capture detector (Whalley *et al.*, 2004). An ancillary measurement of carbon tetrachloride is often gained from using this approach.

8.2.2.3 Sample pre-concentration methods

Whether an atmospheric sample is collected using adsorbent or canister techniques, several stages of sample preparation are required prior to injection to the analytical column. For many species, the size of sample required for sufficient sensitivity precludes any form of direct on-column injection, making sample pre-concentration a vital step.

For volatile species collected in canister or whole air samples, analytes are removed from the canister via either internal canister pressure (for pumped in samples) or by vacuum pump (for atmospheric pressure samples) over a pre-concentration trap. A simple method of pre-concentration is to place a sampling loop either directly into or in the headspace of liquid nitrogen. Liquid argon has also been used since this may reduce the amount of oxygen retained in the refocusing stage. The pre-concentration zone may consist of a packed tube containing either an absorbent such as Tenax™ TA, glass beads or may simply be empty stainless steel tubing. Since the pre-concentration stage is at such low temperatures, the majority of water vapour in the sample must be removed prior to refocusing in order to stop blockage of lines with ice. Once a sufficient volume has been collected on the refocusing trap, the trap is generally flash heated either electrically or using hot water. This results in a very sharp band of compounds being introduced to the head of the analytical column.

With adsorbent tube analysis, the collected analytes are generally thermally desorbed either directly onto the analytical column (in the case of programmed temperature vapourisation (PTV) injection) or onto a refocusing cold trap. The desorption temperature is generally defined by the maximum temperature that the adsorbing material can support. For polymeric adsorbents this may be relatively low (<250°C) whilst carbon-based materials may support desorption at temperatures of 500°C or higher. If a PTV injector is used, desorption may be sufficiently rapid (>16°C s^{-1}) that initial phase ratio refocusing on-column is sufficient to obtain well-resolved peaks. If the desorption from an adsorbent tube is relatively slow then a refocusing step is used, with a similar re-concentration mechanism to that used with canister samples. Once again, water must be removed from the sample since it may affect the column or the detection system.

Automated systems, where a single recycled adsorbent trap is used, may operate by taking air from either a local manifold or canister-collected samples. The addition of a multi-position valve upstream of the instrument can allow for fully automated canister analysis on systems designed for in situ analysis. Fully automated instruments of this kind using either cryogen or Peltier-cooled adsorbent traps are now becoming commercially available.

Whilst the majority of species are thermally desorbed from adsorbent traps onto the analytical column, a few types of compounds require solvent extraction prior to syringe injection. The analysis of some organic nitrates has been described in this way, along with higher molecular weight polycyclic aromatic compounds that may suffer from incomplete or slow release.

Derivatisation can be used to selectively convert species from collected atmospheric samples to analytes more suited to gas chromatography. An example of post-acquisition derivatisation is the analysis of atmospheric alcohols, particularly of interest in Brazil, where ethanol-fuelled cars are used on a large scale (Nguyen *et al.*, 2001). Alcohols are highly polar and can stick to canister walls or can be irreversibly trapped onto absorbants. A novel technique for the analysis of alcohols involves collection of air samples in Pyrex vessels followed by subsequent reaction of alcohols with nitrogen dioxide to form alkyl nitrates. Analysis of adducts by GC-ECD gives detection limits of around 1 ppbV for methanol and ethanol.

Organic compounds associated with the particulate phase require a significant amount of preparation to produce a form of sample that is amenable to GC. The first step is to extract the organic material from the sample matrix and this can be done in a number of ways. The simplest in terms of equipment is Soxhlet extraction, which involves the extraction of material in a refluxing solvent over a period of 24 hours. Following extraction the solvent volume is reduced on a rotary evaporator. Alternative extraction methods that use less organic solvent and are considered more environmentally friendly include supercritical fluid extraction (SFE), accelerated solvent extraction, and microwave assisted extraction. All three techniques require specialised equipment but deliver the extracted organic material in periods of typically 1 hour and into only limited volumes of solvent.

The organic extract from atmospheric particles is often extremely complex with a vast range of functional group types. It is common therefore to perform a sample simplification into aliquots containing non-polar, moderately polar, and highly polar species. This is typically done off-line using disposable solid phase extraction cartridges or hand-packed preparative liquid chromatography columns. The sample is introduced to these typically non-polar phases, and then eluted with successively more polar organic solvents (e.g. hexane, toluene, dichloromethane) to produce three, somewhat simplified, mixtures. Clearly there are problems with this type of crude fraction for species that exist at the margins of each polarity classification. However, for species with well-known properties, for example polycyclic aromatic hydrocarbons, this method of sample clean-up yields a much simplified mixture amenable for GC. In all cases where there are off-line procedures such as separations and solvent volume reductions, there is a potential to lose more volatile species to the atmosphere and to introduce artefacts in the sample. To overcome some of these problems alternative methods whereby the clean-up stage is performed by HPLC (e.g. in coupled LC–GC), or where the sample is thermally extracted direct to GC, have been demonstrated.

8.2.3 Separation of atmospheric samples

Whilst a number of specific applications utilise packed columns (notably in the analysis of methane, CO, and N_2O), current methods for the separation of atmospheric components are performed almost exclusively using capillary column GC. The applications where packed columns are in use generally employ molecular sieve packing (typically 5 Angstrom pores) for the separation of permanent gas species. With the introduction of PLOT columns, where the stationary phase is bonded to the column wall, high resolution analysis of very high volatility species is possible and many applications that previously used packed column GC are now being performed using this type of capillary columns.

In a number of cases only one species is to be isolated by the analytical system, and in these cases simple isothermal separations may be used, often in conjunction with a pre-column backflush step. The analysis of PAN is an example of this where a short backflushing pre-column is used prior to a 10–30 m long analytical column. A simple two-dimensional separation has also been proposed for PAN using heart-cutting, where a small fraction of the eluent from the primary column is diverted to a secondary column, for further separation.

The wide range of analyte volatilities that are encountered in the atmosphere confines each analytical system to only a limited range of species that may be completely resolved on a single column. For the most volatile non-methane hydrocarbons (NMHCs), PLOT columns are used widely for species in the carbon range C_1–C_7. This type of gas-solid chromatography has a thin porous layer of a finely divided solid, usually Al_2O_3 doped with a KCl or Na_2SO_4 salt, deposited on the walls of the tube. Separations on columns of this kind are via adsorption rather than phase partition, the kinetics of which are particularly rapid. As a result theoretical plate numbers (i.e. the number of sites available for adsorption/absorption in a column) in excess of 100 000 are common even for wide bore 0.53 mm internal diameter (i.d.) columns. Whilst in principle suitable for high volatility halogenated compounds, there have been reported hydrogenation and dehydrogenation effects for such species on PLOT columns, and so use has been limited. The retention characteristics of PLOT-type columns are unfavourable for oxygenated compounds, where extremely strong retention, often irreversible, can occur. Because of the strong retention of polar species, water becomes an important interference, and its presence can severely degrade the quality of PLOT separations, manifested in highly variable retention times.

The very strong retention of higher boiling point species on PLOT-type columns leads to extensive peak broadening and very lengthy analysis times. Because of this, the analysis of higher molecular weight species including mono-aromatics, VOCs, CFCs, HCFCs (hydrogen-containing chlorofluorocarbon replacements) and terpenoid compounds is generally performed using wall-coated open tubular (WCOT) columns, where the stationary phase is generally a viscous liquid such as non-polar methylpolysiloxane or slightly polar 5%-phenyl-methylpolysiloxane. Typical column specifications are 0.32 mm i.d., 50 m long with film thickness of 1–5 μm, and a temperature gradient is used to reduce band broadening in later peaks. Wide bore, 0.53 mm i.d. columns are also used where thermal desorption is direct from a pre-concentration trap to the analytical column (Helmig, 1999). The rate of generation of theoretical plates (and hence peak capacity) for a given length, on columns of this type, is lower than for PLOT types, and as a result to obtain full resolution of some species (e.g. HCFC mixtures in the atmosphere), columns as long as 100 m have been reported. This technique is the most widespread method of analysis of mid-range ($\sim C_5$–C_{12}) and semi-volatile organic compounds in atmospheric chemistry. Columns are extremely durable and can be used in both field and laboratory measurements.

To improve the retention and separation of some VOCs (those that fall between standard wall-coated siloxane columns and PLOT columns) without use of sub-ambient cooling, phase thicknesses of up to 15 μm have been reported. Band broadening effects through stationary phase diffusion become significant with films of this thickness and this approach has not been widely adopted. The highest molecular weight gas phase species such as naphthalene, fluorene, and anthracene may be separated efficiently on non-polar columns with film thicknesses of typically 0.25–0.5 μm. For particle phase organic measurements where very high boiling point material is present in solvent extracts, thin film columns must be used, of around 0.1 μm, to allow the elution of the least volatile species. Column bleed, where siloxane fragments break off from the stationary phase and form cyclic products, generally increases with thicker stationary phases. Bleed increases background noise and decreases sensitivity, which is particularly important with an MS detector where very low bleed columns are desirable. Typically for PAHs, coronene is considered the largest species that will successfully elute in routine GC. For particle-bound polychlorinated biphenyls and dioxins the number of isomers is very large and it is very difficult to produce conditions under which all species can be uniquely resolved. To overcome problems of co-elution, chiral columns have found some considerable application in this area.

Using very narrow bore columns and fast temperature ramps (i.e. $40°C\,min^{-1}$), commonly available on modern GC instrumentation, it is possible to perform very fast GC separations. Complex atmospheric samples can be separated in under 10 minutes, allowing significant improvements in sampling frequencies, although the preparation and sampling stages then become the rate-determining step. New stationary phases have been developed to extend the range of compounds amenable to capillary GC analysis. Adding polar functionalities to the methylpolysiloxane backbone can create stationary phases of different polarities and selectivities. For example, 14% cyanopropylphenyl, 86% methyl-polysiloxane columns are moderately polar and have been used in the analysis of hydrocarbons, oxygenates, and CFCs. Chiral cyclodextrin–based stationary phases are also available and can be used for the analysis of enantiomeric mixtures such as biogenic VOCs, especially terpenoids.

In addition, a number of speciality phases have been developed that allow the separation of oxygenated material by capillary GC. Since such species are generally at low concentration in the atmosphere, they often suffer from co-elution with more abundant primary emitted VOCs. Whilst derivatisation followed by HPLC is still the most common method of carbonyl analysis, the use of mixed phase porous layer capillary columns is an emerging technique. An example is the Varian CP-LOWOX column, which uses a 'multi-layer' column coating process to produce an extremely polar column with high stability. Sample pre-concentration can be via standard carbon adsorbent methods, and since no derivatisation stages are required, minimum detectable amounts are greatly improved and sampling volumes drastically reduced. Highly polar WCOT stationary phases have also been developed using polyethylene glycol (wax) and can be used to separate oxygenated analytes such as alcohols, free acids, and esters. This type of column has a lower temperature range and an oxygen trap should be used to extend column lifetime.

Organic nitrate species in the atmosphere may also be determined using capillary GC by either charcoal adsorbent traps, extracted with aromatic organic solvent or via direct cryo-focusing from a canister sample. Lengthy analysis times can result due to the necessity to use combinations of columns to achieve full isolation of target analytes based commonly on moderate polarity 50% phenyl, 50% methyl polysiloxane phases.

Modern GC instrumentation has the ability to perform simultaneous parallel GC separations. The inlet flow can be split and the sample introduced to two columns of different functionality, such as an Al_2O_3 PLOT column and a CP-LOWOX column. Using two detectors, either identical or different, allows additional compositional information to be obtained in a single run.

Using column switching, two-dimensional separations can be achieved where the region of the chromatogram of interest can be diverted to a second column for further separation. An important development in atmospheric analysis by GC since 2000 has been the emergence of comprehensive two-dimensional GC (GC × GC). It is widely accepted that the range of organic compounds found in the atmosphere is vast, driven in large part by the huge complexity of petrochemical makeup, which is a primary source to the urban atmosphere. Even by using the highest resolution capillary columns, a single dimension (e.g. single column) separation may generate no more than a peak capacity of around 200. Given sample complexity running into thousands of compounds, a universal method of analysis is not currently available. As the molecular weight increases, so too does the isomeric complexity, and when a universal detector is used to monitor such a separation, co-elutions occur almost continuously. The serial coupling of two capillary columns of differing selectivities, via a mid-point modulator or injection device, where all material undergoes two separations (distinct from GC heart-cut), produces an analytical system with a peak capacity that approaches the product of the peak capacities of the individual dimensions (Liu *et al.*, 1995). The technique was pioneered by J. Phillips in the early 1990s (Liu & Phillips, 1991; Phillips & Xu, 1995), but was not applied to atmospheric samples until 1999 (Lewis *et al.*, 2000). The separations generated using such a technique are viewed as three-dimensional contour surfaces rather than a two-dimensional data stream, where each spot on the chromatogram corresponds to a compound peak. An example of a GC × GC air separation is shown in Figure 8.10 later in this chapter.

Since the second column separations are very rapid (peak widths of typically 100 ms), at present the universal flame ionisation detector (FID) is the most commonly used detector with a sufficiently fast time-response, but a rapid scanning mass spectrometer can also be used (van Deursen *et al.*, 2000). GC × GC has been shown to be robust enough to be used in the field using either a cryogenic or valve-based modulator (Hamilton *et al.*, 2003b; Xu *et al.*, 2003). The cryogenic modulator gives significant sensitivity enhancements due to the zone compression associated with narrower peak widths over one-dimensional separations (Lee *et al.*, 2001). However, the use of a cryogen in field instruments is avoided wherever possible. The valve modulator requires no cryogen and provides improved transfer of very volatile and polar compounds. However, much of the sample is vented to waste during modulation, and lower sensitivities are achieved (Hamilton *et al.*, 2003a). Whilst GC × GC has some unresolved issues related to quantification of three-dimensional volumes rather than standard two-dimensional areas and the more general problems associated with handling complex and large data sets, it is likely that GC × GC atmospheric separations will be more widespread in the coming years.

8.2.4 Detection

As highlighted in the previous section, even the highest resolution capillary column often has insufficient peak capacity to resolve all components in a typical atmospheric sample. Since the introduction of analyte selectivity in the trapping and preparation stages is not always possible, selectivity in detection is a very useful tool for simplifying atmospheric samples. This may be a detailed structure such as an MS or a simpler detector that gives a universal or group-specific response.

The FID is in general terms by far the most commonly used detector in gas chromatography, since it offers high sensitivity, extremely wide linearity and very good long-term reliability and response. Using well-cleaned fuel gases coupled to low noise electrometer circuitry, it is possible to determine amounts down to as low as 1 pg s^{-1} of eluting peak. With a typical sample volume of 1 L, detection limits for individual species may therefore be in the low, pptV range (10^{-12}). Calibration can be performed with relative ease (and in some cases calculated from standard response characteristics) but the complexity of samples can make peak identification difficult when co-elutions occur. To overcome this lack of selectivity, analytical methods for alkene and aromatic analysis using a selective response from photoionisation detection (PID) and the reduction gas detector (RGD) have been proposed although they are not as widespread as FID techniques. For detectors such as the PID, it is the need to regularly calibrate for decreasing bulb/ionisation intensity that accounts for its limited usage.

The electron capture detector (ECD) was specifically designed for halogen measurements in the atmosphere and although mass spectrometric detection is now the most widespread detection method for these compounds, the ECD still finds usage (Rivett *et al.*, 2003). The ECD offers high sensitivity to electrophilic compounds with almost no response for hydrocarbon species. GC-ECD measurements require careful calibration due to the great variation in response to individual halogenated species, although their high stability allows gas standards to be used over many years. Some halogen-containing species of atmospheric interest (e.g. CH_3Cl, CHF_2Cl, CH_2Cl_2) have a relatively poor ECD response and the use of the oxygen-doped ECD to enhance their response has been successful (Urhahn & Ballschmiter, 2000). The determination of some nitrogen-containing species can also be performed using ECD, notably in the areas of organic nitrate analysis and PAN-type compounds. Organic nitrate analysis using ECD is often complicated by the co-elution of halogenated compounds, so often a nitrogen specific detector such as the chemiluminescence detector is used in parallel.

Detection of CO when separated using GC is generally performed using an RGD where one CO molecule reacts with hot HgO and releases one Hg molecule from the catalytic bed, which is then detected by UV absorption (Kok *et al.*, 1998).

The analysis of sulphur compounds in the atmosphere, in particular dimethyl sulphide (DMS), has often been performed using a combination of GC with sulphur-selective detection to overcome problems of insufficient chromatographic resolution. Flame photometric detection (FPD) has been used extensively in the past although quenching of signal by co-eluting hydrocarbons often results in reduced sensitivity. The *Hall* detector or electrolytic conductivity detector (ELCD) has also been used for atmospheric determinations although it requires regular maintenance making it unattractive for an automated instrument. Emerging methods are now taking advantage of significant improvements in bench-top atomic emission detectors (AED). The multi-elemental nature of the AED offers significant advantages in atmospheric measurements both in terms of sensitivity (sulphur–2 pg s^{-1}), and where concurrent carbon emission line measurements may be used to provide information on empirical formulae of unknowns. The sulphur chemiluminescence detector and the sulphur-specific electron capture detector (SECD) are some of the techniques that offer extremely high sensitivity and selectivity, which may yet find an important role in atmospheric sulphur analysis.

As outlined earlier, oxygenated species in the atmosphere are one of the least studied using gas chromatography. It is an area of fundamental significance since species may be present in the atmosphere as both direct emissions or degradation products of primary VOC emissions. Measurements of species such as acetone have been made using aircraft-portable GC systems in upper tropospheric locations, and there is considerable interest in extending the range of o-VOCs that can be measured since they act as a source of free-radical species when photolysed. Separations on polar wall coated columns (e.g. polyethylene glycol or cyanopropyl-doped phases) or via mixed phase PLOT columns do allow for many aldehydes, ketones and alcohols to be isolated in atmospheric samples. Sensitive and selective detection of oxygenates is, however, still difficult, due to low response in both ECD and FID detectors, and low molecular weight fragmentation in MS. Elemental specific detection such as AED offers some potential in oxygenate analysis although sensitivity is poor at around 100 pg s^{-1}. Detectors such as the helium ionisation detector (HID), which produce a non-selective high sensitivity response to these types of compounds, may in future allow on-line measurements of oxygenates with GC assuming sufficient chromatographic resolution and trapping selectivity can be obtained.

Mass spectrometry offers obvious solutions to problems of compound identification, and bench-top GC-MS technology is now at an advanced stage in terms of reliability, self-calibration, tuning and automation. The eluent from the GC column is transferred to the ion source directly via a gas-tight heated transfer line, making this a simple and convenient hyphenated technique, without the mobile phase evaporation problems associated with LC-MS. A detailed description of the principle and practice of direct mass spectrometric methods for gas and aerosol phase analysis is presented in Chapters 5 and 6, respectively. This chapter will focus on the application of MS detection to GC, allowing structural information of more resolved mixtures to be obtained. The MS has three main components – the ion source, the mass analyser and the detector. Ions can be produced using a number of different ionisation methods, the most popular being electron and chemical ionisation EI and CI respectively.

In EI, an electron beam, operated at about 70 electron volts, is fired at molecules as they enter the ion source. The high energy beam transfers large amounts of energy to molecules, which can break up into smaller ions, releasing excess energy. Fragmentation patterns can give important information about the structure of a molecule, an example being a series of m/z ions that are spaced 14 amu apart ($-CH_2$ group), indicating the presence of a hydrocarbon chain. However, the spectral information obtained from GC-MS of many atmospheric species (in particular hydrocarbon-based compounds) often leads to highly similar fragmentation patterns and assists little in the identification of isomeric species. Similarly, identification of monoterpene species can only be confirmed through a combination of both mass spectral fragmentation information and retention times.

It is often advantageous to obtain the molecular ion, to facilitate identification. In some cases, EI completely fragments the molecule, and the molecular ion is absent from the chromatogram. CI produces ions using an ionised reagent gas (commonly methane, isobutane or ammonia) to transfer protons to (or from) the analyte, resulting in the formation of a pseudo-molecular ion. In the positive ion mode, the reagent gas donates a proton to the analyte to form $[M+H]^+$ and in the negative mode it abstracts a proton to form $[M-H]^-$. This is a softer form of ionisation and produces less fragmentation in the ion source.

The ions formed are accelerated by the accelerating plate into the mass analyser, which separates them according to their mass to charge ratio (m/z). The quadrupole mass spectrometer has now become one of the most widely used mass spectrometers because of its ease of use, small size and relatively low cost. Mass separation in a quadrupole mass filter is based on achieving a stable trajectory for ions of specific m/z values in a hyperbolic electrostatic field (see Chapter 5, Section 5.3.2.1). Operating currently available quadrupole mass spectrometers in full-scan mode often yields insufficient sensitivity for trace-level atmospheric measurements, and selected or single ion monitoring is commonplace.

In order to obtain faster acquisition rates than possible using a quadrupole or ion trap, GC can be coupled to a time-of-flight mass spectrometer (TOF/MS). Ions are accelerated into a field-free region in the flight tube where they are separated according to their m/z ratio. The TOF/MS has a much higher spectral acquisition frequency than most other mass spectrometers, which is of particular importance in fast GC-MS, where peak widths are routinely less than 1 second wide. Whilst coupling of GC × GC to TOF/MS has been reported, this is very much a research rather than fieldwork tool at present. The fast mass spectral acquisition rates of the TOF/MS, up to 500 full mass scans per second, allow adequate peak coverage across the narrow peaks inherent to GC × GC (van Deursen *et al.*, 2000). One of the most interesting new applications of GC × GC–TOF/MS is the analysis of organic aerosol content. GC × GC–TOF/MS has been used to separate over 10 000 individual analytes in an urban PM10 sample, a resolution not possible using any current methodologies (Hamilton *et al.*, 2004).

Where GC-MS is particularly strong is in the measurement of halogenated species in the atmosphere. Whilst full-scan operation of most instruments is insufficient to measure such compounds in clean air, monitoring only selected ions can drastically improve detection limits. Whilst hydrocarbon fragmentation is often very similar with little abundant parent ion, many halogenated compounds give highly unique MS fragmentation with abundant large-*m*/*z* fragments. The combination of favourable fragmentation with selected ion measurements has resulted in GC-MS instruments for atmospheric analysis with extremely low detection limits approaching 0.01 pptV. Long-term measurements of species such as CFCs and their replacements have been performed by GC-MS instruments for many years at a number of locations in the worldwide AGAGE (Advanced Global Atmospheric Gases Experiment) network. Using a triple sector mass analyser (electric–magnetic–electric), with mass resolutions in the order of a few thousand, femtogram sensitivities can be achieved. It is possible to measure CFCs, such as HFC-134a, at highly accurate *m*/*z* ratios (e.g. 83.011 amu–$CF_3CH_2^+$), ensuring there are no background interferences (Oram *et al.*, 1996). This approach has been used also for field measurements of naturally produced trace level iodo-carbon and bromo-carbon compounds.

8.2.5 Calibration and quality control

The quality of data produced in any analytical system is of great importance, particularly if any conclusions are to be drawn from the data, such as air quality statistics and model simulations. In GC systems it is normal to use a number of quality control methods but the most important is the use of calibration standards. A Retention Time Standard, which is a qualitative mix of known components (such as hydrocarbons), is used to determine the chromatographic peak retention time. It can be used during instrument optimisation to ensure maximum peak separation. In addition, it can also be used during routine analysis to ensure there is no retention time shift, which could be due to factors such as changes in water content and flow blockages.

A primary calibration standard is used to determine the response of the GC system to specific analytes and therefore allows quantification. This primary calibration standard is generally a gas standard, made up in zero-grade nitrogen, of the compounds of interest at a known concentration. Standard Reference Materials (SRMs) and Certified Reference Materials (CRMs) are available from speciality gas suppliers and are referenced to a national standard such as the National Institute of Standards and Technology (NIST) in the USA or the National Physics Laboratory (NPL) in the UK. Calibration standards at similar concentrations to those in real air samples are preferable as they reduce any errors associated with dilution during sampling. Gas standards should also be humidified before analysis to ensure they have a similar water concentration as an ambient sample. This takes into account any losses of analytes, in particular polar compounds, to the walls or during the drying process.

The intercomparison of GC instruments around the world is also an important aspect of quality control. The AGAGE instruments are calibrated on-site using alternate analysis of ambient air and a 35 L standard tank. Standard tanks last up to ten months and are calibrated relative to the central laboratory (currently the Scripps Institution of Oceanography, SIO, California, USA) (Prinn *et al.*, 2000). The Accurate Measurement of Hydrocarbons in the Atmosphere (AMOHA) project was developed to evaluate current GC methods of analysis of C_2–C_9 NMHCs across Europe. Calibration gas standards were prepared by NPL, for all laboratories, to reduce variability associated with the standard, and analysis was carried out in 12–14 laboratories over four years. Agreement was generally good for high concentration standard mixtures but disagreement increased as lower concentration, real air samples were analysed. However, some improvements were achieved over the course of the project. A number of inter-comparisons between different analytical measurements including GC-FID/ECD/MD, proton transfer reaction mass spectrometry (PTR-MS) and open path FTIR (OP-FTIR) have been made (Christian *et al.*, 2004). Good agreement between GC and OP-FTIR and between GC and PTR-MS was achieved for a number of compounds. However, differences were found for sticky compounds (e.g. formic acid) and compounds with a low proton affinity (e.g. formaldehyde) with OP-FTIR and PTR-MS having considerable advantages. It is important for real samples to take into account the great difference in time resolution between these methods. In the future, validation of techniques via inter-comparisons of standards will increase.

Typical single column GC-FID instruments may have relative standard deviations of the order of 0.2–0.8% for this type of measurement, with an overall level of uncertainty at the single-figure percentage level arising mainly from gravimetric errors associated with the preparation of standards. The frequency of calibration is dependent on application and detector. An FID is considerably more stable over time than a quadrupole mass spectrometer, which may require calibration every couple of samples. During an intensive field campaign, GC-FID systems should be calibrated at least every 100 samples.

Many of the detectors used in GC analysis can give structure-related responses, allowing a reduced number of analytes to be calibrated. This can reduce calibration time considerably. For example, FIDs give a near-linear increase in response as the carbon number increases in hydrocarbon chains. The addition of hetero-atoms can cause a deviation from linearity which can be compensated for using the Effective Carbon Number (ECN). The ECN relates to an FID response factor as a function of molecular structure, with different hetero-groups causing varying deviations from linearity. The use of ECN can be extremely important for the quantification of oxygenated materials which are usually unstable in gas cylinders, with rapid wall losses and degradation. Carrier gas impurities can lead to a reduction in separation efficiency with increased background signals and damage to detectors and columns, in particular if oxygen is present. It is important to use ultra-high purity gas, and 99.999% helium, or greater, is typically used with both hydrocarbon and oxygen scrubbers.

8.2.6 Instrument deployment in the field

An important aspect of atmospheric chemistry is the ability to deploy instruments outside of the laboratory. In order to make measurements of real air samples, GCs must be easily transported and rapidly installed. The location and accessibility of field sites can vary greatly and the problems associated will be site dependant.

The availability of gases, cryogens and power is one of the main criteria of fieldwork. Gases are an integral part of the GC system and gas cylinders are available in a variety of sizes, usually between 2 and 50 L. In some cases, for example working on an aircraft, weight is an issue and must be kept to a minimum. Cylinders can be made from aluminium making them lighter than conventional steel cylinders. The use of commercially available gas generators can reduce the number of gas cylinders required. This is especially important in locations where gas delivery is not possible. The availability of cryogens, such as liquid nitrogen, in remote locations makes their use limited and systems such as Peltier-cooled PTV are more suitable. The power usage of the entire GC set-up should be kept to a minimum and it is common to use a uninterruptible power supply (UPS) to avoid damaging shutdown of the instrument during power disruptions. Field instrumentation incorporating a PTV trap, sample pump, flow control and GC-FID has a typical power usage of 2 kW (Whalley *et al.*, 2004).

In atmospheric chemistry, the vertical profile of the atmosphere is important. GC systems have been deployed in a number of different ways to obtain information about the composition of the atmosphere at varying altitudes. Ground level and boundary layer measurements can be taken from a number of platforms including conventional and mobile laboratories, research ships and balloons. A lightweight balloon-borne instrument has been developed to measure trace gases in the stratosphere and incorporates a series of Carboxen absorbent tubes which are sampled at different altitudes and analysed on the ground by a GC system (Danis *et al.*, 2000). Long, wide bore sampling manifolds can be used to obtain air above ground level allowing the GC to remain in a laboratory on the ground. For example, the Cape Grim Baseline Air Pollution Station in Tasmania, Australia (http://www.bom.gov.au/inside/cgbaps/index.shtml) has a very tall tower allowing air to be sampled at elevations of 3, 10 and 70 m above ground level. To obtain free tropospheric air without the use of an aircraft, requires a laboratory situated at a high altitude above sea level. The Jungfraujoch High Altitude Research Station (http://www.ifjungo.ch/) is situated on a mountain saddle in Switzerland at 3580 m above sea level. Evidence shows that the Jungfraujoch observatory lies in the free troposphere throughout winter and often in spring and autumn. A number of atmospheric measurements based on GC systems have been made at the Jungfraujoch, including hydrocarbons, PAN and halocarbons (Whalley *et al.*, 2004).

The use of aircraft to obtain a vertical profiles of the atmosphere is becoming increasingly popular. Aircraft allow a large spatial coverage for measurements of atmospheric species. Flight time is limited, however, and measurements therefore only represent snap shots in time. Measurements can be made both in situ with an onboard GC and post analysis with whole air sampling and a ground-based GC system (Blake *et al.*, 1996; Purvis *et al.*, 2003). Examples of in-flight GC systems are ORAC (Organics by Real Time Airborne Chromatograph) (Whalley *et al.*, 2004) and PANAK (PAN-Aldehydes-Alcohols-Ketones) (Singh *et al.*, 2004) which use a series of capillary columns and multiple detectors such as ECD, PID and RGD.

Gas Chromatography is also used extensively in industry to measure stack emissions, landfill emissions and to monitor ambient workplace air quality. In general, grab samples are taken at the emission site using Teflon bags or evacuated containers, which can be analysed in the laboratory. A development in emissions monitoring are portable GCs that can be carried by hand or in a backpack. Portable GCs are available with a variety of detectors including PID, ECD, TCD and MS. These instruments have many applications including hazardous waste site investigation, indoor air quality and emergency response, for example in the detection of chemical weapons (http://www.inficonvocmonitoring.com/).

8.2.7 Examples of applications of chromatography in atmospheric analysis

Table 8.2 shows a number of typical applications of chromatographic techniques in atmospheric analysis. Typical sampling lengths, separation mechanisms, detectors and sensitivity are given, as well as the temporal resolution, indicating the frequency with which measurements can be taken. The following section highlights a number of examples of atmospheric measurements obtained using GC.

Figure 8.3 shows a simple GC chromatogram using a WCOT column and an ECD detector to measure PAN at the Jungfraujoch research station in Switzerland (Whalley *et al.*, 2004). Using a backflush facility through a pre-column, well-retained analytes were diverted, allowing only analytes of interest to enter the analytical column (30 m long × 0.53 mm i.d., 5% phenyl polysilphenylene-siloxane, 1.5 μm film thickness). Subsequent separation results in the isolation of PAN from the large air/water peak every 90 seconds. Calibration of the system was performed using gaseous PAN generated by the photolysis of a mixture of NO_2 with excess acetone, diluted using zero air (Zellweger *et al.*, 2000).

The measurement of hydrocarbons requires a greater separation capacity than the PAN GC, where only one analyte was targeted. Figure 8.4 shows the separation of C_1–C_6 hydrocarbons obtained using a PTV injector, a long PLOT column (50 m long × 0.53 mm i.d., Al_2O_3 modified with Na_2SO_4, 10 μm film thickness) and an FID (Bartle & Lewis, 1996). Automation of the procedure, including the use of a H_2 gas generator to provide carrier and flame gas, allowed the unmanned instrument to produce a chromatogram

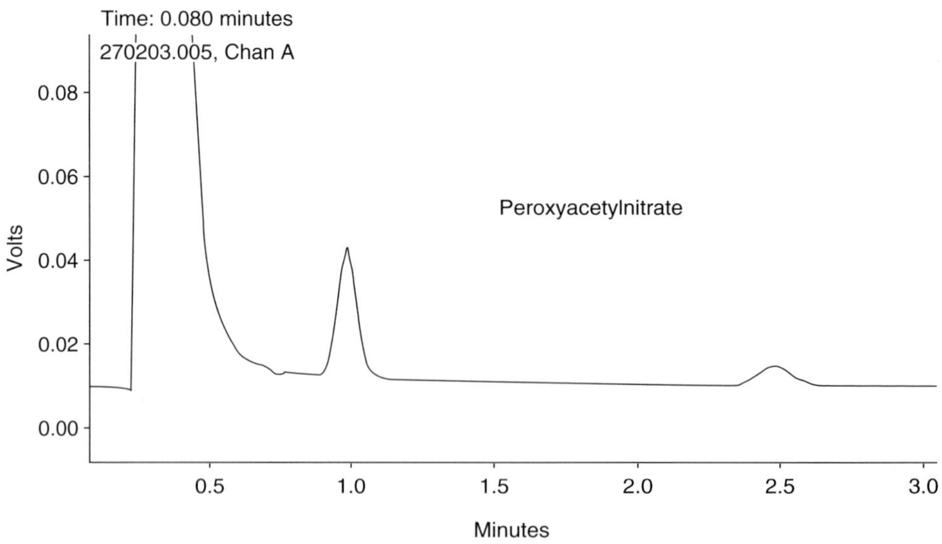

Figure 8.3 PAN chromatogram, with peroxyacetylnitrate highlighted, taken at the Jungfraujoch research station, Switzerland. (Fom Whalley *et al.*, 2004, reproduced by permission of the Royal Society of Chemistry.)

Table 8.2 Typical applications of chromatographic instruments used for gas and condensed phase atmospheric measurements

Description	Separation	Sample size	Detectors	Sensitivity	RDS[a]	Typical resolution	Operation[b]
Permanent gases, CO, CH_4	GC packed columns/molecular sieves	1–10 mL	TCD, FID, HID	Sub-ppm (TCD) to sub-ppb (HID)	Separation	1–10 min	UGS
PAN	GC open or packed columns. With pre column/back flush.	1–10 mL	ECD	10 ppt	Separation	3 min	UGS, AC
Light hydrocarbons C_1- C_6	GC – open tubular PLOT type	100–3000 mL	FID	1–5 ppt	Separation	1 hr	UGS, AC
DMS	GC – WCOT typically methylpolysiloxane	100–3000 mL	FID, FPD	1–10 ppt	Separation	30 min	UGS
Isoprene/ terpenes	GC – WCOT siloxane or cyclodextrin based if chiral	100–3000 mL	FID, MS	1–10 ppt with FID 10–100 ppt full-scan quadrupole MS. Sub-ppt with quad MS in single ion monitoring mode. 1–10 ppt with TOF MS.	Separation	30 min to 1 hr	UGS
CFCs	GC – WCOT typically methylpolysiloxane	100–3000 mL	ECD, MS	1–10 ppt with ECD 10–100 ppt full-scan quadrupole MS. Sub-ppt with quad MS in single ion monitoring mode. 1–10 ppt with TOF MS.	Separation	1 hr	UGS
Reactive halocarbons, organic nitrates	GC – WCOT typically methylpolysiloxane	100–3000 mL	ECD, MS, Mag sector MS, CI-MS	1–10 ppt with ECD, sub-ppt with quad MS in single ion monitoring mode, sub-0.01 ppt with chemical ionisiation, sub 0.01 ppt with magnetic sector MS.	Separation	1 hr	UGS, lab

Table 8.2 (Continued)

Description	Separation	Sample size	Detectors	Sensitivity	RDS[a]	Typical Resolution	Operation[b]
Aromatics	GC – WCOT typically phenyl-methylpolysiloxane	100–3000 mL	PID, MS, FID	1–10 ppt with PID, 1–10 ppt with FID 10–100 ppt full-scan quadrupole MS. Sub-ppt with quad MS in single ion monitoring mode. 1–10 ppt with TOF MS.	Separation	30 min to 1 hr	UGS, AC
Carbonyls (direct)	GC – WCOT or PLOT	100–3000 mL	FID, MS	20–100 ppt with FID, 20–200 ppt full-scan quadrupole MS. 1–10 ppt with quad MS in single-ion monitoring mode. 1–10 ppt with TOF MS.	Separation	1 hr	UGS
Carbonyls (derivatisation)	GC – WCOT typically methylpolysiloxane	5–100 L	ECD, MS	20–200 ppt with ECD, 20–200 ppt full-scan quadrupole MS. 1–10 ppt with quad MS in single-ion monitoring mode. 1–10 ppt with TOF MS.	Sample collection	Several hours	AGS, lab
PAH	LC – normal or reverse phase	1–1000 m^3	UV, MS	Sub-0.01 ppt with large sample size	Sample collection	Hours to days	Lab
Acids	LC – ion exchange column	1–1000 m^3	MS	ppt level	Sample collection	Hours to days	AGS, lab
Peroxides	LC – reverse phase with derivatisation	10–100 L	FL	10–100 ppt depending on sample size	Sample collection	15 min to 1 hr	UGS, lab

[a] RDS – rate determining step in analysis.
[b] Operation: UGS – Unattended ground site, AGS – Attended ground site, AC – Aircraft with attendant ground-laboratory for off-site analysis of Whole Air Sample canisters.

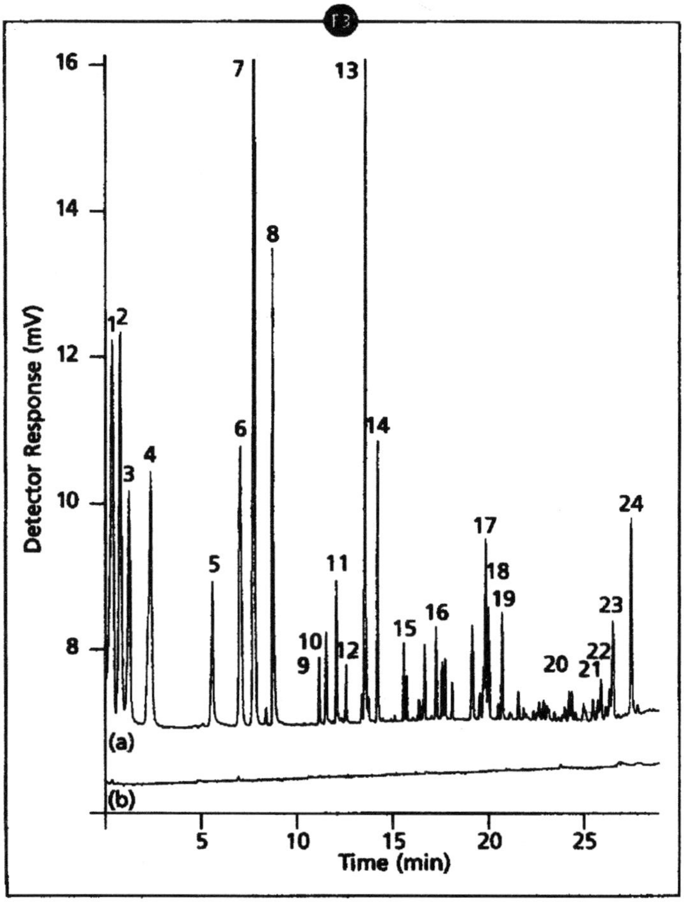

Figure 8.4 Low molecular weight hydrocarbons including methane determined using an on-line activated charcoal adsorbent trap in a programmed temperature vaporisation injector. Column 50 m 0.53 mm i.d. AL_2O_3/Na_2SO_4 PLOT (Chrompack, Netherlands) 10 μm d.f. Desorption temperature was at 16°C s^{-1} from −20 to 400°C, and column temperature programmed from 45 to 200°C. Peaks: 1 = methane, 2 = ethane, 3 = ethene, 4 = propane, 5 = propene, 6 = 2-methylpropane, 7 = ethyne, 8 = n-butane, 9 = trans-2-butene, 10 = 1-butene, 11 = isobutene, 12 = cis-2-butene, 13 = 2-methylbutane, 14 = n-pentane, 15 = 1,3-butadiene, 16 = pentenes, 17 = 2-methylpentane, 18 = 3-methylpentance, 19 = n-hexane, 20 = methyl hexanes and hexenes, 21 = heptane, 22 = methylcyclopentane, 23 = benzene, 24 = toluene. (Reproduced with permission from Bartle and Lewis, 1996.)

every 45 minutes. Using the same instrumentation with a WCOT column (60 m long × 0.53 mm i.d., 100% dimethyl polysiloxane, 3 μm film thickness), which is more suited to the separation of heavier weight analytes, allows the separation of C_5–C_{10} hydrocarbon species in the atmosphere, as shown in Figure 8.5 (Lewis *et al.*, 1996). Using a dual column approach in one instrument facilitates the measurement of the whole range of C_1–C_{10} hydrocarbons in a single analysis.

The analysis of oxidised organics in the atmosphere gives important information on the age and processing of an air mass. Figure 8.6 shows a chromatogram of a C_2–C_5 o-VOCs

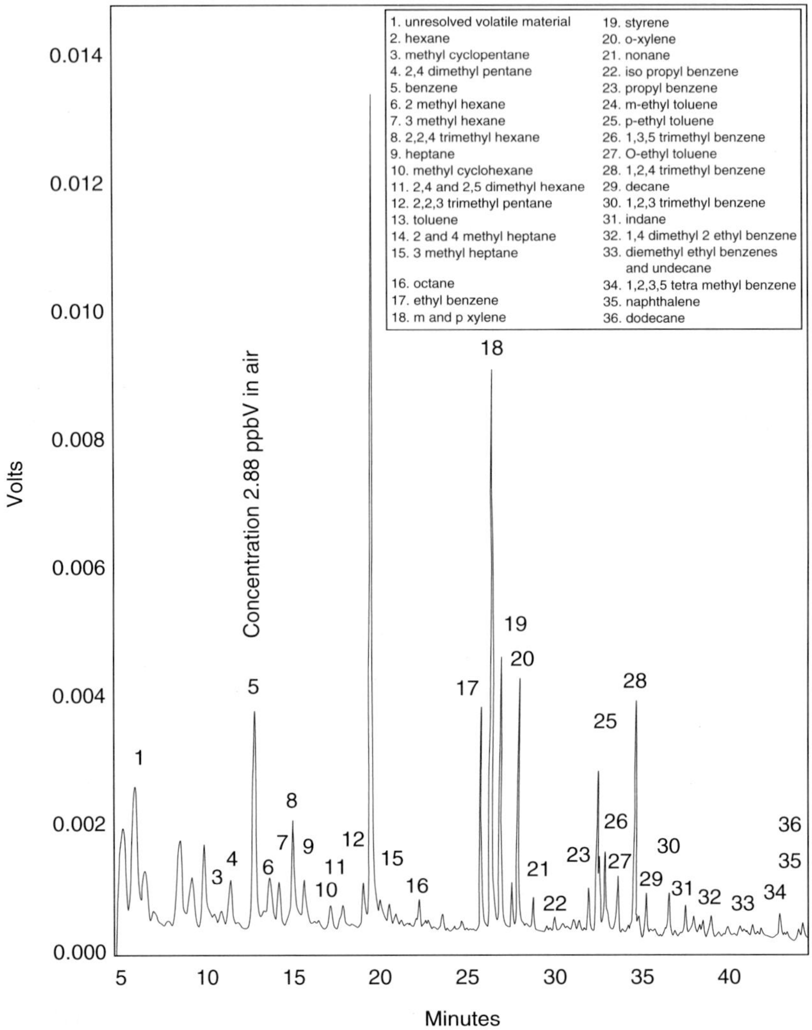

Figure 8.5 Aromatic hydrocarbon species determined using an on-line Tenax™ TA adsorbent trap in a programmed temperature vaporisation injector. Column 60 m 0.53 mm i.d. 100% methyl polysiloxane 3 μm d.f. (Restek RTX-1) Desorption temperature was at 16°C s^{-1} from 0 to 220°C, and column temperature programmed from 35 to 240°C. (Reproduced with permission from Lewis *et al.*, 1996.)

analysis obtained at the Mace Head Atmospheric Research Station (MHARS), situated on the West coast of Ireland (Hopkins *et al.*, 2003). The o-VOCs were collected onto a cooled multi-bed absorbent trap and analysed using a mixed phase capillary column (10 m × 0.53 mm i.d. Varian CP-LOWOX, 10 μm film thickness), designed specifically for trace oxygenate analysis with a highly polar stationary phase. Analytes were calibrated using a combination of liquid permeation tubes and gas standards.

Chlorofluorocarbons (CFCs) have been implicated in the destruction of stratospheric ozone, particularly in the polar regions, leading to the well-known 'hole in the ozone

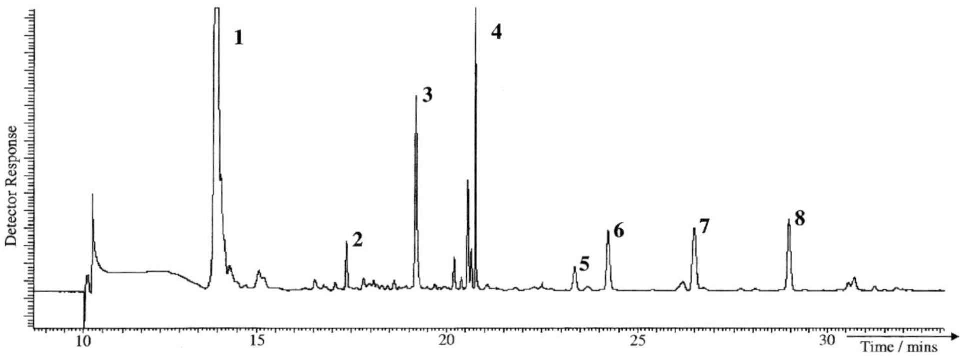

Figure 8.6 Sample chromatograms produced from a LOWOX column where 1 = Aliphatic NMHCs, 2 = benzene, 3 = toluene, 4 = acetaldehyde, 5 = methanol, 6 = acetone, 7 = ethanol, 8 = propanol. (From Hopkins *et al.*, 2003, reproduced by permission of The Royal Society of Chemistry.)

layer' discovered in Antarctica. Figure 8.7 shows a chromatogram of CFCs and Halon species found in clean marine air obtained using both a conventional ECD and an oxygen-doped ECD. The oxygen-doped ECD can enhance the response of certain halogen species (e.g. CH_3Cl, CHF_2Cl, CH_2Cl_2) which have poor responses using conventional ECD detectors. The discovery of the destructive effects of CFCs has led to international protocols limiting their use. A number of CFC replacements have been introduced and their background concentrations have been rising steadily. A GC-magnetic sector MS instrument has been used to measure the concentration of CFCs and Halon species, between 1978 and 1995, at the Cape Grim Observatory in Tasmania. Samples were pre-concentrated onto a cooled glass bead/Tenax™ trap and separated on a PLOT column (50 m × 0.53 mm i.d. HP Alumina PLOT (KCL)). An example chromatogram is shown in Figure 8.8, with peaks detected in Single Ion Mode (SIM) at m/z 69, with concentrations ranging from 80 to 0.1 ppt. Analyte identifications are given and the species detected are long-lived greenhouse gases, with Halon-1301 being an important ozone depleter.

In marine environments, elevated levels of volatile halocarbons are found in the atmosphere, which are emitted from marine sources such as seaweed and phytoplankton. Many of the most important halocarbons have very unique mass spectra, making their analysis particularly suited to GC-MS. Using the mass spectrometer in SIM, over specific retention ranges, gives maximum sensitivity. The GC-MS chromatogram shown in Figure 8.9 was obtained at the Mace Head station, which is situated in a coastal region. The low concentration species have been extracted to show the total speciation achieved. Data of this type can give important information about the air–surface partitioning occurring within the marine boundary layer.

As described earlier, the urban atmosphere contains a highly complex mixture of organic compounds, in both the gas and the aerosol phase. GC × GC with FID detection has been used to separate the C_6–C_{14} organic compounds at a roadside site in Leeds, UK (Hamilton & Lewis, 2003). Using a primary volatility-based column (50 m × 0.32 mm i.d. 100% dimethylpolysiloxane, 4 μm film thickness), followed by a secondary polarity-based separation (2.5 m × 0.10 mm i.d. 50% phenyl polysilphenylene-siloxane, 0.1 μm film thickness) and a valve modulator, gives significant improvement in separation capacity.

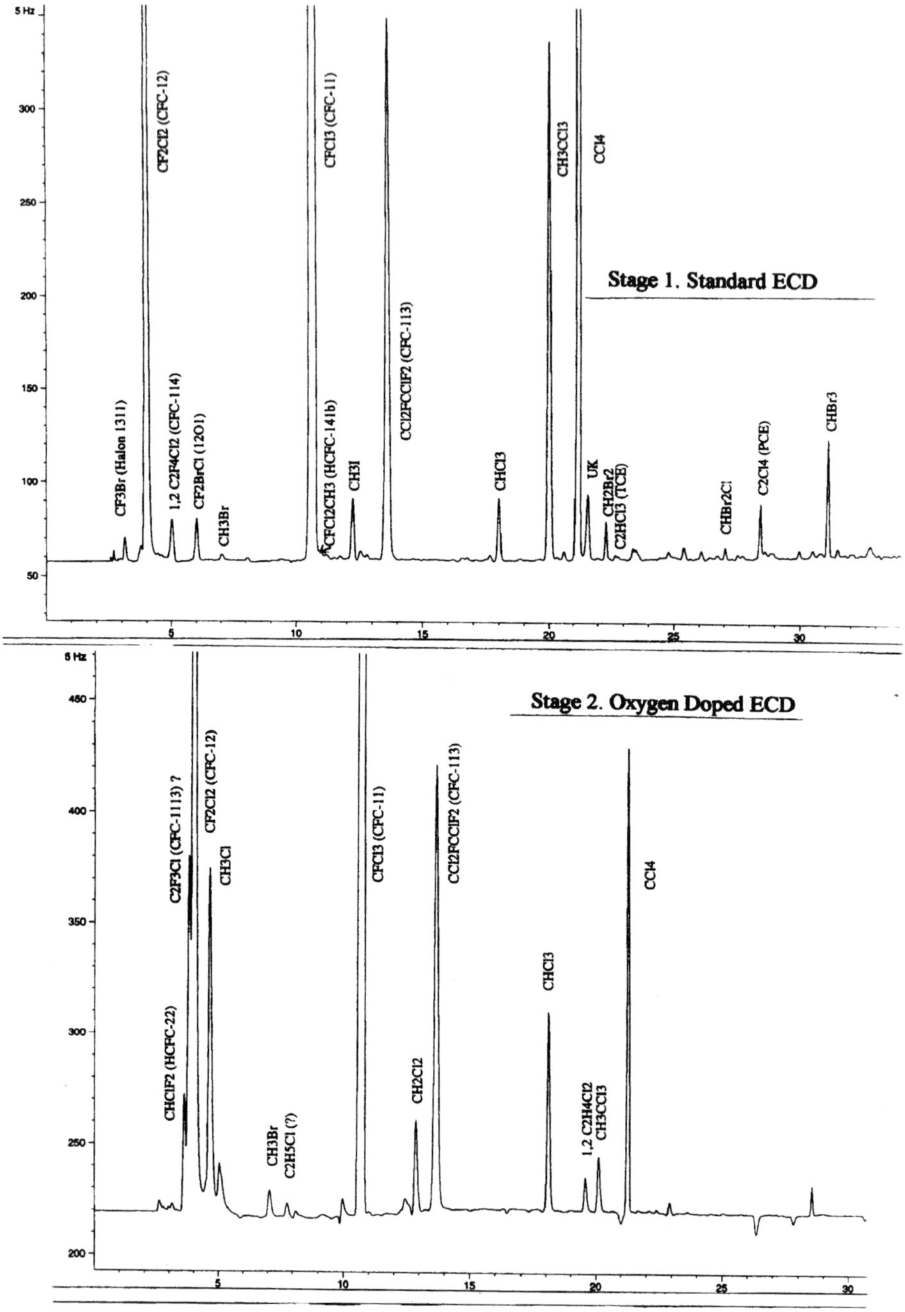

Figure 8.7 CFCs and Halon species in clean marine air determined using an on-line Carbosieve micro adsorbent trap and direct injection to capillary GC. Detection by dual ECD/oxygen doped ECD. 60 m, 0.33 mm i.d., 1 μm film DB-1 (J & W). (By courtesy of M. Bassford and P.G. Simmonds, University of Bristol, Cantocks Place, Bristol.)

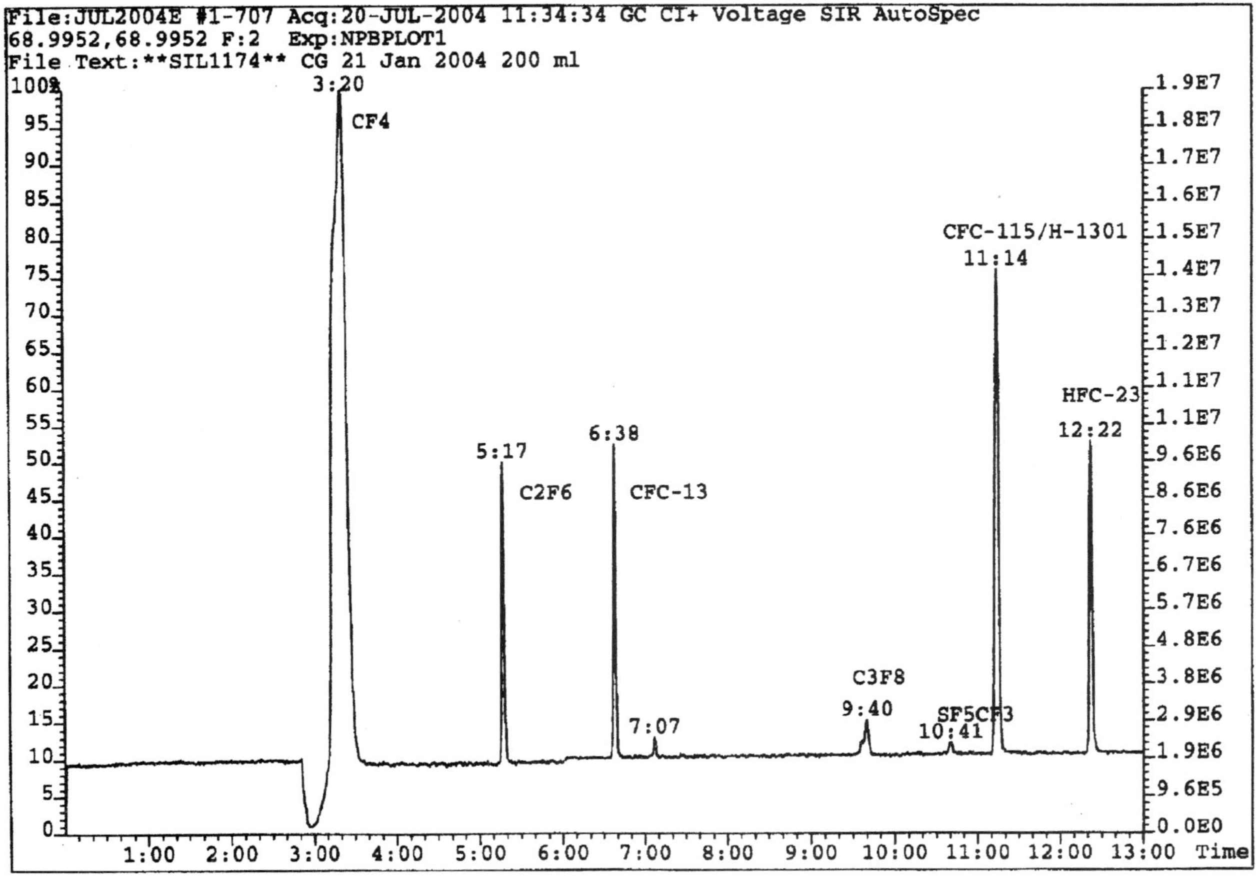

Figure 8.8 Long-lived greenhouse gases observed at the Cape Grim Observatory, Tasmania using a GC/sector MS with SIM at *m*/*z* 69. Samples were pre-concentrated onto Tenax™/glass bead traps and separated using an Al_2O_3/KCl PLOT column. (By courtesy of D. Oram, University of East Anglia, UK.)

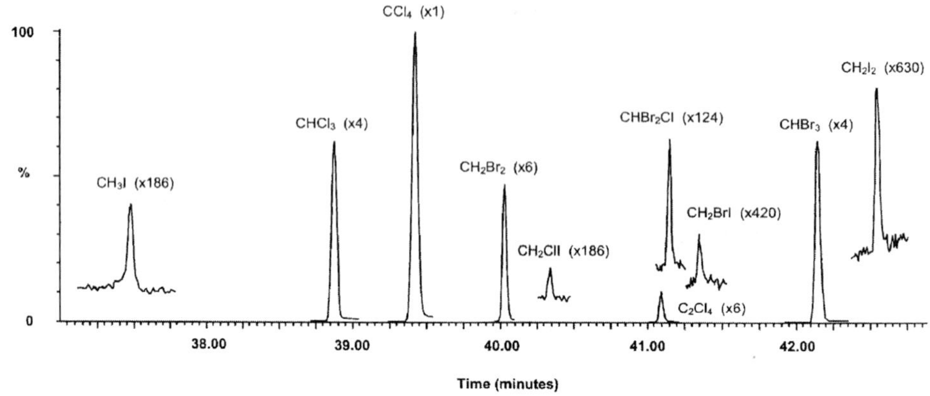

Figure 8.9 Typical ion chromatogram obtained at the Mace Head Atmospheric Research Station in selective recording mode. Identities are (1) CH_3I (*m/z* 127 and 142), (2) $CHCl_3$ (*m/z* 83), (3) C_2H_5I (*m/z* 127 and 156), (5) 2-C_3H_7I (*m/z* 127 and 170), (6) CH_2Br_2 (*m/z* 172 and 174), (7) 1-C_3H_7I (*m/z* 127 and 170), (8) CH_2ClI (*m/z* 176 and 178), (9) $CHBr_2Cl$ (*m/z* 127 and 208), (10) CH_2BrI (*m/z* 220 and 222), (11) $CHBr_3$ (*m/z* 171 and 173), (12) CH_2I_2 (*m/z* 254 and 268). Separation was obtained using a SGE (BPX5) 50 m × 0.32 mm × 3 μm (film) capillary column. (By courtesy of L.J. Carpenter, University of York, UK.)

A GC × GC contour plot is shown in Figure 8.10, where each spot represents an individual analyte. The improvement in resolution between the one-dimensional separation (shown above) and the two-dimensional separation is apparent. The GC × GC chromatogram has been successively expanded to show the greater numbers of isomers at higher molecular weights.

The composition of organic aerosol is even more complex than the gas phase, with considerably more organic compounds in the diesel range. The analysis of this type of sample can prove extremely difficult even with the large separating power of GC × GC. Coupling to a TOF/MS can add an additional separation mechanism, in this case, the *m/z* ratio. Figure 8.11 shows a GC × GC-TOF/MS chromatogram of a PM2.5 particulate samples collected in Augsburg, Germany (Welthagen *et al.*, 2003). In a typical sample, more than 15 000 peaks could be detected. Using an MS as a detector allowed structural information and peak identifications to be obtained, which would have required numerous standards if an FID was used.

The composition of alkyl nitrate compounds has been studied during the TORCH campaign (Tropospheric Organic Chemistry Experiment). This site was situated outside of Chelmsford, UK, and during August 2003 was chosen as a site due to its position relative to the London plume. Air samples were analysed in negative ion chemical ionisation mode (NCI) with methane as the ionisation gas and a sample chromatogram is shown in Figure 8.12.

The use of aircraft in atmospheric measurements is becoming increasingly popular as more research aircraft become available. An on-board GC system has been developed, which uses a cooled absorbant sampling trap followed by GC with HID, which is suitable for use in an aircraft (Whalley *et al.*, 2004). The time resolution of this system is much quicker than conventional GC systems giving a chromatogram every 5 minutes. Figure 8.13 shows a chromatogram collected on 26 March, 2004, on the UK research aircraft, a BAE 146-301 (http://www.faam.ac.uk) during the initial test-flying period. The

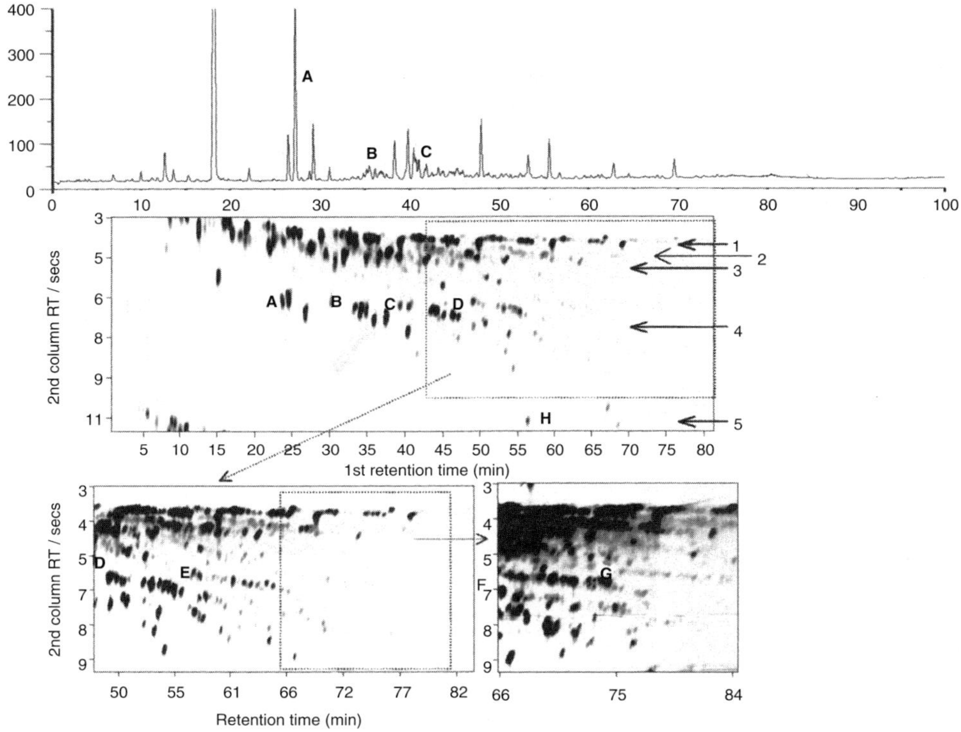

Figure 8.10 Comparison of single column (upper) and GC × GC separations (lower) of a Leeds urban air sample. Areas of the full chromatogram are successively extracted at higher gain to illustrate increasing isomeric complexity at higher boiling points. GC × GC chromatograms are annotated with start of individual C_x isomer band (running right to left) where $A = C_2$, $B = C_3$, $C = C_4$, $D = C_5$, $E = C_6$, $F = C_7$, $G = C_8$, H = naphthalene. Chemical banding assignments, 1; aliphatics, 2; olefinics, 3; oxygenated, 4; mono aromatics, 5; polyaromatics. (From Hamilton & Lewis, 2003, reproduced with permission from Elsevier.) (Reproduced in colour as Plate 4 after page 270.)

aircraft was flying at an altitude of 214.8 m over the Bristol channel (Lat 51.40°N, Lon 3.78°W).

Fieldwork is an integral part of atmospheric analysis and the ability to move equipment from the laboratory to the field site is important. Mobile laboratories housed in vans and shipping containers allow GC instruments to be moved without the need for permanent laboratories facilities at the field site. The University of York, UK, has a GC laboratory fitted within a shipping container, which can be delivered to the site and set-up within a few days. This container was shipped to Mace Head as part of the NAMBLEX experiment described in Section 1.8.2 and Figure 8.14 shows the container on-site and the GC system housed in the interior. Six weeks of almost continuous measurements were made of NMHCs and selected o-VOCs using the dual channel GC-FID system used in Figure 8.6. A region of the concentration *versus* time series, from 30 July to 8 August, 2002, for ethane, acetylene and benzene is shown in Figure 8.15. The directions marked on the graph indicate the origin of the air mass, as calculated by the five-day back trajectory (model used from the European Centre for Medium-Range Weather Forecasts, ECMWF). In the figure, W indicates an air mass that has travelled over the Atlantic Ocean for most

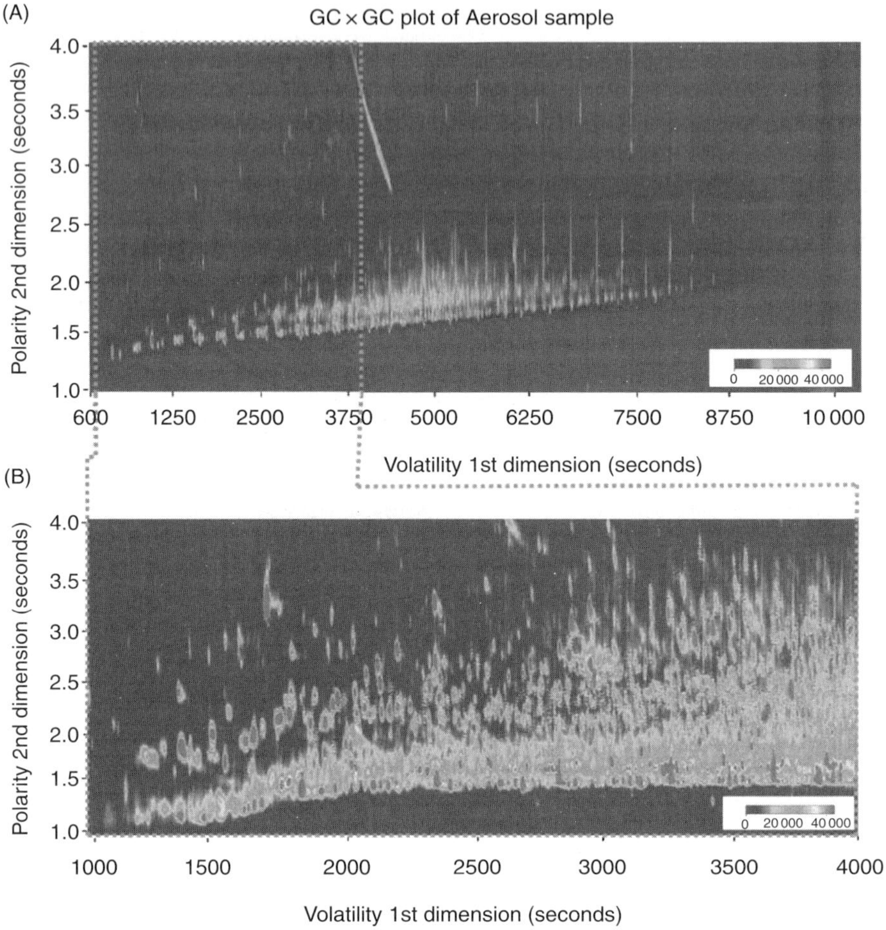

Figure 8.11 Two-dimensional GC × GC-TOFMS total ion current (tic)-plot of an aerosol sample in 2D-contour plot: (A) showing the full chromatogram of the analysed aerosol with (B) the extraction of the selected section for data analysis. (From Welthagen *et al.*, 2003, reproduced with permission from Elsevier.) (Reproduced in colour as Plate 5 after page 270.)

of the last five days; NW indicates an air mass which originated in the northern polar region; and NE five indicates an air mass which has spent the previous five days travelling over northern Europe. From the time series, it is clear to see that the most polluted period is during the NE trajectories. The air mass during this period is travelling over industrial and urban areas of Europe, where the emission of pollution is high. The W and NW air masses are considerably cleaner, having not travelled over any major pollution sources in the previous five days.

In comparison to the short period of measurements in the previous example, the AGAGE network (previously ALE/GAGE) has been operating since 1978 (http://agage.eas.gatech.edu/). One of the primary objectives is to optimally determine from observations the rate of emission and/or chemical destruction (i.e. lifetime) of

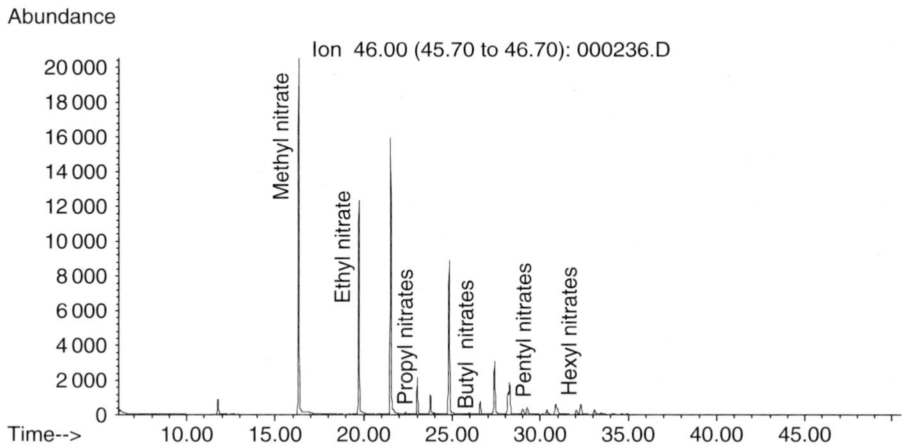

Figure 8.12 Sample alkyl nitrate chromatogram obtained during TORCH 1, at Writtle College, near Chelmsford, UK in August 2003. Agilent Technologies GC (6890)-MS (5973N) running in negative ion chemical ionisation (NCI) mode with methane as chemical ionisation gas. (By courtesy of D. Worton, University of East Anglia, UK.)

the anthropogenic chemicals that contribute most of the reactive chlorine and bromine released into the stratosphere. There are currently five stations around the world measuring a range of important halocarbons. Data collected for the growth of HFC-134a, a CFC replacement, at the Cape Grim site, Tasmania, and over the Atlantic (40°S, 50°W British Antarctic Survey) from 1991 to 1996, is shown in Figure 8.16. The solid data represents the results obtained by a two-dimensional model with 3% of global emissions occurring in the southern hemisphere. Long time-scale continuous measurement networks are of vital importance to determine the concentrations of long-lived species and in understanding their impact on the atmosphere.

8.3 Liquid chromatography

The analyses of the majority of organic compounds in the atmosphere are primarily carried out using GC. However, the complete speciation of organics in the atmosphere is not possible by GC alone. Liquid chromatography (LC) is a complementary technique and is primarily used for the separation of non-volatile substances and highly polar or ionic compounds. High performance liquid chromatography (HPLC) has its origins in classical column chromatography, such as that used extensively in organic synthesis laboratories. However, in theory and practice, it is more similar to GC. Particles as small as 1 μm in diameter can be used as the stationary phase, packed into columns of about 10–25 cm in length. The mobile phase is forced through the packed column by combinations of high pressure pumps (up to 200 bar) to achieve volumetric flow rates of 1–5 $cm^3\ min^{-1}$. Since the mobile phase affects selectivity in HPLC (rather than being inert as in GC), combinations of solvents are dynamically combined to provide binary, ternary or even quaternary mixtures of mobile phase.

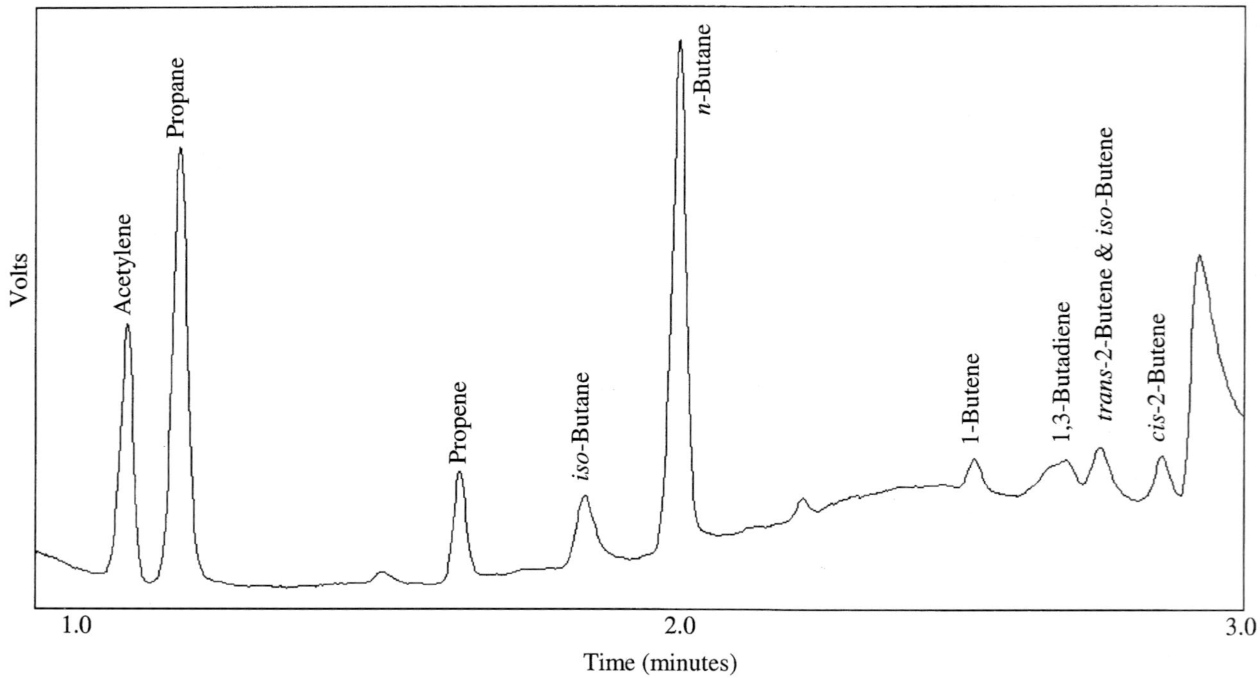

Figure 8.13 ORAC chromatogram collected on 26 March 2004 on the UK research aircraft, a BAE 146–301 during the initial test flying period, altitude of 214.8 m over the Bristol channel (Lat 51.40°, Lon −3.78°). Analyte assignments are given on the chromatogram. (By courtesy of J.B. McQuaid, University of Leeds, UK.)

Figure 8.14 Left – Mace Head Atmospheric Research Station during NAMBLEX, July–September 2002. The GC container is marked. Right – inside of the University of York GC container, showing the GC, sampling and PTV system and a series of canisters.)

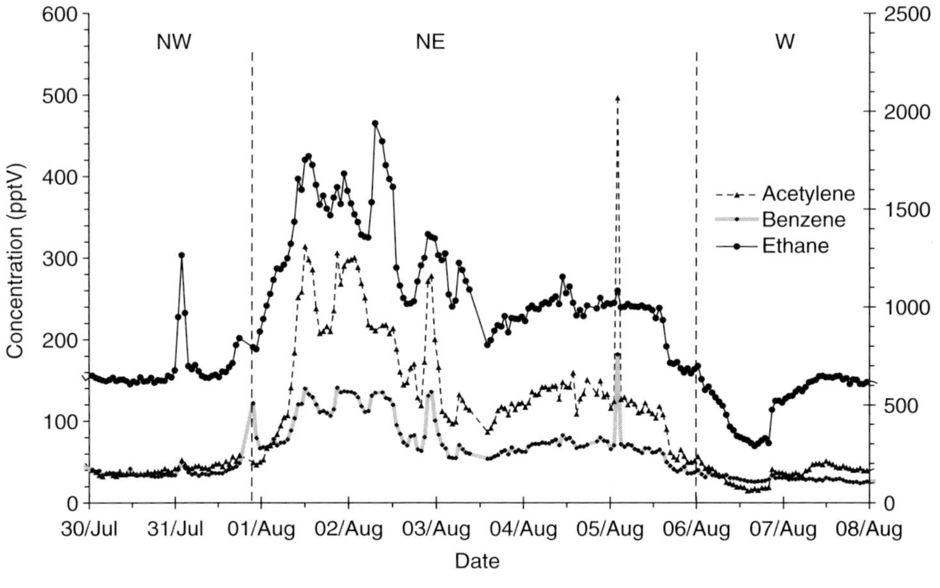

Figure 8.15 Concentrations of acetylene, benzene and ethane observed in air samples at Mace Head during NAMBLEX. NW, NE and W indicate the origin of the air mass arriving at the site as calculated by the 5-day back trajectory. Dotted lines indicate a change in trajectory direction. (By courtesy of J.R. Hopkins, University of York, UK.)

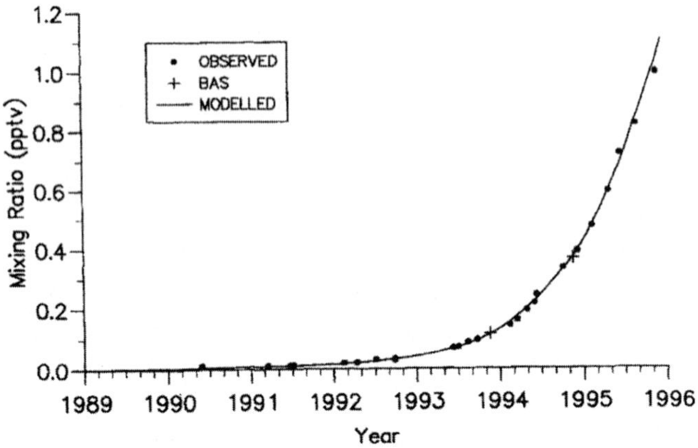

Figure 8.16 HFC-134a concentrations (pptv) observed in air samples collected at Cape Grim (41°S, 145°E) and over the Atlantic Ocean (40°S, 50°W), compared to results from the 2-D model with 3% of global emissions in the Southern Hemisphere. (From Oram *et al.*, 1996, © 1996 American Geophysical Union. Reproduced by permission of American Geophysical Union.)

8.3.1 Sample acquisition and preparation

In atmospheric chemistry the analytes amenable to HPLC are found in the gas and aerosol phase or in atmospheric water, such as fog water. Samples must be introduced to the LC inlet in a liquid solution and a number of trapping methodologies are in use. The primary method of sample acquisition and preparation in atmospheric HPLC analysis uses derivatisation techniques and is one of the most common methods of carbonyl analysis.

In situ derivatisation is commonly used where analytes are derivatised during the collection process. Derivatisation can be carried out in the liquid phase, where air is drawn through an impinger containing the derivatising agent in organic solution (such as *iso*-octane). After collection, the organic fraction is evaporated to dryness and the residue is dissolved in a polar solvent such as methanol. Alternatively, analytes can be derivatised on solid phase absorbents such as SPME, via diffusion or a pumped flow. Derivatised analytes are released from the absorbent using solvent extraction, which can be achieved on-line (Sanchez *et al.*, 2003). Post-acquisition derivatisation is less common in HPLC, although it is used in aerosol analysis and for the detection of atmospheric peroxides (see Chapter 7) (Kok *et al.*, 1995).

Solvent extraction of filters is used extensively in particulate analysis. Much of the aerosol compositional work since 2000 has involved the study of the highly polar organic fraction, which is thought to be the result of numerous heterogeneous reactions in the particulate phase. Solvent extraction using water yields the water-soluble organic content (WSOC), a highly polar mixture that is most suited to LC analysis.

8.3.2 Separation

Early HPLC used columns packed with silica giving a highly polar stationary phase, requiring the use of non-polar mobile phase solvents. This type of separation is known as normal phase high performance liquid chromatography (NP-HPLC) and is limited to the separation of non-polar compounds. In reverse phase (RP-HPLC), the addition of functionalities to the stationary phase has increased the range of polarities that can be analysed by HPLC. The addition of a hydrocarbon chain to the silica, most commonly octadecylsiloxane (C_{18}, ODS), changes the surface characteristics of the packing material to a non-polar stationary phase. Polar mobile phases such as water and acetonitrile can be used to separate hydrocarbon oxidation products, such as carboxylic acids and peroxides, not amenable to GC.

Ion exchange chromatography (IEC), also a form of HPLC, is used for the analysis of extremely polar organic compounds and ionic species, such as organic acids and HONO. Stationary phases in IEC are generally resins, such as PS-DVB, a copolymer of styrene and divinyl-benzene. The polymer is a three-dimensional cross-linked structure that is rigid, porous and highly insoluble. Cation- or anion-exchange properties are introduced to the resin by chemical modification after polymerisation. Metal ions, for example Ca^{2+}, K^+, Mg^{2+}, Na^+, have been measured in aerosol particles using IEC (Lee *et al.*, 2003).

Liquid chromatography can be used as a preparative step in atmospheric analysis to simplify complex mixtures. WSOC in aerosol can be fractionated into three classes of compounds using IEC: (1) neutral/basic compounds; (2) mono- and di-carboxylic acids; and (3) polyacidic compounds. The functional group composition of fractions can then be investigated using proton nuclear magnetic resonance spectroscopy (HNMR).

Multi-dimensional chromatography combinations of HPLC with GC are possible and have been reported where an on-line sample preparation step is desirable. Since the mobile phase from the LC stage must be interfaced to the GC, non-polar solvents that may be easily evaporated are used (e.g. pentane). This limits the technique to normal phase HPLC and water content in the sample becomes a serious problem in achieving full automation.

8.3.3 Detection

The most common detector in routine HPLC analysis is the UV absorbance photometer due to its wide linear range and good sensitivity. Photometers are operated at one or more fixed wavelengths only and require the presence of a suitable chromophore within the molecule. In many atmospheric samples, the molecules of interest have weak absorptions in the UV region. The use of derivatisation to form analytes with strong UV absorbances is widely used in HPLC, and DNPH-carbonyl adducts have maximum UV absorbances at approximately 340–380 nm (Druzik *et al.*, 1990). Diode array detectors can also be coupled to HPLC, offering an extended detection range (190–950 nm), with high spectral resolution and peak identification via UV spectral libraries. Ultraviolet detectors are robust and non-destructive, allowing detected compounds to be collected or further analysed.

Fluorescence detectors are highly selective and among the most sensitive of detectors. Chromatographic analysis of hydrogen and organic peroxides are carried out on a C_{18}-RP-HPLC column followed by post-column reaction with horseradish peroxidase (Kok *et al.*, 1995). One of the reaction by-products is a dimer of p-hydroxyphenyl ethanoic acid that has a strong fluorescence at an excitation of 301 nm and an emission wavelength of 414 nm, giving typical atmospheric detection limits of H_2O_2 and the organic peroxides of 30 pptV.

One of the developments in HPLC is the routine coupling of the column to a mass spectrometer. This is considerably more difficult than in GC-MS, where the eluent from the end of the column flows directly into the ion source. The major difficulty in LC-MS is the removal of the liquid mobile phase, whilst allowing only the analytes to pass into the detector. Several interfaces have been designed for this purpose, but for air analysis the most common are atmospheric pressure chemical ionisation (APCI) and electrospray ionisation (ESI). The former uses a reagent gas such as nitrogen as a nebulising gas. A heated nebuliser is used to vapourise the mobile phase and form reactant ions via a corona discharge. The ions and analyte molecules are accelerated through skimmers into a low-pressure region where the solvent is pumped away. Chemical ionisation (positive or negative) of the analyte molecules occurs via collisions with the excited reagent ions, which then pass into the mass spectrometer, typically a quadrupole-MS, for analysis. LC-APCI/MS is particularly suited to moderately polar and non-polar analytes and has been used to measure aldehydes in both aerosol particles and the gas phase (Grosjean *et al.*, 1999; Van den Bergh *et al.*, 2002). It has also been used to measure nitroaromatic compounds in the atmosphere as an indication of concealed land mines (Sanchez *et al.*, 2003).

In ESI, the eluent from the column flows through a metal capillary, which is at a potential of several kilovolts relative to the surrounding chamber walls. The surfaces of the emerging liquid become charged and the liquid is dispersed in a fine spray via repulsive forces. In this case the analytes are ionised by the makeup gas before the solvent is evaporated and swept away, along with any other non-charged material. Electrospray ionisation is particularly suited to the analysis of charged, polar, and basic compounds, making it a complimentary technique to APCI-MS. An example is the use of both ESI(−) and APCI(+) in terpene aerosol studies to identify the organic acids and carbonyl/alcohol content respectively (Winterhalter *et al.*, 2003).

8.3.4 Examples of the application of liquid chromatography in air analysis

The importance of oxygenated compounds in the atmosphere has already been discussed. Carbonyl species can be emitted from biogenic and anthropogenic sources, such as vehicle emissions, and are also produced via photo-oxidation of hydrocarbons. Using DNPH to convert carbonyls to their 2,4-dinitrophenyl hydrazones produces an analyte ideally suited to HPLC. Figure 8.17 shows a DNPH–carbonyl standard mixture that has been analysed using RP-HPLC column (150 mm × 4.6 mm C18 ODS, 5 μm particle size) with a diode array UV-visible detector (Grosjean *et al.*, 1999). This method has been extended to the analysis of carbonyls in both the gas and particulate phase of the atmosphere. The DNPH derivatives of carbonyls found in a PM sample collected during the Pacific 2001 field study in Fraser Valley, Canada, were analysed by HPLC-UV and a sample chromatogram is shown in Figure 8.18 (Liggio & McLaren, 2003). Simplification of the complex aerosol sample is achieved using a simultaneous extraction/derivatisation technique and an initial guard column is used to trap the hydrazones before flushing onto the analytical column.

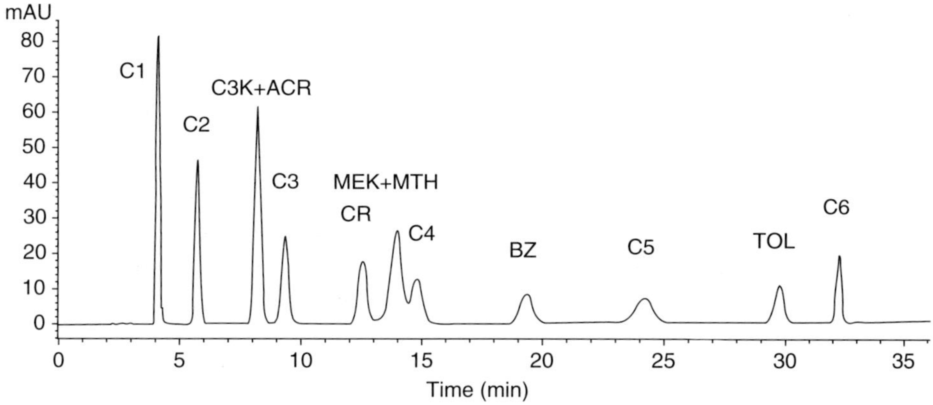

Figure 8.17 Liquid chromatography analysis of a mixture of the DNPH derivatives of 13 carbonyls by ultraviolet absorption at 360 nm (diode array detector). C1, formaldehyde; C2, acetaldehyde; C3K, acetone; ACR, acrolein; C3, propanal; CR, crotonaldehyde; MEK, 2-butanone; MTH, methacrolein; C4, butanal; BZ, benzaldehyde; C5, pentanal; TOL, *m*-tolualdehyde; C6, hexanal. (Reprinted with permission from Grosjean *et al.*, 1999. © 1999 American Chemical Society.)

An important aspect of atmospheric chemistry is the use of smog chambers to simulate the reactions of compounds of interest under controlled pseudo-atmospheric conditions. The production of reaction intermediates can be simulated and the analytes identified, without any background interferences, which would make this process extremely difficult in real air samples. A number of the reaction products of α-pinene ozonolysis have been identified at the EUPHORE smog chamber, Valencia, Spain, using LC-MS (Winterhalter *et al.*, 2003). An example of the ESI(−) ionisation mode chromatogram is shown in Figure 8.19, with the Total Ion Count (TIC) at the top, followed by a series of single ion chromatograms. In the SIM modes, identification and increased sensitivity and resolution are possible. The identification of reaction intermediates can give important information about degradation mechanisms and is vital for accurate modelling studies.

The complexity of compounds in aerosol can be simplified using LC as a fractionation tool. Lewis *et al.* (1995) measured PAH in dust particles using SFE to remove the organic compounds from particulate samples followed by online LC-GC analysis. The initial LC analysis separates the organic content with respect to polarity and removes the aliphatic and polar compounds. Portions of the aromatic fraction were then transferred to the GC for analysis. Figure 8.20 shows both the LC and subsequent GC chromatograms obtained using this technique, with the PAHs indicated by numbers.

8.4 Future work

The use of chromatographic methods in the future for atmospheric measurements seems certain. No other technique can deal so effectively with such a wide range of chemical functionalities and trace species contained within complex mixtures. The science benefits greatly from having increasingly sophisticated and reliable, commercially available

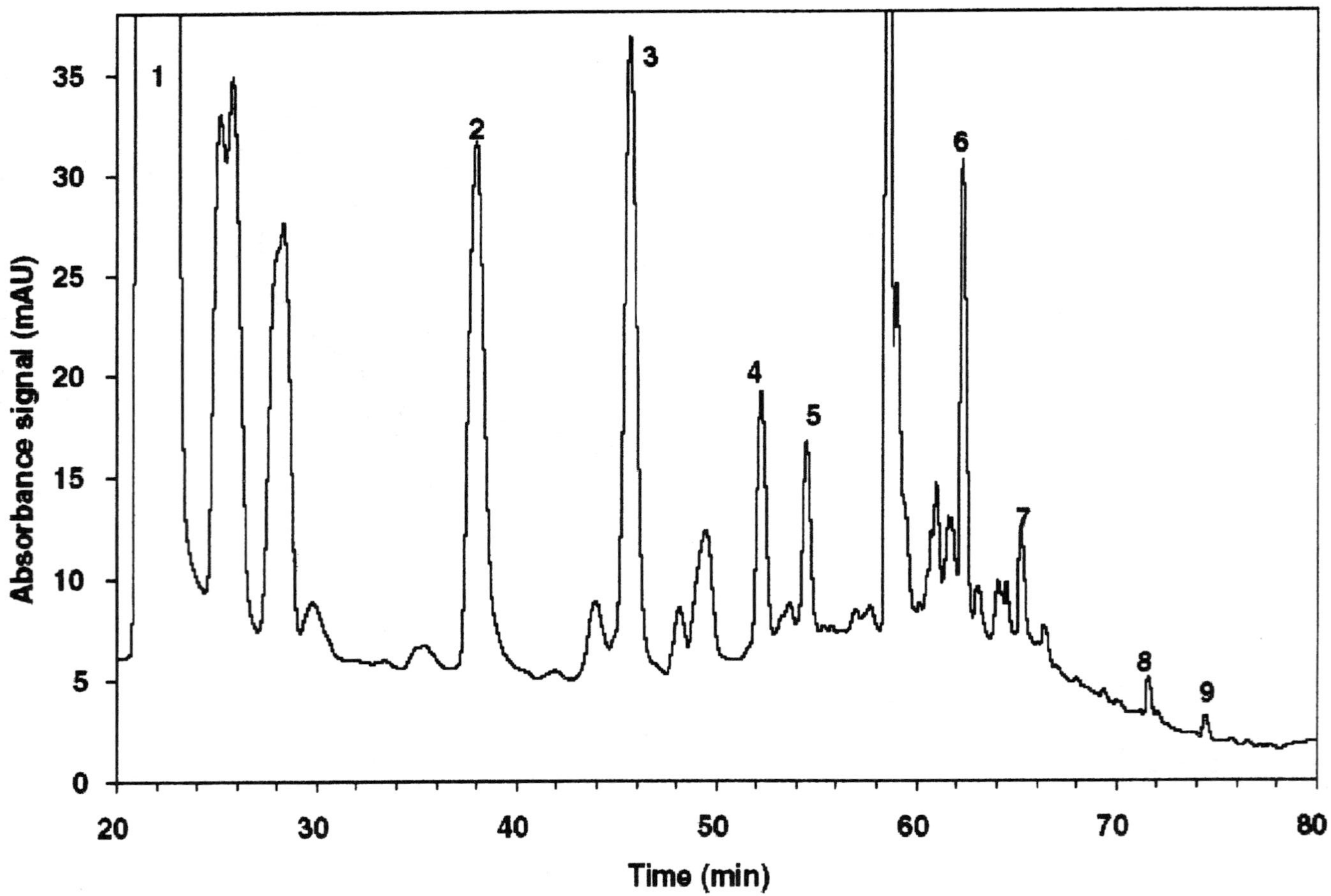

Figure 8.18 Sample chromatogram obtained during the PACIFIC 2001 field study. Peak labels: 1. 2,4-dnph derivatizing agent; 2. formaldehyde-DNPH; 3. acetaldehyde-DNPH; 4. acetone-DNPH; 5. propanal-DNPH; 6. glyoxal-DNPH; 7. methylglyoxal-DNPH; 8. pinonaldehyde-DNPH; 9. nonanal-DNPH. (Reproduced with permission from Liggio & McLaren, 2003), Taylor and Francis Ltd., http://www.tandf.co.uk/journals)

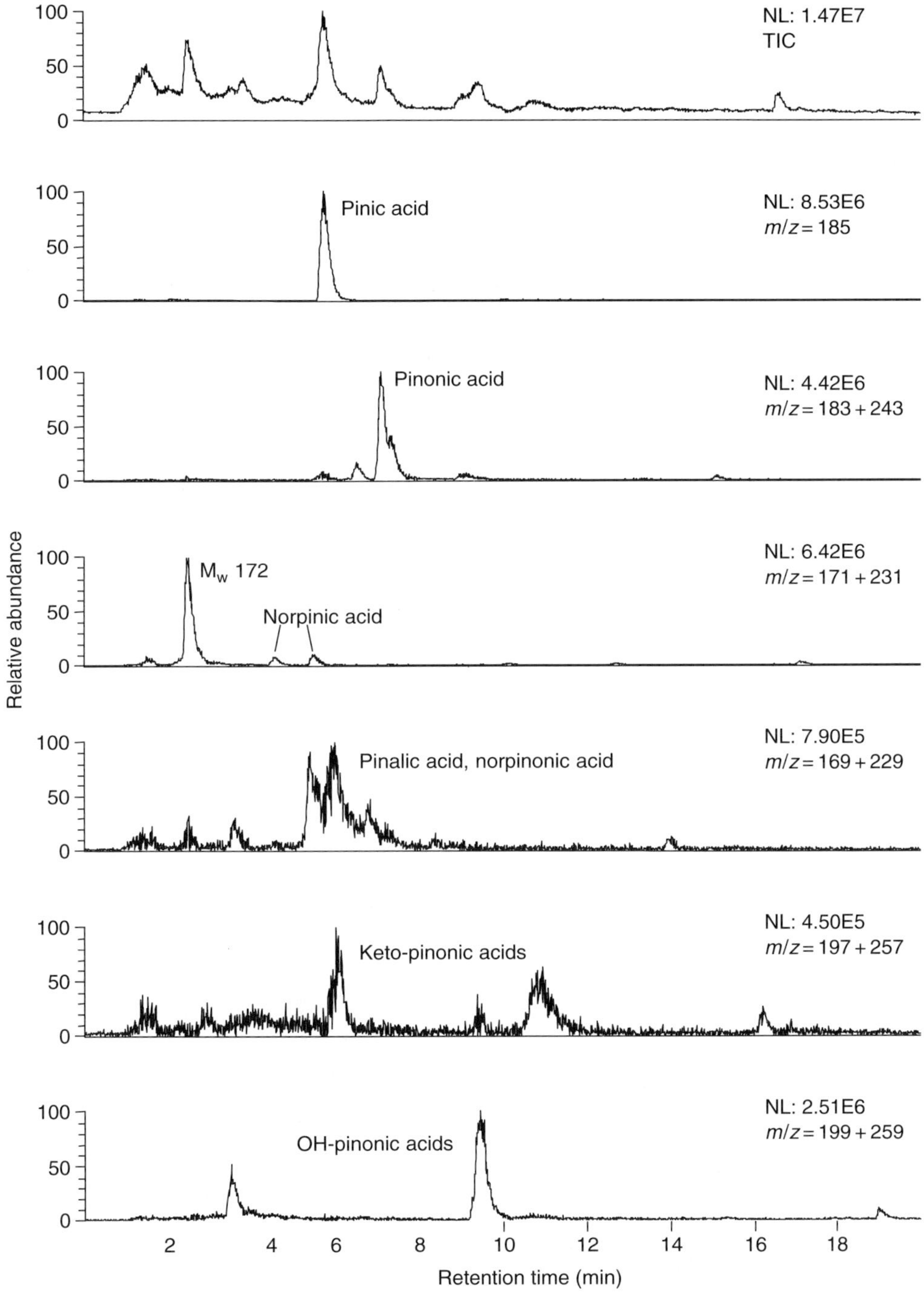

Figure 8.19 TIC and extracted ion chromatogram of a samples from α-pinene-ozonolysis analysed in the ESI(-) mode. (Reproduced with permission from Winterhalter *et al.*, 2003.)

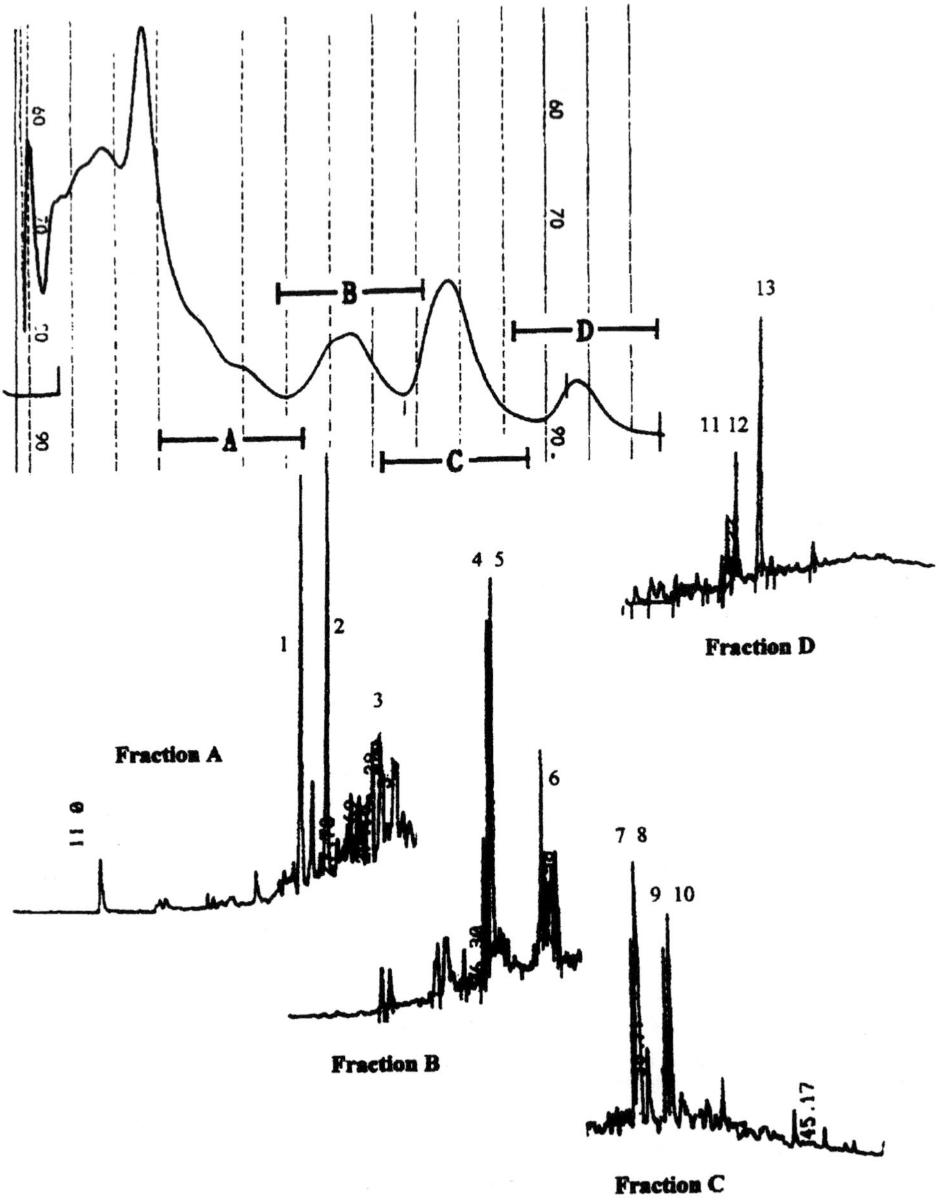

Figure 8.20 LC and GC chromatograms of extracted PAH from urban air particulates collected during October 1994. 1. fluoranthene; 2. pyrene; 3. methyl pyrenes; 4. benz[a]anthracene; 5. chrysene; 6. methyl chrysenes; 7. benzo[b]fluoranthene; 8. benzo[k]fluoranthene; 9. benzo[e]pyrene; 10. benzo[a]pyrene; 11. indeno[1,2,3-cd]pyrene; 12. dibenzo[ah]anthracene; 13. benzo[ghi]perylene. (From Lewis *et al.*, 1995, reproduced with permission from Elsevier.)

equipment at the core of many atmospheric instruments, a spin-off from the enormous industrial and commercial usage of chromatographic techniques.

Since chromatographic techniques are some of the most historically mature in atmospheric chemistry, future developments are likely to be incremental rather than step-change. The reliability and accuracy of GC methods is their key strength and this is only likely to improve as commercially available instrumentation, and in particular GC-MS, which forms the core of many field instruments, develops. Atmospheric measurements may provide some very useful applications for miniaturised GC systems, an area which is beginning to show promise in developmental laboratories. The chemical- and photo-etching technology for production of microfabricated columns onto silica monolith structures has been around for a number of years, but extending into real applications has been limited by detectors. Semiconductor chip–sized thermal conductivity detectors have been the first truly miniaturised detectors to emerge, but it seems only a matter of time before microfabricated ionisation detectors such as PID and HID are produced with the requisite levels of sensitivity for atmospheric applications. The enormous size and electrical power savings that may result from miniaturisation may open up new areas of measurements that have previously only been done using offline chemical methods.

Some of the developments in comprehensive chromatography and the ability to resolve organic compounds in exceedingly high complexity mixtures have been an important step forward. A valuable by-product of comprehensive methods has been the ability to directly introduce heterogeneous samples such as aerosols to the analytical system without sample preparation or clean-up. Saving labour time is an obvious benefit, but importantly the reduction in amount of material required for analysis improves both measurement resolution and detection limits. As high complexity methods reach maturity, however, it is likely that it will be data interpretation that becomes the rate-determining step in the analytical procedure. This may require us to consider adopting data-handling technologies associated with image processing to find ways in which useful and timely information can be extracted from chemical methods with such high levels of detail.

Liquid chromatography methods and LC coupled to mass spectrometry do offer real potential for the future. State-of-the-art coupled LC-MS techniques (using TOF and double or triple quadrupole MS) now offer similar or better detection limits to LC-fluorescence techniques and have improved greatly in terms of reliability, sensitivity and ease of use since the second half of 1990s. This improvement has been driven primarily by the wide uptake of the technique by the bioscience and pharmaceutical communities. Clearly MS as a detector has superior discrimination powers and it seems likely that it may slowly replace non-selective detectors in a number of applications. The use of high resolution LC for the analysis of polar and ionic compounds is perhaps underutilised at present, and there are continuing detection and resolution problems with methods built around industry standard, 4.6 mm i.d. columns. The use of capillary LC (cap-LC) or microbore LC in atmospheric applications has been very limited indeed, but may go some considerable way towards overcoming these limiting factors. The inherently small sample capacity of cap-LC will always be a challenge and reinforces that sample preparation cannot be underestimated in its importance. It is for sample capacity–related reasons that there has yet to be any take-up of capillary electrochromatography (CEC) in atmospheric applications and only very limited use of electrophoresis, but it is in using electrically driven systems, however, that perhaps the most exciting possibilities for very large volume sample pre-concentration may lie.

Further reading

Dewulf, J. & Van Langenhove, H. (2002) Analysis of volatile organic compounds using gas chromatography, *Trac-Trends in Analytical Chemistry*, 21 (9–10), pp. 637–646.

Fifield, F.W. & Kealey, D. (1995) *Principles and Practice of Analytical Chemistry*, CRC Press.

Fowlis, I.A. (1995) *Gas Chromatography*, Second edition, John Wiley & Sons.

Helmig, D. (1999) Air analysis by gas chromatography, *J. Chromatogr. A*, 843 (1–2), pp. 129–146.

Phillips, J.B. & Beens, J. (1999) Comprehensive two-dimensional gas chromatography: A hyphenated method with strong coupling between the two dimensions, *J. Chromatogr. A*, 856 (1–2), pp. 331–347.

Santos, F.J. & Galceran, M.T. (2002) The application of gas chromatography to environmental analysis, *Trac-Trends Anal. Chem.*, 21 (9–10), pp. 672–685.

Skoog, D.A., West, D.M. & Holler, F.J. (2003) *Fundamentals of Analytical Chemistry*, eighth edition, Brooks Cole.

http://www.separationsnow.com/basehtml/SepH

References

Aschmann, S.M., Arey, J. & Atkinson, R. (2003) Kinetics and products of the gas-phase reaction of OH radicals with 5-hydroxy-2-pentanone at 296 +/− 2 K, *J. Atmos. Chem.*, 45 (3), pp. 289–299.

Bartle, K.D. & Lewis, A.C. (1996) A simplified method for the determination of atmospheric hydrocarbons, *LC-GC International*, 9 (5), pp. 297–304.

Bartle, K.D. & Myers, P. (2002) History of gas chromatography, *TRAC-Trends Anal. Chem.*, 21 (9–10), pp. 547–557.

Berresheim, H., Huey, J.W., Thorn, R.P., Eisele, F.L., Tanner, D.J. & Jefferson, A. (1998) Measurements of dimethyl sulfide, dimethyl sulfoxide, dimethyl sulfone, and aerosol ions at Palmer Station, Antarctica, *J. Geophys. Res.-Atmos.*, 103 (D1), pp. 1629–1637.

Blake, D.R., Chen, T.Y., Smith, T.Y., Wang, C.J.L., Wingenter, O.W., Blake, N.J., Rowland, F.S. & Mayer, E.W. (1996) Three-dimensional distribution of nonmenthane hydrocarbons and halocarbons over the northwestern Pacific during the 1991 Pacific Exploratory Mission (PEM-West A), *J. Geophys. Res.-Atmos.*, 101 (D1), pp. 1763–1778.

Cecinato, A., Di Palo, V., Mabilia, R. & Possanzini, M. (2001) Pentafluorophenylhydrazine as a coating reagent for the HRGC-MS determination of semi volatile carbonyl compounds in air, *Chromatographia*, 54 (3–4), pp. 263–269.

Christian, T.J., Kleiss, B., Yokelson, R.J., Holzinger, R., Crutzen, P.J., Hao, W.M., Shirai, T. & Blake, D.R. (2004) Comprehensive laboratory measurements of biomass-burning emissions: 2. First intercomparison of open-path FTIR, PTR-MS, and GC-MS/FID/ECD, *J. Geophys. Res.-Atmos.*, 109 (D2), p. art-D02311.

Danis, F., Harris, N.R.P., Taylor, W.H., McIntyre, J.D., Simmonds, P.G., & Pyle, J.A. (2000) DESCARTES: A novel lightweight balloon-borne instrument for measurement of halocarbons, *Rev. Sci. Instrum.*, 71 (1), pp. 271–280.

Docherty, K.S. & Ziemann, P.J. (2001) On-line, inlet-based trimethylsilyl derivatization for gas chromatography of mono- and dicarboxylic acids, *J. Chromatogr. A*, 921 (2), pp. 265–275.

Druzik, C.M., Grosjean, D., Vanneste, A. & Parmar, S.S. (1990) Sampling of atmospheric carbonyls with small Dnph-coated C18 cartridges and liquid-chromatography analysis with diode-array detection, *Int. J. Environ. Anal. Chem.*, 38 (4), pp. 495–512.

Grosjean, E., Green, P.G. & Grosjean, D. (1999) Liquid chromatography analysis of carbonyl (2,4-dinitrophenyl)hydrazones with detection by diode array ultraviolet spectroscopy and by

atmospheric pressure negative chemical ionization mass spectrometry, *Anal. Chem.*, 71 (9), pp. 1851–1861.
Hamilton, J.F. & Lewis, A.C. (2003) Monoaromatic complexity in urban air and gasoline assessed using comprehensive GC and fast GC-TOF/MS, *Atmos. Environ.*, 37 (5), pp. 589–602.
Hamilton, J.F., Lewis, A.C. & Bartle, K.D. (2003a) Peak amplitude and resolution in comprehensive gas chromatography using valve modulation, *J. Separation Sci.*, 26 (6–7), pp. 578–584.
Hamilton, J.F., Lewis, A.C., Bloss, C., Wagner, V., Henderson, A.P., Golding, B.T., Wirtz, K., Martin-Reviejo, M. & Pilling, M.J. (2003b) Measurements of photo-oxidation products from the reaction of a series of alkyl-benzenes with hydroxyl radicals during EXACT using comprehensive gas chromatography, *Atmos. Chem. Phys.*, 3, pp. 1999–2014.
Hamilton, J.F., Webb, P.J., Lewis, A.C., Hopkins, J.R., Smith, S. & Davy, P. (2004) Partially oxidised organic components in urban aerosol using GCXGC-TOF/MS, *Atmos. Chem. Phys.*, 4, pp. 1279–1290.
Helmig, D. (1999) Air analysis by gas chromatography, *J. Chromatogr. A*, 843 (1–2), pp. 129–146.
Ho, S.S.H. & Yu, J.Z. (2002) Feasibility of collection and analysis of airborne carbonyls by on-sorbent derivatization and thermal desorption, *Anal. Chem.*, 74 (6), pp. 1232–1240.
Ho, S.S.H. & Yu, J.Z. (2004) Determination of airborne carbonyls: Comparison of a thermal desorption/GC method with the standard DNPH/HPLC method, *Environ. Sci. Technol.*, 38 (3), pp. 862–870.
Hopkins, J.R., Lewis, A.C. & Read, K.A. (2003) A two-column method for long-term monitoring of non-methane hydrocarbons (NMHCs) and oxygenated volatile organic compounds (o-VOCs), *J. Environ. Monit.*, 5 (1), pp. 8–13.
Kok, G.L., McLaren, S.E. & Staffelbach, T.A. (1995) Hplc Determination of Atmospheric Organic Hydroperoxides, *J. Atmos. Oceanic Technol.*, 12 (2), pp. 282–289.
Kok, G.L., Prevot, A.S.H., Schillawski, R.D. & Johnson, J.E. (1998) Carbon monoxide measurements from 76 degrees N to 59 degrees S and over the South Tasman Sea, *J. Geophys. Res.-Atmos.*, 103 (D13), pp. 16 731–16 736.
Lee, A.L., Bartle, K.D. & Lewis, A.C. (2001) A model of peak amplitude enhancement in orthogonal two-dimensional gas chromatography, *Anal. Chem.*, 73 (6), pp. 1330–1335.
Lee, Y.N., Weber, R., Ma, Y., Orsini, D., Maxwell-Meier, K., Blake, D., Meinardi, S., Sachse, G., Harward, C., Chen, T.Y., Thornton, D., Tu, F.H. & Bandy, A. (2003) Airborne measurement of inorganic ionic components of fine aerosol particles using the particle-into-liquid sampler coupled to ion chromatography technique during ACE-Asia and TRACE-P, *J. Geophys. Res.-Atmos.*, 108 (D23), p. art-8646.
Lewis, A.C., Bartle, K.D., McQuaid, J.B., Pilling, M.J., Seakins, P.W. & Ridgeon, P. (1996) Atmospheric monitoring of volatile organic compounds using programmed temperature vaporization injection, *Hrc-J. High Resolution Chromatogr.*, 19 (12), pp. 686–690.
Lewis, A.C., Carslaw, N., Marriott, P.J., Kinghorn, R.M., Morrison, P., Lee, A.L., Bartle, K.D. & Pilling, M.J. (2000) A larger pool of ozone-forming carbon compounds in urban atmospheres, *Nature*, 405 (6788), pp. 778–781.
Lewis, A.C., Kupiszewska, D., Bartle, K.D. & Pilling, M.J. (1995) City-Center concentrations of polycyclic aromatic-hydrocarbons using supercritical-fluid extraction, *Atmos. Environ.*, 29 (13), pp. 1531–1542.
Liggio, J. & McLaren, R. (2003) An optimized method for the determination of volatile and semi-volatile aldehydes and ketones in ambient particulate matter, *Int. J. Environ. Anal. Chem.*, 83 (10), pp. 819–835.
Liu, Z.Y., Patterson, D.G. & Lee, M.L. (1995) Geometric approach to factor-analysis for the estimation of orthogonality and practical peak-capacity in comprehensive 2-dimensional separations, *Anal. Chem.*, 67 (21), pp. 3840–3845.
Liu, Z.Y. & Phillips, J.B. (1991) Comprehensive 2-dimensional gas-chromatography using an on-column thermal modulator interface, *J. Chromatogr. Sci.*, 29 (6), pp. 227–231.

Marsella, A.M., Purdham, J.T. & Mabury, S.A. (2000) A new method for the measurement of airborne formaldehyde using derivatization with 3,5-bis (trifluoromethyl) phenylhydrazine and analysis by GC-ECD and GC-MS/SIM, *Int. J. Environ. Anal. Chem.*, 76 (3), pp. 241–256.

Moortgat, G.K., Grossmann, D., Boddenberg, A., Dallmann, G., Ligon, A.P., Turner, W.V., Gab, S., Slemr, F., Wieprecht, W., Acker, K., Kibler, M., Schlomski, S. & Bachmann, K. (2002) Hydrogen peroxide, organic peroxides and higher carbonyl compounds determined during the BERLIOZ campaign, *J. Atmos. Chem.*, 42 (1), pp. 443–463.

Nguyen, H.T.H., Takenaka, N., Bandow, H., Maeda, Y., de Oliva, S.T., Botelho, M.M.F. & Tavares, T.M. (2001) Atmospheric alcohols and aldehydes concentrations measured in Osaka, Japan and in Sao Paulo, Brazil, *Atmos. Environ.*, 35 (18), pp. 3075–3083.

Oram, D.E., Reeves, C.E., Sturges, W.T., Penkett, S.A., Fraser, P.J. & Langenfelds, R.L. (1996) Recent tropospheric growth rate and distribution of HFC-134a (CF3CH2F), *Geophys. Res. Lett.*, 23 (15), pp. 1949–1952.

Phillips, J.B. & Xu, J.Z. (1995) Comprehensive multidimensional gas-chromatography, *J. Chromatogr. A*, 703 (1–2), pp. 327–334.

Prinn, R.G., Weiss, R.F., Fraser, P.J., Simmonds, P.G., Cunnold, D.M., Alyea, F.N., O'Doherty, S., Salameh, P., Miller, B.R., Huang, J., Wang, R.H.J., Hartley, D.E., Harth, C., Steele, L.P., Sturrock, G., Midgley, P.M. & McCulloch, A. (2000) A history of chemically and radiatively important gases in air deduced from ALE/GAGE/AGAGE, *J. Geophys. Res.-Atmos.*, 105 (D14), pp. 17 751–17 792.

Purvis, R.M., Lewis, A.C., Carney, R.A., McQuaid, J.B., Arnold, S.R., Methven, J., Barjat, H., Dewey, K., Kent, J., Monks, P.S., Carpenter, L.J., Brough, N., Penkett, S.A. & Reeves, C.E. (2003) Rapid uplift of nonmethane hydrocarbons in a cold front over central Europe, *J. Geophys. Res.-Atmos.*, 108 (D7), p. art-4224.

Rivett, A.C., Martin, D., Nickless, G., Simmonds, P.G., O'Doherty, S.J., Gray, D.J. & Shallcross, D.E. (2003) In situ gas chromatographic measurements of halocarbons in an urban environment, *Atmos. Environ.*, 37 (16), pp. 2221–2235.

Sanchez, C., Carlsson, H., Colmsjo, A., Crescenzi, C. & Batlle, R. (2003) Determination of nitroaromatic compounds in air samples at femtogram level using C-18 membrane sampling and on-line extraction with LC-MS, *Anal. Chem.*, 75 (17), pp. 4639–4645.

Sin, D.W.M., Wong, Y.C., Sham, W.C. & Wang, D. (2001) Development of an analytical technique and stability evaluation of 143 C-3-C-12 volatile organic compounds in Summa((R)) canisters by gas chromatography-mass spectrometry, *Analyst*, 126 (3), pp. 310–321.

Singh, H.B., Salas, L.J., Chatfield, R.B., Czech, E., Fried, A., Walega, J., Evans, M.J., Field, B.D., Jacob, D.J., Blake, D., Heikes, B., Talbot, R., Sachse, G., Crawford, J.H., Avery, M.A., Sandholm, S. & Fuelberg, H. (2004) Analysis of the atmospheric distribution, sources, and sinks of oxygenated volatile organic chemicals based on measurements over the Pacific during TRACE-P, *J. Geophys. Res.-Atmos.*, 109 (D15), p. art-D15S07.

Urhahn, T. & Ballschmiter, K. (2000) Analysis of halogenated C-1/C-2-trace compounds in marine atmosphere, *Fresenius J. Anal. Chem.*, 366 (4), pp. 365–367.

Vairavamurthy, A., Roberts, J.M. & Newman, L. (1992) Methods for determination of low-molecular-weight carbonyl-compounds in the atmosphere, *Atmos. Environ. Part A – Gen. Top.*, 26 (11), pp. 1965–1993.

Van den Bergh, V., Coeckelberghs, H., Vanhees, I., De Boer, R., Compernolle, F. & Vinckier, C. (2002) HPLC-MS determination of the oxidation products of the reaction between alpha- and beta-pinene and OH radicals, *Anal. Bioanalytical Chem.*, 372 (5–6), pp. 630–638.

van Deursen, M., Beens, J., Reijenga, J., Lipman, P., Cramers, C. & Blomberg, J. (2000) Group-type identification of oil samples using comprehensive two-dimensional gas chromatography coupled to a time-of-flight mass spectrometer (GC × GC-TOF), *HRC-J. High Resolution Chromatogr.*, 23 (7–8), pp. 507–510.

Welthagen, W., Schnelle-Kreis, J. & Zimmermann, R. (2003) Search criteria and rules for comprehensive two-dimensional gas chromatography-time-of-flight mass spectrometry analysis of airborne particulate matter, *J. Chromatogr. A*, 1019 (1–2), pp. 233–249.
Whalley, L.K., Lewis, A.C., McQuaid, J.B., Purvis, R.M., Lee, J.D., Stemmler, K., Zellweger, C. & Ridgeon, P. (2004) Two high-speed, portable GC systems designed for the measurement of non-methane hydrocarbons and PAN: Results from the Jungfraujoch High Altitude Observatory, *J. Environ. Monit.*, 6 (3), pp. 234–241.
Winterhalter, R., Van Dingenen, R., Larsen, B.R., Jensen, N.R. & Hjorth, J. (2003) LC-MS analysis of aerosol particles from the oxidation of a-pinene by ozone and OH-radicals, *Atmos. Chem. Phys. Discuss.*, 3, pp. 1–39.
Xu, X., Williams, J., Plass-Dulmer, C., Berresheim, H., Salisbury, G., Lange, L. & Lelieveld, J. (2003) GC × GC measurements of C-7-C-11 aromatic and n-alkane hydrocarbons on Crete, in air from Eastern Europe during the MINOS campaign, *Atmos. Chem. Phys.*, 3, pp. 1461–1475.
Zellweger, C., Ammann, M., Buchmann, B., Hofer, P., Lugauer, M., Ruttimann, R., Streit, N., Weingartner, E. & Baltensperger, U. (2000) Summertime NOy speciation at the Jungfraujoch, 3580 m above sea level, Switzerland, *J. Geophys. Res.-Atmos.*, 105 (D5), pp. 6655–6667.

Chapter 9

Measurement of Photolysis Frequencies in the Atmosphere

Andreas Hofzumahaus

9.1 Introduction

9.1.1 Photolysis reactions in atmospheric chemistry

The absorption of ultraviolet (UV) and visible solar radiation by atmospheric molecules followed by their dissociation plays a fundamental role in atmospheric chemistry. The photolysis of gaseous compounds contributes to their removal from the atmosphere, but probably the most important role played by these reactions is the generation of highly reactive atoms and radicals. Some examples of relevant reactions are given in Table 9.1, where reactions are listed in the order of their maximum wavelength for which photolysis occurs.

Probably the most important photolysis reactions are the ones that result in the formation of O_3 and OH radicals, the major atmospheric oxidants. In the stratosphere, formation of ozone proceeds via the photodissociation of molecular oxygen into ground-state oxygen atoms $O(^3P)$, which combine with O_2 in a termolecular collision process yielding O_3 molecules:

$$O_2 + h\nu \longrightarrow O(^3P) + O(^3P) \qquad \text{(R 9.1)}$$

$$O(^3P) + O_2 + M \longrightarrow O_3 + M \qquad \text{(R 9.2)}$$

Reaction R 9.1 can occur only at wavelengths less than 243 nm. The threshold corresponds to the minimum photon energy $h\nu = hc/\lambda$, which is required for the dissociation of O_2 into $O(^3P)$ atoms. Here λ and ν denote the wavelength and frequency of the photons, respectively, h is Planck's constant (6.626×10^{-34} J · s), and c is the speed of light (2.9979×10^8 m · s^{-1} in vacuum).

Reaction R 9.1 followed by R 9.2 is the main source of stratospheric ozone which efficiently absorbs all solar radiation below 290 nm when passing into the lower atmosphere. The removal of short-wave UV prevents that ozone formation can take place in the troposphere via R9.1. Rather, photochemical ozone formation in the troposphere

Table 9.1 Examples of photolysis reactions in the atmosphere

Photolysis reaction[a]	Wavelengths[b] (nm)	Photolysis frequency[c] (s^{-1})	
		0 km	25 km
$O_2 + h\nu \longrightarrow O(^3P) + O(^3P)$	<243	4.3×10^{-26}	1.2×10^{-11}
$N_2O + h\nu \longrightarrow O(^1D) + N_2$	<250	1.8×10^{-23}	2.5×10^{-8}
$CCl_3F + h\nu \longrightarrow CCl_2F + Cl$	<250	4.3×10^{-22}	5.1×10^{-7}
$O_3 + h\nu \longrightarrow O(^1D) + O_2(a^1\Delta_g)$	<325[d]	5.1×10^{-5}	1.1×10^{-4}
$CH_3CHO + h\nu \longrightarrow CH_3 + HCO$	<335	6.4×10^{-6}	1.3×10^{-5}
$HCHO + h\nu \longrightarrow H + HCO$	<340	3.3×10^{-5}	6.2×10^{-5}
$CH_3OOH + h\nu \longrightarrow CH_3O + OH$	<350	5.7×10^{-6}	9.9×10^{-6}
$H_2O_2 + h\nu \longrightarrow OH + OH$	<350	7.7×10^{-6}	1.2×10^{-5}
$HNO_3 + h\nu \longrightarrow OH + NO_2$	<350	8.2×10^{-7}	5.6×10^{-6}
$HCHO + h\nu \longrightarrow H_2 + CO$	<360	4.4×10^{-5}	7.6×10^{-5}
$HONO + h\nu \longrightarrow OH + NO$	<400	1.9×10^{-3}	2.8×10^{-3}
$O_3 + h\nu \longrightarrow O(^1D) + O_2(X^3\Sigma_g^-)$	<411[e]	–	–
$NO_2 + h\nu \longrightarrow NO + O(^3P)$	<420	8.8×10^{-3}	1.2×10^{-2}
$HOCl + h\nu \longrightarrow OH + Cl$	<420	2.6×10^{-4}	4.1×10^{-4}
$Cl_2 + h\nu \longrightarrow Cl + Cl$	<490	2.4×10^{-3}	3.5×10^{-3}
$NO_3 + h\nu \longrightarrow NO_2 + O(^3P)$	<640	2.8×10^{-1}	3.0×10^{-1}
$NO_3 + h\nu \longrightarrow NO + O_2$	580–640	2.5×10^{-2}	2.7×10^{-2}
$O_3 + h\nu \longrightarrow O(^3P) + O_2(X^3\Sigma_g^-)$	<1180	4.2×10^{-4}	5.0×10^{-4}
$HNO_4 + h\nu \longrightarrow HO_2 + NO_2$	1450[f]	–	–

[a] Spectral and photochemical data can be found in, e.g., Atkinson *et al.* (2004); Sander *et al.* (2003); Finlayson-Pitts and Pitts, Jr. (2000); Röth *et al.* (1997); Okabe (1978). If not otherwise indicated, photofragments are produced in their electronic ground states.
[b] Wavelengths at which photolysis occurs. Note that in the troposphere, photolysis is not possible below 290 nm, as solar radiation at shorter wavelengths is entirely absorbed in the stratosphere.
[c] From Table B.4 in Jacobson (1999). The photolysis frequencies at 0 and 25 km were modelled for a nonpolluted atmosphere at a solar zenith angle of 0° and a surface albedo of 0.03.
[d] The threshold wavelength for O_3 in the vibrational ground-state is 310 nm. At room temperature internal vibrational energy shifts the threshold to 325 nm (Matsumi *et al.*, 2002).
[e] Spin-forbidden process. The $O(^1D)$ quantum yield for this process is small, about 0.08 ± 0.04 for wavelength up to 340 nm (Matsumi *et al.*, 2002).
[f] Photodissociation by excitation of vibrational overtone transitions, of which the first overtone ($2\nu_1$) at 1.4 μm dominates (Roehl *et al.*, 2002; Donaldson *et al.*, 1997).

is initiated by photolysis of nitrogen dioxide, which has an energetic threshold at a longer wavelength ($\lambda < 420$ nm):

$$NO_2 + h\nu \longrightarrow NO + O(^3P) \quad \text{(R 9.3)}$$

All $O(^3P)$ that is produced in the troposphere is converted into O_3 by reaction with O_2 (R9.2).

Ozone itself can undergo photolysis in the UV and visible range. Most important for the chemistry of the atmosphere is the photolysis by UV radiation, which produces electronically excited oxygen atoms $O(^1D)$:

$$O_3 + h\nu \longrightarrow O(^1D) + O_2(a^1\Delta_g) \quad \text{(R 9.4)}$$

$$\longrightarrow O(^1D) + O_2(X^3\Sigma_g^-) \qquad \text{(R 9.5)}$$

$O(^1D)$ is not only metastable but also highly reactive, and constitutes the main source of atmospheric OH radicals by reaction with water vapour:

$$O(^1D) + H_2O \longrightarrow OH + OH \qquad \text{(R 9.6)}$$

OH is the principal atmospheric oxidant, which reacts with most trace gases and thus prevents the accumulation of chemically reactive, toxic, or climatically active gases in the atmosphere.

Reactions R 9.1–R 9.6 and the processes in Table 9.1 are the prime initiators and drivers of atmospheric chemistry. Details of this chemistry can be found in many text books, for example by Finlayson-Pitts and Pitts, Jr. (2000), Zellner (1999), or Seinfeld and Pandis (1998).

9.1.2 Photolysis frequencies: Kinetic definition

A quantitative description of atmospheric chemistry is not possible without accurate knowledge of the kinetic rates of the relevant photodissociation processes. Given the general photolysis of some compound AB into its products A and B,

$$AB + h\nu \longrightarrow A + B \qquad \text{(R 9.7)}$$

the following kinetic rate coefficient j can be defined:

$$j = -\frac{1}{[AB]}\frac{d[AB]}{dt} \qquad (9.1)$$

Here, the square brackets denote the number density (cm^{-3}) of the respective molecules in the gas phase; and j is a first-order rate coefficient (s^{-1}) that is called *photolysis frequency* or *j-value*. In a given radiation field, it describes the probability per second of the decomposition of the parent compound into its products. Note that j does not have a constant value, but depends on the local radiation field which varies in space and time. As for any first-order decay reaction, the reciprocal rate coefficient is the mean lifetime τ of the decomposing molecules:

$$\tau_{\text{photolysis}} = \frac{1}{j} \qquad (9.2)$$

If the j-value of a photolysis reaction is known, chemical loss and production rates can be calculated. For the general reaction R9.7, it follows:

$$-\frac{d[AB]}{dt} = \frac{d[A]}{dt} = \frac{d[B]}{dt} = j\,[AB] \qquad (9.3)$$

Here, the product $j \times [AB]$ is called the photolysis rate of AB. It is a measure of the number of molecules of AB that decompose per cm^3 and per second, and produce a corresponding number of photofragments A and B in the same volume and time.

9.1.3 Photolysis frequencies: Physical model

The probability of photoabsorption and photodissociation in the bulk gas phase is independent of the direction of incident radiation, because the absorbing molecules are randomly oriented in space. This is of particular importance in the atmosphere, where molecules are exposed to radiation from many different directions. Any volume element in the sunlit atmosphere can receive solar radiation either directly from the sun, or indirectly after scattering by air molecules, aerosol particles and clouds in the atmosphere, or after reflection by the ground (Figure 9.1).

All radiation passing through a volume element can be described by the spectral photon radiance L_λ. This radiometric quantity is defined by the following equation (McCluney, 1994):

$$L_\lambda = \frac{d^2\Phi_\lambda}{d\omega\, dA \cos\vartheta} \tag{9.4}$$

Here, Φ_λ is the spectral photon flux ($s^{-1}nm^{-1}$), that is the number N of photons arriving per time t in a small wavelength interval λ to $\lambda + d\lambda$,

$$\Phi_\lambda = \frac{d^2N}{dt\, d\lambda} = \frac{dN_\lambda}{dt} \tag{9.5}$$

that comes from the solid angle $d\omega$ (sr) along a direction $\vec{\Omega}$ and passes through the surface area dA (m^2) of the volume at angle ϑ (Figure 9.2). The product $dA \cos\vartheta$ appearing in the denominator of Equation 9.4 represents the projected area, dA' of the surface element dA onto a plane that is normal to $\vec{\Omega}$. Thus, the spectral radiance $L_\lambda(s^{-1} \cdot m^{-2} \cdot sr^{-1} \cdot nm^{-1})$ is the photon flux per unit projected area, per unit solid angle and per unit wavelength interval passing through the surface of a volume element in a specified direction.

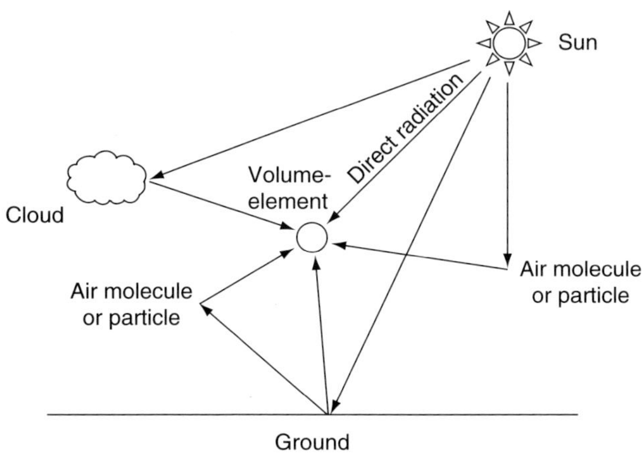

Figure 9.1 Volume element in the sunlit atmosphere which receives solar radiation either directly from the sun, or indirectly after scattering by air molecules, aerosol particles and clouds in the atmosphere, or after reflection by the ground.

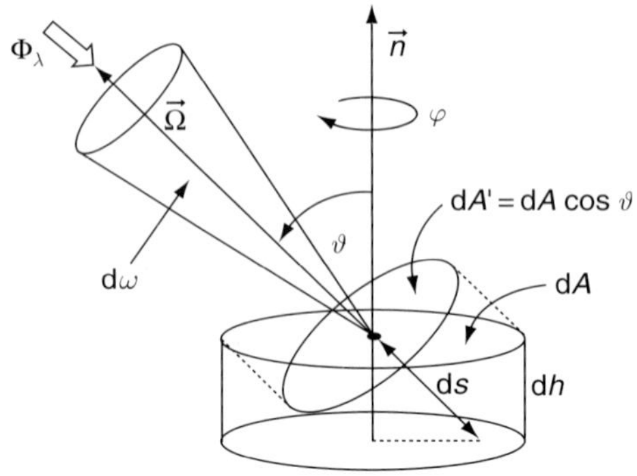

Figure 9.2 Geometry of a volume element that is exposed to a flux of photons Φ_λ from solid angle dω along direction $\vec{\Omega}(\vartheta, \varphi)$. d$A$ and dh are the surface area and height of the volume element, respectively. dA' is the projected area of dA onto a plane normal to $\vec{\Omega}$ and ds is the absorption path length through the volume. $\vec{n}$ is the normal to dA. ϑ and φ are polar and azimuth angles, respectively.

The photoabsorption in the volume element follows Lambert–Beer's law

$$\mathrm{d}L_\lambda = -\sigma\,[\mathrm{AB}]L_\lambda\,\mathrm{d}s \tag{9.6}$$

with σ being the bulk absorption cross section (m^2) of AB, dL_λ represents the decrease of the radiance inside the volume element along the absorption path length $\mathrm{d}s = \mathrm{d}h/\cos\vartheta$, with d$h$ being the height of the volume element $\mathrm{d}V = \mathrm{d}A \times \mathrm{d}h$. Using Equations 9.4–9.6, the decrease of L_λ per path length ds can be expressed as:

$$\frac{\mathrm{d}L_\lambda}{\mathrm{d}s} = \frac{\mathrm{d}^4 N_\lambda}{\mathrm{d}\omega\,(\mathrm{d}A\cos\vartheta)\,\mathrm{d}t\,(\mathrm{d}h/\cos\vartheta)} = \frac{\mathrm{d}^3 N_\lambda}{\mathrm{d}\omega\,\mathrm{d}V\,\mathrm{d}t} = -\sigma[\mathrm{AB}]L_\lambda \tag{9.7}$$

From this equation the spectral number of photons that are absorbed per unit volume and per unit time can be derived as:

$$\frac{\mathrm{d}^3 N_\lambda}{\mathrm{d}V\,\mathrm{d}t} = -\sigma[\mathrm{AB}]\,L_\lambda\,\mathrm{d}\omega \tag{9.8}$$

This equation applies to radiation incident from any single direction $\vec{\Omega}$. For radiation incident from many different directions $\vec{\Omega}(\vartheta, \varphi)$, integration over all solid angles yields:

$$\frac{\mathrm{d}^2 N_\lambda}{\mathrm{d}V\,\mathrm{d}t} = -\sigma[\mathrm{AB}]\int_{4\pi sr} L_\lambda(\vartheta, \varphi)\,\mathrm{d}\omega \tag{9.9}$$

The integral on the right-hand side of Equation 9.9 is called the *spectral actinic flux* F_λ (Madronich, 1987b):

$$F_\lambda = \int_{4\pi sr} L_\lambda(\vartheta, \varphi)\,\mathrm{d}\omega \tag{9.10}$$

It represents the spectral flux of photons per unit area ($\mathrm{m}^{-2}\cdot\mathrm{s}^{-1}\cdot\mathrm{nm}^{-1}$) available to molecules at a particular point in the atmosphere, irrespective of the directions from where the photons arrive.

Assuming that upon absorption of a photon the probability of photodissociation is ϕ, the photolysis rate in the spectral interval λ to $\lambda + d\lambda$ can be written as:

$$\frac{d^2[AB]}{dt\,d\lambda} = \phi\,\frac{d^2N_\lambda}{dV\,dt} = -\sigma\phi F_\lambda [AB] \tag{9.11}$$

ϕ is called *quantum yield* of the photolysis reaction and is, like σ and F_λ, a wavelength-dependent quantity. Furthermore, σ and ϕ can both depend on temperature. Integration of Equation 9.11 over all wavelengths and comparison of the resulting equation with Equation 9.3 then yields the following expression for the photolysis frequency of AB:

$$j = \int \sigma(\lambda, T)\phi(\lambda, T)F_\lambda(\lambda)\,d\lambda \tag{9.12}$$

Note that the integration is to be performed over the wavelength range in which the product $\sigma\phi F_\lambda$ is non-zero and therefore contributes to the j-value.

9.1.4 Actinic flux

Some comments on the terminology and meaning of actinic flux may be in place. The actinic flux F_λ, as defined by Equation 9.10, is a quantitative measure of actinic radiation in the atmosphere. The term *actinic* generally refers to radiation capable of causing photochemical reactions. The term *actinic flux* is not quite exact in the sense that F_λ actually denotes an area density of flux. Actinic flux density would therefore be a more appropriate name, but in fact the standard name in atmospheric-chemistry literature is actinic flux and will be used accordingly in this chapter.

From a chemical kinetic point of view, F_λ can be interpreted as a measure of the spectral photon number-density $[h\nu]_\lambda = dN_\lambda/dV$, with $F_\lambda = c \times [h\nu]_\lambda$. If F_λ is substituted by $[h\nu]_\lambda$, the expression for the spectral photolyis rate (Equation 9.11) assumes the form of a rate equation of a bimolecular chemical reaction:

$$\frac{d^2[AB]}{dt\,d\lambda} = -k\,[h\nu]_\lambda\,[AB] \tag{9.13}$$

Here, by analogy, the photons in the wavelength interval λ to $\lambda + d\lambda$ represent the collision partners of AB and $k = c \times \sigma \times \phi$ can be understood as the effective rate constant of the collision process.

9.1.5 Irradiance

The spectral actinic flux must be distinguished from another radiometric quantity that is widely used in atmospheric physics and is called *spectral irradiance*:

$$E_\lambda = \int_{2\pi sr} L_\lambda(\vartheta, \varphi)\cos\vartheta\,d\omega \tag{9.14}$$

where E_λ is the hemispherically integrated radiance weighted by the cosine of the angle of incidence and represents the photon flux per unit area through a plane surface. It is fundamentally different from the actinic flux, which weights the incident radiation uniformly (Equation 9.10), but has the same physical unit ($m^{-2} \cdot s^{-1} \cdot nm^{-1}$). Note that energy-based quantities can be obtained from the photon-based quantities F_λ and E_λ by multiplication with the corresponding photon energy, for example, $E_\lambda{}^e = hc/\lambda \times E_\lambda$. In this case, the relevant units are $W \cdot m^{-2} \cdot nm^{-1}$.

9.1.6 *Actinic flux in the atmosphere*

Actinic flux in the atmosphere has a spatial and spectral distribution that is highly variable and time dependent. It is mainly influenced by the position of the sun relative to the point of observation, the state and chemical composition of the atmosphere, and the spectral reflectance of the earth's surface.

The actinic flux received at some point in the atmosphere consists of two components, the direct solar radiation F_0 and the diffuse radiation F_{diff}, which originates from scattering in the atmosphere and at the earth's ground (see Figure 9.1):

$$F_\lambda(\lambda) = F_0(\lambda) + F_{diff}(\lambda) \tag{9.15}$$

F_0 represents the integrated, almost unidirectional radiance over the solid angle of the solar disc and F_{diff} denotes the integrated diffuse radiance over the solid angle of the upper and lower hemisphere, excluding the solar disc. We can also write F_{diff} as

$$F_{diff}(\lambda) = F\downarrow(\lambda) + F\uparrow(\lambda) \tag{9.16}$$

where $F\downarrow(\lambda)$ and $F\uparrow(\lambda)$ denote the downwelling and upwelling parts of the diffuse radiation, respectively.

9.1.6.1 *Direct solar radiation*

A beam of direct solar radiation, which passes through the atmosphere (Figure 9.3), is generally attenuated by absorption and scattering. The fraction $\mathcal{T}(\lambda)$ that is transmitted through a slant column of air along a path s can be calculated from Lambert–Beer's law and is given by

$$\mathcal{T}(\lambda) = \exp\{-\int_s \sum_i \sigma_i(h, \lambda)\,[M_i](h)\,ds\} \tag{9.17}$$

where σ_i is the wavelength-dependent cross section for absorption and scattering by an atmospheric component M_i. In general, σ_i and the number density $[M_i]$ vary with altitude h. The summation is performed over all relevant absorbers and scatterers and the integral is calculated along the trajectory s. In the approximation of a non-refractive plane-parallel atmosphere, the trajectory is a straight line in the direction of the sun at solar zenith angle χ. When the zenith angle increases, the path length through the atmosphere becomes larger with $s \approx h/\cos\chi$, and less direct radiation is transmitted according to Equation 9.17. Note that the approximation of a straight line holds only for zenith angles that are not too large, $\chi < 85°$ (Iqbal, 1983).

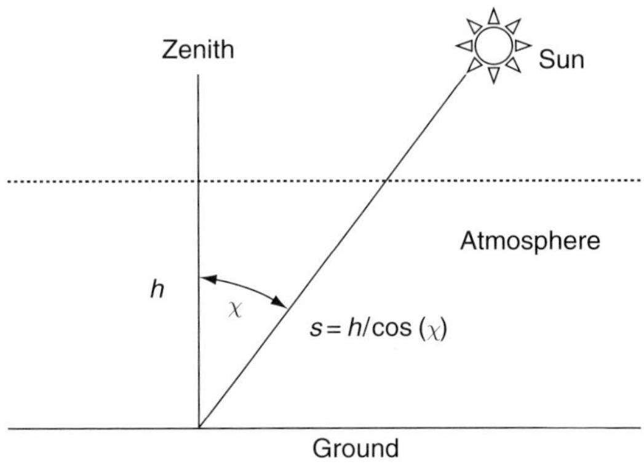

Figure 9.3 Trajectory of direct solar radiation through a non-refractive plane-parallel atmosphere. The solar zenith angle χ specifies the direction of the sun relative to the local vertical direction. *h* is the height of a vertical air column above ground and *s* is the path length of the direct radiation through the corresponding slant air column.

9.1.6.2 Diffuse radiation

Diffuse radiation which is scattered in the atmosphere returns partly into space, while another part reaches the ground where it is absorbed or reflected. Diffuse radiation can also be multiply scattered or absorbed by air components, which comprise gas molecules, particles or cloud droplets. In general the relevance of atmospheric scattering processes increases in air of increasing density or when the path length through the atmosphere becomes longer at increasing solar zenith angle. Scattering becomes more important at shorter wavelengths, according to $\sigma_{\text{scatter}} \propto \lambda^{-n}$ with $n = 4$ for gases (Rayleigh scattering) and $n \approx 0.5$–2.5 for aerosol particles (Mie scattering) that have diameters greater than 100 nm (Iqbal, 1983).

The angular distribution of the diffuse radiance varies with altitude and wavelength. An example for cloud-free conditions is shown in Figure 9.4. At low altitude (0.1 km), most of the diffuse radiation comes from the upper hemisphere. Little radiation returns from

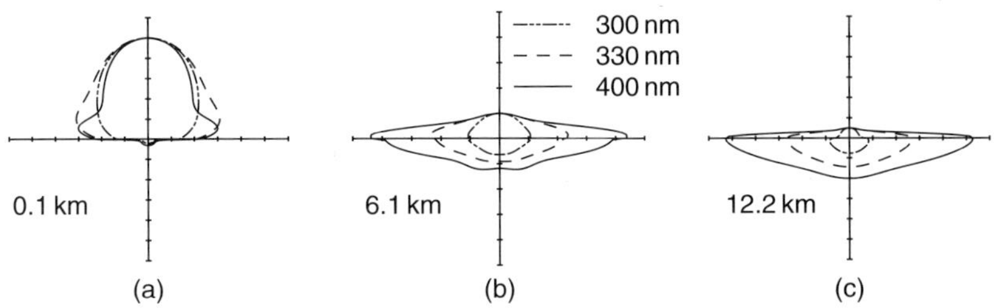

Figure 9.4 Polar diagrams of the relative angular distribution of diffuse radiance, modelled for a cloud-free atmosphere over the Agean Sea in Greece (10 June, 1996, $\chi \sim 20°$, surface albedo 0.03). The radiance data are azimuthally averaged and normalised to unity at the zenith. The data are shown for wavelengths of 300, 330, and 400 nm at altitudes of (a) 0.1, (b) 6.1, and (c) 12.2 km (from Hofzumahaus *et al.*, 2002).

the ground, which, in this example, is assumed to have a small reflectivity. Contrary, at an altitude of ~12 km, the diffuse radiation comes predominantly from the lower hemisphere, where in this case most of the backscattering air mass resides. At longer wavelength (400 nm) a large portion of diffuse radiation comes from the horizon, whereas at shorter wavelength (300 nm) the diffuse radiance is more uniformly distributed. In a rough approximation, the diffuse sky radiance in the UV is often assumed to be isotropically distributed in the lower troposphere. This subject has been discussed in detail, for example, by Ruggaber *et al.* (1993).

Diffuse radiation is also contributed by reflections at the earth's surface. The reflectance (albedo) depends on the composition and structure of the ground, the angle of incidence and wavelength of the radiation. In general, the reflectance of natural surfaces is rather small in the UV and visible regions, typically less than 10% over vegetation or soil, but can reach values close to one over fresh snow. Spectral albedo data for various ground covers can be found, for example, in Iqbal (1983) or McKenzie and Kotkamp (1996). It should be noted that these data are generally specified for solar irradiances. See Appendix A.1 how albedo data can be applied to actinic fluxes.

The partitioning of the actinic flux into direct and diffuse radiation varies with altitude, solar zenith angle and wavelength. In the stratosphere the actinic flux is dominated by direct radiation. Under conditions that favour scattering processes, the diffuse radiation dominates. This is the case in the UV in the lower troposphere, where more than 50% of the actinic flux is contributed by diffuse radiation. The diffuse fraction generally increases with solar zenith angle and, quite naturally, assumes a value of one when the solar disc is hidden, for example, by a thick cloud or a mountain.

9.1.6.3 Spectral composition

The wavelength dependence of absorption and scattering processes gives rise to significant changes in the spectral composition of the solar radiation when it passes through the atmosphere. Figure 9.5 shows the relevant cross sections of O_2 and O_3, the two main absorbers of solar UV radiation, and the Rayleigh-scattering cross section of air as a function of wavelength. The absorption by O_2 and O_3 in the stratosphere has a strong influence on the spectral actinic flux at wavelengths below 320 nm and prevents that any solar radiation reaches the troposphere at $\lambda < 290$ nm (Figure 9.6). The sharp drop of the actinic UV radiation between 290 and 320 nm by many orders of magnitude is called the *UV-B cut-off*. It depends strongly on the vertical column density t of the absorbing ozone in the atmosphere:

$$t_{O_3} = \int_0^\infty [O_3](h)\,dh \tag{9.18}$$

Here t_{O_3} is also called *total ozone* and is specified in Dobson units (DU), with 1 DU = 2.69×10^{16} cm^{-2}. It should be noted that particularly in polluted air, additional absorbers and scatterers, like SO_2 and various kinds of aerosols (sulphate, black carbon, mineral dust, droplets, etc.), can play a significant role for the spatial and spectral distribution of the actinic flux in the atmosphere.

For further reading about the properties of actinic flux in the atmosphere, see, for example, Demerjian *et al.* (1980), Madronich (1987b), Meier *et al.* (1997), Liao *et al.* (1999), Mayer and Madronich (2004).

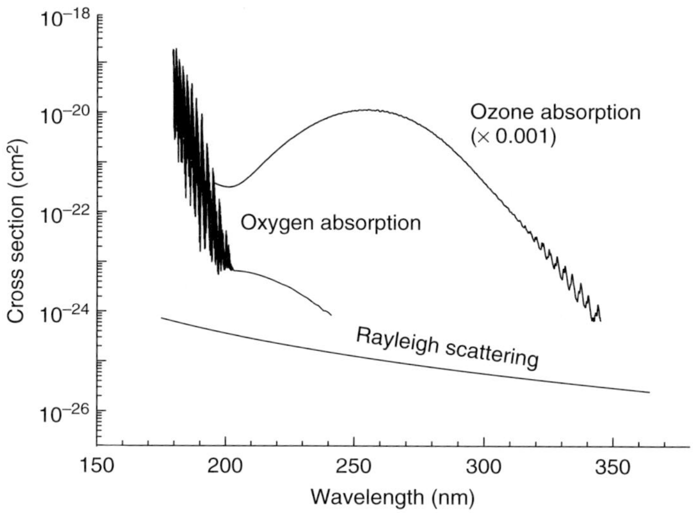

Figure 9.5 Cross sections σ for absorption of solar radiation by atmospheric oxygen and ozone, and for Rayleigh scattering by air. Data are from Yoshino *et al.* (1992), Daumont *et al.* (1992), and Bates (1984), respectively.

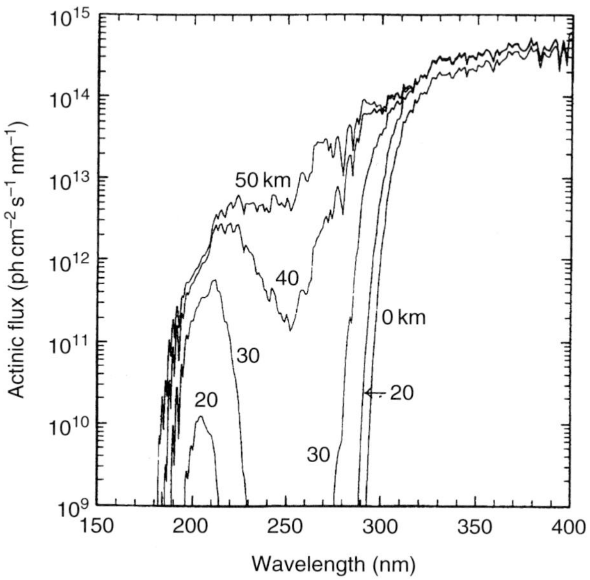

Figure 9.6 Spectral distribution of the solar actinic flux in the earth's atmosphere at different altitudes. The spectra are modelled for the US Standard Atmosphere at a solar zenith angle of 30° and for a ground albedo of 0.3 (from DeMore *et al.*, 1997, courtesy of NASA/JPL/CALTECH).

9.2 Methods for measuring photolysis frequencies

Measurements of actinic flux and photolysis frequencies have important applications in atmospheric sciences. The first one is to provide observational data for field investigations of atmospheric photochemistry. Here, photolysis frequencies are needed besides other chemical and meteorological parameters for the understanding of observed chemical changes in the atmosphere and the subsequent validation of atmospheric chemistry models. The second application is to investigate how the actinic flux depends on the chemical composition and the state of the atmosphere. Such process studies improve the understanding of radiative transport through the atmosphere and help to interprete UV climatologies and trends. The third application is the verification of radiative-transfer models used for the prediction of photolysis frequencies and actinic fluxes in atmospheric chemistry and physics.

9.2.1 Measurement requirements

In order to be useful, techniques for measuring photolysis frequencies must fulfil a number of requirements:

- They must cover the spectral range of wavelengths that contributes to the photolysis processes of interest. For most photolysis reactions, this region includes wavelengths of up to 420 nm (cf. Table 9.1). Only few atmospheric compounds photolyse at longer wavelengths in the visible region and up to 640 nm and even more rare are loosely bound compounds (e.g. O_3 and HNO_4) that photodissociate in the near-infrared. At short wavelengths, a lower limit is generally imposed by the available radiation. As can be seen from Figure 9.6 this limit is around 290 nm in the lower stratosphere and troposphere.
- Given numerous photolysis reactions in the atmosphere, a technique should ideally measure the photolysis frequencies of various chemical compounds with one instrument at the same time.
- The measurements must be sensitive to direct and diffuse radiation arriving from all possible directions in the atmosphere, from the upper and lower hemisphere as well. Ideally, the measurement should give equal weight to radiation from any angle of incidence.
- The measurements require a good temporal and spatial resolution, as the actinic flux can be highly variable. This is particularly the case in situations with broken clouds. Under such conditions light intensity can change over short distances and the local actinic flux can vary in seconds.
- In order to provide spatial coverage, a network of in situ measurement instruments can be used. Alternatively, portable instruments can be mounted on mobile platforms (e.g. aircraft or balloons) for measuring spatial distributions. In the latter case, it is critical that the instruments have a small size and weight, and a low electrical power consumption. No technique is so far available that would measure photolysis frequencies remotely with extended spatial coverage.

9.2.2 General measurement methods

There exist two general methods for measuring atmospheric photolysis frequencies (cf. Figure 9.7).

- *Chemical actinometry* is a method that detects actinic fluxes directly by using the chemical compound of interest. The relevant gas is exposed to solar radiation in a static or flowing gas system and the rate of its photochemical conversion into products is measured. The conversion rate is then used to evaluate a photolysis frequency (Equation 9.1 or 9.3). Chemical actinometry is the most direct way to measure a photolysis frequency and has the particular advantage that it does not require the knowledge of the absorption cross section σ and quantum yield ϕ of the photolysis reaction.
- *Radiometry* is a general method for measuring radiation by optical instruments that utilise photoelectric detectors. In atmospheric chemistry, radiometric methods have been modified such that photolysis frequencies can be derived from the measured intensity of the incident radiation.

Radiometry can be further grouped into methods that measure radiation spectrally resolved (narrowband spectroradiometry) or spectrally integrated (broadband radiometry) (cf. Figure 9.7). Each subgroup can be further divided into methods for measuring actinic fluxes or irradiances. The result is a variety of different types of radiometers that enable a more or less direct determination of photolysis frequencies from measured radiation intensities.

- *Actinic-flux spectroradiometers* provide the most direct way to derive photolysis frequencies from measured radiation. They measure the actinic flux in narrow spectral

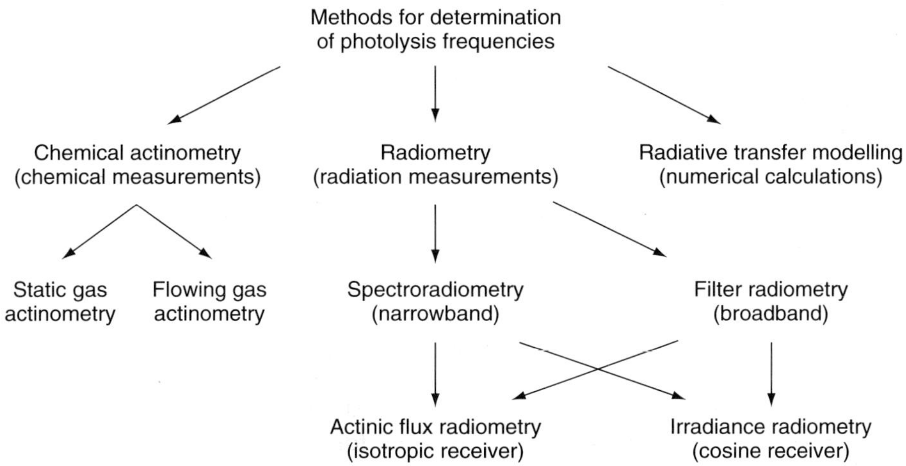

Figure 9.7 Methods for the determination of photolysis frequencies in the atmosphere.

intervals $\Delta\lambda$ over the wavelength range that contributes to the photolysis process of interest. Photolysis frequencies are then calculated from the measured radiation spectra $F_\lambda(\lambda)$ and the relevant molecular cross section σ and quantum yield ϕ, according to Equation 9.12. The main advantage of spectroradiometry is the possibility to determine photolysis frequencies for a whole set of different molecules at the same time.

- *Actinic-flux filter radiometers* have a broad spectral coverage and measure the actinic flux integrated over all wavelengths simultaneously. The optical wavelength selection is chosen to mimic the spectral function $\sigma \times \phi$ of the relevant photolysis frequency by means of an optical transmission filter and a photoelectric sensor. In a first approximation, the filter radiometer signal is proportional to the photolysis frequency and can be calibrated against an absolute reference (e.g. a chemical actinometer). In order to obtain more accurate results, the dependence of the calibration on ambient conditions (e.g. solar zenith angle, temperature) must be taken into account. Filter radiometers are generally useful because of their compact size and fast time response.
- *Irradiance radiometers* (spectroradiometers and broadband radiometers) can be used for the determination of photolysis frequencies in principle like actinic-flux instruments. However, the measured irradiances must be converted into actinic flux data, a step that introduces additional errors in the derived photolysis frequencies. The conversion requires knowledge of the angular distribution of the radiance in the atmosphere. The latter information is difficult to obtain in the field, especially in the presence of clouds and aerosols, and is usually estimated by empirical rules or by numerical models. Irradiance radiometers have the advantage that they have a wide distribution as meteorological instruments and are readily available as standard equipment from various manufacturers. For details, see Section 9.6.

9.2.3 Actinic-flux sensing techniques

For practical purposes, most important are the measurement techniques that are directly sensitive to the actinic flux, that is have the ability to collect radiation with equal sensitivity independent of the angle of incidence (isotropic behaviour). To this class of techniques belong the abovementioned chemical actinometers, actinic-flux spectroradiometers and actinic-flux filter radiometers. The measurement of actinic flux requires the integral detection of the incident radiation from all directions over a solid angle of 4π sr. Spherical or tubular photolysis reactors used in chemical actinometry closely fulfil this requirement (Figure 9.8A). Actinic flux radiometers mostly use hemispherical input optics, such that two instruments must be combined to cover 4π sr (Figure 9.8B). The setup of two radiometers pointing into opposite directions has the additional advantage of providing separate information about the up- and down-welling components of the actinic radiation.

The basic features of actinic-flux sensing techniques are summarized in Table 9.2, while detailed descriptions follow in later sections. It may be noted that chemical actinometry is the oldest actinic-flux method and has been used since the mid-1970s to measure photolysis frequencies in the atmosphere, mainly for NO_2 and O_3 (see Section 9.3). Actinic-flux radiometry was developed later, starting at the end of the 1980s, and is now widely applied in atmospheric chemistry for measurement of photolysis frequencies of

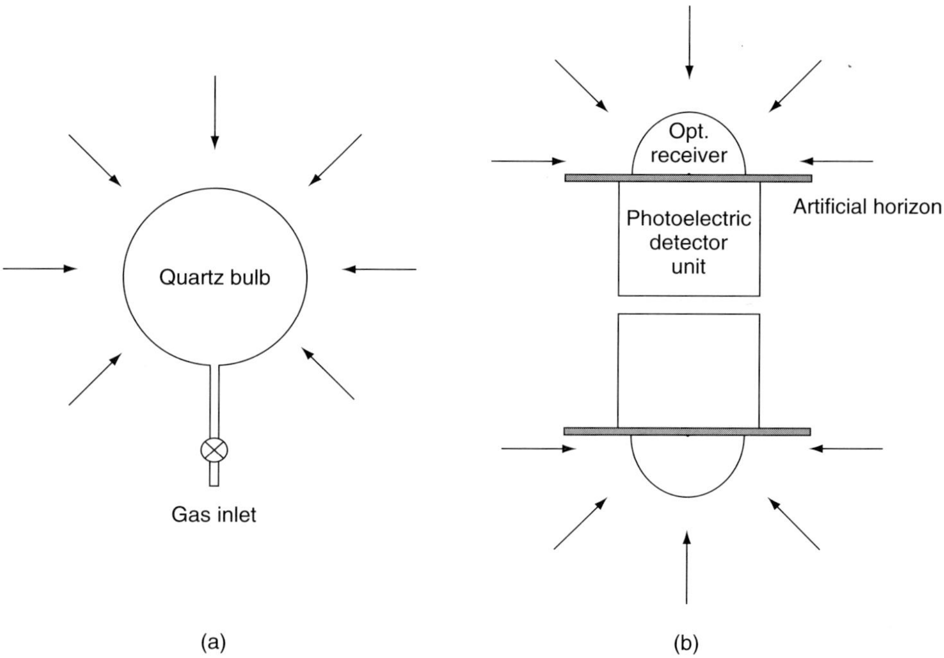

Figure 9.8 Receiver geometries for solar actinic flux. (a) Photolysis reactor of a chemical actinometer with a spherical response. (b) Combination of two hemispheric radiometers facing into opposite directions.

various gases (see Sections 9.4, 9.5, 9.8). The performance and reliability of actinic-flux techniques have been thoroughly tested in several assessments. Examples of results are given in Section 9.8.3.

9.2.4 Radiative transfer models

For completeness, theoretical modelling should be mentioned as a method to determine photolysis frequencies.

Radiative transfer (RT) models calculate first and foremost the spectral actinic flux in the atmosphere. The models use the extraterrestrial solar spectrum as input, as well as optical properties of the atmosphere and the earth's surface albedo. In a second step, photolysis frequencies can be calculated from the simulated radiation spectra $F_\lambda(\lambda)$ and the relevant molecular data σ and ϕ, using Equation 9.12.

Radiative transfer models play an important role for the prediction of photolysis frequencies in atmospheric chemistry. In particular, they provide a broad temporal and spatial coverage that currently cannot be achieved by measurements alone. On the other hand models have difficulties to calculate accurate j-values at local sites where the field-of-view is shadowed by trees, buildings or mountains, or when complex cloudiness prevails. Model results are therefore no good substitute for in situ measurements in local photochemistry field campaigns. However, models are useful to assist in

Table 9.2 In situ measurement techniques for atmospheric photolysis frequencies[a]

Feature	Chemical actinometer	Actinic flux spectroradiometer	Actinic flux filter radiometer
Main components	Spherical or cylindrical quartz reactor + gas analysis	Optical receiver + scanning monochromator or spectrograph + photoelectric detection system	Optical receiver + optical filter + photoelectric detector
Measured quantity	Concentration [AB]	$F_\lambda(\lambda)$	$U = \int D_0 F_\lambda d\lambda$[b]
Signal evaluation	$j = -\frac{1}{\Delta t}\frac{\Delta[\mathrm{AB}]}{[\mathrm{AB}]}$[c]	$j = \sum \sigma\phi F_\lambda \Delta\lambda$[d]	$j = AU$[e]
Versatility	Single specific photolysis reaction	Various photolysis reactions at the same time	Single specific photolysis reaction
Reported applications	O_3, NO_2, $RONO_2$	O_3, NO_2, HONO, HCHO...	O_3, NO_2
Calibration reference	Gas standard	Calibration lamp	Reference instrument[f] or calibration lamp
Sensitivity	Medium to good	Very good	Good
Time resolution	Flow reactors: 1–60 s; Static reactors: 1–60 min	Scanning monochromator systems: 15–120 s; Spectrograph systems: 1–10 s	~1 s
Size	Large to medium; requires pressurised gases	Medium to small	Small

[a] Listed are techniques that collect radiation with isotropic sensitivity.
[b] U is a voltage signal; D_0 represents the wavelength-dependent sensitivity of the filter radiometer; the spectral design of the filter radiometer aims for $D_0(\lambda) \propto \sigma(\lambda)\phi(\lambda)$ and requires knowledge of σ and ϕ.
[c] Equation 9.1, or alternatively equation 9.3 may be applied.
[d] Equation 9.12 is applied; it requires knowledge of $\sigma(\lambda, T)$ and $\phi(\lambda, T)$, and, if necessary, of ambient temperature T.
[e] See equation 9.45; the calibration factor A usually depends on ambient parameters such as solar zenith angle, temperature etc.
[f] Reference instrument can be a chemical actinometer, a spectroradiometer, or another calibrated filter radiometer.

the determination of photolysis frequencies by certain techniques, like filter radiometry or irradiance radiometry. See Section 9.7 for more information about RT modelling.

9.3 Chemical actinometry

Chemical actinometry in its original sense means the quantification of an absolute photon flux by measuring the rate of photoinduced change in a chemical system (Calvert & Pitts, 1966; Pravilov, 1987; Kuhn *et al.*, 2004). In general the chemical system can be a solid, liquid or gas and is called *chemical actinometer*. Actinometry is then based on the principle that the photolytical conversion rate of molecules in an actinometer cell is equal to the absorption rate of photons times the quantum yield of the actinometer. The absorption rate of photons in the actinometer is related to the incident photon flux by Lambert–Beer's law. Application of these fundamental principles allows the calculation of absolute incident photon-fluxes from measured chemical-conversion rates, if the quantum yield of the actinometer is well known (for details, see Calvert & Pitts, 1966).

While classical chemical actinometry has been used in laboratories by photochemists for almost a century (e.g. Warburg, 1912; Vaughan & Noyes, Jr., 1930), the general concept has been adapted, beginning in the 1970s, by atmospheric chemists for the measurement of photolysis frequencies in the atmosphere. Applications have been reported for O_3, NO_2 and alkyl nitrates (Tables 9.3 and 9.4). Corresponding techniques for the photolysis of O_3 and NO_2 will be presented in some detail in Sections 9.3.4 and 9.3.5, respectively.

9.3.1 Principle of chemical actinometry

Chemical actinometers for atmospheric measurements are realised by (1) filling the gas of interest into a photolysis reactor, (2) exposing the actinometer to solar radiation, and (3) measuring the photochemical conversion rate. Here, it is important that the actinometer gas is exposed to the ambient actinic flux without altering the spectral composition and intensity of the radiation significantly. This is achieved by the usage of a transparent quartz cell with a suitable geometrical shape (spherical or cylindrical) and by applying gas concentrations with small optical absorbances (see more about this aspect in Section 9.3.3 and Appendix A.2). If these conditions are fulfilled, the photolysis frequency can be evaluated according to Equation 9.1, which requires measurements of gas concentrations and time but which does not need the knowledge of the absorption spectrum and quantum yield of the primary photodissociation.

There are two possible modes of actinometer operation. In the static batch mode the actinometer gas is filled into the photolysis reactor which is then sealed off by a gas valve and is covered with an opaque hood to exclude sunlight. For measurement, the actinometer is uncovered and exposed to solar radiation for a fixed time interval Δt. Thereafter, the actinometer is covered again and is analysed for change in its chemical composition.

The other mode of operation involves flowing gas. The actinometer gas is passed continuously through a photolysis reactor which is permanently exposed to ambient solar radiation. In this case Δt is equal to the mean residence time of the gas in the illuminated photolysis cell. After having passed the reactor, the gas composition is analysed by an online gas detector. This operational mode allows a continuous monitoring of photolysis frequencies.

9.3.2 Chemical kinetics considerations

The rate of photochemical change for a general reaction $AB + h\nu \rightarrow A + B$ can be determined from either the loss of reactant AB or the increase of its product A (or B).

Table 9.3 $O_3 \rightarrow O(^1D)$ chemical actinometers used in atmospheric field studies

Reference	Photolysis reactor[a]	Gas mixtures[b]	Exposure time	Detection technique	k_a/k_b [c]	Platform
Bahe and Schurath (1978); Bahe *et al.* (1979)	Static and flow types, spher.	O_3/N_2O	0.5–2 h	N_2: gas chromatography	0.62	Ground + balloon ($h = 26$ km)
Dickerson *et al.* (1979, 1982)	Flow type, cyl.	$O_3/O_2/N_2O$	10 s	N_2O_5: conversion into NO + detection by chemiluminescence	0.8	Ground + aircraft ($h = 0.1$–6.8 km)
Junkermann *et al.* (1989)	Static type, spher.	O_3/N_2O	<30 min	N_2: gas chromatography	0.75	Ground
Blackburn *et al.* (1992)	Flow type, cyl.	$O_3/O_2/N_2O$	34 s	$NO_3 + N_2O_5$: transfer into liquid methanol + detection by electrical conductivity	0.79	Ground
Bairai and Stedman (1992)	Flow type, cyl.	$O_3/O_2/N_2O/He$	7 s	NO_2: transfer into luminol solution + fluorescence detection	0.8	Ground
Müller *et al.* (1995)	Static type, spher.	$O_3/O_2/N_2O$	0.8–2 h	N_2: gas chromatography	0.75	Ground
Shetter *et al.* (1996)	Flow type, cyl.	$O_3/O_2/N_2O$	40 s	N_2O_5: transfer into liquid water + detection by electrical conductivity	0.72	Ground

[a] Cyl., cylindrical tube; spher., spherical bulb. Typical reactor volumes are 50–250 cm^3. Walls are made of transparent quartz.
[b] $[N_2O] : [O_3] > 100$; total pressure is about 1 atm in flow systems, and 1–2 atm in static systems.
[c] Branching ratio of the $O(^1D) + N_2O$ reaction channels producing (a) $N_2 + O_2$ and (b) NO + NO, used for the calculation of the $j(O^1D)$-values.

Table 9.4 $NO_2 \rightarrow NO + O$ and $RONO_2 \rightarrow RO + NO_2$ chemical actinometers used in atmospheric field studies

Reference	Photolysis reactor[a]	Gas mixtures[b]	Exposure time	Detection technique[c]	Platform
		$NO_2 \rightarrow NO + O$ actinometer			
Jackson *et al.* (1975); Harvey *et al.* (1977)	Flow type, cyl.	NO_2/Air	1–4 s	NO, NO_2: chemiluminescence	Ground
Zafonte *et al.* (1977)	Flow type, cyl.	NO_2/N_2	7 s	NO, NO_2: chemiluminescence	Ground
Bahe *et al.* (1980)	Flow type, spher. Static type, cyl.	NO_2/N_2 NO_2	10 s 30 s	NO: chemiluminescence NO_2: spectrophotometry	Ground Ground
Dickerson *et al.* (1982)	Flow type, cyl.	NO_2/O_2	1 s	NO, NO_2: chemiluminescence	Ground + aircraft ($h = 0.1$–5.5 km)
Parrish *et al.* (1983)	Flow type, cyl.	NO_2/N_2 (60 hPa, total)	5 s	NO, NO_2: chemiluminescence	Ground
Madronich *et al.* (1983, 1985)	Static type, cyl.	NO_2	5–10 min	Pressure increase	Ground + balloon ($h = 24$, 32 km)
Shetter *et al.* (1992); Lantz *et al.* (1996)	Flow type, cyl.	NO_2/O_2 (67 hPa, total)	0.3–0.5 s	NO, NO_2: chemiluminescence	Ground
Schultz *et al.* (1995)	Static type, cyl.	NO_2	1–2 min	NO_2: photometry	Ground

Table 9.4 (Continued)

Reference	Photolysis reactor[a]	Gas mixtures[b]	Exposure time	Detection technique[c]	Platform
Kelley *et al.* (1995); Dickerson *et al.* (1997)	Flow type, cyl. (dual system, upward and downward looking)	NO_2/Air	1 s	NO, NO_2: chemiluminescence	Aircraft ($h = 0.2$–7.6 km)
Castro *et al.* (1995, 1997)	Flow type, cyl.	NO_2/N_2	1 s	NO, NO_2: chemiluminescence	Ground
Vuilleumier *et al.* (2001)	Flow type, cyl.	NO_2/N_2	1 min	NO, NO_2: chemiluminescence	Ground
Kraus *et al.* (2000)	Flow type, cyl.	NO_2/Air	1 min	NO, NO_2: chemiluminescence	Ground
		$RONO_2 \rightarrow RO + NO_2$ actinometer			
Luke *et al.* (1989)	Flow type, cyl.	$RONO_2$/Air[d]	6–11 s	NO_2: transfer into luminol solution + fluorescence detection	Ground

[a] Cyl., cylindrical tube; spher., spherical bulb. Typical reactor volumes are 7 to 380 cm^3. Walls are made from transparent quartz.
[b] In NO_2 flow actinometers the mixing ratio of NO_2 is typically between 1 and 80 ppmv at total pressures around 1 atm, unless otherwise indicated. In static systems pure NO_2 is used at pressures of 1–5 hPa.
[c] NO_2 can be detected by chemiluminescence after chemical conversion into NO.
[d] R = ethyl, n-propyl, n-butyl, 2-butyl; mixing ratios of $RONO_2$ are between 23–52 ppmv.

If the exposure time Δt is held at less than 1% of the reciprocal j-value, the concentration of AB will change very little and AB can be assumed to have a constant concentration $[AB] \approx [AB]_0$. It is then reasonable to measure the increase of the product A, which can be related to the photolysis frequency as follows:

$$d[A] = j[AB]dt \tag{9.19}$$

Integration over the time interval Δt yields

$$\Delta[A] = [AB]_0 \int_0^{\Delta t} j(t)\,dt. \tag{9.20}$$

In principle the photolysis frequency may vary during the exposure time Δt. It is then useful to define a mean value $\bar{j}$ for this time interval:

$$\bar{j} = \frac{1}{\Delta t} \int_0^{\Delta t} j(t)\,dt \tag{9.21}$$

Combining Equations 9.20 and 9.21 results in

$$\bar{j} = \frac{1}{[AB]_0} \frac{\Delta[A]}{\Delta t} \tag{9.22}$$

This demonstrates that the photolysis frequency derived from the measured increase $\Delta[A]$ is a true mean value over the respective time interval.

Another approach is to let the initial concentration $[AB]_0$ noticeably decay and measure the decrease $\Delta[AB]$ during Δt. In this case we get

$$d[AB] = -j[AB]\,dt \tag{9.23}$$

which after integration yields

$$\bar{j} = -\frac{1}{\Delta t} \ln \frac{[AB]_0 - \Delta[AB]}{[AB]_0} \tag{9.24}$$

Here again, a mean j-value is obtained for the time of exposure.

It should be noted that the photochemistry and its kinetics in real chemical actinometers is often complicated by secondary chemistry. The reason is that the photolytical products, atoms and radicals, are highly reactive. Consecutive reactions in the gase phase and on wall surfaces lead to losses of the products, and possibly also of the initial reactant. For an accurate evaluation of the required photolysis frequency the effect of secondary chemistry must be taken into account. Sometimes scavenger gases are added to the actinometer gas in order to capture the reactive primary photofragments and convert them into species which are more stable or easier to measure. Examples will be given in Sections 9.3.4 and 9.3.5.

9.3.3 *Optical and photochemical considerations*

The advantage of a properly designed chemical actinometer is that the gas molecules inside the photolysis reactor will photodissociate with the same rate coefficient as would molecules do in the surrounding atmosphere. This will only happen if the conditions that have an influence on the photolysis rate are the same inside and outside of the chemical actinometer.

Gas temperature and pressure have a possible influence on the absorption cross section and quantum yield of photolysis reactions. For example, the photodissociation of $O_3 \rightarrow O(^1D)$ exhibits a strong temperature dependence (Dickerson *et al.*, 1982; Bohn *et al.*, 2004), while the NO_2 photolysis is only weakly temperature dependent (Dickerson *et al.*, 1982; Shetter *et al.*, 1988). In few cases, a notable pressure dependence has been observed, such as in the formation of $H_2 + CO$ in the photolysis of HCHO (Moortgat *et al.*, 1983). Whenever a temperature or pressure dependence exists, the operating conditions in the actinometer must be chosen similar to the corresponding ambient conditions in order to provide representative measurements.

Optical effects caused by the design of the chemical actinometer can modify the actinic flux inside the photolysis reactor. Potentially-interfering processes are:

- Strong absorption by the actinometer gas.
- Absorption by the photolysis-cell wall-material.
- Optical reflections at the outer and inner wall surfaces.
- Shielding of solar radiation by the gas inlet/outlet of the photolysis cell.

Gas phase absorption can be a problem at high concentrations of the actinometer gas. If the gas phase is optically thick, for example the absorbance exceeds 1%, an appreciable fraction of radiation is absorbed in the photolysis cell. Correspondingly, the absorption protects some of the actinometer gas from photolysis and the resulting photolysis frequency will be systematically lower than in the surrounding atmosphere. Optically thin conditions can be achieved by a sufficiently low actinometer-gas concentration, which in turn may require the use of highly sensitive analytical instruments for the measurement of correspondingly low reactant and product concentrations.

Chemical-actinometer walls are usually made of quartz and have a typical thickness of a few millimeters. At this thickness the material is highly transparent and has an absorbance of less than 1% between 250 and 1000 nm. Thus, losses of light due to wall absorption are negligible. A considerable amount of incident solar radiation is, however, lost by reflection (I_R) when the light enters the photolysis reactor (Figure 9.9; for details, see Appendix A.2). The reflectance of the quartz wall depends on the angle of incidence (ϑ) and varies for unpolarized UV radiation between ~8% at normal incidence ($\vartheta = 0°$) to 100% at grazing incidence ($\vartheta = 90°$). The attenuated radiation ($I_1 = I_0 - I_R$) entering the photolysis cell is subsequently reflected multiple times at the inner surface of the photolysis cell. These reflections enhance the photon flux inside the chemical actinometer. For a photolysis reactor that has a spherical or infinitely long cylindrical shape and non-absorbing walls, it has been shown in a theoretical study by Zafonte *et al.* (1977) that the reflections at the exterior and interior wall surfaces cancel each other exactly,

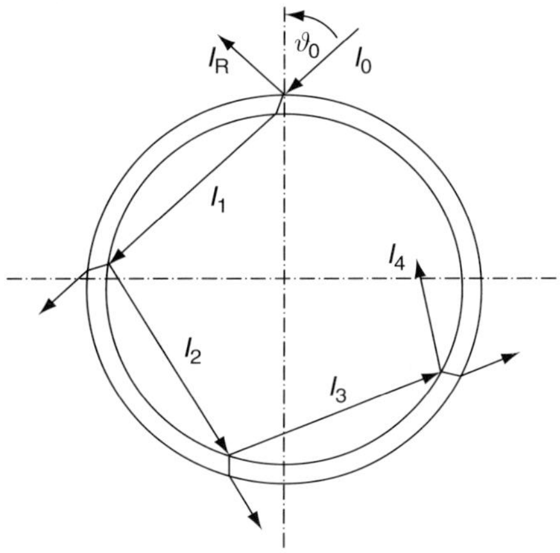

Figure 9.9 Illustration of multiple reflections inside a spherical or cylindrical (infinite length) photolysis reactor. I_0 is the intensity of the incident solar radiation. I_R is the intensity lost by reflection at the outer wall and I_1 is the transmitted intensity. $I_2, I_3, I_4 \ldots$ denote the intensities of the internally reflected light.

that is $I_1 + I_2 + I_3 + \cdots = I_R$ (see Appendix A.2). Therefore, the actinic flux inside the actinometer is expected to be the same as in the surrounding atmosphere. The perfect geometry of the photolysis cell is, however, perturbed by the gas inlet/outlet (usually the ends, in case of cylindrical tubes) which act as a sink for reflected radiation and shield some of the incident solar radiation. It is generally assumed that the corresponding loss of light is small and equal to the ratio of the area of the inlet/outlet to the total surface area of the photolysis reactor (Zafonte *et al.*, 1977).

The theoretical prediction of the cancelling effects of external and internal reflections has been confirmed experimentally (Dickerson & Stedman, 1980; Madronich *et al.*, 1983). In one study, one to three additional cylindrical tubes were slid over the initial photolysis tube in which the photolysis frequency of NO_2 was measured in a field experiment. Within 5% precision of the experiment, the additional tubes had no effect on the measured j-value, suggesting that a single tube caused less than 1.7% loss of radiation inside the actinometer (Dickerson & Stedman, 1980).

9.3.4 Ozone chemical actinometers

The photodissociation of O_3 in the UV and visible region has five energetically possible pathways for which experimental evidence exists (Matsumi & Kawasaki, 2003):

$$O_3 + h\nu(\lambda < 310\,\text{nm}) \longrightarrow O(^1D) + O_2(a^1\Delta_g) \qquad \text{(R 9.4)}$$

$$O_3 + h\nu(\lambda < 411\,\text{nm}) \longrightarrow O(^1D) + O_2(X^3\Sigma_g^-) \qquad \text{(R 9.5)}$$

$$O_3 + h\nu(\lambda < 463\,\text{nm}) \longrightarrow O(^3P) + O_2(b^1\Sigma_g^+) \qquad \text{(R 9.8)}$$

$$O_3 + h\nu(\lambda < 612\,\text{nm}) \longrightarrow O(^3P) + O_2(a^1\Delta_g) \qquad \text{(R 9.9)}$$

$$O_3 + h\nu(\lambda < 1180\,\text{nm}) \longrightarrow O(^3P) + O_2(X^3\Sigma_g^-) \qquad \text{(R 9.10)}$$

When driven by solar radiation, this mechanism produces $O(^3P)$ atoms about five to ten times faster than $O(^1D)$ (Table 9.1). In the troposphere, however, only the formation of $O(^1D)$ is relevant. The reaction of $O(^1D)$ with water vapour produces the important atmospheric OH radicals (see Section 9.1.1), whereas tropospheric $O(^3P)$ atoms are converted back completely to ozone via reaction R 9.2, resulting in a do-nothing cycle.

The photolysis frequency for $O_3 \rightarrow O(^1D)$ is generally called $j(O^1D)$. It has been measured in various tropospheric field experiments by chemical actinometers, which are listed in Table 9.3. The measurement of $j(O^1D)$ by chemical actinometry involves the difficult task of detecting extremely reactive $O(^1D)$ atoms. A suitable scheme is the scavenging of $O(^1D)$ by N_2O and measurement of the resulting reaction products (Bahe and Schurath, 1978; Dickerson *et al.*, 1979):

$$O(^1D) + N_2O \longrightarrow N_2 + O_2 \qquad \text{(R 9.11a)}$$

$$\longrightarrow NO + NO \qquad \text{(R 9.11b)}$$

The resulting NO, however, is not stable and is converted into higher oxides by reactions with excess ozone:

$$NO + O_3 \longrightarrow NO_2 + O_2 \qquad \text{(R 9.12)}$$

$$NO_2 + O_3 \longrightarrow NO_3 + O_2 \qquad \text{(R 9.13)}$$

$$NO_2 + NO_3 + M \rightleftharpoons N_2O_5 + M \qquad \text{(R 9.14)}$$

The $O(^1D)$ production rate can be quantified from measurements either of N_2 or of the nitrogen oxides, $NO_y = NO, NO_2, NO_3, N_2O_5$ (Table 9.3). In any case, the branching ratio k_a/k_b must be accurately known, with k_a and k_b being the bimolecular rate constants for reactions R 9.11a and R 9.11b, respectively. Furthermore, if NO_y is measured, its chemical yield γ per initially formed NO molecule must be known. With this information the photolysis frequency can be determined as

$$j(O^1D) = \left(1 + \frac{k_b}{k_a}\right) \frac{1}{[O_3]} \frac{\Delta[N_2]}{\Delta t} \qquad (9.25)$$

or

$$j(O^1D) = \frac{1}{2}\left(1 + \frac{k_a}{k_b}\right) \frac{1}{\gamma} \frac{1}{[O_3]} \frac{\Delta[NO_y]}{\Delta t} \qquad (9.26)$$

Published measurements of $k_b/(k_a + k_b)$ have been critically evaluated by Cantrell *et al.* (1994), who recommend a value of 0.61 ± 0.08 (95% confidence interval). This value was adopted by IUPAC (Atkinson *et al.*, 2004) and agrees within the uncertainty also with the current recommendation of 0.58 by NASA-JPL (Sander *et al.*, 2003). The value by Cantrell *et al.* (1994) corresponds to $k_a/k_b = 0.64$, which may be compared with the data used by different research groups for their chemical actinometers (Table 9.3). The error of the ratio is usually the dominating uncertainty in the measurement of $j(O^1D)$ by chemical actinometry, which has a total uncertainty of about $\pm(10\text{–}15)\%$.

When $j(O^1D)$ is determined from NO_y measurements, an additional error arises from the determination of γ, which has to be calibrated individually for each chemical actinometer. The calibration can be performed by adding a known amount of NO to the actinometer gas flow at the entrance of the photolysis reactor and measurement of the corresponding increase of the NO_y signal.

The photolysis of ozone to $O(^1D)$ atoms is a process with a pronounced temperature dependence. In order to measure representative $j(O^1D)$ values, it is therefore necessary to operate the chemical actinometer at ambient temperature. If the actinometer gas is not in thermal equilibrium with the surrounding atmosphere, temperature corrections have to be applied to the $j(O^1D)$ data. See Section 9.5.4 for a discussion of the expected temperature dependence.

Static ozone actinometers use the measurement of N_2 as a proxy for $O(^1D)$. This concept has the advantage that nitrogen is a stable compound which undergoes no further chemical reactions. A disadvantage is the limitation of the N_2 detection by spurious background signals which are caused by possible air leaks in the instrumental setup, or by N_2 contaminations in the bottled gases from which the actinometer mixture is prepared. Typical background levels of 1–3 ppmv N_2 have been reported for the chemical actinometers listed in Table 9.3. In order to achieve sufficiently high N_2 concentrations that exceed the background significantly by photolysis, relatively high initial ozone concentrations and long irradiance times have to be applied. For example, given a photolysis frequency of $j(O^1D) = 3 \times 10^{-5}\,s^{-1}$ (a typical clear-sky value at overhead sun and 300 DU of total ozone) and an exposure time of 30 minutes, 2 hPa O_3 in an excess of N_2O will produce approximately 40 ppmv N_2 at 1 atm of total pressure.

An example for a static actinometer is shown in Figure 9.10 (Bahe & Schurath, 1978; Bahe *et al.*, 1979). A quartz bulb (~7 cm dia.) was filled with a mixture of about 13 hPa O_3 and 1.5–3 atm N_2O. After exposure to solar radiation, gas samples were drawn from

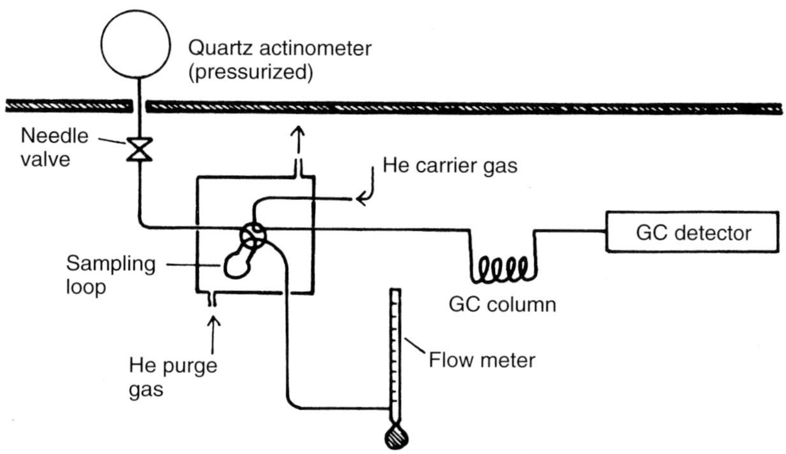

Figure 9.10 Schematic of a chemical actinometer system for measurement of the photolysis frequency of O_3. In the quartz reactor a gas mixture of N_2O and O_3 is exposed to solar radiation. After exposure the amount of photolytically formed N_2 is analysed by gas chromatography (GC) (from Bahe and Schurath, 1978, used with permission of Birkhäuser/Springer).

the reactor for N_2 analysis with a gas chromatographic system. The data were evaluated following Equation 9.25, but corrections (~50%) had to be applied since the actinometer was not optically thin and the ozone concentration was not stable. About 50% O_3 loss within 1 hour of exposure was caused by photolysis and O_3 consumption by the reaction with $O(^3P)$ atoms:

$$O(^3P) + O_3 \longrightarrow O_2 + O_2 \quad \text{(R 9.15)}$$

The concept by Bahe *et al.* (1979) was improved by reducing the initial O_3 concentrations (~2 hPa) to ensure optically thin conditions (Junkermann *et al.*, 1989; Müller *et al.*, 1995). Furthermore, the O_3 losses during sun exposure were reduced by addition of 100 hPa O_2 (Müller *et al.*, 1995). While the O_2 removes a few per cent of $O(^1D)$ by collisional deactivation, it has the advantage to recycle all $O(^3P)$ atoms back to O_3 (via R 9.2), reducing the effective loss of O_3 to about 10% per hour. This technique has been applied successfully for calibration of $j(O^1D)$ filter radiometers (see Section 9.5.4), but owing to its poor time resolution it is less useful for extensive atmospheric field measurements.

Flowing ozone actinometers use online methods with high sensitivity for NO_y detection, which has several advantages. First, the high sensitivity allows to measure much smaller product yields than has been possible with N_2 detection, reducing the necessary measurement time for $j(O^1D)$ to less than a minute (Table 9.3). Second, the online mode enables continuous measurements, which makes them ideal for atmospheric field experiments over extended time periods. Third, small flow tube diameters (typically 1–2 cm) and relative low ozone concentrations provide optically thin conditions which require no corrections in the derived j-values.

An example of a chemical actinometer developed for airborne operation is shown schematically in Figure 9.11 (Dickerson *et al.*, 1982). Two flow reactors, one for O_3 photolysis and one for NO_2 photolysis, are mounted atop the aircraft with free view of the sky. The reactors are supplied with gas mixtures prepared online from stored gases (NO, N_2O, O_2). For the determination of $j(O^1D)$, a flow of oxygen is ozonized and mixed with a larger flow of N_2O, providing about 2000 ppmv of ozone, which is passed through the photolysis reactor. The resulting NO_y is reduced catalytically to NO by passing the gas over heated Palladium at 650°C, which also destroys the remaining ozone. The NO is then measured by a chemiluminescence detector. Care had to be exercised in the choice of the catalyst to avoid conversion of N_2O into NO (Dickerson *et al.*, 1979). The efficiency of the NO retrieval from NO_y is calibrated by injection of a known flow of a standard NO gas mixture upstream of the reactor inlet.

Following the concept by Dickerson *et al.* (1982), other O_3 actinometers were developed, but with major modifications in the NO_y detection. Blackburn *et al.* (1992) passed the gas mixture after its exposure through liquid methanol, where NO_3 and N_2O_5 react with traces of water to form nitric acid. The resulting change in electrical conductivity was calibrated and used to quantify the NO formation in the actinometer. Bairai and Stedman (1992) operated their chemical actinometer with much less ozone (~5 ppmv), resulting in conversion of NO into NO_2 without significant production of higher oxides. The NO_2 was then measured by detection of the chemiluminescence produced during the reaction of NO_2 with a luminol solution. In the actinometer developed by Shetter *et al.* (1996), the NO is converted mostly into N_2O_5, which is then transferred through a permeable

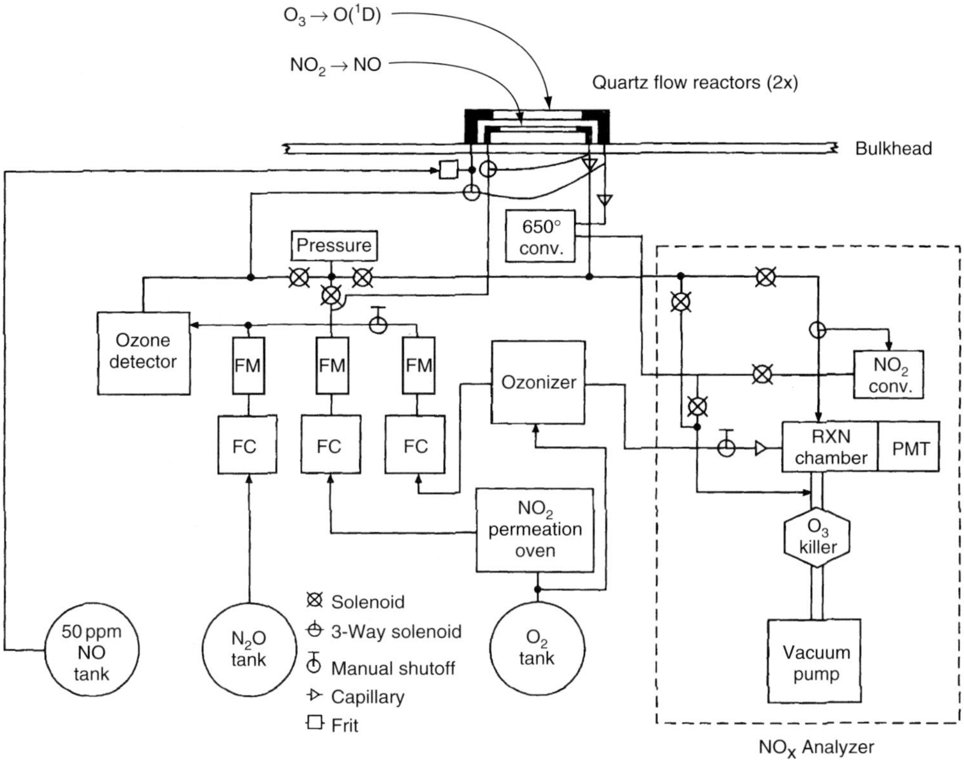

Figure 9.11 Schematic of an airborne chemical actinometer for measurement of photolysis frequencies. Two photolysis flow reactors, one for O_3 and one for NO_2, are mounted atop the aircraft with free view of the sky. NO is photolytically formed in both actinometers and is measured by an NO_x analyser that takes alternatingly samples from one or the other reactor. Abbreviations used: FC, flow controller; FM, flow meter; Conv., conversion of nitrogen oxides into NO; RXN, reaction chamber; PMT, photomultiplier tube (adapted from Dickerson *et al.*, 1982, printed with permission from American Geophysical Union).

Nafion® membrane into liquid water. The increase in electrical water conductivity due to formation of nitrate ions is then measured to quantify the initially formed NO.

Measurement examples are shown in Figures 9.12 and 9.13. The first figure displays a typical diurnal profile, which was measured by a chemical flow actinometer during a field experiment on Mauna Loa in 1992 (Shetter *et al.*, 1996). Smooth $j(O^1D)$ variations under almost cloudless sky can be seen in the morning and late afternoon, demonstrating a high measurement precision. The fluctuations occuring in the early afternoon were caused by clouds that influenced the solar actinic flux. The general shape of the diurnal profile is determined mainly by the $j(O^1D)$ dependence on the slant ozone column $(t_{O_3}/\cos\chi)$, which effectively absorbs solar radiation at wavelengths below 320 nm (see Section 9.1.6 and Figure 9.3). Examples of the functional dependence of $j(O^1D)$ on atmospheric ozone, measured at different ground-based locations, are shown in Figure 9.13. Good agreement of the clear-sky data is found within the errors associated with each measurement technique. For further comparisons of published $j(O^1D)$ data, see Hofzumahaus *et al.* (1992) and Kraus and Hofzumahaus (1998).

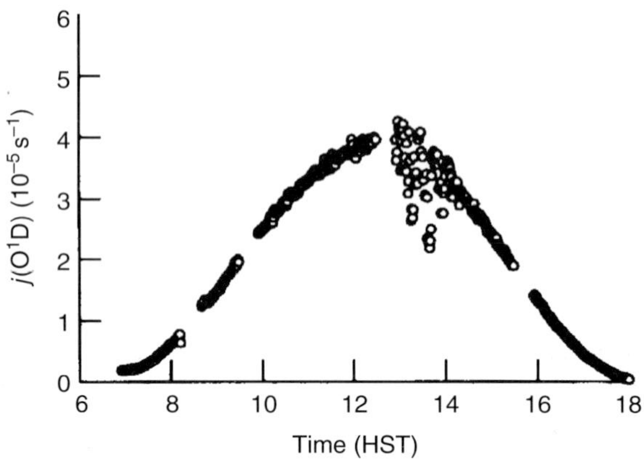

Figure 9.12 Diurnal variation of $j(O^1D)$ measured by a chemical actinometer at the NOAA Mauna Loa Observatory on the island of Hawai, on 3 May, 1992 (adapted from Shetter *et al.*, 1996, printed with permission from American Geophysical Union).

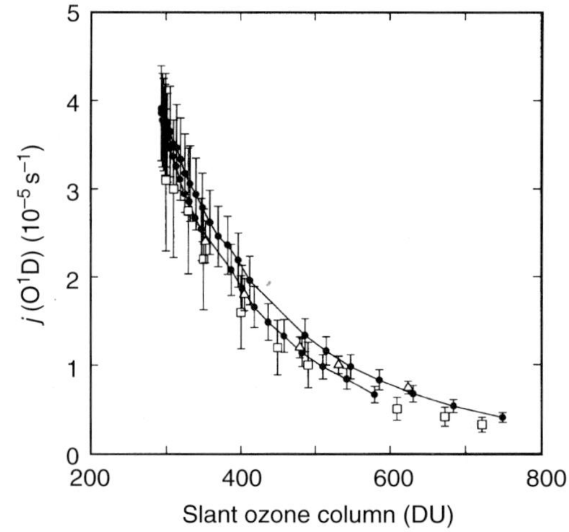

Figure 9.13 Dependence of $j(O^1D)$ on the slant ozone column. Data are from: • Mauna Loa on 6 May, 1992, measured by a chemical actinometer (Shetter *et al.*, 1996); □ Atlantic Ocean (28°N, 30°W) on 24 September, 1988, measured by a calibrated filter radiometer (Hofzumahaus *et al.*, 1992); △ Denver, Colorado, measured by chemical actinometer on 18 May, 1991 (Bairai and Stedman, 1992) (adapted from Shetter *et al.*, 1996, printed with permission from American Geophysical Union).

9.3.5 *Nitrogen-dioxide chemical actinometers*

The photodissociation of NO_2 has one energetically allowed pathway in the troposphere, which exhibits only a weak temperature dependence (Shetter *et al.*, 1988).

$$NO_2 + h\nu(\lambda < 420\,\text{nm}) \longrightarrow NO + O(^3P) \qquad \text{(R 9.3)}$$

Since the photolytical loss of NO_2 is directly related to the primary NO formation, the photolysis frequency $j(NO_2)$ can be determined either from the measured loss rate of NO_2

or from the rate of NO production. However, secondary reactions of the $O(^3P)$ atoms must be taken into account in the evaluation of the $j(NO_2)$ data.

The O-atom chemistry is a function of the operating conditions of the chemical actinometer and must be analysed for each experimental setup. In a system of pure NO_2 (or NO_2 mixed in N_2), $O(^3P)$ will react at low pressure primarily with NO_2 and produce another NO molecule:

$$O(^3P) + NO_2 \longrightarrow NO + O_2 \qquad \text{(R 9.16)}$$

Thus, the effective NO yield per absorbed photon is two. This condition has been realised in a chemical actinometer, for example, by Parrish *et al.* (1983), who photolysed 30 ppmv NO_2 in 60 hPa ultrapure nitrogen.

When the N_2 pressure is increased, side reactions start to compete with R 9.16 and reduce the effective yield of NO:

$$O(^3P) + NO_2 + N_2 \longrightarrow NO_3 + N_2 \qquad \text{(R 9.17)}$$

$$NO_3 + NO \longrightarrow NO_2 + NO_2 \qquad \text{(R 9.18)}$$

$$O(^3P) + NO + N_2 \longrightarrow NO_2 + N_2 \qquad \text{(R 9.19)}$$

At a total pressure of 1 atm N_2, the NO yield reaches a value of (1.61 ± 0.07) for a range of initial concentrations of 1–4 ppmv NO_2 and 0–0.3 ppmv NO (Zafonte *et al.*, 1977). The yield is additionally influenced by trace amounts of O_2, which may be present as a contaminant in the nitrogen supply. A modified value of (1.69 ± 0.08) was estimated for an impurity of 0.5 ppmv O_2 in 1 atm N_2 (Dickerson and Stedman, 1980).

Another approach to perform NO_2 actinometry involves the addition of a large amount of oxygen or air to the actinometer. At sufficiently high pressure, for example 1 atm, nearly all $O(^3P)$ atoms are converted into O_3 by R9.2. When the actinometer conditions are chosen properly, for example when the time between formation and analysis of NO is short, the influence of consecutive O_3 reactions like R9.12 can be neglected and the effective NO yield becomes unity (Harvey *et al.*, 1977; Shetter *et al.*, 1992).

Static nitrogen-dioxide actinometers generally use pure NO_2 (1–5 hPa) as actinometer gas (Table 9.4). During photolysis the NO_2 decay over time is either monitored directly by photometric measurements (Bahe *et al.*, 1980; Schultz *et al.*, 1995) or derived indirectly from the pressure increase associated with the NO_2 photodecomposition in the sealed actinometer cell (Madronich *et al.*, 1983, 1985). The primary photodissociation R9.3 followed by R9.16 results in the net reaction

$$2NO_2 + h\nu \longrightarrow 2NO + O_2 \qquad \text{(R 9.20)}$$

Considering the stoichiometry of this reaction, Equation 9.24 can be modified to calculate $j(NO_2)$ from the NO_2 concentration measured before ($t = 0$) and after ($t = \Delta t$) light exposure:

$$j(NO_2) = -\frac{1}{2\Delta t} \ln \frac{[NO_2]_{\Delta t}}{[NO_2]_{t=0}} \qquad (9.27)$$

Equation 9.27 is only accurate when low NO_2 pressures are used. At high concentrations, NO_2 produces significant amounts of the dimer N_2O_4 in a temperature-dependent equilibrium:

$$NO_2 + NO_2 + M \rightleftharpoons N_2O_4 + M \quad \text{(R 9.21)}$$

When NO_2 is photolysed, the fast equilibrium will replenish some NO_2 by thermal decomposition of N_2O_4 and apparently slows down the NO_2 decay. The N_2O_4 concentration that is present in the actinometer can be calculated from the equilibrium constant $K = p^2_{NO_2}/p_{N_2O_4}$ (147 hPa at 298 K; Sander *et al.*, 2003). At room temperature and 1 hPa NO_2, the fraction of N_2O_4 is 0.7% which can be neglected. At higher NO_2 concentrations or much lower temperatures, N_2O_4 becomes relevant and the equilibrium R9.21 must be explicitly taken into account in the $j(NO_2)$ evaluation (for details, see Madronich *et al.*, 1983).

An example for a static $j(NO_2)$ actinometer, developed for measurements on a stratospheric balloon payload (Madronich *et al.*, 1983, 1985), is shown in Figure 9.14. The inner quartz cylinder serves as a photolysis cell which can be filled with NO_2 from a reservoir, or can be pumped by a 7 litre pre-evacuated tank. The cell is equipped with sensors for pressure and temperature and the NO_2 decay is determined from the total pressure increase associated with R9.20. The photolysis cell is thermally isolated from the ambient air by a second quartz cylinder, with the annulus evacuated. A cylindrical shutter which encloses the photolysis assembly can be driven up and down for exposure of the photolysis cell. This system is the only chemical actinometer that was deployed in the stratosphere. The measurements, which had uncertainties of ±6% to ±15%, were used for tests of radiative-transfer model predictions for this atmospheric regime (Madronich *et al.*, 1985). Other applications of static $j(NO_2)$ actinometers have involved tests and calibrations of actinic-flux filter radiometers for tropospheric use (Schultz *et al.*, 1995; Volz-Thomas *et al.*, 1996; Kraus *et al.*, 1998). In these cases the actinometer measurements had typical uncertainties of about ±5%.

Flowing nitrogen-dioxide actinometers use highly sensitive chemiluminescence instruments for online detection of NO and NO_2, which allows continuous measurements of $j(NO_2)$ with high time resolution (Table 9.4). An example is the airborne actinometer shown schematically in Figure 9.11 (Dickerson *et al.*, 1982). For the determination of $j(NO_2)$, a flow of pure oxygen is doped with about 20 ppmv NO_2 from a permeation device. The mixture is passed through a photolysis reactor mounted atop the aircraft and is analysed after short irradiation ($\Delta t \sim 1$ s) by a chemiluminescence instrument. The $j(NO_2)$ is then derived from the measured increase $\Delta[NO]$ and the initial concentration $[NO_2]_0$:

$$j(NO_2) = \frac{1}{\Delta t}\frac{\Delta[NO]}{[NO_2]_0} \quad (9.28)$$

Other chemical flow actinometers listed in Table 9.4 follow essentially the same principle. Their measurement uncertainties are typically in the range between ±5% and ±12% (e.g., Kelley *et al.*, 1995; Lantz *et al.*, 1996; Shetter *et al.*, 2003).

Measurement examples are shown in Figures 9.15 and 9.16. The first figure displays a typical diurnal profile of $j(NO_2)$, which was measured by a chemical actinometer at

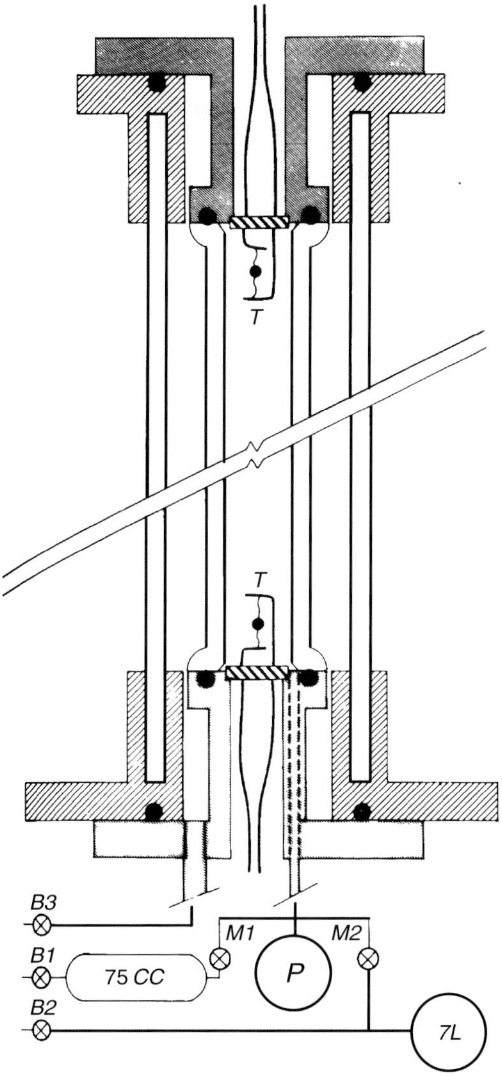

Figure 9.14 Schematic of a balloon-borne chemical actinometer for measurement of the photolysis frequency of NO_2. In the quartz cell, pure NO_2 is photolysed by solar radiation, leading to a measurable increase of total cell pressure. Abbreviations used: P, absolute pressure transducer; T, thermistors; B1, B2, B3 bellows valves (closed during flight); M1, M2, latching solenoid valves; 75 cc, reservoir filled with NO_2; 7 L, volume evacuated before flight (from Madronich *et al.*, 1985, with kind permission of Springer Science and Business Media).

an observatory on Mauna Loa in 1988, and is compared with corresponding theoretical model calculations (Shetter *et al.*, 1992). The maximum $j(NO_2)$ value at noon is around $9 \times 10^{-3}\,s^{-1}$ and corresponds to a photolytical NO_2 lifetime of about 1.9 minutes. The diurnal profile of $j(NO_2)$ may be compared to the one of $j(O^1D)$ measured at the same place and season in a different year (Figure 9.12). The $j(O^1D)$ profile is narrower, because

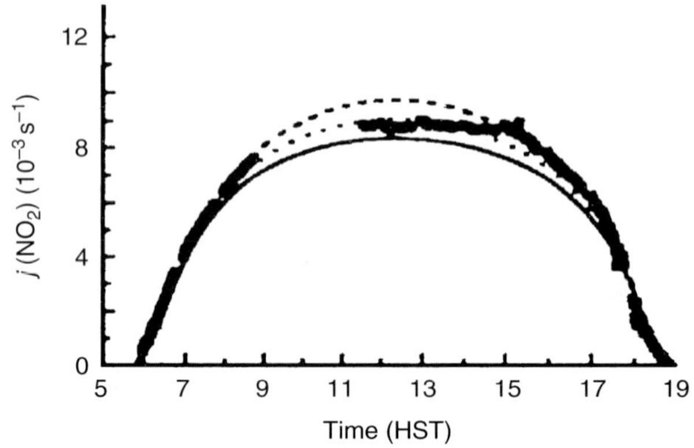

Figure 9.15 Diurnal variation of $j(NO_2)$ measured by chemical actinometry at the Mauna Loa Observatory on the island of Hawai, at an elevation of 3410 m, on 21 May, 1988. Experimental data are shown as points; the curves represent theoretical model calculations assuming slightly different amounts of albedo enhancement by clouds lying in a valley below the observation point (adapted from Shetter *et al.*, 1992, printed with permission from American Geophysical Union).

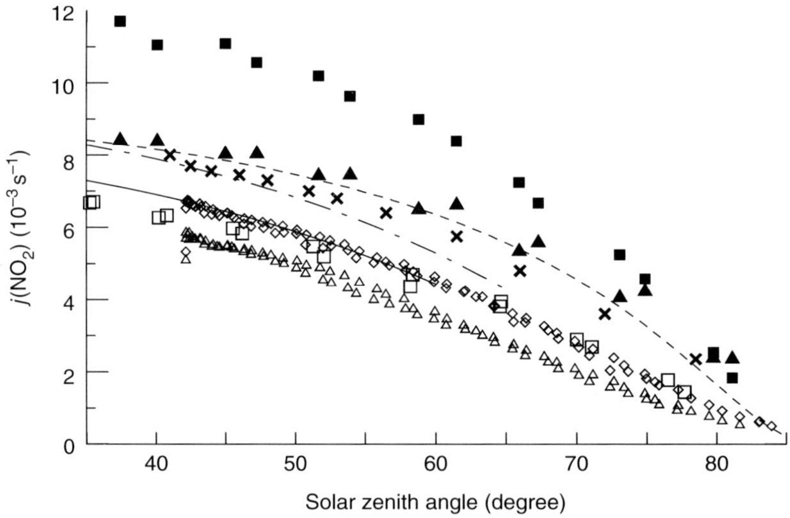

Figure 9.16 Dependence of $j(NO_2)$ on the solar zenith angle under cloud-free conditions. Data obtained by chemical actinometry: — — Dickerson *et al.* (1982), × Madronich *et al.* (1983), - - Parrish *et al.* (1983), — Müller and Schurath (1986), ▲ Shetter *et al.* (1992), ■ Lantz *et al.* (1996); spectroradiometry: ◊ Kraus and Hofzumahaus (1998); filter radiometry: △ Kraus and Hofzumahaus (1998), □ Brauers and Hofzumahaus (1992). Solid and dashed lines are parametrisations from original data (adapted from Kraus and Hofzumahaus, 1998, with kind permission of Springer Science and Business Media).

it depends on radiation that is strongly absorbed by stratospheric ozone at low sun, whereas $j(NO_2)$ is essentially independent on ozone absorption. Examples of the functional dependence of $j(NO_2)$ on the solar zenith angle are given in Figure 9.16 displaying observations at different ground-based locations for clear-sky conditions.

9.4 Actinic-flux spectroradiometry

Spectroradiometry is the measurement of absolute photon fluxes in narrow spectral intervals at selected wavelengths (see e.g. McCluney, 1994). This task requires an optical instrument (spectrometer) that contains three basic elements:

- A receiver optic for radiation incident from outside.
- A device for the spectral dispersion of radiation into different wavelengths.
- A detection system for quantification of the dispersed light.

If a spectrometer is primarily designed and calibrated for the *absolute* measurement of photon-flux densities, it is called a *spectroradiometer*.

A *scanning spectroradiometer* uses a monochromator which selects radiation in a small spectral interval that can be scanned through the wavelength range of interest over time. In this operational mode the radiation spectrum is measured sequentially.

A *spectrograph* exposes a dispersed spectrum of a whole range of wavelengths at once to a many-element detector array, such that radiation of certain wavelengths can be assigned to specific detector elements. In this mode of operation the radiation spectrum is measured simultaneously.

Spectroradiometers can measure different radiometric quantities, usually irradiances or radiances, depending on the design of the reception optics. In atmospheric physics, spectroradiometers have been used for many decades for measuring solar UV and visible irradiances (see Section 9.6). Only since the second half of 1990s, spectroradiometers have been developed and used for the direct measurement of solar actinic fluxes, in order to determine atmospheric photolysis frequencies (e.g. Müller *et al.*, 1995; Kraus & Hofzumahaus, 1998; Shetter & Müller, 1999; Webb *et al.*, 2002b; Eckstein *et al.*, 2003; Edwards & Monks, 2003; Kanaya *et al.*, 2003; Jäkel *et al.*, 2005). Instruments that have been used for field measurements are listed in Table 9.5.

9.4.1 Principle of actinic-flux spectroradiometry

Actinic-flux spectroradiometers collect incident solar radiation with isotropically sensitive receiver optics and record the spectrum of the radiation by either a scanning double-monochromator system or a spectrograph. A photolysis frequency can then be calculated from the measured spectrum (cf. Equation 9.12), when three conditions are fulfilled:

1. The intensities and wavelengths of the measured radiation are absolutely calibrated.
2. The absorption spectrum σ and quantum yield ϕ of the relevant photolysis reaction are accurately known.
3. The measured radiation spectrum covers the full range of wavelengths that contributes to the photodissociation.

This concept has the important advantage that photolysis frequencies of a variety of atmospheric gases can be obtained from the same measurement of radiation obtained by a single instrument.

Table 9.5 Actinic-flux spectroradiometers used in atmospheric field studies

Reference	Inlet Optic[a]	Spectrometer	Detector[b]	Wavelength range (nm)	Bandwidth (nm)	Measur. time	Platform
Hofzumahaus *et al.* (1999, 2002)	$2(4)\pi$ sr[c] Quartz	Double monochromator Bentham DTM300	PMT	280–420	1	80–90 s	Ground + Aircraft
Shetter and Müller (1999)	$2(4)\pi$ sr[d] Quartz	Double monochromator CVI 112	PMT	280–420	1	28 s	Ground + Aircraft
Schmitt (1999)	2π sr Quartz	Spectrograph Carl-Zeiss	PDA	285–700[e]	~ 2	< 5 s	Ground[f]
Webb *et al.* (2002b)	2π sr Quartz	Double monochromator Bentham DTM300	PMT	290–500	0.5–1	–	Ground[g]
Eckstein *et al.* (2003)	4π sr Teflon	Spectrograph Jobin-Yvon Cp200	CCD	300–600[e]	1.8–2.8	30–60 s	Ground
Jäkel *et al.* (2005)	2π sr[d,h] Quartz	Spectrograph Carl-Zeiss	PDA	305–700[e]	2.5	< 1 s	Aircraft

[a] Optical diffusers with isotropic sensitivity; the field-of-view is 2π sr (hemispherical) or 4π sr (spherical).

[b] PMT, photomultiplier tube; PDA, photodiode array; CCD charge-coupled device.

[c] Airborne version uses one double-monochromator with two 2π sr optics for zenith and nadir view.

[d] Airborne version uses two separate 2π sr spectroradiometers for zenith and nadir view.

[e] Instrumental stray light reduces the accuracy of the actinic-flux measurements below 305 nm.

[f] Commercial instrument, used by various research groups (e.g., Kanaya *et al.*, 2003; Edwards and Monks, 2003).

[g] Spectroradiometers equipped with actinic-flux sensing and irradiance-sensing inlet optics.

[h] Zenith and nadir viewing inlet-optics are horizontally stabilised by active control.

When the radiation spectrum is recorded, the photon flux is measured in narrow spectral intervals $\Delta\lambda$ that are centred at discrete equidistant wavelengths λ_i. The step size of the wavelength grid is usually set equal to $\Delta\lambda$, which is chosen to be equal to or less than the spectral bandwidth $\delta\lambda$ of the dispersing spectrometer (see Appendix A.3). Both bandwidth and step size must be sufficiently small to resolve spectral variations of the solar spectrum to allow an accurate determination of photolysis frequencies.

In practice, the calculation of j-values following equation 9.12 is performed as a summation

$$j = \sum_i \langle\sigma\rangle_i \langle\phi\rangle_i \langle F_\lambda\rangle_i \Delta\lambda \tag{9.29}$$

Here, $\langle F_\lambda\rangle_i$, $\langle\sigma\rangle_i$ and $\langle\phi\rangle_i$ denote average values of the quantities over the interval $\Delta\lambda$ at wavelength λ_i (see also annotations on $\langle F_\lambda\rangle_i$ in Appendix A.3). Data for the cross sections and quantum yields, which are functions of wavelength, temperature and possibly pressure, can be taken from the literature (see Section 9.8.4).

9.4.2 Spectroradiometers for actinic-flux measurements

The general outline of a *scanning spectroradiometer* for the measurement of actinic flux (e.g. Hofzumahaus *et al.*, 1999; Shetter & Müller, 1999; Webb *et al.*, 2002b) is illustrated in Figure 9.17. In this example, the actinic radiation is collected by a hemispheric receiver optic that is connected to the entrance slit of a spectrometer by an optical quartz-fibre bundle. The application of the flexible light guide serves two purposes.

1. It allows to set up the spectrometer in a weather-protected, temperature-controlled environment (laboratory container, aircraft cabin, etc.) while the inlet optic is mounted outside, pointing into the sky.
2. In order to enhance light transmission, the fibres in the bundle can be arranged to match one end to the round cross section of the receiver optic and the other end to the rectangular shape of the spectrometer entrance.

Scanning spectroradiometers are generally based on double-monochromators with diffraction gratings turning synchronously with additive dispersion and are usually scanned from 280 to 420 nm for coverage of the most relevant photolysis frequencies. Compared to a single monochromator, the double-monochromator set-up provides double spectral resolution (i.e. half the spectral bandwidth) and several orders of magnitude lower levels of instrumental stray light. These two properties are particularly important for the accurate measurement of solar radiation at wavelengths below 320 nm, where the spectral photon flux changes dramatically over many orders of magnitude (see Figure 9.6).

At the exit of the double-monochromator, the selected radiation (λ_i, $\Delta\lambda$) is detected by a photomultiplier tube (PMT), which needs a high voltage supply for operation. The PMT output signal, a photocurrent, is further amplified and converted into a voltage which is then digitized by an analogue–digital converter and recorded by a computer. The large dynamic range of possible photon fluxes requires good linearity of the detection

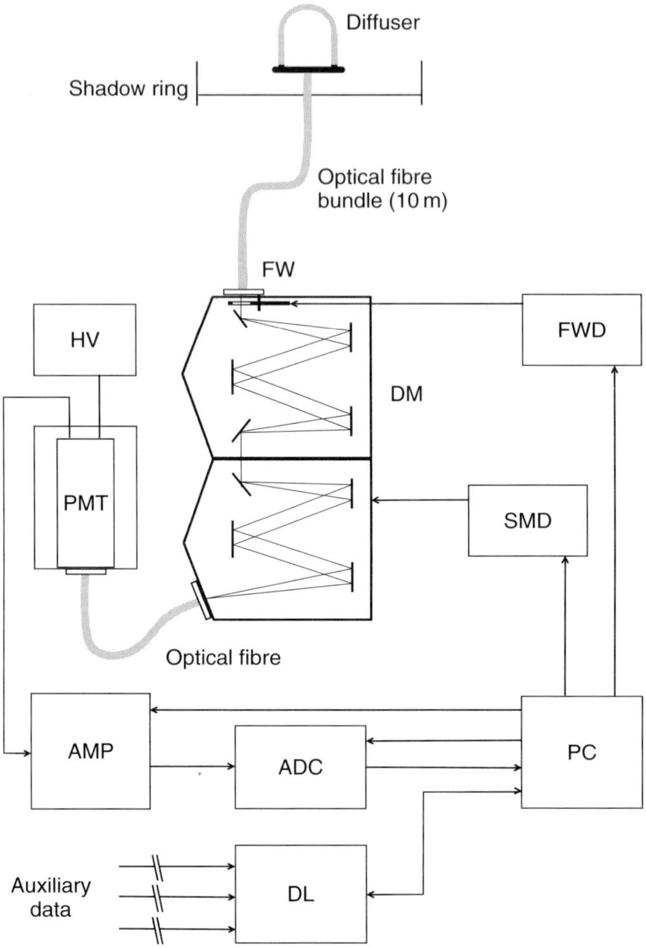

Figure 9.17 Schematic diagram of a scanning spectroradiometer for actinic-flux measurements: DM, double monochromator; HV, high-voltage supply; AMP, current-to-voltage amplifier; ADC, analogue–digital converter; PC, personal computer; DL, data logger; SMD, stepping motor drive; FWD, filter wheel driver; FW, filter wheel with optical shutter (from Hofzumahaus *et al.*, 1999, used with permission of Optical Society of America).

system for at least six decades. This is generally accomplished by selected PMTs with very low dark current and multiple-stage current-to-voltage amplifiers.

A computer collects the measured data and controls the wavelength setting λ_i of the double-monochromator. Some spectrometers, like the one in Figure 9.17, contain an optical shutter that can be closed automatically to measure the dark signal of the PMT. Alternatively, the sum of the dark signal and possible instrumental stray light can be recorded during atmospheric measurements at wavelengths below 290 nm, where no detectable solar radiation is found in the troposphere and lower stratosphere.

Spectrograph systems for the measurement of actinic fluxes use similar inlet optics as scanning spectroradiometers, but make use of a single monochromator in which the exit slit is replaced by a photodiode array (PDA) (Schmitt, 1999; Jäkel *et al.*, 2005) or a charge-coupled device (CCD) (Eckstein *et al.*, 2003). In this way, the complete spectrum can be recorded simultaneously in seconds. An example of a multi-channel spectrometer (manufactured by Zeiss GmbH, Jena, Germany), used, for example, by Schmitt (1999) and Jäkel *et al.*, (2005), is shown schematically in Figure 9.18. The collected light is coupled into the spectrometer via a quartz fibre which is connected to the entrance slit via a cross-section converter (round to rectangular). A flat-field grating disperses the radiation and images the spectrum onto a 512-element detector array. Photoelectric charge generated by the incident photon flux is accumulated on each detector element during a pre-selected exposure time and is then read out by an analogue–digital converter. The digitized spectrum covers usually a usable range from 300 to about 700 nm and is stored in a computer for further data processing.

Detector array–based spectrograph systems have a number of advantages compared to scanning spectrometers, making them particularly useful for field applications (cf. Table 9.5):

- They measure a broader wavelength range in a tenfold to hundredfold shorter time.
- They measure the complete spectrum simultaneously.
- There are no moving parts, making the spectrometer more robust.
- The spectrometer is more compact and lighter.

Probably the most important advantage for the measurement of ambient solar radiation is the fast and simultaneous recording of complete spectra. This capability ensures an accurate representation of the spectra with good time resolution even when the actinic flux changes quickly. In contrast, spectra recorded sequentially by scanning spectroradiometers may be distorted due to temporal intensity fluctuations during a wavelength scan. This is relevant, for example, when measurements are performed under cloudy conditions on fast-flying aircraft.

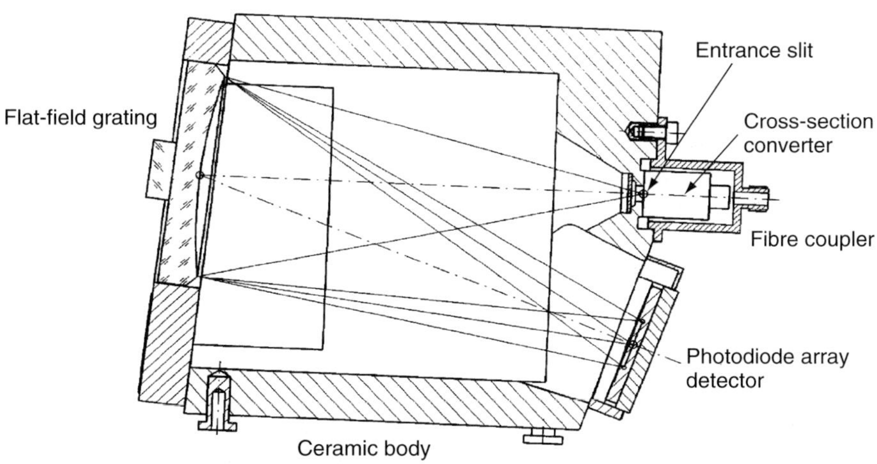

Figure 9.18 Schematic of the optical layout of a spectrograph that uses a photodiode array detector for the detection of the dispersed radiation spectrum (adapted from Zeiss GmbH, 1998, used with permission of Zeiss company).

The main disadvantage of detector-array based spectrometers is the higher instrumental stray-light level compared to double-monochromator systems. Stray light is by definition unwanted light that reaches a detector element from wavelengths that lie outside the spectrometer bandpass. Stray light reaches the detector in an irregular way, for example by random scatter from optical surfaces (mirrors, gratings etc.) and walls inside the spectrometer. The relevance of stray light depends in particular on the spectral distribution of the light source. In case when solar radiation is measured, stray light is contributed mainly by the visible portion of the solar spectrum and produces a background signal on every detector element, reducing the useful dynamic intensity-range of a detector-array based spectrometer by about 2–3 orders of magnitude compared to double-monochromator systems (see e.g. Bais *et al.*, 2003). This presents a particular problem for the measurement of the relatively low photon-fluxes found in the UV-B region. In order to convert the measured signals into reasonable radiation data, sensitive stray-light corrections are needed in the UV-B, which requires a thorough analysis of the instrumental stray-light characteristics (e.g. Edwards & Monks, 2003; Kanaya *et al.*, 2003; Eckstein *et al.*, 2003; Jäkel *et al.*, 2005; Ylianttila *et al.*, 2005).

9.4.3 Actinic-flux receiver optics

Most of the actinic-flux spectroradiometers listed in Table 9.5 use the same type of hemispheric quartz-diffuser for sunlight collection as shown in Figure 9.19. The diffuser is a rigid construction made of a set of concentric frosted quartz-domes, which collect radiation from a single hemisphere and couple all received light with nearly equal efficiency into a quartz light guide. Radiation from the opposite hemisphere is mostly blocked by a shadow ring serving as an artificial horizon. The design of the isotropic receiver is based on a concept that was originally developed by Junkermann *et al.*, (1989). It was further improved by Volz-Thomas *et al.* (1996) and is now being manufactured by Meteorologie Consult GmbH (Glashütten, Germany).

The relative angular response Z_p of the receiver optic must be known for the accurate evaluation of the spectrometer measurements (see Section 9.4.4). It can be measured in the laboratory by exposing the receiver to radiation from a point source (lamp) that is scanned at a constant distance over different polar (ϑ) and azimuth angles (φ) relative to the symmetry axis of the diffuser. An example of results is shown in Figure 9.20 displaying polar-angle scans at different azimuthal angles for a fixed wavelength setting. Note that Z_p is normalised to be unity at the angle of incidence $\vartheta = 0°$. The agreement of the scans at different azimuthal angles shows that Z_p can be considered to be rotationally symmetric. In principle Z_p may exhibit a weak wavelength dependence. A negligible dependence was reported for the 300–420 nm range by Hofzumahaus *et al.* (1999) and Shetter and Müller (1999), while Jäkel *et al.* (2005) found a dependence of up to ±4% from 300 to 700 nm for a similar receiver.

The measured Z_p curves may be compared to the response function of an ideal 2π sr detector (solid lines in Figure 9.20). Owing to its vertical extension, the real receiver collects some radiation from below the artificial horizon line. The limiting polar angle at which Z_p approaches zero can be estimated geometrically from the ratio of the vertical diffuser height and the radius of the shadow ring.

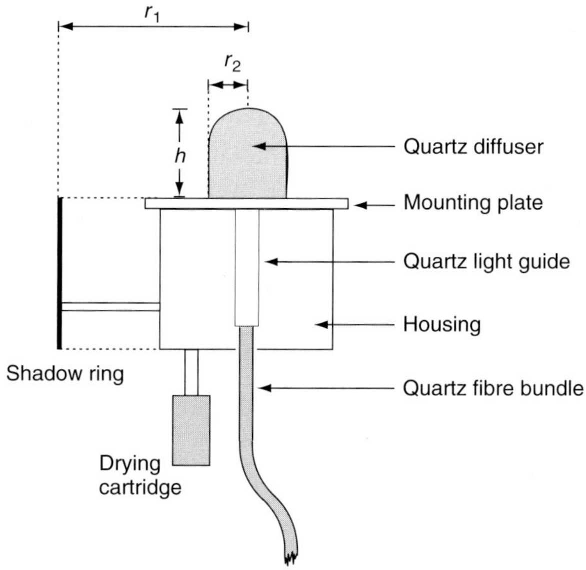

Figure 9.19 Schematic diagram of a receiver optic ($h \approx 35\,\text{mm}$, $r_2 \approx 15\,\text{mm}$) for actinic-flux detection. A shadow ring ($r_1 \approx 75$–$150\,\text{mm}$) serves as an artificial horizon to restrict the field-of-view to one hemisphere (2πsr). A drying cartridge prevents condensation of humidity at cold ambient temperatures. The quartz fiber bundle connects the diffuser to the entrance of the monochromator (adapted from Hofzumahaus *et al.*, 1999, used with permission of Optical Society of America).

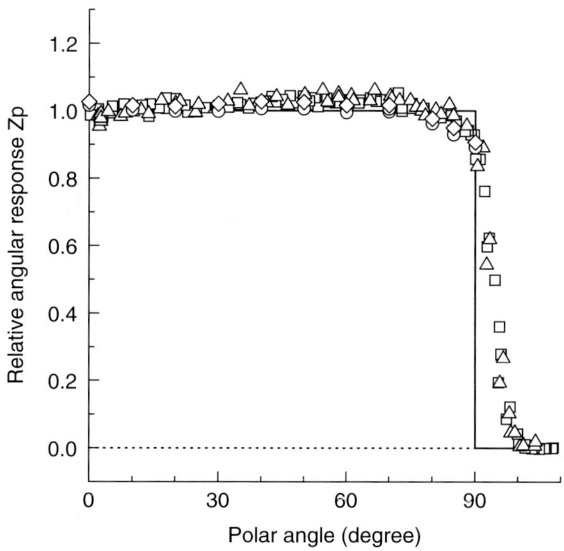

Figure 9.20 Relative angular sensitivity (Z_p) as a function of polar angle (ϑ), measured for the 2πsr inlet-optic of a spectroradiometer operated at Forschungszentrum Jülich. The different symbols denote measurements at four different azimuthal angles ($\varphi = 0°, 90°, 180°$, and $270°$) at 300 nm. The solid line represents the response of an ideal 2π sr actinic-flux detector.

In order to measure 4π sr actinic fluxes, different radiometer concepts have been realised in field experiments:

- One solution is to use two identical spectrometers each of which is equipped with a 2π sr receiver optic. In this case, the inlet optics are oriented into opposite directions (upside and downside) in order to cover the full solid angle of 4π sr. This concept has been applied for airborne measurements by Shetter and Müller (1999) and Jäkel *et al.* (2005). A corresponding approach utilising filter radiometers was realised by Volz-Thomas *et al.* (1996).
- In another aircraft experiment, two 2π sr receiver optics (upside and downside looking) were attached to the same spectroradiometer (Hofzumahaus *et al.*, 2002). Here the inlet optics were connected via fibre bundles to the upper and lower part of the entrance slit of a double-monochromator. Owing to the imaging properties of the spectrometer, the dispersed radiation from the two inlet optics could be separated at the exit slit and measured independently by two photodetectors.
- In a ground-based field study, Eckstein *et al.* (2003) used a receiver-optic with a field-of-view of almost 4π sr. The home-built inlet-optic consisted of a teflon sphere of 30 mm diameter which was connected to the entrance slit of a spectrograph by a thin quartz mono-fibre. Three of such receiver systems were attached in a vertical order over the height of the entrance slit, producing three vertically separated spectra in the focal plane of the spectrograph. The spectra were measured simultaneously by a CCD camera with two-dimensional resolution. This set-up was deployed for measurements at three different heights (4–100 m) on a tower in clouds.

9.4.4 Determination of spectral actinic fluxes

When a spectroradiometer is exposed to solar radiation in the atmosphere, the spectral radiometer signal I_λ can be related to the atmospheric spectral radiance L_λ by the following expression (Hofzumahaus *et al.*, 1999):

$$dI_\lambda(\lambda) = D(\lambda, \vartheta, \varphi)L_\lambda(\lambda, \vartheta, \varphi)d\omega \tag{9.30}$$

Depending on the kind of photodetection the radiometer signal may be a photoelectrical current from a photomultiplier tube or accumulated photoelectric charge from a photodiode array.

D denotes the absolute detection sensitivity of the entire spectroradiometer and may be factorised as:

$$D(\lambda, \vartheta, \varphi) = D_0(\lambda)Z_p(\lambda, \vartheta, \varphi) \tag{9.31}$$

$D_0(\lambda)$ is the wavelength-dependent value of D at $\vartheta = 0°$ and Z_p describes the relative sensitivity of the entrance optic with respect to the direction (ϑ, φ) of the incoming radiation at a given wavelength. Note that Z_p is dimensionless and by definition unity at $\vartheta = 0°$.

In field conditions the entrance optic integrates the atmospheric radiation over all angles of incidence. On the assumption that Z_p has no significant azimuthal dependence (cf. Figure 9.20), the resulting spectroradiometer signal I_λ at a given wavelength λ is given by

$$I_\lambda(\lambda) = D_0(\lambda)\int_{4\pi\,\mathrm{sr}} Z_p(\lambda, \vartheta)L_\lambda(\lambda, \vartheta, \varphi)d\omega \tag{9.32}$$

By introducing the dimensionless function Z_H

$$Z_H = \frac{1}{F_\lambda(\lambda)} \int_{4\pi\,\mathrm{sr}} Z_p(\lambda, \vartheta) L_\lambda(\lambda, \vartheta, \varphi) d\omega \tag{9.33}$$

the following relation between I_λ and the actinic flux F_λ can be derived from Equation 9.32:

$$I_\lambda(\lambda) = D_0(\lambda) Z_H F_\lambda(\lambda) \tag{9.34}$$

In case of an ideal actinic-flux detector, with $Z_p = 1$ for all angles of incidence, Equation 9.33 gives $Z_H = 1$ and equation 9.34 becomes:

$$I_\lambda^{\,\mathrm{ideal}}(\lambda) = D_0(\lambda) F_\lambda(\lambda) \tag{9.35}$$

Comparison with Equation 9.34 shows that Z_H represents a correction factor for the non-ideal angular response of the real detector. Accordingly, the actinic flux can be derived from the measured signal I_λ as

$$F_\lambda(\lambda) = \frac{1}{D_0(\lambda) Z_H} I_\lambda(\lambda) \tag{9.36}$$

Here, the evaluation of $F_\lambda(\lambda)$ requires the absolute detection sensitivity D_0 and the angular-response correction Z_H of the radiometer, both of which can be obtained by calibrations (see Section 9.4.5).

9.4.5 Spectroradiometer calibration

9.4.5.1 Angular-response correction

The correction factor Z_H can be determined by Equation 9.33 from measured Z_p values and the relative angular distribution of the atmospheric radiance $L_\lambda(\lambda)$ estimated by a model. A general procedure to determine Z_H for ground-based upward-looking receivers has been described by Hofzumahaus *et al.* (1999) (see Appendix A.4). The application to ground-based 4π sr measurements was presented by Eckstein *et al.* (2003) and airborne measurements using a pair of upward and downward oriented receivers were discussed by Volz-Thomas *et al.* (1996), Hofzumahaus *et al.* (2002) and Jäkel *et al.* (2005).

The Z_H corrections for upward-looking receivers are usually small (<5%) over ground surfaces with low albedo (<10%). Because the hemispheric receiver optics have some sensitivity to radiation from the opposite hemisphere, the corrections can increase to values of 10–15%, when measurements are performed above strongly reflecting surfaces like snow or clouds (Hofzumahaus *et al.*, 1999; Jäkel *et al.*, 2005).

Downward looking receivers that measure the upwelling actinic radiation need generally larger Z_H corrections (<30%). In this particular case, the unwanted contribution of radiation from the opposite hemisphere is relatively large, because the downward actinic flux is larger than the upward directed component that is to be measured. The Z_H correction applied to the downward oriented receiver has, however, relatively little influence on the determination of the total, upward plus downward, actinic flux (Volz-Thomas *et al.*, 1996; Hofzumahaus *et al.*, 2002; Jäkel *et al.*, 2005).

9.4.5.2 Actinic-flux calibration

The calibration of the absolute detection sensitivity $D_0(\lambda)$ can be achieved by laboratory measurements using certified calibration lamps (Hofzumahaus *et al.*, 1999). The method is based on the equivalence of the spectral actinic flux F_λ and the spectral irradiance E_λ for collimated radiation incident at $\vartheta = 0°$:

$$F_\lambda(\lambda, \vartheta = 0°) = E_\lambda(\lambda, \vartheta = 0°) \tag{9.37}$$

For this condition and considering that $Z_p(\vartheta = 0°) = 1$, integration of Equation 9.32 yields

$$I_\lambda(\lambda) = D_0(\lambda)E_\lambda(\lambda) \tag{9.38}$$

Using this equation, $D_0(\lambda)$ can be directly derived from the measurement of the certified irradiance output of a calibration lamp. This procedure has the advantage that the calibration of the actinic-flux spectroradiometer becomes traceable, via readily available irradiance standards, to national laboratory standards (e.g. from NIST, NPL, PTB).

The application of the concept is complicated by the fact that the surface of the receiver optic is not flat, as is usually the case with irradiance detectors, but has a spatial extension along the optical axis pointing to the calibration lamp (Figure 9.21). Irradiance lamps

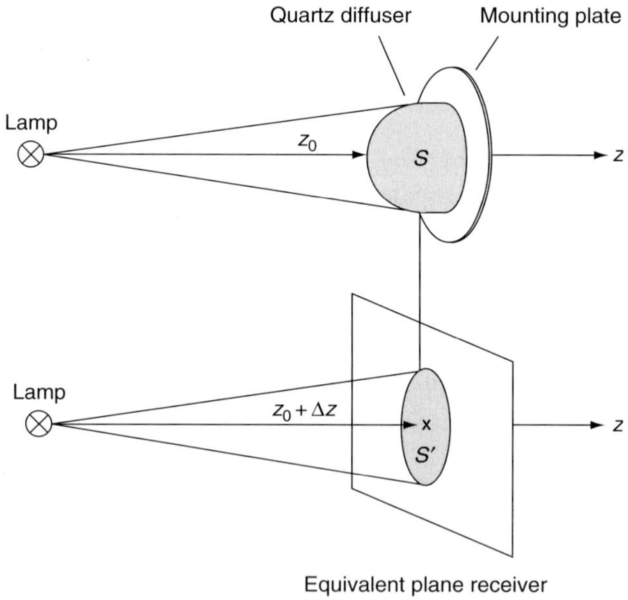

Figure 9.21 Three-dimensional receiver irradiated by a calibration lamp at the working distance z_0 from a reference point (tip of the outer quartz dome) of the receiver. S is the outer surface of the diffuser. The real detector can be thought to behave like a virtual plane receiver with an active surface S', being a projection of S. The displacement Δz of the plane of S' relative to the reference point has a distinct value that is a fixed property of the real receiver (from Hofzumahaus *et al.*, 1999, used with permission of Optical Society of America).

behave as point sources for which the irradiance is proportional to the inverse square of the distance. For this reason the lamp output is certified for a prescribed distance from the lamp (usually 50–70 cm). The longitudinal dimension of an actinic-flux receiver is typically 3 cm (Figure 9.19). Thus, the irradiance from a lamp, operated at a distance of 60 cm, varies about 10% from the top to the bottom of the receiver. The practical problem is to find the correct working distance between the lamp and the receiver that corresponds to the certified irradiance values of the lamp.

The problem is solved by an equivalent plane-receiver (EPR) concept (Hofzumahaus *et al.*, 1999). The three-dimensional receiver can be thought to be replaced by a plane receiver, which is irradiated at normal incidence and has the following properties (Figure 9.21).

1. The EPR and the real detector exhibit the same absolute detection sensitivity $D_0(\lambda)$ with respect to a radiation source located at a working distance z_0 from a reference point of the real receiver. Both receivers yield the same measurement signal I_λ when irradiated by a calibration lamp.
2. The active surface S' of the EPR is a projection of the surface S of the real receiver onto an imaginary plane perpendicular to the optical axis z.
3. The imaginary plane of the EPR has a well-defined position, being displaced relative to the reference point of the quartz diffuser by a fixed length Δz along the optical axis. Δz is a specific property of the real receiver and is independent of z_0.

This concept is valid with a small error of less than 1%, if the longitudinal dimension of the diffuser is less than 6% of the working distance (see Appendix A.5). The validity of the EPR concept has two important implications:

1. Since the real receiver behaves like an EPR, Δz can be determined experimentally from I_λ values, which are measured at different distances z_0 and follow the inverse-square law:

$$I_\lambda = \frac{\text{const.}}{(z_0 + \Delta z)^2} \tag{9.39}$$

   If the measured data are plotted as $I_\lambda^{-1/2}$ versus z_0, Δz can be extrapolated from a linear fit for the condition $I_\lambda^{-1/2} = 0$.
2. Once that Δz is known, the spectroradiometer can be calibrated accurately by using Equation 9.38, provided that the working distance z_0 is set equal to $(z^* - \Delta z)$ where z^* is the prescribed distance for which the irradiance standard is certified.

In practice, the most accurate calibrations are performed in the laboratory using 1000 W tungsten–halogen quartz lamps (FEL type) as irradiance standards. For best accuracy, the lamps need to be operated with highly stable power supplies that keep the lamp current constant within about 10^{-5} of the value specified in the lamp certificate (see, for example, Gardiner & Kirsch, 1995). The error of the absolute calibration obtained in the laboratory is typically around ±5%. It is mainly caused by the uncertainty of the irradiance standard certification (typ. 4%, in the UV, 2σ) and by the determination of the correct working distance (typ. 1–2%) (e.g., Hofzumahaus *et al.*, 1999; Eckstein *et al.*, 2003; Edwards &

Monks, 2003; Jäkel *et al.*, 2005). In the field, varying environmental conditions may influence the stability of the radiometer calibration. Possible sensitivity changes can be monitored by periodical field checks using secondary calibration standards (e.g. 200 W lamps).

9.4.5.3 Wavelength calibration

The wavelength scale of a spectroradiometer is usually calibrated by measuring the spectrum of atomic emission-lines at well-known wavelengths (e.g. from a low-pressure mercury lamp). Monochromator systems can measure the spectra by scanning the monochromator wavelength with small incremental steps, which are much narrower than the spectrometer bandwidth. The resulting spectrum represents the slit function of the monochromator, which normally looks triangular- or Gaussian-shaped and possesses a well-defined maximum (cf. Appendix A.3). The position of the maximum can then be assigned to the known wavelength of the emission line. The resulting wavelength accuracy for the scanning spectrometers in Table 9.5 is generally better than 0.1 nm.

When a spectrograph system measures an atomic emission line, the measured spectrum will display the instrumental line shape sampled with the resolution given by the pixel-to-pixel distance. The maximum of the observed line shape is then assigned to the atomic emission wavelength. If the number of pixels covering the line width is too small, the position of the maximum cannot be determined accurately. In this case, it makes sense to fit a function (e.g. parabola or gaussian) to the measured line spectrum to obtain an improved estimate of the line position. For example, the pixel distance of the spectrograph in Figure 9.18 is only 0.8 nm, but a wavelength accuracy of less than 0.3 nm can be specified (Zeiss GmbH, 1998).

In field applications, the stability of the wavelength calibration is an important instrumental feature. Stability can be provided by a temperature stabilisation of scanning monochromators, which are generally susceptible to temperature effects of their mechanical wavelength-drive. Spectrographs are inherently more stable because they have no moving parts. The instrument in Figure 9.18, for example, has a ceramic body to which all optical components have been cemented, resulting in a negligible temperature drift of $<0.0005\,\mathrm{nm}\cdot\mathrm{K}^{-1}$ (Zeiss GmbH, 1998). The wavelength accuracy of measured atmospheric spectra can be additionally controlled by correlating the Fraunhofer structure in the measured spectra versus reference spectra obtained from the extraterrestrial solar spectrum. Different mathematical procedures have been developed for that purpose by different research groups (see for example McKenzie *et al.*, 1992; Slaper *et al.*, 1995).

9.4.6 Determination of photolysis frequencies

Photolysis frequencies can be calculated from measured actinic-flux spectra according to Equation 9.29. First, the spectral distribution of the photolysis frequency is calculated, $\sigma(\lambda) \times \phi(\lambda) \times F_\lambda(\lambda)$, which is then integrated (summed up) to obtain the respective j-value. The required absorption cross sections, σ, and quantum yields, ϕ, are generally taken from the literature (see references in Section 9.8.4).

9.4.6.1 Examples

Figure 9.22 shows two examples of actinic-flux spectra that were measured with a scanning double-monochromator system on a cloud-free day in Boulder, Colorado, in June 1998. The wavelength dependence between 280 and 420 nm is displayed for two solar zenith angles, $\chi = 16°$ (local noon) and 75° (late afternoon). In both spectra, the solar radiation decreases sharply at wavelengths below 320 nm, caused by atmospheric ozone absorption, with an apparent shift of the UV-B cut-off towards longer wavelength when the solar zenith angle increases. At wavelengths between 320 and 420 nm, the spectra show relative weak wavelength dependencies, with spectral structures caused by solar Fraunhofer lines.

The cross sections and quantum yields for the formation of $O(^1D)$ by ozone photolysis are shown in the upper part of Figure 9.23 together with the solar spectra from the previous figure. The lower part of Figure 9.23 shows the resulting spectral distributions of the photolysis frequency $j(O^1D)$. The photolysis is limited at short wavelength by the available solar radiation, and at long wavelength by the decrease of $\sigma \times \phi$. It is interesting to note that the nearly exponential rise of the O_3 absorption cross section towards shorter wavelength gives high spectral weight to the strongly decreasing photon fluxes. As a result, the radiation in the UV-B cut-off region (290–310 nm) makes a significant contribution to $j(O^1D)$.

Photolysis-frequency distributions of other photochemically relevant molecules are presented in Figure 9.24. It shows the different spectral weighting of the solar radiation by the individual photolysis processes. The different weighting is the main reason for different zenith-angle dependencies of the photolysis frequencies and their different

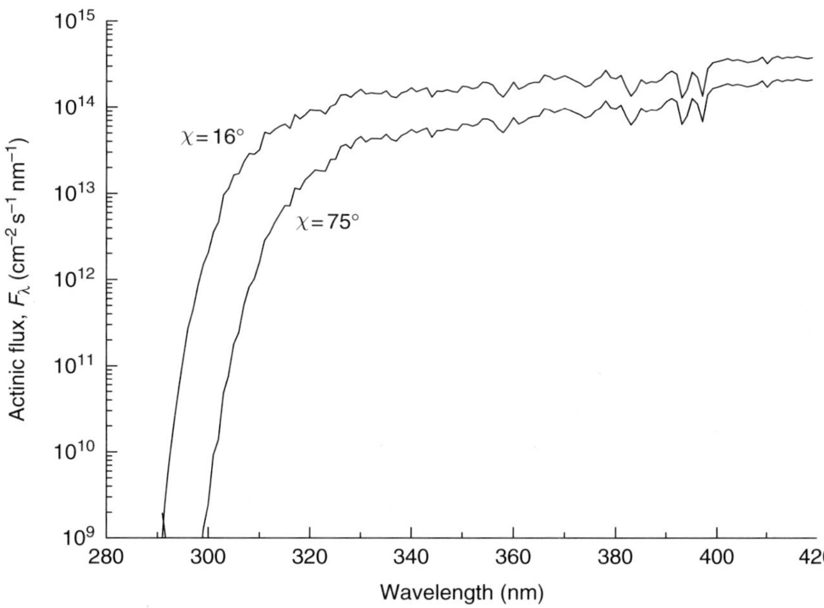

Figure 9.22 Actinic-flux spectra measured on a cloud-free day at two different solar zenith angles. The spectra were recorded with a double-monochromator–based spectroradiometer during the IPMMI field campaign in Boulder, Colorado, on 19 June, 1998, by Forschungszentrum Jülich.

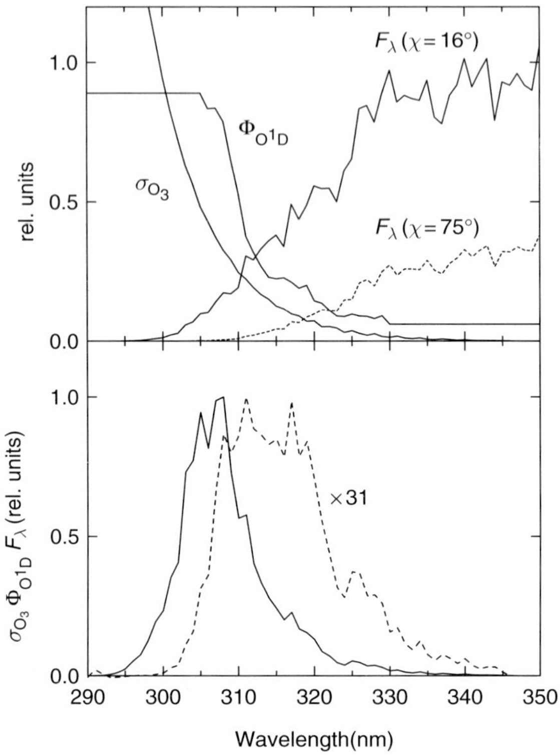

Figure 9.23 (a) Wavelength dependencies of the ozone absorption cross-section (σ_{O_3}) and O(^{1}D) quantum yield (ϕ_{O1D}) at 298 K, and the actinic flux $F_\lambda(\lambda)$ at two different zenith angles. σ_{O_3} (in units of 4.2×10^{-19} cm^2) is from Malicet *et al.* (1995), ϕ_{O^1D} is from Talukdar *et al.* (1998). $F_\lambda(\lambda)$ (in units of 1.67×10^{14} cm^{-2} s^{-1} nm^{-1}) is from Figure 9.22. (b) Spectral distribution of *j*(O^{1}D) (in units of 3×10^{-6} s^{-1} nm^{-1}) at $\chi = 16°$ (solid line) and 75° (dashed line) (adapted from Hofzumahaus *et al.*, 2004, printed with permission from American Geophysical Union).

responses to changes in total ozone, aerosol, and clouds. Figure 9.25 shows as an example the diurnal profiles of photolysis frequencies of O_3, HCHO, and NO_2, derived from the same time series of actinic-flux measurements on a partly cloudy day. The profiles of the three compounds are differently broad and show different temporal modulations caused by clouds. This demonstrates that in general photolysis frequencies of different chemical compounds cannot be scaled to each other, but must be determined individually.

9.4.6.2 Accuracy

Photolysis frequencies derived from spectroradiometer measurements have errors that arise from different sources (Table 9.6). The uncertainty of the absolute detection sensitivity ($D_0 \times Z_H$) is generally in the order of 5–8%, which is more or less wavelength independent. In contrast the spectrometer bandwidth and the wavelength error contribute differently to the uncertainty of *j*-values, depending on the spectral range where the photolysis process occurs. For example, *j*(O^{1}D) values show the largest sensitivity to

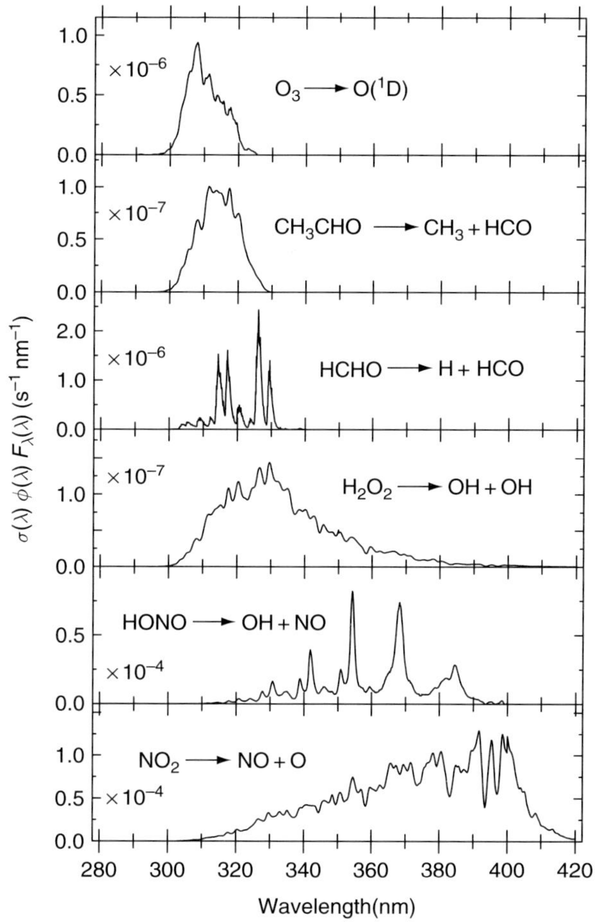

Figure 9.24 Spectral distribution of photolysis frequencies, $\sigma(\lambda) \times \phi(\lambda) \times F_\lambda(\lambda)$, for different photolysis reactions. The actinic flux $F_\lambda(\lambda)$ was measured at ground in North-East Germany at a solar zenith angle of 53° (adapted from Kraus and Hofzumahaus, 1998, with kind permission of Springer Science and Business Media).

the bandwith selection and wavelength error, because the two parameters have a strong influence on how reliable the steep spectral gradient at 290–310 nm can be measured.

Stray light is generally no problem for double-monochromator systems. Spectrograph systems, however, have a high stray-light background for which the measured spectra must be corrected (see Section 9.4.2). The residual error is generally negligible for photolysis frequencies, which are dominated by spectral contributions from above 310 nm, for example $j(NO_2)$. The accurate determination of $j(O^1D)$ values, however, is problematic and residual errors after stray-light corrections can be 10% or higher (Edwards & Monks, 2003; Kanaya *et al.*, 2003; Eckstein *et al.*, 2003; Jäkel *et al.*, 2005). Since spectrograph systems offer a number of advantages compared to scanning systems, more accurate methods for the stray-light correction in the UV-B are highly desirable and remain subject to ongoing research (e.g. Ylianttila *et al.*, 2005).

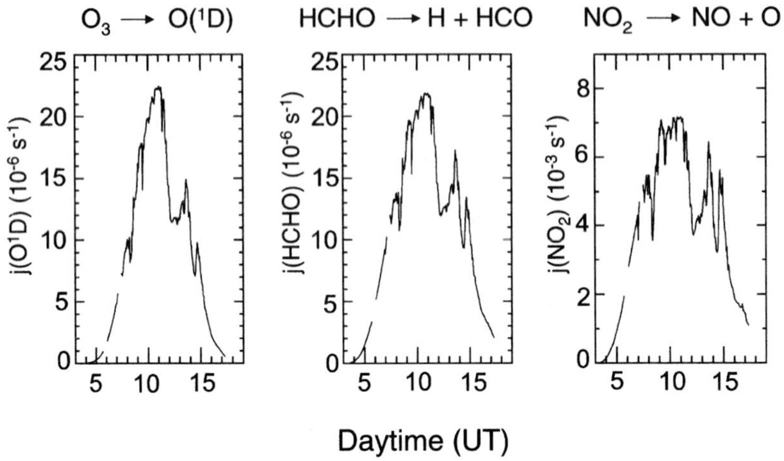

Figure 9.25 Diurnal profiles of the photolysis frequencies for O_3, HCHO, and NO_2. The *j*-values were measured at ground on a cloudy day (10 August, 1994) in North-East Germany. Local noon was at 11:20 UT (adapted from Kraus and Hofzumahaus, 1998, with kind permission of Springer Science and Business Media).

Table 9.6 Error contributions to *j*-values derived from spectral actinic-flux measurements

Source of error	Error contribution (%)		
	$j(O^1D)$	j(HCHO)	$j(NO_2)$
Absolute sensitivity		5–8[a]	
Stray light correction		Instrument dependent[b]	
Spectrometer bandwidth[c]			
1 nm	+0.8	−1.6	−0.08
2 nm	+3.1	−2.9	−0.14
Wavelength accuracy[c]			
±0.1 nm	±2.3	±0.8	±0.16
$\sigma(\lambda)$[a]	±3	±10	±4
$\phi(\lambda)$[a]	±10	±12	±10

[a] See for example Kraus and Hofzumahaus (1998); Shetter *et al.* (2003); Hofzumahaus *et al.* (2004).
[b] Negligible for double-monochromator systems.
[c] See discussion in Hofzumahaus *et al.* (1999); values apply to a solar zenith angle of 30°.

A considerable uncertainty of the photolysis frequencies comes from the errors of the molecular parameters (Kraus & Hofzumahaus, 1998). Although published absorption cross sections have been measured in some cases with high accuracy (e.g. $<3\%$ for O_3), quantum yield data are on principle more difficult to measure and usually not better known than 10%. Care must be exercised about which published data set is the most suitable for the determination of photolysis frequencies. This question has been addressed in field studies where photolysis frequencies from spectroradiometers were compared to measurements by chemical actinometers. For details, see Section 9.8.4.

9.5 Actinic-flux filter radiometry

Filter radiometry is a special case of broadband radiometry, that is the measurement of radiation in an intentionally broad spectral interval (e.g. McCluney, 1994), where an optical transmission filter is used to confine the wavelength range that is measured by a photosensitive detector. Most commercially available broadband radiometers measure irradiances with a spectrally flat response, so that the radiation falling into the sensitivity range of the instrument is measured as an integral quantity in units of $W \cdot m^{-2}$. Examples are pyranometers that are used in meteorology for the measurement of the total irradiance from the sun and the sky. There exists another class of broadband radiometers that apply spectral weighting curves resembling the photoresponse of a specific chemical, physical or biological system. An example is the widely used Robertson-Berger UV radiometers, which measure the solar irradiance with a spectral response proportional to the UV-induced erythemal (sunburning) potential of the human skin (Berger, 1976).

Actinic-flux filter radiometers belong to the class of instruments that produce a signal proportional to a photochemical effect. The concept was first introduced for the measurement of atmospheric photolysis frequencies by Junkermann *et al.* (1989). They combined a receiver optic, which collects radiation with isotropic sensitivity, with combinations of optical filters and photoelectric sensors to build instruments for the specific measurement of $j(O^1D)$ and $j(NO_2)$. The optical wavelength selection was matched closely to the spectral function $\sigma(\lambda) \times \phi(\lambda)$ of the respective photolysis frequency, resulting in an output signal proportional to j. The instruments were then calibrated absolutely by comparison against chemical actinometer measurements. Since the latter half of 1990s, actinic-flux filter radiometers have been more widely used, after they became commercially available from a manufacturer (Meteorologie Consult GmbH, Glashütten, Germany).

Another concept of broadband filter-radiometry uses a cubic array of six, flat radiation sensors to approximate isotropic detection of solar radiation (Lindh *et al.*, 1964; Nader & White, 1969; Van der Hage *et al.*, 1994; Kylling *et al.*, 2003b). Such radiometers have been used in field experiments at ground or suspended under balloons to study the actinic flux in the atmosphere, but were not designed to provide specific photolysis frequencies. Therefore, they will not be further treated here.

In principle, filter radiometers refer specifically to single photolysis reactions. The main advantage of filter-radiometer devices is that they are small and lightweight, and can be operated automatically providing continuous data sets with high time resolution. As will be discussed below, calibration in terms of photolysis frequencies is a laborious task and care has to be exercised in the signal evaluation for accurate results.

9.5.1 Filter radiometer set-up

The general construction of an actinic-flux filter radiometer is illustrated in Figure 9.26. The inlet optic collects solar radiation from one hemisphere with nearly uniform response and couples the light into a light-guide, which directs the radiation through a collimator and an optical transmission filter for detection by a photoelectric sensor. The design of the receiver optic is essentially the same as described in Section 9.4.3 for use in

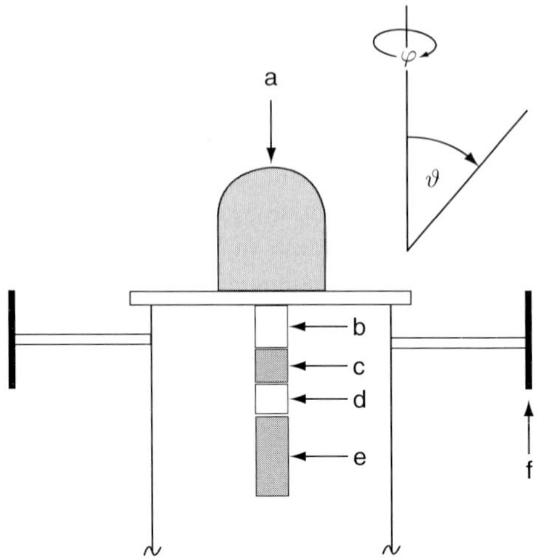

Figure 9.26 Schematic diagram of the instrumental set-up of a $j(O^1D)$ filter radiometer: a, quartz diffuser; b, quartz light-guide; c, collimator; d, optical filter; e, photomultiplier tube; f, shadow ring (from Bohn *et al.*, 2004, printed with permission from American Geophysical Union).

spectroradiometry. The components of the filter–detector combination are chosen to match the photolysis frequency of interest as close as possible. For the measurement of $j(O^1D)$, an interference filter with a centre wavelength of 300 nm and a bandpass of 10 nm (FWHM) is usually combined with a solar-blind photomultiplier (CsTe photocathode; 160–320 nm) to simulate the wavelength dependence of $\sigma_{O_3} \times \phi_{O_1D}$ in the spectral region that is relevant for the troposphere ($\lambda > 300$ nm; see Figure 9.27) (Junkermann *et al.*, 1989; Bohn *et al.*, 2004). In order to avoid a temperature-dependent wavelength-shift of the filter curve, the corresponding radiometer parts are usually kept at constant temperature.

The measurement of $j(NO_2)$ requires an optical filter with a broad bandpass of ~100 nm covering the wavelength range in which NO_2 photodissociates (cf. Figure 9.24). Since the integral photon-flux is high, a vacuum photodiode (Sb-Cs photocathode; 185–650 nm) with a relatively low detection sensitivity can be used, rather than a high-gain photomultiplier that is needed for the determination of $j(O^1D)$. Different combinations of coloured glass filters were tested by Volz-Thomas *et al.* (1996) for their suitability to mimic the spectral behaviour of $j(NO_2)$, including the filter combination (1 mm WG305 + 1 mm UG11) originally developed by Junkermann *et al.* (1989) (Figure 9.28). A significantly improved match for $j(NO_2)$ was achieved with a filter combination (2 mm UG3 + 1 mm UG5) that has been adopted for most $j(NO_2)$ filter radiometers.

The instruments normally include a transimpedance amplifier with a fixed gain that converts the photocurrent of the detector into an analogue voltage signal U which can be recorded automatically by a data logger. The response time of the radiometer can be made fast, 1 s or less, depending on the required time resolution.

The design of the filter radiometers allows measurements from one hemisphere only. Two radiometers must be employed if 4π sr photolysis frequencies are required. In this case, the instruments are oriented in opposite directions, one viewing upward and one downward as illustrated in Figure 9.8.

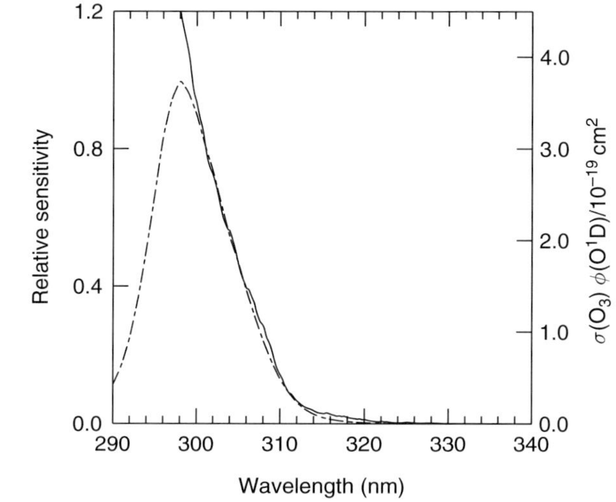

Figure 9.27 Relative spectral sensitivity D_{rel} of a $j(O^1D)$ filter radiometer (dashed line, left axis) and product $\sigma_{O_3} \times \phi_{O^1D}$ at a temperature of 295 K (full line, right axis) (Malicet *et al.*, 1995; Talukdar *et al.*, 1998). The relative wavelength dependencies of the curves agree well in the spectral region that is relevant for the troposphere at $\lambda > 300$ nm (from Bohn *et al.*, 2004).

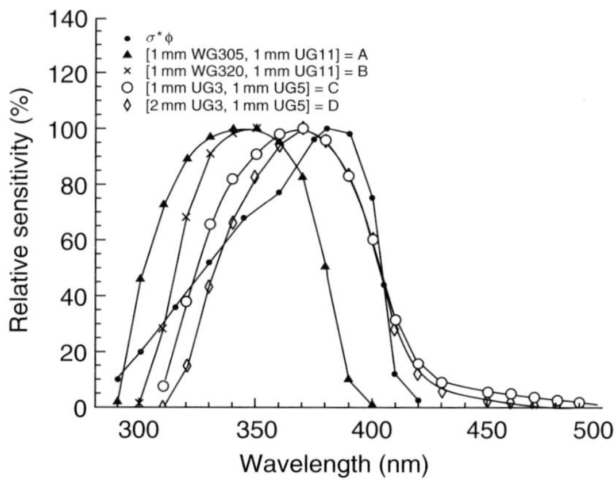

Figure 9.28 Relative spectral sensitivity D_{rel} of a $j(NO_2)$ filter radiometer for different glass filter combinations. Also shown is the product $\sigma_{NO_2} \times \phi_{NO+O}$ (Davidson *et al.*, 1988; Gardner *et al.*, 1987) (from Volz-Thomas *et al.*, 1996).

9.5.2 Determination of photolysis frequencies

The voltage signal of a broadband radiometer can be understood as the spectral signal U_λ integrated over the whole wavelength range that contributes to the signal:

$$U = \int U_\lambda(\lambda)\, d\lambda \tag{9.40}$$

The relation between U_λ and the incident actinic flux can be expressed in analogy to the description of a spectroradiometer signal I_λ (see Section 9.4.4, Equation 9.34) as:

$$U_\lambda(\lambda) = D_0(\lambda) Z_H F_\lambda(\lambda) \tag{9.41}$$

Here, the symbols $D_0(\lambda)$ and Z_H have the same meaning as in Equation 9.34, where $D_0(\lambda)$ is the absolute spectral sensitivity of the radiometer and Z_H is a correction factor for possible deviations from the isotropic response of the receiver optic. It should be noted, however, that in this case $D_0(\lambda)$ refers to a signal which is a voltage.

Assuming that Z_H is nearly wavelength independent over the spectral interval relevant for $j(O^1D)$ and $j(NO_2)$, Equations 9.40 and 9.41 can be combined as follows:

$$U = Z_H \int D_0(\lambda) F_\lambda(\lambda)\, d\lambda \tag{9.42}$$

$D_0(\lambda)$ can be further factorised into a dimensionless function $D_{rel}(\lambda)$, which represents the relative sensitivity of the filter–detector combination as a function of wavelength, and an absolute scaling factor D_{abs}:

$$D_0(\lambda) = D_{abs} D_{rel}(\lambda) \tag{9.43}$$

If this expression is inserted into Equation 9.42, the following relationship between U and the actinc flux is obtained:

$$U = D_{abs} Z_H \int D_{rel}(\lambda) F_\lambda(\lambda) d\lambda \tag{9.44}$$

In this equation the product $D_{abs} \times Z_H$ is a wavelength-independent factor, while the integral represents the total actinic flux weighted by the spectral radiometer response.

Assuming that the photolysis process of interest is driven by the same radiation that is responsible for the radiometer output, the photolysis frequency j will be proportional to the radiometer signal U:

$$j = AU \tag{9.45}$$

If the factor A is known, then j can be derived from U according to this equation. In the most simple and ideal case, A has a constant value that can be obtained by a calibration. However, in practice, A is more or less dependent on ambient parameters (e.g. solar zenith angle or air temperature), which must be considered in the accurate evaluation of the j-values.

In order to understand which quantities influence the value of A, Equations 9.44 and 9.12 can be inserted into Equation 9.45 which yields the following equation for A:

$$A = \frac{j}{U} = \frac{1}{D_{abs} Z_H} \frac{\int \sigma(\lambda)\phi(\lambda) F_\lambda(\lambda)\, d\lambda}{\int D_{rel}(\lambda) F_\lambda(\lambda)\, d\lambda} \tag{9.46}$$

The expression on the right side of the equation will be constant only, if the product $D_{abs} \times Z_H$ has a fixed value and the ratio of the two integrals is invariable. The latter condition applies if

- the radiometer response and the cross section of the photodissociation show the same relative wavelength dependence, that is $D_{rel}(\lambda) \propto \sigma(\lambda)\phi(\lambda)$; or
- the shape of the actinic-flux spectrum $F_\lambda(\lambda)$ remains the same, except for a wavelength-independent scaling factor, for all atmospheric conditions.

Neither of these conditions is fulfilled. In all practical cases, the spectral response of the radiometer does not match exactly the shape of $\sigma(\lambda) \times \phi(\lambda)$. As a result, the ratio of the photolysis frequency and the radiometer signal varies as the spectral distribution of the actinic flux changes with ambient conditions. The most influential parameters that need to be considered are the solar zenith angle χ, the total atmospheric ozone t_{O_3} (if radiation below 320 nm is significantly involved), and the measurement altitude h (Junkermann *et al.*, 1989; Volz-Thomas *et al.*, 1996; Bohn *et al.*, 2004). A further general complication arises, if $\sigma \times \phi$ or $D_{abs} \times D_{rel}$ depends on temperature. Then the corresponding influence of the air temperature, T, and the radiometer temperature, T_R, on the ratio $A = j/U$ must be considered also.

Accordingly, A is a variable quantity that can be generally described by a non-linear function of different parameters p_i:

$$A = A(p_i) \quad \text{with} \quad p_i = \chi, t_{O_3}, T, T_R, h \tag{9.47}$$

For practical reasons A is separated into a constant factor A_0 at selected reference conditions ($p_i^0 = \chi^0, t_{O_3}^0, T^0, T_R^0, h^0$) and a dimensionless function f describing the relative dependence on the parameters p_i:

$$A = A_0 f(p_i) \tag{9.48}$$

with

$$A_0 = A(p_i^0) \quad \text{and} \quad f(p_i^0) = 1.0 \tag{9.49}$$

Using these definitions, j can be related to the radiometer signal U by

$$j = A_0 f(p_i)\, U \tag{9.50}$$

Here, A_0 can be understood as a constant linear calibration factor for selected reference conditions and f as a correction function accounting for the non-linear deviations in the relationship between j and U when the ambient parameters p_i change relative to the reference conditions. As will be shown below, typical f values reported for $j(O^1D)$ radiometer applications in the troposphere deviate from unity by up to a factor 2, but usually less than 10% for $j(NO_2)$ radiometers (Sections 9.5.4 and 9.5.5).

9.5.3 Filter radiometer calibration

The determination of the quantities A_0 and f needed for the evaluation of Equation 9.50 is a laborious task which requires:

- the measurement of the instrumental response functions, $Z_p(\vartheta)$ and $D_{rel}(\lambda)$,
- an absolute calibration of A_0 or D_{abs},
- knowledge of the molecular parameters $\sigma(\lambda)$ and $\phi(\lambda)$, and
- a set of modelled actinic-flux spectra for different solar zenith angles and, where necessary, for a range of values of other atmospheric parameters.

9.5.3.1 Instrumental characteristics

The instrumental characteristics Z_p, D_{rel}, and D_{abs} can be obtained from laboratory measurements. The angular-response correction of a filter radiometer can be determined in a similar way as in the case of a spectroradiometer. The Z_p values are measured by rotating a lamp with broadband emission over the field-of-view of the receiver. The angular-correction factor Z_H is then derived from the measured Z_p values as outlined in Appendix A.4, using a model for the angular distribution of the atmospheric radiance. See Sections 9.4.3 and 9.4.5 for further details.

The spectral response function $D_{rel}(\lambda)$ can be either estimated from the specified wavelength-dependencies of the filter transmission and the detector sensitivity, or measured preferably for the complete radiometer using a tunable narrow-bandwidth light source. The tunable light source can be realised, for example, by a combination of a broadband Xe-arc lamp and a scanning monochromator, which is calibrated against a reference photodetector having a known spectral sensitivity (Bohn *et al.*, 2004).

When required, D_{abs} can be measured by exposing the filter instrument to the radiation of a calibration lamp. The lamp is positioned relative to the inlet optic of the radiometer as described for an actinic-flux spectroradiometer (Section 9.4.5). A value for D_{abs} is then obtained by measuring the radiometer signal U_{cal} when the instrument receives the specified irradiance E_λ (Bohn *et al.*, 2004):

$$D_{abs} = \frac{U_{cal}}{\int D_{rel}(\lambda)\, E_\lambda(\lambda)\, d\lambda} \qquad (9.51)$$

9.5.3.2 Instrumental correction function

The factor f needed for the correction of the filter radiometer data can be determined from Equation 9.48 as:

$$f(\chi, t_{O_3}, T, T_R, h) = \frac{A(\chi, t_{O_3}, T, T_R, h)}{A_0} \qquad (9.52)$$

When A is replaced by Equation 9.46, f can be represented by the product of four factors:

$$f(\chi, t_{O_3}, T, T_R, h) = f_1 f_2 f_3 f_4 \qquad (9.53)$$

where

$$f_1(\chi) = \frac{Z_H(\chi^0)}{Z_H(\chi)} \tag{9.54}$$

$$f_2(T_R) = \frac{D_{abs}(T_R^0)}{D_{abs}(T_R)} \tag{9.55}$$

$$f_3(T, \chi, t_{O_3}, h) = \frac{\int \sigma(\lambda, T)\,\phi(\lambda, T)\,F_\lambda\,d\lambda}{\int \sigma(\lambda, T^0)\,\phi(\lambda, T^0)\,F_\lambda{}^0\,d\lambda} \tag{9.56}$$

$$f_4(T_R, \chi, t_{O_3}, h) = \frac{\int D_{rel}(\lambda, T_R^0)\,F_\lambda{}^0\,d\lambda}{\int D_{rel}(\lambda, T_R)\,F_\lambda\,d\lambda} \tag{9.57}$$

with $F_\lambda = F_\lambda(\lambda; \chi, t_{O_3}, h)$ and $F_\lambda{}^0 = F_\lambda(\lambda; \chi^0, t_{O_3}^0, h^0)$

Here, f_1 represents the relative influence of the solar zenith angle on the angular-response correction; f_2 describes the relative dependence of the absolute detector sensitivity on the radiometer temperature; f_3 considers the influence of the air temperature and ambient parameters (χ, t_{O_3}, h) on the spectral overlap between $\sigma(\lambda) \times \phi(\lambda)$ and $F_\lambda(\lambda)$; and, similarly, f_4 considers the influence of the radiometer temperature and the ambient parameters on the spectral overlap between $D_{rel}(\lambda)$ and $F_\lambda(\lambda)$.

In order to calculate the factors $f_i (i = 1, 2, 3, 4)$, values for all relevant quantities on the right sides of Equations 9.54–9.57 must be available. Values of σ and ϕ may be taken from the literature (see Section 9.8.4), whereas the behaviour of Z_H, D_{rel} and D_{abs} must be derived from laboratory measurements (see p. 458). Furthermore, a set of solar actinic-flux spectra simulated by a radiative transfer model for different ambient conditions is required for the determination of f_3 and f_4. Examples of f_i factors for $j(O^1D)$ and $j(NO_2)$ filter radiometers will be given in Sections 9.5.4 and 9.5.5.

9.5.3.3 *Absolute calibration*

Filter radiometers can be calibrated absolutely in two possible ways: either by a direct field comparison versus another absolute measurement technique, or radiometrically against the intensity of a calibration lamp in the laboratory.

In the first approach, the filter radiometer is operated side-by-side with a reference instrument (e.g. a chemical actinometer) that delivers absolute photolysis frequencies. The calibration factor A_0 is then directly obtained from a linear regression of the measured j-values plotted versus the product $U \times f(p_i)$, where U is the filter-radiometer signal and $f(p_i)$ the correction factor for the ambient conditions during the calibration (cf. Equation 9.50).

In the second approach, A_0 is determined from the quantities obtained in the laboratory (D_{abs}, Z_H, D_{rel}) and the actinic-flux spectrum simulated for the selected standard conditions (cf. Equation 9.46):

$$A_0 = \frac{1}{D_{abs}(T_R^0)\,Z_H(\chi^0)}\,\frac{\int \sigma(\lambda, T^0)\,\phi(\lambda, T^0)\,F_\lambda{}^0\,d\lambda}{\int D_{rel}(\lambda, T_R^0)\,F_\lambda{}^0\,d\lambda} \tag{9.58}$$

This approach was applied for a $j(O^1D)$ filter radiometer (Bohn *et al.*, 2004) and gave good agreement in a field comparison against chemical actinometry and spectroradiometry (Hofzumahaus *et al.*, 2004). The most convenient calibration method for filter radiometers, however, remains the direct field comparison against calibrated reference instruments.

9.5.4 $j(O^1D)$ filter radiometers

The photolysis of ozone leading to the formation of $O(^1D)$ atoms occurs predominantly at wavelengths below 320 nm. In this wavelength region, the solar radiation has a spectral distribution that is strongly dependent on the solar zenith angle and total atmospheric ozone. These two parameters have a strong influence on the calibration of a $j(O^1D)$ filter radiometer, even if the spectral mismatch between D_{rel} and $\sigma \times \phi$ is small. Since the photodissociation of O_3 is strongly temperature dependent, it is even in principle not possible to match $\sigma \times \phi$ exactly with a single filter–detector combination for all conditions. Thus, a significant dependence on the parameters χ, t_{O_3} and T exists for the calibration of a $j(O^1D)$ filter radiometer.

The effect of an imperfect spectral match is demonstrated in Figure 9.29 for a filter radiometer operated at ground (Bohn *et al.*, 2004). Panel (a) reproduces the data from Figure 9.27 on a logarithmic scale, revealing a significant difference in the spectral progression of D_{rel} and $\sigma \times \phi$ at long wavelengths ($\lambda > 315$ nm). The difference is more strongly weighted by the solar spectrum when the zenith angle is high, leading to a substantially different dependence of $j(O^1D)$ and U on χ (see panels (b, c, d)). This behaviour was studied by corresponding field measurements of $j(O^1D)$ and U. Although the two quantities appear to be well linearly correlated (panel (e)), a closer look reveals a significant change of their ratio by a factor 2, when χ varies from 30° to 80° (panel (f)) (Bohn *et al.*, 2004). Similar variations of $j(O^1D)/U$ were found for other $j(O^1D)$ filter radiometers (Junkermann *et al.*, 1989; Hofzumahaus *et al.*, 1992; Shetter *et al.*, 1996; Brauers *et al.*, 1998). The functional dependence on χ, however, looked different for each instrument, because the spectral characteristics of the transmission filters and photosensors were not exactly the same.

In order to understand the variability of $A = j(O^1D)/U$ for a specific filter radiometer, the correction function $f = f_1 f_2 f_3 f_4$ must be determined as discussed in the previous section. For an instrument operated at ground, the factor f_1 has usually a value near unity ($f_1 \approx 1$), because Z_H exhibits only a weak dependence on χ in the UV-B spectral region in the lower troposphere (Hofzumahaus *et al.*, 1999). The factor f_2 is also essentially unity, since $j(O^1D)$ filter radiometers are normally temperature stabilised to avoid unwanted wavelength shifts of the interference-filter transmission curve. The stabilisation also prevents temperature drifts of the photomultiplier sensitivity. Thus, only the correction factors f_3 and f_4 remain to be considered with respect to variations of the air temperature, solar zenith angle and total ozone:

$$f(\chi, t_{O_3}, T) \approx f_3(\chi, t_{O_3}, T) f_4(\chi, t_{O_3}) \tag{9.59}$$

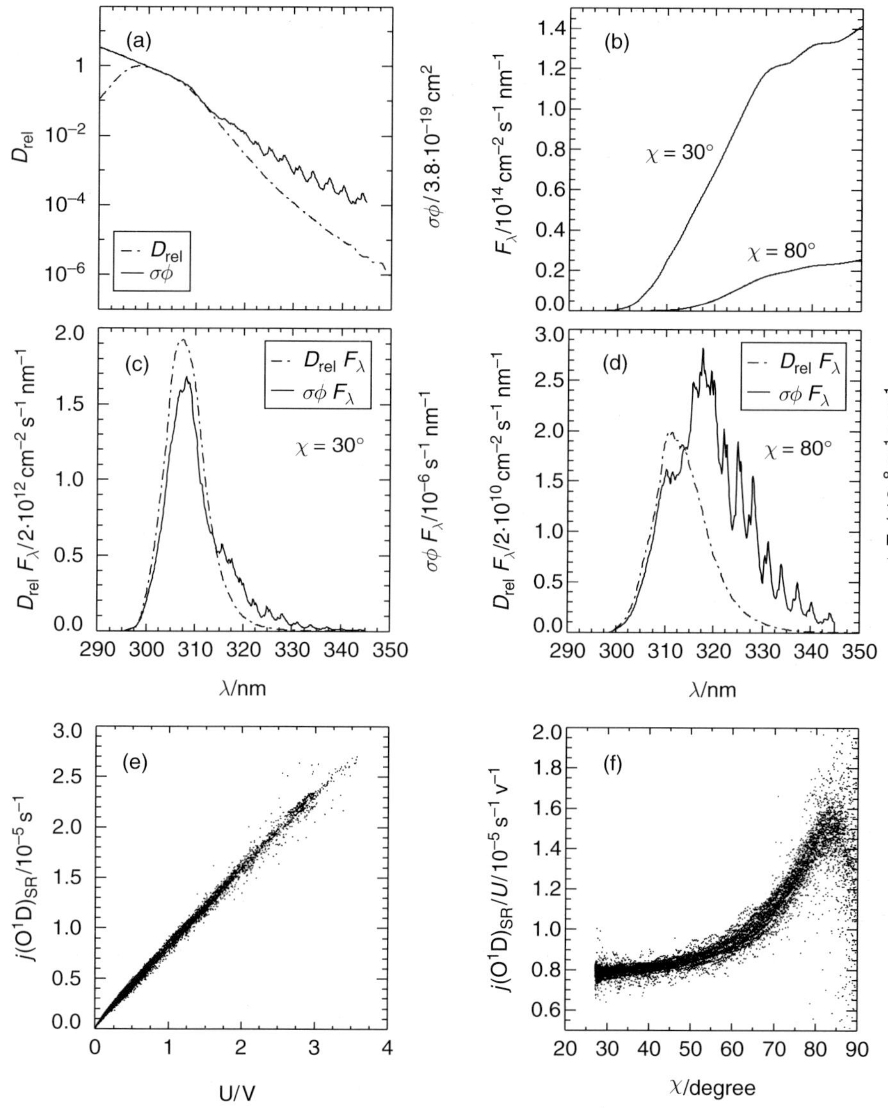

Figure 9.29 Demonstration of the effects caused by an imperfect matching between D_{rel} and $\sigma\phi$ for a $j(O^1D)$ filter radiometer. (a) Wavelength dependence of D_{rel} and $\sigma\phi$ plotted on a logarithmic scale (data from Figure 9.27). (b) Simulated actinic-flux spectra at two different solar zenith angles $\chi = 30°$ and $\chi = 80°$ at a total ozone column $t_{O_3} = 360\,\mathrm{DU}$. (c), (d) Spectral distributions of $D_{rel}F_\lambda$ and $\sigma\phi F_\lambda$ at $\chi = 30°$ and $\chi = 80°$, using the data from panels (a) and (b). (e) Correlation of $j(O^1D)$ field data measured by a spectroradiometer with the simultaneously recorded filter-radiometer signal U. (f) Ratio of $j(O^1D)$ and U as a function of χ, using the data from panel (e) (from Bohn *et al.*, 2004).

The relative temperature dependence of the photodissociation of O_3 is of general interest and can be described separately by the factor b:

$$b = \frac{j(O^1D)(\chi, t_{O_3}, T)}{j(O^1D)(\chi, t_{O_3}, T^0)} \tag{9.60}$$

Thus $j(O^1D)$ can be calculated from the measured voltage U by:

$$j(O^1D) = A_0 f(\chi, t_{O_3}, T^0)\, b(\chi, t_{O_3}, T)\, U \tag{9.61}$$

Figure 9.30 shows as an example $f(\chi, t_{O_3}, T^0)$ values that apply to the spectral radiometer sensitivity shown in Figure 9.29(a). Here f was determined from measured instrumental parameters and modelled actinic-flux spectra for different solar zenith angles ($\chi = 0°–87°$), ozone columns ($t_{O_3} = 240–460$ DU), and a constant air temperature ($T^0 = 295$ K) (Bohn *et al.*, 2004). Figure 9.31 demonstrates that the application of these f values yields good agreement of filter radiometer–derived $j(O^1D)$ values with measurements by a spectroradiometer over a broad range of zenith angles and a highly variable cloud cover. Note that in this example A_0 was absolutely calibrated against a certified irradiance standard using Equations 9.51 and 9.58.

It should be noted that the discussion of $j(O^1D)$ filter radiometers in this section refers to a ground-based operation only. If the instruments are deployed on an aircraft, the correction factors f and b need to be evaluated for the altitude at which measurements

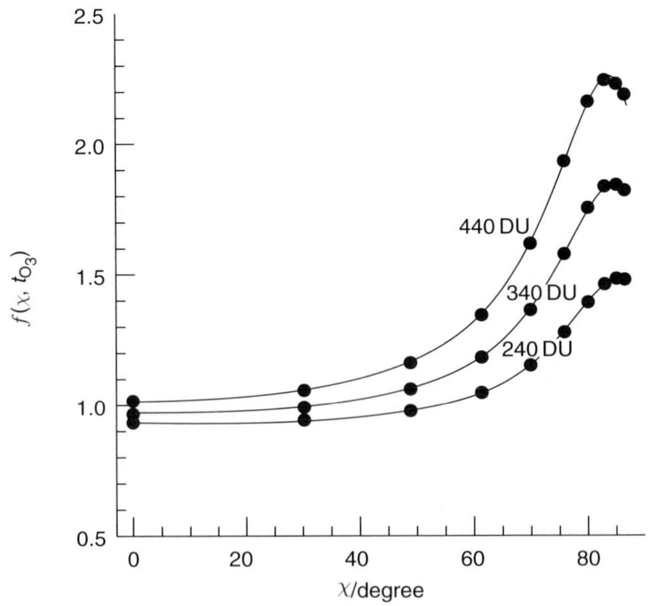

Figure 9.30 Example of correction factors $f(\chi, t_{O_3}, T^0)$ (solid circles) that apply to the spectral filter-radiometer sensitivity shown in Figure 9.29(a). The data refer to an air temperature $T^0 = 295$ K and ground level, and are normalised to the reference conditions $t_{O_3} = 350$ DU and $\chi = 30°$. The full lines show a parametrisation of the data points (from Bohn *et al.*, 2004).

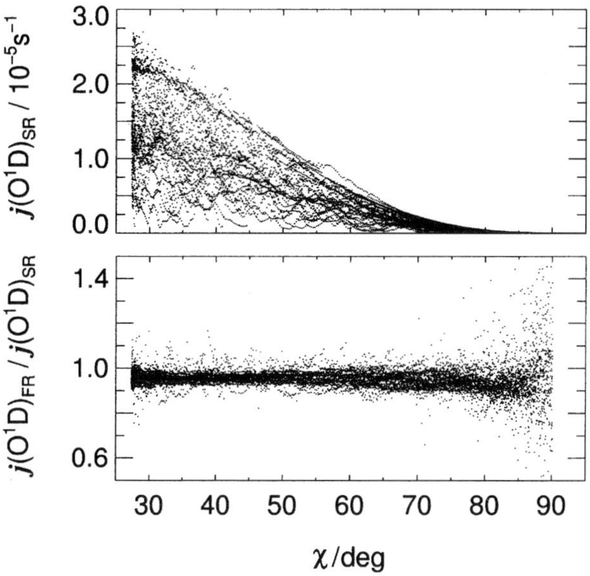

Figure 9.31 Upper panel: Dependence of $j(O^1D)$ on the solar zenith angle, measured by an actinic-flux spectroradiometer (SR) in Jülich, Germany, from 10 to 27 June, 1997. The strong variability of $j(O^1D)$ at given zenith angles is mainly caused by clouds and to some extent by total ozone (315–380 DU). Lower panel: Ratios of simultaneous $j(O^1D)$ measurements obtained from a filter radiometer (FR) and actinic-flux spectroradiometer (upper panel). For both data sets an air temperature of 295 K was assumed (adapted from Bohn *et al.*, 2004).

are performed, requiring an assumption on the partitioning of the total ozone column above and below the measurement altitude.

9.5.4.1 $j(O^1D)$ temperature dependence

The relative dependence of $j(O^1D)$ on the air temperature, represented by the factor b (Equation 9.60), was calculated and parameterised for sea level conditions by Bohn *et al.* (2004) from published ozone absorption cross-sections (Malicet *et al.*, 1995), $O(^1D)$ quantum yields (Talukdar *et al.*, 1998), and simulated actinic-flux spectra for a range of solar zenith angles ($\chi = 0°–87°$), total ozone columns ($t_{O_3} = 240–460$ DU), and temperatures (270–320 K). Figure 9.32 shows two examples of b as a function of T for the maximum and minimum values of total ozone and solar zenith angles covered by the simulated spectra. The two curves represent the minimum and maximum temperature dependence of $j(O^1D)$ within the limit of these conditions and correspond to temperature sensitivities in the order of 0.5–1% K^{-1}. The parametrisation of b and its numerical coefficients can be found in the work of Bohn *et al.* (2004).

9.5.4.2 Accuracy

The accuracy of $j(O^1D)$ derived from filter radiometer measurements is determined by the errors of the factors A_0, f and b entering Equation 9.61. The error of A_0 is mainly

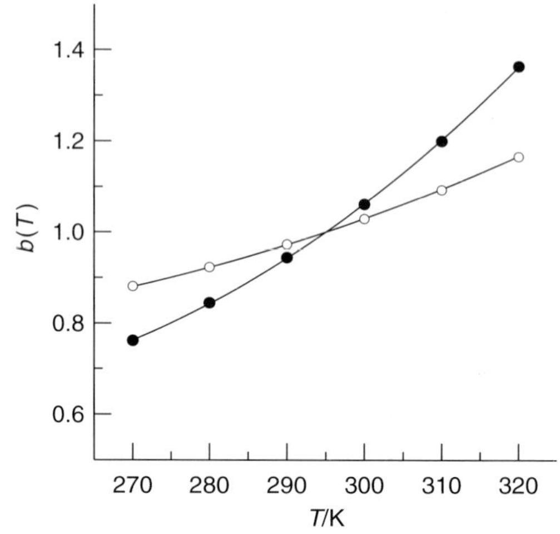

Figure 9.32 Relative temperature dependence $b = j(O^1D)(T)/j(O^1D)(295\,K)$ at the lowest and highest ozone columns considered, combined with low and high zenith angles, respectively, marking the weakest and strongest dependence within the ranges covered. Open symbols: $t_{O_3} = 240\,DU$, $\chi = 0°$; closed symbols: $t_{O_3} = 460\,DU$, $\chi = 85°$. The solid lines are parameterisations of the modelled data points (from Bohn *et al.*, 2004).

caused by the reference technique against which A_0 is calibrated. Possible reference techniques are chemical actinometry (see Section 9.3.4), double monochromator–based spectroradiometry (see Section 9.4.6), or calibration against a certified irradiance standard (see Bohn *et al.*, 2004). In all cases the total uncertainty is quite similar and lies typically in the range of 10–15%. An additional error is contributed by the factors f and b, which correct for non-linearities of the filter-radiometer calibration. If the corrections are ignored and set to unity, errors can be as high as a factor of 2. If the correction factors are applied correctly, remaining uncertainties in the range of 4% at $\chi < 70°$ and 10% at $70° \leq \chi \leq 85°$ are possible (Bohn *et al.*, 2004) and add to the total error.

9.5.5 $j(NO_2)$ filter radiometers

The evaluation of filter-radiometer data for $j(NO_2)$ is less complicated than for $j(O^1D)$. Unlike for O_3, the photolysis of NO_2 depends only weakly on temperature and is dominated by radiation that is not influenced by the total ozone column. In the relevant wavelength range, the variation of the spectral actinic-flux distribution is mainly controlled by the solar zenith angle, but the effect is much weaker than in the UV-B region. Therefore, the spectral mismatch of the radiometer sensitivity to the photodissociation spectrum of NO_2 has a relative small effect on the radiometer calibration. The influence is typically less than 10% and depends mainly on the solar zenith angle (Volz-Thomas *et al.*, 1996).

Figure 9.33 shows as an example the function $f(\chi)$ that is required to correct for the influence of the spectral mismatch at clear-sky conditions. The correction is plotted for the filter combinations A and D from Figure 9.28 as a function of the solar zenith angle,

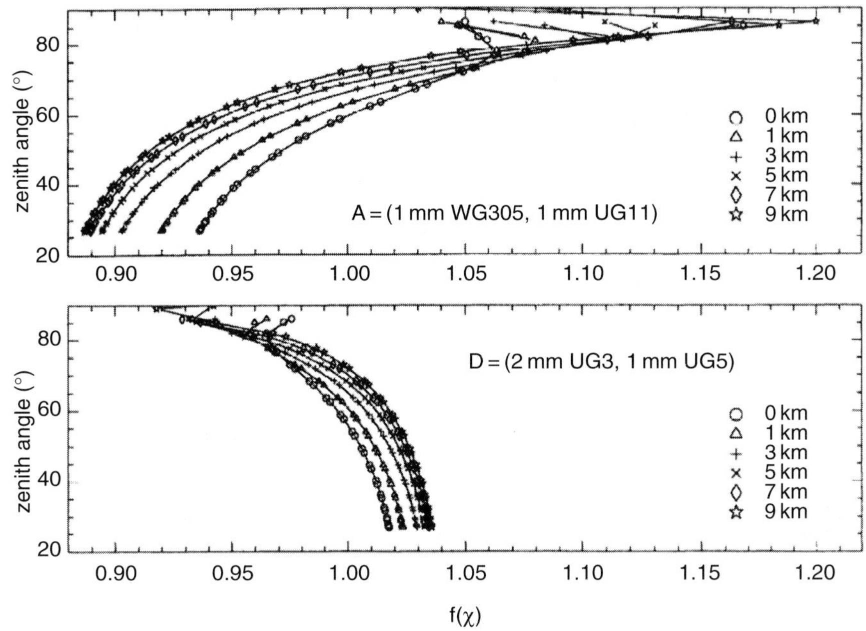

Figure 9.33 Correction factors f as a function of the solar zenith angle at different altitudes for the filter combinations A and D from Figure 9.28. The calculations assume clear-sky conditions (adapted from Volz-Thomas *et al.*, 1996, printed with permission from American Geophysical Union).

and for airborne applications at different altitudes in the troposphere. The correction factors for the preferred filter combination D vary by 6% at ground and by 12% at 9 km altitude when the zenith angle changes between 25° and 90°. The corresponding altitude dependence is small and generally less than 5% at a given zenith angle.

It should be noted that the blocking of direct sunlight by clouds has a relative large influence (3–5%) on the f values of a $j(NO_2)$ radiometer (Volz-Thomas *et al.*, 1996). The reason is that a relative large fraction of direct radiation (up to 50%) contributes to $j(NO_2)$ and has a significantly different spectral composition compared to diffuse radiation. Since it is impractical to correct routinely for such cloud effects, the influence of clouds contributes to the overall uncertainty of the $j(NO_2)$ values.

Temperature has an additional, but small, influence on the $j(NO_2)$ filter-radiometer calibration. In a laboratory experiment, the radiometer sensitivity was found to increase by 2.5% when the temperature was lowered from 298 to 233 K (Volz-Thomas *et al.*, 1996). For most ground-based applications, where temperature variations are much smaller, this effect can be neglected.

Temperature has also a weak influence on the photodissociation of NO_2. This dependence was studied experimentally with chemical actinometers in the laboratory (Dickerson *et al.*, 1982) and in field experiments (Shetter *et al.*, 1988), giving very similar results in both studies. Accordingly, the relative temperature dependence in the range of 227–400K can be described as

$$b = \frac{j(NO_2)(T)}{j(NO_2)(T^0)} = \exp\left(\frac{58\,\mathrm{K}}{T^0} - \frac{(58 \pm 6)\mathrm{K}}{T}\right) \tag{9.62}$$

(Volz-Thomas *et al.*, 1996), corresponding to a sensitivity of about 0.064% K^{-1} at room temperature.

In summary, $j(NO_2)$ values can be derived from measured radiometer voltages U by the following expression

$$j(\mathrm{NO_2}) = A_0 f(\chi, T_\mathrm{R}, T^0, h)\, b(T)\, U \qquad (9.63)$$

where the calibration corrections include dependencies on the solar zenith angle, radiometer temperature, air temperature, and for airborne applications also altitude.

9.5.5.1 Accuracy

The accuracy of $j(NO_2)$ derived from filter-radiometer measurements is determined by the errors of the factors A_0, f and b entering Equation 9.63. The calibration constant A_0 can be derived from field comparisons with chemical actinometers or actinic-flux spectroradiometers. The uncertainty of the chemical actinometer used as a calibration reference for the commercially available filter radiometers is 5–7% (Volz-Thomas *et al.*, 1996; Kraus & Hofzumahaus, 1998), whereas spectroradiometry as a reference has an uncertainty of about 12% (see Section 9.4.6). The uncertainty caused by the correction factors for clear-sky conditions is 2.5% at $\chi < 75°$ and 6.5% at $\chi > 75°$, whereas it is about 5–7% for cloudy conditions (Volz-Thomas *et al.*, 1996).

9.6 Irradiance radiometry

Most observational information about solar UV radiation reaching the earth's surface has been obtained from ground-based radiometers that measure irradiances. Measurements are being performed since decades at different places worldwide, in order to investigate the temporal and spatial variability of solar UV and its potentially harmful impact on biological systems and the environment (see e.g. Urbach, 1969; Zerefos & Bais, 1995, and references therein). The discovery of the stratospheric ozone depletion in the 1980s has been a strong motivation for measuring possible changes in surface UV exposure. Thus, UV-monitoring networks have been established that are providing long-term measurements, for example, via the World Ozone and Ultraviolet Radiation Data Centre (WOUDC) or the European Ultraviolet Database (EUVDB) (Kerr *et al.*, 2002).

Besides UV monitoring, irradiance radiometers have been used also for the determination of photolysis frequencies in the atmosphere since the late 1970s, when actinic-flux sensing radiometers were not readily available (e.g. Sellers & Hanser, 1978; Madronich, 1987a; McElroy *et al.*, 1995). Empirical relationships between actinic fluxes and irradiances were first studied using broadband radiation measurements (Nader & White, 1969; Madronich, 1987a; Van Weele *et al.*, 1995). At the time of writing, new efforts have been made to develop suitable methods for conversion of spectrally resolved UV-irradiance measurements into actinic fluxes (Kazadzis *et al.*, 2000, 2004; McKenzie *et al.*, 2002; Webb *et al.*, 2002a; Kylling *et al.*, 2003a). The main goal of these studies is to explore how databases of surface UV-irradiance measurements can be used to establish climatologies of actinic fluxes and photolysis frequencies that are not available otherwise at present.

9.6.1 Principle of irradiance radiometry

The fundamental difference in the measurement of irradiances and actinic fluxes lies in the angular sensitivity of the receiver-optics for the collection of radiation. Irradiance instruments use flat optical plates that exhibit a cosine-dependent angular response. Photons are detected with maximum sensitivity in the direction perpendicular to the receiver surface and with zero sensitivity in the horizontal direction. Contrary, actinic-flux receivers collect radiation with isotropic sensitivity, that is with equal sensitivity independent of the angle of incidence. The different behaviours are illustrated in Figure 9.34 which shows the relative angular response of an actinic flux (A) and irradiance (B) detector with respect to direct solar radiation.

The effect of the different geometrical response on the measurement of diffuse radiation is shown in Figure 9.35. Assuming an isotropically distributed sky radiance, the upper curve $f(\vartheta_{max})$ shows the fraction of the diffuse hemispheric actinic flux that is observed when the field-of-view is restricted to polar angles between 0° and ϑ_{max}. By definition the fraction becomes unity when ϑ_{max} approaches 90°. The lower curve $e(\vartheta_{max})$ shows the corresponding behaviour of a cosine-dependent receiver. For polar angles limited to less than 30°, the detected irradiance $E(\vartheta_{max})$ is essentially equal to the actinic flux $F(\vartheta_{max})$, but it is a factor of 2 lower when the field-of-view covers the full hemisphere ($\vartheta_{max} = 90°$). Apparently half of the actinic flux arrives from polar angles between 60° and 90°, from where the cosine-dependent detector receives almost no contribution. Thus, changes in actinic flux that are caused by variations of the atmospheric radiance at large polar angles remain largely unnoticed by an irradiance-sensitive radiometer.

Different methods have been developed to derive actinic fluxes and/or photolysis frequencies from measured irradiances and to minimise the inherent uncertainties associated with the angular detector response. The methods can be generally assigned to two groups:

1. *Physical-model based methods.* These methods involve three main steps:

    (a) the measurement of irradiance spectra, $E_\lambda(\lambda)$, by a spectroradiometer;
    (b) the mathematical conversion of the measured spectral irradiances into corresponding actinic fluxes, $E_\lambda(\lambda) \rightarrow F_\lambda(\lambda)$, using a physical radiation model and an assumption of the angular atmospheric radiance distribution; and
    (c) the calculation of j-values from the estimated actinic fluxes, using the cross sections and quantum yields of the photolysis reaction of interest.

    These methods are completely analogous to actinic-flux spectroradiometry (cf. Equation 9.29), but require the additional conversion step (b) in the data evaluation.

2. *Empirical methods.* These methods involve the following steps:

    (a) the simultaneous measurement of j-values and irradiances E;
    (b) the subsequent parametrisation of the relationship between the two correlated observed quantities; and
    (c) the prediction of unknown j-values from measured irradiances by applying the previously derived parametrisation.

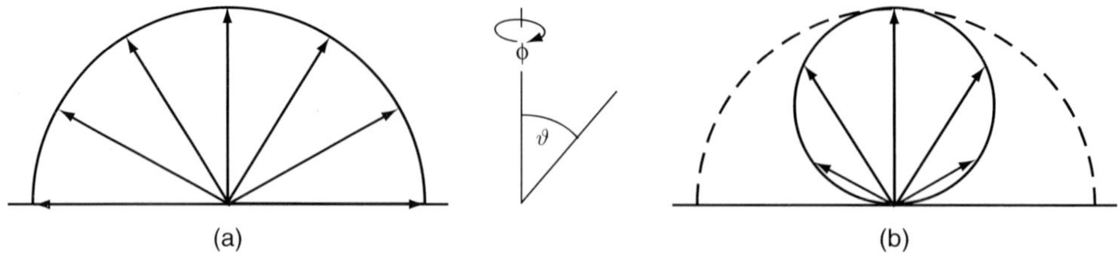

Figure 9.34 Polar diagrams of the relative angular response of radiation detectors for single-hemispheric measurements. The solid curve in (a) (and the dashed curve in (b)) represent the response of an ideal 2π sr detector for actinic flux, while the solid curve in (b) represents the response of an ideal irradiance detector (adapted from Hofzumahaus *et al.*, 1999, used with permission of Optical Society of America).

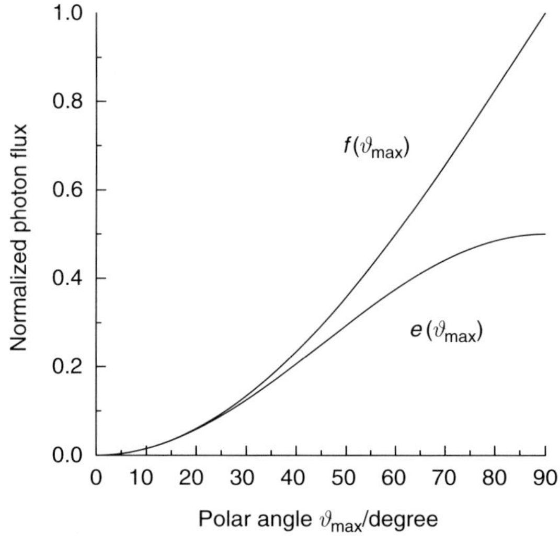

Figure 9.35 Normalised irradiance $e = E(\vartheta_{max})/F(2\pi\,\mathrm{sr})$ and actinic flux $f = F(\vartheta_{max})/F(2\pi\,\mathrm{sr})$ when the field-of-view is restricted to polar angles $\vartheta = 0° - \vartheta_{max}$. $F(2\pi\,\mathrm{sr})$ represents the actinic flux from the full hemisphere ($\vartheta_{max} = 90°$). This example assumes an isotropically distributed sky radiance without direct sunlight (adapted from Hofzumahaus *et al.*, 1999, used with permission of Optical Society of America).

Here the first two steps constitute a calibration which depends on the availability of absolute photolysis-frequency measurements as a reference. The resulting parametrisation is specific to a particular photolysis reaction and usually depends on the individual radiometer properties.

9.6.2 *Physical model–based methods*

The theoretical foundation for the mathematical conversion of atmospheric irradiances into actinic fluxes was outlined by Madronich (1987b) and Van Weele *et al.* (1995). The working equation (see Equation A.10 in Appendix A.1) can be formulated as

$$\frac{F_\lambda}{E_\lambda} = \alpha_E \left(\frac{1}{\cos \chi} - r_{\mathrm{dd}} \right) + (r_{\mathrm{dd}} + r_{\mathrm{du}} A_{\mathrm{s}})$$

where F_λ is the 4π sr actinic flux, E_λ is the downward irradiance, α_E means the fraction of the direct irradiance, r_{dd} is the ratio of the downward diffuse actinic flux to the downward diffuse irradiance, r_{du} is the ratio of the upward diffuse actinic flux to the upward diffuse irradiance, and A_{s} is the ground albedo. If one is mainly interested in the downwelling actinic flux, for example because the ground albedo is small, the equation reduces to:

$$\frac{F_\lambda}{E_\lambda} = \alpha_E \left(\frac{1}{\cos \chi} - r_{\mathrm{dd}} \right) + r_{\mathrm{dd}} \tag{9.64}$$

According to this equation, downwelling actinic fluxes can be derived from measured irradiances, if the parameters α_E and r_{dd} are known. The problem is that the two conversion parameters α_E and r_{dd} have no unique values, but depend on many variables including foremost the solar zenith angle, wavelength, aerosol abundance, and cloud cover. At wavelengths below 320 nm, there also exists a dependence on the absorption by atmospheric ozone.

The direct fraction α_E can be generally calculated as the ratio of the direct to the total downwelling (global) irradiance (see Equation A.9). At some monitoring stations both direct and global irradiance are measured routinely (e.g. Kazadzis *et al.*, 2000). However, when measurements of the direct component are not available, α_E must be obtained from radiative transfer models (e.g. Cotte *et al.*, 1997; McKenzie *et al.*, 2002; Kylling *et al.*, 2003a). The model results are usually calculated as a function of wavelength and solar zenith angle (see examples in Figure 9.36), and require first of all the column density

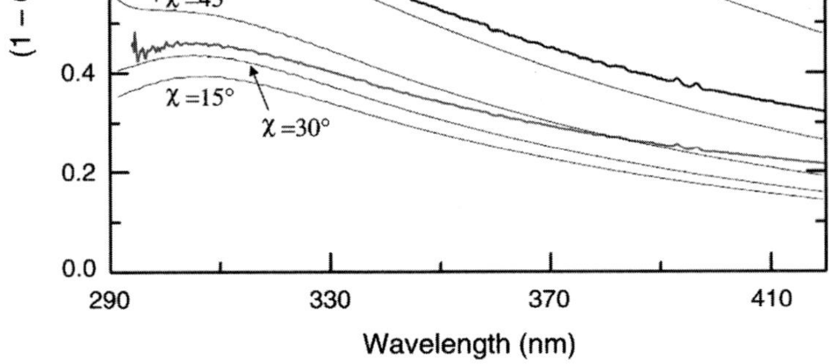

Figure 9.36 Ratio of the diffuse downward irradiance $E\downarrow$ to the global irradiance $E_\lambda = E_0 + E\downarrow$ from measurements during clear-sky and low aerosol conditions in Boulder, Colorado, on 19 June, 1998. Model values for different solar zenith angles are shown for comparison (adapted from McKenzie *et al.*, 2002, printed with permission from American Geophysical Union).

and scattering properties of the atmospheric aerosol as input parameters. When the sun is blocked by clouds, α_E is set to zero. This condition applies in good approximation when the ratio of the measured irradiance to the expected clear-sky value at 330–380 nm is less than 0.8 (McKenzie *et al.*, 2002; Kylling *et al.*, 2003a).

The diffuse ratio r_{dd} depends on the angular distribution of the diffuse downward radiance (cf. Equation A.6; see also Figure 9.4) and is generally difficult to obtain experimentally. In a first approximation it is often assumed that the downward diffuse radiance is isotropically distributed, resulting in a value of $r_{dd} = 2$. Ruggaber *et al.* (1993) investigated theoretically the dependence of r_{dd} on various atmospheric parameters for cloud-free conditions and came to the conclusion that the assumption of $r_{dd} = 2$ results in errors up to 50% for the diffuse actinic flux. They noted that the ratio shows considerable variability as a function of the solar zenith angle, wavelength, aerosol number density, ozone profile, albedo, and height, and that no simple formula exists that can be used to describe r_{dd}. Thus, r_{dd} values are usually derived from one-dimensional radiative transfer models, which can simulate either clear-sky or completely overcast conditions (e.g. Cotte *et al.*, 1997; Kylling *et al.*, 2003a; Kazadzis *et al.*, 2004). Partial cloudiness, however, cannot be treated explicitly owing to the one-dimensional character of the models and the general lack of observed cloud input-data (see Section 9.7). In the particular case of a completely overcast sky, the diffuse ratio was found to lie in the range 1.75 ± 0.15 and to have only a small variability with the solar zenith angle and wavelength for conditions with a small surface albedo (Kazadzis *et al.*, 2000, 2004). This experimental result is in good agreement with corresponding model calculations (Van Weele *et al.*, 1995; Kylling *et al.*, 2003a).

9.6.2.1 Accuracy

The conversion of UV irradiances into actinic fluxes was tested in a number of field experiments, where spectral data of irradiances and actinic fluxes were measured by synchronised spectroradiometers (McKenzie *et al.*, 2002; Webb *et al.*, 2002a; Kylling *et al.*, 2003a; Kazadzis *et al.*, 2004). Equation 9.64 was used to convert the measured irradiances into actinic fluxes, which were then compared to measured data of F_λ. Figure 9.37 shows for example ratios of j-values that were determined from estimated and directly measured actinic-flux spectra. Results are shown for three days with partial cloudiness (15, 16, 18 June, 2002) and one clear-sky day with very little aerosol (19 June, 1998). Values for α_E were obtained from a simple single-layer model (cf. Figure 9.36) and r_{dd} was calculated using a model-based parametrisation (McKenzie *et al.*, 2002). Under clear-sky conditions and at solar zenith angles less than 80°, the photolysis frequencies $j(O^1D)$ and $j(NO_2)$ derived from the estimated actinic-flux data show deviations up to 10% relative to j-values derived from direct actinic-flux measurements (Figure 9.37). Under cloudy conditions, the conversion errors can be larger. They are generally less than 20% for $j(O^1D)$, but can be larger for $j(NO_2)$, which is attributed to the partial cloudiness causing departures from the assumed angular distribution of the sky radiance (McKenzie *et al.*, 2002).

Other field studies have obtained similar results. It was found that the assumption of an isotropic diffuse radiance leads to an overestimation of the actinic flux by 10–15% for clear-sky conditions (Kazadzis *et al.*, 2000; Kylling *et al.*, 2003a). When the diffuse ratio and the fraction of direct radiation are calculated by a detailed one-dimensional

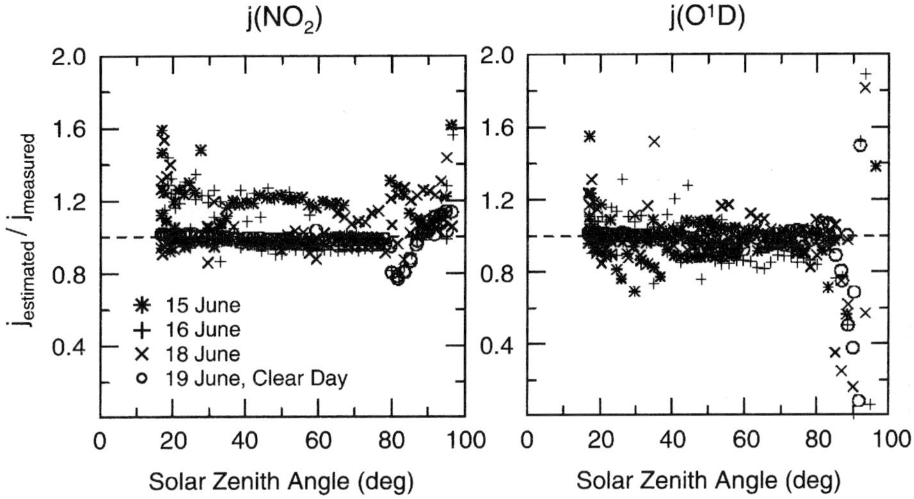

Figure 9.37 Ratios of *j*-values derived from measured spectral irradiances and measured spectral actinic fluxes. The data were obtained at 1.6 km altitude in Boulder, Colorado, from 15 to 19 June, 1998 (adapted from McKenzie *et al.*, 2002, printed with permission from American Geophysical Union).

radiative transfer model, the uncertainty of the irradiance to actinic-flux conversion is generally less than 10% on cloud-free days (Kazadzis *et al.*, 2000; Webb *et al.*, 2002a; Kylling *et al.*, 2003a). The conversion errors for a broad range of conditions, including situations with and without clouds, were studied by Kylling *et al.* (2003a) for different European monitoring sites. The ratios of the estimated to measured actinic fluxes were found to vary around unity with a standard deviation in the order of 10% (1σ) over all conditions, with a tendency to overpredict the actinic flux under clear-sky conditions and to underpredict the actinic flux when clouds were present.

When photolysis frequencies are derived from estimated actinic-flux data, the resulting total error has three main contributions. One is the error of the radiometer calibration and another one is the uncertainty of the absorption cross sections and quantum yields. These two error contributions are essentially the same as for actinic-flux spectroradiometry (see Table 9.6). An additional error is caused by the conversion of the measured irradiance data into actinic fluxes. This error is variable and depends on the ambient conditions, like for example cloudiness, as discussed above.

9.6.3 *Empirical methods*

Different methods have been proposed to predict in a simple way photolysis frequencies from measured irradiances by statistical methods, of which some will be mentioned here. For example, an almost linear relationship was observed in Bonn, Germany, between measured NO_2 photolysis frequencies and broadband (300–3000 nm) irradiances G, which can be approximated by a linear regression (Bahe *et al.*, 1980):

$$j(NO_2)(s^{-1}) = 1.33 \times 10^{-4} + 8.87 \times 10^{-6} G(W \cdot m^{-2}) \qquad (9.65)$$

Because of the constant offset, which has no physical meaning, the equation is useful only for values of $G > 70\,W \cdot m^{-2}$ for estimating $j(NO_2)$ from standard meteorological measurements of broadband radiation. The surprisingly linear correlation is the result of the accidental compensation of different spectral and geometrical dependencies of the $j(NO_2)$ and irradiance measurement instruments. A similar regression was reported for measurements on the Atlantic Ocean (Brauers & Hofzumahaus, 1992).

A nonlinear correlation was observed for $j(NO_2)$ measured by chemical actinometers versus broadband irradiances (295–385 nm) measured by Eppley UV radiometers (Harvey *et al.*, 1977; Zafonte *et al.*, 1977; Dickerson *et al.*, 1982; Parrish *et al.*, 1983; Madronich *et al.*, 1983). The nonlinearity is caused mainly by the different geometrical responses of the instruments and less by the spectral radiometer sensitivity, which matches approximately the wavelength range of the NO_2 photodissociation (Madronich *et al.*, 1983). Based on the physical model that leads to Equation A.12, Madronich (1987b) developed a parameterisation that related measured $j(NO_2)$ values to the corresponding broadband UV irradiances, E, measured by standard Eppley radiometers ($W \cdot cm^{-2}$):

$$\frac{j(NO_2)}{E} = C\left(\frac{1}{\alpha_J \cos\chi + 0.5(1-\alpha_J)} + 2A_s\right) \qquad (9.66)$$

Here, C is an empirical factor ($1.35 \pm 0.05\,cm^2\,J^{-1}$) for clear-sky conditions, χ is the solar zenith angle, A_s is the surface albedo, and α_J is the ratio $J_0/(J_0 + J\downarrow)$, where J_0 and $J\downarrow$ represent the photolysis-frequency components caused by direct and downward diffuse radiation, respectively. The relationship in Equation 9.66 has been used in various field experiments to predict the $j(NO_2)$ values from UV measurements (e.g. Shetter *et al.*, 1992; Lantz *et al.*, 1996). Its application requires α_J values that must be calculated by a model and assumes that the upwelling and downwelling diffuse radiances are isotropic. The parametrisation does not consider possible differences in the spectral sensitivity of individual broadband radiometers and does not correct for the spectral mismatch of the radiometer sensitivity to the photodissociation spectrum of NO_2. The accuracy of the relationship was tested against chemical actinometry for a broad range of conditions at Mauno Loa (Hawai) (Shetter *et al.*, 1992). The estimated $j(NO_2)$ values were found to be accurate to within ±10% for clear skies and independent of the solar zenith angle, and to ±20% for conditions with scattered clouds (Figure 9.38).

Other empirical methods involve spectrally resolved irradiances to estimate the photolysis frequencies of O_3 or NO_2 (McKenzie *et al.*, 2002; Kazadzis *et al.*, 2004). These methods use polynomial statistical fits, but have the disadvantage that they require narrow bandwidth UV measurements by a spectroradiometer for the prediction of j-values. A technically more simple approach uses a multichannel filter radiometer that measures UV irradiances at five different wavelengths with a bandwidth of about 10 nm (Seroji *et al.*, 2004). In this approach the photolysis frequencies of O_3, NO_2, and HCHO can be derived from linear combinations of the spectral measurement channels, which were calibrated against photolysis frequencies determined by actinic-flux spectroradiometry.

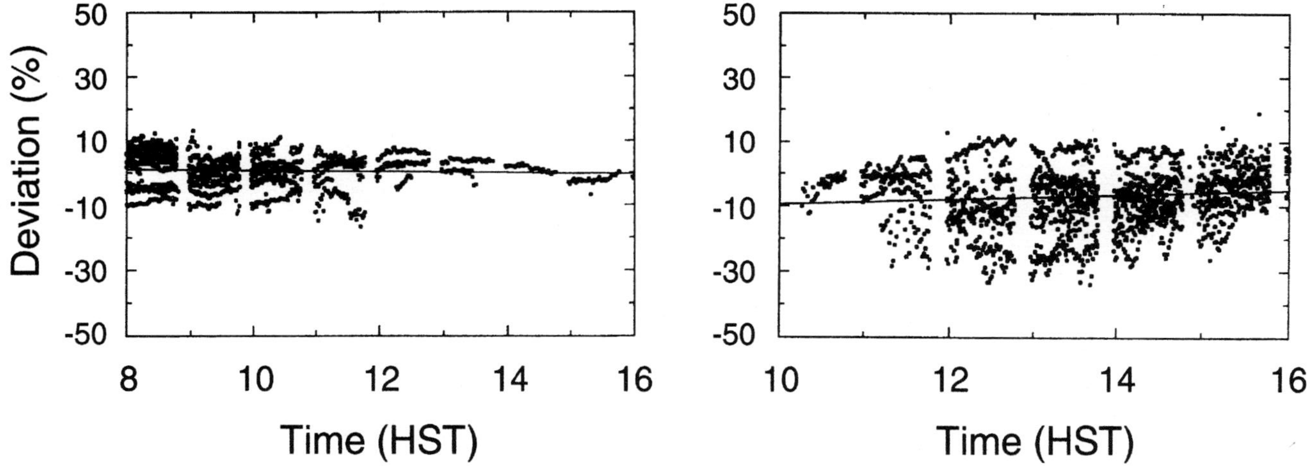

Figure 9.38 Deviation of estimated $j(NO_2)$ values from direct measurements by a chemical actinometer. The estimated values were derived from broadband UV irradiances obtained on Mauna Loa, Hawai, in May 1988. The left panel applies to periods without overhead clouds, but with lower lying valley clouds in many cases. The right panel applies to periods with scattered overhead clouds (adapted from Shetter *et al.*, 1992, printed with permission from American Geophysical Union).

9.7 Modelling of photolysis frequencies

Photolysis frequencies are highly variable in space and time owing to the influences of many varying atmospheric parameters. In situ measurements can provide data only in a few places and times. For this reason, many atmospheric chemistry models rely on photolysis frequencies that are derived from modelled radiation fields.

These are determined by means of radiative transfer (RT) models that use the extraterrestrial solar spectrum as input, as well as vertical distributions of atmospheric parameters (pressure, temperature, aerosol, ozone, clouds) and the earth's surface albedo (see for example Demerjian *et al.* (1980); Madronich (1993); Mayer & Kylling (2005)). The models also require the absorption and scattering cross sections for the interaction of solar radiation with atmospheric gases and aerosols. In a second step, photolysis frequencies are calculated according to Equation 9.12, requiring the knowledge of the relevant parameters σ and ϕ.

Radiative transfer models can differ significantly in the complexity of details describing the transport of radiation in the atmosphere. For example, models can vary in the detailed implementation of the atmospheric structure and composition, the spectral coverage and resolution of the radiation spectrum, or the detailed treatment of the scattered radiation (e.g. Bais *et al.*, 2003). Most operational models are one-dimensional and treat the atmosphere as a vertically inhomogeneous medium that is composed of a finite number of homogeneous layers. This concept requires that absorbers and scatterers are uniformly distributed over horizontal distances that are much wider than the thickness of the respective model layers. In a realistic atmosphere, this is most likely the case under cloud-free conditions, or when stratiform clouds extend across the sky. Despite the limitation to a one-dimensional treatment, such models are widely used in atmospheric chemistry and apply simplified parametrisations to handle broken cloud fields which have horizontally inhomogenous properties (see e.g. Tie *et al.*, 2003).

There have been attempts to simulate the transport of short-wave radiation in cloudy atmospheres with two- and three-dimensional models (e.g. Los *et al.*, 1997; Trautmann *et al.*, 1999). In practice, these models require too much computing time and lack suitable input data, in order to be useful in current atmospheric chemistry models. In some applications like global modelling, even one-dimensional RT models require too much computer resources, because of the many wavelengths needed for the calculation of photolysis frequencies. In these cases atmospheric models often apply parameterised photolysis frequencies that have been precalculated with more sophisticated models as functions of atmospheric input parameters.

Besides its use for predictions in atmospheric chemistry, RT models are also useful as supporting tools for some kind of photolysis-frequency measurement methods, which require some knowledge of the spectral or spatial distribution of the atmospheric radiation. This is the case with actinic-flux filter radiometers, where the spectral composition of the actinic flux is needed for the development of an instrumental calibration function (Section 9.5.3). Another example is the conversion of irradiances into actinic-flux data, which requires the knowledge of the angular distribution of the diffuse radiance (Section 9.6.2).

A detailed discussion of RT models is beyond the scope of this book. Details of the theory of transport of solar radiation in the atmosphere can be found in text books, for example by Goody and Yung (1989), Lenoble (1993), or Thomas and Stamnes (1999). References to currently used models and a discussion of their ability to simulate atmospheric actinic fluxes and photolysis frequencies can be found, for example, in Olson *et al.* (1997); Bais *et al.* (2003); Shetter *et al.* (2003), and Hofzumahaus *et al.* (2004).

9.8 Sample applications

9.8.1 *Photochemistry field experiments*

Photolysis frequency measurements are an important integral part of any photochemistry field experiment. Atmospheric radical concentrations, ozone production rates, or photochemical aging of emitted pollutants can only be understood quantitatively, if the observation of trace gases is accompanied by photolysis-frequency measurements of relevant compounds, like O_3, NO_2, HCHO, etc. The important role of solar radiation in tropospheric chemistry is illustrated in Figure 9.39. It shows the strong dependence of OH radical concentrations on the solar UV, represented here by $j(O^1D)$. The observed close correlation between OH and $j(O^1D)$ is caused by the photochemical OH formation and the short lifetime of the radicals, which respond immediately within seconds to changes of their production rate (Hofzumahaus *et al.*, 1996). Similar observations have been

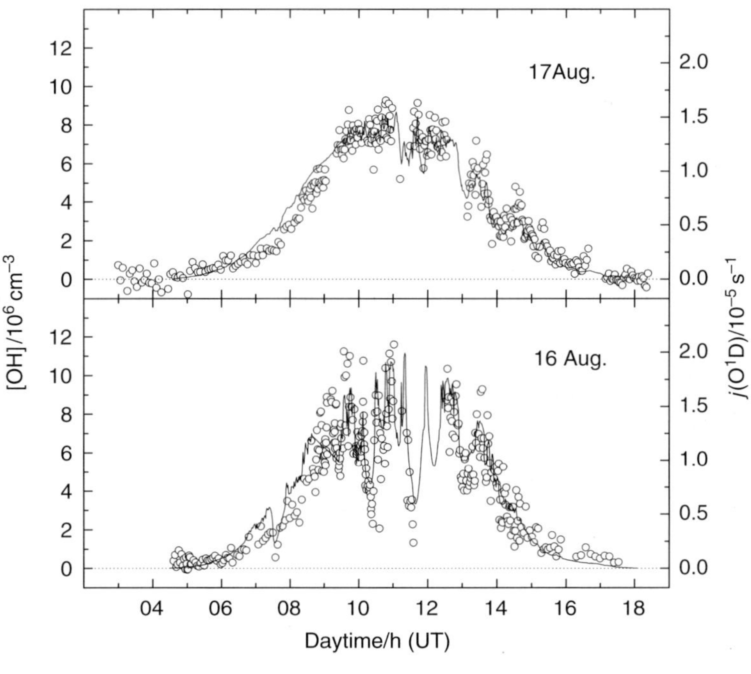

Figure 9.39 Diurnal cycles of OH measured by laser-induced fluorescence and $j(O^1D)$ measured with a filter radiometer. The data were collected at ground in North-East Germany on 16 and 17 August, 1994. The fast variations seen in $j(O^1D)$ were caused by changing broken clouds (adapted from Holland *et al.*, 1998, with kind permission of Springer Science and Business Media).

reported from other field studies for OH (e.g. Eisele *et al.*, 1997; Abram *et al.*, 2000; Brauers *et al.*, 2001; Mauldin III *et al.*, 2001; Berresheim *et al.*, 2003; Holland *et al.*, 2003) and for peroxy radicals, HO_2 and RO_2 (e.g. Cantrell *et al.*, 1992; Penkett *et al.*, 1997; Kanaya *et al.*, 1999). The strong sensitivity of free radical concentrations to solar radiation implies that photolysis frequencies must be measured (or modelled) first and foremost with high accuracy, when photolysis data are to be used as model input for reliable predictions of atmospheric photochemistry.

9.8.2 *Temporal and spatial distributions*

The distribution of photolysis frequencies over different temporal and spatial scales can be simulated by radiative transfer models with different degrees of approximation (see Section 9.7), but corresponding measured data are rather limited, in particular over large scales. The temporal behaviour of photolysis frequencies was investigated mostly for O_3 and NO_2 in the form of diurnal variations that were observed over time periods of weeks or months in field campaigns (e.g. Shetter *et al.*, 1992, 1996). Longer time records are generally not available and may be obtained at present only from the conversion of UV-irradiance data that are recorded at surface monitoring stations (see Section 9.6).

Photolysis frequency measurements over large spatial scales are also rare. Figure 9.40 shows one example of the latitudinal distribution of $j(O^1D)$ that was measured together with $j(NO_2)$ between 50°N and 30°S in early fall 1988 near-equinox conditions (Hofzumahaus *et al.*, 1992; Brauers & Hofzumahaus, 1992). The distribution was recorded on a ship cruise along 30°W and shows superimposed diurnal variations. The diurnal maxima show highest values in the tropics, where the solar zenith angles are smaller and the stratospheric ozone layer is thinner than at high latitudes. A strong reduction of $j(O^1D)$ was observed at 5°N in the innertropical convergence zone (ITCZ), where thick cloud layers attenuated the solar radiation. Also shown are measurements of ozone and water vapour, which were used together with $j(O^1D)$ to calculate the primary production rate of OH, P_{OH}, from the ozone photolysis. The figure indicates that the OH production is largest in the tropics where water vapour and solar actinic flux are most abundant.

Another important aspect is the vertical distribution of atmospheric photolysis frequencies. Figure 9.41 shows as an example the measured actinic flux at selected wavelengths between 120 m and 12 km altitude above the Agean Sea (Greece) for cloudless-sky conditions (Hofzumahaus *et al.*, 2002). The largest vertical gradients were observed at short wavelengths in the lowest 3 km above the sea surface as the result of ozone absorption and aerosol scattering. The observations agree well with model simulations that were constrained by measurements of aerosol and ozone. Similar agreement between measured and modelled radiation has been reported for cloud-free conditions from another aircraft study that was performed in the same altitude range over the western Pacific Ocean (Lefer *et al.*, 2003).

Vertical (and horizontal) distributions of photolysis frequencies are much more complex in the presence of clouds (Junkermann, 1994; Pfister *et al.*, 2000; Früh *et al.*, 2000; Lefer *et al.*, 2003; Kylling *et al.*, 2005). The possible cloud impact can be seen for example

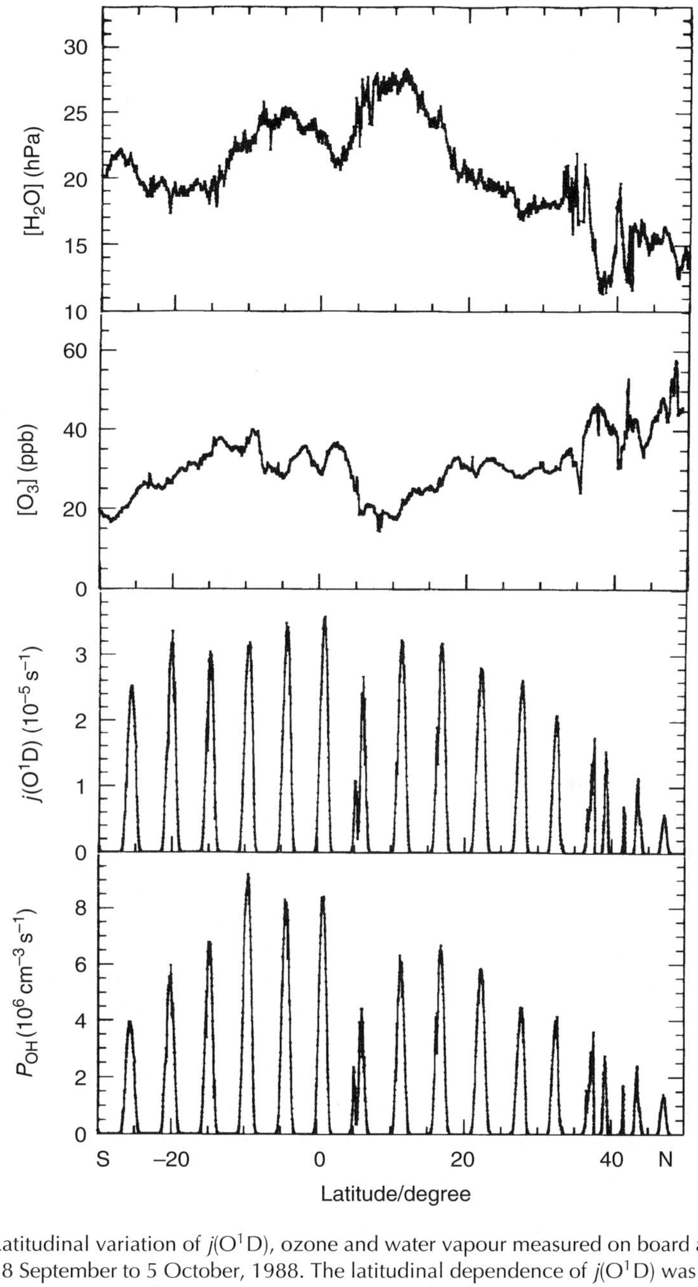

Figure 9.40 Latitudinal variation of $j(O^1D)$, ozone and water vapour measured on board a ship over the Atlantic from 18 September to 5 October, 1988. The latitudinal dependence of $j(O^1D)$ was superimposed by diurnal variations while the ship cruised from North to South. Also shown is the primary OH production rate (P_{OH}) calculated from the ozone photolysis (adapted from Hofzumahaus *et al.*, 1992, with kind permission of Springer Science and Business Media).

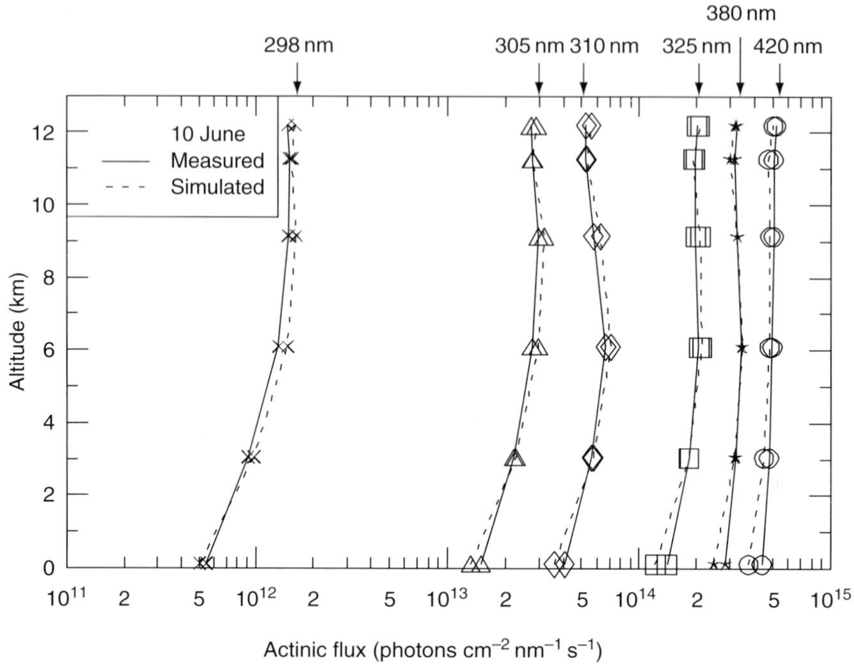

Figure 9.41 Measured and simulated vertical profiles of the actinic flux over the Agean Sea in Greece for selected wavelengths on a cloudless day (10 June, 1996). The solar zenith angle was between 17° and 23° during the flight measurements (adapted from Hofzumahaus *et al.*, 2002, printed with permission from American Geophysical Union).

in Figure 9.42. A strong reduction of measured $j(NO_2)$ values caused by an optically thick cloud layer was observed in the lowest troposphere. The gradual $j(NO_2)$ decrease in the cloud layer is well described by model simulations which used micrometeorological measurements of cloud droplets and particles as model input. When compared to clear-sky conditions, an increase of the actinic flux above clouds and a strong decrease below clouds can be expected (Madronich, 1987b; Van Weele & Duykerke, 1993). Measurements have shown, for example, enhancements above clouds as high as +200% and attenuations below clouds up to −90% (Lefer *et al.*, 2003). While the influence of closed cloud layers on the actinic flux can be described consistently by one-dimensional RT models, the effect of broken clouds cannot be well represented by the same models (Kylling *et al.*, 2005). Here, more research is needed to better understand the three-dimensional effects of clouds on the radiation field.

9.8.3 *Instrumental intercomparisons*

The performance of actinic-flux-sensitive measurement instruments was investigated in various field comparisons. Radiometry-based techniques were tested against chemical actinometers for $j(O^1D)$ (Müller *et al.*, 1995; Hofzumahaus *et al.*, 2004) and $j(NO_2)$ (Kraus *et al.*, 1998, 2000; Shetter *et al.*, 2003). Other investigations involved radiometric

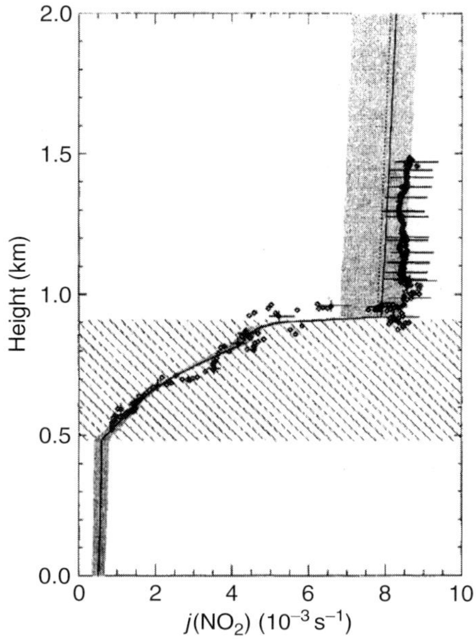

Figure 9.42 Vertical profile of $j(NO_2)$ (symbols) measured by a broadband filter radiometer near Dresden, Germany, on a cloudy day (9 December, 1996) at a solar zenith angle $\chi = 74°$. The solid and dotted lines represent simulations with two different radiative transfer models. The cross-hatched area indicates an optically thick cloud layer (adapted from Früh *et al.*, 2000, printed with permission from American Geophysical Union).

techniques only, comparing spectrally resolved actinic fluxes (e.g. Webb *et al.*, 2002b; Bais *et al.*, 2003; Edwards & Monks, 2003; Eckstein *et al.*, 2003) or results for $j(O^1D)$ and $j(NO_2)$ (e.g. Kraus & Hofzumahaus, 1998; Lefer *et al.*, 2001; Eckstein *et al.*, 2003; Kanaya *et al.*, 2003).

The most comprehensive field comparison so far, the International Photolysis Frequency Measurement and Modeling Comparison (IPMMI), was performed in Boulder, Colorado, on four days with cloudy and clear-sky conditions in summer 1998 (see overview by Cantrell *et al.*, 2003). The assessment was carried out as a blind intercomparison that included chemical actinometers, spectroradiometers, and broadband filter radiometers from eight international research groups. Some examples of diurnal profiles of $j(O^1D)$ and $j(NO_2)$ measured by actinic-flux spectroradiometers and chemical actinometers on a clear day are shown in Figure 9.43.

The main results of the experimental data assessment can be summarised as follows:

- Actinic-flux measurements were compared for three spectroradiometers. Two double-monochromator instruments agreed to within ±6% independent of solar zenith angle and wavelength (300–420 nm) when the measurements were transformed to a common spectral resolution. A third spectroradiometer which was based on a single monochromator showed significant disagreements (up to 30%) with the two other spectroradiometers as a result of instrumental stray light. For further details, see Bais *et al.* (2003); Edwards and Monks (2003).

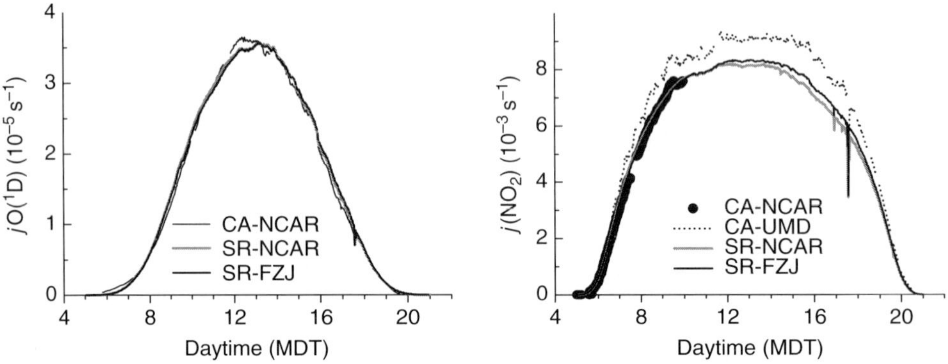

Figure 9.43 Comparison of diurnal profiles of photolysis frequencies measured by different chemical actinometers (CA) and actinic-flux spectroradiometers (SR) during the IPMMI field campaign in Boulder, Colorado, on 19 June 1998 (Shetter *et al.*, 2003; Hofzumahaus *et al.*, 2004). The cross sections and quantum yields used for the evaluation of the spectroradiometer data are in agreement with the IUPAC recommendations from 2004 (Atkinson *et al.*, 2004).

- General good agreement for $j(O^1D)$ within 10% was found among filter radiometers and spectroradiometers at solar zenith angles less than 60°, provided that the instruments used the same cross sections and quantum yields. The deviations were generally larger (up to a factor of 2) when the zenith angles were larger than 60°. Two spectroradiometers and one filter radiometer showed excellent agreement within 5% with the chemical actinometer for zenith angles less than 80°, when recently published $O(^1D)$ quantum yield data were used and when the temperature dependence of $j(O^1D)$ was explicitly taken into account. For further details, see Hofzumahaus *et al.* (2004).
- Measurements of $j(NO_2)$ from filter radiometers, spectroradiometers and two chemical actinometers showed consistent agreement within the stated instrumental uncertainties (from 8 to 20%) at zenith angles below 70°. Spectroradiometers and chemical actinometers agreed within less than 10% when more recent NO_2 absorption cross sections from the literature were used. Filter radiometers appeared to be generally less accurate, with larger discrepancies (20–50%) at larger zenith angles above 70°. For further details, see Shetter *et al.* (2003).

9.8.4 Molecular data assessments

The determination of photolysis frequencies from radiation spectra requires absorption cross sections $\sigma(\lambda, T)$ and quantum yields $\phi(\lambda, T)$ of the photodissociation reactions of interest. In general, there is a choice between published molecular data from different research studies that sometimes show discrepancies larger than their specified uncertainties. This is of considerable relevance in actinic-flux spectroradiometry, because the absorption cross sections and quantum yields make the largest contribution to the error of derived photolysis frequencies (Kraus & Hofzumahaus, 1998; Cantrell *et al.*, 2003) (see also Table 9.6). For a decision on which data are most suitable and accurate, it is generally a good idea to consult the recommendations which are periodically published and updated by institutions like the Jet Propulsion Laboratory (JPL) under contract of the American National Aeronautics and Space Administration (NASA) or the Subcommittee on Gas Kinetic Data Evaluation for Atmospheric Chemistry of the International Union on Pure and Applied Chemistry (IUPAC). Both institutions make their own evaluations available via the World Wide Web (http://jpldataeval.jpl.nasa.gov and www.iupac-kinetic.ch.cam.ac.uk).

For the two most important tropospheric photolysis frequencies, $j(O^1D)$ and $j(NO_2)$, the accuracy of the molecular data was explicitly tested in field experiments by comparing chemical actinometer measurements against photolysis frequency data that were derived from measured actinic-flux spectra and different sets of absorption spectra and quantum yields (Müller *et al.*, 1995; Shetter *et al.*, 1996; Kraus *et al.*, 2000; Cantrell *et al.*, 2003; Shetter *et al.*, 2003; Hofzumahaus *et al.*, 2004).

9.8.4.1 Ozone photodissociation

The most comprehensive study for $j(O^1D)$ was performed in the IPMMI field campaign (see Section 9.8.3) over a temperature range of 5–45°C and zenith angles from 16.5° to 90° (Cantrell *et al.*, 2003; Hofzumahaus *et al.*, 2004). The derived $j(O^1D)$ data were found to have little sensitivity (<3%) to the choice of literature ozone absorption cross sections, which were chosen for example from Bass and Paur (1985), Molina and Molina (1986), and Malicet *et al.* (1995). A large sensitivity, however, was found to be the choice of $O(^1D)$ quantum yields which were taken from different releases of the NASA/JPL recommendations from 1994 to 2003. Compared to chemical actinometry, the differences in the derived $j(O^1D)$ values were generally found to increase with the solar zenith angle. The largest differences were observed in case of the recommendation from 1994 (DeMore *et al.*, 1994), yielding discrepancies over a factor of 2 at low sun ($\chi > 70°$). The application of the most recent recommendation for the $O(^1D)$ quantum yield (Matsumi *et al.*, 2002; Sander *et al.*, 2003; Atkinson *et al.*, 2004) gave the best absolute agreement between chemical actinometry and spectroradiometry, with very small systematic deviations (<5%) at zenith angles $\chi < 80°$ (Figure 9.43). These deviations were smaller than the specified instrumental uncertainty (10%) of the chemical actinometer used for the study.

9.8.4.2 Nitrogen dioxide photodissociation

The cross sections and quantum yields of the NO_2 photolysis were tested in two field studies, JCOM97 (Kraus *et al.*, 2000) and IPMMI (Cantrell *et al.*, 2003; Shetter *et al.*, 2003). The use of absorption cross sections (Davidson *et al.*, 1988) and quantum yields (Gardner *et al.*, 1987; Roehl *et al.*, 1994) recommended by NASA/JPL (DeMore *et al.*, 1997; Sander *et al.*, 2003) led in both studies to spectroradiometer results that were consistently higher than the chemical actinometer measurements by 10–20%. Improved agreement with deviations generally less than 10% was obtained in both studies when IUPAC recommended data (Atkinson *et al.*, 2004) were applied, favouring more recent measurements of the absorption cross sections in accord with the Merienne *et al.* (1995) data and revised quantum yields from Troe (2000).

9.9 Conclusions and outlook

Various experimental techniques are now available for routine measurements of atmospheric photolysis frequencies. The most direct method is chemical actinometry, which has been applied in field campaigns almost exclusively for the photolysis of O_3 and NO_2. It is calibrated by a gas-phase standard and has the particular advantage that it does not require the knowledge of cross sections and quantum yields. The requirement to handle pressurised gases and the instrumental difficulty of detection of reactive photolysis products is probably the reason why it is not more widely used for other photolysis reactions.

The most versatile method is actinic-flux spectroradiometry. It measures the spectrally resolved actinic flux and can be used to determine photolysis frequencies of a variety of chemical compounds from the same measured radiation spectrum. As a further advantage, the absolute calibration is traceable to very accurate radiation standards from national laboratories. For the determination of photolysis frequencies, the method requires knowledge of absorption cross sections and quantum yields. Although the relevant photochemical data have been measured thoroughly, they remain the largest source of uncertainty in the calculation of photolysis frequencies.

The quality of photochemical data for O_3 and NO_2 has been improved greatly owing to new laboratory studies. The improvement is consistent with the outcome of field comparisons between chemical actinometry and actinic-flux spectroradiometry, which agree well when the applied cross sections and quantum yields comply with the most recent IUPAC recommendations (Atkinson *et al.*, 2004). The comparisons also demonstrate that $j(O^1D)$ and $j(NO_2)$ can be measured with spectroradiometers as accurate as by chemical actinometers.

In the future it would be desirable to test critically the photochemical database for other important molecules, like formaldehyde, and perform corresponding field comparisons with chemical actinometers.

Broadband filter radiometers are useful, because they are small and can be run automatically with high time resolution. However, their calibration is laborious and their data evaluation requires careful corrections for their imperfect spectral matching of the respective photolysis processes.

Spectroradiometers developed at the time of writing are based on multichannel-detector spectrograph systems. They combine the advantages of spectroradiometry with the small size and high time resolution known from filter radiometers. Spectrograph instruments offer the largest potential for field studies; however, unsolved stray light problems preclude at present their usefulness in the UV-B region. Their further development will significantly facilitate the fast measurement of actinic radiation and photolysis frequencies, in particular on aircraft and under cloudy conditions.

Acknowledgements

The author gratefully acknowledges fruitful cooperation with present and previous co-workers and colleagues in the field of photolysis frequency measurements, in particular Birger Bohn, Theo Brauers, Alexander Kraus, Martin Müller, and Franz Rohrer;

furthermore Arve Kylling, Christos Zerefos, Rick Shetter and the colleagues who contributed to the IPMMI project, and Rainer Schmitt at Meteorologie Consult GmbH. He also likes to thank Gus Hancock for the stimulating and fruitful collaboration on the photolysis of ozone. Many thanks to Birger Bohn for critical reading of the manuscript and Dwayne Heard for his encouragement in writing this chapter.

Appendix

A.1 Radiation quantities and their relationships

Basic properties of radiance, irradiance, and actinic flux and their relations relevant for the determination of atmospheric photolysis frequencies have been discussed for example by Madronich (1987b) and Van Weele *et al.* (1995). This section summarises some of the important relationships.

The radiance in the atmosphere can be split into three components:

$$L_\lambda = L_0 + L\downarrow + L\uparrow \tag{A.1}$$

where L_0 is the spectral radiance of the direct sunlight, $L\downarrow$ represents the downwelling diffuse sky radiation, and $L\uparrow$ describes the upwelling diffuse radiation. When equation A.1 is inserted into equation 9.10, the actinic flux can be partitioned accordingly into a direct flux F_0, a diffuse downwelling flux $F\downarrow$, and a diffuse upwelling flux $F\uparrow$:

$$F_\lambda = F_0 + F\downarrow + F\uparrow \tag{A.2}$$

In a similar way, the downward irradiance (cf. equation 9.14) can be written as the sum of the direct downward irradiance E_0 and the diffuse downward irradiance $E\downarrow$:

$$E_\lambda = E_0 + E\downarrow \tag{A.3}$$

The upward directed irradiance $E\uparrow$ is related to the downward irradiance by a surface albedo A_s, which is defined as:

$$A_s = \frac{E\uparrow}{E_0 + E\downarrow} \tag{A.4}$$

By definition the direct actinic flux F_0 is related to the direct irradiance E_0 by the cosine of the solar zenith angle χ:

$$\frac{F_0}{E_0} = \frac{1}{cos\chi} \tag{A.5}$$

Ratios can also be defined for the upwelling and downwelling diffuse components:

$$r_{dd} = \frac{F\downarrow}{E\downarrow} = \frac{\int L\downarrow(\vartheta,\varphi)\,d\omega}{\int L\downarrow(\vartheta,\varphi)\,cos(\vartheta)\,d\omega} \tag{A.6}$$

$$r_{du} = \frac{F\uparrow}{E\uparrow} = \frac{\int L\uparrow(\vartheta,\varphi)\,d\omega}{\int L\uparrow(\vartheta,\varphi)\,cos(\vartheta)\,d\omega} \tag{A.7}$$

Here, the ratios r_{dd} and r_{du} can be calculated only if the angular distribution for the corresponding diffuse radiance is known.

Inserting equations A.4–A.7 into equation A.2 allows to express F_λ in terms of the downwelling direct and diffuse irrandiance components:

$$F_\lambda = \frac{E_0}{\cos\chi} + r_{dd}E\downarrow + r_{du}A_s(E_0 + E\downarrow) \tag{A.8}$$

Next the parameter α_E is defined as the direct fraction of the total downward irradiance:

$$\alpha_E = \frac{E_0}{E_0 + E\downarrow} = \frac{E_0}{E_\lambda} \tag{A.9}$$

Combining Eqs. A.8 and A.9, a general relationship between F_λ and E_λ can be derived:

$$\begin{aligned}\frac{F_\lambda}{E_\lambda} &= \frac{\alpha_E}{\cos\chi} + r_{dd}(1-\alpha_E) + r_{du}A_s \\ &= \alpha_E\left(\frac{1}{\cos\chi} - r_{dd}\right) + (r_{dd} + r_{du}A_s)\end{aligned} \tag{A.10}$$

It is often assumed that the reflecting surface giving rise to the upward directed radiance behaves like a Lambertian reflector, that is produces upward radiance without angular dependence. For this special case (isotropy assumption) r_{du} is calculated to be

$$r_{du} = \frac{F\uparrow}{E\uparrow} = 2 \tag{A.11}$$

Then equation A.10 assumes the following form:

$$\frac{F_\lambda}{E_\lambda} = \alpha_E\left(\frac{1}{\cos\chi} - r_{dd}\right) + (r_{dd} + 2A_s) \tag{A.12}$$

This is the basic equation for estimating actinic fluxes from downward irradiances above Lambertian surfaces with albedo A_s.

Another useful equation can be derived from equations A.4 and A.7 for the upward actinic flux $F\uparrow$ above Lambertian surfaces ($r_{du} = 2$):

$$F\uparrow = 2A_s(E_0 + E\downarrow) \tag{A.13}$$

Inserting equations A.5 and A.6, one finds:

$$F\uparrow = 2A_s\left(\cos\chi F_0 + \frac{1}{r_{dd}}F\downarrow\right) \tag{A.14}$$

Assuming a nearly isotropic distribution for the downward radiance $L\downarrow$ yields

$$r_{dd} = \frac{F\downarrow}{E\downarrow} = 2 \tag{A.15}$$

and

$$F\uparrow = A_s(2cos\chi F_0 + F\downarrow) \quad (A.16)$$

When the parameter α_F is defined as the direct fraction of the total downward actinic flux

$$\alpha_F = \frac{F_0}{F_0 + F\downarrow} \quad (A.17)$$

and is inserted into equation A.16, the following relation between the upward and downward actinic flux is found:

$$\beta_F = \frac{F\uparrow}{F_0 + F\downarrow} = A_s[\alpha_F(2cos\chi - 1) + 1] \quad (A.18)$$

Here, the ratio β_F, applicable to actinic fluxes, is the formal counterpart to the albedo A_s which is defined for irradiances and can be used to estimate $F\uparrow$ from the total downwelling actinic flux. Besides an albedo A_s the estimate requires values for the solar zenith angle χ and the fraction α_F of the direct actinic flux. A detailed discussion of reflected actinic fluxes and their relevance above ground and cloud surfaces has been given by Madronich (1987b).

A.2 Effective transmission of solar radiation into a chemical actinometer

We consider the effective transmission of ultraviolet and visible solar radiation into the photolysis cell of a chemical actinometer as discussed in Section 9.3.3. It is assumed that the photolysis cell is either a spherical bulb or an infinitely long cylindrical tube with non-absorbing walls made of quartz. The wall thickness is assumed to be small compared with the radius of the cell. Under the latter assumption, the reactor wall will behave approximately like a plane-parallel plate at the point at which an incident light ray penetrates into the interior of the reactor.

An appreciable amount of incident light is lost by reflections at the two interfaces of the wall, when the radiation enters the photolysis reactor. The reflectance ρ at each interface between air and quartz depends on the refractive indices (n_1 and n_2) of the two media, the angle of incidence (ϑ_1) and the polarisation of light. For the polarisation component parallel $\parallel$ and perpendicular $\perp$ to the plane of incidence (the plane containing a ray incident on the surface and the normal of the surface at the point of incidence), ρ can be calculated according to Fresnel's formulas (McCluney, 1994):

$$\rho_\parallel = \left[\frac{n_1\cos\vartheta_2 - n_2\cos\vartheta_1}{n_1\cos\vartheta_2 + n_2\cos\vartheta_1}\right]^2 \quad \text{and} \quad \rho_\perp = \left[\frac{n_1\cos\vartheta_1 - n_2\cos\vartheta_2}{n_1\cos\vartheta_1 + n_2\cos\vartheta_2}\right]^2 \quad (A.19)$$

Here ϑ_2 is the angle of the refracted light ray in the quartz and is related to ϑ_1 by Snell's law

$$n_1\sin\vartheta_1 = n_2\sin\vartheta_2 \quad (A.20)$$

The reflectances of both polarisation components are equal in the case of normal incidence ($\vartheta_1 = 0°$). With $n_1 = 1$ (air) and $n_2 = 1.46 - 1.5$ (quartz at $\lambda = 650 - 250$ nm), the reflectance is about 3.8% at each interface. For grazing incidence ($\vartheta_1 \rightarrow 90°$), the reflectances both approach 100%.

The effective reflectance R and transmittance T of a plane-parallel plate of quartz surrounded by air is the result of multiple reflections at the two interfaces of the plate (McCluney, 1994). Consider an incident ray of light, with intensity I_0 denoting the flux of photons carried by the ray (Figure 9.44). At the first air–quartz interface, part of the light will be reflected with intensity $I_{\mathrm{R}}^0 = \rho \times I_0$, while the remaining fraction $(1-\rho) \times I_0$ will be refracted into the quartz. Part of the refracted light will be transmitted through the second interface, leaving the quartz as a ray of intensity $I_{\mathrm{T}}^0 = (1-\rho)^2 \times I_0$, with a direction that is collinear to the original light. The radiation that is not transmitted at the second interface will undergo successive internal reflections in the quartz plate, each with reflectance ρ. On each reflection, a fraction $(1-\rho)$ of light will escape from the plate into the air. As a result, an infinite number of rays with intensities $I_{\mathrm{T}}^0, I_{\mathrm{T}}^1, I_{\mathrm{T}}^2, \ldots I_{\mathrm{T}}^i (i \rightarrow \infty)$ will be transmitted through the quartz plate, adding up to the total intensity I_{T}. The intensities can be written as

$$I_{\mathrm{T}}^0 = I_0(1-\rho)^2$$

$$I_{\mathrm{T}}^1 = I_0(1-\rho)^2\rho^2$$

etc.

$$I_{\mathrm{T}}^i = I_0(1-\rho)^2\rho^{2i} \tag{A.21}$$

$$I_{\mathrm{T}} = \sum_{i=0}^{\infty} I_{\mathrm{T}}^i = I_0(1-\rho)^2 \sum_{i=0}^{\infty} \rho^{2i} \tag{A.22}$$

For $x \leq 1$, the following relation is valid

$$\sum_{i=0}^{\infty} x^{2i} = \frac{1}{1-x^2} \tag{A.23}$$

so that

$$T = \frac{I_T}{I_0} = \frac{(1-\rho)^2}{1-\rho^2} = \frac{1-\rho}{1+\rho} \tag{A.24}$$

In a similar way, a succession of rays with the intensities $I_R^0, I_R^1, I_R^2, \ldots I_R^i (i \rightarrow \infty)$ will be reflected with total intensity I_R. For reasons of energy conservation, I_R must be equal to $I_0 - I_T$. It follows that

$$R = \frac{I_R}{I_0} = \frac{2\rho}{1+\rho} \tag{A.25}$$

Equations A.24 and A.25 are applicable to light with parallel or vertical polarisation by setting $\rho = \rho_{\parallel}$ or $\rho = \rho_{\perp}$, respectively. Unpolarised light can be treated as the superimposition of two components with parallel and perpendicular polarisation. The transmitted

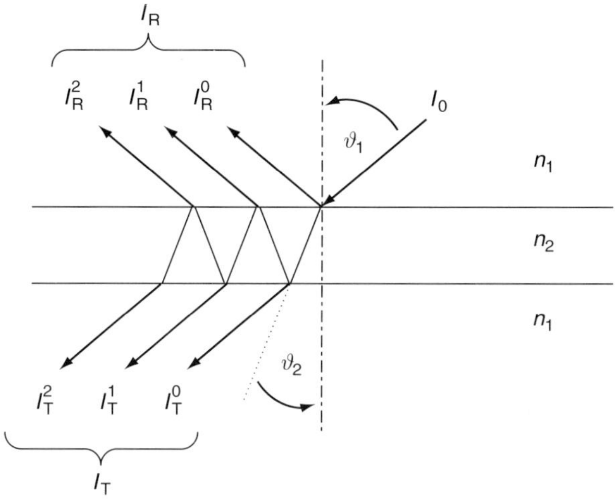

Figure 9.44 Illustration of multiple reflections at the two interfaces of a plane-parallel plate. n_2 denotes the refractive index of the transparent plate material (e.g. quartz) and $n_1 = 1$ is the refractive index of the surrounding air. I_0 is the intensity of the incident light ray, I_R and I_T are the total intensities of the reflected and transmitted rays, respectively. ϑ_1 is the angle of incidence and ϑ_2 the angle of refraction of the original ray.

(reflected) intensities are determined separately for each of the two polarisation components, which finally add up to the total transmitted (reflected) intensity. Note that Equations A.24 and A.25 are valid only for non-coherent light, like solar radiation. The treatment of coherent, monochromatic radiation must take into account interference effects which play no role for broadband radiation.

Next we consider the sun light that has penetrated into the interior of the photolysis reactor. The transmitted ray with intensity $I_1 = (1 - R) \times I_0$ is collinear to the original ray I_0, which has an angle of incidence ϑ_0 (Figure 9.9). The ray I_1 strikes the interior wall surface at an angle which is essentially equal to ϑ_0, if the wall thickness is small, compared with the radius of the reactor. The interior reflection of I_1 produces a succession of an infinite number of reflected rays with intensities I_2, I_3, I_4, etc. Each of these rays has the same angle of incidence at the interior wall and the same path length like ray I_1. The collective contribution of all interior rays to the photolysis in the actinometer is therefore given by

$$I_{\text{internal}} = \sum_{i=1}^{\infty} I_i \tag{A.26}$$

with

$$I_1 = I_0(1 - R)$$
$$I_2 = I_0(1 - R)R$$

etc.

$$I_i = I_0(1 - R)R^i \tag{A.27}$$

and using equation A.23, it follows that

$$I_{\text{internal}} = I_0(1-R)\sum_{i=1}^{\infty} R^i = I_0 \tag{A.28}$$

This means that the initial loss of intensity by reflection at the exterior wall is exactly compensated by multiple reflections on the interior wall surface. Equation A.28 can be derived separately for the vertical and parallel polarisation components of light and is valid for all possible angles of incidence $0 \leq \vartheta_0 \leq 90°$. Thus the result is likewise valid for the solar actinic flux transmitted into the photolysis reactor.

A.3 Spectral spectrometer bandwidth

When monochromatic radiation, for example a narrow atomic emission line, passes through an imaging spectrometer, the transmitted spectrum will exhibit a broadened line profile when the radiation reaches the detector. This broadening is generally caused by the optical imaging of the entrance slit into the plane of detection and to some extent by diffraction of the spectrometer (e.g. Demtröder, 2003). The resulting spectral response of the spectrometer can be represented by an instrumental line-shape function $ILS(\lambda)$ and is characterised by a full-width at half-maximum (FWHM) called the bandwidth ($\delta\lambda$) of the instrument (Figure 9.45).

Any spectrum of light $F_\lambda(\lambda)$ (monochromatic or broadband) passing the spectrometer reaches the plane of detection as a broadened spectrum, which is the spectral convolution of $F_\lambda(\lambda)$ and $ILS(\lambda)$, denoted as $F_\lambda \otimes ILS$:

$$F_\lambda \otimes ILS = \int_0^{\infty} F_\lambda(\lambda')ILS(\lambda-\lambda')d\lambda' \tag{A.29}$$

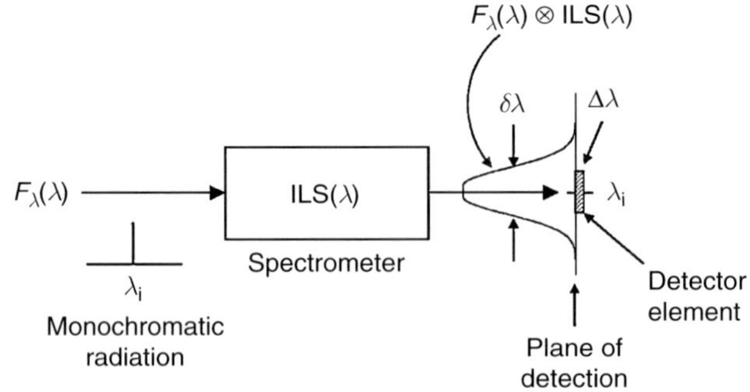

Figure 9.45 Schematic illustration of the instrumental line-broadening of monochromatic radiation transmitted through a spectrometer. $F_\lambda(\lambda)$ is the line spectrum of the incident radiation. $F_\lambda(\lambda) \otimes ILS(\lambda)$ is the broadened line spectrum when the transmitted radiation reaches the plane detection. $\delta\lambda$ is the full-width at half-maximum (FWHM) of the instrumental line shape function (ILS) and $\Delta\lambda$ is the spectral width of the radiation that is actually selected and detected by a detector element.

When this spectrum falls onto a detector element covering a narrow interval $\Delta\lambda$ centred at a specific wavelength λ_i, the mean detected radiation flux $\langle F_\lambda \rangle_i$ is given by:

$$\langle F_\lambda \rangle_i = \frac{1}{\Delta\lambda} \int_{\lambda_i - \frac{1}{2}\Delta\lambda}^{\lambda_i + \frac{1}{2}\Delta\lambda} F_\lambda(\lambda) \otimes ILS(\lambda) d\lambda \tag{A.30}$$

In case of a monochromator, $\langle F_\lambda \rangle_i$ is selected by an exit slit in front of a photoelectric detector. The exit slit width determines $\Delta\lambda$, which is usually chosen to be equal to $\delta\lambda$. In case of a spectrograph, a many-element detector array is exposed to the dispersed spectrum $F_\lambda(\lambda) \otimes \mathrm{ILS}(\lambda)$. Here $\Delta\lambda$ corresponds to the width of a single detector element (pixel). In general, the pixel width is smaller than the spectrograph bandwidth and several adjacent pixel cover the instrumental line profile. It should be noted, however, that the spectral resolution of the spectrograph is determined by $\delta\lambda$, not by the pixel resolution.

The term *instrumental line-shape function* must be distinguished from the term *instrumental slit function*, which is the spectral convolution of the ILS with the response of the exit slit (in case of a monochromator) or a pixel (in case of a spectrograph). The slit function can be recorded as a smooth signal, when the spectrometer receives monochromatic radiation that is tuned continuously over wavelength. If the monochromatic radiation has a fixed wavelength, monochromators can alternatively scan their wavelength setting to obtain the slit function. The FWHM value of the slit function is equal to $\delta\lambda$ if $\Delta\lambda \leq \delta\lambda$ (for further details, see e.g. Demtröder, 2003).

A.4 Correction of the angular response of an actinic-flux receiver (2π sr)

A simple model of the atmospheric radiance L_λ near ground can be used to describe the dependence of Z_H of an upward looking radiometer on atmospheric parameters (Hofzumahaus *et al.*, 1999). It is assumed that the diffuse sky radiance is isotropic, which is often a reasonable approximation for UV radiation in the lower troposphere (cf. Figure 9.4a). The same assumption is made for the upwelling albedo contribution (Lambertian reflector assumption). The radiance L_λ can then be expressed as follows (Madronich, 1987b):

$$L_\lambda = L_0 + L\downarrow + L\uparrow \tag{A.31}$$

$$L_0 = F_0\,\delta(\vartheta_0, \varphi_0) \tag{A.32}$$

$$L\downarrow = \frac{F\downarrow}{2\pi} \tag{A.33}$$

$$L\uparrow = \frac{F\uparrow}{2\pi} \tag{A.34}$$

where the symbols have the meanings as defined in Appendix A.1, and L_0 and F_0 are the spectral radiance and actinic flux of the direct radiation from the direction (ϑ_0, φ_0)

of the sun (δ represents a Dirac delta function). $L\downarrow$ and $F\downarrow$ describe the downwelling diffuse radiation, while $L\uparrow$ and $F\uparrow$ represent the upwelling parts.

If one inserts equations A.31–A.34 into equation 9.33 and makes use of the parameters α_F and β_F (defined in Appendix A.1), the following relationships are obtained:

$$Z_H = \alpha_F Z_p(\chi) + \frac{1-\alpha_F}{2\pi}\int_{\substack{\text{upper}\\\text{hemisphere}}} Z_p\, d\omega + \frac{\beta_F}{2\pi}\int_{\substack{\text{lower}\\\text{hemisphere}}} Z_p\, d\omega \quad (A.35)$$

$$F_\lambda(2\pi\,\text{sr}) = F_0 + F\downarrow \quad (A.36)$$

where $\chi = \vartheta_0$ is the solar zenith angle, α_F is the fraction of the direct actinic flux, β_F is the ratio of the upward to downward actinic flux, and $F_\lambda(2\pi\,\text{sr})$ is the total downward actinic flux that the radiometer aims to measure.

Obviously Z_H has three contributions: the first term is related to the direct solar radiation, the second one to the diffuse sky radiation, and the third one to the ground albedo. The parameters α_F and β_F can be obtained from measurements or must be estimated by a model. Appendix A.1 explains how β_F can be estimated in a simple way from ground-albedo values A_s.

A.5 Equivalent plane-receiver concept

In this part, the validity of the equivalent plane-receiver (EPR) concept is explained following the description by Hofzumahaus *et al.* (1999).

First consider the situation where uniform and well-collimated radiation from a point source at a very large distance is incident onto the surface S of the quartz diffuser (Figure 9.46). In this case, E_λ is constant over the beam cross section and along the optical axis z over the length of the receiver dome. According to equation 9.38, the measured signal I_λ is then related to the incident irradiance E_λ by the detection sensitivity $D_0(\lambda)$ that has been defined by equation 9.31. In order to understand the physical meaning of D

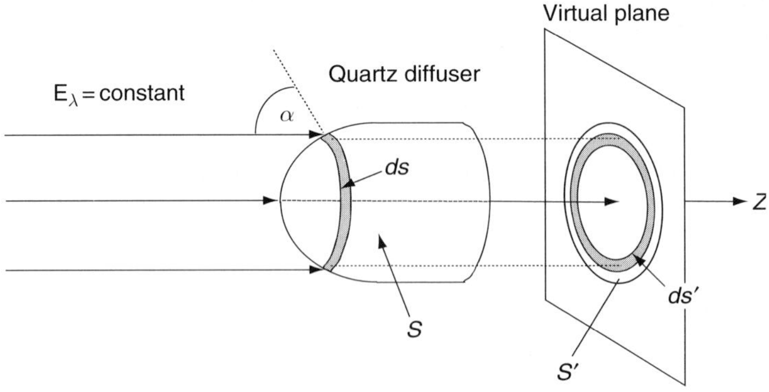

Figure 9.46 Optical receiver irradiated by a collimated beam collinear with the rotational symmetry axis of the optical receiver. S' and ds' are projections of the outer diffuser surfaces, S and ds, onto a plane perpendicular to the optical axis (from Hofzumahaus *et al.*, 1999, used with permission of Optical Society of America).

we next examine the spectral photon flux Φ_λ that is incident onto the detector surface S. Each surface element ds (taken to be a ring element for the rotationally symmetric receiver) receives a specific photon flux $\mathrm{d}\Phi_\lambda$ depending on the angle of incidence α:

$$\mathrm{d}\Phi_\lambda = E_\lambda \, \mathrm{d}s' \tag{A.37}$$

where $\mathrm{d}s'$ is the projection of ds onto a plane perpendicular to the optical axis:

$$\mathrm{d}s' = \mathrm{d}s \cos \alpha \tag{A.38}$$

The contribution $\mathrm{d}I_\lambda$ of the flux $\mathrm{d}\Phi_\lambda$ to the total measurement signal I_λ can be described by:

$$\mathrm{d}I_\lambda = \delta(\vec{r})\mathrm{d}\Phi_\lambda \tag{A.39}$$

Here, $\delta(\vec{r})$ represents the local detection sensitivity of ds at the geometric position $\vec{r}$ on the receiver surface. The term $\delta(\vec{r})$ depends on the efficiency of the transfer of incident radiation from point $\vec{r}$ through the diffuser into the quartz fibre bundle leading to the spectrometer.

Integration of equation A.39 over the whole surface S and substitution of $\mathrm{d}\Phi_\lambda$ (equation A.37) yield:

$$I_\lambda = \int_{S'} \delta(\vec{r}) E_\lambda \, \mathrm{d}s' \tag{A.40}$$

Since E_λ is assumed to be constant, we find by comparison with equation 9.38:

$$D_0 = \int_{S'} \delta(\vec{r}) \, \mathrm{d}s' \tag{A.41}$$

We can interpret this integral to be the detection sensitivity of a virtual plane receiver with the active surface S', where S' is the projection of S of the real receiver onto a plane perpendicular to the optical axis z (see Figure 9.46). Note that the position of the virtual receiver plane along the z axis plays no role as long as the radiation is perfectly collimated.

In the next step we consider the situation in which the receiver optic is exposed to the radiation of a point source at a relatively close distance (Figure 9.21) such that the irradiance E_λ varies along the optical axis from the tip to the end of the receiver optic according to the inverse-square law. With this assumption the irradiance E_λ at any point $\vec{r}$ on the surface S can be described as

$$E_\lambda = E_\lambda(z_0) \left[\frac{z_0}{z_0 + \xi(\vec{r})} \right]^2 \tag{A.42}$$

where z_0 is the working distance from the lamp to the tip of the dome and $z_0 + \xi(\vec{r})$ is the distance to the surface element $\mathrm{d}s(\vec{r})$ (Figure 9.47). We can now apply the approximation

$$\frac{1}{(1+x)^2} \approx 1 - 2x \quad \text{if} \quad |x| \ll 1 \tag{A.43}$$

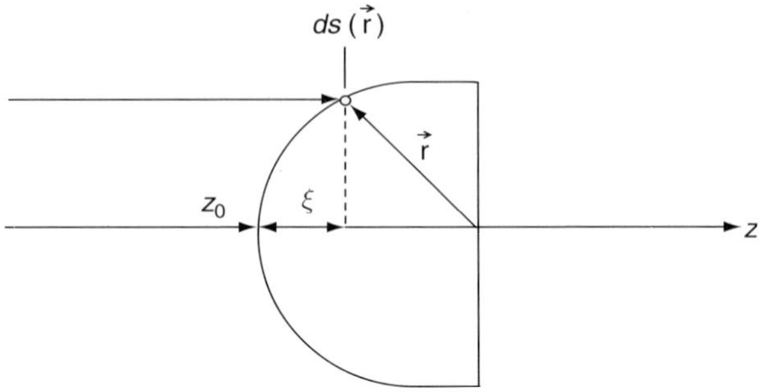

Figure 9.47 Receiver irradiated by a lamp at the working distance z_0 (cf. Figure 9.21). The surface element $ds(\vec{r})$ is displaced relative to the reference point of the receiver by the distance $\xi(\vec{r})$ along the optical axis (from Hofzumahaus *et al.*, 1999, used with permission of Optical Society of America).

which has an error of $\leq 1\%$ for $|x| \leq 0.06$. Upon substitution of $x = (\xi(\vec{r})/z_0)$ we find

$$E_\lambda = E_\lambda(z_0)\left[1 - \frac{2\xi(\vec{r})}{z_0}\right] \tag{A.44}$$

Inserting this relationship in Equation A.40 yields for the measurement signal:

$$I_\lambda = E_\lambda(z_0)\left[\int_{S'} \delta(\vec{r})ds' - \frac{2}{z_0}\int_{S'} \xi(\vec{r})\delta(\vec{r})ds'\right] \tag{A.45}$$

This is the signal that the spectroradiometer measures if the receiver optic is irradiated by a lamp positioned at the distance z_0.

In the next step we define an equivalent distance $(z_0 + \Delta z)$ for the virtual plane receiver (discussed above) such that the plane receiver with the sensitivity D_0 yields exactly the same measurement signal I_λ (equation A.45) as the real detector at the working distance z_0:

$$I_\lambda = D_0 E_\lambda(z_0 + \Delta z) \tag{A.46}$$

$$= D_0 E_\lambda(z_0)\left[\frac{z_0}{z_0 + \Delta z}\right]^2 \tag{A.47}$$

Applying approximation A.43 for small $\Delta z/z_0$ yield:

$$I_\lambda = D_0 E_\lambda(z_0)\left[1 - \frac{2\Delta z}{z_0}\right] \tag{A.48}$$

Inserting equation A.41 and comparison with equation A.45 yields:

$$\Delta z = \frac{\int_{S'} \xi(\vec{r})\delta(\vec{r})\, ds'}{\int_{S'} \delta(\vec{r})\, ds'} \tag{A.49}$$

Thus Δz is the mean displacement of the active surface elements ds along the optical axis weighted by the local sensitivity $\delta(\vec{r})$ and the effective area ds' of each surface element. Note that Δz is independent of z_0 and only depends on properties of the receiver optic.

This demonstrates that the real receiver optic exposed to an irradiance lamp at distance z_0 can indeed be thought to be replaced by an equivalent plane receiver, which has the active surface S', the same detection sensitivity D_0 as the real detector, and follows the inverse-square law if $(z_0 + \Delta z)$ is used as the effective distance from the lamp. Note that the EPR concept is based on an approximation that is valid only if the longitudinal dimension of the receiver is small compared with the working distance z_0.

References

Abram, J.P., Creasey, D.J., Heard, D.E., Lee, J.D., & Pilling, M.J. (2000) Hydroxyl radical and ozone measurements in England during the solar eclipse of 11 August 1999, *Geophys. Res. Lett.*, 27, 3437–3440.

Atkinson, R., Baulch, D.L., Cox, R.A., Crowley, J.N., Hampson, Jr., R.F., Hynes, R.G., Jenkin, M.E., Rossi, M.J., & Troe, J. (2004) Evaluated kinetic and photochemical data for atmospheric chemistry: Volume I gas phase reactions of O_x, HO_x, NO_x and SO_x species, *Atmos. Chem. Phys.*, 4, 1461–1738.

Bahe, F. & Schurath, U. (1978) Measurement of $O(^1D)$ formation by ozone photolysis in the troposphere, *Pure Appl. Geophys.*, 116, 537–544.

Bahe, F.C., Marx, W.N., Schurath, U., & Röth, E.P. (1979) Determination of the absolute photolysis rate of ozone by sunlight, $O_3 + h\nu \rightarrow O(^1D) + O_2(^1\Delta_g)$, at ground level, *Atmos. Environ.*, 13, 1515–1522.

Bahe, F.C., Schurath, U., & Becker, K.H. (1980) The frequency of NO_2 photolysis at ground level as recorded by a continuous actinometer, *Atmos. Environ.*, 14, 711–718.

Bairai, S.T. & Stedman, D.H. (1992) Actinometric measurement of $j < O_3\text{-}O(^1D) >$ using a luminol detector, *Geophys. Res. Lett.*, 19, 2047–2050.

Bais, A.F., Madronich, S., Crawford, J., Hall, S.R., Mayer, B., Van Weele, M., Lenoble, J., Calvert, J.G., Cantrell, C.A., Shetter, R.E., Hofzumahaus, A., Koepke, P., Monks, P.S., Frost, G., McKenzie, R., Krotkov, N., Kylling, A., Swartz, W.H., Lloyd, S., Pfister, G., Martin, T.J., Röth, E.-P., Griffioen, E., Ruggaber, A., Krol, M., Kraus, A., Edwards, G.D., Müller, M., Lefer, B.L., Johnston, P., Schwander, H., Flittner, D., Gardiner, B.G., Barrick, J., & Schmitt, R. (2003) International photolysis frequency measurement and model intercomparison (IPMMI): Spectral actinic solar flux measurements and modeling, *J. Geophys. Res.*, 108(D16), 8543, doi:10.1029/2002JD002891.

Bass, A.M. & Paur, R.J. (1985) The ultraviolet cross sections of ozone: I, II, in C.S. Zerefos and A. Ghazi (eds), *Atmospheric Ozone, Proceedings of the Quadrennial Ozone Symposium, Greece*, pp. 606–616. D. Reidel, Norwell, Mass.

Bates, D.R. (1984) Rayleigh scattering by air. *Planet. Space Sci.*, 32(6), 785–790.

Berger, D.S. (1976) The sunburning ultraviolet meter: Design and performance, *Photochem. Photobiol.*, 24, 587–593.

Berresheim, H., Plass-Dülmer, C., Elste, T., Mihalopoulos, N., & Rohrer, F. (2003) OH in the coastal boundary layer of Crete during MINOS: Measurements and relationship with ozone photolysis, *Atmos. Chem. Phys.*, 3, 639–649.

Blackburn, T.E., Bairai, S.T., & Stedman, D.H. (1992) Solar photolysis of ozone to singlet D oxygen atoms, *J. Geophys. Res.*, 97, 10 109–10 117.

Bohn, B., Kraus, A., Müller, M., & Hofzumahaus, A. (2004) Measurement of atmospheric $O_3 \rightarrow O(^1D)$ photolysis frequencies using filter radiometry, *J. Geophys. Res.*, 109, D10S90, doi:10.1029/2003JD004319.

Brauers, T. & Hofzumahaus, A. (1992) Latitudinal variation of measured NO_2 photolysis frequencies over the Atlantic Ocean between 50°N and 30°S. *J. Atmos. Chem.*, 15, 269–282.

Brauers, T., Dorn, H.-P., Koch, H., Kraus, A.B., & Plass-Dülmer, C. (1998) Meteorological aspects, ozone, and solar radiation measurements during POPCORN 1994, *J. Atmos. Chem.*, 31, 33–52.

Brauers, T., Hausmann, M., Bister, A., Kraus, A., & Dorn, H.-P. (2001) OH radicals in the boundary layer of the Atlantic Ocean 1. Measurements by long-path absorption spectroscopy, *J. Geophys. Res.*, 106, 7399–7414.

Calvert, J.G. & Pitts, J.N. (1966) *Photochemistry*, John Wiley & Sons, London.

Cantrell, C.A., Lind, J.A., Shetter, R.E., Calvert, J.G., Goldan, P.D., Kuster, W., Fehsenfeld, F.C., Montzka, S.A., Parrish, D.D., Westenberg, H.H., Allwine, G., & Martin, R. (1992) Peroxy Radicals in the ROSE Experiment: Measurement and Theory, *J. Geophys. Res.*, 97(D18), 20 671–20 686.

Cantrell, C.A., Shetter, R.E., & Calvert, J.G. (1994) Branching ratio for the $O(^1D) + N_2O$ reaction, *J. Geophys. Res.*, 99(D2), 3739–3743.

Cantrell, C.A., Calvert, J.G., Bais, A., Shetter, R.E., Lefer, B.L., & Edwards, G.D. (2003) Overview and conclusions of the International Photolysis Frequency Measurement and Modeling Intercomparison (IPMMI) study, *J. Geophys. Res.*, 108(D16), 8542, doi:10.1029/2002JD002962.

Castro, T., Ruiz-Suarez, L.G., & Gay, C. (1995) Direct measurements of NO_2 photolysis rates for Mexico City, *Atmosfera*, 8, 137–142.

Castro, T., Ruiz-Suarez, L.G., Molina, J.C., & Montero, M. (1997) Sensitivity analysis of a UV radiation transfer model and experimental photolysis rates of NO_2 in the atmosphere of Mexico City, *Atmos. Environ.*, 31(4), 609–620.

Cotte, H., Devaux, C., & Carlier, P. (1997) Transformation of irradiance measurements into spectral actinic flux for photolysis rates measurements, *J. Atmos. Chem.*, 26, 1–28.

Daumont, D., Charbonnier, J.B.J., & Malicet, J. (1992) Ozone UV spectroscopy I: Absorption cross-sections at room temperature, *J. Atmos. Chem.*, 15, 145–155.

Davidson, J.A., Cantrell, C.A., McDaniel, A.H., Shetter, R.E., Madronich, S., & Calvert, J.G. (1988) Visible-ultraviolet absorption cross sections for NO_2 as a function of temperature, *J. Geophys. Res.*, 93, 7105–7112.

Demerjian, K.L., Schere, K.L., & Peterson, J.T. (1980) Theoretical estimates of actinic (spherically integrated) flux and photolytic rate constants of atmospheric species in the lower troposphere, in J.J.N. Pitts, R.L. Metcalf, & D. Grosjean, (eds), *Advances in Environmental Science and Technology*, volume 10 of NATO ASI Series, pp. 369–459. John Wiley & Sons, New York.

DeMore, W.B., Sander, S.P., Howard, C.J., Ravishankara, A.R., Golden, D.M., Kolb, C.E., Hampson, R.F., Kurylo, M.J., & Molina, M.J. (1994) *Chemical Kinetics and Photochemical Data for Use in Stratospheric Modeling*. Evaluation Number 11, JPL Publ. 94-26. NASA-JPL.

DeMore, W.B., Sander, S.P., Howard, C.J., Ravishankara, A.R., Golden, D.M., Kolb, C.E., Hampson, R.F., Kurylo, M.J., & Molina, M.J. (1997) *Chemical Kinetics and Photochemical Data for Use in Stratospheric Modeling*. Evaluation Number 12, JPL Publ. 97-4. NASA-JPL.

Demtröder, W. (2003) *Laser Spectroscopy*, Third edition, Springer, Berlin.

Dickerson, R.R. & Stedman, D.H. (1980) Precision of NO_2 photolysis rate measurements, *Environ. Sci. Technol.*, 14(10), 1261–1262.

Dickerson, R.R., Stedman, D.H., Chameides, W.L., Crutzen, P.J., & Fishman, J. (1979) Actinometric measurements and theoretical calculations of $j(O_3)$, the rate of photolysis of ozone to $O(^1D)$, *Geophys. Res. Lett.*, 6(11), 833–836.

Dickerson, R.R., Stedman, D.H., & Delany, A.C. (1982) Direct measurements of ozone and nitrogen dioxide photolysis rates in the troposphere, *J. Geophys. Res.*, 87, 4933–4946.

Dickerson, R.R., Kondragunta, S., Stenchikov, G., Civerolo, K.L., Doddridge, B.G., & Holben, B.N. (1997) The impact of aerosols on solar ultraviolet radiation and photochemical smog, *Science*, 278, 827–830.

Donaldson, D.J., Frost, G.J., Rosenlof, K.H., Tuck, A.F., & Vaida, V. (1997) Atmospheric radical production by excitation of vibrational overtones via absorption of visible light, *Geophys. Res. Lett.*, 24(21), 2651–2654.

Eckstein, E., Perner, D., Brühl, C., & Trautmann, T. (2003) A new actinic flux 4π–spectroradiometer: Instrument design and application to clear sky and broken cloud conditions, *Atmos. Chem. Phys.*, 3, 1965–1979.

Edwards, G.D. & Monks, P.S. (2003) Performance of a single monochromator diode array spectroradiometer for the determination of actinic flux and atmospheric photolysis frequencies, *J. Geophys. Res.*, 108(D16), 8546, doi:10.1029/2002JD002844.

Eisele, F.L., Mount, G.H., Tanner, D., Jefferson, A., Shetter, R., Harder, J.W., & Williams, E.J. (1997) Understanding the production and interconversion of the hydroxyl radical during the Tropospheric OH Photochemistry Experiment, *J. Geophys. Res.*, 102, 6457–6465.

Finlayson-Pitts, B.J. & Pitts, Jr., J.N. (2000). *Chemistry of the Upper and Lower Atmosphere: Theory, Experiments and Applications*. Academic Press, San Diego.

Früh, B., Trautmann, T., Wendisch, M., & Keil, A. (2000). Comparison of observed and simulated NO_2 photodissociation frequencies in a cloudless atmosphere and in continental boundary layer clouds, *J. Geophys. Res.*, 105, 9843–9857.

Gardiner, B.G. & Kirsch, P.J. (eds) (1995) *Setting Standards for European Ultraviolet Spectroradiometers*. Air pollution research report 53. European Commission, Brussels.

Gardner, E.P., Sperry, P.D., & Calvert, J.G. (1987) Primary quantum yields of NO_2 photodissociation, *J. Geophys. Res.*, 92, 6642–6652.

Goody, R.M. & Yung, Y.L. (1989) *Atmospheric Radiation. Theoretical Basis*, Second edition, Oxford University Press, New York.

Harvey, R.B., Stedman, D.H., & Chameides, W. (1977) Determination of the absolute rate of solar photolysis of NO_2, *J. Air Poll. Contr. Assoc.*, 27, 663–666.

Hofzumahaus, A., Brauers, T., Platt, U., & Callies, J. (1992) Latitudinal variation of measured O_3 photolysis frequencies $J(O^1D)$ and primary OH production rates over the Atlantic Ocean between 50°N and 30°S, *J. Atmos. Chem.*, 15, 283–298.

Hofzumahaus, A., Aschmutat, U., Heßling, M., Holland, F., & Ehhalt, D.H. (1996) The measurement of tropospheric OH radicals by laser-induced fluorescence spectroscopy during the POPCORN field campaign, *Geophys. Res. Lett.*, 23, 2541–2544.

Hofzumahaus, A., Kraus, A., & Müller, M. (1999) Solar actinic flux spectroradiometry: A technique for measuring photolysis frequencies in the atmosphere, *Appl. Opt.*, 38(21), 4443–4460.

Hofzumahaus, A., Kraus, A., Kylling, A., & Zerefos, C. S. (2002) Solar actinic radiation (280–420 nm) in the cloud-free troposphere between ground and 12 km altitude: Measurements and model results, *J. Geophys. Res.*, 107(D18), 8139, doi:10.1029/2001JD900142.

Hofzumahaus, A., Lefer, B.L., Monks, P.S., Hall, S.R., Kylling, A., Mayer, B., Shetter, R.E., Junkermann, W., Bais, A., Calvert, J.G., Cantrell, C.A., Madronich, S., Edwards, G.D., Kraus, A., Müller, M., Bohn, B., Schmitt, R., Johnston, P., McKenzie, R., Frost, G.J., Griffioen, E., Krol, M., Martin, T., Pfister, G., Röth, E.P., Ruggaber, A., Swartz, W.H., Lloyd, S.A., & Van Weele, M. (2004). Photolysis frequency of O_3 to $O(^1D)$: Measurements and modeling during the International Photolysis Frequency Measurement and Modeling Intercomparison (IPMMI), *J. Geophys. Res.*, 109, D08S90, doi:10.1029/2003JD004333.

Holland, F., Aschmutat, U., Heßling, M., Hofzumahaus, A., & Ehhalt, D.H. (1998) Highly time resolved measurements of OH during POPCORN using laser-induced fluorescence spectroscopy, *J. Atmos. Chem.*, 31, 205–225.

Holland, F., Hofzumahaus, A., Schäfer, J., Kraus, A., & Pätz, H.-W. (2003) Measurements of OH and HO_2 radical concentrations and photolysis frequencies during BERLIOZ, *J. Geophys. Res.*, 108(D4), 8246, doi:10.1029/2001.

Iqbal, M. (1983) *An Introduction to Solar Radiation*. Academic Press, Toronto.

Jackson, J.O., Stedman, D.H., Smith, R.G., Hecker, L.H., & Warner, P.O. (1975) Direct NO_2 photolysis rate monitor, *Rev. Sci. Instrum.*, 46(4), 376–378.

Jacobson, M.Z. (1999) *Fundamentals of Atmospheric Modeling*. Cambridge University Press, Cambridge.

Jäkel, E., Wendisch, M., Kniffka, A., & Trautmann, T. (2005) Airborne system for fast measurements of upwelling and downwelling spectral actinic flux densities, *Appl. Opt.*, 44(3), 434–444.

Junkermann, W. (1994) Measurements of the $J(O^1D)$ actinic flux within and above stratiform clouds and above snow surfaces, *Geophys. Res. Lett.*, 21(9), 793–796.

Junkermann, W., Platt, U., & Volz-Thomas, A. (1989) A photoelectric detector for the measurement of photolysis frequencies of ozone and other atmospheric molecules, *J. Atmos. Chem.*, 8, 203–227.

Kanaya, Y., Sadanaga, Y., Matsumoto, J., Sharma, U.K., Hirokawa, J., Kajii, Y., & Akimoto, H. (1999) Nighttime observation of the HO_2 radical by an LIF Instrument at Oki Island, Japan, and its possible origins, *Geophys. Res. Lett.*, 26(14), 2179–2182.

Kanaya, Y., Kajii, Y., & Akimoto, H. (2003) Solar actinic flux and photolysis frequency determinations by radiometers and a radiative transfer model at Rishiri Island: Comparisons, cloud effects, and detection of an aerosol plume from Russian forest fires, *Atmos. Environ.*, 37, 2463–2475.

Kazadzis, S., Bais, A.F., Balis, D., Zerefos, C.S., & Blumenthaler, M. (2000) Retrieval of downwelling UV actinic flux density spectra from spectral measurements of global and direct solar UV irradiances, *J. Geophys. Res.*, 105(D4), 4857–4864.

Kazadzis, S., Topaloglou, C., Bais, A.F., Blumenthaler, M., Balis, D., Kazantzidis, A., & Schallhart, B. (2004) Actinic flux and O^1D photolysis frequencies retrieved from spectral measurements of irradiance at Thessaloniki, Greece, *Atmos. Chem. Phys.*, 4, 2215–2226.

Kelley, P., Russel, R.R., Luke, W.T., & Kok, G.L. (1995) Rate of NO_2 photolysis from the surface to 7.6 km altitude in clear sky and clouds, *Geophys. Res. Lett.*, 22(19), 2621–2624.

Kerr, J.B., Seckmeyer, G., Bais, A.F., Bernhard, G., Blumthaler, M., Diaz, S.B., Krotkov, N., Lubin, D., McKenzie, R.L., Sabziparvar, A.A., & Verdebout, J. (2002) Surface ultraviolet radiation: Past and future, Chapter 5 in *Scientific Assessment of Ozone Depletion: 2002, Global Ozone Research and Monitoring*, Project Report No. 47, Geneva, 2003. WMO (World Meteorological Organization).

Kraus, A. & Hofzumahaus, A. (1998) Field measurements of atmospheric photolysis frequencies for O_3, NO_2, HCHO, H_2O_2 and HONO by UV spectroradiometry, *J. Atmos. Chem.*, 31(1–2), 161–180.

Kraus, A., Brauers, T., Brüning, D., A. Hofzumahaus, Rohrer, F., Houben, N., Pätz, H.W., & Volz-Thomas, A. (1998) Results of the NO_2-photolysis frequency intercomparison JCOM97 (in German). Reports of Forschungszentrums Jülich No. 3578, ISSN 0944–2952, Forschungszentrum Jülich GmbH, D-52425 Jülich, Germany.

Kraus, A., Rohrer, F., & Hofzumahaus, A. (2000) Intercomparison of NO_2 photolysis frequency measurements by actinic flux spectroradiometry and chemical actinometry during JCOM97, *Geophys. Res. Lett.*, 27, 1115–1118.

Kuhn, H.J., Braslavsky, S.E., & Schmidt, R. (2004) Chemical Actinometry – A IUPAC Technical Report, *Pure Appl. Chem.*, 76(12), 2105–2146.

Kylling, A., Webb, A.R., Bais, A.F., Blumenthaler, M., Schmitt, R., Thiel, S., Kazantzidis, A., Kift, R., Misslbeck, M., Schallhart, B., Schreder, J., Topaloglou, C., Kazadzis, S., & Rimmer, J. (2003a) Actinic flux determination from measurements of irradiance, *J. Geophys. Res.*, 108(D16), doi:10.1029/2002JD003236.

Kylling, A., Danielsen, T., Blumenthaler, M., Schreder, J., & Johnsen, B. (2003b) Twilight tropospheric and stratospheric photodissociation rates derived from balloon borne radiation measurements, *Atmos. Chem. Phys.*, 3, 377–385.

Kylling, A., Webb, A.R., Kift, R., Gobbi, G.P., Ammannato, L., Barnaba, F., Bais, A., Kazadzis, S., Wendisch, M., Jäkel, E., Schmidt, S., Kniffka, A., Thiel, S., Junkermann, W., Blumthaler, M., Silbernagl, R., Schallhart, B., Schmitt, R., Kjeldstad, B., Thorseth, T.M., Scheirer, R., & Mayer, B. (2005) Spectral actinic flux in the lower troposphere: Measurement and 1-D simulations for cloudless, broken cloud and overcast situations, *Atmos. Chem. Phys.*, 5, 1975–1997.

Lantz, K.O., Shetter, R.E., Cantrell, C.A., Flocke, S.J., Calvert, J.G., & Madronich, S. (1996) Theoretical, actinometric, and radiometric determinations of the photolysis rate coefficient of

NO_2 during the Mauna Loa Observatory Photochemistry Experiment 2, *J. Geophys. Res.*, 101, 14 613–14 630.

Lefer, B.L., Hall, S.R., Cinquini, L., Shetter, R.E., Barrick, J.D., & Crawford, J.H. (2001) Comparison of airborne NO_2 photolysis frequency measurements during PEM-Tropics B, *J. Geophys. Res.*, 106(D23), 32 645–32 656.

Lefer, B.L., Shetter, R.E., Hall, S.R., Crawford, J.H., & Olson, J.R. (2003) Impact of clouds and aerosols on photolysis frequencies and photochemistry during TRACE-P: 1. Analysis using radiative transfer and photochemical box models, *J. Geophys. Res.*, 108(D21), 8821, doi:10.1029/2002JD003171.

Lenoble, J. (1993) *Atmospheric Radiative Transfer*. A. Deepak Publishing, Hampton, Virginia.

Liao, H., Yung, Y.L., & Seinfeld, J.H. (1999) Effects of aerosols on tropospheric photolysis rates in clear and cloudy atmospheres, *J. Geophys. Res.*, 104(D19), 23 697–23 707.

Lindh, K.G., Buchberg, H., & Wilson, K.W. (1964) Omni-directional ultraviolet radiometer, *Solar energy*, 8(4), 112–116.

Los, A., Van Weele, M., & Dynkerke, P.G. (1997) Actinic fluxes in broken cloud fields, *J. Geophys. Res.*, 102(D4), 4257–4266.

Luke, W.T., Dickerson, R.R., & Nunnermacker, L.J. (1989) Direct measurements of the photolysis rate coefficients and Henry's Law constants of several alky nitrates, *J. Geophys. Res.*, 94(12), 14 905–14 921.

Madronich, S. (1987a) Intercomparison of NO_2 photodissociation and U.V. radiometer measurements, *Atmos. Environ.*, 21(3), 569–578.

Madronich, S. (1987b) Photodissociation in the atmosphere, 1. Actinic flux and the effects of ground reflection and clouds, *J. Geophys. Res.*, 92, 9740–9752.

Madronich, S. (1993) UV Radiation in the natural and perturbed atmosphere, Chapter 2 in *UV-B Radiation and Ozone Depletion*, pp. 17–69. Lewis Publishers, Boca Raton.

Madronich, S., Hastie, D.R., Ridley, B.A., & Schiff, H.I. (1983) Measurement of the photodissociation coefficient of NO_2 in the atmosphere: I. Method and surface measurements, *J. Atmos. Chem.*, 1, 3–25.

Madronich, S., Hastie, D.R., Schiff, H.I., & Ridley, B.A. (1985) Measurement of the photodissociation coefficient of NO_2 in the atmosphere: II. Stratospheric measurements, *J. Atmos. Chem.*, 3, 233–245.

Malicet, J., Daumont, D., Charbonnier, J., Parisse, C., Chakir, A., & Brion, J. (1995) Ozone UV spectroscopy. II. Absorption cross-sections and temperature dependence, *J. Atmos. Chem.*, 21, 263–273.

Matsumi, Y. & Kawasaki, M. (2003) Photolysis of atmospheric ozone in the ultraviolet region, *Chem. Rev.*, 103, 4767–4781.

Matsumi, Y., Comes, F.J., Hancock, G., Hofzumahaus, A., Hynes, A.J., Kawasaki, M., & Ravishankara, A.R. (2002) Quantum yields for production of $O(^1D)$ in the ultraviolet photolysis of ozone: Recommendation based on evaluation of laboratory data, *J. Geophys. Res.*, 107(D3), doi:10.1029/2001JD000510.

Mauldin III, R.L., Eisele, F.L., Cantrell, C.A., Kosciuch, E., Ridley, B.A., & Lefer, B. (2001) Measurements of OH aboard the NASA P-3 during PEM-Tropics B, *J. Geophys. Res.*, 106(D23), 32 657–32 666.

Mayer, B. & Kylling, A. (2005) Technical note: The libRadtran software package for radiative transfer calculations description and examples of use, *Atmos. Chem. Phys.*, 5, 1855–1877.

Mayer, B. & Madronich, S. (2004) Actinic flux and photolysis in water droplets: Mie calculations and geometrical optics limit, *Atmos. Chem. Phys.*, 4, 2241–2250.

McCluney, W.R. (1994). *Introduction to Radiometry and Photometry*. Artech House, Boston.

McElroy, C.T., Midwinter, C., Barton, D.V., & Hall, R.B. (1995) A comparison of J-values from the composition and photodissociative flux measurement with model calculations, *Geophys. Res. Lett.*, 22(11), 1365–1368.

McKenzie, R., Johnston, P., Hofzumahaus, A., Kraus, A., Madronich, S., Cantrell, C.A., Calvert, J.G., & Shetter, R.E. (2002) Relationship between photolysis frequencies derived from spectroscopic measurements of actinic fluxes and irradiances during the IPMMI campaign, *J. Geophys. Res.*, 107(D5), doi:10.1029/2001JD000601.

McKenzie, R.L. & Kotkamp, M. (1996) Upwelling UV spectral irradiances and surface albedo measurements at Lauder, New Zealand. *Geophys. Res. Lett.*, 23(14), 1757–1760.

McKenzie, R.L., Johnston, P.V., Kotkamp, M., Bittar, A., & Hamlin, J.D. (1992) Solar ultraviolet spectroradiometry in New Zealand: instrumentation and sample results from 1990, *Appl. Opt.*, 31(30), 6501–6509.

Meier, R.R., Anderson, G.P., Cantrell, C.A., Hall, L.A., Lean, J., Minschwaner, K., Shetter, R.E., Shettle, E.P., & Stamnes, K. (1997) Actinic radiation in the terrestrial atmosphere, *J. Atmos. and Solar Terr. Phys.*, 59(17), 2111–2157.

Merienne, M.F., Jenouvrier, A., & Coquart, B. (1995) The NO_2 absorption spectrum: 1. Absorption cross-sections at ambient temperature in the 300–500 nm region, *J. Atmos. Chem.*, 20, 281–297.

Molina, L.T. & Molina, M.J. (1986) Absolute absorption cross sections of ozone in the 185- to 350-nm wavelength range, *J. Geophys. Res.*, 91(D13), 14 501–14 508.

Moortgat, G.K., Seiler, W., & Warneck, P. (1983) Photodissociation of HCHO in air: CO and H_2 quantum yields at 220 and 300 K, *J. Chem. Phys.*, 78(3), 1185–1190.

Müller, R. and Schurath, U. (1986) Entwicklung eines Gerätes zur kontinuierlichen Messung der Photodissoziations Geschwindigkeit von Aldehyden in der Atmosphäre durch Nachweis des erzeugten CO, Gesellschaft für Strahlenund Umweltforschung mbHMünchen, Abschlussbericht Vorhaben KBF 53.

Müller, M., Kraus, A., & Hofzumahaus, A. (1995) $O_3 \rightarrow O(^1D)$ photolysis frequencies determined from spectroradiometric measurements of solar actinic UV-radiation: Comparison with chemical actinometer measurements, *Geophys. Res. Lett.*, 22, 679–682.

Nader, J.S. & White, N. (1969) Volumetric measurement of ultraviolet energy in an urban atmosphere, *Environ. Science Technol.*, 3(9), 848–854.

Okabe, H. (1978) *Photochemistry of Small Molecules.* John Wiley and Sons.

Olson, J., Prather, M., Berntsen, T., Carmichael, G., Chatfield, R., Connell, P., Derwent, R., Horowitz, L., Jin, S., Kanakidou, M., Kasibhatla, P., Kotamarthi, R., Kuhn, M., Law, K., Penner, J., Perliski, L., Sillman, S., Stordal, F., Thompson, A., and Wildt, O. (1997) Results from the Intergovernmental Panel on Climatic Change photochemical model intercomparison (PhotoComp), *J. Geophys. Res.*, 102, 5979–5991.

Parrish, D.D., Murphy, P.C., Albritton, D.L., & Fehsenfeld, F.C. (1983) The measurement of the photodissociation rate of NO_2 in the atmosphere, *Atmos. Environ.*, 17, 1365–1379.

Penkett, S.A., Monks, P.S., Carpenter, L.J., Clemitshaw, K.C., Ayers, G.P., Gillett, R.W., Galbally, I.E., & Meyer, C.P. (1997) Relationships between ozone photolysis rates and peroxy radical concentrations in clean marine air over the Southern Ocean, *J. Geophys. Res.*, 102(D11), 12 805–12 817.

Pfister, G., Baumgartner, D., Maderbacher, R., & Putz, E. (2000) Aircraft measurements of photolysis rate coefficients for ozone and nitrogen dioxide under cloudy conditions, *Atmos. Environ.*, 34, 4019–4029.

Pravilov, A.M. (1987). Gas-phase actinometry for UV and vacuum UV spectral regions (Review). *High Energy Chem. USSR*, 21, 243–255.

Roehl, C.M., Orlando, J.J., Tyndall, G.S., Shetter, R.E., Vazquez, G.J., Cantrell, C.A., & Calvert, J.G. (1994) Temperature dependence of the quantum yields for the photolysis of NO_2 near the dissociation limit, *J. Phys. Chem.*, 98, 7837–7843.

Roehl, C.M., Nizkiorodov, S.A., Zhang, H., Blake, G.A., & Wennberg, P.O. (2002) Photodissociation of peroxynitric acid in the near-IR, *J. Phys. Chem. A*, 106, 3766–3772.

Röth, E.-P., Ruhnke, R., Moortgat, G., Meller, R., & Schneider, W. (1997) UV/VIS-Absorption cross sections and quantum yields for use in photochemistry and atmospheric modeling. Reports of Forschungszentrums Jülich No. 3341, ISSN 0944–2952, Forschungszentrum Jülich GmbH, D-52425 Jülich, Germany.

Ruggaber, A., Forkel, R., & Dlugi, R. (1993) Spectral actinic flux and its ratio to spectral irradiance by radiation transfer calculations, *J. Geophys. Res.*, 98, 1151–1162.

Sander, S.P., Friedl, R.R., Golden, D.M., Kurylo, M.J., Huie, R.E., Orkin, V.L., Moortgat, G.K., Ravishankara, A.R., Kolb, C.E., Molina, M.J., & Finlayson-Pitts, B.J. (2003) *Chemical Kinetics and Photochemical Data for Use in Atmospheric Studies.* Evaluation Number 14, JPL Publ. 02-25. NASA-JPL.

Schmitt, R. (1999) *Fast Monolithic Single Monochromator – Spectral Radiometer With Diode Array Detector For Atmospheric Research.* Metcon Inc, Boulder, USA and Meteorologie Consult GmbH, Glashütten, Germany. Technical Data Sheet.

Schultz, M., Houben, N., Mihelcic, D., Pätz, H.-W., & Volz-Thomas, A. (1995) A chemical actinometer for the calibration of photoelectric detectors for the measurement of jNO_2 (in German). Reports of Forschungszentrum Jülich No. 3135, ISSN 0944–2952, Forschungszentrum Jülich GmbH, D-52425 Jülich, Germany.

Seinfeld, J.H. & Pandis, S.N. (1998) *Atmospheric Chemistry and Physics, From Air Pollution to Climate Change,* John Wiley and Sons, New York.

Sellers, B. & Hanser, F.A. (1978) Measurement of tropospheric 300 nm solar ultraviolet flux for determination of $O(^1D)$ photoproduction rate, *J. Atmos. Sci.*, 35, 912–918.

Seroji, A.R., Webb, A.R., Coe, H., Monks, P.S., & Rickard, A.R. (2004) Derivation and validation of photolysis rates of O_3, NO_2, and CH_2O from a GUV-541 radiometer, *J. Geophys. Res.*, 109(D21307), doi:10.1029/2004JD004674.

Shetter, R.E. & Müller, M. (1999) Photolysis frequency measurements using actinic flux spectroradiometry during the PEM-Tropics Mission: Instrumentation description and some results, *J. Geophys. Res.*, 104, 5647–5661.

Shetter, R.E., Davidson, J.A., Cantrell, C.A., Burzynski, N.J., & Calvert, J.G. (1988) Temperature dependence of the atmospheric photolysis rate coefficient for NO_2, *J. Geophys. Res.*, 93, 7113–7118.

Shetter, R.E., McDaniel, A.H., Cantrell, C.A., Madronich, S., & Calvert, J.G. (1992) Actinometer and Eppley radiometer measurements of the NO_2 photolysis rate coefficient during the Mauna-Loa-Observatory Photochemistry Experiment, *J. Geophys. Res.*, 97, 10 349–10 359.

Shetter, R.E., Cantrell, C.A., Lantz, K.O., Flocke, S.J., Orlando, J.J., Tyndall, G.S., Gilpin, T.M., Fischer, C.A., Madronich, S., Calvert, J.G., & Junkermann, W. (1996) Actinometric and radiometric measurement and modeling of the photolysis rate coefficient of ozone to $O(^1D)$ during the Mauna Loa Observatory Photochemistry Experiment 2, *J. Geophys. Res.*, 101, 14 631–14 642.

Shetter, R.E., Junkermann, W., Swartz, W.H., Frost, G.J., Crawford, J.H., Lefer, B.L., Barrick, J.D., Hall, S.R., Hofzumahaus, A., Bais, A., Calvert, J.G., Cantrell, C.A., Madronich, S., Müller, M., Kraus, A., Monks, P.S., Edwards, G.D., McKenzie, R., Johnston, P., Schmitt, R., Griffioen, E., Krol, M., Kylling, A., Dickerson, R.R., Lloyd, S.A., Martin, T., Gardiner, B., Mayer, B., Pfister, G., Röth, E.P., Köpke, P., Ruggaber, A., Schwander, H., & Weele, M.V. (2003) Photolysis frequency of NO_2: Measurement and modeling during the International Photolysis Frequency Measurement and Modeling Intercomparison (IPMMI), *J. Geophys. Res.*, 108(D16), 8544, doi:10.1029/2002JD002932.

Slaper, H., Reinen, H.A.J.M., Blumenthaler, M., Huber, M., & Kuik, F. (1995) Comparing ground-level spectrally resolved solar UV measurements using various instruments: A technique resolving effects of wavelength shift and slit width, *Geophys. Res. Lett.*, 22(20), 2721–2724.

Talukdar, R.K., Longfellow, C.A., Gilles, M.K., & Ravishankara, A.R. (1998) Quantum yields of $O(^1D)$ in the photolysis of ozone between 289 and 329 nm as a function of temperature, *Geophys. Res. Lett.*, 25(2), 143–146.

Thomas, G.E. & Stamnes, K. (1999) *Radiative Transfer in the Atmosphere and Ocean,* Cambridge University Press.

Tie, X., Madronich, S., Walters, S., Zhang, R., Rasch, P., & Collins, W. (2003) Effect of clouds on photolysis and oxidants in the troposphere, *J. Geophys. Res.*, 108(D20), 4642, doi:10.1029/2003JD003659.

Trautmann, T., Podgorny, I., Landgraf, J., & Crutzen, P.J. (1999) Actinic fluxes and photodissociation coefficients in cloud fields embedded in realistic atmospheres, *J. Geophys. Res.*, 104, 30, 173–30 192.

Troe, J. (2000) Are primary quantum yields of NO_2 photolysis at $\lambda \leq 398$ nm smaller than unity?, *Zeit Phys. Chem.*, 214(5), 573–581.

Urbach, F. (ed.) (1969) *The Biologic Effects of Ultraviolet Radiation (With Emphasis on the Skin)*, Proceedings of the First International Conference, sponsored jointly by the Skin and Cancer Hospital, Temple University Health Sciences Center and The International Society of Biometeorology. Pergamon Press, Oxford.

Van der Hage, J.C.H., Boot, W., Van Dop, H., Duynkerke, P.G., & Vila-Guerau de Arellano, J. (1994) A photo-electric detector suspended under a balloon for actinic flux measurements, *J. Atmos. Oceanic Technol.*, 11(3), 674–679.

Van Weele, M. & Duykerke, P.G. (1993) Effect of clouds on the photodissociation of NO_2: Observations and modelling, *J. Atmos. Chem.*, 16, 231–255.

Van Weele, M., Vila-Guerau De Arellano, J., & Kuik, F. (1995) Combined measurements of UV-A actinic flux, UV-A irradiance and global radiation in relation to photodissociation rates, *Tellus*, 47B, 353–364.

Vaughan, W.E. & Noyes, Jr., W.A. (1930) Photochemical studies. XI. The quantum efficiency of ozone formation in the fluorite region, *J. Am. Chem. Soc.*, 52, 559–568.

Volz-Thomas, A., Lerner, A., Pätz, H.-W., Schultz, M., McKenna, D.S., Schmitt, R., Madronich, S., & Röth, E.P. (1996) Airborne measurements of the photolysis frequency of NO_2, *J. Geophys. Res.*, 101(D13), 18 613–18 627.

Vuilleumier, L., Bamer, J.T., Harley, R.A., & Brown, N.J. (2001) Evaluation of nitrogen dioxide photolysis rates in an urban area using data from the 1997 Southern California Ozone Study, *Atmos. Environ.*, 35, 6525–6537.

Warburg, E. (1912) Über den Energieumsatz bei photochemischen Vorgängen in Gasen II. *Sitzungsberichte der Preussischen Akademie der Wissenschaften Berlin*, pp. 216–225.

Webb, A.R., Kift, R., Thiel, S., & Blumenthaler, M. (2002a) An empirical method for the conversion of spectral UV irradiance measurements to actinic flux data, *Atmos. Environ.*, 36, 4397–4404.

Webb, A.R., Bais, A.F., Blumenthaler, M., Gobbi, G.-P., Kylling, A., Schmitt, R., Thiel, S., Barnaba, F., Danielsen, T., Junkermann, W., Kazantzidis, A., Kelly, P., Kift, R., Liberti, G.L., Misslbeck, M., Schallhart, B., Schreder, J., & Topaloglou, C. (2002b) Measuring spectral actinic flux and irradiance: Experimental results from the Actinic Flux Determination from Measurements of Irradiance (ADMIRA) Project, *J. Atmos. Ocean. Technol.*, 19, 1049–1062.

Ylianttila, L., Visuri, R., Huurto, L., & Jokela, K. (2005) Evaluation of a single-monochromator diode array spectroradiometer for sunbed UV-radiation measurements, *Photochem. Photobiol.*, 81, 333–341.

Yoshino, K., Esmond, J.R., Cheung, A.S.-C., Freeman, D.E., & Parkinson, W.H. (1992) High resolution absorption cross sections in the transmission window region of the Schumann-Runge Bands and Herzberg Continuum of O_2, *Planet. Space Sci.*, 40(2–3), 185–192.

Zafonte, L., Rieger, P.L., & Holmes, J.R. (1977) Nitrogen dioxide photolysis in the Los Angeles atmosphere, *Environ. Sci. Technol.*, 11(5), 483–487.

Zeiss GmbH (1998) *Product Information MCS Multichannel Spectrometer*, Zeiss GmbH, Jena, Germany.

Zellner, R. (guest ed.) (1999) *Topics in Physical Chemistry: Global Aspects of Atmospheric Chemistry*, Vol. 6, Deutsche-Bunsengesellschaft für Physikalische, Chemic e.v.(ed.) Steinkopff-Verlag, Darmstadt.

Zerefos, C.S. & Bais, A.F. (ed.) (1995) *Solar Ultraviolet Radiation: Modelling, Measurements and Effects.*, volume 52 of NATO ASI Series I: Global Environmental Change, Springer Verlag, Berlin.

Index